REVIEWS in MINERALOGY and GEOCHEMISTRY

Volume 52   2003

URANIUM-SERIES GEOCHEMISTRY

EDITORS:

Bernard Bourdon	*Institut de Physique du Globe de Paris* *Paris, France*
Gideon M. Henderson	*University of Oxford* *Oxford, United Kingdom*
Craig C. Lundstrom	*University of Illinois, Urbana-Champaign* *Urbana, Illinois, USA*
Simon P. Turner	*University of Bristol* *Bristol, United Kingdom*

FRONT COVER: The background of the front cover image is an exert from a paper published exactly 100 years before this volume. That paper was the first to calculate the half life for ^{234}Th, or U-X as this newly discovered radioactive substance was then called. Further details of the early history of U-series science can be found in the second preface to this volume. Superimposed on this background is the full series of radioactive nuclides produced by the initial decay of ^{238}U, and finally resulting in the formation of ^{206}Pb. Nuclides with half lives longer than a year are shown in red, while those with half lives between one day and one year are shown in blue.

Series Editors: **Jodi J. Rosso &**

GEOCHEMICAL SOCIETY
MINERALOGICAL SOCIETY of AMERICA

COPYRIGHT 2003

MINERALOGICAL SOCIETY OF AMERICA

The appearance of the code at the bottom of the first page of each chapter in this volume indicates the copyright owner's consent that copies of the article can be made for personal use or internal use or for the personal use or internal use of specific clients, provided the original publication is cited. The consent is given on the condition, however, that the copier pay the stated per-copy fee through the Copyright Clearance Center, Inc. for copying beyond that permitted by Sections 107 or 108 of the U.S. Copyright Law. This consent does not extend to other types of copying for general distribution, for advertising or promotional purposes, for creating new collective works, or for resale. For permission to reprint entire articles in these cases and the like, consult the Administrator of the Mineralogical Society of America as to the royalty due to the Society.

REVIEWS IN MINERALOGY AND GEOCHEMISTRY

(Formerly: REVIEWS IN MINERALOGY)

ISSN 1529-6466

Volume 52

U-series Geochemistry

ISBN 0-939950-54-5

Additional copies of this volume as well as others in this series may be obtained at moderate cost from:

THE MINERALOGICAL SOCIETY OF AMERICA
1015 EIGHTEENTH STREET, NW, SUITE 601
WASHINGTON, DC 20036 U.S.A.

URANIUM–SERIES GEOCHEMISTRY
52   *Reviews in Mineralogy and Geochemistry*   52

FOREWORD

Anniversaries always cause us to reflect upon where we have been and where we are going. Exactly 100 years before the publication of this volume, the first paper which calculated the half-life for the newly discovered radioactive substance U-X (now called ^{234}Th), was published. Now, in this volume, the editors Bernard Bourdon, Gideon Henderson, Craig Lundstrom and Simon Turner have integrated a group of contributors who update our knowledge of U-series geochemistry, offer an opportunity for non-specialists to understand its basic principles, and give us a view of the future of this active field of research. It was prepared in advance of a two-day short course (April 3-4, 2003) on U-series geochemistry, jointly sponsored by GS and MSA and presented in Paris, France prior to the joint EGS/AGU/EUG meeting in Nice.

As Series Editor, I thank Bernard, Gideon, Craig and Simon for the considerable time and effort that they put into the preparation and organization of this volume. I also thank the many authors who contributed to this volume for their timely thoroughness during the preparation and review process. And, as always, I thank my infinitely patient and supportive family, Kevin, Ethan and Natalie.

Jodi J. Rosso, Series Editor
West Richland, Washington
January 31, 2003

ACKNOWLEDGMENTS

We would like to thank all the individuals and organizations who have made possible the publication of this volume including the Board of directors of the Geochemical Society and the Mineralogical Society of America. Special thanks are due to Scott Wood and Jodi Rosso for handling the production of this volume. Karl K. Turekian is thanked for writing the preface and all the authors are thanked for their prompt and comprehensive contributions. We would like to thank Seth Davis (GS) and Alex Speer (MS) who have helped organize and muster support for the short course associated with this volume (Paris, April 2003). We would like to thank the US Department of Energy, the Commissariat à l'Energie Atomique (Atomic Energy Commission, France), The French Agence National des Déchets Radioactifs (Radioactive Waste National Agency) and the Thermo-Finnigan Company who have all provided financial support for this short-course. We would also like to thank Claude Jaupart, director of IPGP for provision of the venue. Finally, we would also like to thank all the reviewers who significantly helped to improve the quality and inclusiveness of the chapters in this volume.

Bernard Bourdon
Gideon Henderson
Craig Lundstrom
Simon Turner

January 2003

PREFACE

URANIUM DECAY SERIES

Karl K. Turekian

The discovery of the ^{238}U decay chain, of course, started with the seminal work of Marie Curie in identifying and separating ^{226}Ra. Through the work of the Curies and others, all the members of the ^{238}U decay chain were identified. An important milestone for geochronometrists was the discovery of ^{230}Th (called Ionium) by Bertram Boltwood, the Yale scientist who also made the first age determinations on minerals using the U-Pb dating method (Boltwood in 1906 established the antiquity of rocks and even identified a mineral from Sri Lanka-then Ceylon as having an age of 2.1 billion years!)

The application of the ^{238}U decay chain to the dating of deep sea sediments was by Piggott and Urry in 1942 using the "Ionium" method of dating. Actually they measured ^{226}Ra (itself through ^{222}Rn) assuming secular equilibrium had been established between ^{230}Th and ^{226}Ra.

Although ^{230}Th was measured in deep sea sediments by Picciotto and Gilvain in 1954 using photographic emulsions, it was not until alpha spectrometry was developed in the late 1950's that ^{230}Th was routinely measured in marine deposits. Alpha spectrometry and gamma spectrometry became the work horses for the study of the uranium and thorium decay chains in a variety of Earth materials. These ranged from ^{222}Rn and its daughters in the atmosphere, to the uranium decay chain nuclides in the oceanic water column, and volcanic rocks and many other systems in which either chronometry or element partitioning, were explored.

Much of what we learned about the ^{238}U, ^{235}U and ^{232}Th decay chain nuclides as chronometers and process indicators we owe to these seminal studies based on the measurement of radioactivity.

The discovery that mass spectrometry would soon usurp many of the tasks performed by radioactive counting was in itself serendipitous. It came about because a fundamental issue in cosmochemistry was at stake. Although variation in ^{235}U/^{238}U had been reported for meteorites the results were easily discredited as due to analytical difficulties. One set of results, however, was published by a credible laboratory long involved in quality measurements of high mass isotopes such as the lead isotopes. The purported discovery of ^{235}U/^{238}U variations in meteorites, if true, would have consequences in defining the early history of the formation of the elements and the development of inhomogeneity of uranium isotopes in the accumulation of the protoplanetary materials of the Solar System.

Clearly the result was too important to escape the scrutiny of falsification implicit in the way we do science. The Lunatic Asylum at Caltech under the leadership of Jerry Wasserburg took on that task. Jerry Wasserburg and Jim Chen clearly established the constancy and Earth-likeness of ^{235}U/^{238}U in the samplable universe. In the hands of another member of the Lunatic Asylum, Larry Edwards, the methodology was transformed into a tool for the study of the ^{238}U decay chain in marine systems. Thus the mass spectrometric techniques developed provided an approach to measuring the U and Th isotopes in geological materials as well as cosmic materials with the same refinement and accommodation for small sample size.

Soon after this discovery the harnessing of the technique to the measurement of all the U isotopes and all the Th isotopes with great precision immediately opened up the entire field of uranium and thorium decay chain studies. This area of study was formerly the poaching ground for radioactive measurements alone but now became part of the wonderful world of mass spectrometric measurements. (The same transformation took place for radiocarbon from the various radioactive counting schemes to accelerator mass spectrometry.)

No Earth material was protected from this assault. The refinement of dating corals, analyzing volcanic rocks for partitioning and chronometer studies and extensions far and wide into ground waters and ocean bottom dwelling organisms has been the consequence of this innovation.

Although Ra isotopes, ^{210}Pb and ^{210}Po remain an active pursuit of those doing radioactive measurements, many of these nuclides have also become subject to the mass spectrometric approach.

In this volume, for the first time, all the methods for determining the uranium and thorium decay chain nuclides in Earth materials are discussed. The range of problems solvable with this approach is remarkable—a fitting, tribute to the Curies and the early workers who discovered them for us to use.

ONE HUNDRED YEARS AGO: THE BIRTH OF URANIUM-SERIES SCIENCE

Gideon M. Henderson

One hundred years ago: the date is 1903, and it is an auspicious year in the history of radiochemistry. 1903 witnessed the first published version of a radioactive decay chain; the submission of Marie Curie's doctoral thesis; the award of the Nobel prize for physics to Becquerel and the Curies; and the recognition that radioactivity released both heat and He, with important implications for the age of the Earth and for absolute dating. These events were part of the rapid development of a new science that followed the discovery of radioactivity in 1896. For those geochemists familiar with U-series geochemistry, the early history of the field can make fascinating reading. In these early years, armed only with simple chemistry (the mass spectrometer, for instance, was not to be invented until 1918), the pioneers of the field were able to piece together an almost complete picture of the three naturally occurring decay series. This preface provides a brief introduction to this period of discovery—discovery that underlies all the geochemical applications detailed in the chapters that follow.

As the end of the 19th century approached, several workers were investigating the recently discovered X-rays. One of these, Henri Becquerel, discovered that phosphorescent uranium salts released penetrating rays, distinct from X-rays, which were capable of exposing photographic plates (Becquerel 1896b). In a key, but somewhat fortuitous experiment, Becquerel demonstrated that the rays from the uranium salts did not require light in order to be emitted and were therefore independent of the phosphorescence (Becquerel 1896a). Becquerel had discovered radioactivity, although it was two years before this name was coined (by Marie Curie) and the phenomenon was initially termed "Becquerel" radiation, or "uranic" radiation. This discovery was pursued by Marie Curie who checked the radioactivity of many compounds and minerals. She demonstrated that radioactivity came particularly from uranium and thorium (Curie 1898). And she provided the first indication of its atomic, rather than molecular nature because natural compounds emitted radioactivity proportionally to their U or Th content, regardless of their chemical form.

One curious observation, however, was that pure U actually had a lower radioactivity than natural U compounds. To investigate this, Curie synthesized one of these compounds from pure reagents and found that the synthetic compound had a lower radioactivity than the identical natural example. This led her to believe that there was an impurity in the natural compound which was more radioactive than U (Curie 1898). Since she had already tested all the other elements, this impurity seemed to be a new element. In fact, it turned out to be two new elements—polonium and radium—which the Curies were successfully able to isolate from pitchblende (Curie and Curie 1898; Curie et al. 1898). For radium, the presence of a new element was confirmed by the observation of new spectral lines not attributable to any other element. This caused a considerable stir and the curious new elements, together with their discoverers, achieved rapid public fame. The Curies were duly awarded the 1903 Nobel prize in Physics for studies into "radiation phenomena," along with Becquerel for his discovery of "spontaneous radioactivity." Marie Curie would, in 1911, also be awarded the Nobel prize in chemistry for her part in the discovery of Ra and Po.

Shortly after the discovery of these new radioactive materials it was recognized that there were two different forms of radiation. All radiation caused ionization of air so that it would conduct electricity, but only some radiation was capable of passing through material, such as the shielding paper which was placed on photographic plates to prevent them being exposed by light (Rutherford 1899). Rutherford named the non-penetrating form α-rays, and the penetrating form β-rays.

The discovery of two new elements started a frenetic race to find more. Actinium was soon unearthed (Debierne 1900) and many other substances were isolated from U and Th which also seemed to be new elements. One of these was discovered somewhat fortuitously. Several workers had noticed that the radioactivity of Th salts seemed to vary randomly with time and they noticed that the variation correlated with drafts in the lab, appearing to reflect a radioactive emanation which could be blown away from the surface of the Th. This "Th-emanation" was not attracted by charge and appeared to be a gas, ^{220}Rn, as it turns out, although Rutherford at first speculated that it was Th vapor. Rutherford swept some of the Th-emanation into a jar and repeatedly measured its ability to ionize air in order to assess its radioactivity. He was therefore the first to report an exponential decrease in radioactivity with time, and his 1900 paper on the subject introduced the familiar equation $dN/dt = -\lambda N$, as well as the concept of half-lives (Rutherford 1900a). His measured half-life for the Th emanation of 60 seconds was remarkably close to our present assessment of 55.6 seconds for ^{220}Rn.

Rutherford also noticed that the walls of the vessels in which Th emanation was investigated became radioactive during the experiment. This "excited activity" lasted longer than the activity of the Th emanation, but itself decayed away with a half-life of about 11 hours (Rutherford 1900b). Unwittingly, he was working down the Th decay chain and was measuring the decay of ^{212}Pb, the grand-daughter of ^{220}Rn, formed when it decayed.

At about the same time, solid substances with strong radioactivity were separated chemically from U and Th. That from U was named U-X (Crookes 1900) and turned out to be ^{234}Th, while that from Th was named Th-X (Rutherford and Soddy 1902) and was ^{224}Ra. Becquerel, returning to radioactivity research, noted that U, once stripped of its U-X, had a dramatically lower radioactivity and that this radioactivity seemed to return to the uranium if it was left for a sufficiently long time (Becquerel 1901). Rutherford and Soddy pursued this idea using the more rapidly decaying Th-X. They separated Th-X from Th and made a series of measurements that demonstrated an exact correspondence between the return of the radioactivity to the Th, and the decay of radioactivity in the Th-X. On this basis they deduced that much of the, "radioactivity of thorium is not due to thorium itself but to the presence of a non-thorium substance in minute amount which is being continuously produced." And they went on to give the first description of secular equilibrium:

> "The normal or constant radioactivity possessed by thorium is an equilibrium value, where the rate of increase of radioactivity due to the production of fresh active material is balanced by the rate of decay of radioactivity of that already formed" (Rutherford and Soddy 1902).

In the same paper, Rutherford and Soddy also suggested that elements were undergoing "spontaneous transformation." The use of the word transformation smacked of alchemy and Rutherford was loath to use it, but by then it seemed clear that elements were really changing and that "radioactivity may therefore be considered as a manifestation of subatomic change" (Rutherford and Soddy 1902).

The recognition of element transformation allowed the idea of a series of elements forming sequentially from the decay of a parent element and led to the first published set of "U-series" in 1903 (Rutherford 1903):

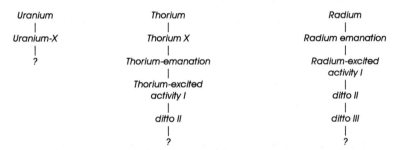

By painstaking chemical separations, and careful study of the style and longevity of radioactivity from the resulting separates, these series were rapidly added to and only a year later more than 15 discrete radioactive substances were known, each with measured half-lives, and all arranged into four decay series from U, Th, Ac, and Ra (Rutherford 1904).

Were all of these newly discovered substances also new elements? This question would not be answered for some years but there was a flurry of other major discoveries to keep the protagonists occupied. Pierre Curie discovered that radioactivity released large quantities of heat (Curie and Laborde 1903) which appeared mysterious—as if the heat was coming from nowhere. This discovery provided an extra heat source for the Earth and reconciled the estimates of a very old Earth, based on geological estimates, with the young age calculated by Lord Kelvin from cooling rates. The year 1903 also witnessed the first demonstration that α-decay released He (Ramsay and Soddy 1903). The build up of He was soon put to use to date geological materials, initially by Rutherford in 1905 who calculated the first ever radiometric age of ≈500 Myr for a pitchblende sample, and then by Strutt who examined a wide variety of minerals (Strutt

U-series Geochemistry – Preface

1905). Shortly thereafter, Boltwood recognized that the Pb content of minerals increases with age and it became clear that Pb was the final product of radioactivity. Boltwood was also responsible for adding another substance to the decay series through his discovery of ionium (^{230}Th), and therefore for linking the U and Ra decay chains (Boltwood 1907). The discovery of ^{234}U, initially known as UrII, followed in 1912.

Increasingly, new attempts to use basic chemistry to separate substances from radioactive material were meeting with failure. In many cases, two substances which were known to have different radioactive properties and molecular masses simply could not be separated from one another and appeared chemically identical. By 1910, this problem led Soddy to speculate that there were different forms of the same element (Soddy 1910). By 1913 he was confident of this interpretation and coined the term "isotope" to describe the various types of each element, recognizing that each isotope had a distinct mass and half-life (Soddy 1913b). In the same year he wrote that "radiothorium, ionium, thorium, U-X, and radioactinium are a group of isotopic elements, the calculated atomic masses of which vary from 228-234" (a completely accurate statement- we now call these isotopes ^{228}Th, ^{230}Th, ^{232}Th, ^{234}Th, ^{227}Th respectively). Soddy received the Nobel prize for chemistry in 1921 for his work on isotopes.

The various U series were by now all but complete. Branched decays were understood, and a daughter of U-X was discovered—^{234}Pa as it is now known, but initially named "brevium" to reflect it's short half-life (Fajans and Gohring 1913). By 1913, a published ^{238}U decay series (Fajans 1913) was remarkably close to that in use today, differing only in the absence of some of the branched decays after Ra-A (^{218}Po), and in the precise values of some of the half-lives:

$$\text{UrI} \xrightarrow{\alpha} \text{UrX} \xrightarrow{\beta} \text{UrX}_2 \xrightarrow{\beta} \text{UrII} \xrightarrow{\alpha} \text{I}_o \xrightarrow{\alpha} \text{Ra} \xrightarrow{\alpha} \text{RaEm} \xrightarrow{\alpha} \text{RaA} \xrightarrow{\alpha} \text{RaB} \xrightarrow{\beta}$$
$\quad$ 5×10^9 yrs $\quad$ 24.6 days $\qquad\qquad$ 10^6 yrs $\quad$ 10^6 yrs $\quad$ 2000 yrs $\quad$ 3.86 days $\quad$ 3 min $\quad$ 26.7 min

$$\xrightarrow{} \text{RaC}_2 \xrightarrow{} 1.4 \text{ min}$$
$$\text{RaC}_1 \quad 19.5 \text{ min}$$
$$\xrightarrow{\beta} \text{RaC}' \xrightarrow{\alpha} \text{RaD} \xrightarrow{\beta} \text{RaE} \xrightarrow{\beta} \text{RaF} \xrightarrow{\alpha} \text{Pb}$$
$\qquad\qquad$ 10^{-6} sec $\quad$ 16 yrs $\quad$ 5 days $\quad$ 136 days

Working independently of one another, Fajans and Soddy also deduced the displacement rule (Soddy 1913a) (Fajans 1913). Based on the chemical behavior of the isotopes in the decay chains, and on their molecular masses, they realized that each time an element changed by emitting an α-ray, the resulting element belonged to a group in the periodic table shifted two to the left of the initial isotope. Similarly, each time an element changed by emitting a β-ray, the resulting element was shifted one group to the right. This enabled the decay series to be plotted on a figure of mass against atomic number, as shown in Figure 1, and to look even more familiar to the modern U-series geochemist.

The fundamental work to establish the sequence of isotopes in the U and Th decay chains was therefore almost complete by 1913, only 17 years after the first discovery of radioactivity. It would be another 40 years before techniques for the routine measurement of some of these isotopes were developed (as detailed in Edwards et al. 2003) and the U-series isotopes started to see their widespread application to questions in the earth sciences.

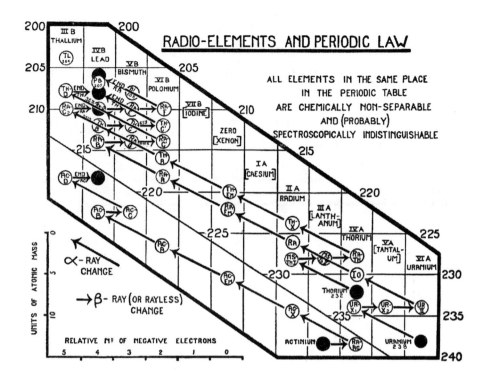

Figure 1. The three decay series from uranium, thorium, and actinium as published by Soddy in 1913 (Soddy 1913b).

REFERENCES

Becquerel AH (1896a) On the invisible rays emitted by phosphorescent bodies. Comptes Rendus de Seances de l'academie de Sciences 122:501-503
Becquerel AH (1896b) On the rays emitted by phosphoresence. Comptes Rendus de Seances de l'academie de Sciences 122:420-421
Becquerel H (1901) Sur la radioactivitie de l'uranium. Comptes Rendus de Seances de l'academie de Sciences 83:977-978
Boltwood BB (1907) Note on a new radioactive element. Amer J Sci 24:370-372
Crookes W (1900) Radio-activity of Uranium. Proc R Soc London 66:409-422
Curie M (1898) Rays emitted by compounds of uranium and thorium. Comptes Rendus de Seances de l'academie de Sciences 126:1101-1103
Curie P, Curie M (1898) Sur une nouvelle substance radioactive, contenue dans la pechblende. Comptes Rendus de Seances de l'academie de Sciences 127:175-178
Curie P, Curie M, Bemont G (1898) Sur une nouvelle substance fortement radioactive, contenue dans la pechblende. Comptes Rendus de Seances de l'academie de Sciences 127:1215-1217
Curie P, Laborde A (1903) On the heat spontaneously released by the salts of radium. Comptes Rendus de Seances de l'academie de Sciences 86:673
Debierne A (1900) Sur un nouvel element radio-actif: l'actinium. Comptes Rendus de Seances de l'academie de Sciences 130:906-908
Edwards RL, Gallup CD, Cheng H (2003) Uranium-series dating of marine and lacustrine carbonates. Rev Mineral Geochem 52:363-405
Fajans K (1913) Radioactive transformations and the periodic system of the elements. Berichte der Dautschen Chemischen Gesellschaft 46:422-439

Fajans K, Gohring O (1913) Uber die komplexe Natur des UrX. Naturwissenschaften 1:339
Ramsay W, Soddy F (1903) Experiments in radioactivity and the production of helium from radium. Proc R Soc London 72:204-207
Rutherford E (1899) Uranium radiation and the electrical conduction produced by it. Philos Mag 47:109-163
Rutherford E (1900a) A radioactive substance emitted from thorium compounds. Philos Mag 49:1-14
Rutherford E (1900b) Radioactivity produced in substances by the action of thorium compounds. Philos Mag 49:161-192
Rutherford E (1903) Radioactive change. Philos Mag 5:576-591
Rutherford E (1904) The succession of changes in radioactive bodies. Philos Trans R Soc 204:169-219
Rutherford E, Soddy F (1902) The cause and nature of radioactivity Part 1. Philos Mag 4:370-396
Soddy F (1910) Radioactivity. *In*: Annual Reports on the Progress of Chemistry, Vol. 7. The Chemical Society, London, p 257-286
Soddy F (1913a) The radio-elements and the periodic law. Chemical News 107:97-99
Soddy F (1913b) Radioactivity. *In*: Annual Reports on the Progress of Chemistry. The Chemical Society, London, p 262-288
Strutt RJ (1905) On the radio-active minerals. Proc R Soc London 76:88-101

U-series Geochemistry

Table of Contents

1   Introduction to U-series Geochemistry
Bernard Bourdon, Simon Turner,
Gideon M. Henderson and Craig C. Lundstrom

1. NEW DEVELOPMENTS IN U-SERIES GEOCHEMISTRY 1
2. U AND TH RADIOACTIVE DECAY SERIES 2
   2.1. Basic concepts 2
   2.2. Disequilibrium between U-series nuclides 6
   2.3. Processes creating disequilibria between U-series nuclides 7
3. CHEMISTRY AND GEOCHEMISTRY OF THE U-SERIES NUCLIDES 10
4. DETERMINATION OF THE HALF-LIVES OF U-SERIES NUCLIDES 12
   4.1. Methods for measurement of half-life 13
   4.2. Recommended half-lives for key nuclides 15
5. OUTLINE OF THE VOLUME 16
REFERENCES 17
APPENDIX: GENERAL SOLUTIONS OF U-SERIES DECAY EQUATION USING LAPLACE TRANSFORMS 20

2   Techniques for Measuring Uranium-series Nuclides: 1992-2002
Steven J. Goldstein and Claudine H. Stirling

1. INTRODUCTION 23
2. SAMPLE PREPARATION 24
   2.1. Microwave digestion of solids 24
   2.2. Tracer addition and tracer/sample equilibration 24
3. CHEMICAL SEPARATIONS 26
   3.1. Extraction chromatography resins/disks 26
4. INSTRUMENTAL ANALYSIS METHODS 27
   4.1. Alpha spectrometry 27
   4.2. Gamma spectrometry 29
   4.3. Thermal ionization mass spectrometry (TIMS) 30
   4.4. Secondary ion mass spectrometry (SIMS) 36
   4.5. ICPMS and MC-ICPMS 37
5. COMPARISON OF ANALYTICAL METHODS FOR U-SERIES NUCLIDES 48
6. FUTURE DEVELOPMENTS 52
ACKNOWLEDGMENTS 53
REFERENCES 53

3   Mineral-Melt Partitioning of Uranium, Thorium and Their Daughters
Jonathan Blundy and Bernard Wood

1. INTRODUCTION 59
2. PARTITIONING PRELIMINARIES 60
3. SOURCES OF PARTITIONING DATA 61
4. THE IMPORTANCE OF PARTITIONING IN INTERPRETING
    U-SERIES DATA 62
5. THE LATTICE STRAIN MODEL 65
    5.1. Proxies.. 77
    5.2. Derivation of proxy relationships 81
6. ADDITIONAL CONSIDERATIONS 81
    6.1. Henry's Law.. 82
    6.2. Ingrowth 83
7. MINERAL-MELT PARTITION COEFFICIENTS 84
    7.1. Clinopyroxene.. 84
    7.2. Orthopyroxene 90
    7.3. Olivine.. 92
    7.4. Garnet... 94
    7.5. Amphibole.. 100
    7.6. Plagioclase 102
    7.7. Alkali-feldspar 106
    7.8. Phlogopite (biotite) 108
    7.9. Oxide minerals 111
    7.10. Zircon... 114
    7.11. Other accessory phases 116
8. CONCLUSIONS.. 116
ACKNOWLEDGMENTS 118
REFERENCES 118

4   Timescales of Magma Chamber Processes and Dating of Young Volcanic Rocks
Michel Condomines, Pierre-Jean Gauthier, and Olgeir Sigmarsson

1. INTRODUCTION 125
2. GENERAL PRINCIPLES AND USEFUL DIAGRAMS 126
    2.1. Radioactive disequilibria as dating tools 126
    2.2. Isochron diagrams.. 126
    2.3. Diagrams showing the evolution of disequilibria vs. time 128
3. TIMESCALES OF MAGMA TRANSFER AND MAGMA CHAMBER
    PROCESSES... 129
    3.1. Preliminary comments and definitions 129
    3.2. Disequilibria in single volcanic samples.. 131
    3.3. Comparison of disequilibria in comagmatic lavas:
           timescales of differentiation 133

U-series Geochemistry – Table of Contents

    3.4. Evolution of disequilibria during the eruptive history of a volcano:
          magma storage or residence times 135
    3.5. Studies on minerals and crystallization ages:
          timescales of crystallization 140
    3.6. Timescales of magma degassing 153
4. DATING YOUNG VOLCANIC ROCKS 160
    4.1. Mineral isochrons 160
    4.2. Dating of whole-rocks 163
5. SUMMARY AND CONCLUSIONS 167
ACKNOWLEDGMENTS 169
REFERENCES 169

5   Uranium-series Disequilibria in Mid-ocean Ridge Basalts: Observations and Models of Basalt Genesis

Craig C. Lundstrom

1. INTRODUCTION 175
    1.1. A primer on U-series disequilibria 175
2. OBSERVED U-SERIES DISEQUILIBRIA IN MORB 176
    2.1. Analysis of U-series nuclides 176
    2.2. Observed U-series disequilibria in MORB relative to other
          tectonic settings 177
    2.3. Observed U-series disequilibria at specific ridge locations 179
    2.4. Applications: dating of MORB using U-series disequilibria 188
    2.5. Assessing secondary contamination in creating U-series disequilibria 189
3. MODELS OF U-SERIES DISEQUILIBRIA GENERATION DURING
    MELTING 190
    3.1. Constraints on element partitioning 191
    3.2. Time-independent melting models 192
    3.3. Ingrowth melting models 193
    3.4 A review of U-series melting models 196
4. RELATING MODELS TO OBSERVATIONS 200
5. INTEGRATING U-SERIES DISEQUILIBRIA MODELS WITH
    CONSTRAINTS FROM OTHER GEOCHEMICAL TRACERS 203
    5.1. Comparison of melting at ridges to other tectonic settings 203
    5.2. The relationship between Th isotopes and long-lived isotope systems 203
    5.3. Reconciling U-series interpretations with other geochemical observations 204
6. CONCLUDING REMARKS 207
ACKNOWLEDGMENTS 207
REFERENCES 207
APPENDIX: TRANSPORT-BASED MODELS FOR CREATING U-SERIES
    DISEQUILIBRIA 212

6 U-series Constraints on Intraplate Basaltic Magmatism
Bernard Bourdon and Kenneth W. W. Sims

1. INTRODUCTION 215
2. DIFFICULTIES IN CONSTRAINING HOTSPOT MELTING PROCESSES 215
   2.1. Source composition and source heterogeneities 215
   2.2. Role of the lithosphere 217
   2.3. Complexities in the melting region 218
3. THE ROLE OF SOURCE HETEROGENEITIES ON U-SERIES FRACTIONATION IN HOTSPOT MAGMATISM 219
   3.1. Identifying residual mineral phases 219
   3.2. Role of source heterogeneities on melting processes 226
   3.3. Tracing mantle sources 227
4. MELTING PROCESSES AND RELATION TO CONVECTIVE STRUCTURE OF PLUME 230
   4.1. Time dependent melting models for hotspot magmatism 230
   4.2. Sources of uncertainty in these models 236
   4.3. Observational constraints 237
5. SUMMARY AND PERSPECTIVES 244
ACKNOWLEDGMENTS 244
REFERENCES 244
APPENDIX: ANALYTICAL SOLUTIONS FOR TIME DEPENDENT MELTING MODELS 249
   A.1. Dynamic melt transport 249
   A.2. Chromatographic melt transport 249
   A.3. Box-model for equilibrium melting 253

7 Insights into Magma Genesis at Convergent Margins from U-series Isotopes
Simon Turner, Bernard Bourdon and Jim Gill

1. INTRODUCTION 255
2. CONVERGENT MARGIN MAGMATISM 256
   2.1. Geochemical signatures of source components 256
   2.2. The enriched component—sediment or OIB? 256
   2.3. The fluid component 258
   2.4. Relative depletion of the mantle wedge 260
3. U-SERIES ISOTOPES IN ARC LAVAS 260
4. BEHAVIOUR OF THE U-SERIES NUCLIDES IN AQUEOUS FLUIDS 261
   4.1. General empirical evidence 263
   4.2. Empirical observations from arc lavas 263
   4.3. Experimental constraints on mineral/fluid partitioning 264
   4.4. Composition of fluids released from the subducting plate 268
   4.5. Chromatographic interaction with the mantle wedge 268
5. SEDIMENT ADDITION, MASS BALANCE FOR Th AND TIME SCALES 269
   5.1. Mass balance for Th content and $^{230}Th/^{232}Th$ ratios 270

    5.2. Mechanism of sediment transfer—implications for the temperature
        of the wedge? 271
    5.3. Time scale of sediment transfer 272
6. FLUID ADDITION TIME SCALES 273
    6.1. U addition time scales 273
    6.2. Ra addition time scales 276
    6.3. Reconciling the U and Ra time scales 278
    6.4. Single-stage fluid addition 278
    6.5. Two-stage fluid addition 280
    6.6. Continuous fluid addition 281
    6.7. Mechanisms of fluid addition 282
7. PARTIAL MELTING AND MELT ASCENT RATES 283
    7.1. ^{231}Pa- and ^{230}Th-excess evidence for a partial melting
        signature in the wedge 284
    7.2. Ra evidence for melt ascent rates 284
    7.3. Models to reconcile the Pa, Th and Ra data 286
    7.4. Batch and equilibrium porous flow melting models 286
    7.5. Dynamic melting 286
    7.6. Flux melting 290
    7.7. Partial melting of the subducting oceanic crust 293
8. DISCUSSION OF U-SERIES TIME SCALE IMPLICATIONS FOR
    ARC LAVAS 293
9. REAR ARC LAVAS 295
10. MODIFYING PROCESSES 297
    10.1. Time since eruption 297
    10.2. Alteration and seawater interaction 297
    10.3. Magma chambers processes 298
    10.4. Radioactive decay 298
    10.5. Crystal fractionation or accumulation 299
    10.6. Magma recharge 299
    10.7. Crustal contamination 299
11. TIME-INTEGRATED U-Th-Pb EVOLUTION OF THE CRUST-MANTLE 301
12. FUTURE WORK 303
ACKNOWLEDGMENTS 303
REFERENCES 303
APPENDIX: EQUATIONS FOR SIMULATING MELTING AND
    DEHYDRATION MODELS IN ARCS 311
    A1. Single-stage model 311
    A2. Two stage-model for fluid addition 311
    A3. Continuous dehydration and melting 313

8   The Behavior of U- and Th-series Nuclides in Groundwater

Donald Porcelli and Peter W. Swarzenski

1. INTRODUCTION 317
2. NUCLIDE TRANSPORT IN AQUIFERS 320
    2.1. General modeling considerations 320
    2.2. Radon and the recoil rate of U-series nuclides 331

2.3. Ra isotopes 334
  2.4. Th isotopes 339
  2.5. U isotopes 343
  2.6. ^{210}Pb 348
3. GROUNDWATER DISCHARGE INTO ESTUARIES 349
  3.1. Background 349
  3.2. Tracing groundwater using ^{222}Rn and the Ra quartet 351
4. OPEN ISSUES 353
  4.1. The effects of well construction and sampling 353
  4.2. Quantification of model parameters 354
  4.3. Interpreting model-derived information 354
  4.4. Inputs at the water table 355
  4.5. Applications to pollutant radionuclide migration studies 355
  4.6. Tracing groundwater discharges 355
ACKNOWLEDGMENTS 355
REFERENCES 356

9 Uranium-series Dating of Marine and Lacustrine Carbonates
R.L. Edwards, C.D. Gallup, and H. Cheng

1. HISTORICAL CONSIDERATIONS 363
2. THEORY 365
  2.1. Decay chains 365
  2.2. Secular equilibrium and uranium-series dating 366
  2.3. ^{230}Th, ^{231}Pa, and ^{230}Th/^{231}Pa age equations 366
  2.4. Tests for ^{231}Pa-^{230}Th age concordancy 373
3. TESTS OF DATING ASSUMPTIONS 376
  3.1. Are initial ^{230}Th/^{238}U and ^{231}Pa/^{235}U values equal to zero? 376
  3.2. Tests of the closed-system assumption 378
4. SOURCES OF ERROR IN AGE 387
  4.1. Errors in half-lives and decay constants 387
  4.2. Errors in measurement of isotope ratios 389
  4.3. Error in initial ^{230}Th/^{232}Th 391
5. LATE QUATERNARY SEA LEVELS FROM CORAL DATING 391
  5.1. Deglacial sea level 391
  5.2. Sea level during the last interglacial/glacial cycle and earlier 393
6. DATING OF OTHER MARINE AND LACUSTRINE MATERIALS 394
  6.1. Deep sea corals 394
  6.2. Carbonate bank sediments 396
  6.3. Mollusks and foraminifera 397
  6.4. Lacustrine carbonates 398
7. CONCLUSIONS 399
ACKNOWLEDGMENTS 399
REFERENCES 400

10 Uranium-series Chronology and Environmental Applications of Speleothems
David A. Richards and Jeffrey A. Dorale

1. INTRODUCTION	407
2. BASIC GEOCHRONOLICAL PRINCIPLES AND ASSUMPTIONS	410
2.1. General principles of ^{230}Th-^{234}U-^{238}U and ^{231}Pa-^{235}U dating	410
2.2. Initial conditions	412
2.3. Closed system decay	419
2.4. ^{234}U/^{238}U dating methodology	423
2.5. U-Th-Pb dating of secondary carbonates of Quaternary age	424
3. SPELEOTHEM GEOCHRONOLOGY IN PRACTICE	428
3.1. Speleothem sampling strategy	428
3.2. Treatment of U-series ages	430
4. SPELEOTHEM CHRONOLOGY AND ENVIRONMENTAL CHANGE	431
4.1. Applications based on the presence/absence or growth rate of speleothems	431
4.2. Applications based on proxy evidence for environmental change contained within speleothems	439
5. CONCLUDING REMARKS	449
ACKNOWLEDGMENTS	450
REFERENCES	450

11 Short-lived U/Th Series Radionuclides in the Ocean: Tracers for Scavenging Rates, Export Fluxes and Particle Dynamics
J. K. Cochran and P. Masqué

1. INTRODUCTION	461
2. MEASUREMENT TECHNIQUES	462
3. SCAVENGING FROM SEAWATER	465
3.1. Early observations of Th scavenging	465
3.2. Development of scavenging models based on Th	467
3.3. The role of colloids in Th scavenging	468
3.4. Scavenging of Po	469
4. THORIUM AND POLONIUM AS TRACERS FOR ORGANIC CARBON CYCLING IN THE OCEANS	469
4.1. Basis, approach and early results	469
4.2. Results of the past decade: JGOFS and other studies	472
4.3. Unresolved issues	476
5. ^{234}Th AS A TRACER FOR PARTICLE TRANSPORT AND SEDIMENT PROCESSES IN THE COASTAL OCEAN	482
5.1. Sediment mixing rates	482
5.2. ^{234}Th as a tracer of particle transport in shelf and estuarine environments	484
6. CONCLUDING REMARKS	486
ACKNOWLEDGMENTS	487
REFERENCES	487

12   The U-series Toolbox for Paleoceanography
Gideon M. Henderson and Robert F. Anderson

1. INTRODUCTION 493
2. U-SERIES ISOTOPES IN THE OCEAN ENVIRONMENT 493
   2.1. The ocean uranium budget 493
   2.2. Chemical behavior of U-series nuclides in the oceans 496
3. HISTORY OF WEATHERING – (^{234}U/^{238}U) 497
4. SEDIMENTATION RATE – ^{230}Th$_{xs}$ 499
   4.1. The downward flux of ^{230}Th 499
   4.2. Seafloor sediments 505
   4.3. Mn crusts 508
5. PAST EXPORT PRODUCTIVITY – (^{231}Pa$_{xs}$/^{230}Th$_{xs}$) 508
   5.1. Chemical fractionation and boundary scavenging 510
   5.2. (^{231}Pa$_{xs}$/^{230}Th$_{xs}$) ratios as a paleoproductivity proxy 511
   5.3. The role of particle composition 513
   5.4. Prospects for future use 514
6. RATES OF PAST OCEAN CIRCULATION – (^{231}Pa$_{xs}$/^{230}Th$_{xs}$) 517
7. HOLOCENE SEDIMENT CHRONOLOGY – ^{226}Ra 518
8. SEDIMENT MIXING – ^{210}Pb 520
9. CONCLUDING REMARKS 522
ACKNOWLEDGMENTS 523
REFERENCES 523
APPENDIX 530

13   U-Th-Ra Fractionation During Weathering and River Transport
F. Chabaux, J. Riotte and O. Dequincey

1. INTRODUCTION 533
2. ORIGIN OF RADIONUCLIDE FRACTIONATION DURING
    WEATHERING AND TRANSFERS INTO SURFACE WATERS 534
   2.1. Chemical fractionation and mobilization factors 534
   2.2. Alpha recoil 542
3. RADIOACTIVE DISEQUILIBRIA IN WEATHERING PROFILES:
    DATING AND TRACING OF CHEMICAL MOBILITY 542
   3.1. Dating of pedogenic concretions 543
   3.2. Characterization and time scale of chemical mobility
        in weathering profiles 543
4. TRANSPORT OF U-Th-Ra ISOTOPES IN RIVER WATERS 553
   4.1. Transport of uranium isotopes in river waters 553
   4.2. Transport of thorium and radium isotopes 558
5. ESTIMATES OF WEATHERING MASS BALANCE FROM U-SERIES
    DISEQUILIBRIA IN RIVER WATERS 565
6. CONCLUDING REMARK 568
ACKNOWLEDGMENTS 569
REFERENCES 569

14 The Behavior of U- and Th-series Nuclides in the Estuarine Environment
Peter W. Swarzenski, Donald Porcelli, Per S. Andersson and Joseph M. Smoak

1. INTRODUCTION 577
    1.1. Estuarine mixing 578
    1.2. Estuary fluxes 579
    1.3. Colloids 581
2. URANIUM 583
    2.1. U in seawater 584
    2.2. River water U inputs 584
    2.3. U behavior in estuaries 584
    2.4. Uranium removal to anoxic sediments 586
    2.5. Importance of particles and colloids for controlling estuarine uranium ... 587
    2.6. The (^{234}U/^{238}U) activity ratios in estuaries 588
3. THORIUM 590
    3.1. ^{234}Th 590
    3.2. ^{228}Th 592
    3.3. Long-lived Th isotopes – ^{232}Th and ^{230}Th 592
4. RADIUM 593
5. RADON-222 597
6. LEAD AND POLONIUM 597
7. CONCLUSIONS 599
REFERENCES 600

15 U-series Dating and Human Evolution
A. W. G. Pike and P. B. Pettitt

1. INTRODUCTION 607
2. U-SERIES DATING OF BONES AND TEETH 608
    2.1. The diffusion-adsorption (D-A) model 610
    2.2. U-series combined with electron spin resonance dating 615
    2.3. Non-destructive U-series dating by gamma spectrometry 617
    2.4. Future developments 617
3. APPLICATIONS 618
    3.1. The issue of chronology in hominid evolution 618
    3.2. Neanderthals and modern humans in Israel 619
    3.3. *Homo erectus* and *Homo sapiens* in Java 620
    3.4. *Homo erectus* and *Homo sapiens* in China 621
    3.5. The "Pit of the Bones" and a new species of hominid in Spain 622
    3.6. The earliest Australian human remains 624
4. CONCLUSION 625
ACKNOWLEDGMENTS 626
REFERENCES 626
APPENDIX: FURTHER DETAILS OF THE D-A MODEL 630

16 Mathematical–Statistical Treatment of Data and Errors for ^{230}Th/U Geochronology

K. R. Ludwig

1. INTRODUCTION	631
2. WHY ERROR ESTIMATION IS IMPORTANT	631
3. ERRORS OF THE MEASURED ISOTOPIC RATIOS	632
4. ERROR CORRELATIONS	633
5. FIRST ORDER ESTIMATION OF ERRORS	634
6. WHEN FIRST ORDER ERROR ESTIMATION IS INADEQUATE	635
6.1. Improving the first order estimate analytically	636
6.2. Error estimation by Monte Carlo	636
7. CORRECTING A SINGLE ANALYSIS FOR DETRITAL THORIUM AND URANIUM	639
8. ISOCHRONS	641
8.1. Isochron representations for the general ^{230}Th/U system	642
8.2. Error-weighted regressions and isochrons for x-y data	644
8.3. 3-dimensional error-weighted regressions and isochrons	646
8.4. Isochrons with excess scatter	647
8.5. Beyond error-weighted least-squares isochrons	648
8.6. Robust and resistant isochrons	648
9. PITFALLS IN DATA PRESENTATION	650
10. NOTE ON IMPLEMENTATION OF ALGORITHMS	651
ACKNOWLEDGMENTS	651
REFERENCES	651
APPENDIX I: ESTIMATING ERROR CORRELATIONS	653
APPENDIX II: WORKED EXAMPLE OF DETRITAL CORRECTION AND ERROR PROPAGATION	655
APPENDIX III: FUNCTIONS/ROUTINES FOR ^{230}Th/U DATING PROVIDED BY *ISOPLOT*	656

1 Introduction to U-series Geochemistry

Bernard Bourdon
Laboratoire de Géochimie et Cosmochimie
IPGP-CNRS UMR 7579
4 Place Jussieu
75252 Paris cedex 05, France

Simon Turner
Department of Earth Sciences
Wills Memorial Building
University of Bristol
Bristol, BS8 1RJ, United Kingdom

Gideon M. Henderson
Department of Earth Sciences
University of Oxford
Parks Road
Oxford, OX1 3PR, United Kingdom

Craig C. Lundstrom
Department of Geology
University of Illinois at Urbana Champaign
245 NHB, 1301 W. Green St.
Urbana, Illinois, 61801, U.S.A.

1. NEW DEVELOPMENTS IN U-SERIES GEOCHEMISTRY

During the last century, the Earth Sciences underwent two major revolutions in understanding. The first was the recognition of the great antiquity of the Earth and the second was the development of plate tectonic theory. These leaps in knowledge moved geology from its largely descriptive origins and established the modern, quantitative, Earth Sciences. For any science, and particularly for the Earth Sciences, time scales are of central importance. Until recently, however, the study of time scales in the Earth Sciences was largely restricted to the unraveling of the ancient history of our planet. For several decades, Earth scientists have used a variety of isotope chronometers to unravel the long-term evolution of the planet. A fuller understanding of the physical and chemical processes driving this evolution often remained elusive because such processes occur on time scales (1-10^5 years) which are simply not resolvable by most conventional chronometers. The U-series isotopes, however, do provide tools with sufficient time resolution to study these Earth processes. During the last decade, the Earth Sciences have become increasingly focused on fundamental processes and U-series geochemistry has witnessed a renaissance, with widespread application in disciplines as diverse as modern oceanography and igneous petrology.

The uranium and thorium decay-series contain radioactive isotopes of many elements (in particular, U, Th, Pa, Ra and Rn). The varied geochemical properties of these elements cause nuclides within the chain to be fractionated in different geological environments. while the varied half-lives of the nuclides allows investigation of processes occurring on time scales from days to 10^5 years. U-series measurements have therefore revolutionized the Earth Sciences by offering some of the only quantitative constraints on time scales applicable to the physical processes that take place on the Earth.

The application of U-series geochemistry to the Earth Sciences was thoroughly summarized in 1982 and again in 1992 with the two editions of "Uranium-series Disequilibrium, Applications to Earth, Marine and Environmental Sciences," edited by M. Ivanovich and R. S. Harmon. It is now over a decade since the publication of the second of those volumes, and a great deal of new U-series work has been conducted. Much of this new work has relied on the development of new analytical techniques. These advances began in the late 1980's with the development of thermal ionization mass spectrometric techniques for the measurement of U, Th, Pa and Ra including the ability to measure isotope ratios greater than 10^5 (see Goldstein and Stirling 2003). More recently, multi-collector inductively coupled plasma mass spectrometry (MC-ICP-MS) has further improved the sensitivity, speed, and possibly precision of U-series measurements. Advances have also been made in the chemical separation techniques for U-series nuclides. The analytical chemistry of these elements has been known since the 1950's but adapting these techniques to geological samples with a complex matrix occurred more recently. Efficient techniques involving actinide-specific resin have also only been implemented in the last 10 years.

The need to understand the processes operating on Earth, coupled to recent analytical advances, have ensured that the U-series nuclides have seen widespread application since the last Ivanovich and Harmon book (1992). This volume does not set out to repeat material in that book, but is an attempt to bring together the advances in the subject over the last ten years, highlighting the excitement and rapid expansion of U-series research. The scope of the various chapters in this book is laid out at the end of this introduction. The remainder of this chapter introduces some of the basic concepts of U-series geochemistry, the chemical behavior of the elements involved, and the half-lives of the U- and Th-series nuclides. This chapter is not intended to be an exhaustive summary of the nuclear or radio-chemistry of the U-series nuclides and for additional information, the reader is referred to Ivanovich (1992).

2. U AND Th RADIOACTIVE DECAY SERIES

2.1. Basic concepts

There are three naturally occurring radioactive decay chains; each starts with an actinide nuclide (^{238}U, ^{235}U and ^{232}Th) having a long half-life ($t_{1/2} > 0.7$ Gyr) and ends with a stable isotope of lead (Fig. 1). In between is a series of nuclides with half-lives ranging from microseconds to hundreds of thousands of years. U-series disequilibrium refers to any fractionation between different members within a decay chain resulting in a non-steady state condition (steady state is known as secular equilibrium as explained below). The members of the decay chain most commonly studied are ^{234}U ($t_{1/2} = 245$ ka), ^{230}Th ($t_{1/2} = 76$ ka) and ^{226}Ra ($t_{1/2} = 1599$ yr) in the ^{238}U decay chain and ^{231}Pa ($t_{1/2} = 33$ ka) in the ^{235}U decay chain. The relatively long half-lives make these nuclides particularly suited to investigating many geological processes that occur over time scales similar to their decay period. An important characteristic of these decay-chains is that the ultimate parent isotope (e.g., ^{238}U) is radioactive and has a much longer half-life than all the intermediate nuclides.

The equation of decay for a given number of atoms, N, of a given nuclide can be written as follows:

For the parent nuclide:

$$\frac{dN_1}{dt} = -\lambda_1 N_1 \tag{1}$$

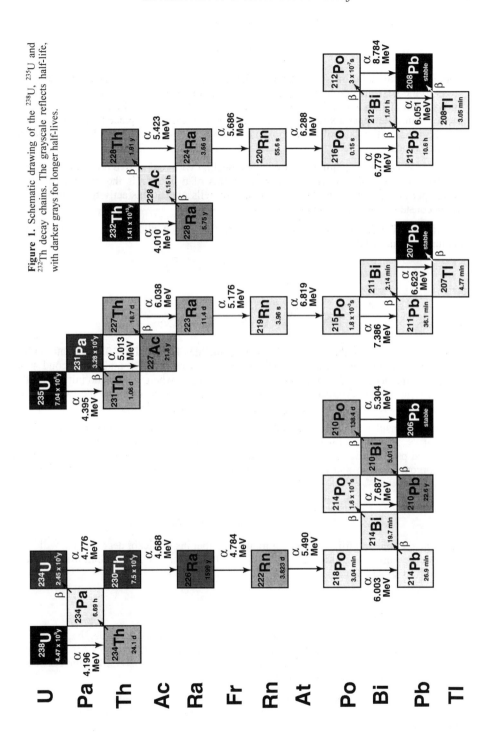

Figure 1. Schematic drawing of the ^{238}U, ^{235}U and ^{232}Th decay chains. The grayscale reflects half-life, with darker grays for longer half-lives.

For all the intermediate nuclides of interest:

$$\frac{dN_i}{dt} = -\lambda_i N_i + \lambda_{i-1} N_{i-1} \tag{2}$$

where the decay constant, λ is related to the half-life by $\lambda = \ln2/t_{1/2}$ and subscript $i-1$ refers to the next nuclide higher up in the decay chain. These equations were first solved by Bateman (1910) and are illustrated here with reference to the first two nuclides of a decay chain. The appendix provides a general solution using a Laplace transform that can also be used in the case of linear box models which track changes in U-series nuclide concentrations.

The concept most commonly used when dealing with radioactive nuclides is activity. By definition, the activity of a number of atoms of a nuclide is the number of decay events per unit of time. The law of radioactivity tells us that this activity is equal to the decay constant times the number of atoms.

If we solve Equations (1) and (2) for the first two nuclides in a decay chain, we obtain (Bateman 1910):

$$N_1 = N_1^0 e^{-\lambda_1 t} \tag{3a}$$

and

$$N_2 = \frac{\lambda_1}{\lambda_2 - \lambda_1} N_1^0 \left(e^{-\lambda_1 t} - e^{-\lambda_2 t} \right) + N_2^0 e^{-\lambda_2 t} \tag{3b}$$

where N_1^0 and N_2^0 are the number of atoms of nuclides 1 and 2 at $t = 0$. In Equation (3b), the first term represents the ingrowth of N_2 by decay of N_1 and the second term represents the decay of N_2 initially present. In situations where there is no initial N_2, the second term can be dropped. These equations enable one to estimate the evolution of activity as a function of time starting with an initial activity ratio $\lambda_2 N_2^0 / \lambda_1 N_1^0$.

If we consider this pair of radioactive isotopes for time scales greater than six half-lives of N_2, Equation (3b) can be simplified. Because each decay series starts with a long-lived parent, it is commonly the case that $\lambda_1 \ll \lambda_2$. In this case, after six half lives, $e^{-\lambda_2 t}$ approaches zero and can be removed from the equation. For time scales such that $6T_2 < t \ll T_1$ then

$$N_2 \approx \frac{\lambda_1}{\lambda_2 - \lambda_1} N_1^0 e^{-\lambda_1 t} \approx \frac{\lambda_1}{\lambda_2 - \lambda_1} N_1 \tag{4}$$

With the second simplification reflecting the fact that insignificant decay of N_1 will have occurred because $t \ll T_1$. Because $\lambda_1 \gg \lambda_2$, Equation (4) can be further simplified because $\lambda_2 - \lambda_1 \approx \lambda_2$ so that:

$$N_2 \approx \frac{\lambda_1}{\lambda_2} N_1 \tag{5}$$

This situation, when the activity of the higher atomic number nuclide, the "parent," is equal to the activity in the next step in the chain, the "daughter," is known as radioactive equilibrium (also referred to as secular equilibrium). Thus, secular equilibrium between a parent and a daughter implies an activity ratio of 1.

A useful analogy for understanding secular equilibrium is visualizing a decay chain as a series of pools of water (Fig. 2). These pools eventually lead to a continuously filling pool representing a stable isotope of lead (either ^{206}Pb, ^{207}Pb or ^{208}Pb). Over the timescale

Introduction to U-series Geochemistry

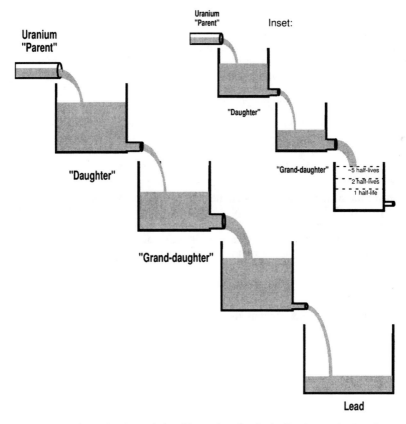

Figure 2. Analogy of U-series decay chain with a series of tanks feeding into each other. See text for description.

relevant to U-series disequilibria studies, the parent nuclides ^{238}U, ^{235}U, and ^{232}Th all have half-lives much longer than any of the other isotopes in the chain. Therefore, the flux of radioactive decay (the activity) coming from the top of the chain remains essentially constant over the time relevant to study of the short-lived members. Thus, we can assume the flux of water coming out of the top pipe in Figure 2 is constant. In this schematic model, each tank represents an individual nuclide and has a spout of size proportional to the decay constant of that nuclide. In steady state, the flux of water through each tank in the system will be constant when the hydraulic head within a tank increases to equalize the flux in and out of the tank. Thus, the flux of water through a set of tanks is analogous to the activity (in Bq/g or dpm/g) of all nuclides in the chain in secular equilibrium. In both the decay chain and water analogy, flow occurs only in one direction and the change in any pool's depth (or nuclide concentration) does not affect the pool above it. Because the activity of any nuclide is equal to λN, the amount of water in a given tank is proportional to the number of atoms of a nuclide at secular equilibrium. In other words, a tank with a small spout having a large volume of water is analogous to a long-lived nuclide that has a relatively high concentration of atoms in secular equilibrium. In contrast, a tank with a large spout (which results in small steady-state amount of water) is analogous to a short-lived nuclide that has a lower concentration of atoms in secular equilibrium.

2.2. Disequilibrium between U-series nuclides

Processes that fractionate nuclides within a chain produce parent-daughter disequilibrium; the return to equilibrium then allows quantification of time. Because of the prescribed decay behavior, U-series disequilibria can be used for geochronology or for examining the rates and time scales of any dynamic processes which induces fractionation. In many cases, the direction of disequilibrium (activity ratios above or below one) provides a powerful means of tracing specific processes.

Based on Equation (3), in the case of a system where there is an initial disequilibrium in the chain (namely: $\lambda_1 N_1 \neq \lambda_2 N_2$), it is generally stated that the system returns to secular equilibrium after ≈six half-lives of the daughter. The wide variety of parent-daughter pairs allows disequilibria to provide temporal constraints over a wide range in time scales (Fig. 3).

One of the behaviors of the system not easy to grasp is why the return to equilibrium is mostly controlled by the half-life of the daughter nuclide? This can be investigated by considering the ^{226}Ra/^{230}Th system (^{230}Th decays to form ^{226}Ra with a half-life of 1599 years). If fractionation by some process results in an activity ratio greater than 1 at time t = 0, the equation describing the return to equilibrium, as shown above, is:

$$^{226}\text{Ra} = \frac{\lambda_{\text{Th}}}{\lambda_{\text{Ra}} - \lambda_{\text{Th}}} {}^{230}\text{Th}_0 \left(e^{-\lambda_{\text{Th}} t} - e^{-\lambda_{\text{Ra}} t} \right) + {}^{226}\text{Ra}_0 e^{-\lambda_{\text{Ra}} t} \tag{6}$$

If we now take into account the fact that since $\lambda_{\text{Ra}} \gg \lambda_{\text{Th}}$, this equation can be rewritten using activity ratios as:

$$\left[\left(\frac{^{226}\text{Ra}}{^{230}\text{Th}} \right) - 1 \right] = \left[\left(\frac{^{226}\text{Ra}}{^{230}\text{Th}} \right)_i - 1 \right] e^{-\lambda_{\text{Ra}} t} \tag{7}$$

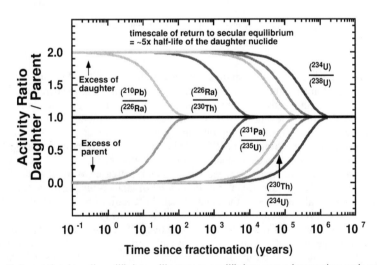

Figure 3. Parent daughter disequilibrium will return to equilibrium over a known time scale related to the half-life of the daughter nuclide. To return to within 5% of an activity ratio of 1 requires a time period equal to five times the half-life of the daughter nuclide. Because of the wide variety of half-lives within the U-decay-series, these systems can be used to constrain the time scales of processes from single years up to 1 Ma.

The quantity in brackets represents the excess of ^{226}Ra relative to ^{230}Th. It follows that after six half-lives of ^{226}Ra, the term $e^{-\lambda t}$ will be equal to $1/2^6 = 1/64$ which means that ^{226}Ra/^{230}Th activity ratios will be equal to 1 within approximately 1%. Now consider the case of a system where there is no initial ^{226}Ra. The same equation applies except that in this case $(^{226}$Ra/^{230}Th$)_i = 0$. Since there is no initial ^{226}Ra, how is it possible that the half-life of ^{226}Ra controls the return to secular equilibrium for this system? In fact, the system is also controlled by the half-life of ^{230}Th and the issue boils down to the concept of activity. In the case of a system with no initial ^{226}Ra, there are $\lambda_{Th} N_{Th} \Delta t$ atoms of ^{226}Ra produced during Δt. If we assume that ^{226}Ra is not decaying, it will take $\Delta t = 1/\lambda_{Ra}$ to accumulate enough ^{226}Ra atoms to reach the condition for secular equilibrium $N_{Ra} = \lambda_{Th} N_{Th}/\lambda_{Ra}$. Because ^{226}Ra is decaying, it effectively takes more time than $1/\lambda_{Ra}$ to return to secular equilibrium (see Fig. 4). This illustrates that the return to secular equilibrium is limited by the decay of ^{226}Ra.

2.3. Processes creating disequilibria between U-series nuclides

The previous section showed that if the decay chain remains undisturbed for a period of approximately 6 times the longest half-lived intermediate nuclide then the chain will be in a state of secular equilibrium (i.e., equal activities for all the nuclides). The key to the utility of the U-series is that several natural processes are capable of disrupting this state of equilibrium.

Two types of mechanisms need to be distinguished here. Firstly, each element has distinct chemical properties and thus, the U-series nuclides can become fractionated during processes that discriminate chemical behavior: phase change, partial melting, crystallization, partitioning, dissolution, adsorption, degassing, oxidation/reduction, complexation. For example, during crystallization of a mineral from a melt, the $(^{230}$Th/^{238}U$)$ activity ratio in the mineral will be:

$$\left(\frac{^{230}Th}{^{238}U}\right)_{mineral} = \frac{D_{Th}}{D_U} \left(\frac{^{230}Th}{^{238}U}\right)_{melt} \tag{8}$$

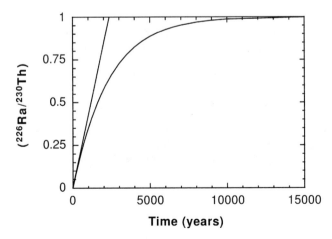

Figure 4. Return to secular equilibrium of ^{226}Ra/^{230}Th activity ratio with no initial ^{226}Ra. The return to secular equilibrium essentially depends on the half-life of the daughter isotope. If the decay of ^{226}Ra is "turned off," the return to secular equilibrium takes $1/\lambda_{Ra}$ years.

where D_{Th} and D_U are the mineral/melt partition coefficients. Provided that the partition coefficients are different, the ^{230}Th-^{238}U activity ratio in the mineral will be distinct from that of the melt. This process is generally called fractionation but for the U-series nuclides, the result is transitory disequilibrium.

Secondly, fractionation can also take place as a result of radioactive decay, especially in the low-temperature environment, and these effects are generally described as recoil effects. To illustrate the physics of recoil, we choose for example the decay of ^{238}U:

$$^{238}U \rightarrow {}^{234}Th + {}^4He + Q \tag{9}$$

The resulting particles (^{234}Th and ^{4}He) are charged and emitted with finite kinetic energy. In order to estimate this energy, it is necessary to undertake an energy balance for decay that is analogous to the energy balance for a chemical reaction. During decay, we assume that both total momentum and kinetic energy are conservative. If we also assume that the nucleus was initially at rest:

$$Q = \Delta Mc^2 = E_c^{Th} + E_c^\alpha \tag{10a}$$

$$M_\alpha V_\alpha = M_{Th} V_{Th} \tag{10b}$$

where E_c, M and V are the kinetic energy, the mass and the velocity of the daughter nuclide and the helium atom. From these equations, the kinetic energy of the daughter nuclide can be shown to be a small fraction of the total energy release Q:

$$E_c^{Th} = Q \frac{M_\alpha}{M_\alpha + M_{Th}} \tag{11}$$

The effect of recoil is three-fold: firstly, the recoil atom is displaced from the site where it was located. It can thus be ejected directly into an adjacent phase. The displacement distance is approximately 40 nm, depending on the substrate (Harvey 1962) and is known as the range. It can be estimated using the following equation:

$$\delta = \frac{\left(M_{Th} + M_a\right)\left(M_a E_c^{Th} K\right)\left(Z_{Th}^{2/3} + Z_a^{2/3}\right)^{1/2}}{\left(M_{Th} Z_{Th} Z_a \rho\right)} \tag{12}$$

where Z_{Th} is the atomic number of ^{234}Th, and Z_a and M_a denotes the atomic number and mass of the absorber, K is a constant (6.02) and ρ is the density. E is given here in keV while δ is in nanometers. Secondly, the site is damaged by the α-particle, which makes the daughter more prone to subsequent mobilization. Lastly, the atom is displaced from its original site and can also be more easily removed. It is difficult to determine which of these processes control the extent to which a daughter nuclide is more readily mobilized than its parent. Nevertheless, it is possible to estimate the fraction of daughter nuclides that would be directly ejected by recoil. This calculation was made by Fleischer and Raabe (1975) for plutonium particles and is described below. For a spherical particle with radius r and for a range δ, the fraction of volume that will be affected by recoil into another phase is:

$$f_v = \frac{r^3 - (r-\delta)^3}{r^3} \tag{13}$$

Out of this volume, only half of the particles will be ejected towards the rim of the grain and only a half of those will have trajectories which actually cross the grain boundary (Fig. 5). Thus, for a porous media with porosity ϕ and density ρ_s the number of daughter

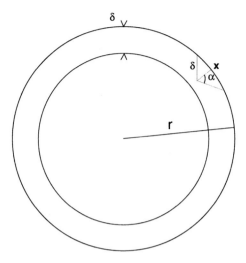

Figure 5. Ejection of daughter nuclide out of a grain due to recoil. Only a fraction of the nuclides located in a cone will be ejected for a given range δ. As shown on the diagram, at a distance x from the surface of the grain, the only nuclide to escape by recoil will be located in a cone with an angle α.

nuclides ejected per unit of time per unit of mass is:

$$R_N = \rho_s (1-\phi) \frac{r^3 - (r-\delta)^3}{4r^3} \lambda N \tag{14}$$

where λ and N are the decay constant and number of atoms per unit of mass of the parent nuclide.

Such alpha-recoil plays a fundamental role in fractionating the nuclides from one another in the low-temperature environment. During igneous processes, on the other hand, alpha recoil is probably not important in the generation of disequilibria (^{230}Th, ^{226}Ra, and ^{231}Pa). Beattie (1993) pointed out that the time scale of annealing of alpha decay damage at high temperatures was much shorter than the time scale of decay of these nuclides.

One of the important theoretical advances in the last decade was the development of models of disequilibria generation based on the dynamic process of a fluid moving relative to a solid. McKenzie (1985) was the first to show how radioactive disequilibrium can be created if U-series nuclides have different residence times within a melting column during two-phase flow. This model was developed to simulate mantle melting beneath mid-ocean ridges but in reality applies to any earth process where U-series nuclides move at different velocities due to exchange between the fluid and the solid. In this model, different elements will have different effective velocities based on how they partition between the moving fluid and a solid. This situation could be a magma exchanging with mantle minerals or could be groundwater exchanging with the surfaces of minerals

Consider a steady-state situation of a parent-daughter pair within a fluid flowing through, and chemically exchanging with, mineral grains in a column of length x. Assume the mineral grains and the fluid are homogenous and in chemical equilibrium with each other and the U-series nuclides in both phases are in secular equilibrium ($A_{parent} = A_{daughter}$). If a continuous stream of the fluid is fed into the column, elements that do not partition into the solid ($D^{solid/liquid} = 0$) will move at the velocity of the fluid whereas elements that partition into the solid, even slightly, will travel more slowly. Thus, if an element favors the fluid phase, it will have a shorter residence time in the column than elements which favor the solid phase. Once it has achieved steady state this system will have no effect on the concentration of non-radioactive elements in the fluid leaving the column and they will equal the inputs to the column. Nevertheless, this situation can have profound effects on short-lived nuclides within a decay chain, even once steady state has been achieved.

If the parent nuclide, for instance, has a greater preference for the solid phase than the daughter it will have a correspondingly greater residence time in the column. The daughter will, therefore, effectively "see" more parent than in a static situation where there is no differential movement of nuclides. For instance the total amount of the parent in the column at a given time will be $A_{parent} \times / w_{eff\text{-}parent}$ where $w_{eff\text{-}parent}$ is the effective velocity of the parent. In contrast, the number of daughter atoms in the column will be $A_{daughter} \times / w_{eff\text{-}daughter}$. If $w_{eff\text{-}daughter} > w_{eff\text{-}parent}$, then $A_{parent} \times / w_{eff\text{-}parent}$ will be greater than $A_{daughter} \times / w_{eff\text{-}daughter}$ and the daughter will be effectively supported by a greater amount of parent than that in secular equilibrium. These dynamical effects will result in greater U-series fractionation than expected in static systems.

3. CHEMISTRY AND GEOCHEMISTRY OF THE U-SERIES NUCLIDES

Knowledge of the chemical properties of the U-series nuclides is essential to any understanding of fractionation within the U-series chains. The nuclides of particular geochemical interest are ^{238}U, ^{234}U, ^{234}Th, ^{230}Th, ^{226}Ra, ^{231}Pa, ^{222}Rn and ^{210}Pb (Table 1). In this section, we review some basic chemical properties of these nuclides. Much of this material can also be found in Ivanovich and Harmon (1992).

Table 1. Chemical properties of the main U-series nuclides

	Z	Electronic Configuration	Oxidation States*	Geochemical Properties	Ionization Potential (kJ/mol)	Melting point (K)
U	92	$[Rn]5f^36d^17s^2$	0, 3, 4, **5**, **6**	Lithophile Incompatible Soluble (+6) Insoluble (+4)	584	1405.5
Pa	91	$[Rn]5f^26d^17s^2$	3, 4, **5**	Lithophile Incompatible Insoluble	568	2113
Th	90	$[Rn]6d^27s^2$	0, 3, **4**	Lithophile Incompatible Insoluble	587	2023
Ra	89	$[Rn]7s^2$	**2**	Lithophile Incompatible Sl. soluble	509.3	973
Rn	86	$[Rn]$	**0**, 2	Volatile Soluble	1037	202
Po	84	$[Xe]4f^{14}5d^{10}6s^26p^4$	-2, +2, **+4**, +6	Volatile	812	527
Pb	82	$[Xe]4f^{14}5d^{10}6s^26p^2$	0, **2**, 4	Chalcophile Incompatible M. Volatile Insoluble	715.5	600.65

*bold indicates preferred oxidation state in terrestrial material
Source material: Emsley (1989)

Most of the U-series nuclides are metals. Five of them belong to the actinide family corresponding to the filling of the internal orbitals while the orbitals 7s are filled. A sixth, Ra is an alkali earth and shares some chemical properties with other alkali earths, particularly the heavier ones (Sr and Ba), while a seventh, Rn, is a noble gas. The filling of the orbitals prescribes the possible oxidation states of these elements. Their preferred oxidation state is obtained when the electronic configuration is that of the closest rare gas (Rn).

An important chemical property of relevance to geochemistry is the ionic radii in different coordination. In general, Th has a larger ionic radius than U despite U having the larger atomic number; this is a phenomenon known as the actinide contraction and is similar to the well-known lanthanide contraction. This means that in general the heavier actinide (e.g., U) should be more easily be accommodated in minerals than the lighter ones (e.g., Th) at a given oxidation state. There are exceptions to this as explained by Blundy and Wood (2003).

The diffusion of U and Th within a solid is, in general, very slow due to their large size and charge (Van Orman et al. 1998). Even at mantle temperatures, it is expected that a solid will not fully equilibrate with the surrounding phases (fluid, melt or other solid phases) if solid diffusion controls the equilibration. As yet, there have been no direct determinations of diffusion coefficients for any other decay chain element.

From a geochemical viewpoint, U is an incompatible lithophile and refractory element. U exists in three distinct oxidation states in nature (Calas 1979) but the most common are U^{4+} ([Rn] $5f^2$) and U^{6+} ([Rn]). The most reduced form (metal) is never found in natural environments. At the surface of the earth, U is dominantly in the U^{6+} form. However, in a reducing environment, it will be in the U^{4+} state where it is insoluble and therefore generally far less mobile than U(VI). In the mantle, U is thought to occur in the U^{4+} state, except in the subduction zone environment where the oxygen fugacity is thought to be higher (as discussed by Turner et al. 2003).

In aqueous solutions, the chemistry of U and Th is highly dependent on their ability to form complexes with other ions in solution. In non-complexing acid media, they generally exists as M^{n+}. At higher oxidation states (e.g., U^{6+}), the cations will react with water, a reaction known as hydrolysis:

$$M^{n+} + xH_2O \rightarrow MO_x^{(n-2x)+} + 2xH^+ \tag{15}$$

U^{6+} in the aqueous solutions is thus present as the uranyl ion $(UO_2)^{2+}$. As pH increases, the form of these ions evolves to $M(OH)^{(n-1)+}$ or $MO_xOH^{(n-2x-1)+}$. In a non-complexing media, the limitation to solubility is generally the hydroxide form (e.g., $Th(OH)_4$ or $UO_2(OH)_2$). The mobility of U and Th is largely controlled by their ability to form complexes both with inorganic (F^-, Cl^-, PO_4^{3-}, CO_3^{2-}) and organic ligands. This has been described in detail for Th, U and Pa (Langmuir 1978; Langmuir and Herman 1980; Guillaumont et al. 1968). In natural environments, depending on the ratio of organic acids to inorganic ligands, the dominant species might be variable. For inorganic ligands, the strongest affinity is obtained with those ligands with the highest charge. The order of affinity for single and double-charge species is:

$$F^- > H_2PO_4^{2-} > SCN_2^- > NO_3^- > Cl^- \text{ for } M^{4+}$$

$$CO_3^{2-} > HPO_4^{2-} > SO_4^{2-}$$

In general, however, organic ligands such as humic acids have the strongest affinities.

Radium has a distinctive chemistry compared with the actinides since it is an alkali earth. Its large ionic radius (Table 1) makes it incompatible in most minerals although minerals which accommodate Ba generally take up large amounts of Ra (e.g., barite, phlogopite or celsiane, a Ba-rich feldspar). In general, Ra is only slightly soluble compared with Ca or Sr and its solubility is controlled by the concentrations of sulfate or carbonate ions. In many cases, its concentration in natural waters is controlled by recoil from the host rock (see above).

Protactinium is an element with complex behavior. It is part of the actinide family with atomic number (Z = 91) between those of U and Th. It generally occurs in the +5 oxidation state such that the +5 ion is slightly smaller than U or Th. The +4 oxidation state also exists but is generally thought to be metastable, at least in solutions. The +5 oxidation state of protactinium makes its properties in solutions similar in behavior to high field strength element such as Nb and Ta but its ionic radii is very different. This similarity of properties has mostly been noted in the case of aqueous solutions but their respective behavior in the mantle could be quite different because their ionic radii which controls mineral/melt partitioning are quite different. Pa is insoluble and forms hydroxide complexes or polymers similar to Th (see above). Its affinity for inorganic ligands follows the following order:

$$F^- > SO_4^{2-} > NO_3^- > Cl^- > ClO_4^-$$

Pa has strong affinities for organic complexing ligands and so it is expected that Pa might be more soluble in natural waters where such ligands are present compared with solutions containing only inorganic ligands. For example, it has been shown that the residence time of Pa in seawater is longer than that of Th (Henderson and Anderson 2003). At high pressure and temperature, the behavior of Pa is far less well known that for U and Th because of the difficulties in running experiments. Predictions of its partitioning behavior are given in Blundy and Wood (2003).

Radon is a noble gas and is therefore not readily ionized or chemically reactive. Its properties in terrestrial material will be controlled by its solubility in melt and fluid as well as its diffusion coefficients. Compared with the lighter noble gases, Rn diffuses more slowly and has a lower solubility in water. It will also more readily adsorb onto surface that the lighter rare gases. It can, however be lost by degassing in magmatic systems (Condomines et al. 2003). More information about the behavior of Rn can be found in Ivanovich and Harmon (1992).

Lead is a chalcophile element that is also slightly volatile at high temperature. During mantle melting, in the absence of sulfide phase, it tends to partition into the melt relative to the solid. The solubility of lead in aqueous fluids is very low. However, in the presence of complexing agent such as chloride, sulfide, carbonate or organic ligands, its solubility is enhanced. This is particularly true at higher temperature where lead can be transported in fluids and latter precipitate as galena (PbS). With respect to the short-lived ^{210}Pb, The short-lived lead nuclide, ^{210}Pb, is often found in secular equilibrium with ^{226}Ra. Exceptions to this are environments where ^{222}Rn is lost from the system by degassing (e.g., see Condomines et al. 2003), or aqueous systems where the insoluble nature of ^{210}Pb leads to its preferential removal. The speciation of Pb in natural waters is rather complex and heavily depends on the availability of organic complexing agents for which Pb has the highest affinity. In the oceans, Pb has a very short residence (30-150 yrs) and is rapidly scavenged by particles.

4. DETERMINATION OF THE HALF-LIVES OF U-SERIES NUCLIDES

Application of the U-series theory outlined above relies on accurate knowledge of the half-lives of the various nuclides, especially when U-series based chronologies are compared with other chronologies. Considerable analytical effort has gone into careful measurement of the half-lives and most are now known to better than 1%. Recommended half-lives for all U-series nuclides are provided in Table 2. These half-lives have been assessed using one of five techniques as follows:

4.1. Methods for measurement of half life

Decay of the nuclide itself. The conceptually simplest approach is to take a known quantity of the nuclide of interest, P, and repeatedly measure it over a sufficiently long period. The observed decrease in activity with time provides the half-life to an acceptable precision and it was this technique that was originally used to establish the concept of half-lives (Rutherford 1900). Most early attempts to assess half lives, such as that for ^{234}Th depicted on the front cover of this volume, followed this method (Rutherford and Soddy 1903). This approach may use measurement of either the activity of P, or the number of atoms of P, although the former is more commonly used. Care must be taken that the nuclide is sufficiently pure so that, for instance, no parent of P is admixed allowing continued production of P during the experiment. The technique is obviously limited to those nuclides with sufficiently short half-lives that decay can readily be measured in a realistic timeframe. In practice, the longest-lived isotopes which can be assessed in this way have half-lives of a few decades (e.g., ^{210}Pb; Merritt et al. 1957).

Ingrowth of a daughter nuclide. For longer-lived nuclides where reduction of P itself is not observable, an alternative is to start with a quantity of P which is initially entirely clean of its daughter, D, and to observe the ingrowth of D with time. If D is stable, calculation of the half-life is then straightforward. A stable daughter is not normally the case for U-series nuclides; however α-decay also produces ^{4}He which is stable. Ingrowth of ^{4}He has been used, for instance, to assess the half-life ^{226}Ra (Kohman et al. 1949), although care must be taken to allow for additional ^{4}He production from the decay of ^{222}Rn and subsequent daughters when these start to grow in from ^{226}Ra. Where D is not stable, its ingrowth will be dependent on the product of the mean lives of P and D. If the half-life of D is well known, the ingrowth of D can therefore still be used to assess the half-life of P. This approach was used for most early attempts to assess the half-life of ^{230}Th, via the ingrowth of ^{226}Ra.

Measurement of specific activity. The half-life of a nuclide can be readily calculated if both the number of atoms and their rate of decay can be measured, i.e., if the activity A and the number of atoms of P can be measured, then λ is known from $A = \lambda P$. As instrumentation for both atom counting and decay counting has improved in recent decades, this approach has become the dominant method of assessing half-lives. Potential problems with this technique include the accurate and precise calibration of decay-counter efficiency; and ensuring sufficient purity of the nuclide of interest. This technique provides the presently used half-lives for many nuclides, including those for the parents of the three decay chains, ^{238}U, ^{235}U (Jaffey et al. 1971), and ^{232}Th.

Calorimetry. Radioactive decay produces heat and the rate of heat production can be used to calculate half-life. If the heat production from a known quantity of a pure parent, P, is measured by calorimetry, and the energy released by each decay is also known, the half-life can be calculated in a manner similar to that of the specific activity approach. Calorimetry has been widely used to assess half-lives and works particularly well for pure α-emitters (Attree et al. 1962). As with the specific activity approach, calibration of the measurement technique and purity of the nuclide are the two biggest problems to overcome. Calorimetry provides the best estimates of the half lives of several U-series nuclides including ^{231}Pa, ^{226}Ra, ^{227}Ac, and ^{210}Po (Holden 1990).

Table 2. Half-lives for the U- and Th- decay series nuclides, with decay modes.

^{238}U decay chain

Nuclide	Half-life	Ref.
^{238}U α	4.4683 ±0.0048 Byrs	1
^{234}Th β	24.1 days	2
^{234}Pa β	6.69 hours	2
^{234}U α	245,250 ± 490 yrs	3
^{230}Th α	75,690 ± 230 yrs	3
^{226}Ra α	1599 ± 4 yrs	4
^{222}Rn α	3.823 ± 0.004 days	4
^{218}Po α,β	3.04 min	2
^{218}At α,β	1.6 sec	2
^{218}Rn α	35 msec	2
^{214}Pb β	26.9 min	2
^{214}Bi α,β	19.7 min	2
^{214}Po α	0.1637 msec	2
^{210}Tl β	1.3 min	2
^{210}Pb α,β	22.6 ± 0.1 yrs	4
^{210}Bi α,β	5.01 days	2
^{210}Po α	138.4 ± 0.1 days	4
^{206}Hg β	8.2 min	2
^{206}Tl β	4.2 min	2
^{206}Pb	stable	

^{235}U decay chain

Nuclide	Half-life	Ref.
^{235}U α	0.70381 ± 0.00096 Byrs	1
^{231}Th β	1.063 days	2
^{231}Pa α	32,760 ± 220 yrs	5
^{227}Ac α,β	21.77 ± 0.02 yrs	4
^{227}Th α	18.72 days	2
^{223}Fr α,β	22 min.	2
^{223}Ra α	11.435 days	2
^{219}At α,β	50 sec	2
^{219}Rn α	3.96 sec	2
^{215}Bi β	7.7 min	2
^{215}Po α,β	1.78 msec	2
^{215}At α	0.1 msec	2
^{211}Pb β	36.1 min	2
^{211}Bi α,β	2.14 min	2
^{211}Po α	0.516 sec	2
^{207}Tl β	4.77 min	2
^{207}Pb	stable	

^{232}Th decay chain

Nuclide	Half-life	Ref.
^{232}Th α	14.0100 Byrs	4
^{228}Ra β	5.75 ± 0.03 yrs	4
^{228}Ac β	6.15 hours	2
^{228}Th α	1.912 ± 0.002 yrs	4
^{224}Ra α	3.66 days	2
^{220}Rn α	55.6 sec	2
^{216}Po α	0.145 sec	2
^{212}Pb β	10.64 hours	2
^{212}Bi α,β	1.009 hours	2
^{212}Po β	0.298 μsec	2
^{208}Tl β	3.053 min	2
^{208}Pb	stable	

*References: 1. Jaffey et al. (1971); 2. Lide (1998); 3. Cheng et al. (2000); 4. Holden (1990); 5. Robert et al. (1969)

Introduction to U-series Geochemistry 15

Secular equilibrium materials. For materials that have remained a closed system for sufficient time that secular equilibrium has been achieved, the half-lives of nuclides within the decay chain can be calculated from the relationship $\lambda_P P = \lambda_D D$. If the atom ratio P/D is measured, and one of the decay constants is well known, then the other can be readily calculated. Limitations on this approach are the ability to measure the atom ratios to sufficient precision, and finding samples that have remained closed systems for a sufficient length of time. This approach has been used to derive the present recommended half lives for ^{230}Th and ^{234}U (Cheng et al. 2000; Ludwig et al. 1992).

4.2 Recommended half-lives for key nuclides

In general, the half lives of the U-series nuclides are known with sufficient precision and accuracy that the geological uncertainty in the behavior of the nuclides far outweighs any uncertainty due to poorly known half-lives. The one area where this is not entirely true is in some applications of absolute chronology where analytical precisions are now sufficiently high (typically at the permil level), and the geochemical system sufficiently well understood, that half-life uncertainty can become a significant fraction of final age uncertainty. The key nuclides presently used for such chronology are ^{234}U, ^{230}Th, and ^{231}Pa and a brief word follows about the recommended half lives for each of these three nuclides.

^{234}U. The history of ^{234}U half-life measurements has recently been summarized by Cheng et al. (2000). For many years a value of 244,500 ± 1000 kyr was used, based on the close agreement between two assessments: 244,600 ± 730 yrs (De Bievre et al. 1971) and 244,400 ± 1200 yrs (Lounsbury and Durham 1971). The Commission on Radiochemistry and Nuclear Techniques assessed the half-lives of several nuclides and, for ^{234}U, reassessed the data of both of these studies together with five other studies and suggested a half-life of 245,500 ± 1,200 yrs (Holden 1989). This differs by 4‰ from the commonly used value, a significant difference given present analytical precisions of ≈1‰. Ludwig et al. (1992) used a secular equilibrium approach using a single uraninite material and measured a half-life of 245,290 ± 140 yrs. Using a similar approach, but applying it to a wide variety of secular equilibrium materials, Cheng et al. (2000) derived a half-life of 245,250 ± 490 yrs, in good agreement with the values of Ludwig et al. (1992) and Holden (1989) This value is therefore the present best estimate of the ^{234}U half live (Table 2). Given the relative ease of measurement of ^{234}U/^{238}U atom ratios with modern technology, other laboratories have already repeated this secular equilibrium approach (e.g., Bernal et al. 2002) and further refinement of the half-life is possible. Nevertheless, at the time of publication of this volume, the broad agreement between various laboratories suggests that the true value lies within the uncertainty quoted by Cheng et al. (2000)

^{230}Th. The early history of ^{230}Th half-life measurements was summarized by Meadows et al. (1980). These measurements relied on a variety of techniques including daughter ingrowth, calorimetry, and specific activity. The study of Meadows et al. itself used specific activity to derive a value of 75,381 ± 590 yrs and a very similar value was recommended by the Commission on Radiochemistry and Nuclear Techniques (Holden, 1989). The uncertainty on this value of some 8‰ is significant given present analytical precisions of ≈1‰. This has led to a recent re-appraisal of the ^{230}Th half-life using a variety of secular equilibrium materials (Cheng et al. 2000). The Cheng et al value of 75,690 ± 230 yrs is within error of the Meadows et al. (1980) value but with a two-fold improvement in uncertainty. This value is therefore recommended (Table 2).

^{231}Pa. The most recent measurement of the ^{231}Pa half-life established a value of 32,760 ± 220 yrs based on calorimetry (Robert et al. 1969). This value has been accepted as the commonly used value by most laboratories (e.g., Picket et al. 1994; Edwards et al. 1997). The Commission on Radiochemistry and Nuclear Techniques suggest a value of

32,500 ± 200 yrs based on a straightforward average of five studies. The difference between these two values, and indeed the uncertainty on either value, is comparable to the best presently achievable measurement precision and so the choice of half-life is not particularly significant. With future improvements of measurement precision, the uncertainty on the ^{231}Pa may become a dominant source of error and a reassessment will be required. Until that time, we recommend a value of 32,760 ± 220 (Robert et al. 1969) to enable straightforward comparison with published data.

5. OUTLINE OF THE VOLUME

The chapters of this volume provide detailed reviews of the current understanding in a range of U-series isotope applications with emphasis on advances made since 1992. Although each chapter has been written to stand alone, cross referencing to other chapters is included where appropriate.

The second chapter (Goldstein and Stirling) gives an overview of developments in analytical techniques over the last decade including, thermal ionization and MC-ICP mass spectrometry. This chapter shows how the development of these new techniques have improved the sensitivity and accuracy of U-series measurements and how new analytical schemes have been developed for measuring Ra and Pa by mass spectrometry. These breakthroughs have greatly aided many of the exciting new advances discussed in subsequent chapters.

In the third chapter, Blundy and Wood discuss the advances made in our understanding of U-series partitioning between minerals and their melts. In the second edition of "Uranium-series disequilibria" (Ivanovich and Harmon 1992), Gascoyne wrote "At present, little information is available on the partitioning of uranium and thorium between crystals and silicate melts." This picture has radically changed over the past ten years. This has been made possible by both the introduction of ion-microprobe measurements of partition coefficients and by the development of the lattice-strain model allowing prediction of mineral/melt partition coefficients for a wide range of U-series nuclides in the major igneous rock forming minerals.

The next four chapters are concerned with the application of U-series isotopes to magmatic processes within the Earth. Chapter four reviews recent developments in the dating of young volcanic rocks and in the rapidly expanding study of magma residence times and magma chamber processes. These authors emphasize some of the difficulties in interpreting mineral ages but also how these may often provide different, but complimentary, information to groundmass ages. New methods for estimating the rates of degassing have also emerged providing the potential for closer links to volcanology and hazard prediction. Chapters five to seven deal with the study of magmatism at mid-ocean ridges, above hotspots and at convergent margins respectively. Over the last decade many investigations of melting processes have been made and our view of this field has completely changed due to both a greater number of high-quality analyses and to improved theoretical modeling. Thus, it is now possible to build a much more detailed picture of how melt generation, migration and modification occurs within the Earth—aspects fundamental to models for Earth differentiation and dynamics. Central to these advances is the notion that knowledge of time scale constrains the physical mechanism of a process.

From Chapter 8 onwards, the focus of the volume shifts to lower temperature geochemistry, starting with a chapter on the behavior of the U-series nuclides in groundwaters. This subject merited a chapter on its own in the Ivanovich and Harmon (1992) volume and its continued interest has led to significant advances in understanding and in application over the last decade. The next two chapters deal with the dating of carbonates: in ocean and lakes in Chapter 9; and in caves in Chapter 10. The U/Th chronometer is one of the few tools that can be applied to date important climate and sea-level events in the Quaternary and this area has seen some of the more high-profile applications of the U-series in the past decade.

Chapters 11 and 12 focus on the oceans. The first of these describes the use of U-series nuclides in the modern ocean, where they have been particularly useful during the last decade to study the downward flux of carbon. The second ocean chapter looks at the paleoceanographic uses of U-series nuclides, which include assessment of sedimentation rates, ocean circulation rates, and paleoproductivity. Both of these ocean chapters demonstrate that knowledge of the behavior of the U-series is now sufficiently well developed that their measurement provides useful quantitative information about much more than just the geochemistry of these elements.

The emphasis shifts to the continental domain in Chapter 13 which addresses the fractionation of the U-series nuclides during continental weathering and subsequent riverine transport. This subject area also merited a chapter in the volume by Ivanovich and Harmon (1992) and, to some extent, the community has been slow to capitalize on the understanding of terrestrial fractionation processes during the last decade. Recent studies of soil and riverine U-series chemistry look set to change this omission, however, and these are detailed in this chapter. Having seen how the U-series nuclides are mobilized and transported in surface waters, Chapter 14 summarizes their behavior during transfer to the oceans at estuaries. This is a complicated field, with diverse aspects of the river chemistry and environment controlling the precise behavior. But our developing understanding of the possible range of behavior is providing ever more accurate oceanic budgets for the U-series elements.

Chapter 15 brings the emphasis to the human scale as it describes the use of U-series chronometers to date the history of human evolution. U/Th dating provides one of the few chronometers for a crucial period of such evolution which saw the emergence of modern humans, the extinction of both Neanderthals and *H. erectus*, and a great deal of environmental change. Recently developed tools, described in this chapter, look set to put the time scales for such change onto a much more secure footing.

Finally, Chapter 16 provides information about the handling of U-series data, with a particular focus on the appropriate propagation of errors. Such error propagation can be complex, especially in the multi-dimensional space required for ^{238}U-^{234}U-^{230}Th-^{232}Th chronology. All too often, short cuts are taken during data analysis which are not statistically justified and this chapter sets out some more appropriate ways of handling U-series data.

REFERENCES

Attree RW, Cabell MJ, Cushing RL, Pieroni JJ (1962) A calorimetric determination of the half-life of thorium-230 and a consequent revision to its neutron capture cross section. Can J Phys 40:194-201

Bateman H (1910) Solution of a system of differential equations occurring in the theory of radioactive transformations. Proc Cambridge Phil Soc 15:423-427

Beattie PD (1993) The generation of uranium series disequilibria by partial melting of spinel peridotite; constraints from partitioning studies. Earth Planet Sci Lett 117:379-391

Bernal JP, McCulloch MT, Mortimer GE, Esat T (2002) Strategies for the determination of the isotopic composition of natural Uranium. Geochim Cosmochim Acta 66(S1):A72

Blundy J, Wood B (2003) Mineral-melt partitioning of uranium, thorium and their daughters. Rev Mineral Geochem 52:59-123

Calas G (1979) Etude expérimentale du comportement de l'uranium dans les magmas: états d'oxydation et coordinance. Geochim Cosmochim Acta 43:1521-1531

Chabaux F, Ben Othman D, Birck J-L (1994) A new Ra-Ba chromatographic separation and its application to Ra mass-spectrometric measurements in volcanic rocks. Chem Geol 114:191-197

Cheng H, Edwards RL, Hoff J, Gallup CD, Richards DA, Asmerom Y (2000) The half lives of uranium-234 and thorium-230. Chem Geol 169:17-33

Condomines M, Gauthier P-J, Sigmarsson O (2003) Timescales of magma chamber processes and dating of young volcanic rocks. Rev Mineral Geochem 52:125-174

De Bievre P, Lauer KF, Le Duigon Y, Moret H, Muschenborn G, Spaepen J, Spernol A, Vaninbroukx R, Verdingh V (1971) The half life of 234-U. In Proc. Int. Conf. Chem. Nucl. Data, Measurement and applications (ed. M. L. Hurrell), pp. 221-225. Canterbury. Inst. Civil Engineers.

Edwards RL, Chen JH, Wasserburg GJ (1986) ^{238}U-^{234}U-^{230}Th-^{232}Th systematics and the precise measurement of time over the past 500,000 years. Earth Planet Sci Lett 81:175-192.

Edwards RL, Cheng H, Murrell MT, Goldstein SJ (1997) Protactinium-231 dating of carbonates by thermal ionization mass spectrometry: implications for Quaternary climate change. Science 276:782-785

Emsley J (1989) The Elements. Clarendon Press, Oxford

Faure G (1986) Principles of Isotope Geology, Second Edition. John Wiley and Sons, New York

Fleischer RL, Raabe OG (1975) Recoiling alpha-emitting nuclei. Mechanisms for uranium-series disequilibrium. Geochim Cosmochim Acta 42:973-978

Goldstein SJ, Murrell MT, Williams RW (1993) ^{231}Pa and ^{230}Th chronology of mid-ocean ridge basalts. Earth Planet Sci Lett 115:151-159

Goldstein SJ, Murrell, MT, Williams RW (1993) ^{231}Pa and ^{230}Th chronology of mid-ocean ridge basalts. Earth Planet Sci Lett 115:151-159

Goldstein SJ, Stirling CH (2003) Techniques for measuring uranium-series nuclides: 1992-2002. Rev Mineral Geochem 52:23-57

Guillaumont R, Bouissières G, Muxart R (1968) Chimie du Protactinium. I. Solutions aqueuses de protactinium penta- et tétravalent. Actinides Rev 1:135-163

Harvey BG (1962) Introduction to Nuclear Physics and Chemistry. Prentice Hall Inc, New Jersey

Henderson GM, Anderson RF (2003) The U-series toolbox for paleoceanography. Rev Mineral Geochem 52:493-531

Holden NE (1989) Total and spontaneous fission half-lives for uranium, plutonium, americium and curium nuclides. Pure Appl Chem 61(8):1483-1504

Holden NE (1990) Total half-lives for selected nuclides. Pure Appl Chem 62(5):941-958

Ivanovich M (1992) The phenomenon of radioactivity. In: Uranium-series Disequilibrium: Applications to Earth, Marine, and Environmental Sciences. Ivanovich M, Harmon RS (eds) Clarendon Press, Oxford, p 1-33

Ivanovich M, Harmon RS (1992) Uranium-series Disequilibrium: Applications to Earth, Marine, and Environmental Sciences. Clarendon Press, Oxford

Jaffey AH, Flynn KF, Glendenin LE, Bentley WC, Essling AM (1971) Precision measurement of half-lives and specific acitivities of ^{235}U and ^{238}U. Phys Rev C4:1889

Kohman TP, Ames DP, Sedlet J (1949) The transuranium elements. National Nuclear Energy Series IV(14B):1675

Langmuir D (1978) Uranium-solution-mineral equilibria at low temperatures with applications to sedimentary ore deposits. Geochim Cosmochim Acta 42:547-569.

Langmuir D, Herman JS (1980) The mobility of thorium in natural waters at low temperatures. Geochim Cosmochim Acta 44:1753-1766

Le Roux LJ, Glendenin LE (1963) Half-life of ^{232}Th. Proceedings of the National Meeting on Nuclear Energy, Pretoria, South Africa, 83-94

Lounsbury M, Durham RW (1971) The alpha half-life of ^{234}U. Proceedings of the International Conference of Chemical Nuclear Data Measurements and Applications

Ludwig KR, Simmons KR, Szabo BJ, Winograd IJ, Landwehr JM, Riggs AC, Hoffman RJ (1992) Mass-spectrometric ^{230}Th-^{234}U-^{238}U dating of the Devils Hole calcite vein. Science 258:284-287

Meadows JW, Armani RJ, Callis EL, Essling AM (1980) Half-life of ^{230}Th. Phys Rev C 22(2):750-754

McKenzie D (1985) ^{230}Th-^{238}U disequilibrium and the melting processes beneath ridge. Earth Planet Sci Lett 72:149-157

Merritt WR, Champion PJ, Hawkings RC (1957) The half-life of ^{210}Pb. Can J Phys 35:16

Pickett DA, Murrell MT, Williams R.W (1994) Determination of femtogram quantities of protactinium in geological samples by thermal ionization mass spectrometry. Anal Chem 66:1044-1049

Robert J, Miranda CF, Muxart R (1969) Mesure de la periode du protactinium-231 par microcalorimetrie. Radiochim Acta 11:104-108

Rutherford E (1900) A radioactive substance emitted from thorium compounds. Philosoph Mag 49:1-14

Rutherford E, Soddy F (1903) The radioactivity of uranium. Philosoph Mag 5:441-445

Turner S, Bourdon B, Gill J (2003) Insights into magma genesis at convergent margins from U-series isotopes. Rev Mineral Geochem 52:255-315

Van Orman JA, Grove TL, Shimizu N (1998) Uranium and thorium diffusion in diopside. Earth Planet Sci Lett 160:505-519

APPENDIX:
GENERAL SOLUTIONS OF U-SERIES DECAY EQUATION USING LAPLACE TRANSFORMS

The solution to the general decay equations is often given in textbooks (e.g., Faure 1986). However, this solution is given for initial abundances of the daughter nuclides that are equal to zero. In the most general cases, the initial abundances of the daughter nuclides are not equal to zero. For example, in many geological examples, we make the assumptions that the decay chain is in secular equilibrium. The solutions of these equations can also be used to solve simple box models of U-series nuclides where first order kinetics are assumed.

The Laplace transform of a function f is defined as:

$$\tilde{f}(s) = \int_0^\infty f(t)e^{-st}dt \tag{A1}$$

By applying the Laplace transform to the U-series decay equation, one obtains simple linear equations that can be solved for the Laplace transforms of N_i (the number of nuclei i in the system). By inverting the Laplace transforms using tables, the time-dependent solutions are directly obtained. The Laplace transform for Equation (1) is:

$$s\tilde{N}_1 - \tilde{N}_1^0 = -\lambda_1 \tilde{N}_1 \tag{A2}$$

For Equation (2), the Laplace transform is:

$$s\tilde{N}_i - \tilde{N}_i^0 = -\lambda_i \tilde{N}_i + \lambda_{i-1}\tilde{N}_{i-1} \tag{A3}$$

We now give a solution for the first four nuclides in the chain corresponding for example to ^{238}U-^{234}U-^{230}Th-^{226}Ra which has relevance for application in the study of magmatic processes or weathering.

$$\tilde{N}_1 = \frac{N_1^0}{s+\lambda_1} \tag{A4a}$$

$$\tilde{N}_2 = \frac{N_2^0}{s+\lambda_2} + \frac{\lambda_1 N_1^0}{(s+\lambda_1)(s+\lambda_2)} \tag{A4b}$$

$$\tilde{N}_3 = \frac{N_3^0}{s+\lambda_3} + \frac{\lambda_2 N_2^0}{(s+\lambda_2)(s+\lambda_3)} + \frac{\lambda_1 \lambda_2 N_1^0}{(s+\lambda_1)(s+\lambda_2)(s+\lambda_3)} \tag{A4c}$$

$$\tilde{N}_4 = \frac{N_4^0}{s+\lambda_4} + \frac{\lambda_3 N_3^0}{(s+\lambda_3)(s+\lambda_4)} + \frac{\lambda_2 \lambda_3 N_2^0}{(s+\lambda_2)(s+\lambda_3)(s+\lambda_4)} + \frac{\lambda_1 \lambda_2 \lambda_3 N_1^0}{(s+\lambda_1)(s+\lambda_2)(s+\lambda_3)(s+\lambda_4)} \tag{A4d}$$

Each of these equations can then be inverted to time-space once the fractions are decomposed as follows:

$$\frac{1}{(s+\lambda_1)(s+\lambda_2)\ldots(s+\lambda_n)} = \sum_{i=1}^{n} \frac{a_i}{s+\lambda_i} \tag{A5}$$

where $a_i = \prod_{j \neq i} \dfrac{1}{\lambda_j - \lambda_i}$

By remembering that the inverse of the function $1/(s+\lambda_i)$ is $e^{-\lambda_i t}$, the general solutions can be written in a rather compact form as:

$$N_n(t) = N_n^0 e^{-\lambda_n t} + \sum_{i=1}^{n-1}\left(\left(\prod_{j=i}^{n-1}\lambda_j\right) N_i^0 \sum_{k=i}^{n} a_{ki} e^{-\lambda_k t}\right) \tag{A6}$$

where N_i^0 is the initial number of atoms of nuclide i and $a_{ki} = \prod_{j \neq k, j=i} \dfrac{1}{\lambda_j - \lambda_k}$.

2   Techniques for Measuring Uranium-series Nuclides: 1992-2002

Steven J. Goldstein
Isotope and Nuclear Chemistry Group
Los Alamos National Laboratory MS K484
Los Alamos, New Mexico, USA 87545
sgoldstein@lanl.gov

Claudine H. Stirling
Institute for Isotope Geology and Mineral Resources
Department of Earth Sciences
ETH Zentrum, Sonneggstrasse 5, CH-8092, Zürich, Switzerland
stirling@erdw.ethz.ch

1. INTRODUCTION

Advances in geochemistry and geochronology are often closely linked to development of new technologies for improved measurement of elemental and isotopic abundance. At the beginning of the past decade, thermal ionization mass spectrometric (TIMS) methods were just beginning to be applied for long-lived uranium-series nuclide measurement (Edwards et al. 1987; Goldstein et al. 1989; Bard et al. 1990), with considerable advances in measurement speed, precision, and sensitivity over decay-counting methods. This opened up a vast number of applications in uranium-series geochronology and geochemistry of young sediments, volcanic rocks, and aqueous systems. Over the past decade there have continued to be advances in thermal ionization techniques, and the advent of alternative mass spectrometric methods, particularly multi-collector inductively coupled plasma mass spectrometry (MC-ICPMS), has continued to improve the quality of uranium-series studies.

So the past decade has been a particularly dynamic time for not only development of mass spectrometric techniques, but initiation of other methods related to long-lived uranium-series nuclide measurement. In the area of sample preparation, further development of microwave digestion methods had led to advances in speed and cost of analysis. In chemical separations, development of extraction chromatographic resins for isolating specific elements have simplified many separation problems and consequently improved analytical characteristics including sensitivity, speed of analysis, waste generation, and cost. With regard to instrumental analysis, advances in both decay-counting and mass spectrometry instrumentation have improved either measurement sensitivity or precision, speed of analysis, or analytical cost. One could argue that instrumental developments will continue to drive scientific breakthroughs in the application of uranium-series nuclides as tracers and chronometers in the earth and other sciences.

In this chapter we discuss improvements documented in the literature over the past decade in these areas and others. Chemical procedures, decay-counting spectroscopy, and mass spectrometric techniques published prior to 1992 were previously discussed by Lally (1992), Ivanovich and Murray (1992), and Chen et al. (1992). Because ICPMS methods were not discussed in preceding reviews and have become more commonly used in the past decade, we also include some theoretical discussion of ICPMS techniques and their variants. We also primarily focus our discussion of analytical developments on the longer-lived isotopes of uranium, thorium, protactinium, and radium in the uranium and thorium decay series, as these have been more widely applied in geochemistry and geochronology.

2. SAMPLE PREPARATION

2.1. Microwave digestion of solids

Sample digestion/dissolution is generally required for all forms of uranium-series analysis excluding gamma spectrometry, and for all sample matrices excluding the dissolved components of aqueous solutions. Development of microwave digestion techniques for solids predates the 1990's, and numerous reviews have documented use of these systems for sample digestion (e.g., Kuss 1992; Lamble and Hill 1998). Prior to the past decade, most of these digestion systems were closed system, in which both pressure and temperature increases in the vessel during the microwave process were used to aid the digestion process. In the past decade, systems with more precise temperature and pressure control have been developed to dissolve samples under more specifically predetermined conditions. These methods have been applied to measurement of elemental U and Th in geologic standards (e.g., Totland et al. 1992; Sen Gupta and Bertrand 1995), and excellent agreement has been obtained with consensus values for these elements using open vessel-hot plate or fusion digestions.

In addition, open system microwave processes where pressure is maintained at atmospheric have also been developed. These systems provide additional options for sample digestion and avoid safety issues with pressurized closed-system digestion. Along with open system methods, automated techniques for addition of acid solutions to reaction vessels have also been implemented, providing options for on-line elemental or isotopic analysis. The size range of samples to be digested has also been extended to 1 gram or greater through use of larger sample vessels in the open and closed system configurations, which also provides more suitability for uranium-series analysis. Although microwave methods have been mainly applied to determination of elemental concentrations, these techniques may provide distinct advantages for uranium-series isotopic analysis in terms of reduced acid requirements and improved speed of analysis. In addition, microwave energy may also promote more complete sample-tracer equilibration in isotope dilution analysis than would be achieved using only conventional digestion methods.

2.2. Tracer addition and tracer/sample equilibration

In most alpha and mass spectrometric methods for which sample preparation is extensive and chemical recoveries can vary considerably from sample to sample, precise elemental concentrations are determined by isotope dilution methods (e.g., Faure 1977). This method is based on the determination of the isotopic composition of an element in a mixture of a known quantity of a tracer with an unknown quantity of the normal element. The tracer is a solution containing a known concentration of a particular element or elements for which isotopic composition has been changed by enrichment of one or more of its isotopes.

In these situations, addition of a tracer of unique isotopic composition is required, and the nature of the tracers added depends on the measurement technique. For example, short-lived ^{232}U and ^{228}Th (with respective half-lives of 70 and 1.9 years) are commonly used as a tracer for alpha spectrometric analysis of U and Th, whereas longer-lived ^{236}U, ^{233}U and ^{229}Th (with half-lives of 23 million, 160 thousand and 7900 years, respectively) are commonly used for mass spectrometric isotope dilution analysis. For thorium and uranium, measurement of isotopic composition (i.e., ^{230}Th/^{232}Th and ^{234}U/^{238}U) can be combined with measurement of abundance, and in these cases the tracer must be of high-purity and appropriate corrections must be made for the low levels of natural U and Th isotopes that are inevitably present. For isotope dilution measurements only, it is also advantageous to have high-purity tracers to minimize errors associated with measurement of the isotopic composition of the tracer. When a mixed spike (e.g., Wasserburg et al. 1981) is used, such as ^{233}U-^{229}Th, the relative abundances of U-series isotopes can be determined more precisely, leading to improvements in dating precision. Normally, tracers are standardized by mixing them with either a gravimetric standard or a well-characterized secular equilibrium standard for which U-series isotope abundances are well known (e.g., Ludwig et al. 1992). Analysis of secular equilibrium rock standards such as Table Mountain Latite or TML (Williams et al. 1992) and Harwell Uraninite or HU-1 (e.g., Ludwig et al. 1992), ideally matching the matrix to be analysed, also allow one to evaluate U-series measurement accuracy.

Methods for preparing U-series tracer isotopes vary according to element, but generally involve nuclear reactions to provide the radioactive tracers of interest. Preparation of the tracer is particularly an issue for protactinium, since the only option for mass spectrometric measurement (^{233}Pa) has a short half-life (27 d) and must be replenished frequently (Pickett et al. 1994; Bourdon et al. 1999). ^{233}Pa can be prepared by milking a solution of ^{237}Np or by neutron activation of ^{232}Th. Both methods are somewhat limiting, as the ^{237}Np method involves handling of considerable alpha activity, and the ^{232}Th method requires frequent access to a nuclear reactor. For radium, enriched ^{228}Ra solutions can be prepared by milking a solution of ^{232}Th (Volpe et al. 1991; Cohen and O'Nions 1991). ^{229}Th is commonly prepared by milking a ^{233}U supply, whereas ^{233}U and ^{236}U can be prepared by neutron and alpha activation of ^{232}Th, respectively. These nuclear reactions are commonly followed by chemical separations to purify the element of interest.

Regardless of the tracers added, proper tracer-sample equilibration must be obtained for accurate isotope dilution analysis. For chemically unreactive elements such as Th and Pa, it is generally desirable to ensure tracer-sample equilibration with acids that have a high boiling point such as nitric and perchloric acids, for which tracer and sample are converted to the same chemical form upon increasing temperature and dry-down. Some laboratories have reported good results for Th (Stirling et al. 1995; Pietruszka et al. 2002) and Pa (Bourdon et al. 1999) using only nitric acid during dry-down, whereas earlier work has generally found perchloric acid to be necessary for accurate results for these elements (Goldstein et al. 1989; Pickett et al. 1994; Lundstrom et al. 1998). Other laboratories use boric acid rather than perchloric acid to eliminate fluorides and help promote complete equilibration. For more reactive elements such as U and Ra, such steps are less critical due to the tendency for these elements to chemically exchange between different compound forms, hence nitric or hydrochloric acid dry-downs are thought to be sufficient.

For precise measurement of isotopic composition by mass spectrometry, it is also common to use either a natural, known isotopic ratio to correct for instrumental mass fractionation (e.g., internal normalization) or to add a tracer for this purpose. For example for natural uranium samples, one can use the natural ^{238}U/^{235}U of 137.88 to correct for fractionation. Alternatively, one can use an added ^{233}U-^{236}U double spike of ratio ~unity for the most precise fractionation correction and/or non-natural samples (e.g., Chen et al. 1986; Goldstein et al. 1989).

In contrast to thermal ionization methods, where the tracer added must be of the same element as the analyte, tracers of different elemental composition but similar ionization efficiency can be utilized for inductively coupled plasma mass spectrometry (ICPMS) analysis. Hence, for ICPMS work, uranium can be added to thorium or radium samples as a way of correcting for instrumental mass bias (e.g., Luo et al. 1997; Stirling et al. 2001; Pietruszka et al. 2002). The only drawback of this approach is that small inter-element (e.g., U vs. Th) biases may be present during ionization or detection that need to be considered and evaluated (e.g., Pietruszka et al. 2002).

3. CHEMICAL SEPARATIONS

Chemical separations to purify the element of interest are required for most forms of decay-counting excluding gross gamma spectrometry and for most mass spectrometric methods excluding some types of ICPMS analysis. This is mainly due to the necessity for removal of unwanted interferences in the mass or decay-energy spectrum of uranium-series elements. In many cases in mass spectrometry, chemical separations are also required to provide optimal ionization efficiency or measurement signal to noise ratio. While most standard separation methods of anion or cation exchange chemistry, solvent extraction, phase separation, and/or coprecipitation were developed prior to the last decade, recently developed extraction chromatography materials have improved many aspects of uranium-series analysis. Automated separation systems based on these materials and others have also been developed which further increase analytical throughput (e.g., Hollenbach et al. 1994).

3.1. Extraction chromatography resins/disks

Extraction chromatography resins consist of inert beads that are coated by an organic extractant that can be selective for concentration of any of a number of elements from aqueous solutions. These materials were largely developed and basic separation capabilities determined at Argonne National Laboratory by P. Horwitz and colleagues and have been applied for uranium-series measurement by investigators around the world. These materials have been applied in two geometric configurations: as extraction chromatography beds for standard gravimetric column chromatography, and as disks for more rapid extraction of elements from aqueous solutions (e.g., Joannon and Pin 2001).

One of the first bed materials was based on the extractant diamyl amylphosphonate (DAAP; marketed under the name U-TEVA-Spec) and was designed for purification of the tetravalent actinides (U (IV), Th (IV), Pu (IV)) and hexavalent uranium (U(VI)). This material is characterized by high (>10-100) distribution coefficients for U and Th in significant (>3 M) concentrations of both nitric and hydrochloric acids, and so is useful for both U and Th purification (Horwitz et al. 1992; Goldstein et al. 1997; Eikenberg et al. 2001a).

An additional material based on the extractant octyl-phenyl-N,N-diisobutyl-carbamoylmethylphosphine oxide, or CMPO, (marketed under the name TRU-Spec) has also been widely utilized for separations of transuranic actinides (Horwitz et al. 1993a) but is also useful for uranium-series separations (e.g., Burnett and Yeh 1995; Luo et al. 1997; Bourdon et al. 1999; Layne and Sims 2000). This material has even greater distribution coefficients for the uranium-series elements U (>1000), Th (>10000), and Pa. As shown in Figure 1, use of this material allows for sequential separations of Ra, Th, U, and Pa from a single aliquot on a single column. Separations of protactinium using this material (Bourdon et al. 1999) provide an alternative to liquid-liquid extractions documented in Pickett et al. (1994).

Another material based on the crown ether extractant 4,4'(5')-bis(t-butyl-cyclohexano)-18 crown-6, marketed under the name Sr-Spec, is useful for separations involving divalent cations including Pb, Ba, and Ra (Horwitz et al. 1991). For ^{226}Ra analysis by TIMS, Ra-Ba separations are required because the presence of Ba drastically decreases the ionization efficiency of fg Ra samples from 10% to <1%. This material has been widely used for separations of Ra from Ba (e.g., Chabaux et al. 1994; Lundstrom et al. 1998; Rihs et al. 2000; Joannon and Pin 2001; Pietruszka et al. 2002) and is a complement or alternative to cation exchange separations for EDTA complexes of these elements (Volpe et al. 1991; Cohen and O'Nions 1991). Sr-Spec material would also be useful for ^{210}Pb analysis, since Pb has a greater distribution coefficient than Sr with this extractant.

Separation of U, Th, Pa, and Ra on a TRU™ Spec Resin Column

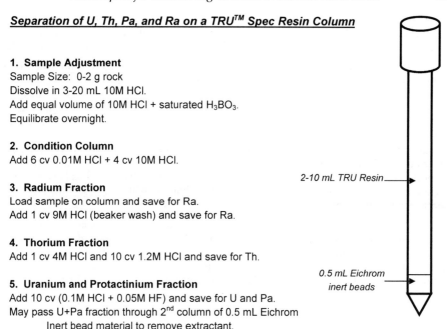

1. Sample Adjustment
Sample Size: 0-2 g rock
Dissolve in 3-20 mL 10M HCl.
Add equal volume of 10M HCl + saturated H_3BO_3.
Equilibrate overnight.

2. Condition Column
Add 6 cv 0.01M HCl + 4 cv 10M HCl.

3. Radium Fraction
Load sample on column and save for Ra.
Add 1 cv 9M HCl (beaker wash) and save for Ra.

4. Thorium Fraction
Add 1 cv 4M HCl and 10 cv 1.2M HCl and save for Th.

5. Uranium and Protactinium Fraction
Add 10 cv (0.1M HCl + 0.05M HF) and save for U and Pa.
May pass U+Pa fraction through 2nd column of 0.5 mL Eichrom
   Inert bead material to remove extractant.

Figure 1. Schematic diagram showing a TRU-spec extraction chromatography method for separation of uranium, thorium, protactinium, and radium from a single rock aliquot. Further purification for each element is normally necessary for mass spectrometric analysis. Analysis of a single aliquot reduces sample size requirements and facilitates evaluation of uranium-series dating concordance for volcanic rocks and carbonates. For TIMS work where ionization is negatively influenced by the presence of residual extractant, inert beads are used to help remove dissolved extractant from the eluant.

However, there are some negative side-effects of these extractant materials, both in TIMS and ICPMS. For TIMS work, residual extractant in the eluant may negatively influence ionization behavior of uranium-series elements. This may occur by raising ionization temperature due to formation of refractory organic compounds on the filament, or by contributing to undesirable organic isobars in the mass spectrum. These effects can be minimized by using a lower bed of uncoated resin beads to help absorb the excess extractant (Pietruszka et al. 2002) and/or use of appropriate clean-up anion or cation exchange column chemistry. Similar effects have generally not been reported for ICPMS, presumably due to effective breakdown of the extractants at higher temperature in the ICP source, although in a few cases organic interferences in the ICPMS spectrum may be attributed to residual resins or extractants (Pietruszka et al. 2002). The presence of organic extractants in final ICPMS sample solutions may also contribute to significantly increased washout times between sample analysis.

4. INSTRUMENTAL ANALYSIS METHODS

4.1. Alpha spectrometry

Sources. It is well known that the optimal geometry for source preparation for alpha spectrometry is as an "infinitely thin" and uniform deposit on a flat surface. In practice, "infinitely thin" can correspond to up to 100 µg of material deposited on a planchet of 1-2 cm in diameter. Both electrodeposition and co-precipitation (NdF$_3$) methods have traditionally been utilized for preparation of alpha spectrometric sources (e.g., Lally 1992).

However, during the past decade some innovative methods for preparation of radium, thorium, and uranium sources for aqueous samples have been reported (e.g., Surbeck 2000; Morvan et al. 2001; Eikenberg et al. 2001b). In these methods, radionuclides in water samples are adsorbed onto a thin layer (20 um) of manganese oxide, followed by direct alpha counting of the oxide layer. As shown in Figure 2, energy resolution obtained from alpha counting of these materials can be nearly as good as for electroplated sources. Uranium adsorbing films (Surbeck 2000) have also been prepared utilizing a commercially available cation exchanger containing diphosphonate and sufonate groups (Horwitz et al. 1993b). Energy resolution with these films is poorer than for the manganese oxide films, but is better than obtained with liquid scintillation alpha spectrometry (see below).

Instrumentation. Traditional methods of alpha and beta spectrometry instrumentation have changed little over the past decade. Alpha spectrometric methods typically rely on semi-conductor or lithium-drifted silicon detectors (Si(Li)), or more historically gridded ion chambers, and these detection systems are still widely used in various types of uranium-series nuclide measurement for health, environmental, and

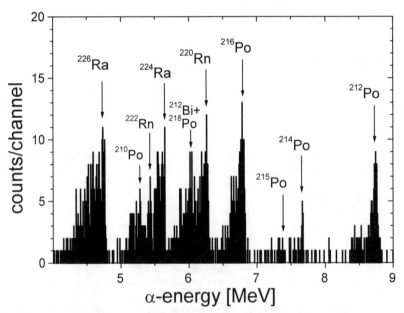

Figure 2. Alpha spectrum for a radium adsorbing manganese-oxide thin film exposed to a groundwater sample, after Surbeck (2000) and Eikenberg et al. (2001b). A 2×2 cm sheet is exposed to 0.1-1.0 L of sample for 2 days, capturing nearly all of the radium in the sample. These sample discs can be used directly for low-level alpha spectrometry without the need for further separation and preparation methods to produce planar sample sources. Energy resolution is nearly as good as for electroplated sources, and detection limits are typically 0.2 mBq/L (6 fg ^{226}Ra/L) for ^{226}Ra and ^{224}Ra for a one-week counting period. These sensitivities are comparable to traditional methods of alpha spectrometry. [Used by permission of Elsevier Science, from Eikenberg et al. (2001), *J Environ Radioact*, Vol. 54, Fig. 4, p. 117]

geological applications. For many short-lived isotopes in the uranium and thorium decay series, traditional alpha spectrometry with semi-conductor detection is still either the method of choice or currently the only realistic alternative for low-level detection (e.g., Po-210 (Rubin et al. 1994) and Ac-227 (Martin et al. 1995)).

However, new techniques for detection of alpha-emitting isotopes via alpha liquid scintillation spectrometry have recently been developed (PERALS, or photon-electron rejecting alpha-liquid scintillation; e.g., Dacheux and Aupiais 1997). These methods combine chemical separation by liquid-liquid extraction with measurement of alpha activity by liquid scintillation. Scintillation detection is obtained because the organic extractant also contains an energy transfer reagent and a light-emitting fluor. A variety of combinations of aqueous phases and organic phases can be utilized, depending on the particular isotopes to be measured. For instrumentation, the method uses pulse shape discrimination to nearly eliminate the background from beta-emitting isotopes and to reduce background from gamma-ray activity. Detection limits comparable to alpha spectrometry are obtained, but with much less sample preparation for many matrices. Although isotope dilution is required for analysis of samples with complex matrices, aqueous solutions have more consistent recoveries, resulting in the option of direct analysis without tracer addition.

4.2. Gamma spectrometry

Over the past decade, traditional instrumentation for highest resolution gamma spectrometry has continued to rely on solid-state germanium detectors (e.g., Knoll 1989). Although there have been developments in the geometry of such detectors to improve counting efficiency, from extension of flat-planar detectors to larger sizes and to different geometries (e.g., well-type), their basic characteristics have not changed (El-Daoushy and Garcia-Tenorio 1995).

Minimization of background from cosmic rays or earth's natural radioactivity continues to be a major focus and relies on development of improved shielding or shielding materials (e.g., Semkow et al. 2002). Ground-level locations may use active rejection of cosmic-ray muons, either using NaI active shields, plastic scintillator, or proportional counters, and these anti-coincidence techniques provide reduced background. However, lowest backgrounds are achieved at deep-underground locations (e.g., Neder et al. 2000) for which cosmic rays are greatly reduced or not present.

Compton suppression gamma spectrometry has been applied for ^{228}Ra determination (James et al. 1998), but is particularly applicable for high-activity measurements where background is created by Compton scattering of gamma radiation originating from the sample itself. For low-activity measurements the Compton suppressor may act as a guard against cosmic radiation, although surrounding a Ge detector with the Compton suppressor may actually increase background from natural radioactivity present in the components of the suppressor.

In addition to instrumental improvements, various approaches have been used to improve the purity or geometry of sources of natural samples for gamma spectrometric measurement. For example, improvements in source preparation for ^{234}Th measurement in water and sediment samples by gamma spectrometry are discussed in Cochran and Masque (2003). It should be emphasized that one of the main advantages of gamma spectrometry is ease of use, since in many cases samples may be analyzed directly or with significantly reduced sample preparation compared to alpha, beta, or mass spectrometric techniques.

4.3. Thermal ionization mass spectrometry (TIMS)

The end of the 1980's saw the application of TIMS to U-series measurement (Chen et al. 1986; Edwards et al. 1987; Goldstein et al. 1989). This represented a major technological advance. Analysis time was reduced from one week to several hours, sample sizes for many carbonate or volcanic rock samples decreased from ~10-100 µg to 0.1-1 µg U or Th, measurement precision improved from percent to permil levels, and for the ^{238}U-^{234}U-^{230}Th decay series, the dating range was extended from 350,000 years to 500,000 years, compared with earlier alpha-spectrometric techniques. The first TIMS instruments utilized a single-collector and "peak jumping" analysis routines (e.g., Edwards et al. 1987; Goldstein et al. 1989; Bard et al. 1990). Despite the sequential measurement of isotopes, peak-tailing corrections, and limited dynamic range, analytical uncertainties of ~5‰ could be attained for U and Th. As a result, ^{230}Th-age uncertainties improved from ±10,000 to ±2,000 years in 120,000 year-old carbonate samples, and ^{230}Th-ages could be determined with uncertainties of ±10,000 years in 300,000 year-old samples. The improved precision and sensitivity opened up a vast number of applications in paleoclimatology and geochemistry.

Subsequently, a wide array of developments in TIMS methods for uranium-series measurement occurred during the past decade including initiation of methods for measurement of long-lived radium (Volpe et al. 1991; Cohen and O'Nions 1991) and protactinium isotopes (Pickett et al. 1994; Bourdon et al. 1999), development of improved sources or ionization methods for TIMS analysis, and introduction of commercially available multi-collector TIMS instruments designed specifically for uranium and thorium isotopic measurement.

Sources. The ionization efficiency in a TIMS source depends on a number of factors including the difference between filament work function and elemental ionization potential, as well as the temperature of the filament, elemental oxidation state and volatility, distribution of sample on the filament surface, etc. Ionization efficiency is defined as the number of atoms ionized in the source relative to the total number of atoms introduced to the instrument. In TIMS, transmission of ionized material through the mass spectrometer is normally assumed to be close to 100%, and ionization efficiency is given as the number of ions detected as a proportion of the number of atoms loaded onto the filament. Although effects of ionization potential, filament work function and temperature are theoretically given by the Saha-Langmuir equation (e.g., Wayne et al. 2002), in practice other chemical, physical, and geometrical effects often exert a more significant influence on ionization behavior and efficiency. In addition, evaporation and ionization of an element off the filament surface under vacuum causes an increasing depletion of the lighter isotopes over the heavier isotopes as progressively more of the sample is volatilized. This gives rise to a time-dependent instrumental mass fractionation, which varies with ionization method and may be corrected by either internal normalization or external standardization techniques.

Over the past decade, a number of improvements in source preparation have been utilized for uranium-series nuclide analysis by TIMS. These generally either involve methods to increase the work function of the filament material, addition of enhancers to optimize elemental volatility and oxidation state, geometrical improvements in filament configuration or sample loading, or instrumental improvements in source mechanism or design. Early TIMS uranium and thorium work utilized single Re filaments for analysis of small Th and U samples (ng or smaller; Chen et al. 1986; Edwards et al. 1987), and multiple filament assemblies for more effective ionization of larger samples (Goldstein et al. 1989). For single filament measurements of Th isotopics in low-^{232}Th carbonate or water samples, ^{232}Th filament blanks can be significant and need to be carefully controlled or monitored (Edwards et al. 1987). Samples for single filaments were typically loaded as dilute HCl or HNO_3 solutions between layers of colloidal graphite, called the graphite sandwich method. In this case, the graphite provides a reduced environment and samples are analyzed as the U^+ and Th^+ metal species at temperatures around 1700°C and 1900°C, respectively. Because ionization efficiency for this method is quite negatively correlated with sample size loaded, particularly for Th, triple filament configurations provide better ionization efficiency for larger U and Th samples (Goldstein et al. 1989). In multiple filament techniques for U and Th, the side filaments are used to control sample volatility at lower temperatures than the center, ionizing filament which is typically maintained at about 2100°C to produce U^+ and Th^+ metal species more efficiently. For U and Th, ionization efficiencies are typically a few epsilon to a few permil, but in exceptional cases, have been reported at the percent-level (e.g., Edwards et al. 1987; Esat 1995). More recently, Yokoyama et al. (2001) used a silicic acid–dilute phosphoric acid activator to measure $^{234}U/^{238}U$ by TIMS using UO_2^+ ions. This approach yielded analytical uncertainties of 1‰ (2σ) on small samples of uranium (10-100 ng).

Radium samples for TIMS work can be prepared in a number of ways, due to the low ionization temperature for this element (~1300°C). For single filament analysis, samples can be loaded with a silica-gel mixture on single Re or Pt filaments, in either standard or reduced length (Delmore) configurations (Volpe et al. 1991). Alternatively, samples can be loaded onto the center W filament of a triple Ta-W-Ta filament assembly with a Ta-HF-H_3PO_4 activator solution (Cohen and O'Nions 1991). In this method, the side filaments are used to preheat the center filament and gently remove hydrocarbon interferences from the sample filament prior to Ra ionization. Lundstrom et al. (1998) modified these procedures slightly and used a $TaCl_5$ activator on single Re filaments.

Protactinium samples for TIMS analysis are typically prepared by the graphite sandwich technique on single Re filaments (Pickett et al. 1994; Lundstrom et al. 1998), similar to methods for small Th and U samples (Chen et al. 1986; Edwards et al. 1987). An alternative method is to analyze protactinium with a silica-gel enhancer on tungsten filaments as the double-oxide at filament temperatures around 1400°C (Bourdon et al. 1999). For the graphite sandwich methods, identifying the appropriate amount of graphite is critical, since it is desirable to let uranium ionization complete prior to protactinium isotopic measurement to remove the ^{233}U isobar produced from ^{233}Pa decay from the $^{231}Pa/^{233}Pa$ measurement. With too much graphite, uranium and protactinium ionization overlap, resulting in more complicated data reduction or rejection of data. However, it is desirable for some graphite to be present at the higher temperatures of Pa ionization (1900°C), so proper adjustment is a somewhat tricky aspect of these methods. The ^{233}U isobar can also be minimized by ensuring a clean separation of Pa from U during ion exchange column chemistry, although some ^{233}U isobar is inevitably present due to ^{233}Pa decay between separation and analysis.

Carburization of rhenium filaments has been used to optimize Th and Pa ionization efficiency for TIMS analysis on single filaments (Esat 1995). ReC has a greater work function than Re metal, and elemental oxidation state is maintained in the reduced or metal state by the presence of carbon in the filament. Using this method and a mass spectrometer with improved ion optics, Esat (1995) was able to improve Th transmission and ionization efficiency by about a factor of 30 over conventional methods. Using more conventional mass spectrometry, Murrell et al. (personal communication) were able to improve ionization efficiency for Pa and Th by a factor of 5-10 over conventional graphite sandwich loads on Re filaments (Goldstein et al. 1989; Pickett et al. 1994). For Pa analysis, one drawback is that Pa and U ionization commonly overlap using this procedure, and so a deconvolution procedure, where ^{238}U and ^{233}U peak intensities are approximately correlated and involving the isotopes ^{238}U, ^{233}U + ^{233}Pa, and ^{231}Pa, should be applied to utilize more data and obtain accurate ^{231}Pa/^{233}Pa ratios.

Resonance ionization methods (RIMS) have also been explored for improving Th ionization efficiency for mass spectrometric measurement (Johnson and Fearey 1993). As shown in Figure 3, two lasers are required, a continuous resonant dye laser for resonance of thorium atoms, and a continuous UV argon laser for transition from resonance to ionization. Consequently, sophisticated laser instrumentation is required for these methods,

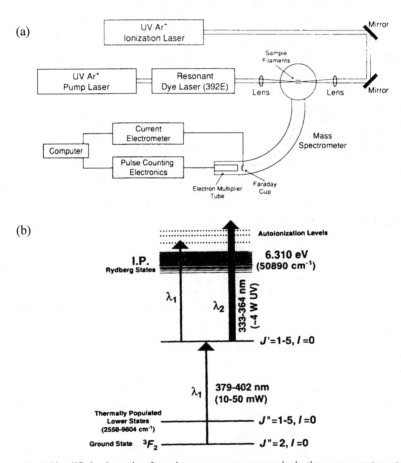

Figure 3. a) Simplified schematic of continuous wave resonance ionization mass spectrometry for thorium isotopic analysis, after Johnson and Fearey (1993). Two lasers are required for ionization, a narrow band laser for the resonant step and a high-powered ultraviolet Ar$^+$ laser for the ionization step, and are coupled with a standard (NBS-1290) mass spectrometer. b) Thorium energy level diagram illustrating the two-step ionization process. Ionization/detection efficiency is improved by at least an order of magnitude (to 0.4%) relative to standard TIMS methods for large thorium samples. Internal precision is comparable to TIMS techniques (~0.5%), although a small bias of 2-4% is introduced due to differences in the transition strengths of ^{230}Th vs. ^{232}Th. However, the bias is reproducible and can be corrected using standards. [Used by permission of Elsevier Science, from Johnson and Fearey (1993), *Spectrochim Acta Part B*, Vol. 48, Figs. 1 & 2, p. 1066 & 1067]

and laser stability needs to be maintained over the measurement period for these methods to be utilized. However, ionization efficiency was increased by greater than an order of magnitude (to ~0.4%) relative to standard TIMS techniques. Consequently, volcanic rock samples with as little 1-5 ng Th may potentially be analyzed using this method.

Ion guns or cavity sources have also been developed for improving elemental ionization for TIMS measurement (Duan et al. 1997; Wayne et al. 2002). As shown in Figure 4, one ion source is based on a tungsten crucible with a cavity (0.02 cm diameter, 1 cm depth) into which sample is loaded. The crucible is heated by electron

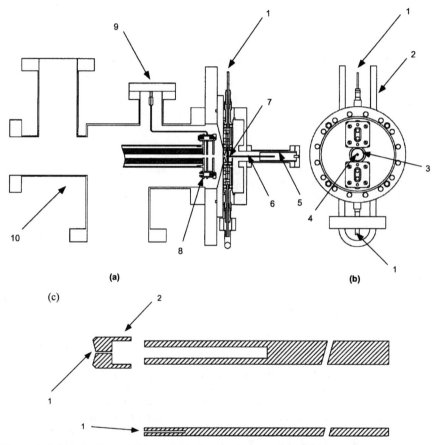

Figure 4. Schematic diagram of a cavity ionization source designed for TIMS or quadrupole mass spectrometry, after Duan et al. (1997). a) the ion source mounted on a mass spectrometer: 1, electrodes for power supply; 2, water or air cooling tubing; 3, electron bombardment shielding can; 4, the orifice of the shielding can; 5, crucible holder; 6, crucible; 7, electron bombarding filament; 8, ion lens; 9-10, quadrupole mass spectrometer; b) a coaxial view of the source, with same components as in a), c) tungsten crucibles used for the thermal ionization source. Top diagram illustrates a crucible for an isotope separator, bottom is the crucible for a quadrupole or magnetic sector spectrometer. 1, ionization channel; 2, crucible cap. Dimensions for the smaller thermal ionization crucible are 0.02 cm diameter and 1 cm depth. Ionization efficiency for many elements (ng to ug) is improved by at least an order of magnitude relative to standard filaments, although backgrounds can also be elevated relative to standard methods and need to be carefully evaluated. [Used by permission of Elsevier Science, from Duan et al. (1997), *Int J Mass Spectrom Ion Processes*, Vol. 161, Figs. 1 & 2, p. 29 & 30]

bombardment from a filament surrounding the crucible that can be modified to control the crucible temperature. Ionization efficiency for large samples (ng to ug) of many elements is enhanced by factors of 10-100 relative to standard TIMS sources due to the increased interaction of gaseous analyte atoms with the greater surface area of the crucible walls (Duan et al. 1997). Wayne et al. (2002) adapted this source to a time-of-flight mass spectrometer and obtained a maximum ionization of efficiency of 1-3% for thorium samples of ~0.1-20 ng size, representing a factor of 10-30 improvement over conventional methods. However, blanks from the W cavity source are a particular issue for uranium analysis and also need to be carefully evaluated for thorium and other uranium-series measurements.

Detection systems. Prior to the past decade, most instruments used for uranium-series analysis were single-collector instruments, for which ion beams of the various isotopes are cycled onto a single low-intensity detector, usually with electronics operating in pulse counting mode (Chen et al. 1986; Edwards et al. 1987; Bard et al. 1990; Goldstein et al. 1989; Volpe et al. 1991; Pickett et al. 1994), in order to measure the low-intensity ion beams of ^{230}Th, ^{234}U, ^{231}Pa, ^{233}Pa, ^{226}Ra and ^{228}Ra. Daly detectors and discrete-dynode multipliers provide the best combination of linearity and dynamic range for isotope ratio measurement, although precise measurement (<1%) is usually limited to ratios below a few thousand (Cheng et al. 2000; Richter et al. 2001). Thus, the measurement of extremely large isotope ratios of uranium and thorium (^{238}U/^{234}U = 20,000 and ^{232}Th/^{230}Th = 200,000) is problematic using single-collector instruments. A "bridge" isotope must instead be used (e.g., ^{235}U and ^{229}Th) and two isotope ratio measurements of about 100-1000 in magnitude are combined (e.g., ^{238}U/^{235}U × ^{235}U/^{234}U and ^{232}Th/^{229}Th × ^{229}Th/^{230}Th) to yield the desired measurement (Chen et al. 1986; Goldstein et al. 1989).

An alternative to the bridge technique was recently reported for thorium analysis in silicate rocks for which both ^{230}Th and ^{232}Th are measured on a single ion-counting detector (Rubin 2001). With careful chemistry and mass spectrometry, ^{230}Th/^{232}Th ratios of igneous rocks can be measured with this technique with a precision that is similar to the bridge method. The disadvantage of this technique is that ^{230}Th ion-count rates are extremely low (around 10 cps) with normal silicate thorium ratios and are therefore subject to perturbations from background variation and low-level isobaric interferences in "normal" samples.

A double-focusing magnetic sector TIMS was initially used for thorium isotope measurements in silicate rocks to provide adequate abundance sensitivity (Goldstein et al. 1989). However in the late 1980's, commercially available multi-collector instruments became available that provide more options for uranium-series analysis. These instruments generally consist of an array of Faraday cups situated after the magnet, with a Faraday cup and/or low level counting system (Daly knob or electron multiplier) located after an electrostatic analyzer (ESA) or retarding potential filter (e.g., Palacz et al. 1992; Cohen et al. 1992; Rubin 2001). The energy filter improves abundance sensitivity. Abundance sensitivity is defined as the peak tailing contribution from one mass to an adjacent mass and is normally determined by measuring the contribution of mass 238 to mass 237 using a pure U standard solution. By using an energy filter, abundance sensitivity is typically <0.5 ppm at one amu, enabling the measurement of low abundance ^{230}Th in the presence of high-abundance ^{232}Th, typical of silicate rocks. Without an energy filter, abundance sensitivity is about an order of magnitude worse (~5 ppm at one amu), and in silicate rocks, the ^{230}Th peak may be of the same order of magnitude to the contribution from ^{232}Th tailing. Tailing corrections based on mathematical subtraction can be made (McDermott et al. 1993), but these can also be difficult to determine precisely. With an energy filter, more time can also be spent measuring peak count rates rather than counting backgrounds, increasing the efficiency of the analysis. An energy filter may also be desirable to enhance the abundance sensitivity on ^{234}U in the presence of peak tailing from ^{238}U.

The first multiple-collector TIMS instruments permitted simultaneous measurement of the U-series isotopes in static or multi-static mode. The main advantage of these techniques is that multi-collection permits simultaneous collection of the major and minor isotope beams, improving the efficiency of data collection. In addition, greater variations in ion beam intensity can be accommodated than with single-collector techniques. In static collection, the main disadvantage is that the Faraday-ion counting gain must be stable during the course of the analysis and must be measured either during or before analysis. In multi-static modes of collection for uranium, the gain can be obtained during analysis by switching the ^{235}U beam between Faraday and ion-counting detectors, and so ^{234}U/^{238}U ratios with extremely high precision can be obtained. This, coupled with high abundance sensitivity capabilities, permits reduced analysis times, smaller sample sizes and analytical precisions of up to 1-2‰ for the U-Th isotopes for carbonates (e.g., Stirling et al. 1995). Analysis is possible on high-^{232}Th silicate rocks, as well as low-^{232}Th carbonate samples.

Despite the application of multiple-collector arrays to TIMS U-series measurement, it is usually still necessary to cycle the low-level Th isotopes (^{230}Th, ^{229}Th and carbonate ^{232}Th) through a single low-level detector, so that measurement precision is still limited by beam intensity fluctuations and poor data acquisition efficiency. To overcome these limitations, Esat (1995) applied charge-collection TIMS (CC-TIMS) to thorium isotope measurement. In "charge-collection" mode, a Faraday collector is still utilized, but the usual high-value feed-back resistor (10^{11} Ω) in the electrometer amplifier is replaced with a 20 pico-Farad air-core capacitor (Fig. 5). Provided that the ion beam is stable, charge builds up on the capacitor with time in a linear fashion and is displayed in analogue mode as a monotonic increase in voltage and the rate of voltage accumulation is related to the ion current. This technique allows the measurement range to be extended below 10^{-14} A, while preserving the advantages of Faraday cup arrays, allowing all low-level ion currents to be monitored simultaneously, in multiple channels. Charge collection achieves very high levels of precision; the uncertainty in ^{229}Th/^{230}Th is routinely 0.6‰ for sample sizes corresponding to 30 pg of ^{230}Th.

In static collection for untraced thorium samples, the gain is typically measured during the warm-up stage of analysis for which the ^{232}Th beam is switched between Faraday and ion-counting detectors, and variations in this gain during a run have sometimes been noted (Rubin 2001). In multi-static collection for traced Th analysis, ^{229}Th can be switched between Faraday and ion-counting detectors to provide an in-run measurement of Faraday-ion counting gain. This method can therefore provide very high precision results, similar to uranium analysis by this technique.

Accuracy for all thorium measurements by TIMS is limited by the absence of an appropriate normalization isotope ratio for internal correction of instrumental mass fractionation. However, external mass fractionation correction factors may be obtained via analysis of suitable thorium standards, such as the UC-Santa Cruz and IRMM standards (Raptis et al. 1998) for ^{230}Th/^{232}Th, and these corrections are usually small but significant (< few ‰/amu). For very high precision analysis, the inability to perform an internal mass fractionation correction is probably the major limitation of all of the methods for thorium isotope analysis discussed above. For this reason, MC-ICPMS techniques where various methods for external mass fractionation correction are available, provide improved accuracy and precision for Th isotope determinations (Luo et al. 1997; Pietruszka et al. 2002).

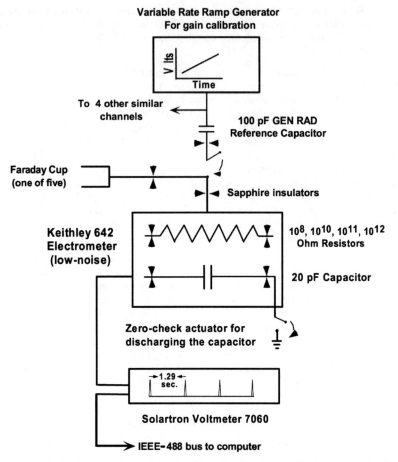

Figure 5. Schematic diagram of one of the five channels of a charge-collection system for thermal ionization mass spectrometry of thorium, after Esat (1995). In "charge-collection" mode, a Faraday collector is still utilized, but the usual high-value feed-back resistor (10^{11} Ω) in the electrometer amplifier is replaced with a 20 pico-Farad air-core capacitor. Charge builds up on the capacitor with time and is displayed in analogue mode as a monotonic increase in voltage, with the rate of voltage accumulation related to the ion current. This technique allows the measurement range to be extended below 10^{-14} A, while preserving the advantages of Faraday cup arrays, allowing all low-level ion currents to be monitored simultaneously, in multiple channels. Charge collection achieves very high levels of precision; the uncertainty in ^{229}Th/^{230}Th is routinely 0.6‰ for sample sizes corresponding to 30 pg of ^{230}Th. [Used by permission of Elsevier Science, from Esat (1995), *Int J Mass Spectrom Ion Processes*, Vol. 148, Fig. 1, p. 163]

4.4. Secondary ion mass spectrometry (SIMS)

SIMS has also been successfully applied for thorium isotopic measurement during the past decade. This technique has been applied for both chemically separated thorium samples (England et al. 1992; Bourdon et al. 1994; Layne and Sims, 2000), as well as in-situ analysis of minerals with high thorium content such as zircons (Reid et al. 1997).

Although detailed methods depend on the instrument used, sources and ionization method are the major difference between SIMS and TIMS methods. SIMS methods employ either primary Ar^+ or O^- beams which bombard spectrographically pure carbon planchets containing precipitated thorium sample. Ion separation is similar to TIMS methods, obtained with either large radius magnetic sectors and/or electrostatic analyzers, which can provide adequate abundance sensitivity for measurement of ^{230}Th in the presence of a large ^{232}Th beam (<0.1 ppm for a change in one mass unit at mass 232). Ion detection is similar to TIMS methods, with sequential or simultaneous collection of ^{232}Th on a Faraday cup and ^{230}Th on an axial electron multiplier, and the use of standards to determine the Faraday-multiplier gain. The main advantage of SIMS techniques for chemically separated samples is that ionization efficiency for Th (~2%) is substantially greater than for TIMS methods. Hence sample size requirements for Th analysis in volcanic rocks are substantially lower with SIMS work (10 ng or less), while overall measurement reproducibility is comparable to TIMS methods (0.5 to 1%). However, the main disadvantage of SIMS is equipment cost, as commercially available SIMS instrumentation is considerably more expensive than TIMS instruments.

4.5. ICPMS and MC-ICPMS

Inductively coupled plasma mass spectrometry (ICPMS) techniques have been widely applied for uranium-series nuclide measurement during the past decade. The major strengths of ICPMS are five-fold. First, the high ~6000 K temperature attained in the plasma source produces efficient ionization (>90%) of nearly all elements (Gray 1985; Jarvis et al. 1992). This includes those with high first ionization potentials that are difficult to ionize by TIMS, thorium being just one example. Second, in contrast to TIMS, ionization efficiency is not a function of load size and signal intensity can be increased simply by using a more concentrated solution. Third, the mass discrimination of the plasma source, although large, is essentially constant during an analysis and can be reliably corrected at high levels of precision (Walder and Freedman 1992, Walder et al. 1993; Halliday et al. 1995; 1998). Fourth, the mass discrimination is widely considered to be independent of the chemical properties of the element, and to a first order, is a function of mass only with relatively minor inter-element biases during ionization and/or detection. Therefore, two elements with overlapping mass ranges can be admixed, and the mass bias in the isotopic composition of one can be used to correct for mass discrimination in the isotopic composition of the other (e.g., Halliday et al. 1995; Luo et al. 1997; Marechal et al. 1999; Rehkamper and Halliday 1999). Fifth, sample throughput is typically faster than can be achieved using TIMS. Less sample preparation may be required in certain applications such as direct analysis of aqueous solutions, although care must be taken to eliminate matrix effects. Matrix effects refer to the influence of the major element composition on measured isotopic composition, due to mass discrimination, isobaric interferences, or other phenomena in the source region.

Techniques in "conventional" ICPMS are comprised of quadrupole ICPMS (Q-ICPMS), high-resolution sector-field ICPMS (HR-ICPMS) and time-of-flight ICPMS (TOF-ICPMS). For low-level ion counting, all are equipped with a single detector, typically an electron multiplier operating in pulse counting mode. These techniques share the same ICP source but utilize different methods of ion separation and focusing (quadrupole, magnetic sector, and time-of-flight). Further comparisons between these methods are discussed in Section 5.

Multiple-collector inductively coupled plasma mass spectrometry (MC-ICPMS) combines sector-field ICPMS with a multiple collector detector system and has recently emerged as an alternative to TIMS for precise U-Th isotope measurement. The full potential of MC-ICPMS has yet to be realized. Yet despite this, its performance in high precision isotope measurement already challenges and, in some cases, surpasses that ever achieved by TIMS (e.g., Lee and Halliday 1995; Blichert-Toft and Albarède 1997).

The first isotopic measurements of U using MC-ICPMS were published by Walder and Freedman (1992) and Taylor et al. (1995). This work was followed by a series of combined U and Th isotopic studies at the University of Michigan using a VG Elemental Plasma 54, the first commercially available MC-ICPMS instrument (Fig. 6). The results of these studies demonstrated that uncertainties of ~1‰ ($2\sigma_M$) were achievable for $^{234}U/^{238}U$, $^{230}Th/^{238}U$ and $^{230}Th/^{232}Th$ in both standard reference materials and natural samples containing ~200-600 ng of U and silicate Th (Luo et al. 1997; Stirling et al. 2000 and Stirling et al. 2001). This is competitive with even the highest quality data acquired by TIMS.

The weakness of MC-ICPMS lies in the inefficiency by which ions are transferred from the plasma source into the mass spectrometer. Therefore, despite very high ionization efficiencies for nearly all elements, the overall sensitivity (defined as ionization plus transmission efficiencies) of first generation MC-ICPMS instruments is of the order of one to a few permil for the U-series nuclides. For most, this is comparable to what can be attained using TIMS.

Several second-generation MC-ICPMS instruments have been developed since this initial work. With one exception, all are double-focusing instruments, and in contrast to the Plasma 54, all are equipped with fast laminated magnets to allow rapid switching between masses. Some have variable mass resolution (400 to 10,000). While moderate to high mass resolution is critical for some elements to avoid spectral interferences, these are not considered to be problematic in the U-Th mass range and the standard resolution of 400 normally suffices for U-series measurements. All second-generation instruments have benefited from design improvements at the interface between the plasma source and the mass spectrometer, resulting in significantly enhanced sensitivities of up to an order of magnitude compared with the Plasma 54. This has allowed precise data to be acquired

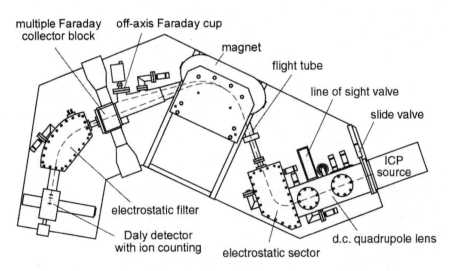

Figure 6. Schematic outline of the first commercially available multiple collector ICPMS, the "Plasma 54," after Halliday et al. (1995). This instrument uses Nier-Johnson double-focusing and is equipped with eight independently adjustable Faraday collectors. The axial collector can be wound down to provide access to a Daly detector equipped with ion counting capabilities and a second-stage energy filter for high abundance sensitivity measurements. The sample may be introduced to the plasma source by either solution aspiration or laser ablation.

on smaller sample sizes with U and Th contents at the 100-10 ng level (e.g., Nakai et al. 2001; Turner et al. 2001; Robinson et al. 2002). Multiple-collector ICPMS is thus fast becoming a widely accepted means for obtaining high-quality data for the U-series isotope systems. The following discussion mainly focuses on features of MC-ICPMS analysis, although both MC-ICPMS and ICPMS methods share basic characteristics in source and general design.

Sources/sample introduction. Solution aspiration. In both ICPMS and MC-ICPMS, the standard means of sample introduction is to aspirate the chemically separated analyte into the plasma source as an aerosol. It is usual to first dissolve the sample in a solution of dilute acid. For U-Th analysis, solutions of dilute HNO_3 are normally used for sample introduction, although it may be desirable to add a trace amount of HF to the thorium fraction to prevent precipitation. Molecular interferences and/or memory effects may also determine the acid medium for sample introduction. Conventional glass aspiration systems use a Meinhard cross-flow nebulizer to aspirate the sample into droplets, coupled to a cooled spray chamber, which removes the larger droplets and excess solution by condensation. Solution uptake rates are typically ~500 µl/min. Conventional aspiration produces a "wet" plasma containing some polyatomic oxides and hydrides, which may create molecular interferences that require correction. Such interferences can often be reduced to negligible proportions by injecting the sample as a "dry" aerosol using a micro-concentric nebulizer or MCN (with solution uptake rates of 50 to 150 µl/min; Vanhaecke et al. 1996) coupled to a heated spray chamber and membrane desolvator. Molecular interferences can also be controlled by chemically separating the analyte from the sample matrix so that only pure solutions are introduced to the system. As an added benefit, desolvating nebulization systems are highly efficient, typically resulting in significantly better sample utilization compared with conventional nebulizers. Other high-efficiency systems are available (e.g., Liu and Montaser 1994; Becker et al. 1999; Huang et al. 2000; McLean et al. 2001), but in general, all are wet nebulizers and as such, are still prone to hydride and oxide interference formation. Between sample runs, the nebulizer must be flushed with a cleaning solution, typically dilute acid in order to minimize sample memory effects.

Laser ablation. A laser ablation system can be used in place of solution nebulization as the sample introduction system. During laser ablation sampling, the high-energy laser beam produces a micro-plasma from the sample and nebulizer gas that ablates the sample surface. The sample is enclosed by an optical cell, through which the laser beam may pass undisturbed. Typically, the cell is mounted on a stage that can be rotated or moved in the x, y and z directions by computer control, and several samples can be loaded simultaneously to minimize sample changeover time. Sample preparation is very easy; provided the sample fits into the sample chamber, and sits a minimum distance below the top of the sample cell, all that is required is to attach the sample to a mount with an adhesive. Polishing of the sample surface may also be desirable to enhance any surface textures and to promote stable ion beams. During ablation, it is usual to view the sample and pit morphology with transmitted (plane or polarized) or reflected light using a binocular microscope or an on-line video camera and monitor. Laser ablation sampling for U-series isotopic analysis has been used in both conventional ICPMS and MC-ICPMS (e.g., Becker et al. 2000; Stirling et al. 2000; Guillong and Günther 2002).

Ion extraction. The aspirated or laser ablated sample is transported from the sample introduction system into the center of the torch by a ~1 l/min flow of Ar carrier gas where it is immediately dissociated and ionized by energy transfer with the hot ~6000 K temperature of the surrounding Ar plasma. Ionization efficiencies are >95% for U and Th (Jarvis et al., 1992). For laser ablation sampling, helium may be employed as the carrier gas instead of Ar because ablation in a pure He atmosphere has been found to significantly enhance sensitivity for some elements (e.g., Eggins et al. 1998; Günther and Heinrich 1999; Horn et al. 2000).

For most instrument configurations, the ions are then accelerated into a rotary-pumped "expansion chamber" situated between the first water-cooled Ni "sample" cone and second Ni "skimmer" cone by a large accelerating voltage of up to 10,000 volts. The expansion chamber is held at a vacuum of ~1 mbar, enabling the ions to undergo supersonic expansion and a small proportion (several permil) are extracted behind the skimmer cone. It is during this extraction process that most of the sample is lost. Ion sampling in this interface region is associated with the preferential extraction of heavy isotopes over lighter ones. This is generally ascribed to steady-state "space-charge effects," in which heavy isotopes are less deflected from the optical axis than lighter ones, and creates an essentially constant instrumental mass bias that is largely a function of mass alone. The MC-ICPMS mass bias is large— ~0.5 to 1% per a.m.u in the mass range of the U-Th isotopes—and is in stark contrast to TIMS, where instrumental bias is approximately an order of magnitude smaller but time-dependent due to the progressive evaporation of the sample from the filament surface. Once behind the skimmer cone, the ions are optically focused onto the entrance slit into the mass analyzer by a series of ion lenses. This transfer lens region is held at a vacuum of $\sim 10^{-5}$ to 10^{-8} mbar by turbomolecular pumps.

Ion focusing. Collision/reaction cells. Collision or reaction cells are commonly employed to convert molecular interferences into either neutral species or species that are different from the analyte mass, which enables the analyte ions to emerge from the cell free of interferences. The analyte ions can then be directed into the mass analyzer (quadrupole or sector) for normal mass separation. Single-focusing instruments often utilize a quadrupole or hexapole collision cell/reaction cell, positioned between the mass spectrometer interface and the magnetic sector analyzer. As analyte ions are directed into the collision/reaction cell, they collide with a "collision gas" fed into the cell. Energy is transferred during collision-induced reactions, which thermalizes or cools the sample ions, reducing their energy spread, and dissociates molecular ions creating potential interference masses. In Q-ICPMS, the application of a collision/reaction cell has been found to significantly enhance ion transmission, sensitivity and measurement precision, compared to Q-ICPMS without a collision cell (e.g., Becker and Dietze 2000). Collision/reaction cells have also been utilized in MC-ICPMS, most notably with the second-generation Micromass IsoProbe.

Nier Johnson double focusing. Most MC-ICPMS instruments utilize double-focusing with Nier Johnson geometry, in which ions passing through the mass spectrometer entrance slit are energy- then direction-focussed. The ICP sector magnet provides double dispersion and direction-focusing and is similar to that used in TIMS. Energy-focusing is combined with the magnet because the ion beams produced during plasma ionization show a large ion energy spread, significantly larger than the range of kinetic energies produced during thermal ionization. This must be compensated for to prevent a blurring of the image, which occurs because an ion beam passing through a magnetic field is deflected by an amount dependent on its ion energy as well as its mass and charge. Normally energy-focusing is achieved using an electrostatic analyzer. The double-focusing configuration provides the "flat-topped" peak shapes that are a requisite in high-precision isotope measurement. Nier Johnson (or reversed Nier Johnson) geometry is also used in double-focusing sector field HR-ICPMS.

The quadrupole mass analyzer utilized by Q-ICPMS is comprised of quadrupole rods with combined DC and RF potentials that can be set to allow analyte ions with a specific mass-to-charge ratio access to the detector. All other ions will collide with the quadrupole rods and will not reach the detector. In TOF-ICPMS, the analyte ions are accelerated into the flight tube with the same kinetic energy, so that ions of a different mass will also have different velocities. The lightest ions will arrive at the detector first. The entire mass spectrum will reach the detector within 50 µs, providing ~20,000 spectra/s.

Ion detection. Only moderately high precision (0.5-10%) can normally be obtained using conventional ICPMS instruments and "peak-hopping" because the ionization conditions in the plasma source are highly unstable. This causes short-term fluctuations in ion beam intensity at the 1-second level (referred to as "plasma flicker") upon which longer period intensity oscillations may be superimposed due to irregularities in nebuliser uptake and sample introduction rates. MC-ICPMS instruments alleviate these difficulties by the simultaneous collection of ion beams in multiple detectors. Isotope ratios are monitored instead of ion beam intensities (as is often the approach in TIMS and ICPMS), in order to cancel out plasma source instability, thereby increasing measurement precision dramatically. High levels of precision have also been reported for HR-ICPMS instruments employing fast electrostatic mass scanning (by adjusting the acceleration voltage while keeping the magnetic field constant) to smooth out signal fluctuations due to plasma flicker (e.g., Quetel et al. 2000a; Choi et al. 2001; Shen et al. 2002).

All MC-ICPMS instruments are equipped with a multiple Faraday collector array oriented perpendicular to the optic axis, enabling the simultaneous "static" or "multi-static" measurement of up to twelve ion beams. Most instruments use Faraday cups mounted on motorized detector carriers that can be adjusted independently to alter the mass dispersion and obtain coincident ion beams, as is the approach adopted for MC-TIMS measurement. However, some instruments instead employ a fixed collector array and zoom optics to achieve the required mass dispersion and peak coincidences (e.g., Belshaw et al. 1998).

For low-level ion detection, the first-generation Plasma 54 instrument is equipped with a single Daly detector behind a second stage ESA for the sequential measurement of low-level isotopes at high abundance sensitivity. The second-generation instruments are each equipped with multiple ion-counting channels, one operating at high abundance sensitivity, for the simultaneous measurement of multiple low-level ion beams (Figs. 7-9). This feature may greatly improve the precision and sensitivity of radium and protactinium isotope measurement, where both tracer and analyte isotopes are typically measured on low-level ion detectors.

Multiple-collection techniques. *Uranium.* Table 1 shows a typical protocol used by multi-collector instruments (equipped with one ion counting channel) both in MC-TIMS, MC-ICPMS and LA-MC-ICPMS (e.g., Cohen et al. 1992; Stirling et al. 1995; Luo et al. 1997; Stirling et al. 2000; Pietruszka et al. 2002). A first sequence monitors the atomic ratios between ^{234}U, ^{235}U and ^{238}U by aligning Faraday collectors for masses ^{235}U (10^{-13} A) and ^{238}U (10^{-11} A), while the low intensity ^{234}U ion beam (10^{-15} to 10^{-16} A) is measured simultaneously in the low-level detector. A second sequence shifts all masses to monitor ^{235}U in the ion counting channel relative to ^{238}U. A comparison of the two sequential ^{235}U/^{238}U measurements using different collector configurations provides an estimate of the drift in the relative gain between the ion counter and Faraday cup, for which the ^{234}U ion beam intensity must be corrected at the end of each two-sequence cycle. In TIMS, where the ion beams are relatively stable, either ^{235}U ion beam intensities, or more precisely ^{235}U/^{238}U ratios, can be used to monitor the gain. A second correction is also applied at the cycle level: Using empirically derived linear, power or exponential laws (e.g., Russell 1978; Wasserburg et al. 1981; Hart and Zindler 1989; Habfast 1998), all ratios are corrected for instrumental mass fractionation using ^{238}U/^{235}U normalized to the assumed "true" value of 137.88 (Steiger and Jäger 1977) for natural

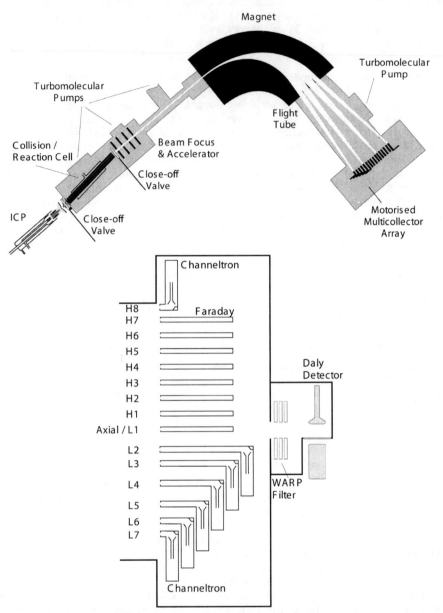

Figure 7. Schematic outline of a second-generation MC-ICPMS instrument (Micromass IsoProbe 2) (top) and its collector array (bottom). This instrument is single focusing and utilizes a collision cell to minimize the energy spread of ions entering the magnet. A schematic of the collector array also illustrates one possible configuration for multi-collection for either TIMS or MC-ICPMS, with a motorized multiple-collector array of moveable Faraday cups on the high mass side and moveable channeltrons for ion counting on the low mass side. The axial position is occupied by a removable Faraday, with a Daly detector located behind the retarding potential filter for high abundance sensitivity measurements. [Used with permission of Micromass.]

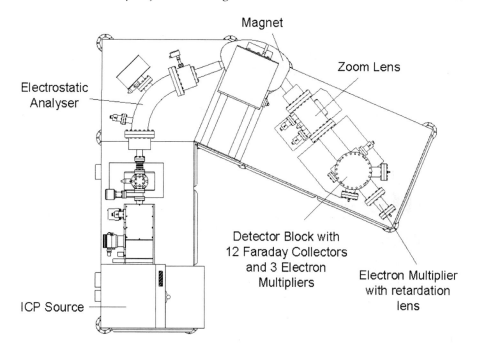
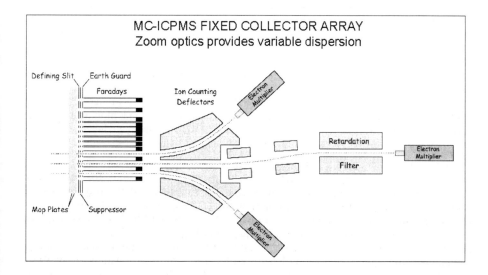

Figure 8. Schematic outline of a second-generation MC-ICPMS instrument (Nu Instruments Nu Plasma), equipped with a multiple-Faraday collector block for the simultaneous measurement of up to 12 ion beams, and three electron multipliers (one operating at high-abundance sensitivity) for simultaneous low-intensity isotope measurement. This instrument uses zoom optics to obtain the required mass dispersion and peak coincidences in place of motorized detector carriers. [Used with permission of Nu Instruments Ltd.]

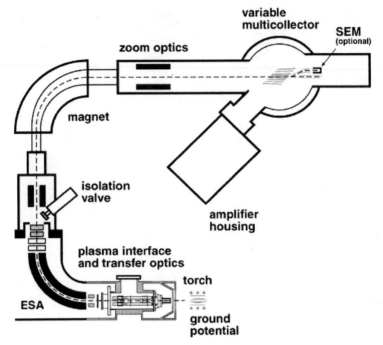

Figure 9. Schematic diagram showing a second-generation MC-ICPMS instrument (ThermoFinnigan Neptune). This instrument utilizes double-focusing and is equipped with a motorized multiple-Faraday collector block with two channels that can be operated in high-resolution mode. Optional multiple-ion counting channels are also available for the simultaneous measurement of low-intensity ion beams. [Used with permission of Thermo Finnigan.]

samples. This practice of normalizing the measured ratio to an invariant ratio of the same element is referred to as "internal normalization". The large instrumental mass fractionations inherent to MC-ICPMS (up to 1% per amu in the mass range of the U-Th isotopes for all signal intensities) necessitates careful control and monitoring of the correction factors. Applying a mass fractionation correction at the cycle level also allows the small but time dependent TIMS correction to be reliably made.

For solution nebulization MC-ICPMS and MC-TIMS, typical runs consist of up to 100 ratios collected over ~30 minutes to 1 hour. This includes time for background measurement (typically monitored at half mass on either side of the peak) and peak centering, but excludes filament warm-up time for TIMS and nebulizer cleaning for MC-ICPMS. Sample sizes of ~1000-10 ng of total U for TIMS and MC-ICPMS are required to attain 2σ measurement uncertainties of a permil or better on $^{234}U/^{238}U$ (e.g., Stirling et al. 1995; Luo et al. 1997; Yokoyama et al. 2001; Robinson et al. 2002). Smaller samples containing only 1 ng-levels of ^{238}U have been measured by laser ablation MC-ICPMS, over shorter acquisition times of ~10 minutes, yielding $^{234}U/^{238}U$ measurements at 100 μm scale resolution, with 2σ uncertainties of ±3-4‰ (Stirling et al. 2000).

Isotope dilution measurements using a ^{233}U and/or ^{236}U spike tracer can be performed separately, or combined with the isotopic composition run. Spike tracers can be measured on an ion counting channel, or alternatively, it may be preferable to concentrate the tracer (provided the spike corrections are small) and measure it on a

Techniques for Measuring U-series Nuclides: 1992-2002

Table 1. Examples of collector configurations for U-series measurements using a multiple-Faraday array coupled to a single ion counter.

		Far-1	IC	Far-2	Far-3	Far-4	Far-5
URANIUM							
Uranium Multi-Static, Internal Normalization using MC-TIMS or MC-ICPMS							
Sample	S1	(^{233}U)	^{234}U	^{235}U	(^{236}U)		^{238}U
	S2		^{235}U			^{238}U	
Uranium Static, External Standardization using MC-ICPMS							
Standard	S1	(^{233}U)	^{234}U	^{235}U	(^{236}U)		^{238}U
Sample	S1	(^{233}U)	^{234}U	^{235}U	(^{236}U)		^{238}U
Standard	S1	(^{233}U)	^{234}U	^{235}U	(^{236}U)		^{238}U
THORIUM							
Thorium Static, External Standardization (no internal mass fractionation correction) using TIMS							
Faraday-Ion Counter gain determined at the start of the session							
Sample	S1		^{230}Th		^{232}Th		
Thorium Multi-Static, Traced analysis using TIMS							
Faraday-Ion Counter gain determined during analysis							
Sample	S1	(^{229}Th)	^{230}Th		^{232}Th		
	S2		(^{229}Th)			^{232}Th	
Thorium Multi-Static, External Normalization using MC-ICPMS							
Sample	S1	(^{229}Th)	^{230}Th		^{232}Th		
	S2	(^{233}U)	^{234}U	^{235}U	(^{236}U)		^{238}U
Thorium Static, External Standardization using MC-ICPMS							
Standard	S1	(^{233}U)	^{234}U	^{235}U	(^{236}U)		^{238}U
Sample	S1	(^{229}Th)	^{230}Th		^{232}Th		
Standard	S1	(^{233}U)	^{234}U	^{235}U	(^{236}U)		^{238}U
or							
Standard	S1	(^{229}Th)	^{230}Th		^{232}Th		
Sample	S1	(^{229}Th)	^{230}Th		^{232}Th		
Standard	S1	(^{229}Th)	^{230}Th		^{232}Th		
Thorium Multi-Static, External Standardization using MC-ICPMS							
Standard	S1		^{230}Th		^{232}Th		
	S2		(^{229}Th)			^{232}Th	
Sample	S1		^{230}Th		^{232}Th		
	S2		(^{229}Th)			^{232}Th	
Standard	S1		^{230}Th		^{232}Th		
	S2		(^{229}Th)			^{232}Th	
URANIUM-THORIUM							
U-Th Multi-Static, Internal Normalization using MC-ICPMS or LA-MC-ICPMS							
Sample	S1	(^{229}Th)	^{230}Th		^{232}Th		
	S2	(^{233}U)	^{234}U	^{235}U	(^{236}U)		^{238}U
	S3		^{235}U		^{238}U		

Notes: S1 through S3 refer to respective sequences 1 through 3 in a single cycle of data acquisition. Internal normalization protocols for uranium monitor ^{235}U sequentially in a Faraday collector and in the ion counter (IC). The ion counter to Faraday relative gain is then determined by comparing (^{238}U/^{235}U)$_{Faraday}$ with (^{238}U)$_{Faraday}$/(^{235}U)$_{IC}$. Internal normalization also corrects each ratio for instrumental mass fractionation at the end of each cycle using (^{238}U/^{235}U)$_{Faraday}$ normalized to the true value of 137.88. For traced Th analysis by TIMS, the ion counter to Faraday relative gain can be determined by comparing (^{229}Th/^{232}Th)$_{Faraday}$ with (^{229}Th)$_{IC}$/(^{232}Th)$_{Faraday}$. External standardization corrects for the ion counter/Faraday gain based on normalization to bracketing measurements for well-characterized standards. The mass fractionation correction may also be applied using sample-standard bracketing, or alternatively, may be applied at the cycle level, as is the approach for internal normalization protocols. Spike tracers, ^{233}U or ^{236}U and ^{229}Th, shown in parentheses, can also be monitored during combined isotopic composition and isotope dilution runs. During LA-MC-ICPMS, ^{230}Th and ^{232}Th are monitored during the same sequence as ^{238}U (monitored on a high-mass Faraday collector; not shown) to obtain reliable measurements of ^{230}Th/^{238}U and ^{232}Th/^{238}U without isotope dilution.

Faraday collector, simultaneously with ^{238}U, ^{235}U and ^{234}U during the first sequence. This shortens the analysis routine, consuming less sample. Ion beam intensities are typically larger in MC-ICPMS than in TIMS due to the ease with which signal size can be increased by introducing a more concentrated solution. While this yields more precise data, non-linearity of the low-level detector response and uncertainties in its dead-time correction become more important for larger beam intensities, and must be carefully monitored (Cheng et al. 2000; Richter et al. 2001).

In MC-ICPMS, external standardization procedures have also been implemented (Table 1), whereby sample–standard bracketing is used to cancel out the drift in the Faraday to ion counter gain (Stirling et al. 2000; Robinson et al. 2002). In this case, U isotopic composition is determined as a static measurement in one cycle only, by normalizing mass fractionation-corrected ratios for the unknown sample to those determined on a well-constrained standard. It may also be desirable to monitor instrumental mass fractionation using external standardization techniques. A typical approach is to analyze the standard before and after the sample so that drift in the Faraday to ion counter gain can be cancelled out by linear interpolation. External standardization protocols have two main advantages. First, data acquisition times are shorter than those for multi-static protocols, so that less sample is consumed. Second, ion beam intensities are not restricted by the necessity to keep ^{235}U ion counts below 1,000,000 cps. The main disadvantages of external standardization are also two-fold. First, it is not possible to correct for short-term non-linear perturbations in the Faraday to ion counter gain, which may ultimately limit the precision and accuracy of the measurement. Second, matrix differences between the sample and standard may compromise the reliability of the mass bias correction, although during solution nebulization, these may not be significant at the permil-levels of precision attainable in U-Th isotopic measurement. Matrix-dependent effects during laser ablation MC-ICPMS, on the other hand, can be extremely large and at the 10 to 100% level (e.g., Stirling et al. 2000). External standardization was tested for solution nebulization measurements by Luo et al. (1997) using a Plasma 54 instrument, but was found to create greater variability in the data than observed for multi-static protocols, which was presumed to be due to non-linear short-term variability in the gain correction. For this reason, internal monitoring of the gain at the cycle level is often considered preferable, particularly for U where sample size is generally not limited. Nevertheless, provided the gain is stable, and care is taken to matrix-match the bracketing standard with the sample, external standardization can yield results that are of comparable precision and accuracy to those determined using internal normalization techniques (e.g., Robinson et al. 2002).

Thorium. Multiple-collector measurement protocols by TIMS for thorium isotopic analysis typically involve the simultaneous measurement of ^{232}Th and ^{230}Th (for silicate rocks), or ^{229}Th and ^{230}Th, then ^{229}Th and ^{232}Th (for low-^{232}Th samples), using an axial ion counter and off-axis Faraday collector (Table 1). Various methods are used to correct for the relative gain between the low-level and Faraday detectors and 2σ-uncertainties of 1-5‰ are typically obtained (Palacz et al. 1992; Cohen et al. 1992; McDermott et al. 1993; Rubin 2001). Charge-collection TIMS protocols enable ^{229}Th, ^{230}Th and ^{232}Th to be monitored simultaneously on a multiple-Faraday array and can achieve measurement uncertainties at the sub-permil level (Esat et al. 1995; Stirling et al. 1995).

However, thorium has only two naturally occurring long-lived isotopes, and all Th measurements by TIMS are limited by the absence of a well-constrained isotope ratio that can be used for internal normalization purposes to correct for instrumental mass fractionation. In this regard, one of the most important advantages of MC-ICPMS over MC-TIMS is the ability to admix two elements with overlapping mass ranges and use the instrumental mass discrimination factor determined for one to calculate mass fractionation corrections for the other. This approach, referred to as external normalization, relies on the mass fractionation factor being independent of the chemical properties of the analyte, but means that U chemically separated from the same sample can be admixed with the Th fraction, allowing all thorium ratios to be corrected for mass fractionation using ^{238}U/^{235}U normalized to the 137.88 true value. A two-sequence multi-static routine can then be utilized (Table 1), in which thorium isotopic composition is measured in one sequence, and uranium isotopic composition in the other (Luo et al. 1997; Stirling et al. 2001; Pietruszka et al. 2002). The U sequence provides both the mass fractionation and Faraday–ion counter gain corrections, the latter of which can be determined by monitoring ^{234}U in the ion counter and comparing the fractionation-corrected ^{234}U/^{238}U to the true ratio determined earlier from the U run. For isotope dilution runs, the protocols can be modified to include ^{229}Th measurement, either in an ion counter or Faraday cup. As is the case for U, 100 thorium ratios can be collected over 30 minutes to one hour, although Th measurement is normally significantly faster by MC-ICPMS than by MC-TIMS because thermal ionization often requires very long warm-up times of up to several hours. Very precise and accurate data can be acquired using these MC-ICPMS protocols; measurement precision is ~1‰ on ^{230}Th/^{232}Th and can be better than 1‰ on ^{230}Th/^{238}U. Typical sample sizes are ~10-100 ng of total Th for silicate rocks and 10 pg-levels of ^{230}Th for low-^{232}Th measurements. For TIMS, Th ionization efficiency is a function of load size. Therefore TIMS, with lower overall background count rates, may be best suited for small Th samples, whereas MC-ICPMS, with significantly greater ionization efficiency, may provide significantly improved results for large Th samples. In MC-ICPMS, thorium isotopic composition can be measured at the same time as U in a combined U-Th analysis. Although this consumes more Th than a two-sequence routine, analytical precision is comparable and accuracy may be better when Th and U are monitored simultaneously (Stirling et al. 2001).

External standardization was applied to silicate Th isotopic measurement using MC-ICPMS by Nakai et al. (2001). These authors determined Faraday-ion counter gain and mass fractionation corrections for thorium based on bracketing measurements for standard solutions of natural uranium. The advantage of this approach is that Th isotopic compositions can be acquired on small samples (~10 ng of ^{232}Th) as a single static measurement over a short 10-15 minute analysis period. However, because instrumental mass fractionation (and Faraday-ion counter gain) can vary between solution runs, as a function of both time and solvent loading, it is advantageous to check the veracity of the instrument calibrations using additional, matrix-matched Th standards if using techniques in external standardization. Turner et al. (2001) determined ^{230}Th/^{232}Th to a precision of 1% or better on ng levels of silicate thorium, and Robinson et al. (2002) obtained measurement uncertainties at the 1‰-level on ~250 ng of carbonate thorium using bracketing measurements for a well-characterized Th standard.

Without isotope dilution, the simultaneous measurement of U and Th is essential in LA-MC-ICPMS, although large (10-100% level) elemental, matrix-dependent fractionation effects can still be observed between U and Th (e.g., Stirling et al. 2000). As a result, Th/U ratios can be systematically lower, and apparent ^{238}U-^{234}U-^{230}Th-ages systematically younger than the true values. The source of this U-Th fractionation appears to be incomplete vaporization and ionization of laser-generated particles within the plasma (Günther 2002; Guillong and Günther 2002) and the effect becomes increasingly problematic for larger particle sizes.

Radium and Protactinium. TIMS protocols for both radium and protactinium currently involve cycling the isotopes ^{226}Ra and ^{228}Ra (tracer or normal), and ^{231}Pa and

^{233}Pa (tracer), through a single low-intensity detector (Cohen and O'Nions 1991; Volpe et al. 1991; Pickett et al. 1994; Bourdon et al. 1999). These two-isotope TIMS methods do not permit internal correction for instrumental mass fractionation, which is generally thought to be small relative to overall measurement precision (5-10‰). MC-ICPMS techniques can provide more precise corrections for mass fractionation utilizing external normalization. For example, multiple-collector measurement protocols using MC-ICPMS for ~100 fg load sizes of silicate radium have been reported by Pietruszka et al. (2002). Natural uranium was admixed to the Ra sample fractions to allow instrumental mass fractionation to be monitored using ^{238}U/^{235}U, and the isotopes ^{226}Ra and ^{228}Ra were cycled through a single Daly detector. Using this approach analytical precision to ~3‰ on ^{226}Ra/^{230}Th was obtained. However, unlike uranium and thorium, overall collection efficiency for MC-ICPMS of 0.5% is still lower than typically achieved by TIMS (1-10%).

5. COMPARISON OF ANALYTICAL METHODS FOR U-SERIES NUCLIDES

A variety of parameters for various analytical methods for uranium, thorium, radium, and protactinium isotopes are summarized in Table 2. Estimated detection limits for various techniques are also presented in Figure 10. Alpha spectrometric methods are the traditional method of measurement for these isotopes, and still provide one of the more cost effective methods of analysis. Hence they are well suited for many applications where higher detection limits and lower precision are acceptable and analytical cost is paramount (e.g., environmental monitoring). They are also in many cases the only option for analysis of very short-lived isotopes of the uranium and thorium decay series. However, for the longer-lived isotopes in terms of basic parameters including sample size requirements, detection limit, analytical precision, and time of analysis, alpha spectrometry currently lags behind other methods based on atom counting. Gamma spectrometry also continues to be a viable option for measurement of beta-emitting uranium-series nuclides with short half-lives, such as Pb-210, Ac-227, Ra-228, and Th-234 (Fig. 10). Although this method has higher detection limits than other techniques, it generally has simpler and faster sample preparation compared to alpha, beta, or mass spectrometry.

TIMS methods emerged in the late 1980's and early 1990's as an alternative to decay counting techniques. Order of magnitude improvements in sample size requirements, analytical precision, and time of analysis were achieved for uranium, thorium, protactinium, and radium analysis, with a corresponding penalty in terms of analytical cost and throughput. Although analysis times are smaller for TIMS, sample throughput is normally lower than for decay counting, for which chemical separations are not as extensive, clean laboratory chemistry space may not be required, and multiple counters can be operated simultaneously at less cost. However application of TIMS methods has led to many scientific breakthroughs in uranium-series geochronology and geochemistry.

In the past three years, MC-ICPMS has emerged as an alternative to TIMS for precise measurement of the U-series isotopes with comparable or better precision. U-Th isotopes can now be routinely measured at the sub-permil level. Previously, this had only been demonstrated using charge-collection TIMS applied to thorium isotope measurement. Data collection efficiency, sample size requirements, and detection limits can also be greatly improved over TIMS. For the ^{238}U-^{234}U-^{230}Th system applied to carbonate samples, this has extended the dating range beyond 600,000 years, and ^{230}Th-age uncertainties of ±2000 years are now attainable on 300,000 year-old samples (e.g., Stirling et al. 2001).

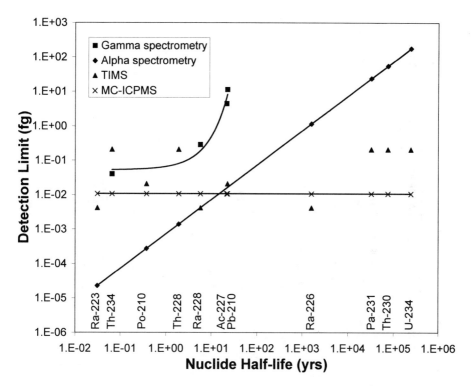

Figure 10. Estimates of detection limits as a function of uranium-series nuclide half-life for various decay and atom counting techniques. Gamma spectrometry has the highest detection limits but also is the simplest and least expensive method of uranium-series analysis in terms of sample preparation. Alpha spectrometry has higher detection limits for longer-lived nuclides but provides optimal detection for short-lived (<10 yr half-life) nuclides. TIMS sensitivities vary with element analyzed and ionization efficiency, but TIMS provides improved detection limits (~0.1 fg) for longer lived nuclides. MC-ICPMS detection limits are relatively element independent (~0.01 fg). Detection limits are generally lower than for TIMS due to improved signal collection utilization (ion multi-collection) and ionization/collection efficiency, although for some elements (e.g., radium) TIMS would provide optimal detection. Detection limits are based on counting statistics only (three times the standard deviation of the background) and are estimated for situations where interferences or isobars are absent. In cases of longer-lived nuclides (^{235}U, ^{238}U, ^{232}Th), detection limits are blank limited. Parameters used for each method are: 1) Gamma spectrometry: 7 day count period, 3% detector collection efficiency, 10 count per day background; 2) Alpha spectrometry: 7 day count period, 30% detector collection efficiency, 1 count per day background; 3) TIMS: ionization/collection efficiency ranges from 0.1 to 5%, 1 count per second background, 60 min total measurement period with 10% collection utilization 4) MC-ICPMS: ionization/collection efficiency is 0.5%, 1 count per second background, 20 min total measurement period with 50% collection utilization.

Conventional ICPMS has been widely applied to U-series isotope measurement over the past decade (e.g., Shaw and Francois 1991; Heumann et al. 1998; Becker et al. 1999; Hinrichs and Schnetger 1999; Platzner et al. 1999; Becker et al. 2000; Becker and Dietze 2000; Halicz et al. 2000; Quetel et al. 2000a; Quetel et al. 2000b; Choi et al. 2001; McLean et al. 2001; Shen et al. 2002). Isotope ratio measurements are based on the successive "peak-jumping" measurement of isotopes as they are cycled through a single electron multiplier. Using Q-ICPMS, analytical precision (2σ) approaches 3-10% for

Table 2. Comparison of analytical methods for U-series nuclides

		Alpha Spec.	TIMS	MC-TIMS	ICPMS	MC-ICPMS	LA-ICPMS	LA-MC-ICPMS	SIMS
Sample Size	U	1-100 µg	100-2000 ng	10-1000 ng	10-450 ng	10-450 ng	1 ng	1 ng	---
	Th (silicates)	1-100 µg	100-2000 ng	10-1000 ng	10-600 ng	10-600 ng	1 ng	1 ng	10-100 ng
	Pa	10-50 pg	30-1000 fg	---	10-1000 fg	---	---	---	---
	Ra	10-20 pg	1-1000 fg	---	20-1000 fg	10-3600 fg	---	---	---
Precision (2σ)	^{238}U	2-10%	0.1%	0.1%	0.1-5%	<0.1%	5%	<0.2%	---
	^{235}U/^{238}U	2-10%	0.5%	0.1%	0.5%	0.01%	4%	<0.2%	---
	^{234}U/^{238}U	2-10%	0.5%	0.1-0.2%	0.1-10%	0.05-0.2%	1-2%	0.4%	---
	^{232}Th	2-10%	0.3%	0.3%	0.5-5%	<0.1%	10%	---	---
	^{230}Th/^{232}Th	2-10%	0.5%	0.1-0.4%	0.5-10%	0.1-0.3%	1-12%	0.8%	0.5-1.0%
	^{231}Pa/^{235}U	2-10%	0.3-1.0%	---	1-5%	---	---	---	---
	^{226}Ra	2-10%	0.5-1.5%	---	2-6%	0.3%	---	---	---
Meas. Time	U	1-28 day	4 hrs	1-2 hrs	1-10 min	15-30 min	1-5 min	5-10 min	---
	Th	1-28 day	4 hrs	2 hrs	1-10 min	15-60 min	1-5 min	5-10 min	1 hr
	Pa	1-28 day	3 hrs	---	3 min	---	---	---	---
	Ra	1-28 day	3 hrs	---	3 min	30 min	---	---	---
Detector Type	U, Th, Pa, Ra	Semiconductor	Single ion counter	Multiple-Faraday + single or multiple-ion counters	Single ion counter	Multiple-Faraday + single or multiple-ion counters	Single ion counter	Multiple-Faraday + single or multiple counters	Multiple-Faraday + single ion counter

		Alpha Spec.	TIMS	MC-TIMS	ICPMS	MC-ICPMS	LA-ICPMS	LA-MC-ICPMS	SIMS
Sensitivity (Collection Efficiency)	U	N/A	0.01-1% load size dependent	0.01-1% load size dependent	0.01-1% nebulizer & instrument dependent (HR > Q)	0.01-1% nebulizer & instrument dependent ($1^{st} < 2^{nd}$)	0.01-0.1% matrix & instrument dependent (HR > Q)	0.1% matrix & instrument dependent ($1^{st} < 2^{nd}$)	---
	Th	N/A	0.01-1% load size dependent	0.01-1% load size dependent	0.01-1% nebulizer & instrument dependent (HR > Q)	0.01-1% nebulizer & instrument dependent ($1^{st} < 2^{nd}$)	0.01-0.1% matrix & instrument dependent (HR > Q)	0.01-0.1% matrix & instrument dependent ($1^{st} < 2^{nd}$)	2%
	Pa	N/A	0.1-1.0%	0.1-1.0%	0.1% nebulizer & instrument dependent (HR > Q)	---	---	---	---
	Ra	N/A	1-10%	1-10%	0.01-1% nebulizer & instrument dependent (HR > Q)	0.5% nebulizer & instrument dependent ($1^{st} < 2^{nd}$)	---	---	---
Instrumental Mass Fractionation	U	N/A	<0.3%/AMU Time-dependent	<0.3%/AMU Time-dependent	<0.5%/AMU Time-independent	<1%/AMU Time-independent	<0.5%/AMU Time-independent	<1%/AMU Time-independent	<0.1%/AMU
	Th	N/A	<0.3%/AMU Time-dependent	<0.3%/AMU Time-dependent	<0.5%/AMU Time-independent	<1%/AMU Time-independent	<0.5%/AMU Time-independent	<1%/AMU Time-independent	---
	Pa	N/A	<0.3%/AMU Time-dependent	<0.3%/AMU Time-dependent	<0.5%/AMU Time-independent	---	---	---	---
	Ra	N/A	<0.3%/AMU Time-dependent	<0.3%/AMU Time-dependent	---	<1%/AMU Time-independent	---	---	---

Notes: HR = High Resolution HR-ICPMS, Q = Quadrupole Q-ICPMS, 1st = First generation MC-ICPMS, 2nd = Second generation MC-ICPMS.

various isotope ratios and is limited by instability of the plasma during sequential peak measurement and difficulties in obtaining reliable peak shapes. In general, higher precision measurements to the permil-level can be obtained from HR-ICPMS because the magnetic sector design produces flat-topped peak shapes, significantly higher sensitivities and lower instrumental background, resulting in higher signal-to-noise ratios and lower detection limits. Low abundance sensitivity (at the ~5 ppm level) can affect the accuracy of ^{230}Th/^{232}Th measurements on silicate samples, and accuracy is typically monitored using a well-characterized matrix-matched standard, and corrected by employing a tailing correction (e.g., Shen et al. 2002). Corrections for instrumental mass fractionation are normally obtained using external standardization procedures, although internal normalization has also been applied (e.g., Shen et al. 2002). HR-ICPMS instruments can be operated in high-resolution mode (normal mass resolution M/ΔM is 300) in order to resolve molecular interferences and/or to improve abundance sensitivity (e.g., Hinrichs and Schnetger 1999), but this results in a decrease in sensitivity and in precision. Techniques in Q-ICPMS can rival those of HR-ICPMS if collision cells are installed (e.g., Becker and Dietze 2000). TOF-ICPMS allows measurements of up to 20,000 spectra per second but to date, has not been extensively applied to the U-series systems.

None of the above-mentioned ICPMS techniques can rival MC-TIMS and MC-ICPMS in terms of analytical precision, but the advantage of conventional ICPMS lies in the speed and ease with which data can be acquired. Analysis times are typically less than 10 minutes, and results can be obtained on solid, liquid or gas samples directly, without chemical preparation. Direct analysis will, however, give rise to high levels of molecular ion formation (e.g., ^{207}Pb^{16}O^{12}C$^+$ or ^{203}Tl^{16}O$_2^+$ on ^{235}U$^+$) and matrix effects, which limit precision and are difficult to correct.

SIMS techniques have occupied somewhat of a narrower niche in uranium-series analysis, but have significantly improved Th isotope analysis relative to TIMS for chemically separated samples. The major improvement relative to TIMS is an improvement by about an order of magnitude in efficiency or sample size requirements for silicates. For uranium and/or thorium rich minerals such as carbonates and zircons, both SIMS and laser-ablation MC-ICPMS have been used for the direct *in situ* analysis of U and Th isotopes (Reid et al. 1997; Stirling et al. 2000) on very small (pg to ng levels of total U and Th) samples, at 10-100 μm scale resolution.

6. FUTURE DEVELOPMENTS

MC-ICPMS instruments have only recently been applied for uranium-series analysis, and so there is much potential for further development of measurement techniques using their unique capabilities. Second-generation instruments are now equipped with multi-collecting ion counting capabilities, and such capabilities are also available for TIMS instruments. In particular, application of low-level multi-collection to uranium, thorium, radium, and protactinium analysis may greatly improve analytical sensitivity for these measurements, thereby extending analysis to materials with lower U and Th concentrations such as mineral separates for internal mineral isochron dating applications. Low-level multi-collection may also significantly improve many aspects of direct uranium and thorium analysis of minerals using laser-ablation techniques. Finally, multi-collection ion counting may extend the range of analytes to uranium-series isotopes with shorter half-lives, such as lead-210 and actinium-227.

For both TIMS and MC-ICPMS, improvements in sources are likely to play a major role in enhancing measurement sensitivity. This includes further development of cavity sources for TIMS and newer ICP sources and nebulizers with reduced isobaric interferences or background. With regard to measurement sensitivity, one of the current weaknesses of MC-ICPMS lies in the inefficiency by which ions are transferred from the plasma source into the mass spectrometer. The challenge of future instrumentation will be to improve the overall sensitivity in this "interface" region, resulting in additional improvements in detection limits for many elements. However, continued application of MC-ICPMS techniques should almost certainly lead to enhanced analytical throughput and decreased analytical costs. For example, it may be possible to reduce costly and time-intensive chemical separations for many types of uranium-series analysis, with particular emphasis on aqueous systems. This may lead to larger uranium-series data sets than are now possible, improving both the quality and quantity of uranium-series applications and studies in geochemistry and geochronology.

ACKNOWLEDGMENTS

We thank G. Henderson, A. Pietruszka, T. Elliott, M. Murrell and M. Rehkämper for insightful reviews. SJG thanks the Department of Energy, Office of Basic Energy Sciences, Geosciences Research Program and the LANL Laboratory Directed Research and Development (LDRD) Program, for geochemistry support. CHS was primarily supported by Swiss NF grants.

REFERENCES

Bard E, Hamelin B, Fairbanks RG, Zindler A, Mathieu G, Arnold M (1990) U/Th and ^{14}C ages of corals from Barbados and their use for calibrating the ^{14}C time scale beyond 9000 years BP. Nucl Instrum Methods Phys Res Sect B 52:461-468

Becker JS, Dietze H-J, McClean JA, Montaser A (1999) Ultratrace and isotope analysis of long-lived radionuclides by inductively coupled plasma quadrupole mass spectrometry using a direct injection high efficiency nebulizer. Anal Chem 71:3077-3984

Becker JS, Dietze H-J (2000) Precise and accurate isotope ratio measurements by ICP-MS. Fresenius J Anal Chem 368:23-30

Becker JS, Pickhardt C, Dietze H-J (2000) Laser ablation inductively coupled plasma mass spectrometry for the trace, ultratrace and isotope analysis of long-lived radionuclides in solid samples. Intl J Mass Spectrom 202:283-297

Belshaw NS, Freedman PA, O'Nions RK, Frank M, Guo Y (1998) A new variable dispersion double focusing plasma mass spectrometer with performance illustrated for Pb isotopes. Intl J Mass Spectrom 181:51-58

Blichert-Toft J, Alberede F (1997) The Lu-Hf isotope geochemistry of chondrites and the evolution of the mantle-crust system. Earth Planet Sci Lett 148:243-258

Bourdon B, Zindler A, Worner G (1994) Evolution of the Laacher See magma chamber: evidence from SIMS and TIMS measurements of U-Th disequilibria in minerals and glasses. Earth Planet Sci Lett 126:75-90

Bourdon B, Joron J-L, Allegre CJ (1999) A method for ^{231}Pa analysis by thermal ionization mass spectrometry in silicate rocks. Chem Geol 157:147-151

Burnett WC, Yeh CC (1995) Separation of protactinium from geochemical materials via extraction chromatography. Radioact Radiochem 6:22-32

Chabaux F, Ben Othman D, Birck JL (1994) A new Ra-Ba chromatographic separation and its application to Ra mass spectrometric measurement in volcanic rocks. Chem Geol 114:191-197

Chen JH, Edwards RL, Wasserburg GJ (1986) ^{238}U, ^{234}U, and ^{232}Th in seawater. Earth Planet Sci Lett 80:241-251

Chen JH, Edwards RL, Wasserburg GJ (1992) Mass spectrometry and applications to uranium-series disequilibrium. *In*: Uranium-Series Disequilibrium: Applications to Earth, Marine, and Environmental Sciences, 2nd Ed. Ivanovich M, Harmon RS (eds) Oxford Univ. Press, Oxford

Cheng H, Edwards RL, Hoff J, Gallup CD, Richards DA, and Asmerom Y (2000) The half-lives of uranium-234 and thorium-230. Chem Geol 169:17-33

Choi MS, Francois R, Sims K, Bacon MP, Brown-Leger S, Fleer AP, Ball L, Schneider D, Pichat S (2001) Rapid determination of ^{230}Th and ^{231}Pa in seawater by desolved micro-nebulization Inductively Coupled Plasma magnetic sector mass spectrometry. Mar Chem 76:99-112

Cochran JK, Masqué P (2003) Short-lived U/Th-series radionuclides in the ocean: tracers for scavenging rates, export fluxes and particle dynamics. Rev Mineral Geochem 52:461-492

Cohen AS, O'Nions RK (1991) Precise determination of femtogram quantities of radium by thermal ionization mass spectrometry. Anal Chem 63:2705-2708

Cohen AS, Belshaw NS, O'Nions RK (1992) High precision uranium, thorium, and radium isotope ratio measurements by high dynamic range thermal ionization mass spectrometry. Intl J Mass Spectrom Ion Processes 116:71-81

Dacheux N, Aupiais J (1997) Determination of uranium, thorium, plutonium, americium, and curium ultratraces by photon electron rejecting alpha liquid scintillation. Anal Chem 69:2275-2282

Duan YX, Chamberlin EP, Olivares JA (1997) Development of a new high-efficiency thermal ionization source for mass spectrometry. Intl J Mass Spectrom Ion Processes 161:27-39

Edwards RL, Chen JH, Wasserburg GJ (1987) ^{238}U-^{234}U-^{230}Th-^{232}Th systematics and the precise measurement of time over the past 500,000 years. Earth Planet Sci Lett 81:175-192

Eggins SM, Kinsley LPJ, Shelley JMG (1998) Deposition and element fractionation processes occurring during atmospheric pressure sampling for analysis by ICPMS. Appl Surf Sci 129:278-286

Eikenberg J, Vezzu G, Zumsteg I, Bajo S, Ruethi M, Wyssling G (2001a) Precise two chronometer dating of Pleistocene travertine: The $^{230}Th/^{234}U$ and $^{226}Ra_{ex}/^{226}Ra(0)$ approach. Quat Sci Rev 20:1935-1953

Eikenberg J, Tricca A, Vezzu G, Bajo S, Ruethi M, Surbeck H (2001b) Determination of ^{228}Ra, ^{226}Ra, and ^{224}Ra in natural water via adsorption on MnO_2-coated discs. J Environ Radioact 54:109-131

El-Daoushy F, Garcia-Tenorio R (1995) Well Ge and semi-planar Ge (HP) detectors for low-level gamma spectrometry. Nucl Instrum Methods Phys Res Sect A 356:376-384

England JG, Zindler A, Reisberg LC, Rubenstone JL, Salters V, Marcantonio F, Bourdon B, Brueckner H, Turner PJ, Weaver S, Read P (1992) The Lamont-Doherty Geological Observatory Isolab-54 isotope ratio mass spectrometer. Intl J Mass Spectrom Ion Processes 121:201-240

Esat TM (1995) Charge collection thermal ion mass spectrometry of thorium. Intl J Mass Spectrom Ion Processes 148:159-170

Faure G (1977) Principles of isotope geology. J. Wiley and Sons, New York

Goldstein SJ, Murrell MT, Janecky DR (1989) Th and U isotopic systematics of basalts from the Juan de Fuca and Gorda Ridges by mass spectrometry. Earth Planet Sci Lett 96:134-146

Goldstein SJ, Rodriguez JM, Lujan N (1997) Measurement and application of uranium isotopes for human and environmental monitoring. Health Phys 72:10-18

Gray AL (1985). Solid sample introduction by laser ablation for inductively coupled plasma source mass spectrometry. Analyst 110:551-556

Guillong M, Günther D (2002) Effect of particle size distribution on ICP-induced elemental fractionation in laser ablation-inductively coupled plasma-mass spectrometry. J Anal At Spectrom 7:831-837

Günther D (2002) Laser-ablation inductively coupled plasma mass spectrometry. Anal Bioanal Chem 372:31-32

Günther D, Heinrich CA (1999) Enhanced sensitivity in laser ablation-ICP mass spectrometry using helium-argon mixtures as aerosol carrier. J Anal At Spectrom 14:1363-1368

Habfast K (1998) Fractionation correction and multiple collectors in thermal ionization isotope ratio mass spectrometry. Intl J Mass Spectrom 176:133-148

Halicz L, Segal I, Gavrieli I, Lorber A, Karpas Z (2000) Determination of the $^{234}U/^{238}U$ ratio in water samples by inductively coupled plasma mass spectrometry. Anal Chim Acta 422:203-208

Halliday AN, Lee D-C, Christensen JN, Walder AJ, Freedman PA, Jones CE, Hall CM, Yi W, and Teagle D. (1995) Recent developments in inductively coupled plasma magnetic sector multiple collector mass spectrometry. Intl J Mass Spec Ion Proc 146/147:21-33

Halliday AN, Lee D-C, Christensen JN, Rehkamper M, Yi W, Luo X-Z, Hall CM, Ballentine CJ, Pettke T, Stirling C (1998) Applications of multiple collector-ICPMS to cosmochemistry, geochemistry, and paleoceanography. Geochim Cosmochim Acta 62:919-940

Hart SR, Zindler A (1989) Isotope fractionation laws: A test using calcium. Intl J Mass Spectrom Ion Processes 89:287-301

Heumann KG, Gallus SM, Rädlinger G, Vogl J (1998) Precision and accuracy in isotope ratio measurements by plasma source mass spectrometry. J Anal At Spectrom 13:1001-1008

Hinrichs J, Schnetger B (1999) A fast method for the simultaneous determination of ^{230}Th, ^{234}U and ^{235}U with isotope dilution sector field ICP-MS. Analyst 124:927-932

Hollenbach M, Grohs J, Mamich S, Kroft M, Denoyer ER (1994) Determination of ^{99}Tc, ^{230}Th, and ^{234}U in soils by inductively-coupled plasma-mass spectrometry using flow-injection preconcentration. J Anal At Spectrom 9:927-933

Horn I, Rudnick RL, McDonough WF (2000) Precise elemental and isotope ratio determination by simultaneous solution nebulization and laser ablation-ICP-MS: Application to U-Pb geochronology. Chem Geol 164:281-301

Horwitz EP, Dietz ML, Fisher DE (1991) Separation and preconcentration of strontium from biological, environmental, and nuclear waste samples by extraction chromatography using a crown ether. Anal Chem 63:522-525

Horwitz EP, Dietz ML, Chiarizia R, Diamond H, Essling AM, Graczyk D (1992) Separation and preconcentration of uranium from acidic media by extraction chromatography. Anal Chim Acta 266:25-37

Horwitz EP, Chiarizia R, Dietz ML, Diamond H, Nelson DM (1993a) Separation and preconcentration of actinides from acidic media by extraction chromatography. Anal Chim Acta 281:361-372

Horwitz EP, Chiarizia, R., Diamond H, Gatrone RC, Alexandratos SD, Trochimzuk AQ, Crick DW (1993b) Uptake of metal ions by a new chelating ion exchange resin. 1. Acid dependencies of actinide ions. Solvent Extr Ion Exch 11:943-966

Huang M, Hirabayashi A, Shirasaki T, Koizumi H (2000) A multimicrospray nebulizer for microwave-induced plasma mass spectrometry. Anal Chem 72:2463-2467

Ivanovich M, Murray A (1992) Spectroscopic methods. *In:* Uranium-Series Disequilibrium: Applications to Earth, Marine, and Environmental Sciences, 2nd Ed. Ivanovich M, Harmon RS (eds) Oxford Univ. Press, Oxford

James WD, Boothe PN, Presley BJ (1998) Compton suppression gamma-spectroscopy in the analysis of radium and lead isotopes in ocean sediments. J Radioanal Nucl Chem 236:261-265

Jarvis KE, Gray AL, Houk RS (1992) Handbook of Inductively Coupled Plasma Mass Spectrometry, Blackie, Glasgow

Joannon S, Pin C (2001) Ultra-trace determination of ^{226}Ra in thermal waters by high sensitivity quadrupole ICP-mass spectrometry following selective extraction and concentration using radium-specific membrane disks. J Anal At Spectrom 16:32-37

Johnson SG, Fearey BL (1993) Spectroscopic study of thorium using continuous-wave resonance ionization mass-spectrometry with ultraviolet ionization. Spectrochim Acta Part B 48:1065-1077

Knoll GF (1989) Radiation Detection and Measurement. J. Wiley and Sons, New York

Kuss HM (1992) Applications of microwave digestion technique for elemental analyses. Fresenius J Anal Chem 343:788-793

Lally AE (1992) Chemical procedures. In: Uranium-Series Disequilibrium: Applications to Earth, Marine, and Environmental Sciences, 2nd Ed. Ivanovich M, Harmon RS (eds) Oxford Univ. Press, Oxford

Lamble KJ, Hill SJ (1998) Microwave digestion procedures for environmental matrices. Analyst 123:R103-R133

Layne GD, Sims KW (2000) Secondary ion mass spectrometry for the measurement of ^{232}Th/^{230}Th in volcanic rocks. Intl J Mass Spectrom 203:187-198

Lee D-C, Halliday AN (1995) Precise determinations of the isotopic compositions and atomic weights of molybdenum, tellurium, tin and tungsten using ICP source magnetic sector multiple collector mass spectrometry. Intl J Mass Spectrom Ion Processes 146/147:35-46

Liu H, Montaser A (1994) Phase-Doppler diagnostic studies of primary and tertiary aerosols produced by a new high-efficiency nebulizer. Anal Chem 66:3233-3242

Ludwig KR, Simmons KR, Szabo BJ, Winograd IJ, Landwehr JM, Riggs AC, Hoffman RJ (1992) Mass spectrometric ^{230}Th-^{234}U-^{238}U dating of the Devils Hole calcite vein. Science 258:284-287

Lundstrom CC, Gill JB, Williams Q, Hanan BB (1998) Investigating solid mantle upwelling rates beneath mid-ocean ridges using U-series disequilibria, II. A local study at 33°S Mid-Atlantic Ridge. Earth Planet Sci Lett 157:167-181

Luo X, Rehkämper M, Lee D-C, Halliday AN (1997) High precision ^{230}Th/^{232}Th and ^{234}U/^{238}U measurements using energy-filtered ICP magnetic sector multiple collector mass spectrometry. Intl J Mass Spectrom Ion Processes 171:105-117

McLean JA, Becker JS, Boulyga SF, Dietze H-J, Montaser A (2001) Ultratrace and isotopic analysis of long-lived radionuclides by double-focusing sector field inductively coupled plasma mass spectrometry using direct liquid sample introduction. Intl J Mass Spectrom 208:193-204

McDermott F, Elliott TR, van Calsteren P, Hawkesworth CJ (1993) Measurement of Th-230/Th-232 ratios in young volcanic rocks by single-sector thermal ionization mass-spectrometry. Chem Geol 103:283-292

Marechal CN, Telouk P, Alberede F (1999) Precise analysis of copper and zinc isotopic compositions by plasma-source mass spectrometry. Chem Geol 156:251-273

Martin P, Hancock GJ, Paulka S, Akber RA (1995) Determination of Ac-227 by alpha-particle spectrometry. Appl Radiat Isot 46:1065-1070

Morvan K, Andres Y, Mokili B, Abbe JC (2001) Determination of radium-226 in aqueous solutions by alpha-spectrometry. Anal Chem 73:4218-4224

Nakai S, Fukuda S, Nakada S (2001) Thorium isotopic measurements on silicate rock samples with a multi-collector inductively coupled plasma mass spectrometer. Analyst 126:1707-1710

Neder H, Heusser G, Laubenstein M (2000) Low-level γ-ray germanium-spectrometer to measure very low primordial radionuclide concentrations. Appl Radiat Isot 53:191-195

Palacz ZA, Freedman PA, Walder AJ (1992) Thorium isotope ratio measurements at high abundance sensitivity using a VG 54-30, an energy-filtered thermal ionization mass spectrometer. Chem Geol 101:157-165

Pickett DA, Murrell MT, Williams RW (1994) Determination of femtogram quantities of protactinium in geologic samples by thermal ionization mass spectrometry. Anal Chem 66:1044-1049

Pietruszka AJ, Carlson RW, Hauri EH (2002) Precise and accurate measurement of ^{226}Ra-^{230}Th-^{238}U disequilibria in volcanic rocks using plasma ionization multicollector mass spectrometry. Chem Geol 188:171-191

Platzner IT, Becker JS, Dietze H-J (1999) Stability study of isotope ratio measurements for uranium and thorium by ICP-QMS. At Spectrosc 20:6-12

Quétel CR, Vogl J, Prohaska T, Nelms S, Taylor PDP, De Bievre P (2000a) Comparative performance study of ICP mass spectrometers by means of U "isotopic measurements." Fresenius J Anal Chem 368:148-155

Quétel CR, Prohaska T, Hamester M, Kerl W, Taylor PDP (2000b) Examination of the performance exhibited by a single detector double focusing magnetic sector ICP-MS instrument for uranium isotope abundance ratio measurements over almost three orders of magnitude and down to pg g-1 concentration levels. J Anal At Spectrom 15:353-358

Raptis K, Mayer K, Hendrickx F, De Bievre P (1998) Preparation and certification of new thorium isotopic reference materials. Fresenius J Anal Chem 361:400-403

Rehkämper M, Halliday AN (1999) The precise measurement of Tl isotopic compositions by MC-ICP-MS: Application to the analysis of geological materials and meteorites. Geochim Cosmochim Acta 63:935-944

Reid MR, Coath CD, Harrison TM, McKeegan KD (1997) Prolonged residence time for the youngest rhyolites associated with Long Valley Caldera, ^{230}Th-^{238}U ion microprobe dating of young zircons. Earth Planet Sci Lett 150:27-39

Richter S, Goldberg SA, Mason PB, Traina AJ, Schwieters JB (2001) Linearity tests for secondary electron multipliers used in isotope ratio mass spectrometry. Intl J Mass Spectrom 206:105-127

Rihs S, Condomines M, Sigmarsson O (2000) U, Ra, and Ba incorporation during precipitation of hydrothermal carbonates: implications for ^{226}Ra-Ba dating of impure travertines. Geochim Cosmochim Acta 64:661-671

Robinson LF, Henderson GM, Slowey NC (2002) U-Th dating of marine isotope stage 7 in Bahamas slope sediments. Earth Planet Sci Lett 196:175-187

Rubin KH, Macdougall JD, Perfit MR (1994) ^{210}Po-^{210}Pb dating of recent volcanic eruptions on the sea floor. Nature 368:841-844

Rubin KH (2001) Analysis of ^{232}Th/^{230}Th in volcanic rocks: A comparison of thermal ionization mass spectrometry and other methodologies. Chem Geol. 175:723-750

Russell WA, Papanastassiou DA, Tombrello TA (1978) Ca isotope fractionation on the Earth and other Solar System materials. Geochim Cosmochim Acta 42:1075-1090

Semkow TM, Parekh PP, Schwenker CD, Khan AJ, Bari A, Colaresi JF, Tench OK, David G, Guryn W (2002) Low-background gamma spectrometry for environmental radioactivity. Appl Radiat Isot 57:213-223

Sen Gupta JG, Bertrand NB (1995) Direct ICP-MS determination of trace and ultratrace elements in geological materials after decomposition in a microwave oven I. Quantitation of Y, Th, U, and the lanthanides. Talanta 42:1595-1607

Shaw TJ, Francois R (1991) A fast and sensitive ICP-MS assay for the determination of ^{230}Th in marine sediments. Geochim Cosmochim Acta 55:2075-2078

Shen CC, Edwards RL, Cheng H, Dorale JA, Thomas RB, Moran SB, Weinstein SE, Edmonds HN (2002) Uranium and thorium isotopic and concentration measurements by magnetic sector inductively coupled plasma mass spectrometry. Chem Geol 185:165-178

Steiger RH, Jäger E (1977) Subcommission on geochronology: Convention on the use of decay constants in geo- and cosmochronology. Earth Planet Sci Lett 36:359-362

Stirling CH, Esat TM, McCulloch MT, Lambeck K (1995) High-precision U-series dating of corals from Western Australia and implications for the timing and duration of the Last Interglacial. Earth Planet Sci Lett 135: 115-130.

Stirling CH, Lee D-C, Christensen JN, Halliday AN (2000) High-precision in situ ^{238}U-^{234}U-^{230}Th isotopic analysis using laser ablation multiple-collector ICPMS. Geochim Cosmochim Acta 64:3737-3750

Stirling CH, Esat TM, Lambeck K, McCulloch MT, Blake SG, Lee D-C, Halliday AN (2001) Orbital forcing of the Marine Isotope Stage 9 interglacial. Science 291:290-293

Surbeck H (2000) Alpha spectrometry sample preparation using selectively adsorbing thin films. Appl Radiat Isot 53:97-100

Taylor PDP, De Bievre P, Walder AJ, Entwistle A (1995) Validation of the analytical linearity and mass discrimination correction model exhibited by a Multiple Collector Inductively Coupled Plasma Mass Spectrometer by means of a set of synthetic uranium isotope mixtures. J Anal At Spectrom 10:395-398

Totland M, Jarvis I, Jarvis KE (1992) An assessment of dissolution techniques for the analysis of geological samples by plasma spectrometry. Chem Geol 95:35-62

Turner S, van Calsteren P, Vigier N, Thomas L (2001) Determination of thorium and uranium isotope ratios in low concentration geological materials using a fixed multi-collector-ICP-MS. J Anal At Spectrom 16:612-615

Vanhaecke F, Van Holderbeke M, Moens L, Dams R (1996) Evaluation of a commercially available microconcentric nebulizer for inductively coupled plasma mass spectrometry. J Anal At Spectrom 11:543-548

Volpe AM, Olivares JA, Murrell MT (1991) Determination of radium isotope ratios and abundances in geologic samples by thermal ionization mass spectrometry. Anal Chem 63:913-916

Walder AJ, Freedman PA (1992) Isotopic ratio measurement using a double focusing magnetic sector mass analyser with an inductively coupled plasma as an ion source. J Anal At Spectrom 7:571-575

Walder AJ, Koller D, Reed NM, Hutton RC, Freedman PA (1993) Isotope ratio measurement by inductively coupled plasma multiple collector mass spectrometry incorporating a high efficiency nebulization system. J Anal At Spectrom 8:1037-1041

Wasserburg GJ, Jacobsen SB, DePaolo DJ, McCulloch MT, and Wen T (1981) Precise determination of Sm/Nd ratios, Sm and Nd isotopic abundances in standard solutions. Geochim Cosmochim Acta 45:2311-2323

Wayne DM, Hang W, McDaniel DK, Fields RE, Rios E, Majidi V (2002) The thermal ionization cavity (TIC) source: elucidation of possible mechanisms for enhanced ionization efficiency. Intl J Mass Spectrom 216:41-57

Williams RW, Collerson KD, Gill JB, Deniel C (1992) High Th/U ratios in subcontinental lithospheric mantle: mass spectrometric measurement of Th isotopes in Gaussberg lamproites. Earth Planet Sci Lett 111:257-268

Yokoyama T, Makishima A, Nakamura E (2001) Precise analysis of $^{234}U/^{238}U$ ratio using UO_2^+ ion with thermal ionization mass spectrometry for natural samples. Chem Geol 181:1-12

3 Mineral-Melt Partitioning of Uranium, Thorium and Their Daughters

Jonathan Blundy and Bernard Wood

CETSEI, Department of Earth Sciences
University of Bristol
Wills Memorial Building
Bristol, BS8 1RJ, United Kingdom

1. INTRODUCTION

The uranium and thorium decay series (hereafter "U-series") include the nuclides of ten elements, all of which can be found at trace levels in rocks and minerals. The relatively short half-lives of the U-series nuclides give them considerable potential to decipher a wide variety of natural processes. The common observation of secular radioactive disequilibrium between parent and daughter nuclides provides a time dimension that is not possible with the more commonly used trace elements. However, just like conventional trace elements, the behavior of U-series elements depends on their partitioning between coexisting phases, such as minerals and melts. Interpreting radioactive disequilibrium behavior of the U-series critically requires an understanding of how parent and daughter nuclides of these elements are fractionated one from another under the conditions of interest. Without appropriate partition coefficients (D) it is difficult to separate that part of any disequilibrium signal that is due to process and that part which is due to time. This problem is minimized, but by no means eliminated, by the use of activity ratios rather than concentration ratios, as conventionally used for trace elements. But still, there is very little point in determining isotopic concentrations at the sub-femtogram level, if the data themselves cannot be interpreted or modeled with comparable precision and accuracy. Unfortunately, with the partial exception of U, Th and Pb, our knowledge of partitioning of the U-series elements lags well behind our ability to measure them. The problem is exacerbated by the fact that, because nearly all of the elements of interest are highly incompatible ($D \ll 1$) in all common silicate and oxide minerals, and many lack stable or long half-life isotopes, there are serious technical difficulties associated with determining their partition coefficients experimentally

The purpose of this chapter is to first establish a case for the importance of using the correct partition coefficients. Second we review experimental techniques for measuring partitioning and highlight any new advances that may be suitable for the U-series. Third we describe and implement a simple lattice strain model that can be used to estimate one partition coefficient from that of a "proxy" with the same charge and similar ionic radius. Finally we propose a set of proxies for U-series elements in the key mineral phases in magmatic environments. The partitioning behavior of the proxies is well understood in a wide range of conditions, and can be used to derive partition coefficients for those U-series elements for which there are presently no experimental determinations. The elements to be considered include nine of the ten U-series members: uranium, thorium, protactinium, radium, radon, actinium, polonium, bismuth and lead. In view of the very short half-lives (minutes) of both thallium nuclides (^{208}Tl and ^{207}Tl) in the U-series this element will not be considered further.

Although our principal objective is to develop an understanding of U-series partitioning in magmatic situations, it is worth noting that partitioning of U-series elements is not confined to natural geochemical problems. All of the elements in the natural U and Th decay series, together with a number of chemically similar transuranic elements, occur in nuclear wastes. Many strategies of radioactive waste encapsulation rely on assumptions about element solubility and partitioning (e.g., Lumpkin et al. 1995). Devising containment strategies that involve both melt and solid phases requires an understanding of how trace radionuclides will partition between them. This is an important area of applied geochemistry, though beyond the scope of this chapter.

We make no attempt to discuss the partitioning behavior of U-series elements between aqueous fluids and minerals at ambient conditions. Examples where this behavior is important include uptake of U-series elements by calcite in speleothems or by bone apatite. Also we do not consider U-series behavior in hydrothermal solutions at high temperatures, such as during dehydration of subducted crust. In both cases complexation behavior in the fluid may play an important role, and at low temperatures kinetic controls may dominate. These are fruitful areas for future experimental study.

2. PARTITIONING PRELIMINARIES

Partition coefficients are conventionally expressed as the weight concentration ratio of an element in one phase to that in another. In the case of magmatic processes it is conventional to express the partition coefficient as the trace element concentration in the mineral over that in the melt. An element, i, is said to be incompatible if the mineral-melt partition coefficient (D_i) is less than 1 and compatible if $D_i > 1$. All of the U-series elements are incompatible in the major rock-forming minerals: quartz, olivine, pyroxenes, feldspars, amphiboles, micas and garnets. They are, however, less incompatible in some trace or accessory phases, such as zircon and titanite, and may even become compatible in some phases, e.g., Th in chevkinite and allanite. Consequently, tiny inclusions of accessory phases (or glass) in many natural minerals, mean that determining partition coefficients by bulk analysis of, for example, coexisting phenocryst-groundmass pairs from volcanic rocks is highly unreliable (Michael 1988). For these reasons we will only discuss data obtained experimentally, or by microbeam analysis of phenocrysts and glass in volcanic rocks. Because of the low concentrations involved, data from the latter are very scarce.

Another issue of concern is diffusion. During melting or crystallization trace elements must be transferred between solid and melt phases. The rate at which this happens is ultimately controlled by solid-state diffusion. If melting or crystallization rates are slow relative to diffusion, then surface equilibrium, or even bulk crystal equilibrium, will be maintained. As most partition coefficients and trace element models pertain to the case of surface or bulk equilibrium then the behavior of fast-diffusing species in slow-moving processes is easily modeled. However, many of the U-series elements form large cations that are unlikely to diffuse sufficiently rapidly that surface equilibrium can be assumed *a priori*. Van Orman et al. (1998), for example, have shown that U and Th diffusivities are too low for mantle melts to be modeled using equilibrium partition coefficients. They show that large lanthanides, such as La and Ce, are similarly vulnerable to disequilibrium. Although diffusivities have not been measured, it seems likely that large cations such as Ac^{3+}, Ra^{2+}, Po^{4+}, and large noble gases such as Rn, will not always show equilibrium partitioning. In the rest of this chapter, which is concerned exclusively with the equilibrium case, it is important to bear these disequilibrium issues in mind. The optimum approach would be one that modifies equilibrium partition coefficients for the effects of diffusion, so that the rate dependence of the trace element chemistry of melts can be modeled. That approach, which is discussed in some detail by Van Orman et al. (2002), is beyond the scope of this review. Of course, slow diffusing species are also less likely to achieve equilibrium during experimental studies of mineral melt partitioning.

3. SOURCES OF PARTITIONING DATA

There are two principal sources of reliable partitioning data for any trace element: glassy volcanic rocks and high temperature experiments. For the reasons outlined above, both sources rely on analytical techniques with high spatial resolution. Typically these are microbeam techniques, such as electron-microprobe (EMPA), laser ablation ICP-MS, ion-microprobe secondary ion mass spectrometry (SIMS) or proton-induced X-ray emission (PIXE).

Microbeam analysis of coexisting phenocrysts and glassy matrix can be reliable provided that the phenocryst and matrix glass can be shown to be in equilibrium. Because of their incompatible nature and relatively low diffusivities, U-series elements are likely to undergo large changes in concentration during melting or crystallization, with the result that phenocrysts may be zoned in trace elements, even if they are homogeneous in major elements. For this reason, it is important that the phenocrysts analyzed are either shown to be unzoned in trace elements, or have rims in contact with glass that are suitably broad to be analyzed. The relatively large size of natural crystals, versus those produced experimentally, means that this is often the case. Conversely, the greatest disadvantage of the natural approach versus the experimental approach is that the intensive variables pressure, temperature and redox state are never as well constrained in nature as they are in the laboratory. Moreover, in nature one is limited by natural concentrations, which for most U-series elements, are too low to be measured by any current microbeam technique. To date, reliable phenocryst-matrix partitioning data are only available for U, Th and Pb (Table 1a). Advances in microbeam technology, notably lower detection limits, could lengthen this list.

The second source of partitioning data is experimental equilibration of crystals and liquids followed by microbeam analysis of quenched run products. Starting materials can be natural rocks, or synthetic analogues. In either case it is customary to dope the starting material with the U-series element(s) of interest, in order to enhance analytical precision. Of course, doping levels should not be so high as to trigger trace phase saturation (e.g.,

Table 1a. Sources of partitioning data: Microbeam or track-counting analyses of coexisting crystals and matrix (or glass) in natural rocks.

Source	Minerals	Rock type	Technique	U	Th	Pb
LaTourrette et al. (1991)	Mt	Rhyolite	Fission Track	√		
Stimac & Hickmott (1994)	Ilm, Opx, Pyrrhotite	Rhyolite	PIXE	√	√	√
Ewart & Griffin (1994)	Cpx, Amph, Phlog, Plag, Ilm, Kfs[+]	Various	PIXE			√
Jeffries et al. (1995)	Cpx	Basalt	LA-ICP-MS	√	√	
Foley et al. (1996)	Cpx, Phlog	Lamprophyre	LA-ICP-MS		√	
Chazot et al. (1996)	Cpx	Lherzolite	SIMS		√	
Marshall et al. (1998)	Fluorite	Rhyolite	SIMS	√	√	
Thompson & Malpas (2000)	Cpx	Basalt	LA-ICP-MS		√	
Macdonald et al. (2002)	Chevkinite	Rhyolite	SIMS	√	√	

[+]Also includes Pb, U and Th data for accessory phases allanite, leucite and aenigmatite

Mineral abbreviations:
Amph = amphibole   Ilm = ilmenite   Oliv = olivine   Phlog = phlogopite
Cpx = clinopyroxene Kfs = alkali feldspar Opx = orthopyroxene Plag = plagioclase
Gt = garnet   Mt = magnetite

with thorite in Th-doped experiments) or violate Henry's Law (see below). Experimental studies can be designed either to directly reproduce a particular natural process, e.g., initiation of melting at the peridotite solidus, or to investigate a sufficient range of pressures, temperatures and compositions, so that a model can be derived for extrapolation and interpolation to other conditions. The principal difficulty with experimental partitioning studies is growing crystals large enough for microbeam analysis. The conventional method is to isobarically undercool the sample slowly (a few °C per hour) from a super-liquidus or near-liquidus temperature to the run temperature. This technique is usually successful, even for phases such as plagioclase, but care must be taken to avoid rapid disequilibrium uptake of trace elements, especially for slow diffusing species such as the large U-series ions. Once again, experimental partitioning data are limited to U, Th and Pb (Table 1b). This is largely because the doping levels required to precisely measure partition coefficients for other U-series elements, which lack stable or long-lived isotopes, may require introducing highly radioactive materials into sensitive mass spectrometers. This need not be a problem for cases where the partition coefficient is close to one, but for highly incompatible elements ($D < 10^{-2}$) the doping levels required to get detectable concentrations in the crystal phase may be prohibitive.

Because U-series elements are radioactive, it is also possible to use fission-, gamma-, beta- or alpha-track counting techniques for the determination of partition coefficients. This offers the significant advantage that it does not require placing radioactive materials into mass spectrometers. For track counting to provide a viable alternative to microbeam techniques, however, it is important that full account (or advantage) can be taken of *in situ* daughter nuclides (i.e., those produced after the experiment is quenched), and that the huge concentration difference between crystal and liquid does not compromise analytical precision. This problem is particularly acute where track images of mixed crystal and glasses are used, such that some tracks originating in the glass actually appear to have originated in the crystal. Possible techniques include: the use of silver nitrate coated photographic film for the detection of beta-tracks (e.g., Mysen and Seitz 1973); NaI detectors for counting gamma-rays (e.g., Watson et al. 1987); fission-track counting of neutron-irradiated samples using mica detectors (La Tourrette et al. 1991); and the use of alpha-sensitive plastics, such as CR-39 Tastrak (Henshaw 1989), for counting alpha-tracks. All of these techniques offer the huge advantage of very low (ppb to sub-ppb) detection limits, potentially enabling experiments to be conducted at much lower (and safer) overall concentrations than is possible with microbeam techniques. Recent U-series partitioning studies that use these techniques include, Benjamin et al. (1980), La Tourrette et al. (1991) and La Tourrette and Burnett (1992). For elements that have no stable or very long-lived isotopes (e.g., Pa, Ra), track counting remains one of the most promising methods of partition coefficient determination. Similar techniques may also prove useful in analysing volcanic phenocrysts for U-series nuclides, providing only that the concentration is sufficient to generate a statistically significant number of tracks during a reasonable counting period.

4. THE IMPORTANCE OF PARTITIONING IN INTERPRETING U-SERIES DATA

In the introduction we asserted that it was important to use the correct partition coefficients when interpreting U-series data. Both the ratio of daughter and parent partition coefficients and their absolute values are important. Small errors in the ratio can propagate to quite large errors in predictions of activity ratios even when the source material is assumed to have a parent-daughter ratio of unity (i.e., in radioactive

Table 1b. Sources of partitioning data: Experimental studies of natural and synthetic starting materials.

Source	Minerals	P /GPa	T /°C	U	Th	Pb
Benjamin et al. (1980)	Cpx	2.0	1415-1450	√	√	
Watson et al. (1987)	Cpx	10^{-4}	1275	√		√
LaTourette & Burnett (1992)	Cpx	10^{-4}	1160-1273	√	√	
Beattie (1993a)	Cpx, Opx, Oliv	10^{-4}-2.9	1190-1390	√	√	√
Beattie (1993b)	Gt	3.0-3.6	1300-1565	√	√	√
Hart & Dunn (1993)	Cpx	3.0	1380			√
Kennedy et al. (1993)	Opx, Oliv	10^{-4}	1150-1525	√	√	
LaTourette et al. (1993)	Gt	2.7	1307-1313	√	√	
Hauri et al. (1994)	Cpx, Gt	1.7-2.5	1405-1430	√	√	√
Lundstrom et al. (1994)	Cpx	10^{-4}	1285	√	√	√
Kennedy et al. (1994)	Perovskite	10^{-4}	1150-1525	√	√	
Nielsen et al. (1994)	Mt	10^{-4}	1080-1129	√	√	
Brenan et al. (1995)	Amph	1.5	1000	√	√	√
LaTourette et al. (1995)	Amph, Phlog	1.5-2.0	1150-1165	√	√	√
Andreessen et al. (1996)	Amph	0.5	940-980		√	
Bindeman et al. (1998)	Plag	10^{-4}	1153-1297			√
Lundstrom et al. (1998)	Cpx	10^{-4}	1275		√	√
Salters & Longhi (1999)	Cpx, Gt, Opx	1.2-2.8	1375-1595	√		
Schmidt et al. (1999)	Cpx, Phlog	1.5	1040-1175	√	√	√
Van Westrenen et al. (1999)	Gt	3.0	1530-1565	√	√	
Wood et al. (1999)	Cpx, Opx, Oliv	1.5-1.9	1260-1320	√	√	
Bindeman & Davis (2000)	Plag	10^{-4}	1153-1297	√		
Blundy & Dalton (2000)	Cpx	0.8-3.0	1375-1640			√
Dalpé et al. (2000)	Amph	1.5-2.5	1000-1100	√	√	
Foley et al. (2000)	Rutile	1.8	900			√
Law et al. (2000)	Wollastonite	3.0	1420	√	√	√
Van Westrenen et al. (2000)	Gt, Cpx	2.9-3.0	1538-1540	√	√	
Landwehr et al. (2001)	Cpx, Gt	1-8	1330-1735	√	√	
Wood & Trigila (2001)	Cpx	10^{-4}-0.2	1042-1140	√	√	
Taura et al. (2001)	Perovskite	25	2350	√	√	
Barth et al. (2002)	Gt, Cpx	1.8	1000-1400			√
Salters et al. (2002)	Cpx, Opx, Oliv, Gt	1.0-3.4	1350-1660	√	√	√
Klemme et al. (2002)	Gt, Cpx	3.0	1400	√	√	√
Pertermann & Hirschmann (2002)	Cpx	3.0	1335-1365	√	√	
Hermann (2002)	Allanite	2.0	900	√	√	
Corgne & Wood (2003)	Perovskite	25	2300	√	√	
Bennett et al. (2003)	Cpx, Gt	3.0	1298-1402	√	√	
McDade et al. (2003a)	Cpx, Opx, Gt	3.0	1495-1505	√	√	
McDade et al. (2003b)	Cpx, Opx, Oliv	1.5	1315	√	√	

Mineral abbreviations:
Amph = amphibole   Ilm = ilmenite   Oliv = olivine   Phlog = phlogopite
Cpx = clinopyroxene   Kfs = alkali feldspar   Opx = orthopyroxene   Plag = plagioclase
Gt = garnet   Mt = magnetite

equilibrium). Conversely the absolute values of the partition coefficients contains information about the porosity of the source region at the point of melt extraction, and the upwelling rate (e.g., Beattie 1993a; Asmerom et al. 2000). Before going on to derive partition coefficients for the U-series elements, we will illustrate the importance of partition coefficients by reference to the case of U-Th radioactive disequilibrium in mid-ocean ridge basalts.

Mid-ocean ridge basalts (MORB) are widely held to form by decompression melting of adiabatically upwelling mantle at divergent plate margins. The depth beneath ridges at which melting initiates, however, is still a topic of debate and one that bears on the extractability of mantle melts, the extent to which upwelling is passive or active, and even on the water content of the sub-ridge mantle. Thus at the East-Pacific Rise (EPR) seismic tomography (Toomey et al. 2002) reveals a partially molten region down to depths of over 150 km, yet it is impossible to tell from seismology alone whether the melt forms an interconnected network (and is therefore extractable) and whether the melt contains water (which would reduce the melting temperature). U-series analyses provide a potentially powerful tool to resolve these questions in a less ambiguous way than is possible from conventional trace element geochemistry.

Young MORB (<350 ka) from the EPR and elsewhere commonly display secular disequilibrium in the ^{238}U-decay series, with excesses of (^{230}Th) over (^{238}U) most frequently observed (the parentheses denote activity). Providing that the mantle source region is in secular equilibrium for ^{238}U and ^{230}Th at the onset of melting (i.e., more than c. 500 ka has elapsed since the region was last perturbed), then U-Th disequilibrium is generally taken as evidence that (a) U and Th are chemically fractionated from each during the melting process; and (b) melt transport is rapid relative to the half-life of ^{230}Th. (See Chapter of Lundstrom and Spiegelman for discussion of more complex transport models). As the bulk solid partition coefficients for U and Th are very small (<5 × 10^{-3} for spinel or garnet lherzolite) this fractionation must occur during the earliest stages of melting such that the bulk solid partition coefficient for U, D_U^{Bulk}, is of the same order as the threshold porosity at which melt is extracted. Although not well known, the latter is likely to be greater than 10^{-5} (McKenzie 1989) and possibly as large as 10^{-3} or 10^{-2} (Faul 2001). If the chemical fractionation of U from Th is an equilibrium (rather than disequilibrium) process, then the presence of U-Th radioactive disequilibrium can be used to constrain the source mineralogy and hence depth of onset of melting. Other variables that influence the extent of radioactive disequilibrium include the rate of melt extraction, grainsize and the threshold porosity at which melt is extracted.

A global dataset of (^{230}Th)/(^{238}U) shows an inverse correlation between (^{230}Th) excess and the depth to the ridge axis (Bourdon et al. 1996). Shallower ridges are thought to result from greater cumulative amounts of melting, initiated at greater depth, while deeper ridges are thought to result from less cumulative melting, initiated at shallower depth. Significant excess of (^{230}Th) in MORB from shallow ridges constrains melting to begin in a source rock whose bulk D_{Th} is less than D_U. Conversely, smaller excess of (^{230}Th) and occasional excess of (^{238}U) in MORB from deep ridges do not constrain melting to begin in a source rock with bulk D_U greater than D_{Th} and may indicate the opposite sense of fractionation. Until recently, available experimental studies for olivine, orthopyroxene and clinopyroxene, the three major upper mantle minerals, showed that olivine and orthopyroxene have $D_U/D_{Th} > 1$, but very low absolute D_U values of about 10^{-4} to 10^{-5} (Beattie 1993a). Almost all available clinopyroxene data suggested $D_U/D_{Th} < 1$ at absolute D_U of 10^{-2} to 10^{-3} (Beattie 1993a; La Tourette and Burnett 1992; Hauri et al. 1994). When these data are used to calculate bulk partition coefficients relevant to mantle melting the value for clinopyroxene dominates. Therefore, in the absence of an additional phase with high D_U values and $D_U/D_{Th} > 1$, excesses of ^{230}Th can not be generated during melting in the spinel lherzolite stability field. In contrast the pyrope-rich garnets typical of mantle peridotite were found to have high D_U/D_{Th} values of about 4 and absolute D_U values similar to those of clinopyroxene (Beattie 1993b; La Tourrette et al. 1993; Van Westrenen et al. 1999; Salters and Longhi 1999; Landwehr et al. 2001; Salters et al. 2002). For this reason the generation of ^{230}Th-^{238}U disequilibrium during mantle melting is almost universally attributed to the initiation of melting in the garnet lherzolite stability field, i.e., at depths greater than ~90 km (Robinson and Wood 1998).

A major shortcoming of most of the available experimental clinopyroxene partitioning data on which the above conclusions were based is the discrepancy between experimental and natural conditions. These include both temperature and pressure and, of particular importance for the partitioning of highly charged ions, the crystal composition. For example Lundstrom et al. (1994) have demonstrated that clinopyroxene D_{Th} is a strong function of the concentration of Al in tetrahedral coordination in clinopyroxene. Experimental studies of the melting behavior of lherzolite in natural and synthetic systems (Takahashi and Kushiro 1983; Falloon and Green 1987; Walter and Presnall 1994; Gudfinnsson and Presnall 1996, 2000; Hirose and Kushiro 1998; Robinson et al. 1998) show that the composition of solidus clinopyroxene at the onset of melting is sensitive to changes in pressure and temperature. With increasing pressure and temperature along the lherzolite solidus, clinopyroxenes become progressively enriched in the enstatite, jadeite and calcium-tschermaks components and depleted in the diopside component. Wood et al. (1999) argued that the observation of $D_U/D_{Th} < 1$ in clinopyroxene was a consequence of the diopside-rich nature of the experimental clinopyroxenes, which had large, Ca-rich M2 lattice sites, more favourable to incorporation of the larger Th^{4+} ion (1.05 Å in VIII-fold coordination) than U^{4+} (1.00 Å). Wood et al. (1999) performed clinopyroxene-melt partitioning experiments on compositions and under conditions more characteristic of the mantle solidus. They showed that a smaller, Ca-poor M2 lattice site produces the opposite sense of U-Th fractionation, with $D_U/D_{Th} > 1$ (Fig. 1). These findings were confirmed by the more exhaustive study of Landwehr et al. (2001) who showed that D_U/D_{Th} for clinopyroxene increases roughly linearly with increasing depth (Fig. 2), such that even in the absence of garnet, the greater the pressure of onset of melting, the greater the likely ^{230}Th excess. Although this finding does not rule out garnet in the source region of mid-ocean ridge basalts, it removes the *requirement* for this phase. To demonstrate that melting did begin in the presence of garnet, *additional* evidence must be mustered, such as lanthanide concentration patterns (McKenzie and O'Nions 1991) or Lu-Hf isotope systematics (Salters and Hart 1989).

We will return in more detail to U-Th partitioning between clinopyroxene and melt later. The MORB example simply demonstrates the care required when selecting partition coefficients, which are sensitive to changes in pressure, temperature and composition. Without detailed partitioning data at the appropriate conditions, an erroneous inference could be drawn from U-series disequilibrium data, namely that excess of (^{230}Th) over (^{238}U) *proves* a garnetiferous source. One can envisage countless other examples where unanticipated fractionations between daughter-parent pairs can lead to surprising conclusions.

5. THE LATTICE STRAIN MODEL

Two important observations are clear from the foregoing. The first is that, although there are partitioning data for U, Th and Pb (Table 1), they are sufficiently variable to present a problem for the geochemist: which value to choose for a particular problem.

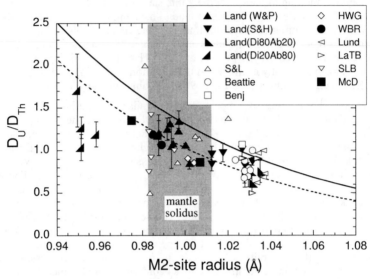

Figure 1. Variation in D_U/D_{Th} with M2-site radius in clinopyroxene for all available experimental data. Data sources are: Land (Landwehr et al. 2001); HWG (Hauri et al. 1994); LaTB (La Tourrette and Burnett 1992); SLB (Salters et al. 2002); McD (McDade et al. 2003a,b); WBR (Wood et al. 1999); Beattie (1993a); Benj (Benjamin et al. 1980); S&L (Salters and Longhi 1999); Lund (Lundstrom et al. 1994). In each case the M2-site radius is calculated from the experimental clinopyroxene composition using Equation (12a). Error bars (1 s.d.) are only shown for experiments carried out in Bristol, where D_U/D_{Th} is taken from the ratio $(U/Th)_{cpx}/(U/Th)_{melt}$, and is consequently more precise (Wood et al. 1999). Error bars for all other data are larger, sometimes up to 50% of D_U/D_{Th}. They have been omitted for clarity. The four different categories of Landwehr et al. (2001) denote different starting mixes, designed to bracket a wide range of clinopyroxene composition and hence M2-site radius. The deviation for four of the points labelled "Di20Ab80", which derive from 5.6-8.1 GPa experiments in the system diopside$_{20}$-albite$_{80}$, is probably due to their unusually high Na$_2$O content (9-11 wt%), which is not adequately accounted for in the M2-site radius algorithm (Eqn. 12a). The curved lines denote the lattice strain model (see text) with ionic radii from Shannon (1976), shown as solid line, and from Table 2 (broken line). The range of M2-site radii on the mantle solidus, from experimental studies, is shown as a shaded box. It is evident from this plot that the D_U/D_{Th} ratio is very sensitive to clinopyroxene composition.

Evidently one reason for the variability of the data stems from the fact that they were acquired under variable conditions of pressure, temperature and composition. A method of parameterising the data for U, Th and Pb is required, so that full account can be taken of pressure, temperature and composition. The second observation is that there are no experimental partitioning data for the other U-series elements Pa, Ra, Rn, Ac, Po, or Bi. What partition coefficients should be used for these elements? It is certainly not always sufficient to simply assume that an element, such as Pa or Ra, is totally incompatible ($D = 0$) in all mineral phases, because this removes the opportunity to chemically fractionate these elements from each other by processes such as melting, crystallization or degassing. Clearly what is needed is a means of estimating partition coefficients for U-series elements for which there are no experimental values, and using these estimates to gauge the senses and likely magnitudes of chemical fractionations between them. Conventionally, this has involved the use of proxies: elements whose chemistry is similar (though rarely identical) to the U-series element of interest, and for which there are

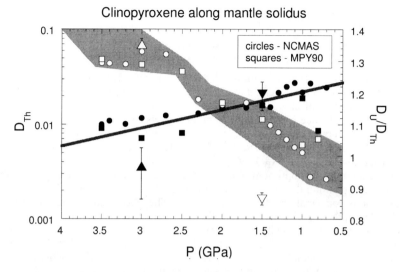

Figure 2. Calculated variation in D_U/D_{Th} (open symbols) and D_{Th} (solid symbols) along mantle solidus using model of Landwehr et al. (2001). Solidus clinopyroxene compositions were taken from experiments on natural compositions (MPY90 – Falloon and Green 1987) and synthetic systems (NCMAS – Walter and Presnall 1994). They both describe similar trends of increasing D_U/D_{Th} and decreasing D_{Th} with increasing pressure on the solidus. The shaded area and solid line are designed to guide the eye; they have no statistical significance. Also shown on this plot are two experimental determinations of D_U/D_{Th} and D_{Th} (McDade et al. 2003a,b), on the mantle solidus at 1.5 and 3 GPa, that post-date the Landwehr et al. (2001) model. Errors bars are 1 s.d. Note that the model reproduces D_{Th} at both pressures, but overestimates D_U/D_{Th} at 1.5 GPa. This is partly because the 1.5 GPa experiment of McDade et al. (2003b) was performed on a slightly more refractory bulk composition than MPY90. The Wood et al. (1999) determination of D_U/D_{Th} for MPY90 composition at 1.5 GPa is 1.07 ± 0.04.

partitioning data. The best example of a proxy is Ba, generally used as a stable analogue for Ra, which lacks a stable reference isotope. For other U-series elements, the choice of proxy is less straightforward. Furthermore, it is not clear what effect small physicochemical differences between a U-series element and its proxy have on the partition coefficients. To employ proxies effectively, these issues must be resolved. Here we describe a simple technique for doing this that takes account of the principal controls on trace element incorporation and partitioning into minerals.

Since the pioneering work of Goldschmidt (1937) it has been generally accepted that, at a given pressure (P) and temperature (T), the principal controls on trace element partitioning are the mismatch in valence and ionic radius between the substituent and substituted ion. Because of their regular, rigid structures (relative to melt or fluid phases) minerals can readily accept onto their lattice sites only ions of similar radius and charge to the "host" ion normally resident at that site. Substitutions at extended or intrinsic defects may also be important, but these have not yet been subject to systematic study. Because trace element partitioning is a thermodynamic process, P and T and, in the case of variable valence ions, oxygen fugacity (fO_2), also exert an important control on partition coefficients. Goldschmidt proposed three "rules" for trace element uptake into minerals from melts (Mason 1966, p. 132):

1. If two ions have the same radius and the same charge, they will enter a given crystal lattice with equal facility.

2. If two ions have similar radii and the same charge, the smaller ion will enter a given lattice more readily.

3. If two ions have similar radii and different charge, the ion with the higher charge will enter a given crystal lattice more readily.

Goldschmidt's second and third rules are oversimplifications. Still, his first "rule" provided the basis for a theoretical treatment of the problem, which came with the advent of atomistic models of crystal lattices. These "lattice strain" models were based on the simple premise of trace ions as charged point defects in a dielectric, elastic continuum. Consequent disruption of the lattice around the defect is minimised by relocating (or *relaxing*) the neighbouring ions and distributing the surplus elastic or electrostatic energy through the lattice (Nagasawa 1966; Brice 1975). The elastic strain energy is roughly symmetrical about an optimum ionic size. In other words the energy penalty for accommodating an ion that is fractionally too large for a site is about the same as that for an ion that is too small by the same amount. The elastic strain energy is inversely correlated with the logarithm of the partition coefficient, which, for a specific lattice site, should vary near-parabolically with ionic radius as first observed from analyses of volcanic phenocrysts and their glassy groundmass by Onuma et al. (1968). The optimum radius, i.e., that for which D is largest, is close to the radius of the host cation in the crystal of interest. Where substitution occurs at more than one lattice site a series of overlapping parabolae are observed for each cation valence (Onuma et al. 1968; Purton et al. 1996), with maxima corresponding to the optimum radii of the various sites. Note that cations that have ionic radii intermediate between two sites (i.e., too small for one site and too large for the other) can be disordered across these sites and therefore will have higher overall partition coefficients than would be expected from their size mismatch on either site alone. Mn^{2+} incorporation in clinopyroxene is one such example.

The advent of trace element microbeam techniques revolutionised the experimental determination of partition coefficients and provided the means to quantify the lattice strain models. Two papers, both published in 1994, successfully applied lattice strain models to experimental partition data for olivine (Beattie 1994) and for clinopyroxene and plagioclase (Blundy and Wood 1994). Beattie used the formulation of Nagasawa (1966), while Blundy and Wood used that of Brice (1975). The two formulations are very similar and can be applied with equal facility, differing by at most 10% in lattice strain energy for the largest misfit cations. For simplicity we shall use the Brice (1975) formulation here, as adapted by Blundy and Wood (1994), largely because that formulation is mathematically more tractable. However, we recognise that the Brice equation is a slightly less complete description of the lattice strain process than the Nagasawa equation, which explicitly takes account of both radial and tangential strains around the substituent cation.

According to the Brice model, for an isovalent series of ions with charge $n+$ and radius r_i entering crystal lattice site M, the partition coefficient, $D_{i(M)}$, can be described in terms of three parameters (Fig. 3): $r_{0(M)}^{n+}$, the radius of that site; E_M^{n+} the elastic response of that site to lattice strain (as measured by its *effective* Young's Modulus) caused by ions that are bigger or smaller than $r_{0(M)}^{n+}$; and $D_{0(M)}^{n+}$, the "strain-compensated" partition coefficient for a (fictive) ion with radius $r_{0(M)}^{n+}$, according to the expression (Blundy and Wood 1994):

$$D_i = D_{0(M)}^{n+} \times \exp\left\{ \frac{-4\pi N_A E_M^{n+} \left[\frac{1}{2} r_{0(M)}^{n+} \left(r_i - r_{0(M)}^{n+}\right)^2 + \frac{1}{3}\left(r_i - r_{0(M)}^{n+}\right)^3 \right]}{RT} \right\} \quad (1)$$

Lattice strain model

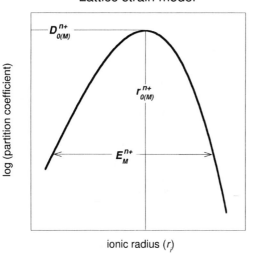

Figure 3. Cartoon illustrating the *lattice strain model* of trace element partitioning. For an isovalent series of ions with charge $n+$ and radius r_i entering crystal lattice site M, the partition coefficient, $D_{i(M)}$, can be described in terms of three parameters: $r_{0(M)}^{n+}$, the radius of that site; the elastic response of that site (as measured by its apparent Young's Modulus, E_M^{n+}) to lattice strain caused by ions that are bigger or smaller than $r_{0(M)}^{n+}$; and $D_{0(M)}^{n+}$, the partition coefficient for a (fictive) ion with radius $r_{0(M)}^{n+}$. The mathematical description of the lattice strain model is given in Equation (1).

where N_A is Avogadro's Number (6.02214×10^{23}), R is the universal gas constant (8.31451 J K^{-1} mol^{-1}) and T is in K. We can simplify Equation (1) by explicitly evaluating the constant terms, adopting E in GPa and ionic radii in Å, as follows:

$$D_i = D_{0(M)}^{n+} \times \exp\left\{\frac{-910.17 E_M^{n+}\left[\frac{1}{2}r_{0(M)}^{n+}\left(r_i - r_{0(M)}^{n+}\right)^2 + \frac{1}{3}\left(r_i - r_{0(M)}^{n+}\right)^3\right]}{T}\right\} \quad (2)$$

Blundy and Wood (1994) showed that Equation (2) could be used to fit the partition coefficients for series of isovalent cations entering the large cation sites in plagioclase and clinopyroxene (Fig. 4). Significantly they were able to show that, in the case of homovalent substitution, i.e., where trace ion and host ion have the same charge, the fit parameters E and r_0 have physical significance. Thus E_M^{2+} and E_M^{1+} in plagioclase were found to lie close to the bulk crystal Young's Moduli for anorthite and albite (Fig. 5), respectively, while E_{M2}^{2+} in clinopyroxene was similar to the bulk crystal Young's Modulus of diopside, the dominant end-member component in their synthetic crystals. This suggested that the elastic properties of the crystal are strongly influenced by those of the large cation site, rather than by the relatively rigid tetrahedral Si-O(-Al) framework. Crystals with higher bulk elastic moduli (e.g., garnet) are, in general, characterised by tighter partitioning parabolae than crystals with lower bulk elastic moduli (e.g., plagioclase), which accommodate misfit cations more easily. Evidently the ability of a crystal lattice to accommodate cations of a different ionic radius to the "host" or "optimum" cation is determined to a large extent by the elastic response of the crystal to strain. As crystals are characterised by a regular arrangement of ions in relatively rigid lattices, it is likely that they will be less tolerant of misfit cations than silicate melts with their essentially disordered and flexible structures. Consequently, it is not surprising that strain of the crystal lattice dominates the energetics of partitioning. It is for that reason that no melt composition-related term appears in Equation (1). However, the pre-exponential term, D_0, the strain-compensated partition coefficient, will depend on activity-composition relationships in both crystal and melt phases, and may show some complex sensitivity to melt composition (e.g., Blundy et al. 1996), as well as to P and T.

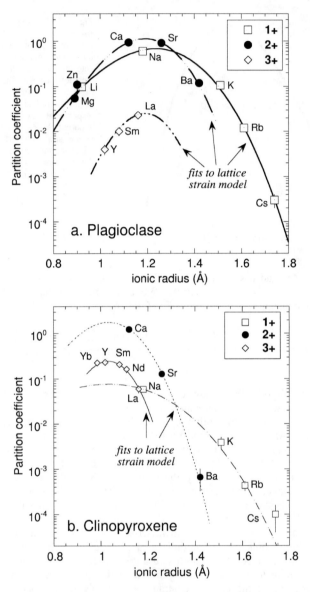

Figure 4. Fits of lattice strain model to experimental mineral-melt partition coefficients for (a) plagioclase (run 90-6 of Blundy and Wood 1994) and (b) clinopyroxene (run DC23 of Blundy and Dalton 2000). Different valence cations, entering the large cation site of each mineral, are denoted by different symbols. The curves are non-linear least squares fits of Equation (1) to the data for each valence. Errors bars, when larger than symbol, are 1 s.d. Ionic radii in VIII-fold coordination are taken from Shannon (1976).

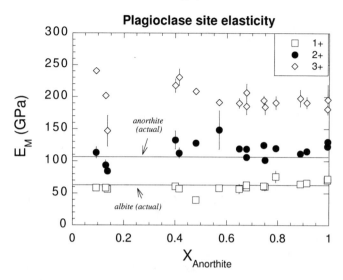

Figure 5. Variation in apparent lattice-site Young's Modulus (E_M^{n+}) with molar anorthite content of the host crystal for the plagioclase partitioning experiments of Blundy and Wood (1994). E_M^{1+}, E_M^{2+} and E_M^{3+} were derived by fitting Equation (1) to 1+, 2+ and 3+ cations, respectively, for each experiment in the system diopside-albite-anorthite. Horizontal lines labeled *anorthite* and *albite* denote measured Young's Moduli of the bulk crystals. Note the similarity between E_M^{1+} and E_{albite}, and E_M^{2+} and $E_{anorthite}$ suggesting that, in this case, it is the elasticity of the large cation site that controls the elasticity of the bulk crystal. Error bars are 1 s.d.

Blundy and Wood (1994) found that values for the optimum site radius, $r_{0(M)}^{n+}$, were not necessarily the same as the radius of the cation normally resident at the site of interest. Thus $r_{0(M2)}^{2+}$ in clinopyroxene is slightly smaller that the radius of Ca^{2+}, the ion which normally occupies that site (Wood and Blundy 1997). Instead they observed that $r_{0(M)}^{n+}$ was closely related to the known metal-oxygen bond lengths (d_{M-O}) in the host minerals, provided that allowance was made for the ionic radius of co-ordinating O^{2-} ions, taken to be 1.38Å (Shannon 1976), i.e.,

$$d_{M-O} = r_{0(M)}^{n+} + 1.38 \text{ Å} \quad (3)$$

Moreover, for solid solution series, such as pyrope-grossular garnets (Fig. 6a; Van Westrenen et al. 1999) or albite-anorthite plagioclases (Fig. 6b; Blundy and Wood 1994), the variation in $r_{0(M)}^{n+}$ across the solid solution, obtained by fitting the partitioning data, is in agreement with known variations in d_{M-O}. This observation has since been confirmed by Tiepolo et al. (1998) and Bottazzi et al. (1997) in studies of lanthanide partitioning between amphibole and melt. These authors were able to measure d_{M-O} in synthetic amphiboles using single crystal X-ray techniques and compare the values to $r_{0(M4)}^{3+}$ determined experimentally from the same crystals. The strong correlation between the two, and offset of 1.41 ± 0.04 Å in d_{M-O} (cf. Eqn. 3) confirmed a strong link between crystal structure and trace element incorporation.

A further feature of the fit parameters obtained by Blundy and Wood (1994) for plagioclase and clinopyroxene was that the partitioning parabolae become tighter (E_M^{n+} increases) and displaced to lower $r_{0(M)}^{n+}$ as charge increases (Law et al. 2000; Blundy and Dalton 2001). These observations have since been confirmed by a large number of

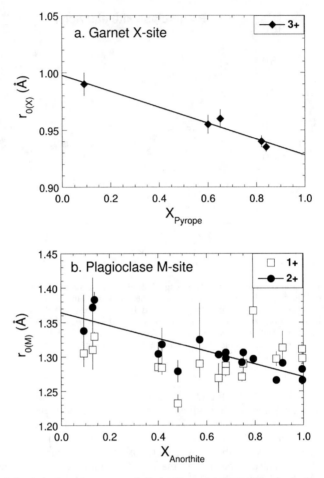

Figure 6. Variation in lattice size parameter ($r_{0(M)}^{n+}$) with crystal composition for the large cations sites in (a) garnet and (b) plagioclase. Garnet data ($r_{0(X)}^{3+}$) were derived from 3+ partitioning data from the 3 GPa CMAS experiments of Van Westrenen et al. (1999) and are plotted against molar fraction pyrope. Plagioclase data ($r_{0(M)}^{3+}$ and $r_{0(M)}^{2+}$) were derived from 1+ and 2+ partitioning data from the 0.1 MPa diopside-albite-anorthite experiments of Blundy and Wood (1994) and are plotted against molar fraction anorthite. In both cases the observed decrease in site dimension with increasing $X_{Anorthite}$ or X_{Pyrope} is quantitatively consistent with measured changes in metal-oxygen bond lengths in these minerals. Error bars are 1 s.d.

partitioning studies on other mineral phases, including wollastonite, amphibole, orthopyroxene, diopside, perovskite and garnet. The observations are interesting because once again they appear to have physical significance. In the case of E_M^{n+} the observed behavior is analogous to that of ionic crystals, wherein for a given class of compound (e.g., oxides) the *bulk* modulus, K, is a function of the ratio of cation charge (Z_c) to molecular volume (d_{M-O}^{-3}). Anderson and Anderson (1970) showed that, for an ionic crystal with electrostatic (Coulombic) attractive forces and a Born power-law repulsive potential, K is linearly dependent on cation charge for fixed structure type and molecular volume. This finding was extended by Hazen and Finger (1979) to individual cation-anion coordination polyhedra in a large number of simple and compound oxides, leading to the following relationship:

$$K = 750(\pm 20)\, Z_c\, d_{M-O}^{-3}\ \text{GPa} \qquad (4)$$

where d_{M-O} (in Å) is given by Equation (3). K is therefore a linear function of cation charge in oxygen polyhedra of the kind we are concerned with in silicate minerals. The observed correlation can be extended to Young's modulus, E, by considering the identity relating E and K in isotropic materials:

$$E = 3K(1-2\sigma) \qquad (5)$$

where σ is Poisson's ratio. Most minerals approximate Poisson solids ($\sigma \approx 0.25$), giving:

$$E \approx 1.5\, K \qquad (6)$$

Strictly speaking this identity applies only to bulk crystal elastic properties, and cannot be expected to hold in the case of apparent lattice site elastic properties. However, we find that the elastic properties and dimensions of the large (non-tetrahedral) cation sites in a large number of silicate minerals, as obtained from trace element partitioning data, are in very good agreement with Equation (4) (Fig. 7). In detail the similarity between oxides and lattice sites breaks down at large charges and small bond-lengths: lattice sites appear to be stiffer than oxide minerals with the same value of $Z_c d_{M-O}^{-3}$. This may relate to the tendency of minerals with small coordination polyhedra to deform by twisting or torqueing of the polyhedra rather than by bulk compression, or is perhaps a reflection of the gross simplifications used in comparing the two datasets, such as the conversion from E to K (Eqn. 6). In a recent lattice statics (i.e., 0 K) computer simulation of divalent cation incorporation into pyrope-grossular garnets, Van Westrenen et al. (2003) showed that the configuration of Ca and Mg as first, second and third nearest X-site neighbours to the substituent cation can profoundly influence the energetics of trace element incorporation. In some intermediate Gr-Py solid solutions this leads to an apparent "softening" of the lattice, as measured by the curvature of the partitioning parabolae, which is in fact simply a consequence of ordering between trace substituent and host Ca and Mg. In any event it is apparent that small lattice sites are very resistant to the accommodation of highly charged cations, even with small size misfits. This has implications for the incorporation of the highly charged U-series cations (notably Pa^{5+}), onto the smaller octahedral sites in silicates, because partition coefficients for these elements will be very sensitive to small changes in r_0 and E (e.g., Hill et al. 2000). Note also that extrapolation of the lattice site trend to zero charge (Fig. 7) shows that E_M^{0+} will be very low, suggesting that there should be very little discrimination between the noble gases on the basis of their atomic radii (Wood and Blundy 2001).

The decrease in $r_{0(M)}^{n+}$ with charge suggests that displacement of the coordinating oxygens around the lattice site plays a role (Fig. 8). Thus a highly charged cation (with net positive charge at the site of interest) draws in the oxygen anions reducing the optimum radius of the site. Conversely a low-charged cation (with net negative charge) repels the oxygens, thereby increasing the optimum radius of the site. There is a limit to how far the oxygen ions can be drawn in before repulsive interactions between adjacent oxygens becomes significant. As a result $r_{0(M)}^{n+}$ is nearly constant at high charge (Fig. 8). This inference is borne out by the variation in ionic radius with charge for several polyvalent elements (Am, U, Np, Pu) in VI-, VII- and VIII-fold coordination (Fig. 8) as derived from various actinide compounds by Shannon (1976). As the cation charge increases so the M-O bond length in these compounds decreases. The rate of decrease in M-O is less at high charge than at low charge, due to repulsion between the coordinating oxygens as they are drawn closer to the more highly charged central atom. This trend is

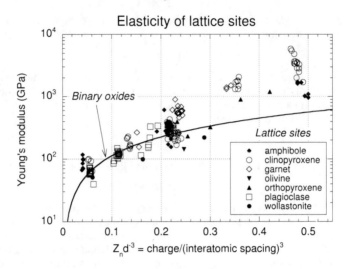

Figure 7. Variation in apparent site elasticity (Young's Modulus, E_M^{n+}) with the ratio Z_n/d^3, where Z_n is the ionic charge and d is the metal-oxygen bond length (in Å) at the site of interest, given as $r_{0(M)}^{n+}$ + the radius of IV-coordinate O^{2-} ions (= 1.38 Å; Shannon 1976). Data are presented for (large) non-tetrahedral cation sites only. The solid line shows the variation in bulk E with Z_n/d^3 for binary oxides (Hazen and Finger 1979), assuming that $E \equiv 1.5 \times$ bulk modulus (K). Note the agreement between oxides and lattice sites at values of Z_n/d^3 less than 0.3 Å^{-3}. At higher values lattice sites appear to be considerably stiffer than oxides. Evidently small lattice sites are very resistant to the accommodation of highly charged cations, even with small size misfits. This has implications for the incorporation of, for example, Pa^{5+} onto M1 in clinopyroxene. Extrapolation of the lattice site trend to zero charge shows that E_M^{0+} will be very low (near zero), suggesting that there should be very little discrimination between the noble gases on the basis of their atomic radii. This is consistent with the experimental results of Brooker et al. (2003) on Ne-Ar-Kr-Xe partitioning between clinopyroxene and melt. Data in this figure are compiled from a large number of experimental sources and over a wide of pressure and temperature. All were obtained by fitting Equation (1) to cation partitioning data.

very similar to that observed from the partitioning data, although the two trends are displaced by approximately 2 charge units.

The physical plausibility of $r_{0(M)}^{n+}$ and E_M^{n+} means that (a) for homovalent cations both parameters can be estimated from known M-O distances and bulk-crystal elastic moduli and (b) values of $r_{0(M)}^{n+}$ and E_M^{n+} for higher and lower charged substituents (i.e., heterovalent substituents) can be reliably estimated using the trends in Figures 7 and 8. These two facts form the basis for a quantitative model of mineral-melt partitioning. Wood and Blundy (1997) used this approach to rationalize a large data set of lanthanide partition coefficients for clinopyroxene. Their objective was to obtain values of $D_{0(M2)}^{3+}$ for each of several hundred experiments and then parameterize them as a function of P and T. Van Westrenen et al. (2001a) took a similar approach to the partitioning of lanthanides and Sc between garnet and melt. With the exception of garnet-melt and clinopyroxene-melt partitioning of U and Th (Landwehr et al. 2001), there are insufficient U-series partitioning data to make such an exercise worthwhile. Instead, for most elements we will use the lattice strain approach as a means of estimating the partition coefficient of one element based on the partition coefficient of a proxy.

In all cases considered below, the proxy and the U-series element have the same ionic charge. However, in some cases there are no partitioning data even for the proxy at

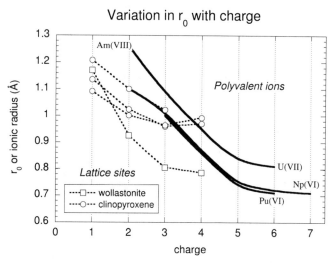

Figure 8. Variation in $r_{0(M)}^{n+}$ with cation charge (Z_n) for lattice sites in clinopyroxene (Blundy and Dalton 2000; Brooker et al. 2003) and wollastonite (Law et al. 2000), compared to change in ionic radius with charge for some polyvalent cations (Am, U, Np and Pu in VI-, VII- and VIII-fold coordination; Shannon 1976). In both cases there is a decrease in ionic radius or $r_{0(M)}^{n+}$ with increasing charge. Although they are displaced along the abscissa, the form of the curves is strikingly similar suggesting that it is electronic polarisation of the co-ordinating oxygens around the lattice site that is responsible for the change in $r_{0(M)}^{n+}$ with charge. This variation must be taken into account when attempting to extrapolate partitioning data from one valence to another.

the conditions of interest, and its partition coefficient must be estimated from the partition coefficients of other cations with different charge. To effect this we use the relationship between the strain-compensated partition coefficient, D_0, and cation charge, Z_c (Wood and Blundy 2001). Typically, D_0 is larger for homovalent substitution than for heterovalent substitution. In fact when plotted against Z_c, D_0 shows another parabolic dependence (Fig. 9), similar to that of D_i versus r_i (Fig. 4) The form of this relationship relates to the electrostatic work that is done when the substituent ion has a different charge to that of the cation normally resident at a lattice site. Such *heterovalent* substitution requires the presence of a charge-compensating defect (cation or anion) elsewhere in the lattice (e.g., $Ln^{3+} + Na^{1+} = 2Ca^{2+}$). At high temperatures there is likely to be considerable disorder between the trace ion and its compensating defect (Purton et al. 1997), and the two need not sit on adjacent sites. In that case the electrostatic work is that required to dissipate the excess (or deficit) charge, which in turn is controlled by the size of the charged region and the dielectric constant of the lattice.

Wood and Blundy (2001) developed an *electrostatic model* to describe this process. In essence this is a continuum approach, analogous to the lattice strain model, wherein the crystal lattice is viewed as an isotropic dielectric medium. For a series of ions with the optimum ionic radius at site M, ($r_{0(M)}^{n+}$), partitioning is then controlled by the charge on the substituent (Z_n) relative to the optimum charge at the site of interest, $Z_{0(M)}$ (Fig. 10):

$$D_{0(M)}^{n+} = D_{00(M)} \times \exp\left\{\frac{-N_A e_0^2 (Z_n - Z_{0(M)})^2}{(2\varepsilon\rho)RT}\right\} = D_{00(M)} \times \exp\left\{\frac{-83594(Z_n - Z_{0(M)})^2}{(\varepsilon\rho)T}\right\} \quad (7)$$

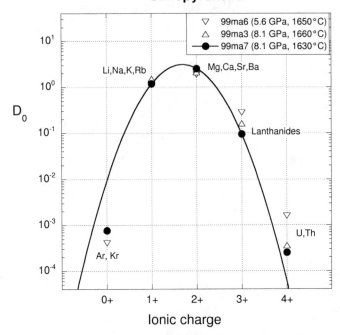

Figure 9. Variation in the strain-free partition coefficient for the clinopyroxene M2-site { $D_{0(M2)}^{n+}$ } with ionic charge for three experimental runs of Brooker et al. (2003). For 1+, 2+ and 3+ charges in each experiment $D_{0(M2)}^{n+}$ was obtained by fitting Equation (1) to the partitioning data, for the cations shown. For 4+ cations, we have adopted the D_U or D_{Th} (whichever is larger), because there are insufficient 4+ partitioning data to fit a parabola. For the noble gases ("0+") either D_{Ar} or D_{Kr} is plotted. Note the near-parabolic variation of $D_{0(M2)}^{n+}$ with charge, such that noble gases (Ne-Rn), with effective charge of approximately −2 at M2, are expected to have similar clinopyroxene-melt partition coefficient to U^{4+} and Th^{4+} with effective charge +2 at the same site. This forms the basis of the electrostatic model (Eqn. 7 and Fig. 10). The solid curve shows the fit of the model to experiment 99ma7. The fit parameters, in this case, are: $Z_{0(M2)}$ = 1.69 and $\rho\varepsilon$ = 21.6 Å. The low value of $Z_{0(M2)}$ reflects the sodium-rich nature of this clinopyroxene (10.9 wt% Na_2O).

where e_0 is the charge on the electron and N_A, R and T are defined above. The tightness of the parabola is inversely proportional to the radius (ρ) of the region over which the excess charge is distributed, and the dielectric constant (ε) of that region. The partition coefficient for a (fictive) ion that can enter a lattice site without causing either elastic strain or electrostatic charging is denoted $D_{00(M)}$.

Experimental clinopyroxene-melt (Brooker et al. 2003) and wollastonite-melt (Law et al. 2000) partitioning data can be fitted to the electrostatic model (see Fig. 9). Only species entering the large cation site (e.g., M2 in clinopyroxene) are considered. For each isovalent group (actinides, lanthanides etc.) $D_{0(M)}^{n+}$, derived as shown in Figure 4, is plotted against charge. The resulting parabola is consistent with the simple theory in Equation (7), and can be fit with $Z_{0(M)}$ of 1.8-2.0 and $\varepsilon\rho$ of 18-26 Å. This relationship can be used to predict the strain-free partition coefficient for a particular valence based only on the partition coefficients of other valences at the same lattice site. For example, we can use strain-free partition coefficients for 1+, 2+ and 3+ cations on a specific lattice site to estimate the strain-free partition coefficient for 4+ cations, even if there are no 4+

Electrostatic model

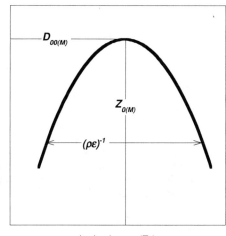

Figure 10. Cartoon illustrating the *electrostatic model* of trace element partitioning. For a series of ions with the optimum ionic radius at site M ($r^{n+}_{0(M)}$), partitioning is controlled by the ionic charge (Z_n) relative to the optimum charge at the site of interest, $Z_{0(M)}$. The larger the charge mismatch, the smaller the partition coefficient. The tightness of the parabola is inversely proportional to the product of the radius (ρ) of the region over which the excess charge is distributed and the dielectric constant (ε) of that region. The partition coefficient for a (fictive) ion that can enter a lattice site without causing elastic strain or electrostatic charging is denoted $D_{00(M)}$. The mathematical description of the electrostatic model is given in Equation (7).

partitioning data available. $D^{1+}_{0(M)}$, $D^{2+}_{0(M)}$ and $D^{3+}_{0(M)}$ can be derived by fitting the available 1+, 2+ and 3+ partitioning data to Equation (2). $D^{4+}_{0(M)}$ can then be obtained by extrapolation using Equation (7). Next, $r^{4+}_{0(M)}$ and $E^{4+}_{(M)}$ must be estimated: a reasonable assumption, based on the foregoing discussion, is that $r^{4+}_{0(M)} = r^{3+}_{0(M)}$ (Fig. 8) and $E^{4+}_{(M)} = \frac{4}{3} E^{3+}_{(M)}$ (Fig. 7). Finally, the partition coefficient for a 4+ cation of interest (e.g., Po^{4+}) can then be obtained from $D^{4+}_{0(M)}$ through Equation (2) using $r^{4+}_{0(M)}$ and $E^{4+}_{(M)}$.

A particular value of the electrostatic model is that it enables predictions to be made of the partitioning behavior of the noble gases. We have previously noted that there should be very little discrimination between the noble gases on the basis of their atomic radii, because E^{0+}_M is very small. Equation (7) predicts that if the noble gases enter lattice sites with a negative effective charge, say −2 at a Ca site, then they would be expected to have similar partition coefficients to U^{4+} and Th^{4+}, with effective charge of +2 at the same site. Recent determination of clinopyroxene-melt partition coefficients for the larger noble gases (Ne-Xe) show very little variation with atomic radius and an overall similarity to D_U and D_{Th} in the same crystals (Fig. 9, Brooker et al. 2003), which is consistent with noble gas incorporation at the M2 lattice site. Computer simulation studies of noble gas incorporation in clinopyroxene (Brooker et al. 2003; Allan et al. 2003) support this contention, showing that the energy penalty for inserting the heavier noble gases at lattice sites, charge balanced by other species elsewhere in the lattice, is much lower than that for inserting them interstitially, or at grain boundaries. We shall use this approach to estimate partition coefficients for Rn.

5.1. Proxies

Proxies are widely used in U-series geochemistry in the study of Ra isotopes. Because Ra has no stable (or long-lived) isotope against which to be ratioed, the common practice is to ratio Ra isotope activity to Ba concentration. The assumption is that Ba and Ra are chemically similar (i.e., both form large divalent cations) and behave coherently during magmatic processes. Ba can be therefore described as a "proxy" for Ra. The conventional approach, however, is to assume that the partitioning behavior of Ba and Ra are identical. In fact, as is apparent from Equation (1), the slightly different ionic radii of

Ba^{2+} and Ra^{2+} (1.42 Å and 1.48 Å, respectively, in VIII-fold coordination; Table 2) mean that in most cases their partition coefficients will not be same (Blundy and Wood 1994; Purton et al. 1996). Under certain circumstances these differences could lead to significant fractionation of Ra from Ba. Without correcting for the small partition coefficient differences, Ra isotopic data could be misinterpreted, as recently emphasized by Cooper et al. (2001) and Cooper and Reid (2003).

The approach taken here is to use the lattice strain model to derive the partition coefficient of a U-series element (such as Ra) from the partition coefficient of its proxy (such as Ba) under the appropriate conditions. Clearly the proxy needs to be an element that forms ions of the same charge and similar ionic radius to the U-series element of interest, so that the pair are not significantly fractionated from each other by changes in phase composition, pressure or temperature. Also the partitioning behavior of the proxy must be reasonably well constrained under the conditions of interest. Having established a suitable partition coefficient for the proxy, the partition coefficient for the U-series element can then be obtained via rearrangement of Equation (2) (Blundy and Wood 1994):

$$D_a = D_b \exp\left[\frac{-910.17 E_M^{n+}}{T}\left\{\frac{r_{0(M)}^{n+}}{2}(r_b^2 - r_a^2) - \frac{1}{3}(r_b^3 - r_a^3)\right\}\right] \quad (8)$$

where subscript a denotes the U-series element and subscript b denotes the proxy. Where an element enters two sites then Equation (8) must be evaluated for both sites, ideally with a different proxy on each, and the two partition coefficients summed (Purton et al. 1996). Account must be taken of the different coordination numbers and elastic properties of the two sites.

Table 2. Ionic radii of U-series elements and their proxies

U-series ion	Coordination				Proxy ion	Coordination			
	VI	VIII	X	XII		VI	VIII	X	XII
*U^{4+}	0.875	0.983	1.08	1.17	na				
*Th^{4+}	0.919	1.041	1.13	1.21	na				
Pb^{2+}	1.19	1.29	1.40	1.49	Sr^{2+}	1.18	1.26	1.36	1.44
Ra^{2+}	1.40	1.48	1.59	1.70	Ba^{2+}	1.35	1.42	1.52	1.61
Pa^{5+}	0.78	0.91	na	na	†Nb^{5+}	0.640	0.74	na	na
					Ta^{5+}	0.660	0.76		
Bi^{3+}	1.03	1.17	1.28	1.36	La^{3+}	1.032	1.032	1.27	1.36
Ac^{3+}	1.12	1.24	1.34	1.42	La^{3+}	1.032	1.032	1.27	1.36
Po^{4+}	0.94	1.08	1.17	1.25	Th^{4+}	0.919	1.041	1.13	1.21
					U^{4+}	0.875	0.983	1.08	1.17
‡Rn^0					Ar^0				
					Xe^0				

Ionic radii in Ångstroms (10^{-10} m). Values in italics are estimated from Fig. 2 of Shannon (1976). All others taken directly from Table 1 of Shannon (1976), except as noted below.

na is not applicable

†Radius of Ta^{5+} ions assumed 0.020 Å larger than Nb^{5+} ions following Tiepolo et al. (2000b)

‡It is assumed that there is no size fractionation between noble gases. In the absence of partition coefficients for other noble gases, D_{Rn} estimated from electrostatic model.

*U and Th ionic radii in VIII-coordination from Wood et al. (1999). Those in VI-coordination are introduced here; X and XII are from Shannon (1976).

We will use the proxy approach for all of the U-series elements, except U and Th for which there are already sufficient experimental data (Table 1b). For both of these elements we will discuss 4+ cations only. U also forms 5+ and 6+ cations in oxidizing environments. These are more relevant to aqueous and hydrothermal settings and will not be considered further here. However, in some experiments run at atmospheric pressure in air (e.g., Beattie 1993b), U will occur in one of its oxidized forms. These data are excluded from discussion, although it should be noted that in all minerals discussed here, the uranium partition coefficient will be considerably smaller when uranium is dominantly 5+ and 6+ compared to when it is dominantly 4+.

Selecting suitable proxies involves knowledge of the redox state of the U-series element and its ionic radius for the appropriate coordination number. The latter are known for almost all cations from the compilation of Shannon (1976). However, some care is required when using these data because not all of the ionic radii were obtained from oxides and some problems may occur when applying such radii to oxygen-coordinated sites in silicate minerals. For example, Wood et al. (1999) noted that the Shannon ionic radii for U^{4+} and Th^{4+}, which were derived from fluorides, were slightly larger than those derived from the corresponding oxides, UO_2 and ThO_2. In addition, some ionic radii in Shannon are only quoted to two significant figures, while a third figure is required to explain some of the observed fractionations. Nb^{5+} and Ta^{5+} are examples: Tiepolo et al. (2000) noted that the VI-fold radius of Nb^{5+} must be ~0.015 Å larger than Ta^{5+}, even though Shannon (1976) quotes 0.64 Å for both. With these reservations in mind, however, we have adopted the Shannon (1976) values for most U-series elements (Table 2), in the absence of alternative data on their compounds or their partitioning behavior. If no value is given by Shannon (1976), we have used his correlations between radius and coordination number (Fig. 2 in Shannon 1976) to extrapolate or interpolate an appropriate value (Table 2).

The simplest proxy to identify is that of Ba^{2+} for Ra^{2+}. As noted above, both are heavy alkaline earths, which form large divalent cations. They exclusively enter large cation sites with at least VIII-fold coordination. There are a large number of Ba partition coefficients for the major rock-forming minerals over a wide range of conditions, which make it an ideal proxy.

There is a relatively large set of partitioning data for Pb in most rock-forming minerals (Table 1). As far as possible we will use these data. Where there are no data, the proxy approach will be used. In most magmatic environments Pb forms divalent ions with a similar ionic radius to Sr^{2+} (Table 2), which will be used as a proxy. However, Pb^{2+} is unusual in having a lone pair of electrons ($6s^2$), which gives a tendency to covalent bonding. Shannon (1976) draws attention to the variable ionic radius of Pb^{2+}, depending on the extent to which the lone pair participates in bonding, which in turn may be linked to site distortion. This may partly explain the apparent anomalous partitioning behavior of Pb in some minerals. In that case, the lattice strain approach, using the Shannon ionic radii alone, is inadequate. One possible remedy is to modify the radii of cations with lone pairs of electrons (e.g., Pb^{2+}, Tl^{1+}, Bi^{3+}) according to their specific coordination environment. Such an exercise is beyond the scope of this review. We will however note when the lattice strain approach does not appear to be valid for the Pb-Sr pair, and, where possible, suggest a more appropriate "effective" ionic radius for Pb^{2+}. No consideration will be given to Pb^{4+}, which is unlikely to be common in magmatic environments, although like U^{5+} and U^{6+}, it may occur in some atmospheric pressure experiments conducted in air. In all minerals discussed, with the likely exception of zircon, Pb^{4+} will be more incompatible than Pb^{2+}.

Polonium forms both 4+ and 6+ ions, the former being considerably more stable in most natural environments, although PoO_6^{6-} complexes may occur in water (cf. telluric acid). The ionic radii of Po^{4+} are 0.94 Å (VI) and 1.08 Å (VIII), which are only slightly larger than Th^{4+} and U^{4+} (Table 2). We shall therefore use the latter two elements as proxies.

Actinium is similarly easy to find a proxy for. It forms large trivalent cations with an ionic radius of 1.12 Å in VI-fold coordination. This is somewhat larger than La^{3+} (1.032 Å), which we will adopt as a proxy. The partitioning behavior of the lanthanides (denoted collectively Ln) is sufficiently well understood to make this a prudent choice. In some minerals, however, the larger size of Ac^{3+}, may place it onto a larger lattice site than the lanthanides. This possibility should be considered for minerals with very large cation sites, such as amphiboles and micas.

Bismuth forms both 3+ and 5+ cations, although the former are by far the more common in nature. The ionic radius of Bi^{3+} is even closer to that of La^{3+}, than Ac^{3+}, so again La^{3+} is taken as the proxy. As noted above, Bi^{3+} has the same electronic configuration as Pb^{2+}, with a lone pair. It is unlikely therefore that the Shannon (1976) radius for Bi^{3+} is universally applicable. Unfortunately, there is too little known about the magmatic geochemistry of Bi, to use its partitioning behavior to validate the proxy relationship, or propose a revised effective radius for Bi^{3+}. The values of D_{Bi}/D_{La} derived here should be viewed in the light of this uncertainty.

Protactinium forms nominally 5+ cations with an ionic radius of 0.78 Å in VI-fold coordination and 0.91 Å in VIII-fold coordination (Table 2). The only other 5+ cations with comparable radii are Nb^{5+} and Ta^{5+} (nominally 0.64 and 0.74 Å in VI and VIII-coordination, respectively) and U^{5+} with 0.76 Å in VI-fold (no value is given by Shannon (1976) for VIII-fold). There are almost no partitioning data for U^{5+}, which makes it unsuitable as a proxy. We have therefore chosen Nb^{5+} and Ta^{5+} as Pa^{5+} proxies. We have adopted the observation of Tiepolo et al. (2000b) that Nb^{5+} is slightly larger than Ta^{5+} (Table 2). In our fitting of Nb-Ta fractionation in a large number of minerals, we have found that VI-fold radii of 0.640 Å for Ta^{5+} and 0.660 Å for Nb^{5+} provide the best fits, and will be adopted here. (VIII-fold radii are assumed to show the same 0.02 Å difference between Ta and Nb; Table 2). In some minerals, the ionic radius of Pa^{5+} falls between that of two sites, and the partition coefficients for each site must be summed. In the absence of a suitably large 5+ cation proxy, we are obliged to use the electrostatic model to estimate D_0^{5+} on the larger cation site. It is worth mentioning here a useful rule of thumb in predicting Pa partitioning. The relative ionic radii of Pa^{5+}, Ta^{5+} and the revised Nb^{5+} radius indicate that only where the appropriate cation site is larger than Nb^{5+} will appreciable Pa^{5+} be incorporated. Minerals with sites larger than Nb^{5+} will exhibit strong Nb-Ta fractionation, with $D_{Nb} \gg D_{Ta}$. Consequently only minerals with large D_{Nb}/D_{Ta} ratios and large absolute values of D_{Nb} will have appreciable D_{Pa}. We note also that it is at high charge and small radius where the lattice strain model is most sensitive to small changes in the parameters r_0 and E. This makes estimates of D_{Pa} particularly challenging and potentially subject to larger errors than other U-series elements.

Radon is the largest noble gas, although its exact atomic radius is not constrained. We will assume that, like other heavy noble gases, it enters the large cation sites in minerals (Brooker et al. 2003). However, we recognise that the larger noble gases (viz. Xe) do have a tendency to migrate to mineral surfaces and grain boundaries, and this possibility should be considered. There are insufficient mineral-melt partitioning data on the other noble gases to use them as proxies, with the exception of olivine and clinopyroxene, for which data are now emerging (Brooker et al. 1998; Chamorro et al. 2001). Instead will use the electrostatic model to estimate D_{Rn}, by extrapolating $D_{0(M)}^{n+}$ to zero charge, using Equation (7). For minerals with divalent host cations this will tend to give values of D_{Rn} comparable to D_U or D_{Th} within an order of magnitude (Fig. 9). This more circuitous route to obtaining D_{Rn} makes the values considerably less reliable than for the other U-series elements.

5.2. Derivation of proxy relationships

Having selected the optimum proxies, the problem now reduces to one of first determining the ratio of the partition coefficient of the U-series element to that of its proxy, via Equation (8), for each rock-forming mineral, and then choosing the most suitable partition coefficient for the proxy under the conditions of interest. A comprehensive review of the latter is beyond the scope of this paper and the reader is referred to compilations of partition coefficients for this purpose (Jones 1994; Green 1994; Nielsen 2002; Blundy and Wood 2003; Wood and Blundy 2003). Where appropriate we will, however, suggest suitable proxy partition coefficients for those minerals and conditions for which we have reliable data. Ideally, the user should attempt to determine the proxy partition coefficient by analysis of co-existing phases in the sample of interest. This is straightforward in volcanic systems where, for example, Ba in crystal and glass can be analyzed to obtain D_{Ba} (e.g., Cooper and Reid 2003). In total we require partition coefficients for the following: U, Th and Pb, and the proxies, La, Sr, Ba and Nb (or Ta). This is a reduction from 8 to 6 or 7 in the number of elements whose partitioning needs to be characterised. As discussed above, the procedure for Rn is more circuitous.

Equation (8) contains three variables. The first is temperature, which can be fixed by the user as appropriate to the problem at hand. The two unknown variables are $r_{0(M)}^{n+}$ and E_M^{n+}, which vary from mineral to mineral and, in some cases, along solid solution series. In this section we will use existing partitioning data to derive $r_{0(M)}^{n+}$ and E_M^{n+}. Where possible we will obtain $r_{0(M)}^{n+}$ and E_M^{n+} for cations with the same charge as the proxy. Where that is not possible we will use the observed relationships between $r_{0(M)}^{n+}$ or E_M^{n+} and charge to extrapolate values to the appropriate charge. The desired outcome is a set of $r_{0(M)}^{n+}$ and E_M^{n+} values for each rock-forming mineral for each charge and for each site at which U-series substitution occurs. Where there are sufficient data we will provide expressions that relate $r_{0(M)}^{n+}$ and E_M^{n+} to changing composition along a solid solution. Otherwise a single value of $r_{0(M)}^{n+}$ and E_M^{n+} will be used for all compositions. As more partitioning data become available, and as more rock-forming minerals are subject to careful X-ray structure determinations, it is hoped to refine the $r_{0(M)}^{n+}$ and E_M^{n+} values.

An additional advantage of the proxy approach is that the relationship between a U-series element and its proxy is unlikely to be significantly modified by the presence of water. Wood and Blundy (2002) have shown that water can have the effect of either increasing or decreasing partition coefficients due to the combined effect of water on melting temperatures and component activities in melts. For the same reason water can fractionate one valence group from another. It will not, however, produce fractionation between different-sized ions of the same valence entering a specific lattice site. The principal effect of water on the proxy relationship lies in the lower temperature at which hydrous processes tend to occur, relative to anhydrous processes. This is readily accounted for by the presence of temperature in the denominator of Equation (8).

6. ADDITIONAL CONSIDERATIONS

Before discussing mineral-melt partition coefficients in detail, it is useful to consider other factors that may influence partition coefficients for the U-series elements. Such factors arise both because the U-series elements typically occur at very low abundances in nature, and because they are radioactive. The first feature introduces the possibility of deviations from Henry's Law, at very low concentrations. The second feature raises questions about the ingrowth of highly incompatible daughter nuclides with time. We discuss these issues separately.

6.1. Henry's Law

Many U-series elements occur in nature at much lower concentrations than the more commonly used trace elements, such as the lanthanides. This fact, coupled with the highly incompatible nature of many U-series elements, means that in some cases factors other than lattice strain may influence the uptake of cations into a mineral, and so affect the partition coefficient.

The effect of concentration on partitioning is normally discussed in the context of Henry's Law. Henry's Law pertains to a region of activity-composition space where the thermodynamic activity of a species is related directly to its concentration, via the Henry's Law constant. We can define a Henry's Law constant for both melt and mineral phases. If either phase fails to show Henrian behavior over a particular concentration range, then a relationship will emerge between the partition coefficient and concentration. This may take the form of a steady change in partition coefficient with concentration (in melt or crystal), or a sudden change from one value to another as the Henry's Law threshold is crossed. Early studies of Henry's Law behavior focussed on possible deviations at relatively high concentrations, when an element changes from being a passive trace species to a minor structural constituent. Although an early study by Mysen (1979) of Ni partitioning between olivine and melt appeared to show deviations from Henry's Law as the Ni content of the melt exceeded 1000 ppm, there have been countless subsequent experimental studies (e.g., Beattie 1993c) to show that trace element partition coefficients do not change with concentration from a few ppm to a few weight per cent trace element in the melt. The conventional way of establishing that Henry's Law is obeyed is to determine the partition coefficient of a trace element at various concentration levels, while maintaining all other variables (pressure, temperature, major element phase compositions) constant. Good illustrations of Henrian behavior can be found in Beattie's (1993a) study of U-series element partitioning between clinopyroxene and melt. For a ten-fold change in crystal concentration of thorium (0.3-3 ppm), the thorium partition coefficient is constant within analytical error.

We find further evidence of Henrian behavior in our own studies of lanthanide partitioning between minerals and melts in synthetic systems where the doping level varies dramatically from one lanthanide to another. For example in the clinopyroxene-melt partitioning study of Blundy and Dalton (2000), partition coefficients were determined for La, Nd, Sm ,Y, and Yb. All five partition coefficients define a smooth parabola (Fig. 4b), consistent with the lattice strain model introduced above, despite the fact that the melt concentration levels varies from 100 ppm (Yb) to 4000 ppm (Sm). Any deviation in Henry's Law over this concentration range would be manifest in a deviation from the partitioning parabola for the element at highest concentration. This is not evident from Figure 4b. In contrast, Bindeman and Davis (2000), in their study of lanthanide partitioning between plagioclase and melt, do detect a decrease in partition coefficients from runs where lanthanides were present in plagioclase at natural (1-10 ppm) concentrations to runs where lanthanides were highly doped and occur at 100-10,000 ppm levels in plagioclase. The effect is relatively small (25-50% decrease), but consistent from run to run, and from lanthanide to lanthanide. Significantly, however, the curvature of the partitioning parabola, when plotted against ionic radius, is not influenced by doping level. In other words, the same structural controls operate irrespective of the doping level. Bindeman and Davis (2000) suggest that the lower partition coefficients at higher concentration reflects changes in the available charge balancing mechanisms as concentration increases. In particular, substitutions involving vacancies or other charge-balancing cations are saturated at relatively low concentrations. At higher concentrations fewer substitution mechanisms are available, and there is a consequent decrease in the partition coefficient. Significantly, the concentration effect is only seen for heterovalent cations, that require a charge-balancing mechanism. The homovalent cations Sr and Ba obey Henry's Law over a very wide concentration range (e.g., 800-11,000 ppm Sr; Bindeman and Davis 2000).

The Bindeman and Davis (2000) example illustrates the importance of considering the substitution mechanism of heterovalent cations. At high doping levels fewer substitution mechanisms may be available than at very low concentrations. The effect noticed by Bindeman and Davis is relatively small, and certainly well within the uncertainty of the approach we will use here to estimate U-series partition coefficients. However, in the case of heterovalent cations which occur at very low concentration in nature it is possible that a number of substitution mechanisms may aid their incorporation into minerals, causing appreciable deviation from the values predicted by the lattice strain approach. Pa^{5+} provides a good example. The concentration of Pa in most basaltic rocks is of the order 30-700 fg g^{-1} (or $0.3\text{-}7 \times 10^{-4}$ ppb). This is several orders of magnitude lower concentration than that of all conventional trace elements. At the same time, the Pa partition coefficient for most minerals, calculated on the basis of its ionic radius and charge via the lattice strain model, is very low ($< 10^{-6}$, see below). Thus if Pa incorporation is controlled by substitution mechanisms that do not apply to other highly charged cations, or at least make a negligible contribution to their overall partition coefficient, then D_{Pa} as estimated by lattice strain, may be a considerable underestimate.

The kinds of substitution mechanisms that may be relevant to super-low concentration elements such as Pa involve intrinsic defects, such as lattice vacancies or interstitials. Vacancy defects can potentially provide a low energy mechanism for heterovalent cation substitution, in that they remove or minimise the need for additional charge balancing substitutions. Formation of a vacancy *per se* is energetically unfavourable (e.g., Purton et al. 1997), and the trace element must rely instead on the thermal defect concentration in the mineral of interest, at the conditions of interest. Extended defects, such as dislocations or grain boundaries, may also play a key role, but as these are essentially non-equilibrium features, they will not be considered further here.

Urusov and Dudnikova (1998) discuss the problem of Henry's Law in the context of heterovalent cations. They present numerical expressions that describe the transition from vacancy-dominated control at very low concentration, to coupled-substitution control at higher concentration in terms of the equilibrium constants for the relevant equilibria. They conclude that for Schottky (vacancy) defect concentrations of 10^{-7}, as observed in many natural minerals, the increase in partition coefficient at very low concentrations is unlikely to be more than an order of magnitude, and typically no more than a factor of two. This is in keeping with the experimental studies of Harrison and Wood (1980) and Bindeman and Davis (2000). We conclude that, although an increase in partition coefficient is expected at very low concentration, especially for highly incompatible, heterovalent elements, this effect is extremely unlikely to exceed an order of magnitude, and for most elements may be negligible.

6.2. Ingrowth

U-series elements are unusual in that, although they are trapped as one species at the time of crystal growth, they will decay to a different species with time. Consequently, the possibility exists for a highly compatible parent to decay to a highly incompatible daughter nuclide. With time the daughter will be expelled from the lattice, typically by diffusion. This creates a balance between uptake and decay that is not a consideration for stable trace elements, and so deserves brief mention here.

During equilibrium crystal growth from a melt a U-series parent and daughter will be incorporated according to their equilibrium partition coefficients, D_p and D_d, respectively:

$$\frac{[p]_x}{[d]_x} = \frac{[p]_l}{[d]_l} \times \frac{D_p^{x/l}}{D_d^{x/l}} \qquad (9)$$

where the square brackets denote concentration, and subscripts x and l denote crystal and melt, respectively.

Immediately that the parent is incorporated into the crystal it will start to decay to the daughter. At secular equilibrium the (radio)activity of parent and daughter, will be equal, such that:

$$\lambda_p n_p = \lambda_d n_d \qquad (10)$$

where λ is the decay constant and n the number of atoms in the crystal. The latter is directly related to concentration, such that, after rearrangement:

$$\frac{[p]_x}{[d]_x} = \frac{\lambda_d}{\lambda_p} \qquad (11)$$

If the diffusion rate is negligible compared to the decay rate of the parent, then with time the concentration ratio of parent to daughter will evolve from (9) to (11) and secular equilibrium will be established. If diffusion is very rapid compared to the decay rate of the parent, then the concentration ratio will remain as given by (9) as all daughter is diffusively expelled from the crystal. In reality the situation is likely to be somewhere between these two extremes, probably in a steady state in which diffusive fluxes are balanced by radioactive production and decay (Van Orman et al. 2002a). This has some important consequences for the development and interpretation of radioactive excesses. For example, if the concentration ratio given by (9) is significantly smaller than that given by (11), then the crystal will accumulate ingrown daughter, which will then diffuse out of the crystal into a melt or other coexisting phase so generating an unsupported daughter excess. Clearly diffusion may play an important role in fractionating the U-series isotopes, a process termed "diffusive fractionation" by Van Orman et al. (2002b). Specific examples of this complexity have been discussed briefly by Feineman et al. (2002) and Van Orman et al. (2002). As diffusivity data for U-series radionuclides become better constrained, the full implications of ingrowth for U-series geochemistry will become apparent. The diffusion model of Van Orman et al. (2001), based upon principles similar to the lattice strain model of partitioning presented here, represents an important step in that direction.

We will not go into greater detail about ingrowth here, other than to note that the daughter-parent partition coefficient ratio has some significance, even if extremely low. For this reason we attempt to constrain partition coefficients for some highly incompatible values in the major rock-forming minerals, even when they fall below 10^{-6}. For simple melting and crystallization calculations these values can be assumed zero. However, for ingrowth modelling, the parent-daughter partition coefficient ratio may still be finite and significant.

7. MINERAL-MELT PARTITION COEFFICIENTS

7.1. Clinopyroxene

Clinopyroxene is the mineral for which most partitioning data are available. It has two sites at which the U-series elements substitute: the large distorted VIII-fold M2 site, normally occupied by Ca, Na, Fe or Mg, and the octahedral M1 site occupied by Mg, Fe, Al, Cr and Ti. All of the U-series cations are expected to substitute at M2, with the exception of Pa^{5+}, which may potentially enter both sites. The potential proxies, Nb^{5+} and Ta^{5+}, are smaller than Pa^{5+}, and exclusively located at the M1 site.

Wood and Blundy (1997) adapted the lattice strain model to describe lanthanide partitioning between clinopyroxene and melt as a function of crystal composition, pressure and temperature. In developing the model, they arrived at relationships between $r^{3+}_{0(M2)}$ and E^{3+}_{M2} and, respectively, crystal composition, and pressure and temperature:

$$r^{3+}_{0(M2)} = 0.974 + 0.067 X^{M2}_{Ca} - 0.051 X^{M1}_{Al} \text{ (Å)} \tag{12a}$$

$$E^{3+}_{M2} = 318.6 + 6.9P - 0.036T \text{ (GPa)} \tag{12b}$$

where X^{M2}_{Ca} and X^{M1}_{Al} refer to the atomic fractions of Ca and Al on the clinopyroxene M2 and M1 sites, P is in GPa and T in K. For E^{3+}_{M2} the dependencies on pressure and temperature are derived from measurements of the elastic properties of diopside (Wood and Blundy 1997). Equations (12a) and (12b) can be used to relate D_{Ac} and D_{Bi} to D_{La}. For both diopside-rich clinopyroxenes at low pressure and CaTs-rich clinopyroxenes at 3 GPa D_{Ac}/D_{La} is 0.06 and D_{Bi}/D_{La} is 0.75, as calculated from Equations (8) and (9).

Landwehr et al. (2001) extended the model of Wood and Blundy (1997) to include U^{4+} and Th^{4+}. They measured experimentally D_U and D_{Th} in a wide variety of synthetic clinopyroxene compositions in order to evaluate the crystal compositional dependence of U-Th fractionation. Their observations confirm the predictions of Wood et al. (1999), namely that as the M2 site becomes smaller, so D_{Th} becomes smaller than D_U (Figs. 1 and 11). The M2 site becomes smaller as the enstatite component of the clinopyroxene increases and Ca on M2 is replaced by Mg. Enstatite solubility in clinopyroxene increases with increasing temperature, consequently clinopyroxene coexisting with orthopyroxene will show higher D_U/D_{Th} at higher temperature. For this reason, D_U/D_{Th} increases with increasing pressure along the mantle solidus, as discussed above.

The curves in Figure 1 show the predicted variation in D_U/D_{Th} from Equation (8). In both cases E^{4+}_{M2} is assumed to be $\frac{4}{3} E^{3+}_{M2}$ as determined by Equation (12b) and $r^{4+}_{0(M2)} = r^{3+}_{0(M2)}$ as derived from Equation (12a). The solid curves in Figure 1 show the variation with the ionic radii of Shannon (1976), the broken curves show the predicted variation using the revised ionic radii of Wood et al. (1999), obtained from ThO_2 and UO_2 (Table 2). The latter better describe the experimentally measured U-Th fractionation. To some extent this may be an artifact of the assumption that $r^{4+}_{0(M2)} = r^{3+}_{0(M2)}$, when in fact there is strong evidence to suggest that $r^{4+}_{0(M2)} < r^{3+}_{0(M2)}$ (Fig. 8). However, the difference between the revised ionic radii of U^{4+} and Th^{4+} is not the same as the difference between the Shannon (1976) values. Consequently, a simple reduction in $r^{4+}_{0(M2)}$ relative to $r^{3+}_{0(M2)}$ will not, in itself, produce the same result. For this reason, we have adopted the Wood et al. (1999) radii throughout this study (Table 2).

Having successfully described the variation in D_U/D_{Th} with crystal composition, pressure and temperature, Landwehr et al. (2001) went on to parameterize D_{Th} from their own and published studies in terms of a simple partitioning equilibrium:

$$ThMgAl_2O_6 \text{ (pyroxene)} = ThMgAl_2O_6 \text{ (melt)} \tag{13}$$

The equilibrium constant for reaction (13) K_{13} is related to the entropy (ΔS_f), enthalpy (ΔH_f) and volume (ΔV_f) of fusion of fictive $ThMgAl_2O_6$ pyroxene by:

$$-RT \ln K_{13} = \Delta H_f - T \Delta S_f + P \Delta V_f \tag{14}$$

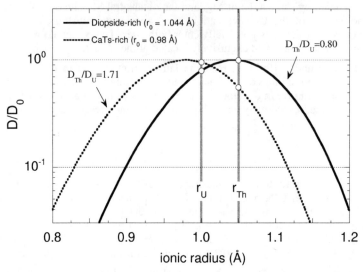

Figure 11. The clinopyroxene-liquid partition coefficient for 4+ ions entering the M2-site shown as a function of the ionic radius of the trace cation. Changes in clinopyroxene composition along a solid solution lead to small changes in the dimensions of M2 ($r_{0(M2)}^{4+}$), which can lead in turn to changes in the relative fractionation between 4+ ions of similar ionic radii, such as U^{4+} and Th^{4+} (shown as vertical lines). We contrast the partitioning behavior of a diopside-rich clinopyroxene ($r_{0(M2)}^{4+}$ = 1.044 Å) and a Ca-Tschermaks-rich clinopyroxene ($r_{0(M2)}^{4+}$ = 0.980 Å). In the first case $r_{0(M2)}^{4+}$ is closer to the ionic radius of Th^{4+} and D_U/D_{Th} = 0.8; in the second case $r_{0(M2)}^{4+}$ is closer to the ionic radius of U^{4+} and D_U/D_{Th} = 1.17. This is the origin of the behavior observed in Figure 1.

In order to derive activity–composition relationships for $ThMgAl_2O_6$ in clinopyroxene, Landwehr et al. (2001) used a mixing-on-sites model assuming complete short-range order between M and T sites. The activity of $ThMgAl_2O_6$ in clinopyroxene is thus given by:

$$a_{ThMgAl_2O_6}^{Cpx} = X_{Th}^{M2}\gamma_{Th}^{M2}X_{Mg}^{M1}\gamma_{Mg}^{M1} \quad (15)$$

where the X and γ terms denote mole fraction and activity coefficients, respectively, for the species and site of interest. γ_{Mg}^{M1} is given by $\exp(W_{Mg-Al}/RT)$ where W_{Mg-Al} is an Mg-Al Margules binary interaction parameter for the M1 site, estimated by Landwehr et al. (2001) to be 7.5 kJmol^{-1}. The value of γ_{Th}^{M2} can be readily calculated from the lattice strain model, in terms of the energy required to insert Th into a lattice site of radius r_0:

$$\gamma_{Th}^{M2} = \exp\left(\frac{\Delta G_{strain}^{Th-M2}}{RT}\right) = \exp\left[\frac{4\pi E_{M2}^{4+} N_A}{RT}\left[\frac{r_0}{2}(r_{Th}-r_0)^2 + \frac{1}{3}(r_{Th}-r_0)^3\right]\right] \quad (16)$$

VIII-fold ionic radius for Th^{4+} is taken from Table 2, and E_{M2}^{4+} and $r_{0(M2)}^{4+}$ are defined above.

For the melt, Landwehr et al. (2001) adopted a model of ideal mixing of six-oxygen $[Al_2O_6]^{6-}$ units, similar to that of Wood and Blundy (1999) and Blundy et al. (1995):

$$a_{ThMgAl_2O_6}^{melt} = X_{Th}^{6ox} X_{Mg}^{6ox} \quad (17)$$

where the X terms denote cation molar fractions on a six-oxygen basis.

Fitting 71 D_{Th} values (27 from their study and 44 from the literature) and subsequently excluding 5 outliers, Landwehr et al. (2001) derived the following expression for Th partitioning between clinopyroxene and silicate melt as a function of temperature, pressure, crystal chemistry and the molar Mg_{M1} partition coefficient:

$$RT \ln\left(\frac{D_{Th}\gamma_{Th}^{M2} X_{Mg}^{M1} \gamma_{Mg}^{M1}}{X_{Mg}^{melt}}\right) = 214.79 - 0.1757T + 16.42P - 1.50P^2 \quad (18)$$

where R is the gas constant (0.008314 kJmol^{-1}) and X_{Mg}^{X1} and X_{Mg}^{melt} are the mole fractions of Mg on the clinopyroxene M1 position and in the liquid, respectively, calculated on a 6 oxygen basis. (Note that this expression differs slightly from the original Landwehr et al. (2001) equation (their Eqn. 10), which contains a typographical error in ΔS_f). Equation (18) faithfully reproduces D_{Th} over 2 orders of magnitude. The fit is best at high pressures relevant to mantle melting. D_U can be calculated from D_{Th} using Equation (8).

Salters et al. (2002) have recently proposed an alternative expression for calculating D_U and D_{Th} in clinopyroxene as a function of crystal and melt composition. The expressions are calibrated on over 40 experimental determinations of D_U and D_{Th}. Salters et al. (2002) do not give values for the average absolute deviation. The full expressions (with 1 s.d. uncertainties in brackets) are:

$$\ln\left(\frac{D_U^*}{X_{Ca}^{melt}\left(X_{Al}^{melt}\right)^2}\right) = -5.65(\pm 1.84) + 6.04(\pm 2.53)\left(1 - X_{Mg}^{melt}\right)^2 + 5.00(\pm 0.55)\left(1 - X_{Di}^{cpx}\right)^2 \quad (19a)$$

$$\ln\left(\frac{D_{Th}^*}{X_{Ca}^{melt}\left(X_{Al}^{melt}\right)^2}\right) = -3.07(\pm 1.11) + 3.29(\pm 1.27)\left(1 - X_{En}^{cpx}\right)^2 + 5.8(\pm 1.17)\left(1 - X_{Di}^{cpx}\right)^2 \quad (19b)$$

The * denotes molar, rather than weight fraction, partition coefficient. The X^{melt} terms denote the cation molar fractions of Ca, Al and Mg, while the X^{cpx} terms denote the molar fraction of diopside and enstatite components in clinopyroxene. Note that this expression contains no explicit P-T dependence.

A full comparison of Equations (19a) and (19b), and (18) and (8) for calculating D_U and D_{Th} will not be presented here. However, it is useful to assess their ability to retrieve partition coefficients from experiments not used in the original calibration of the models. To this end we have used the recent experimental studies of McDade et al. (2003a,b) on clinopyroxene-melt partitioning close to the solidus of MORB-pyrolite at 3 GPa and Tinaquillo Lherzolite at 1.5 GPa. These experiments post-date the Landwehr et al. (2001) and Salters et al. (2002) studies, and therefore provide a fair evaluation. Results are presented in Table 3. The Landwehr et al. (2001) model retrieves the correct sense of change in D_U and D_{Th} with changing pressure and temperature. Their absolute D_U and D_{Th} values at 1.5 GPa are in excellent agreement with the experiments, while the 3 GPa values are underestimated by a factor 2. The Salters et al. model predicts the opposite sense of change with P and T, and underestimates D_U and D_{Th} at 1.5 GPa by a factor of ~3. However, the ratios D_U/D_{Th} are slightly more closely predicted by the Salters et al. model than by Landwehr et al. (2001), but, importantly, they both predict that D_U/D_{Th} increases with increasing temperature (and pressure). We conclude that both models can be applied with some confidence.

Partition coefficients for Po can be derived from those for Th using Equation (8) and the ionic radii in Table 2. We have not modified the ionic radius for Po^{4+} from the value

Table 3. Comparison of models for calculation of clinopyroxene D_U and D_{Th}.

	Sample	D_U	D_{Th}	D_U/D_{Th}
Experiments	R84-19[a]	0.0049(27)[c]	0.0036(20)	1.353(18)
	R64-11[b]	0.018(6)	0.021(7)	0.863(20)
Landwehr et al. (2001) model[d]	R84-19	0.0117	0.0078	1.502
	R64-11	0.0173	0.0162	1.066
Salters et al. (2002) model[e]	R84-19	0.0152	0.0119	1.272
	R64-11	0.0053	0.0056	0.940

a. 3.0 GPa, 1495°C, MORB-pyrolite source (McDade et al. 2003a)
b. 1.5 GPa, 1315°C, Tinaquillo Lherzolite source (McDade et al. 2003b)
c. Uncertainties in the experimental values are 1 s.d. expressed in terms of least significant figures.
d. Equations (18) and (8), with lattice parameters give by Equations (12a,b)
e. Equations (19a,b)

given in Shannon (1976), even though his value, like those for $r_{U^{4+}}$ and $r_{Th^{4+}}$, is derived from fluorine-bearing compounds. Such a modification will not be possible until Po partitioning data become available. In the meantime, values of D_{Po} for clinopyroxene calculated using the Shannon (1976) radius, should be seen as minima, i.e., if $r_{Po^{4+}}$ is reduced slightly it will enter M2 more readily. Typical calculated D_{Po}/D_{Th} are 0.6 for low-pressure diopside-rich clinopyroxenes and 0.7 for higher pressure CaTs-rich clinopyroxenes.

There are 10 published experimental studies of Pb partitioning between clinopyroxene and melt, covering a wide range of pressure, temperature and composition (Table 1b). There are a total of 20 D_{Pb} determinations, with values varying from 0.8 to 0.05. There is no obvious correlation with either composition or intensive parameters. For 16 of the 20 runs, D_{Sr} was also determined, allowing an evaluation of the proxy relationship between Pb and Sr. D_{Pb}/D_{Sr} ranges from 0.04-6.3 and is strongly positively correlated with D_{Pb}, suggesting that it is variability in this parameter which is responsible for the extreme variation in D_{Pb}/D_{Sr}. Certainly this variation in D_{Pb}/D_{Sr} is not consistent with the similarity in ionic radii of Pb^{2+} and Sr^{2+} (Table 2), as previously noted by Watson et al. (1987). For example, if we assume, following Wood and Blundy (1997) that $E_{M2}^{2+} = \frac{2}{3} E_{M2}^{3+}$, as determined by Equation (12b), and that $r_{0(M2)}^{2+} = r_{0(M2)}^{3+}$, D_{Pb}/D_{Sr} should be in the range 0.23-0.45 for the various experiments. Increasing $r_{0(M2)}^{2+}$ slightly relative to $r_{0(M2)}^{3+}$ only increases the calculated D_{Pb}/D_{Sr} ratio. It is difficult to account for this discrepancy between observed and estimated D_{Pb}. It is unlikely that significant Pb is present as Pb^{4+} in the melt, as some of the runs with low D_{Pb}/D_{Sr} were run in graphite capsules, which ensure low oxygen fugacity. Moreover some 60% of Pb would need to be present as Pb^{4+} to explain the lowest D_{Pb} values, which seems highly unlikely. Pb loss to the capsule materials cannot account for the low partition coefficients, although it may in part explain the anomalously high D_{Pb} in the runs of Blundy and Dalton (2000) and Klemme et al. (2002), which were both run in unlined platinum capsules. The possibility remains that the electronic structure of Pb^{2+}, with its lone pair of electrons, reduces the compatibility of Pb in clinopyroxene. In this regard, it is interesting to note that increasing the ionic radius of Pb^{2+} to 1.32Å faithfully reproduces the observed D_{Pb}/D_{Sr} ratios of Beattie (1993a) and Hauri et al. (1994). We conclude that, for clinopyroxene, Sr does not represent a straightforward proxy for Pb, and that this approach should only be used with great caution. Further investigation of the systematics of Pb partitioning between clinopyroxene and melt are clearly required. Until that has been done, the only reasonable approach is to adopt a value for D_{Pb} from the sources compiled in Table 1b.

Ra and Ba are also divalent cations that enter the M2 site. For experiments which report both D_{Sr} and D_{Ba} (i.e., Beattie 1993a; Hart and Dunn 1993; Hauri et al. 1994; Lundstrom et al. 1994 1998; Klemme et al. 2002; Blundy and Dalton 2000; Green et al. 2000; Blundy and Brooker 2003; Blundy, unpublished data; Chamorro, unpublished data, reported in Brooker et al. 2003; Wood and Trigila 2001) it is possible to calibrate E_{M2}^{2+} and $r_{0(M2)}^{2+}$. To facilitate this exercise (fitted E_{M2}^{n+} and $r_{0(M2)}^{n+}$ are statistically highly correlated) we have assumed that $E_{M2}^{2+} = \frac{2}{3}E_{M2}^{3+}$ and then calculated the best-fit value of $r_{0(M2)}^{2+}$. We find that using $r_{0(M2)}^{2+} = r_{0(M2)}^{3+} + 0.06$ Å reproduces 72% of 47 experimental D_{Ba}/D_{Sr} ratios to within 3 s.d. We consider that these lattice strain parameters can be used to derive D_{Ra} from D_{Ba}, or even D_{Sr}, although that correction would be inaccurately large. Typical calculated D_{Ra}/D_{Ba} ratios are 0.01-0.07.

Protactinium differs from all of the preceding U-series elements in that it may enter both the M1 and M2 sites. M1 is known to behave in a considerably stiffer fashion than M2 (Fig. 7; Hill et al. 2000), as befits its small size. The lack of a suite of small 5+ cations (unlike the lanthanides) makes it difficult to constrain $r_{0(M1)}^{5+}$ and E_{M1}^{5+} by fitting a partitioning parabola. However, it is possible to estimate $r_{0(M1)}^{5+}$ and E_{M1}^{5+} from the observed Nb-Ta fractionation in clinopyroxene. Hill et al. (2000) have shown that the clinopyroxene M1 site has $r_{0(M1)}^{4+}$ of ~0.65 Å and E_{M1}^{4+} of 3000 GPa. $r_{0(M1)}^{5+}$ should be slightly smaller and E_{M1}^{5+} larger. Assuming that $r_{0(M1)}^{5+}$ is on the order 0.62 Å, then $Z_c d_{M1-O^{-3}}$ is approximately 0.6, which is consistent with E_{M1}^{5+} of 4000 GPa (Fig. 7). We have chosen to adopt this fixed value, and then calculate the value of $r_{0(M1)}^{5+}$ which is required to produce the observed D_{Nb}/D_{Ta} in a large number of clinopyroxene-melt experiments using the modified ionic radii in Table 2. We have used 38 experimental partitioning experiments in which D_{Nb} and D_{Ta} were determined with reasonable precision (Skulski et al. 1994; Forsyth et al. 1994; Blundy et al. 1998; Lundstrom et al. 1998; Klein et al. 2000; Green et al. 2000; Hill et al. 2000; McDade et al. 2003a,b; Bennett et al. 2003; Blundy and Brooker 2003). Calculated $r_{0(M1)}^{5+}$ values are in the range 0.56 to 0.66 Å. Regression of the best fit $r_{0(M1)}^{5+}$ (in Å) to crystal composition produces the following empirical expression:

$$r_{0(M1)}^{5+} = 0.576 + 4.71 \times 10^{-5}T - 0.013 Al^{M1} - 0.017 Ca - 0.020 Mg - 0.033 Ti - 0.001 Na \quad (20)$$

Where all cations are expressed as atoms per six oxygen formula unit and T is in K. This expression reproduces $r_{0(M1)}^{5+}$ to within 0.011 Å. The predicted versus observed D_{Nb}/D_{Ta} are plotted in Figure 12. Given that the experimental values are rarely measured to better than 30% relative, the fit is considered acceptable. Significantly, using the derived $r_{0(M1)}^{5+}$ and E_{M1}^{5+} parameters to estimate D_{Pa}, gives values of D_{Pa}/D_{Nb} in the range 10^{-8} to 10^{-10}. We conclude that it is very unlikely that any Pa enters the clinopyroxene M1 site.

On the clinopyroxene M2 site, Pa^{5+} will be discriminated against both on account of its small size and high charge, relative to the site. If we assume that $r_{0(M2)}^{5+} = r_{0(M2)}^{4+} = r_{0(M2)}^{3+}$, from Equation (12a), and set $E_{M2}^{5+} = \frac{5}{3}E_{M2}^{3+}$ (Eqn. 12b), we can calculate $D_{Pa}/D_{0(M2)}^{5+}$. Values are in the range 10^{-4} to 10^{-2}. $D_{0(M2)}^{5+}$ is, of course, not constrained experimentally. However, if we extrapolate the electrostatic model for the clinopyroxene in Figure 9, we estimate that $D_{0(M2)}^{5+} \approx 10^{-3} D_{0(M2)}^{4+}$, which in turn can be estimated from D_U and D_{Th}. For most natural clinopyroxenes, $D_{0(M2)}^{4+}$ is larger than D_U and D_{Th} by a factor of at most 2. Consequently, $D_{Pa(M2)}$ is a factor of 2×10^{-5} to 2×10^{-7} lower than D_U or D_{Th}. Although very small, this is greater than D_{Pa} on M1. We conclude that D_{Pa} is vanishingly small in clinopyroxene, consistent with $D_{Nb}/D_{Ta} < 1$, and Pa is considerably more incompatible than its parent U and Th.

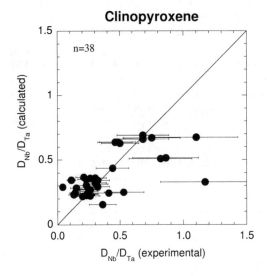

Figure 12. Application of the lattice strain model to Nb-Ta fractionation by clinopyroxene. The calculated D_{Nb}/D_{Ta} is based on the lattice strain parameters for the M1-site given in the text (E_{M1}^{5+} = 4000 GPa; $r_{0(M1)}^{5+}$ from Eqn. 20). The 38 experimental values are taken from the sources listed in the text. Error bars on the experimental measurements are 1 s.d. This model is used to estimate $D_{Pa(M1)}$.

Radon partitioning is better constrained for clinopyroxene than for any other phase because reliable partition coefficients for other noble gases are available. Chamorro et al. (2000), in a study of synthetic starting materials at 1-8 GPa, found D_{Ar} in the range 7×10^{-5} to 4×10^{-4}. A recent experimental determination of D_{Ne}, D_{Ar}, D_{Kr} and D_{Xe} in a diopside-rich synthetic composition at 0.1 GPa and 1200°C (Brooker et al. 2003) has shown that, consistent with lattice strain and electrostatic theory (see above), there is negligible difference between partition coefficients of different noble gases on the basis of their atomic radii. All 4 gases have approximately the same partition coefficient (10^{-3}), lending support to the proposal that Rn will also have the same partition coefficient, whatever its exact atomic radius. Significantly noble gas partition coefficients are roughly the same as D_{Th} (3.2×10^{-3}) and D_U (1.5×10^{-3}) in the same experiments (Fig. 9), in accord with the electrostatic theory above. If the mean charge on M2 decreases, for example due to increased jadeite component, we would expect noble gas partition coefficients to increase relative to D_U and D_{Th}. The effect may be even more pronounced if the M2 contains significant vacancies, as in the case of the eskolaite component, $\square_{0.5}Ca_{0.5}AlSi_2O_6$. Pertermann and Hirschmann (2002) have shown that clinopyroxenes on the eclogite solidus at 3 GPa contain appreciable eskolaite component, and they propose that this will facilitate noble gas incorporation, probably leading to higher values of D_{Rn} than those estimated above.

7.2. Orthopyroxene

Orthopyroxene has a VI-fold M1 site and a VI-fold M2 site. Both are predominantly filled by Mg and Fe. The smaller M1 site shares many characteristics with the clinopyroxene M1 site. It is therefore reasonable to assume that no U-series cations, including Pa^{5+} (see above) enter that site. We will confine our discussion to the octahedral M2, which is smaller than the equivalent (VIII-fold) clinopyroxene site, even after allowing for the different coordination number. Consequently most of the U-series elements have very low orthopyroxene-melt partition coefficients.

Lanthanides are more incompatible in orthopyroxene than clinopyroxene, typically by a factor of 4-8 on the mantle solidus (Blundy and Wood 2003; Salters et al. 2002; McDade et al. 2003a,b). Not surprisingly there are fewer lanthanide partitioning data for orthopyroxene to be fitted to the lattice strain model. We have identified a total of 16 runs from the studies of Kennedy et al. (1993), Wood et al. (1999), Blundy and Wood (2003), Blundy and Brooker (2003), Salters and Longhi (1999), Green et al. (2000) and Salters et al. (2002). (Only the two slowest cooling-rate runs of Kennedy et al. were used). Preliminary fits of Equation (2) to the lanthanide partition coefficients indicate that a single value of E_{M2}^{3+} will suffice for all runs. We have selected a value of 360 GPa, irrespective of P and T. (There are insufficient data to justify a more complex expression along the lines of Eqn. 12b). The derived values of $r_{0(M1)}^{3+}$ are in the range 0.758 to 0.819 Å, and correlate strongly with both total Al content and Ca content of the orthopyroxene. The following expression reproduces $r_{0(M1)}^{3+}$ with an average absolute deviation of 0.012 Å for the 16 experiments fitted:

$$r_{0(M2)}^{3+} = 0.753 + 0.118\text{Al}^{\text{tot}} + 0.114\text{Ca} \tag{21}$$

Where Al^{tot} and Ca denote atoms per 6-oxygen formula unit. These values of $r_{0(M1)}^{3+}$ combined with E_{M2}^{3+} of 360 GPa allow the ratios D_{Ac}/D_{La} and D_{Bi}/D_{La} to be calculated. Typical values for mantle solidus orthopyroxenes are in the range 0.001-0.009 and 1.1-1.2, respectively.

Uranium and thorium partitioning into orthopyroxene have been studied experimentally by Beattie (1993a), Kennedy et al. (1993), Salters and Longhi (1999), Wood et al. (1999), Salters et al. (2002) and McDade et al. (2003a,b). All of these studies involved broadly basaltic liquid compositions. D_{Th} on the mantle solidus is around 0.001. In all cases $D_U \geq D_{Th}$ as befits the small M2 site radius. The M2 site radius will increase as diopside component is dissolved in orthopyroxene, which is favoured at high temperatures. Thus, in contrast to clinopyroxene, we would expect D_U/D_{Th} to decrease with increasing temperature (and pressure) along the mantle solidus. This is just about apparent from the studies of McDade et al. (2003a,b) where D_U/D_{Th} changes from 0.98 ± 0.66 at 3 GPa/1505°C to 1.49 ± 0.03 at 1.5 GPa/1315°C. However, elsewhere in the dataset there is considerable variability in D_U/D_{Th}, which is not obviously correlated with crystal chemistry. This is partly a result of analytical precision, which is rarely better than $\pm 30\%$ due to the very low levels of U and Th in the experimental pyroxenes, and partly the result of not all experimental orthopyroxenes coexisting with clinopyroxene, which buffers Ca content and hence M2 site radius. The most precise estimate of D_U/D_{Th} available is 2.52 ± 0.24, at 1.5 GPa and 1268°C (Wood et al. 1999). We have modeled U and Th partitioning data by assuming that $E_{M2}^{4+} = \frac{4}{3}E_{M2}^{3+} = 480$ GPa and $r_{0(M2)}^{4+} = r_{0(M2)}^{3+}$. We have revised the VI-fold ionic radii of U and Th downwards in approximate proportion to the revised VIII-fold ionic radii for these elements based on their oxides. The preferred values are 0.875 Å and 0.919 Å, respectively (Table 2). These values reproduce the D_U/D_{Th} ratios of the Wood et al. (1999) and Beattie (1993a) data very closely, with values consistently around 2.6 at the mantle solidus. Polonium partition coefficients can be derived from D_{Th} using the same lattice strain parameters. D_{Po}/D_{Th}, so calculated, is typically 0.4-0.5.

The large alkaline earths (Sr-Ba) are highly incompatible in orthopyroxene, and it is clear that Ra will be even more so. At atmospheric pressure D_{Sr} is in the range $9\text{-}14 \times 10^{-4}$ (Beattie 1993a; Kennedy et al. 1993). On the mantle solidus D_{Sr} is 3.7×10^{-3} at 3 GPa, 1500°C (McDade et al. 2003a) and 0.074 at 1.5 GPa, 1315°C (McDade et al. 2003b), suggesting a dependence on both temperature and pressure. Green et al. (2000) report $D_{Sr} = 0.012$ under hydrous conditions at 2 GPa. D_{Ba} is much less well constrained due to its very low partition coefficient, and consequent low analytical precision. The most precise studies of D_{Ba} in orthopyroxene (Beattie 1993a) suggest that D_{Ba} is 50-100 times less than D_{Sr}. We have used D_{Ca}, D_{Sr} and D_{Ba} from the experiments of Beattie (1993a) and Kennedy et al. (1993) to derive $r_{0(M2)}^{2+}$, assuming that $E_{M2}^{2+} = \frac{2}{3}E_{M2}^{3+} =$

240 GPa. In the four experiments fitted $r_{0(M2)}^{2+}$ is found to be approximately 0.08 Å larger than $r_{0(M2)}^{3+}$, as calculated from Equation (18), an increment consistent with that obtained above for clinopyroxene (Fig. 8). These values can be used to estimate D_{Ra} from D_{Ba}. Typical D_{Ra}/D_{Ba} ratios calculated along the mantle solidus are 0.01 to 0.02.

There is only one determination of D_{Pb} in orthopyroxene, that of Salters et al. (2002) at the mantle solidus at 2.8 GPa. This value (0.009 ± 0.006) is within error of that calculated from the D_{Sr} value of McDade et al. (2003a) under similar conditions, using the lattice strain model, i.e., 0.0024 ± 0.0012. However, the uncertainties on both measurements should not be taken as strong support for the potential of Sr as a proxy for Pb. Still, there is no evidence for the anomalously low D_{Pb} values observed in clinopyroxene.

By analogy with clinopyroxene it is likely that Pa enters the orthopyroxene M2 site. In light of the fact that D_U and D_{Th} in orthopyroxene are approximately ten times lower than in clinopyroxene, it is likely that D_{Pa} is also lower in orthopyroxene. However, this effect is offset to some extent by the smaller M2 site in orthopyroxene, which will tend to be more favourable to Pa^{5+} than the M2 site in clinopyroxene. We have used the electrostatic model, applied to the two orthopyroxene-melt partitioning experiments of McDade et al. (2003a,b) to derive $D_{0(M2)}^{5+}$ (Fig. 13). Both datasets, at 1.5 and 3 GPa, are well fitted by Equation (7), giving fit parameters (see Fig. 13) for $\varepsilon\rho$ (37-39 Å) and Z_0 (1.78-1.98) that are in reasonable agreement with those obtained for clinopyroxene. In both cases the extrapolated value of $D_{0(M2)}^{5+}$ is 2-4 × 10^{-6}. After correcting for lattice strain, assuming $E_{M2}^{5+} = \frac{5}{3}E_{M2}^{3+} = 600$ GPa and $r_{0(M2)}^{5+} = r_{0(M2)}^{4+} = r_{0(M2)}^{3+}$ (from Eqn. 21), we get values of D_{Pa} of 1.8-3.6 × 10^{-6} and D_{Pa}/D_U of approximately 1.1-1.7 × 10^{-3}, i.e., 2 to 4 orders of magnitude larger than D_{Pa}/D_U in clinopyroxene.

To our knowledge there are no noble gas partition coefficients for orthopyroxene. We can make an estimate using the relationship between D_0 and charge in Figure 13. Because Z_0 is slightly less than 2 for the two orthopyroxenes plotted, noble gas partition coefficients should be slightly higher than D_U. We estimate D_{Rn} in the range 0.004-0.02 under upper mantle conditions. This is slightly higher than the estimate for clinopyroxene.

7.3. Olivine

All of the U-series elements are highly incompatible in olivine, with partition coefficients consistently lower than for either of the pyroxenes. Kennedy et al. (1993) and Beattie (1994) have experimentally investigated lanthanide partitioning between olivine and melt at atmospheric pressure. Salters et al. (2002) and McDade et al. (2003b) present a single set of olivine-melt lanthanide partition coefficients in multiply-saturated mantle melts at 1 and 1.5 GPa, respectively. Taura et al. (1998) have studied the partitioning of a large number of elements in addition to the lanthanides at pressures of 3-14 GPa. Lanthanide partition coefficients from all four studies can be fitted to the lattice strain model to derive best-fit parameters for E_M^{3+} and $r_{0(M)}^{3+}$. Although olivine has two octahedral (VI-fold) sites (M1 and M2), they are sufficiently close in size and geometry that only one site was used for fitting (cf. Beattie 1994). We included both Sc and octahedral Al in the fitting, to better resolve the parabolae at low ionic radii. Like orthopyroxene, we found that a single value of E_M^{3+} of 360 GPa can be used to fit all experiments. Fitted $r_{0(M)}^{3+}$ values lie in the range 0.70 to 0.73 Å. There is a hint of a slight increase in $r_{0(M)}^{3+}$ with increasing Fe content, consistent with the large M2-site in fayalite versus forsterite. However, there are insufficient data to quantify this relationship. For forsterite-rich olivines (>90 mol%) a single value of $r_{0(M)}^{3+} = 0.710$ Å is adequate to fit all experiments, with the exception of Taura et al. (1998) data which are too imprecise to be useful. Using these fixed values of E_M^{3+} and $r_{0(M)}^{3+}$, we calculate D_{Ac}/D_{La} and D_{Bi}/D_{La} at the mantle solidus of 0.01 and 1.1, respectively.

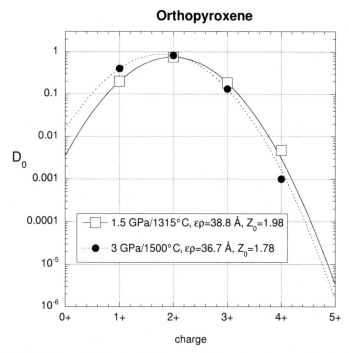

Figure 13. Electrostatic model fitted to partition coefficients for cations entering the M2-site in orthopyroxene, based on the experiments of McDade et al. (2003a,b). The curves are fits to Equation (7) and can be used to estimate D_{Rn} and $D^{5+}_{0(M2)}$, from which $D_{Pa(M2)}$ can be calculated via the lattice strain model. The fit parameters are given in the legend.

There are four published studies of olivine-melt partitioning of U and Th: Beattie (1993a) $D_U = 6(\pm 1) \times 10^{-6}$ at 2.9 GPa; Salters et al. (2002) $D_U = 5(\pm 7) \times 10^{-4}$ at 1 GPa; Wood et al. (1999) $D_U = 18(\pm 1) \times 10^{-6}$ at 1.5 GPa; McDade et al. (2003b) $D_U = 5.9 \times 10^{-5}$ at 1.5 GPa. The data are too sparse to establish whether these differences in D_U are real. However, in all four cases $D_U > D_{Th}$, with D_U/D_{Th} values of 3.3 ± 2.7 at 1 atmosphere, 12 ± 5 at 1 GPa and 4.3 at 1.5 GPa. We have modeled U-Th fractionation using the lattice strain models by taking $E^{4+}_{M2} = \frac{4}{3}E^{3+}_{M2} = 480$ GPa and $r^{4+}_{0(M)} = r^{3+}_{0(M)} = 0.710$ Å (cf. orthopyroxene). We have used the same VI-fold ionic radii for U^{4+} and Th^{4+} as for orthopyroxene (Table 2), which gives D_U/D_{Th} of 6.3, in broad agreement with the two experimental values. D_{Po}/D_U calculated from the same lattice strain parameters is 0.03-0.10.

Lattice strain parameters for 2+ cations can be derived by fitting the partitioning data for Fe^{2+}, Mn, Ca, Sr and Ba from the experiments of Beattie (1993a) and Kennedy et al. (1993). In both cases $r^{2+}_{0(M)}$ is found to be 0.078 Å larger than $r^{3+}_{0(M)}$ as previously observed for the pyroxenes. Taking $E^{2+}_{M2} = \frac{2}{3}E^{3+}_{M2} = 240$ GPa this gives a D_{Ra}/D_{Ba} ratio at the mantle solidus of 0.004. D_{Pb}/D_{Sr} at the mantle solidus calculated in the same way gives 0.53. There is only one published determination of D_{Pb}: 0.0035 ± 0.0011 at 1 GPa from Salters et al. (2002). If Sr can be used as a crude proxy for Pb in olivine, then this value would imply $D_{Sr} \approx 0.007$ at the mantle solidus. The best values for comparison are the two 3 GPa experiments of Taura et al. (1998), which give D_{Sr} in the range 0.006-0.091, the 2.9 GPa $D_{Sr} = 1.5 \times 10^{-5}$ of Beattie (1993b) and the 1.5 GPa $D_{Sr} = 1.8 \times 10^{-4}$

of McDade et al. (2003b). Thus D_{Pb} appears to be slightly higher for olivine than would be predicted from lattice strain arguments, although the data are too sparse to draw any firm conclusion. Using the same 2+ parameters, D_{Ra}/D_{Ba} is calculated to be 0.003-0.02.

Only McDade et al. (2003b) report experimental data on the partitioning of Nb and Ta between olivine and melt. At 1.5 GPa they find $D_{Nb} = 1 \times 10^{-4}$ and $D_{Ta} = 6 \times 10^{-4}$, which indicates that $r_{0(M)}^{5+}$ is smaller than r_{Nb} (0.660 Å). This will serve to exclude Pa (r_{Pa} = 0.78 Å) from the M-sites in much the same way as it is excluded from M1 in clinopyroxene. Furthermore D_{Nb} and D_{Ta} are themselves two orders of magnitude lower in olivine than coexisting clinopyroxene (McDade et al. 2003b). For these reasons we suggest that Pa is substantially more incompatible in olivine than clinopyroxene.

Argon partition coefficients for olivine have recently been determined by Brooker et al. (1998). Their measured D_{Ar} values are around 10^{-2}, which is higher than measured for clinopyroxene, but similar to the estimated values for orthopyroxene. We propose a value of 10^{-2} for D_{Rn}.

7.4. Garnet

Garnet has two sites onto which U-series elements partition: the dodecahedral (VIII-fold) X-site, occupied principally by Ca, Mg and Fe^{2+}, and the octahedral (VI) Y-site, occupied by Al, with lesser Fe^{3+}, Cr, Ti and, at high pressure, Si and Mg via the majorite substitution ($Al_2 = MgSi$). With the exception of Pa^{5+}, which is likely to enter the Y-site, all U-series cations are expected to occur on the X-site. An unusual feature of garnets is the extreme variation in X-site size due to the complete, but non-ideal, solid solution between the grossular, pyrope and almandine end-members. This is in marked contrast to the pyroxene, where the large difference in size between Ca^{2+} and Mg^{2+} leads to a pronounced miscibility gap at most temperatures and pressures. The variable dimension of garnet X-sites, leads to complexities in application of the lattice strain model, which are not apparent for olivine or the pyroxenes. One of the most striking features about garnet-melt partitioning to emerge from experimental studies is the non-linear variation in partition coefficients with garnet composition. Thus, along the grossular-pyrope join partition coefficients for some elements are appreciably higher in mixed garnets than in either end-member (Van Westrenen et al. 1999, 2000a, 2001, 2003).

Van Westrenen et al. (2001a) present a model of lanthanide and Sc partitioning between the garnet X-site and melt. The model is a variant of the lattice strain model of clinopyroxene-melt partitioning of Wood and Blundy (1997), and is based on 160 experimental garnet-melt pairs in the pressure-temperature range 2.5-7.5 GPa and 1450-1930°C. The model includes composition-sensitive expressions for $r_{0(X)}^{3+}$ and accounts for the non-linear variation in E_X^{3+} with composition, as follows:

$$r_{0(X)}^{3+}(\text{Å}) = 0.930 X_{Py} + 0.993 X_{Gr} + \\ 0.916 X_{Alm} + 0.946 X_{Sp} + 1.05 (X_{And} + X_{Uv}) - 0.005(P-3) \quad (22a)$$

$$E_X^{3+}(\text{GPa}) = 3.5 \times 10^{12} \left(1.38 + r_{0(X)}^{3+}\right)^{-26.7} \quad (22b)$$

where the X terms denote the molar fractions of the garnet components grossular (Gr), pyrope (Py), almandine (Alm), spessartine (Sp), andradite (And) and uvarovite (Uv), and P is pressure in GPa. $D_{0(X)}^{3+}$ is described in terms of pressure and temperature via the fusion equilibrium of a fictive lanthanide-bearing pyrope, $LnMg_2Al_2Si_3O_{12}$, with analogous activity-composition relationships to those adopted for clinopyroxene by Wood and Blundy (1997). The full expression for $D_{0(X)}^{3+}$ is:

$$D^{3+}_{0(X)} = \frac{\exp\left(\dfrac{406+12.1P-0.224T}{RT}\right)}{\left(\gamma^{gt}_{Mg}D_{Mg}\right)^2} \tag{23}$$

where P is in GPa, T in Kelvin, and D_{Mg} is the Mg partition coefficient under the conditions of interest. The term γ^{gt}_{Mg} accounts for non-ideal mixing between Ca^{2+} and Mg^{2+} on the garnet X-site:

$$\gamma^{gt}_{Mg} = \exp\left[\frac{33(X_{Ca})^2}{RT}\right] \tag{24}$$

where X_{Ca} is the molar fraction Ca on the X site (i.e., $X_{Gr}+X_{And}$). Equations (22)-(24) can be used to derive the proxy relationships between Ac, Bi and La. For both pyrope-rich garnets at the mantle solidus and more calcic garnets typical of eclogites, calculated D_{Ac}/D_{La} and D_{Bi}/D_{La} are 2×10^{-5} to 7×10^{-4} and 0.3 to 0.5, respectively.

Uranium and thorium partitioning between garnet and melt has been studied by Beattie (1993b), La Tourrette et al. (1993), Salters and Longhi (1999), Van Westrenen et al. (1999, 2000), Salters et al. (2002), Klemme et al. (2002) and McDade et al. (2003a). These data show that not only are D_U and D_{Th} highly variable ($0.001 < D_{Th} < 0.3$), but so too is the D_U/D_{Th} ratio. This variability in D_U/D_{Th} is in large part due to the large variation in the size of the X site. If we use Equation (19a) to estimate $r^{3+}_{0(X)}$ and assume $r^{4+}_{0(X)} = r^{3+}_{0(X)}$ we see that the variation in D_U/D_{Th} with $r^{4+}_{0(X)}$ (Fig. 14) is remarkably similar to that for clinopyroxene (Fig. 1). The fact that the plotted garnet partitioning data derive from a wide range of pressures, temperatures and compositions is particularly encouraging.

The variation in D_U/D_{Th} is broadly consistent with $E^{4+}_{(X)} = \frac{4}{3}E^{3+}_{(X)}$ (as calculated from Eqn. 22b), and the revised VIII-fold ionic radii for U^{4+} and Th^{4+} in Table 2 (Fig. 14). For pyrope-rich garnets characteristic of the mantle solidus garnet D_U/D_{Th} is in the range 2.4 to 7.5. Note that the uvarovite component appears in the equation for $r^{3+}_{0(X)}$ (22a). For the Cr-rich mantle solidus garnets of Salters and Longhi (1999) and Salters et al. (2002) the presence of 2-4 mol% uvarovite has a significant effect on $r^{3+}_{0(X)}$, increasing the size of the X-site and so reducing U/Th fractionation (Fig. 14). Evidently the Cr content of mantle garnets must be accounted for when estimating U-Th fractionation. For the more grossular-rich garnets characteristic of eclogites, D_U/D_{Th} has an experimental value of ~3.2 (Klemme et al. 2002)

D_{Th} (and D_U) vary inversely with reciprocal temperature (Fig. 15). For mantle solidus garnets the correlation is reasonably good and can be used to make a first-order estimate of D_{Th}. A more comprehensive model for D_U and D_{Th}, as a function of pressure, temperature and melt composition is provided by Salters et al. (2002). Their full expressions (for the molar partition coefficients, D^*) are:

$$\tfrac{1}{2}\ln\left(\frac{D^*_U}{\left(X^{melt}_{FM}\right)^4 \left(X^{melt}_{Si}\right)^2}\right) = \\ 7.85 - \frac{24300}{T} + 4.01\left(1-X^{melt}_{Al}\right)^2 + 10.84\left(1-X^{melt}_{FM}\right)^2 - 1.88\left(1-X^{gt}_{Gr}\right)^2 \tag{25a}$$

and

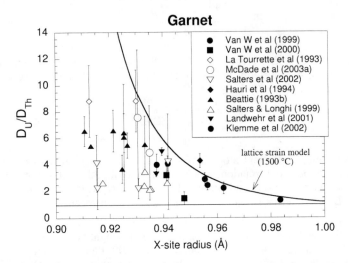

Figure 14. U-Th fractionation by garnet as a function of X-site dimension ($r_{0(X)}^{4+}$) for 33 experimental garnet-melt pairs from the sources listed in the legend. $r_{0(X)}^{4+}$ is assumed equal to $r_{0(X)}^{3+}$ as given by Equation (22a). Note the much larger U-Th fractionation produced by garnet relative to clinopyroxene (Fig. 1). The curved line shows the prediction of the lattice strain model (at fixed temperature of 1500°C) using the VIII-fold ionic radii in Table 2 and $E_X^{4+} = \frac{4}{3} E_X^{3+}$, as given by Equation (22b). Errors bars are 1 s.d.

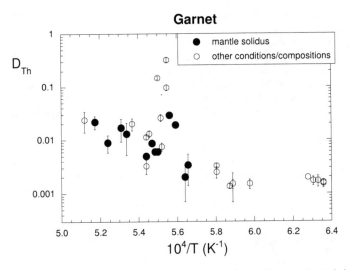

Figure 15. Variation in D_{Th} for garnet versus reciprocal temperature for experimental data sources listed in Table 1b at a variety of pressures ($n = 33$). A distinction is made between mantle solidus partition coefficients (Salters and Longhi 1999; Salters et al. 2002; McDade et al. 2003a,b) and the rest. Note the strong temperature dependence, which is qualitatively similar to that incorporated in Equation (25b). The scatter is due to additional compositional controls.

$$\tfrac{1}{2}\ln\left(\frac{D^*_{Th}}{\left(X^{melt}_{FM}\right)^4\left(X^{melt}_{Si}\right)^2}\right)=11.46-\frac{24200}{T}+8.6\left(1-X^{melt}_{FM}\right)^2-2.08\left(1-X^{gt}_{Gr}\right)^2 \quad (25b)$$

where X_{FM} is the molar cation fraction of Fe+Mg in the melt, X_{Si} is the molar cation fraction Si in the melt, and X_{Gr} is the molar fraction grossular in the garnet. As an independent test of these expressions we have calculated D_U and D_{Th} for the two garnet-melt partitioning runs of McDade et al. (2003a), which post-date the Salters et al. (2002) calibration. A comparison of observed and calculated partition coefficients (Table 4) shows remarkable agreement, suggesting that Equations (25a,b) can be applied with confidence to garnet-melt partitioning on the mantle solidus. Under other conditions, an alternative approach is to adopt an appropriate D_U value from one of the studies in Table 1b and then estimate D_{Th} from the lattice strain equation using the parameters above. D_{Po}/D_{Th} calculated using these parameters is in the range 0.04 to 0.1.

Table 4. Test of Salters et al. (2002) model for garnet-melt D_U and D_{Th}

	Sample	D_U	D_{Th}	D_U/D_{Th}
Experiments	R84-9[a]	0.015(3)[c]	0.002(13)	5.0(9)
	R64-19[b]	0.018(7)	0.0046(19)	3.9(22)
Salters et al. (2002) model[d]	R84-9	0.012	0.0034	3.4
	R84-19	0.019	0.0047	4.0

a. 3.0 GPa, 1495°C, MORB-pyrolite source (McDade et al. 2003a)
b. 3.0 GPa, 1500°C, MORB-pyrolite source (McDade et al. 2003a)
c. Uncertainties in the experimental values are 1 s.d. expressed in terms of least significant figures.
d. Equation (25)

Large alkaline earths (Sr,Ba) are highly incompatible in all garnets, and it is likely that D_{Ra} will be similarly small. For the limited experimental datasets where it is possible to estimate both $r^{3+}_{0(X)}$ and $r^{2+}_{0(X)}$ (Van Westrenen et al. 2000; Klemme et al. 2002; McDade et al. 2003a) we find that $r^{3+}_{0(X)}$ is 0.053 ± 0.004 Å smaller than $r^{2+}_{0(X)}$. This is within the range observed previously for pyroxenes and olivine. We recommend using a fixed value of 0.053 Å, such that $r^{2+}_{0(X)} = 0.053 + r^{3+}_{0(X)}$ (as derived from Eqn. 22a). Combining this value with $E^{2+}_{(X)} = \tfrac{2}{3}E^{3+}_{(X)}$ from Equation (22b) we can calculate D_{Ra}/D_{Ba}. Along the mantle solidus this ratio is approximately 10^{-4}. D_{Ba} itself is not well constrained for garnets because of difficulties in analysing such trace quantities. The most robust estimates are probably those of Beattie (1993b), which are around 10^{-5}. Suffice to say D_{Ra} in garnet is vanishingly small.

There are a number of D_{Pb} determinations for garnet (Beattie 1993b; Salters et al. 2002; Klemme et al. 2002). As noted for clinopyroxene, the Klemme et al. (2002) D_{Pb} value (0.18 ± 0.03) appears remarkably high, possibly due to an experimental artefact. Salters et al. (2002) give two values around 5×10^{-3}, while Beattie (1993b) gives 1-67 × 10^{-5} under similar conditions. The latter values are about twenty times lower than D_{Sr} in the same experiments (5-13 × 10^{-4}). The lattice strain model, using the 2+ parameters derived above, gives a consistent D_{Pb}/D_{Sr} of 0.02-0.09, suggesting that Sr may be a reasonable proxy for Pb in garnet.

Perhaps the biggest challenge in estimating U-series partition coefficients is the case of protactinium partitioning into garnet. The difficulty arises because the ionic radius of

Pa^{5+} in VI-fold (0.78 Å) and VIII-fold (0.91 Å) coordination places it between the optimum radii of the X and Y sites, i.e., it is slightly too large for the Y-site and too small for the X-site. Consequently, as for clinopyroxene, it is necessary to evaluate Pa partitioning onto both sites. In the analogous case of garnet-melt partitioning of Zr^{4+} and its slightly smaller twin, Hf^{4+}, Van Westrenen et al. (2001b) have shown that both cations can enter both sites, with $D_{Zr}>D_{Hf}$ on the larger X-site ($r_{0(X)}^{4+}$ = 0.9-1.0 Å) and $D_{Hf}>D_{Zr}$ on the smaller Y-site ($r_{0(Y)}^{4+} \approx$ 0.67Å). The bulk garnet D_{Zr}/D_{Hf} is therefore a complex function of garnet composition. As we are concerned here with the relationship between D_{Pa} and D_{Nb} (or D_{Ta}) it is important to evaluate whether similar complexity occurs for 5+ cations. The problem is compounded by the fact that Nb and Ta are the only 5+ cations for which there are partitioning data, whereas Ti^{4+}, Zr^{4+} and Hf^{4+} can all be used to constrain the problem for 4+ cations (Van Westrenen et al. 2001b). We will begin by evaluating $D_{0(X)}^{5+}$ using the electrostatic model in order to estimate $D_{Pa(X)}$, $D_{Nb(X)}$ and $D_{Ta(X)}$. We will use these estimates to derive $D_{Nb(Y)}$ and $D_{Ta(Y)}$ from D_{Nb} and D_{Ta}, respectively which can in turn be used to calibrate a lattice strain model for the Y-site. This model will be based around the 4+ partitioning model of Van Westrenen et al. (2001b), in terms of $r_{0(Y)}^{5+}$ and $E_{(Y)}^{5+}$. From these parameters it is possible to estimate $D_{Pa(Y)}$, which can be combined with $D_{Pa(X)}$ to give D_{Pa} and hence D_{Pa}/D_{Nb}.

In order to derive an electrostatic model for the garnet X-site it is necessary to have experimental data on 4+, 3+, 2+, 1+ cation partitioning onto this site. Only the study of Klemme et al. (2002) presents data for all four valences. These data show that $D_{0(X)}^{1+}$ is sufficiently close to D_{Li} that it is reasonable to use D_{Li} instead. $D_{0(X)}^{4+}$ is derived from the D_U and D_{Th} lattice strain model presented above. By these means we can increase to ten the number of experimental data that can be fitted (Van Westrenen et al. 1999, 2000; Klemme et al. 2002; McDade et al. 2003a). For each experiment we plot $D_{0(X)}^{n+}$ vs. Z_n and fit Equation (7) to obtain $D_{00(X)}$, $\varepsilon\rho$ and $Z_{0(X)}$. Typical fits are shown in Figure 16 and the results for all runs given in Table 5. All three parameters are broadly consistent between runs, especially $Z_{0(X)}$, which is in the range 2.6-2.9 for all runs except Klemme et al. (2002), where it is 2.36. $\varepsilon\rho$ is in the range 12-39 Å, which is similar to that previously obtained for clinopyroxene and wollastonite (Blundy and Wood 2003). For all ten runs we have used the parameters in Table 5 to estimate $D_{0(X)}^{5+}$, from which we can calculate $D_{Nb(X)}$, $D_{Ta(X)}$ and $D_{Pa(X)}$ using $r_{0(X)}^{5+} = r_{0(X)}^{3+}$ and $E_{(Y)}^{5+} = \frac{5}{3}E_{(Y)}^{3+}$ (from Eqn. 22a,b), as before. For all ten garnets there is negligible incorporation of Nb or Ta onto the X-site, however $D_{Pa(X)}$ is significant, though variable (10^{-4} to 6 × 10^{-3}). We conclude, as expected, that Nb and Ta predominantly enter the Y-site, while Pa enters both X and Y.

Y-site partitioning of 5+ cations is difficult to constrain because there are data for only two cations, Nb^{5+} and Ta^{5+}, and yet three parameters need to be constrained: $r_{0(Y)}^{5+}$, $E_{(Y)}^{5+}$ and $D_{0(Y)}^{5+}$. We will make the assumption, following Van Westrenen et al. (2001b) that the dimensions of the Y-site do not vary significantly with garnet composition. This is likely to be true provided that there is very little andradite, uvarovite or majorite component to replace Al with Fe^{3+}, Cr^{3+} or Mg+Si, respectively, as the dominant cation on Y-site. This is true for all of our calibrant experiments, but as noted above, the relatively high uvarovite component on the mantle solidus may cause complications. We will assume that $r_{0(Y)}^{5+}$ and $E_{(Y)}^{5+}$ are the same for all ten runs. We will further assume that $r_{0(Y)}^{5+} \leq r_{0(Y)}^{4+}$ and $E_{(Y)}^{5+} \geq E_{(Y)}^{4+}$ as estimated by Van Westrenen et al. (2001b). We can then find the values of these parameters that best model the observed $D_{Nb(Y)}/D_{Ta(Y)}$, i.e., after allowing for the negligible $D_{Nb(X)}$ and $D_{Ta(X)}$. For 12 published and unpublished anhydrous garnet-melt partition experiments the weighted mean value of D_{Nb}/D_{Ta} is 0.81 ± 0.05. This ratio can be reproduced with $r_{0(Y)}^{5+}$ = 0.67 and $E_{(Y)}^{5+}$ = 1500 GPa. We can use these parameters to calculate $D_{Pa(Y)}$, which can be added to the above estimates of $D_{Pa(X)}$ to give D_{Pa}. In this way we calculate a very consistent set of D_{Pa}/D_{Nb} values of

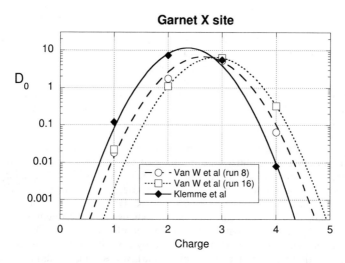

Figure 16. Electrostatic model fitted to partition coefficients for cations entering the X-site in garnet, based on the experiments of Van Westrenen et al. (1999, 2000) and Klemme et al. (2002). The curves are fits to Equation (7) and can be used to estimate D_{Rn} and $D_{0(X)}^{5+}$, from which $D_{Pa(X)}$ can be calculated via the lattice strain model. The fit parameters are given in Table 5.

Table 5. Electrostatic model fit parameters for garnet and estimates of D_{Pa}/D_{Nb}

Reference	Run	T (°C)	P (GPa)	$D_{00(X)}$	$\varepsilon\rho$ (Å)	$Z_{0(X)}$	D_{Pa}/D_{Nb}
Van Westrenen et al. (1999)	11	1565	3	5.62(19)	16.0(4)	2.56(1)	0.022
Van Westrenen et al. (1999)	8	1560	3	6.79(26)	19.9(5)	2.64(1)	0.022
Van Westrenen et al. (1999)	13	1530	3	2.58(13)	38.6(26)	2.69(5)	0.026
Van Westrenen et al. (1999)	12	1545	3	3.96(29)	20.6(14)	2.61(3)	0.022
Van Westrenen et al. (1999)	14	1530	3	2.96(14)	38.6(35)	2.80(3)	0.073
Van Westrenen et al. (2000)	16	1540	3	6.49(36)	19.7(12)	2.87(2)	0.028
Van Westrenen et al. (2000)	18	1538	3	7.51(39)	14.3(5)	2.72(1)	0.021
McDade et al. (in press)	R84-19	1495	3	8.81(70)	12.5(5)	2.73(1)	0.019
McDade et al. (in press)	R84-9	1500	3	10.1(7)	12.4(5)	2.71(1)	0.020
Klemme et al (2002)	BS21	1400	3	11.6(13)	18.8(8)	2.36(3)	0.015

Partitioning data, in terms of D_0^{n+}, fitted to Equation (7) by non-linear least squares.
Uncertainties in brackets given in terms of least significant figures.

0.02-0.07 for all ten garnets considered. As garnet D_{Nb} on the mantle solidus is in the range 0.007-0.051 (Salters et al. 2002; McDade et al. 2003a), D_{Pa} is likely to be in the range 0.001-0.01, which is about an order of magnitude less than D_{Th} for garnet under the same conditions, but considerably larger than for pyroxenes or olivine.

There are no published noble gas partition coefficients for garnet. However, we can make a crude estimate by extrapolation of the electrostatic model (Table 5; Fig. 16) to zero charge. Calculated values are extremely variable, in the range 10^{-4}-10^{-12}, with the lowest values obtained for the mantle solidus, and 4×10^{-6} for eclogitic garnets (Klemme et al. 2002). These values are several orders of magnitude lower than for olivines and pyroxenes, suggesting that garnet is an insignificant host for noble gases (including radon).

7.5. Amphibole

Amphibole adds further complexity to the problem of estimating U-series partition coefficients because it contains a multiplicity of cation sites at which substitution may occur: three structurally distinct octahedral (VI) sites, M1, M2 and M3; a larger VIII-fold M4 site, which in clino-amphiboles is occupied by Ca and Na; and a XII-fold A-site, which may be vacant, or occupied by Na and K. Fortunately, Tiepolo and co-workers (1998, 2000a,b, 2001) have carried out an exhaustive experimental investigation of amphibole-melt partitioning, which sheds considerable light on the likely site occupancy of the U-series elements. The lanthanides (and by association bismuth) and actinides are thought to substitute at M4 (Tiepolo et al. 2000a), while lead, barium (and radium) enter the A-site (Tiepolo et al. 2000a; Dalpé and Baker 2000). Actinium may partition between A and M4 sites, although its 3+ charge suggests that it will favour M4. Nb and Ta are known to enter M1 (Tiepolo et al. 2000b). Pa may also enter this site, although its large size relative to Nb and Ta means that some Pa may enter M4. The site occupancy of the noble gases is not known.

Tiepolo et al. (2000a) have studied lanthanide partitioning at 1.4 GPa for a wide range of amphibole compositions. They propose that there are two symmetrically distinct M4 sites (M4 and M4') and that the lanthanides distribute themselves between these two, with the larger lanthanides (La-Ce) predominantly on M4 and the smaller lanthanides (Dy-Lu) exclusively on M4' (Botazzi et al. 1999). The relationship between the sizes of the M4' and M4 sites varies with amphibole composition. In kaersutites M4' and M4 are similar in size (≤ 0.07 Å difference) and both play a role in controlling lanthanide partitioning. In richterites, however, M4' is appreciably smaller than M4 (by ~0.5 Å) and influences only the partitioning of Gd-Lu. As La is the proxy of interest for Ac and Bi, we need only be concerned with the M4 site. We have obtained lattice strain parameters for the light lanthanides from the experimental data of Brenan et al. (1995), La Tourrette et al. (1995), Andreeßen et al. (1996), Klein et al. (2000), Dalpé and Baker (2000), Tiepolo et al. (2000a) and Blundy and Brooker (2003). $E^{3+}_{(M4)}$ ranges from 156-445 GPa (mean = 311 GPa), with no obvious correlation with pressure, temperature or composition. $r^{3+}_{0(M4)}$ is in the range 1.01-1.04 Å (mean 1.033 Å), again with no obvious compositional dependence. A comprehensive study of crystal-chemical controls on lattice strain parameters for the amphibole M4 site, although very worthwhile, is beyond the scope of this study. We will therefore opt for fixed values of 330 GPa and 1.033 Å, which results in mean calculated values of D_{Ac}/D_{La} and D_{Bi}/D_{La} of 0.024 and 0.71, respectively, for all the experimental amphiboles studied.

The partitioning of La itself was found by Tiepolo et al. (2000a) to be a linear function of melt silica content:

$$\ln D^{amph}_{La(M4)} = -7.8(\pm 0.6) + 10(\pm 1)\frac{X^{melt}_{nf}}{X^{melt}_{total}} \tag{26}$$

where X_{nf} and X_{total} are the molar fraction of network-forming cations and total cations, respectively, in the melt. Tiepolo et al. (2000a) take network-forming cations to be Si plus all Al that is charge-balanced by alkalis. We note that equally strong correlations exist between D_{Ln} and D_{Ca} for the same dataset, suggesting that Ca exchange between amphibole and melt may be equally as important as melt SiO_2 content. However, for simplicity here we will adopt Tiepolo et al.'s (2000a) empirical expressions for D_{La} and several other proxy elements.

Uranium and thorium partitioning into amphibole were also studied experimentally by Tiepolo et al. (2000a). Under the redox conditions of their experiments (FMQ-2 log units) U was dominantly tetravalent. They find no correlation between D_U, D_{Th} and crystal composition, and, but again find a linear correlation with silica content:

$$\ln D^{amph}_{U(M4)} = -11(\pm 1) + 11(\pm 2) \frac{X^{melt}_{nf}}{X^{melt}_{total}} \tag{27a}$$

$$\ln D^{amph}_{Th(M4)} = -11(\pm 1) + 10(\pm 1) \frac{X^{melt}_{nf}}{X^{melt}_{total}} \tag{27b}$$

These data clearly indicate that U and Th are not fractionated from each other by amphibole. This is supported by a compilation of all available data, which show that D_U/D_{Th} is within error of unity over a range in D_U values from 0.004 to 0.034. Tiepolo et al. (2000a) conclude that amphibole plays no role in fractionation of uranium from thorium in magmas or in the mantle. By the same token we predict $D_{Po} \approx D_{Th}$.

Radium and barium both enter the large (XII-fold) A-site in amphibole. There are a large number of experimental D_{Ba} and D_{Sr} determinations with which to calibrate the parameters $r^{2+}_{0(A)}$ and $E^{2+}_{(A)}$. To perform this exercise we have assigned Sr to both M4 and A. $D_{Sr(M4)}$ is estimated from D_{Ca} using the elastic strain parameters for 3+ cations on M4, modified in the usual way to account for the change in charge. We calculate that approximately 2-16% of the Sr in the amphiboles studied resides on M4, with the rest on A. $D_{Sr(A)}$ (i.e., $D_{Sr} - D_{Sr(M4)}$) is then combined with D_{Ba} to arrive at optimum values for $E^{2+}_{(A)}$ and $r^{2+}_{0(A)}$. From the experiments of Brenan et al. (1995), La Tourrette et al. (1995) and Dalpé and Baker (2000) we find that setting $E^{2+}_{(A)}$ = 160 GPa and $r^{2+}_{0(A)}$ = 1.504 Å satisfactorily describes all D_{Ba}/D_{Sr} ratios. Using these parameters we calculate consistent D_{Ra}/D_{Ba} = 0.080 ± 0.007 for all of the amphiboles studied. The value of D_{Ba} itself varies from 0.10 to 0.72 in these three studies. There is no obvious correlation with pressure, temperature or composition. A slightly lower D_{Ba} (0.05) is obtained by Andreeßen et al. (1996) for andesitic and basaltic andesite melts. For amphibole phenocrysts in acid volcanic rocks, Ewart and Griffin (1994) find D_{Ba} = 0.16-0.30.

Tiepolo et al. (2000a) also studied lead partitioning between amphibole and melt, again finding a linear correlation with melt SiO_2 content:

$$\ln D^{amph}_{Pb(A)} = -7.6(\pm 0.7) + 8.4(\pm 1.2) \frac{X^{melt}_{nf}}{X^{melt}_{total}} \tag{28}$$

They ascribe the higher value of D_{Pb} compared to D_U (and D_{Th}) to its incorporation in the larger amphibole A site. The A site is normally occupied by Na^{1+}, consequently Pb^{2+} incorporation requires charge balancing elsewhere in the structure. Tiepolo et al. (2000) suggest that tetrahedral Al is the most likely charge compensating species. They argue that in richterites D_{Pb} will be much lower than the values predicted by Equation (28) due to their much lower Al contents. Tiepolo et al. (2000a) do not report D_{Sr} for their experiments, so no test of the proxy relationship is possible. Such a relationship is complicated by the possibility that Sr^{2+}, which is slightly smaller than Pb^{2+}, may also enter the smaller M4 site (see above). La Tourrette et al. (1995) and Brenan et al. (1995) measure both D_{Pb} and D_{Sr} and find very variable D_{Pb}/D_{Sr} ratios (0.13-0.55) suggestive of some site decoupling between this element pair.

Protactinium may enter either the M1 site, in the company of Nb^{5+} and Ta^{5+}, or the M4 site, along with the actinides. Tiepolo et al. (2000b) present an exhaustive study of Nb-Ta fractionation by amphibole, using the same pargasite and kaersutite dataset described above. They find that the ratio D_{Nb}/D_{Ta} correlates positively with M1-O bond length (d_{M1-O}) as determined by X-ray diffraction on the same synthetic crystals. d_{M1-O} in turn correlates inversely with amphibole mg# (molar $Mg/[Mg+Fe^{tot}]$). The higher compatibility of Nb at longer d_{M1-O} strongly suggests that the ionic radius of Nb is greater than that of Ta by 0.01-0.02 Å. The exact value of each radius is impossible to determine from the partitioning data alone, but if we assume that Ta^{5+} has a radius of 0.640 Å (as listed by Shannon 1976), then we can use the partitioning data to estimate what value of the Nb^{5+} radius best accounts for the observed fractionation. We use the observed relationship between d_{M1-O} and D_{Nb}/D_{Ta} to estimate $r^{5+}_{0(M1)}$. We find that simply subtracting 1.38 Å from d_{M1-O} gives values of $r^{5+}_{0(M1)}$ that are implausibly large for the observed D_{Nb}/D_{Ta}. This may in part be due to our adoption of 0.640 Å for the radius of Ta^{5+}, or it may be due to the distortion of the M1 site leading to slightly larger effective radii for the coordinating O^{2-}. Whatever the explanation, we find that the following expression gives the best results:

$$r^{5+}_{0(M1)}(\text{Å}) = d_{M1-O} - 1.426 \tag{29a}$$

d_{M1-O} is given by Equation (1) of Tiepolo et al. (2000b):

$$d_{M1-O} = 2.10 - 0.023 mg\# - 0.014 Ti_{tot} \tag{29b}$$

where Ti_{tot} is the total Ti content of the amphibole in atoms per 23 oxygen formula unit. Using Equation (29) we can reproduce all of the available experimental D_{Nb}/D_{Ta} data to within 2 standard deviations with a constant $E^{5+}_{(M1)}$ = 3500 GPa and Nb^{5+} ionic radius of 0.660 Å (as previously used for clinopyroxene and garnet). With these parameters we calculate that $D_{Pa(M1)}/D_{Nb(M1)}$ is negligibly small, suggesting that Pa does not enter the M1 site.

We can estimate $D_{Pa(M4)}$ only by using the electrostatic model with $D^{4+}_{0(M4)}$ derived from D_U and D_{Th}, $D^{3+}_{0(M4)}$ from the lanthanides and $D^{2+}_{0(M4)}$ from D_{Ca} and $D_{Sr(M4)}$. For these purposes we will not take into account the added complexity of M4 and M4' sites. A fit of Equation (7) to the $D_{0(M4)}$ values reported by of Brenan et al. (1995) and La Tourrette et al. (1995) gives electrostatic fit parameters of 2.3 and 1.9, respectively, for $Z_{\rho(M4)}$ and ~40 Å for $\varepsilon\rho$, in both cases (Fig. 17). We estimate that $D_{Pa(M4)}/D_U$ is $1-4 \times 10^{-4}$ in these amphiboles. D_U itself can be estimated from Equation (27a). (Note that as Pa and Nb now enter different sites (cf. Pb and Sr) it is no longer useful to use Nb as a proxy for Pa.) The very low values of D_{Pa} (= $D_{Pa(M4)} + D_{Pa(M1)}$) suggests that amphibole exercises little control over the behavior of this element.

There are no noble gas partitioning data for amphiboles. Given the multiplicity of cation and anion sites in this mineral it seems likely that D_{Rn} will be higher than in other silicate minerals.

7.6. Plagioclase

Plagioclase has a single large cation site (M) into which all U-series elements partition. This site is normally occupied by Ca and Na, with coordination number increasing with increasing Na content. For simplicity we will assume VIII coordination

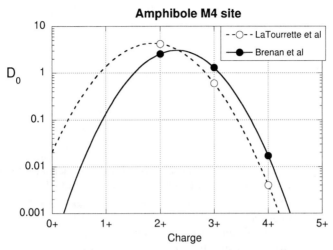

Figure 17. Electrostatic model fitted to partition coefficients for cations entering the M4-site in amphibole, based on the experiments of Brenan et al. (1995) and La Tourrette et al. (1995). A single M4-site is assumed, rather than M4 and M4', as proposed by Bottazzi et al. (1999). The curves are fits to Equation (7) and can be used to estimate $D^{5+}_{0(M4)}$, from which $D_{Pa(M4)}$ can be calculated via the lattice strain model. Because of the multiplicity of sites in amphibole, it is unlikely that extrapolation of the curves to zero charge gives a reliable estimate for D_{Rn}. The fit parameters are $Z_{0(M2)} = 1.87$ and $\rho\varepsilon = 38.1$ Å (La Tourrette et al. 1995), and 2.31, 36.1 Å (Brenan et al. 1995).

across the entire solid solution. As the difference between VIII and X co-ordinated ionic radii is approximately the same for all U-series cations (Shannon 1976), this assumption has very little bearing on our findings. Experimental studies of plagioclase-melt partitioning have been carried out by Drake (1972), Blundy and Wood (1994), Bindeman et al. (1998), Bindeman and Davis (2000) and Blundy (unpublished data). All of these studies confirm the findings of Blundy and Wood (1991) that partition coefficients are strongly dependent on plagioclase molar anorthite content (X_{An}). Blundy and Wood (1991) derived linear relationships between $RT\ln D_{Sr}$ and $RT\ln D_{Ba}$ and X_{An}. Similar relationships, but with different slopes, have been found for a large number of other elements, including U and some U-series proxies, by Blundy and Wood (1994), Bindeman et al. (1998), and Bindeman and Davis (2000). These correlations are useful in predicting U-series partition coefficients.

Bindeman et al. (1998) and Bindeman and Davis (2000) present SIMS analyses of Drake's (1972) experimental run products. These were doped with selected lanthanides, Sr and Ba to derive partition coefficients that could be determined by electron-microprobe analysis. However, a large number of other trace elements occur at natural levels in the starting materials, and these were measured by Bindeman et al. (1998) and Bindeman and Davis (2000) to derive partition coefficients. The experiments crystallised plagioclase in the composition range An_{80}-An_{40}, which covers most terrestrial magmatic plagioclases. In the case of lanthanides Bindeman et al. (1998) find a positive correlation between $RT\ln D_{Ln}$ and X_{An}. For La the relationship is:

$$RT \ln D_{La}\left(\text{kJmol}^{-1}\right) = -10.8(\pm 2.6) X_{An} - 12.4(\pm 1.8) \qquad (30)$$

Lattice strain parameters for 3+ cations entering plagioclase are difficult to derive because $r^{3+}_{0(M)}$ is clearly larger than La^{3+}, meaning that one limb of the partitioning parabola is not defined. Blundy and Wood (1994), fitting D_Y, D_{Sm} and D_{La} from 1 atmosphere experiments in the system diopside-albite-anorthite, assumed $r_{0(M)}^{3+} = r_{0(M)}^{2+}$, which was itself derived from fitting a larger number of 2+ cations. $r_{0(M)}^{2+}$ was found to increase linearly with decreasing X_{An} according to relationship (Fig. 6b; Blundy and Wood 1994):

$$r_{0(M)}^{2+}(\text{Å}) = 1.258 - 0.057 X_{An} \tag{31}$$

The values for $E_{(M)}^{3+}$ obtained in this way are approximately constant at 210 ± 1 GPa across the solid solution series (Fig. 5). We have fitted the lanthanide partitioning data of Bindeman et al. (1998) and Bindeman and Davis (2000) using this value of $E_{(M)}^{3+}$ and find a consistent decrease in $r_{0(M)}^{3+}$ relative to $r_{0(M)}^{2+}$ (as calculated from Eqn. 26) of ~0.03Å, i.e., slightly smaller than was found for ferromagnesian minerals. Using $r_{0(M)}^{3+} = r_{0(M)}^{2+} - 0.03$Å we can calculate D_{Ac}/D_{La} and D_{Bi}/D_{La}, which vary from 1.0 and 1.1, respectively, at An_{40} (at 900°C) to 0.74 and 1.03, respectively, at An_{80} (1200°C).

Bindeman and Davis (2000) analyzed uranium in Drake's (1972) plagioclases and found a correlation between D_U and X_{An}, with values of 0.01 to 0.08. However Drake's experiments were run in air and therefore more likely pertain to U^{6+} rather than U^{4+}. Nonetheless, the values of D_U are still surprisingly large, given that D_{La} in the same runs is only about ten times larger. Fitting the electrostatic model to the 1+, 2+ and 3+ data of Bindeman and Davis (2000) suggests that $D_{U^{4+}}$ should be of the order 10^{-5} (Fig. 18) i.e.,

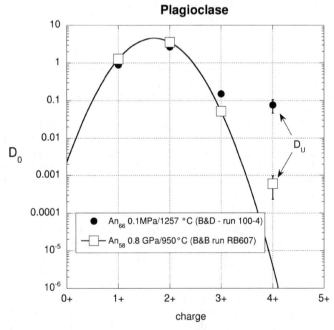

Figure 18. Electrostatic model fitted to partition coefficients for cations entering the large M-site in plagioclase, based on the experimental results of Bindeman and Davis (2000) and Blundy and Brooker (2003). The solid curve is a fit of the $D_{0(M)}^{1+}$, $D_{0(M)}^{2+}$ and $D_{0(M)}^{3+}$ data of Blundy and Brooker (2003) to Equation (7) giving $Z_{0(M2)} = 1.70$ and $\rho\varepsilon = 25.9$ Å. Also plotted is D_U from both experiments, with 1 s.d. error bar. In both cases, D_U is higher than the extrapolated $D_{0(M4)}^{4+}$. This is especially true for the Bindeman and Davis (2000) data. The cause of the discrepancy is not known, and further experimental studies on uranium partitioning into plagioclase are required. D_{Rn} can be estimated as ~2×10^{-3}.

considerably lower than the reported values. The ratio D_U/D_{Zr} measured by Bindeman et al. (1998) is also surprisingly low given the size difference between U^{4+} and Zr^{4+} and their high charge. A single experimental determination of plagioclase-melt partitioning of uranium and thorium in a hydrous dacite melt at 0.8 GPa and 950°C (Blundy and Brooker 2003) gives $D_U = 6.0 \pm 3.7 \times 10^{-4}$ and $D_{Th} = 4.6 \pm 2.6 \times 10^{-4}$ for An_{58}. Once again, however, the value for D_U (and D_{Th}) seems higher than would be expected from the electrostatic model (Fig. 18). Cooper and Reid (2003) provide ion-microprobe measurements on Th in plagioclase phenocrysts, from Mount St. Helens lavas, which can be combined with Th concentrations in the corresponding groundmass or bulk rock to give D_{Th} in the range $7-53 \times 10^{-4}$ (7 samples, An_{43-60}). We can estimate D_U/D_{Th} using the lattice strain model by assuming that $r_{0(M)}^{4+} = r_{0(M)}^{2+} - 0.06$ and that $E_{(M)}^{4+} = \frac{4}{3}E_{(M)}^{3+} = 280$ GPa. This gives values for D_U/D_{Th} of 0.18 in An_{40} (900°C) and 0.30 in An_{80} (1200°C). The corresponding D_U for the Cooper et al. (2003) plagioclases (at 900°C) is $1-12 \times 10^{-4}$, in reasonable agreement with Blundy and Brooker (2003) and the measurements shown in Figure 18. D_{Po}/D_{Th} calculated using the same parameters is 2.3 in An_{40} and 1.7 in An_{80}. If these estimates of D_{Th}/D_U and D_{Po}/D_{Th} are combined with Bindeman and Davis' (2000) measured D_U, then polonium appears to become compatible in sodic plagioclase ($An_{\leq 30}$) at low temperatures. We emphasise that conclusions regarding U, Th and Po need to be re-evaluated in the light of further experimental partitioning studies on plagioclase.

Strontium and barium partitioning into plagioclase are well constrained from a large number of experiments (Blundy and Wood 1991 and references therein; Bindeman et al. 1998; Bindeman and Davis 2000). Both D_{Ba} and D_{Sr} are strongly correlated with X_{An}. Blundy and Wood derived relationships of the form $RT\ln D = aX_{An} + b$ for both elements. Bindeman et al. (1998) revisited these relationships in the light of their new data and found very similar relationships. Although the expressions for $RT\ln D_{Ba}$ and $RT\ln D_{Sr}$ given by Bindeman et al. (1998) differ slightly from those of Blundy and Wood (1991), the calculated the D_{Sr} and D_{Ba} are very similar in the composition range An_{40-70} and 750-1000°C. For that reason the Blundy and Wood (1991) relationships are retained here:

$$RT\ln D_{Ba}\left(\text{kJmol}^{-1}\right) = -38.2(\pm 3.2)X_{An} + 10.2(\pm 1.8) \quad (32)$$

$$RT\ln D_{Sr}\left(\text{kJmol}^{-1}\right) = -26.7(\pm 1.9)X_{An} + 26.8(\pm 1.2) \quad (33)$$

Equation (32) can be used to estimate D_{Ra}, using the lattice strain approach. $E_{(M)}^{2+}$ is taken as 116 GPa (Fig. 5; Blundy and Wood 1994), while $r_{0(M)}^{2+}$ is taken from Equation (31). This gives D_{Ra}/D_{Ba} of 0.043 at An_{80} (1200°C) and 0.19 at An_{40} (900°C). Cooper et al. (2002) and Cooper and Reid (2003) used a similar method to estimate D_{Ra} from D_{Ba}. Note that as plagioclase becomes more sodic D_{Ba} and D_{Ra}/D_{Ba} increase, although D_{Ra} never becomes compatible ($D_{Ra} > 1$) in plagioclase. The maximum D_{Ra}/D_{Ba} occurs for sodic plagioclases at high temperatures.

Lead partition coefficients are reported by Bindeman et al. (1998), who present the following relationship between D_{Pb} and X_{An}:

$$RT\ln D_{Pb}\left(\text{kJmol}^{-1}\right) = -60.5(\pm 11.8)X_{An} + 25.3(\pm 7.8) \quad (34)$$

This expression is relatively imprecise because of the scarcity of data. Also, the oxidation state of Pb in these experiments is not known. However, it is interesting that for those experiments in which both D_{Pb} and D_{Sr} have been determined, the D_{Pb}/D_{Sr} ratio is consistently less than would be predicted from the 2+ lattice strain model using parameters presented above. As in the case of clinopyroxene, increasing the effective VIII-fold ionic radius of Pb^{2+} in plagioclase, to 1.38 Å, does retrieve the observed ratios. Thus one can potentially use the relationship between D_{Sr} and X_{An} (Equation 33) to derive D_{Pb}, assuming an effective ionic radius of 1.38 Å, with greater precision than Equation (34).

Niobium and tantalum are highly incompatible in plagioclase, largely because their high charge is at odds with the predominantly 1+ and 2+ cations in the large cation site. It is possible that some Nb and Ta enter the tetrahedral site, normally occupied by Si and Al. In experiments with hydrous dacitic melts at 0.8-1.3 GPa and 950-1025°C, Blundy and Brooker (2003) measure D_{Nb} in the range $1.3\text{-}3.6 \times 10^{-4}$ for $An_{51\text{-}58}$. The larger VIII-fold ionic radius of Pa compared to Nb (and Ta) will result in $D_{Pa} \gg D_{Nb}$. Assuming that $E_{(M)}^{5+} = \frac{5}{3} E_{(M)}^{3+} = 336$ GPa and $r_{0(M)}^{5+} = r_{0(M)}^{2+} - 0.06$, D_{Pa}/D_{Nb} is calculated to be ~5000, which gives a surprisingly large value of $D_{Pa} = 0.65\text{-}1.8$. We suspect that this is a result of Nb and Ta incorporation on the plagioclase tetrahedral site, making the measured D_{Nb} an overestimate of $D_{Nb(M)}$. Alternatively, we can place an upper bound on D_{Pa} by estimating $D_{0(M)}^{5+}$ for plagioclase using the electrostatic model (Fig. 18). For the Bindeman and Davis (2000) experiment $Z_{0(M)}$ is 1.77, as befits an M site occupied by a mixture of Na^{1+} and Ca^{2+}, and $\varepsilon\rho$ (28 ± 1 Å) is comparable to values for garnet (Table 5) and other silicate minerals. The corresponding values for the Blundy and Brooker (2003) experiment are 1.70 and 26 ± 2 Å. In both cases $D_{0(M)}^{5+}$ is $<10^{-6}$, which is at odds with the value of D_{Nb} quoted above. We conclude that although D_{Pa} is likely to very small, further experimental data on plagioclase-melt partitioning of highly-charged cations are required before it can be reliably estimated.

The electrostatic model can also be used to estimate noble gas partition coefficients for plagioclase. It seems reasonable to assume that the noble gases enter the large M site. As $Z_{0(M)}$ for plagioclase is less than 2, the noble gases are less incompatible in plagioclase than U and Th. For example, D_{Rn} is estimated to be approximately 0.003-0.006 for the two plagioclases plotted in Figure 18.

7.7. Alkali-feldspar

All U-series elements partition onto the large (X-fold) M-site in alkali-feldspar. However, there are very few experimental data on trace element partitioning between alkali feldspar and melt, other than for the alkalis and alkaline earths. There are no published data for U, Th, the lanthanides or Nb, which makes estimating the partition coefficients for most U-series elements extremely difficult. As alkali feldspar may be an important fractionating phase in evolved rocks at low temperatures, when partition coefficients are generally at their highest, it could potentially play an important role in the evolution of some U-series elements. Clearly new experimental studies of alkali feldspar-melt partitioning should be an urgent priority. Until such data become available, we will confine ourselves here to estimates of radium and lead partitioning.

Barium and strontium are well known to be compatible in alkali feldspar. Several experimental studies demonstrate this over a wide range of pressures and temperatures (Long 1978; Guo and Green 1989; Icenhower and London 1996). In their PIXE study of trace element partitioning between alkali feldspar phenocrysts and matrix glass, Ewart and Griffin (1994), obtain values of D_{Ba} in the range 1.3 to >20. The Icenhower and London (1996) study provides data for D_{Mg}, D_{Ca}, D_{Sr} and D_{Ba}, such that lattice strain parameters for 2+ cations can be derived. Preliminary fits (using X-fold ionic radii from Shannon 1976) indicates that $E_{(M)}^{2+}$ is in the range 55-150 GPa, We have elected for a fixed $E_{(M)}^{2+}$ of 91 GPa, which provides an adequate fit to all of the data as well as being broadly consistent with $E_{(M)}^{2+}$ for plagioclase (Fig. 5). Fitted $r_{0(M)}^{2+}$ values using $E_{(M)}^{2+} = 91$ GPa for the Icenhower and London (1996) data give values of $r_{0(M)}^{2+}$ which increase linearly with the molar fraction of orthoclase component (X_{Or}) (Fig. 19). A weighted fit gives:

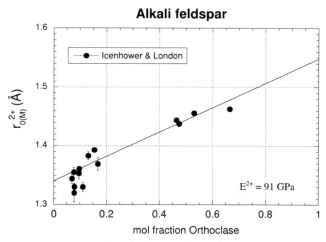

Figure 19. Variation in dimension of the X-fold cation site in alkali-feldspar for 2+ cations ($r^{2+}_{0(M)}$), obtained by fitting the experimental data of Icenhower and London (1996) for D_{Mg}, D_{Ca}, D_{Sr} and D_{Ba} at 0.2 GPa and 650-750°C. In performing the fits E^{2+}_M was set at 91 GPa for all runs. Error bars are 1 s.d. The positive slope is consistent with measured changes in metal-oxygen bond length from albite to orthoclase (cf. Fig. 6). The solid line shows the best-fit linear regression given in Equation (35).

$$r^{2+}_{0(M)}(\text{Å}) = 1.341(\pm 0.002) + 0.207(\pm 0.005) X_{Or} \qquad (35)$$

From these values it is straightforward to calculate D_{Ra}/D_{Ba} as a function of temperature and X_{Or}. For all three datasets we find a linear correlation between D_{Ra}/D_{Ba} and X_{Or} (Fig. 20). For example, for the Icenhower and London (1996) data at 0.2 GPa and 650-750°C the relationships are:

$$D_{Ba} = 1.919 + 23.92 X_{Or} \qquad (36a)$$

$$\frac{D_{Ra}}{D_{Ba}} = 0.104 + 0.545 X_{Or} \qquad (36b)$$

Under these conditions our calculations show that Ra becomes compatible ($D_{Ra} \geq 1$) at $X_{Or} \geq 0.13$. For the 0.8 GPa experiments of Long (1976) Ra is calculated to be compatible at all temperatures from 740-770°C. In the Guo and Green (1989) experiments at 1-2.5 GPa and 900-1100°C, Ra is calculated to be compatible at all but the highest pressure (2.5 GPa) and temperatures (≥ 1000°C). Evidently alkali-feldspar has a dominant influence over the behavior of Ra in evolved silicic systems.

The above 2+ fit parameters can also be used to derive D_{Pb} from D_{Sr}, assuming that the proxy relationship is valid for alkali feldspar. For the Icenhower and London (1996) experiments we obtain a further linear correlation with X_{Or}:

$$\frac{D_{Pb}}{D_{Sr}} = 0.801 + 1.124 X_{Or} \qquad (37)$$

As Sr is highly compatible in alkali feldspar under all conditions, it is likely that so too is Pb, whatever the validity of the Sr-Pb proxy relationship. Leeman (1979) determines D_{Pb} of 0.84-1.37 for separated sanidine phenocrysts (Or$_{\sim 50}$) from various acid volcanic rocks, in broad agreement with Equation (37).

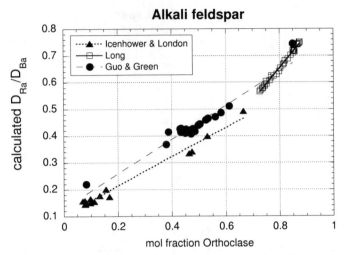

Figure 20. Variation in calculated D_{Ra}/D_{Ba} for three experimental alkali-feldspar-melt partitioning studies (Long 1978; Icenhower and London 1996; Guo and Green 1989) as a function of molar fraction orthoclase content. The lines denote linear best-fits to the different data sets; the fit for the Icenhower and London data is given as Equation (36b). The slopes appear to vary with pressure and temperature, but are broadly consistent between the three studies.

We can extend the relationships for 2+ cations to 3+ cations, by assuming that $E^{3+}_{(M)}$ = $\frac{3}{2} E^{2+}_{(M)}$ and $r^{3+}_{0(M)} = r^{2+}_{0(M)} - 0.03$ Å (cf. plagioclase). For the pressures and temperatures of the Icenhower and London's (1996) experiments (0.2 GPa, 650-750°C) both D_{Bi}/D_{La} and D_{Ac}/D_{La} increase linearly with X_{Or} from 1.1 and 1.8, respectively, at Or_7 to 1.4 and 7.5, respectively, at Or_{66}. Unfortunately there are no experimentally-determined values of D_{La} on which to base these proxy relationships. Leeman and Phelps (1981), working with separated sanidine phenocrysts (Or_{50-67}) and glasses from Yellowstone rhyolites, derive a consistent set of D_{La} values in the range 0.10-0.15. However, Mahood and Stimac (1980), again working with separates, measure D_{La} in the range 1.5-4.6 × 10^{-3}. It therefore seems unlikely that Ac and Bi are compatible in alkali-feldspar. Experimental studies of lanthanide partitioning between alkali-feldspar and melt are urgently required to evaluate these conclusions.

7.8. Phlogopite (biotite)

Trioctahedral micas, such as phlogopite or biotite, are characterised by four distinct cation lattice sites: tetrahedral (Z) sites occupied by Si and Al; octahedral (Y) sites, denoted M1 and M2, occupied by Al, Cr, Fe^{3+}, Ti, Fe^{2+}, Mg and Mn; and a large XII-fold co-ordinated interlayer X-site occupied by K, Na and Ca. Schmidt et al. (1999) argue that only the X-site is suitable for incorporation of large trace cations. The low partition coefficients for lanthanides, U and Th in phlogopite (<2 × 10^{-4}) indicate, moreover, that large highly charged cations cannot be incorporated at the X-site, presumably because of a lack of suitable charge-balancing mechanism. We will therefore not consider further the partitioning of U, Th, Ac or Bi. To estimate the partitioning behavior of the other U-series elements we have used the following experimental phlogopite (or biotite)-melt partition coefficients: La Tourrette et al. (1995), Icenhower and London (1995), Schmidt et al. (1999) and Green et al. (2000). These data cover a wide range in pressure (0.2-3 GPa), temperature (650-1140°C) and Al content (2-4 atoms pfu).

2+ cations, such as Sr and Ba, readily enter the X-site, with the excess positive charge balanced by Li substitution on an M-site or Al on a Z-site. In fact, partition coefficients for barium in phlogopite can be even larger even than in alkali-feldspar. Phlogopite D_{Ba} is consistently larger than D_{Sr}, indicative of a site with very large $r^{2+}_{0(X)}$. This observation indicates that phlogopite will be a significant host for radium and possibly lead.

Partition coefficients for radium can be obtained by fitting the 2+ partitioning data to the lattice strain model (using XII-fold ionic radii) to obtain $r^{2+}_{0(X)}$ and $E^{2+}_{(X)}$. Unfortunately, $r^{2+}_{0(X)}$ appears slightly larger than the radius of Ba^{2+}, which makes one limb of the parabola poorly constrained. Moreover, typically only three divalent partition coefficients are measured (Ba, Sr and Ca), and D_{Ca}, when analyzed by EMPA, is normally imprecise. We have taken an alternative approach, using 1+ partitioning data to constrain $r^{1+}_{0(X)}$ and $E^{1+}_{(X)}$, which are then converted to $r^{2+}_{0(X)}$ and $E^{2+}_{(X)}$, using some of the simple relationships described above. The data of Icenhower and London (1995) are ideal for this purpose as they report partition coefficients for Na, K, Rb and Cs, which span the size of the X site. For 15 experiments at 650-750°C and 0.2 GPa, we obtain a very tight cluster of $r^{1+}_{0(X)}$ (1.650-1.673 Å) and $E^{1+}_{(X)}$ (47-56 GPa), with mean values of 1.665 ± 0.007 Å and 50 ± 2 GPa. A typical fit is shown in Figure 21. There are no higher pressure partitioning data for all four alkali metals, and we are forced to fix $E^{1+}_{(X)}$ = 50 GPa in order to fit the 1.5 GPa experiments of Schmidt et al. (1999). Significantly we derive much larger values of $r^{1+}_{0(X)}$ in the range 1.71-1.73 Å. The Schmidt et al. (1999)

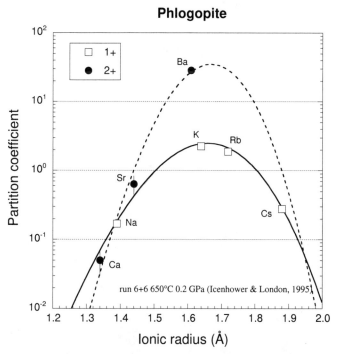

Figure 21. Fits of lattice strain model to experimental phlogopite-melt partition coefficients for 1+ and 2+ cations on the large X-site for run 6+6 of Icenhower and London (1995) at 0.2 GPa and 650°C. 1+ data can be readily fitted; 2+ data were fitted by assuming that $E^{2+}_X = 2E^{1+}_X$ = 100 GPa. Note that $r^{2+}_{0(X)}$ is slightly larger than r_{Ba}, making this site ideal for incorporation of Ra. Errors bars, when larger than symbol, are 1 s.d. Ionic radii in XII-fold coordination are taken from Shannon (1976).

phlogopites are considerably less aluminous than those of Icenhower and London (1995), it therefore is likely that this change in $r_{0(X)}^{1+}$ is related to crystal chemistry, notably the extent of eastonite solid solution. We do not, as yet, have sufficient data to quantify this possibility. However, as we shall show, variations in $r_{0(X)}^{1+}$ and $r_{0(X)}^{2+}$ do not have a significant impact on calculated D_{Ra}/D_{Ba}.

There are only five experiments that report D_{Ca}, D_{Sr} and D_{Ba}: three from Schmidt et al. (1999) and two from Icenhower and London (1995). We have fitted these data using the simplest reasonable approximation, $E_{(X)}^{2+} = 2E_{(X)}^{1+} = 100$ GPa. In each experiment we find that the calculated $r_{0(X)}^{2+}$ is 0.01-0.06 Å smaller than $r_{0(X)}^{1+}$, consistent with observations elsewhere in this chapter. The difference in $r_{0(X)}^{2+}$ between the low-Al Schmidt et al. (1999) biotites and the high-Al Icenhower and London (1995) biotites persists. However, when we use the fitted $r_{0(X)}^{2+}$ and $E_{(X)}^{2+} = 100$ GPa to calculate D_{Ra}/D_{Ba}, the values for all five experiments are remarkably consistent, from 0.7 to 1.8 (mean = 1.0 ± 0.3), despite the wide range in pressure, temperature and crystal and melt chemistry. We suggest that, uniquely, in phlogopite $D_{Ra} = D_{Ba}$.

D_{Ba} can itself be parameterized from the above experimental data, supplemented by the data of Guo and Green (1990), who confined themselves to Ba partitioning only, in the pressure range 1-3 GPa. This provides a total of 33 data points. From Figure 22 it is clear that lnD_{Ba} is inversely correlated with temperature. However, there is a discrepancy between D_{Ba} determined by La Tourrette et al. (1995) and that determined by Guo and Green (1990) at almost identical pressure and temperature. This difference must be compositional in origin, although there are no consistent melt of crystal compositional

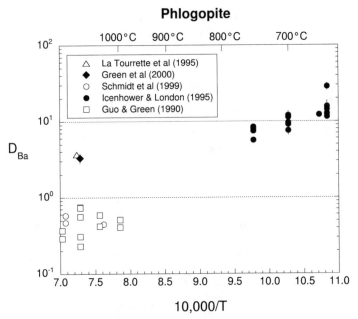

Figure 22. Variation in phlogopite-melt D_{Ba} with reciprocal temperature. The low temperature data of Icenhower and London (1995) define a strong linear trend. At higher temperature, no trend is apparent, and the data fall into two distinct clusters: the high D_{Ba} points of Green et al. (2000) and La Tourrette et al. (1995) and the low D_{Ba} points of Schmidt et al. (1999) and Guo and Green (1990). The cause of these differences is unresolved.

differences between the low and high D_{Ba} phlogopites. Until $r_{0(X)}^{2+}$ can be better parameterised in terms of biotite composition, the relationship between D_{Ba} and $D_{0(X)}^{2+}$, and hence with pressure and temperature, will remain elusive. In the meantime we suggest using Figure 22 to estimate D_{Ba}. At temperatures less than ~1000°C, we calculate that Ra will become compatible in phlogopite.

Schmidt et al. (1999) report D_{Pb} of 0.034-0.045 for two experiments with leucite lamproite melt composition; for a basanitic melt composition La Tourrette et al. (1995) give D_{Pb} = 0.10. In all three cases D_{Pb} consistently falls below, by a factor of ~3, the parabola defined by the other 2+ cations, as previously noted for several other minerals. Here the implication is that the effective XII-fold ionic radius of Pb^{2+} is slightly *smaller* than the value given in Table 2, i.e., closer in size to r_{Sr}. D_{Pb}/D_{Sr} is between 0.6 and 1.2, in these experiments. In the PIXE partition study of Ewart and Griffin (1994) for acid volcanic rocks, D_{Pb} ranges from 0.21 to 2.1 (3 samples), with D_{Pb}/D_{Sr} of 0.29 to 2.9. Until there are further experimental determinations of D_{Pb}, or better constraints on its ionic radius, we suggest that $D_{Pb} = D_{Sr}$.

Phlogopites show high values for D_{Ti}, in the range 0.7-28 in the experimental dataset introduced above. Ti enters the M-sites, charge balanced either by Al-Si exchange on the Z-sites or by the formation of octahedral site vacancies. It seems possible, therefore, that other highly charged cations, such as Pa^{5+}, may also enter this site. However, the very large values of D_{Ti}/D_{Zr} (54-104 for the experiments listed above) show that the M-site dimension is considerably smaller than the VI-fold ionic radius of Zr (0.72 Å), so that cations as large as Pa^{5+} (0.78 Å) will be excluded. The partition coefficient for Nb, whose ionic radius is significantly smaller than Pa, is 0.02-0.09. Green et al. (2000) report D_{Nb}/D_{Ta} = 0.191 ± 0.013, indicating that $r_{0(M)}^{5+}$ is smaller than the ionic radius of Nb (0.66 Å). By analogy with 5+ cation partitioning in other minerals, we suggest that D_{Pa} will be vanishingly small in phlogopite.

There are no noble gas partition coefficients for phlogopite. The large size and low mean charge of the large X-site ($Z_{0(X)}$) suggest that noble gases could be readily incorporated into phlogopite.

7.9. Oxide minerals

The most important magmatic oxide minerals are spinel (including magnetite and chromite), ilmenite and rutile, and we will confine our discussion to these. Because all of these oxides are dominated by relatively small cations, it is unlikely that any of the larger U-series cations will be appreciably incorporated. Due to the relatively low modal abundance of oxides in rocks (typically 2 wt% or less), these minerals will only exert significant leverage on trace element fractionation for elements whose partition coefficients are greater than approximately 10^{-3}. We will further confine our discussion to the more highly charged and/or smaller U-series cations.

Spinels. There are limited experimental data on uranium and thorium partitioning between magnetite and melt (Nielsen et al. 1994; Blundy and Brooker 2003). Both studies find U and Th to be moderately incompatible. Blundy and Brooker's results for a hydrous dacitic melt at 1 GPa and 1025°C give D_U and D_{Th} of approximately 0.004. The accuracy of these values is compromised by the very low concentrations in the crystals and the lack of suitable SIMS secondary standards for these elements in oxide minerals. Nonetheless, these values are within the range of D_{Th} of magnetites at atmospheric pressure: 0.003-0.025 (Nielsen et al. 1994). It is difficult to place these values within the context of the lattice strain model, firstly because there are so few systematic experimental studies of trace element partitioning into oxides and secondly because of the compositional diversity of the spinels and their complex intersite cation ordering.

Nonetheless, we can gain some insight from the atmospheric pressure experimental study of Horn et al. (1994). Their run #44/3 (1275°C; logfO_2 = −0.68) yielded partition coefficients for ten different trace elements of differing charge and size (Fig. 23). Assuming that at high temperatures there is complete disorder across the tetrahedral and octahedral sites, i.e., they can be defined by single parabolae for each charge, we can sketch lattice strain curves through 2+, 3+ and 4+ data. Although our assumption is a gross oversimplification, we can nonetheless see a systematic increase in E and decrease in r_0 with charge, as observed for silicate minerals. The curves for 4+ cations suggest that D_U will be <10^{-5}, and substantially higher than D_{Th}. These estimates are much lower than the experimental values noted above. However, D_{Ti} in the Horn et al. (1994) experiment is about an order of magnitude less than in those experiments of Nielsen et al. (1994) for which D_{Th} was measured, and two orders of magnitude less than in the Blundy and Brooker (2003) experiment. It is therefore possible that D_0^{4+} is larger in the experiments where D_{Th} was measured, than in the Horn et al. (1994) experiment shown in Figure 23. We suggest that the D_{Th} values given above are reasonable estimates for magnetite, but that they will be very sensitive to temperature and D_{Ti}. D_U/D_{Th} will be very large (>10). It is not possible to gauge the compositional dependence of D_{Th} and D_U at this stage.

Protactinium partition coefficients cannot be estimated with any degree of accuracy from the available data. However, from a compilation of 40 magnetite-melt pairs where

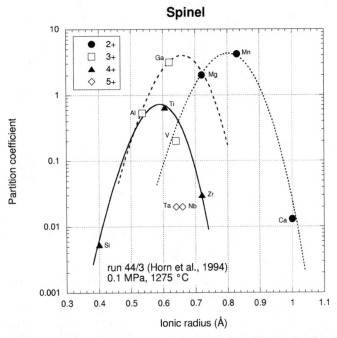

Figure 23. Lattice strain model applied to experimental spinel-melt partitioning data (run 44/3 of Horn et al. 1994), assuming a single site with VI-fold coordination. The spinel in this run is a magnesioferrite-hercynite solid solution with composition $Mg_{0.78}Fe^{2+}_{0.26}Fe^{3+}_{1.58}Al_{0.32}Ti_{0.04}O_4$. Curves show approximate fits to the experimental data; for 3+ cations there are insufficient data to define a parabola. Note the apparent increase in r_0 and E with cation charge, as observed for silicates. The low D_V suggests mixed 3+, 4+ and 5+ valence at the experimental fO_2 ($10^{-0.68}$). The low D_0^{4+} in this spinel may account for the low extrapolated values of D_U and D_{Th}, compared to the experimentally measured D_{Th} of Nielsen et al. (1994) and D_U and D_{Th} of Blundy and Brooker (2003).

D_{Nb} and D_{Ta} have been measured, all but 5 have D_{Nb}/D_{Ta} <1, with an average value of 0.89 ± 0.53, consistent with the Horn et al. (1994) data in Figure 23. Taking $r^{5+}_{0(oct)} \approx r^{4+}_{0(oct)}$ and $E^{5+}_{oct} \approx \frac{5}{4} E^{4+}_{oct}$ (as suggested by Fig. 23), gives D_{Nb}/D_{Ta} of 0.54, which is within1 s.d. of the experimental mean. The corresponding value of D_{Pa}/D_{Nb} is on the order 10^{-4}. If we adjust $r^{5+}_{0(oct)} \approx r^{4+}_{0(oct)}$ and $E^{5+}_{oct} \approx \frac{5}{4} E^{4+}_{oct}$ so as to reproduce the mean experimental D_{Nb}/D_{Ta}, then D_{Pa}/D_{Nb} increases to 10^{-2}. As D_{Nb} itself can be 1 or more in silicic volcanic rocks (Ewart and Griffin 1994), then magnetite D_{Pa} may be considerably larger than that for any of the silicate minerals discussed above and may play an important role in controlling Pa fractionation in evolved silicic rocks. Significantly, magnetites may have broadly similar D_U and D_{Pa}, and so effect relatively little fractionation between U and Pa.

The lack of systematic lanthanide partitioning data for spinels precludes estimating values for D_{Ac}, D_{Bi} etc. Suffice to say that they will be vanishingly small, as will D_{Ra} and D_{Pb}, given the low partition coefficients for Ca and Sr in spinels.

Ilmenite. There are currently no experimentally-determined ilmenite-melt partition coefficients. We must rely instead on phenocryst-groundmass data analyzed by PIXE (Stimac and Hickmott 1994; Ewart and Griffin 1994). Like magnetite, ilmenite is characterised by two relatively small sites, and is therefore unlikely to take up appreciable amounts of lanthanides or large alkaline earths. We will confine ourselves again to U, Th and Pa.

Neither of the phenocryst-groundmass studies analyzed for U or Th. Consequently we must try to estimate D_U and D_{Th} from the behavior of other 4+ cations, viz. Ti, and Zr, for which there are data. D_{Ti}/D_{Zr} ranges from 100-220 in the study of Stimac and Hickmott (1994), with absolute D_{Ti} in the range 150-225. With such a large fractionation between Ti (VI-fold ionic radius 0.605 Å) and Zr (0.72 Å) it is unlikely that there will be significant incorporation of even larger cations such as U and Th onto the Ti site. There are insufficient data for 1+ and 3+ cations entering the large Fe site to estimate D_0^{4+} using the electrostatic model. We estimate that $D_U < 10^{-4}$ for ilmenite and that it is considerably larger (>100 times) than D_{Th}.

Ilmenite is characterised by very large partition coefficients for Nb and Ta, in the range 7-150, raising the possibility of appreciable D_{Pa}. Stimac and Hickmott (1994) show that D_{Nb}/D_{Ta} ranges from 1 to 0.8, similar to magnetite. Assuming that the site characteristics are therefore similar in these two minerals, we suggest that D_{Pa}/D_{Nb} will have a similar, or slightly higher, value (10^{-2}-10^{-4}). However, because D_{Nb} is itself 25-60 times larger in ilmenite than coexisting magnetite (e.g., Ewart and Griffin 1994), D_{Pa} will be correspondingly larger. Like magnetite, ilmenite may also play an important role in controlling the Pa evolution of silicic magmas. Experimental data on ilmenite-melt partitioning are required to better constrain this possibility.

Rutile. It is unlikely that rutile will incorporate any U-series cations other than U, Th and Pa. D_U and D_{Th} can be estimated from D_{Ti}, D_{Zr} and D_{Hf}, using the experimental data of Foley et al. (2000). In three experiments on a doped natural tonalite under hydrous conditions at 1.8 to 2.5 GPa, Foley et al. (2000) measure D_{Zr} and D_{Hf} by LA-ICP-MS. Unfortunately they do not report D_{Ti}. However, we can estimate the TiO_2 content of the melt from the glass analysis in Jenner et al. (1993) for the same tonalite starting material at similar P and T, which gives D_{Ti} = 84. The D_{Ti}/D_{Zr} and D_{Hf}/D_{Zr} ratios for the three experiments in Foley et al. (2000) can be fitted (using VI-fold ionic radii) with r_0^{4+} = 0.585 Å and E^{4+} = 700 GPa. This gives D_U of ~4×10^{-5} and D_{Th} some three orders of magnitude smaller.

Foley et al. (2000) report a single value for D_{Pb} of 0.015 at 2.5 GPa and 1100°C.

Experimental D_{Nb} and D_{Ta} are provided by Jenner et al. (1993) and Bennett et al. (2003). Foley et al. (2000) determine D_{Nb}, but place only lower bounds on D_{Ta}. The first two studies give D_{Nb}/D_{Ta} of 0.53 and 0.78 ± 0.14, respectively. This is remarkably similar to both magnetite and ilmenite, suggesting that the site characteristics are broadly similar in all three minerals. As a consequence we propose that D_{Pa}/D_{Nb} is similar in rutile to the other oxides. D_{Nb} appears to be very sensitive to both temperature and, possibly, melt composition. For example in the 3 GPa Ti-CMAS experiments of Bennett et al. (2003), D_{Nb} is 28 ± 3, while it is 53 in the Jenner et al. (1993) experiments and >100 in the Foley et al. (2000) experiments, both on hydrous tonalite. This range is comparable to D_{Nb} for ilmenite and we suggest that these two minerals will have a comparably strong influence on Pa behavior in evolved liquids. The 5+ cation partitioning in ilmenite and rutile could be better constrained experimentally by also adding vanadium and running at elevated fO_2 where V is all 5+.

7.10. Zircon

There are surprisingly few microbeam studies of zircon-melt partitioning in natural systems and none in experimental systems. Recently Thomas et al. (2002) have derived zircon-melt partition coefficients from rehomogenised glass inclusions in zircons from an intrusive tonalite, while Hinton et al. (R. Hinton, S. Marshall and R. Macdonald, written comm.) have used an ion-microprobe to measure zircon-melt partition coefficients from a Kenyan peralkaline rhyolite, with an estimated eruption temperature of 700°C (Scaillet and Macdonald 2001). We have used the lanthanide partition coefficients from these two studies to derive best-fit values for r_0^{3+} and E^{3+} for the large VIII-co-ordinated site. In total there are 13 individual sets of partition coefficients. All of these yield broadly consistent values of r_0^{3+}, in the range 0.968-1.018 Å, but very variable E^{3+}, in the range 373-1575 GPa. Because Lu^{3+} is comparable in size to r_0^{3+}, E^{3+} cannot be well constrained from the lanthanides alone. Only the Hinton et al. (written comm.) study has determined D_{Sc}, which better constrains E^{3+} to be 620 GPa and r_0^{3+} to be 0.962 Å (Fig. 24). These values have been adopted here, for the calculation of D_{Ac}/D_{La} and D_{Bi}/D_{La}, which at 700°C are 2×10^{-6} and 0.04 respectively.

Only Hinton et al. (written comm.) have determined D_U and D_{Th} for zircon. From one sample they obtain $D_U = 130$ and $D_{Th} = 20$ and from the other $D_U = 97$ and $D_{Th} = 15$. These values are very similar to those obtained by Charlier and Zellmer (2000) for zircons separated from a rhyolite: $D_U = 125$, $D_{Th} = 21.2$. The high D_U/D_{Th} ratios, ~6 in all three cases, are consistent with r_0^{4+} being smaller than r_{U4+}, while $D_{Zr}/D_{Hf} \approx 2$, indicates that r_0^{4+} is slightly larger than r_{Zr}. However, it is not possible to fit these data (and D_{Ti}) satisfactorily using the lattice strain model. The curve drawn through the 4+ cations in Figure 23, with $r_0^{4+} = 0.912$ Å and $E^{4+} = 750$ GPa, only crudely reproduces the data; D_U/D_{Th} is grossly overestimated. One possible explanation for this mismatch is that some U^{6+} is present in the melt, so reducing D_U. This is unlikely as the magmatic fO_2 under which the Kenyan zircons crystallised is thought to be at or below FMQ (Scaillet and Macdonald 2001). More plausible is the proposal of Linnen and Keppler (2002), on the basis of the different solubilities of zircon and hafnon in silicate melts, that the D_{Zr}/D_{Hf} ratio should be very sensitive to melt composition, especially in evolved silicic compositions. Thus, the curvature of the 4+ parabola that passes through D_{Zr} and D_{Hf} should vary systematically with melt composition. This is clearly a possibility that requires experimental testing for other 4+ cations, including Ti, Th and U. Until then, we can only conclude, based on the very limited data available, that $D_U \approx 100$ and $D_U/D_{Th} \approx 6$.

One of the striking features about r_0^{3+} as constrained above, is that it is almost identical to the ionic radius of Pa^{5+} in VIII-fold coordination, suggesting that Pa will readily partition into zircon. However, until there are experimental data with which to

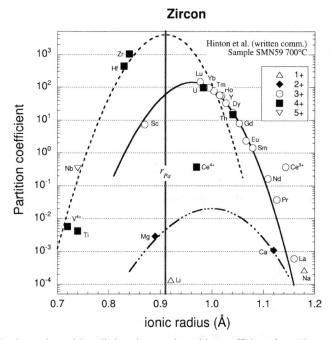

Figure 24. Lattice strain model applied to zircon-melt partition coefficients from Hinton et al. (written comm.) for a zircon phenocryst in peralkaline rhyolite SMN59 from Kenya. Ionic radii are for VIII-fold coordination (Shannon 1976). The curves are fits to Equation (1) at an estimated eruption temperature of 700°C (Scaillet and Macdonald 2001). Note the excellent fit of the trivalent lanthanides, with the exception of Ce, whose elevated partition coefficient is due to the presence of both Ce^{3+} and Ce^{4+} in the melt, with the latter having a much higher partition coefficient into zircon. The 4+ parabola crudely fits the data from D_{Ti} and D_V, through D_{Zr} to D_{Th}, but does not reproduce the observed D_U/D_{Th} ratio. We speculate that this is due to melt compositional effects on D_{Zr} and D_{Hf} (Linnen and Keppler 2002), and possibly other 4+ cations, in very silicic melts. Because of its VIII-fold ionic radius of 0.91 Å (vertical line), D_{Pa} is likely to be at least as high as D_{Nb}, and probably considerably higher.

calibrate the lattice strain model, and any sensitivity to melt composition, we cannot quantify the partitioning behavior of Pa relative to other 5+ cations, such as Nb. It is reasonable to conclude, however, that D_{Pa} is likely to be larger than D_{Nb}, based on simple ionic radius considerations. Unfortunately, there are very few D_{Nb} data for zircon. The problem is exacerbated in ion-microprobe studies by the isobaric interference of $^1H^{92}Zr$ on ^{93}Nb, which is not negligible in zircon itself. This problem, if not fully addressed will tend to overestimates of D_{Nb}. Hinton et al. (written comm.) measured $D_{Nb} = 0.34$ for one zircon-melt pair (Fig. 23), which we consider as a minimum estimate for D_{Pa}. It would seem that zircon may have the highest Pa partition coefficient of any mineral, possibly by several orders of magnitude. Further experimental determinations of the partitioning of high-field strength elements (and even Pa itself) between zircon and a variety of silicic melts are urgently needed to gauge the extent to which this mineral can control the entire Pa budget of silicic magmas.

Divalent cations do not readily enter zircon. Thomas et al. (2002) give D_{Sr} of 0.014-0.043 and D_{Ba} of 0.003-0.005. On this basis we estimate that D_{Ra}/D_{Ba} will be on the order 10^{-6}. There are no data for D_{Pb}, which will presumably be somewhat lower than D_{Sr}, based on our experiences with other minerals.

There are no constraints on D_{Rn} at this time, but given the high mean site charge ($Z_0 \approx 4$), it is unlikely that any noble gases dissolve in zircon.

7.11. Other accessory phases

Other accessories that may play an important role in the fractionation of some U-series elements, include, monazite, apatite, allanite, titanite, thorite and chevkinite. Hermann (2002) has recently determined experimentally the partitioning of U, Th and lanthanides between allanite and granitic melt at 2.0 GPa and 900°C. He finds $D_U = 20$ and $D_{Th} = 60$, confirming that allanite can play an important role in controlling U-Th budgets in silicic melts. The very high D_{La} in the same experiment (~200), indicates that allanite will also be an important host for Bi and Ac.

We are not aware of any experimental U-series partitioning data for other accessory phases. Macdonald et al. (2002) determined $D_{Th} = 87$ and 158, and $D_U = 7$ and 11 for two chevkinite phenocrysts in peralkaline rhyolites from Kenya.

8. CONCLUSIONS

Experimentally-determined partition coefficients exist for U, Th and Pb in most of the major rock-forming minerals. Through careful consideration of experimental and analytical techniques these data can be filtered for use in U-series studies. In some cases, e.g., U and Th in garnet and clinopyroxene, there are sufficient partitioning data to be parameterised as functions of pressure, temperature and composition. This allows polybaric, polythermal natural processes to be modeled with greater precision. For the rest of the U-series elements, however, there are no experimentally-determined partition coefficients, and values must be estimated. One method for doing this is the lattice strain model of Blundy and Wood (1994), which allows the partition coefficient of one cation to be derived from that of another cation with the same charge entering the same lattice site. We have developed this model for application to the U-series elements by identifying a proxy element for each U-series element. We then derive input parameters for the lattice strain model based on available experimental partition coefficients, and so quantify the relationship with the proxy. In those cases where there are no experimental partitioning data for isovalent cations, we apply the electrostatic model of Wood and Blundy (2001), which enables the partition coefficients for a variety of cation charges to be parameterised in terms of a simple parabolic expression. This approach is especially useful in the case of Pa partitioning, which is compromised by a lack of other 5+ cations of similar size.

Our approach has enabled, for the first time, estimates to be made of the partition coefficients for nine of the ten U-series elements in twelve rock-forming minerals. We summarize our estimates of the proxy relationships in the form $D_{U\text{-series}}/D_{proxy}$ (Table 6). These ratios can be combined with estimates of D_{proxy} from the literature. We hope that these data will provide a useful source until such time as direct experimental data become available. In the meantime, new partition coefficient determinations for other elements can be used to refine and modify the estimates presented here. For the U-series elements and their proxies, we have shown that partitioning data are presently lacking, or contradictory, in the cases of plagioclase, alkali feldspar and zircon. The oxide mineral dataset is also rather sparse. For many accessory phases there are no partitioning data at all. The effect of oxidation state on U and Pb partitioning would also represent a worthwhile experimental endeavour. This is especially true in the case of fluid-mineral partitioning where there are presently very few partitioning data and where elevated fO_2 is more likely than in magmatic systems. It is not yet clear to what extent the lattice strain model can be extended to cover mineral-fluid partitioning.

Table 6. Summary of partition coefficients for U-series elements or their proxies.

ion	Cpx	Opx	Olivine	Garnet	Amphibole
U^{4+}	(18) and (8) or (19b)	$2.6 \times D_{Th}$	6×10^{-5}	(25a)	(27a)
Th^{4+}	(18) or (19a)	0.001-0.003	$D_U/6.3$	(25b)	(27b)
Po^{4+}	$0.7 \times D_{Th}$	$0.5 \times D_{Th}$	$0.2\text{-}0.6 \times D_{Th}$	$0.04\text{-}0.1 \times D_{Th}$	$= D_{Th}$
Ra^{2+}	$0.01\text{-}0.07 \times D_{Ba}$	$0.01\text{-}0.02 \times D_{Ba}$	$0.003\text{-}0.02 \times D_{Ba}$	$10^{-4} \times D_{Ba}$	$0.08 \times D_{Ba}$
Pb^{2+}	(8) with $r_{Pb} = 1.32$Å	0.009	0.004	$0.02\text{-}0.09 \times D_{Sr}$	(28)
Ac^{3+}	$0.06 \times D_{La}$	$1\text{-}9 \times 10^{-3} \times D_{La}$	$0.01 \times D_{La}$	$0.2\text{-}7 \times 10^{-5} \times D_{La}$	$0.02 \times D_{La}$
Bi^{3+}	$0.75 \times D_{La}$	$1.2 \times D_{La}$	$1.1 \times D_{La}$	$0.3\text{-}0.5 \times D_{La}$	$0.71 \times D_{La}$
Pa^{5+}	$10^{-7} \times D_U$	$10^{-3} \times D_U$	$10^{-3} \times D_U$	Table 4	$1\text{-}4 \times 10^{-4} \times D_U$
Rn^0	10^{-3}	0.004-0.02	10^{-2}	$< 10^{-4}$	

ion	Plagioclase	K-feldspar	Phlogopite	Oxides	Zircon
U^{4+}	6×10^{-4}			$\geq 10 \times D_{Th}$	≈ 100
Th^{4+}	$5.6 \times D_U$			$\leq 10^{-3} \times D_{Ti}$	$\approx D_U/6$
Po^{4+}	$2 \times D_{Th}$				
Ra^{2+}	(32) and (8)	(36a,b)	$\approx D_{Ba}$	$\ll$	$10^{-6} \times D_{Ba}$
Pb^{2+}	(34)	0.8-1.4 or (37)	$\approx D_{Sr}$	≈ 0.02	
Ac^{3+}	$0.02 \times D_{La}$	$2\text{-}8 \times D_{La}$		$\ll$	$2 \times 10^{-6} \times D_{La}$
Bi^{3+}	$\approx D_{La}$	$1\text{-}2 \times D_{La}$		$\ll$	$0.04 \times D_{La}$
Pa^{5+}	$< 10^{-6}$?		$< 10^{-8}$	$10^{-2}\text{-}10^{-4} \times D_{Nb}$	$\geq D_{Nb}$
Rn^0	$3\text{-}6 \times 10^{-3}$			$\ll$	$\ll$

Partition coefficients are expressed either as values, or as ratios to the proxy element indicated.
(#) indicates Equation (#) in the text.
$\ll$ indicates vanishingly small.
Blank space indicates that the value is presently unconstrained.
To some extent values in this table are generic. The reader is urged to consult the text for more details on how to modify the values for different melting environments.

Lead partitioning is difficult to estimate using the lattice strain approach, because D_{Pb} tends to deviate from parabolae defined by other divalent cations. We suggest that the ionic radius of Pb^{2+} is affected by its lone-pair of electrons and may vary from structure to structure. The effective ionic radius of Pb^{2+} appears inconsistent with the values tabulated by Shannon (1976) for different coordination numbers. This is probably a consequence of divalent lead's electronic configuration, a problem that also afflicts Bi^{3+}. Nonetheless, a systematic study of Pb partitioning between melts and cation sites of varying coordination number and geometry could be used to derive an effective set of ionic radii for Pb^{2+}, which could be extrapolated to Bi^{3+} and any other ions with a lone pair of electrons. Until this issue is resolved, the estimated mineral-melt partition coefficients for bismuth, which are consistently close to D_{La}, should be viewed as crude.

In most minerals the U-series elements are highly incompatible. There are, however, several exceptions. We calculate that radium will be compatible in phlogopite and alkali-feldspar and that these two minerals will dominate the magmatic radium budget in cases where they are present. Protactinium has elevated partition coefficients in magnetite, ilmenite and rutile, and may be higher still in zircon. Protactinium abundance and ^{231}Pa-^{235}U disequilibria in magmatic rocks may therefore be very sensitive to saturation by accessory phases, particularly zircon. Although actinium is not compatible in any of the minerals studied, its principal host will be sodic plagioclase, and possibly alkali feldspar, at low magmatic temperatures, where D_{Ac} may be as high as 0.3. Polonium partitioning is likely to be very similar to Th, which has a slightly smaller ionic radius. D_{Po} will only exceed D_{Th} in minerals with large cation sites that accommodate 4+ cations, such as sodic plagioclase, where polonium may become compatible at low temperatures. There is a lack of noble gas partitioning data for most rock-forming minerals, and no estimate of the atomic radius of radon.

Finally, the experimental community should not despair of directly determining partition coefficients for key U-series elements. Obviously this will not be possible for highly incompatible elements because the doping levels would involve hazardous levels of radioactivity. However, for elements with partition coefficients close to unity in some phases, partitioning could be studied experimentally at reduced doping levels. One objective of this review is to highlight such phases and elements. Radium partitioning between melt and phlogopite, alkali-feldspar or sodic plagioclase should be experimentally tractable, as should protactinium partitioning between melts and oxides or zircon. Careful experimental design, including choice of analytical technique (microbeam or track-counting), is essential if the simple lattice strain model and its predictions are to be tested, and a consensual set of U-series partition coefficients derived.

ACKNOWLEDGMENTS

This chapter represents the culmination of an 8-year research fellowship from The Royal Society to JDB, which he gratefully acknowledges. We would also like to thank NERC for several research grants, and the contributions of Bristol PhD students W. Van Westrenen, E. Hill, K. Law and S. Bennett over the past 5 years. JDB is grateful to N. Allan for a useful discussion about the physical chemistry of Pb^{2+}, to P. Asimow, J. Brenan and T. Green for constructive reviews, and to R. Macdonald, M. Feineman and K. Cooper for informal discussions. R. Hinton, S. Marshall and R. Macdonald kindly provided unpublished zircon-melt partitioning data. We thank Simon Turner for his patience in editing this chapter.

REFERENCES

Allan NL, Du Z, Lavrentiev MY, Blundy JD, Purton JA (2003) Similarity in solid state chemistry: trace element incorporation in ceramics minerals and melts. Proceedings of the Gerona Conference on Molecular Similarity (in press)

Anderson DL, Anderson OL (1970) The bulk modulus-volume relationship for oxides. Jour Geophys Res 75:3494-3500

Andreeßen T, Bottazzi P, Vannucci R, Mengel K, Johannes W (1996) Experimental determination of trace element partitioning between amphibole and melt. 6th Annual Goldschmidt Conf Heidelberg, Germany, J Conf Abstr 1:17

Asmeron Y, Cheng H, Thomas R, Hirschmann M, Edwards RL (2000) Melting of the Earth's lithospheric mantle inferred from protactinium-thorium-uranium isotopic data. Nature 406:293-296

Barth MG, Foley SF, Horn I (2002) Partial melting in Archean subduction zones: constraints from experimentally determined trace element partition coefficients between eclogitic minerals and tonalitic melts under upper mantle conditions. Precamb Res 113:323-340

Beattie P (1993a) The generation of uranium series disequilibria by partial melting of spinel peridotite: Constraints from partitioning studies. Earth Planet Sci Lett 117:379-391

Beattie P (1993b) Uranium-thorium disequilibria and partitioning on melting of garnet peridotite. Nature 363:63-65

Beattie P (1993c) On the apparent non-Henry's Law behavior in experimental studies. Geochim Cosmochim Acta 57:47-55

Beattie P (1994) Systematics and energetics of trace-element partitioning between olivine and silicate melts: Implications for the nature of mineral/melt partitioning. Chem Geol 117:57-71

Benjamin T, Heuser WR, Burnett DS, Seitz MG (1980) Actinide crystal-liquid partitioning for clinopyroxene and $Ca_3(PO_4)_2$. Geochim Cosmochim Acta 44:1251-1264

Bennett S, Blundy J, Elliott T (2003) The effect of sodium and titanium on trace element partitioning. Geochim Cosmochim Acta (submitted)

Bindeman IN, Davis AM (2000) Trace element partitioning between plagioclase and melt: investigation of dopant influence on partition behavior. Geochim Cosmochim Acta 64:2863-2878

Bindeman IN, Davis AM, Drake MJ (1998) Ion microprobe study of plagioclase-basalt partition experiments at natural concentration levels of trace elements. Geochim Cosmochim Acta 62:1175-1193

Blundy JD, Wood BJ (1991) Crystal-chemical controls on the partitioning of Sr and Ba between plagioclase feldspar silicate melts and hydrothermal solutions. Geochim Cosmochim Acta 55:193-209

Blundy JD, Wood BJ (1994) Prediction of crystal-melt partition coefficients from elastic moduli. Nature 372:452-454

Blundy JD, Falloon TJ, Wood BJ, Dalton JA (1995) Sodium partitioning between clinopyroxene and silicate melts. J Geophys Res 100:15501-15515

Blundy JD, Wood BJ, Davies A (1996) Thermodynamics of rare earth element partitioning between clinopyroxene and melt in the system $CaO-MgO-Al_2O_3-SiO_2$. Geochim Cosmochim Acta 60:359-364

Blundy JD, Robinson JAC, Wood BJ (1998) Heavy REE are compatible in clinopyroxene on the spinel lherzolite solidus. Earth Planet Sci Lett 160:493-504

Blundy JD, Dalton JA (2000) Experimental comparison of trace element partitioning between clinopyroxene and melt in carbonate and silicate systems and implications for mantle metasomatism. Contrib Mineral Petrol 139:356-371

Blundy J, Wood B (2003) Partitioning of trace elements between crystals and melts. Earth Planet Sci Lett (in press)

Blundy JD, Brooker RA (2003) Trace element partitioning during melting and crystallization of mafic rocks in the lower crust. Contrib Mineral Petrol (submitted).

Bottazzi P, Tiepolo M, Vannucci R, Zanetti A, Brumm R, Foley SF, Oberti R (1999) Distinct site preferences for heavy and light REE in amphibole and the prediction of $^{Amph/L}D_{REE}$. Contrib Mineral Petrol 137:36-45

Bourdon B, Zindler A, Elliott T, Langmuir C H (1996) Constraints on mantle melting at mid-ocean ridges from global $^{238}U-^{230}Th$ disequilibrium data. Nature 384:231-235

Brenan JM, Shaw HF, Ryerson FJ, Phinney DL (1995) Experimental determination of trace element partitioning between pargasitic amphibole and synthetic hydrous melt. Earth Planet Sci Lett 135:1-11

Brice JC (1975) Some thermodynamic aspects of the growth of strained crystals. J Cryst Growth 28:249-253

Brooker RA, Wartho J-A, Carroll MR, Kelley SP, Draper DS (1998) Preliminary UVLAMP determinations of argon partition coefficients for olivine and clinopyroxene grown from silicate melts. Chem Geol 147:185-200

Brooker RA, Du Z, Blundy JD, Kelley SP, Allan NL, Wood BJ, Chamorro EM, Wartho J-A, Purton JA (2003) The "zero charge" partitioning behavior of noble gases during mantle melting. Nature (submitted)

Chamorro EM, Wartho JA, Brooker RA, Wood BJ, Kelley SP, Blundy JD (2002) Ar and K partitioning between clinopyroxene and silicate melt to 8 GPa. Geochim Cosmochim Acta 66:507-519

Charlier B, Zellmer G (2000) Some remarks on U-Th mineral ages from igneous rocks with prolonged crystallization histories. Earth Planet Sci Lett 183:457-469

Chazot G, Menzies MA, Harte B (1996) Determination of partition coefficients between apatite clinopyroxene amphibole and melt in natural spinel lherzolites from Yemen: implications for wet melting of the lithospheric mantle. Geochim Cosmochim Acta 60:423-437

Cooper KM, Reid MR, Murrell MT, Clague DA (2001) Crystal and magma residence at Kilauea Volcano Hawaii: $^{230}Th-^{226}Ra$ dating of the 1995 east rift eruption. Earth Planet Sci Lett 184:703-718

Cooper KM, Reid MR (2003) Re-examination of crystal ages in recent Mount St Helens lavas: implications for magma reservoir processes. Earth Planet Sci Lett (submitted)

Corgne A, Wood B (2003) Trace element partitioning between silicate melt and Ca-perovskites ($CaTiO_3$ and $CaSiO_4$)—implications for mantle differentiation. Geophys Res Lett 29(19):1903, doi:10.1029/2001GL014398

Dalpé C, Baker DR (2000) Experimental investigation of large-ion lithophile-element-high-field-strength-element- and rare-earth-element-partitioning between calcic amphibole and basaltic melt: the effects of pressure and oxygen fugacity. Contrib Mineral Petrol: 140 233-250

Drake MJ (1972) The distribution of major and trace elements between plagioclase feldspar and magmatic silicate liquid: an experimental study. PhD thesis (unpubl), University of Oregon, Eugene, Oregon

Ewart A, Griffin WL (1994) Application of proton-microprobe data to trace-element partitioning in volcanic rocks. Chem Geol 117:251-284

Falloon TJ, Green DH (1987) Anhydrous partial melting of MORB pyrolite and other peridotite compositions at 10 kbar implications for the origin of primitive MORB glasses. Mineral Petrol 37:181-219

Faul UH (2001) Melt retention and segregation beneath mid-ocean ridges. Nature 410:920-923

Feineman MD, DePaolo DJ, Ryerson FJ (2002) Steady-state ^{226}Ra/^{230}Th disequilibrium in hydrous mantle minerals. Geochim Cosmochim Acta 66:A345 (abstr)

Foley SF, Jackson SE, Fryer BJ, Greenough JD, Jenner GA (1996) Trace element partition coefficients for clinopyroxene and phlogopite in an alkaline lamprophyre from Newfoundland by LAM-ICP-MS. Geochim Cosmochim Acta 60:629-638

Foley SF, Barth MG, Jenner GA (2000) Rutile/melt partition coefficients for trace elements and an assessment of the influence of rutile on the trace element characteristics of subduction zone magma. Geochim Cosmochim Acta 64:933-938

Forsythe LM, Nielsen RL, Fisk MR (1994) High-field-strength element partitioning between pyroxene and basaltic to dacitic magmas. Chem Geol 117:107-125

Goldschmidt VM (1937) The principles of the distribution of chemical elements in minerals and rocks. J Chem Soc London 140:655-673

Green TH (1994) Experimental studies of trace element partitioning applicable to igneous petrogenesis--Sedona 16 years later. Chem Geol 117:1-36

Green TH, Blundy JD, Adam J, Yaxley GM (2000) SIMS determination of trace element partition coefficients between garnet clinopyroxene and hydrous basaltic liquids at 2-7.5 GPa and 1080-1200°C. Lithos 53:165-187

Gudfinnsson GH, Presnall DC (1996) Melting relations of model lherzolite in the system CaO-MgO-Al$_2$O$_3$-SiO$_2$ at 2.4-3.4 GPa and the generation of komatiites. J Geophys Res 101:27701-27709

Gudfinnsson GH, Presnall DC (2000) Melting behavior of model lherzolite in the system CaO-MgO-Al$_2$O$_3$-SiO$_2$-FeO at 0.7-2.8 GPa. J Petrol 41:1241-1269

Guo J, Green TH (1989) Barium partitioning between alkali feldspar and silicate melts at high temperature and pressure. Contrib Mineral Petrol 102:328-335

Guo J, Green TH (1990) Experimental study of barium partitioning between phlogopite and silicate liquid at upper-mantle pressure and temperature. Lithos 24:83-96

Harrison WJ, Wood BJ (1980) An experimental investigation of the partitioning of REE between garnet and liquid with reference to the role of defects. Contrib Mineral Petrol 72:145-155

Hart SR, Dunn T (1993) Experimental cpx/melt partitioning of 24 trace elements. Contrib Mineral Petrol 113:1-8

Hauri EH, Wagner TP, Grove TL (1994) Experimental and natural partitioning of Th U Pb and other trace elements between garnet clinopyroxene and basaltic melts. Chem Geol 117:149-166

Hazen RM, Finger LW (1979) Bulk Modulus-volume relationship for cation-anion polyhedra. J Geophys Res 84:6723-6728

Henshaw DL (1989) Application of solid state nuclear track detectors to measurements of natural alpha-radioactivity in human body tissues. Nucl Tracks Radiat Meas 16(4):253-270 Int J Radiat Appl Instrum Part D

Hermann J (2002) Allanite: thorium and light rare earth element carrier in subducted crust. Chem Geol 192:289-306

Hill E, Wood BJ, Blundy JD (2000) The effect of Ca-Tschermaks component on trace element partitioning between clinopyroxene and silicate melt. Lithos 53:205-217

Hilyard M, Nielsen RL, Beard JS, Patiño-Douce A, Blencoe J (2000) Experimental determination of the partitioning behavior of rare earth element and high field strength elemtns between pargasitic amphibole and natural silicate melts. Geochim Cosmochim Acta 64:1103-1120

Hirose K, Kushiro I (1998) The effect of melt segregation on polybaric mantle melting: Estimation from incremental melting experiments. Phys Earth Planet Int 107:111-118

Horn I, Foley SF, Jackson SE, Jenner GA (1994) Experimentally determined partitioning of high field strength- and selected transition elements between spinel and basaltic melt. Chem Geol 117:193-218

Icenhower J, London D (1995) An experimental study of element partitioning among biotite muscovite and coexisting peraluminous silicic melt at 200 MPa (H$_2$O). Am Mineral 80:1229-1251

Icenhower J, London D (1996) Experimental partitioning of Rb, Cs, Sr, and Ba between alkali feldspar and peraluminous melt. Am Mineral 81:719-734

Jeffries TE, Perkins WT, Pearce NJG (1995) Measurements of trace elements in basalts and their phenocrysts by laser probe microanalysis inductively coupled plasma mass spectrometry. Chem Geol 121:131-144

Jenner GA, Foley SF, Jackson SE, Green TH, Fryer BJ, Longerich HP (1994) Determination of partition coefficients for trace elements in high pressure-temperature experimental run products by laser ablation microprobe-inductively coupled plasma-mass spectrometry (LAM-ICP-MS). Geochim Cosmochim Acta 58:5099-5103

Jones JH (1995) Experimental trace element partitioning. In: Ahrens TJ (ed) Rock physics and phase relations: A handbook of physical constants Am Geophys Union Reference Shelf 3:73-104

Kennedy AK, Lofgren GE, Wasserburg GJ (1993) An experimental study of trace element partitioning between olivine orthopyroxene and melt in chondrules: equilibrium values and kinetic effects. Earth Planet Sci Lett 115:177-195

Kennedy AK, Lofgren GE, Wasserburg GJ (1994) Trace-element partition-coefficients for perovskite and hibonite in meteorite compositions. Chem Geol 117:379-390

Klein M, Stosch HG, Seck A (1997) Partitioning of high field-strength and rare earth elements between amphibole and quartz-dioritic to tonalitic melts: an experimental study. Chem Geol 138:257-271

Klein M, Stosch HG, Seck HA, Shimizu N (2000) Experimental partitioning of high field strength elements between clinopyroxene and garnet in andesitic to tonalitic systems. Geochim Comsochim Acta 64:99-115

Klemme S, Blundy JD, Wood BJ (2002) Experimental constraints on major and trace element partitioning during partial melting of eclogite. Geochim Comsochim Acta 66:3109-3123

Landwehr D, Blundy J, Chamorro-Perez EM, Hill E, Wood BJ (2001) U-series disequilibria generated by partial melting of spinel lherzolite. Earth Planet Sci Lett 188:329-348

La Tourrette TZ, Burnett DS, Bacon CR (1991) Uranium and minor-element partititioning in Fe-Ti oxides and zircon from partially melted granoditorite. Crater Lake Oregon Geochim Comsochim Acta 55:457-469

La Tourrette TZ, Burnett DS (1992) Experimental determination of U-partitioning and Th-partitioning between clinopyroxene and natural and synthetic basaltic liquid. Earth Planet Sci Lett 110:227-244

La Tourrette TZ, Kennedy AK, Wasserburg GJ (1993) Thorium-Uranium fractionation by garnet: Evidence for a deep source and rapid rise of oceanic basalts. Science 261:739-742

La Tourrette T, Hervig RL, Holloway JR (1995) Trace element partitioning between amphibole phlogopite and basanite melt. Earth Planet Sci Lett 135:13-30

Law KM, Blundy JD, Wood BJ, Ragnarsdottir KV (2000) Trace element partitioning between wollastonite and carbonate-silicate melt. Mineral Mag 64:155-165

Leeman WP (1979) Partitioning of Pb between volcanic glass and coexisting sanidine and plagioclase feldspars. Geochim Cosmochim Acta 43:171-175

Leeman WP, Phelps DW (1981) Partitioning of rare earths and other trace elements between sanidine and coexisting volcanic glass. J Geophys Res 86:10193-10199

Linnen RL, Keppler H (2002) Melt composition control of Zr/Hf fractionation in magmatic processes. Geochim Comsochim Acta 66:3293-3301

Long PE (1978) Experimental determination pf partition coefficients for Rb Sr and Ba between alkali feldspar and silicate liquid. Geochim Cosmochim Acta 42:833-846

Lundstrom CC, Shaw HF, Ryerson FJ, Phinney DL, Gill JB, Williams Q (1994) Compositional controls on the partitioning of U, Th, Ba, Pb, Sr, and Zr between clinopyroxene and haplobasaltic melts: implications for uranium series disequilibria in basalts. Earth Planet Sci Lett 128:407-423

Lundstrom CC, Shaw HF, Ryerson FJ, Williams Q, Gill J (1998) Crystal chemical control of clinopyroxene-melt partitioning in the Di-Ab-An system: Implications for elemental fractionations in the depleted mantle. Geochim Cosmochim Acta 62:2849-2862

McDade P, Wood BJ, Blundy JD, Dalton JA (2003a) Trace element partitioning at 3 GPa on the anhydrous garnet peridotite solidus. J Petrol (submitted)

McDade P, Blundy J, Wood B (2003b) Trace element partitioning on the Tinaquillo Lherzolite solidus at 1.5 GPa. Phys Planet Earth Int (submitted)

Macdonald R, Marshall AS, Dawson JB, Hinton RW, Hill PG (2002) Chevkinite-group minerals from salic volcanic rocks of the East African Rift. Mineral Mag 66:287-299

McKenzie D (1989) Some remarks on the movement of small melt fractions in the mantle. Earth Planet Sci Lett 95:53-72

McKenzie D, O'Nions RK (1991) Partial melt distributions from inversion of rare earth element concentrations. J Petrol 32:1021-1091

Mahood GA, Stimac JA (1990) Trace-element partitioning in pantellerites and trachytes. Geochim Cosmochim Acta 54:2257-2276

Marshall AS, Hinton RW, Macdonald R (1998) Phenocrystic fluorite in peralkaline rhyolites, Olkaria, Kenya Rift Valley. Mineral Mag 62:477-486

Mason B (1966) Principles of Geochemistry (3rd Ed). Wiley, New York

Michael PJ (1988) Partition coefficients for rare earth elements in mafic minerals of high silica rhyolites: the importance of accessory mineral inclusions. Geochim Cosmochim Acta 52:275-282

Mysen BO (1979) Nickel partitioning between olivine and silicate melt: Henry's Law revisited. Am Mineral 64:1107-1114

Mysen BO, Seitz MG (1973) Trace element partitioning determined with beta-track mapping: an experimental study using carbon and samarium as examples. J Geophys Res 80:2627-2635

Nagasawa H (1966) Trace element partition coefficient in ionic crystals. Science 152:767-769

Nielsen R (2002) Trace element partitioning. *http://earthreforg/GERM/tools/tephtm*

Nielsen RL, Forsythe LM, Gallahan WE, Fisk MR (1994) Major and trace element magnetite-melt equilibria. Chem Geol 117:167-191

Nielsen RL, Beard JS (2000) Magnetite-melt HFSE partitioning. Chem Geol 164:21-34

Onuma N, Higuchi H, Wakita H, Nagasawa H (1968) Trace element partition between two pyroxenes and the host lava. Earth Planet Sci Lett 5:47-51

Pertermann M, Hirschmann MM (2002) Trace element partitioning between vacancy-rich eclogitic clinopyroxene and silicate melt. Am Mineral 87:1365-1376

Purton JA, Allan NL, Blundy JD, Wasserman EA (1996) Isovalent trace element partitioning between minerals and melts - a computer simulation model. Geochim Cosmochim Acta 60:4977-4987

Purton JA, Allan NL, Blundy J D (1997) Calculated solution energies of heterovalent cations in forsterite and diopside: Implications for trace element partitioning. Geochim Cosmochim Acta 61:3927-3936

Robinson JAC, Wood BJ (1998) The depth of the garnet/spinel transition in fractionally melting peridotite. Earth Planet Sci Lett 164:277-284

Robinson JAC, Wood BJ, Blundy J D (1998) The beginning of melting of fertile and depleted peridotite at 1.5 GPa. Earth Planet Sci Lett 155:97-111

Salters VJM, Hart SR (1989) The hafnium paradox and the role of garnet in the source of mid-ocean-ridge basalts. Nature 342:420-422

Salters VJM, Longhi J (1999) Trace element partitioning during the initial stages of melting beneath mid-ocean ridges. Earth Planet Sci Lett 166:15-30

Salters VJM Longhi JE Bizimis M (2002) Near mantle solidus trace element partitioning at pressures up to 3.4 GPa. Geochem Geophys Geosys 3:2001GC000148

Scaillet B, Macdonald R (2001) Phase relations of peralkaline silicic magmas and petrogenetic implications. J Petrol 42:825-845

Schmidt KH, Bottazzi P, Vannucci R, Mengel K (1999) Trace element partitioning between phlogopite, clinopyroxene, and leucite lamproite melt. Earth Planet Sci Lett 168:287-299

Shannon RD (1976) Revised effetive ionic radii and systematic studies of interatomic distances in halides and chalcogenides. Acta Cryst A32:751-767

Skulski T, Minarik W, Watson EB (1994) High-pressure experimental trace-element partitioning between clinopyroxene and basaltic melts. Chem Geol 117:127-147

Stimac J, Hickmott D (1994) Trace-element partition coefficients for ilmenite, orthopyroxene, and pyrrhotite in rhyolite determined by micro-PIXE analysis. Chem Geol 117:313-330

Takahashi E, Kushiro I (1983) Melting of a dry peridotite at high pressures and temperatures and basalt magma genesis. Am Mineral 68:859-879

Taura H, Yurimoto H, Kurita K, Sueno S (1998) Pressure dependence on partition coefficients for trace elements between olivine and the coexisting melts. Phys Chem Min 25:469-484

Taura H, Yurimoto H, Kato T, Sueno S (2001) Trace element partitioning between silicate perovskites and ultracalcic melt. Phys Planet Earth Int 124:25-32

Tiepolo M, Vannucci R, Zanetti A, Brumm R, Foley SF, Bottazzi P, Oberti R (1998) Fine-scale structural control of REE site-preference: the case of amphibole. Mineral Mag 62A: 1517-1518

Tiepolo M, Vannucci R, Oberti R, Foley SF, Botazzi P, Zanetti A (2000a) Nb and Ta incorporation and fractionation in titanian pargasite and kaersutite: crystal-chemical constraints and implications for natural systems. Earth Planet Sci Lett 176:185-201

Tiepolo M, Vannucci R, Bottazzi P, Oberti R, Zanetti A, Foley S (2000b) Partitioning of rare earth elements, Y, Th, U, and Pb between pargasite, kaersutite, and basanite to trachyte melts: implications for percolated and veined mantle. Geochem Geophys Geosys 1:2000GC000064

Tiepolo M, Bottazzi P, Foley SF, Oberti R, Vannucci R, Zanetti A (2001) Fractionation of Nb and Ta from Zr and Hf at mantle depths: the role of titanian pargasite and kaersutite. J Petrol 42:221-232

Thomas JB, Bodnar RJ, Shimizu N, Sinha AK (2002) Determination of zircon/melt trace element partition coefficients from SIMS analysis of melt inclusions in zircon. Geochim Cosmochim Acta 66:2887-2901

Thompson GM, Malpas J (2000) Mineral/melt partition coefficients of oceanic alkali basalts determined on natural samples using laser ablation-inductively couple plasma-mass spectrometry (LAM-ICP-MS). Mineral Mag 64:85-94

Toomey DR, Wilcock WSD, Conder JA, Forsyth DW, Blundy JD, Parmentier EM, Hammond WC () Asymmetric mantle dynamics in the MELT region of the East Pacific Rise. Earth Planet Sci Lett 200:287-295

Urusov VS, Dudnikova VB (1998) The trace component trapping effect: experimental evidence, theoretical interpretation, and geochemical applications. Geochim Cosmochim Acta 62:1233-1240

Van Orman JA, Grove TL, Shimizu N (1998) Uranium and thorium diffusion in diopside. Earth Planet Sci Lett 160:505-519

Van Orman JA, Grove TL, Shimizu N (2001) Rare earth element diffusion in diopside: influence of temperature, pressure and ionic radius, and an elastic model for diffusion in silicates. Contrib Mineral Petrol 141:687-703

Van Orman J, Saal A, Bourdon B, Hauri E (2002a) A new model for U-series isotope fractionation during igneous processes with finite diffusion and multiple solid phases. EOS Trans, Am Geophys Union 83(47) Fall Meet Suppl Abstract V71C-02

Van Orman JA, Grove TL, Shimizu N (2002b) Diffusive fractionation of trace elements during production and transport of melt in Earth's upper mantle. Earth Planet Sci Lett 198:93-112

Van Westrenen W Blundy JD Wood BJ (1999) Crystal-chemical controls on trace element partitioning between garnet and anhydrous silicate melt. Am Mineral 84:838-847

Van Westrenen W, Blundy JD, Wood BJ (2000a) Effect of Fe^{2+} on garnet-melt trace element partitioning: Experiments in FCMAS and quantification of crystal-chemical controls in natural systems. Lithos 53:191-203

Van Westrenen W, Blundy JD, Wood BJ (2000b) HFSE/REE fractionation during partial melting in the presence of garnet: implications for identification of mantle heterogeneities. Geochem Geophys Geosys 2:2000GC000133

Van Westrenen W, Wood BJ, Blundy JD (2001) A predictive thermodynamic model of garnet-melt trace element partitioning. Contrib Mineral Petrol 142:219-234

Van Westrenen W, Allan NL, Blundy JD, Purton JA, Wood BJ (2003) Trace element incorporation into pyrope-grossular solid solutions: an atomistic simulation study. Phys Chem Minerals (in revision)

Walter MJ, Presnall DC (1994) Melting behavior of simplified lherzolite in the system $CaO-MgO-Al_2O_3-SiO_2-Na_2O$ from 7 to 35 kbar. J Petrology 35:329-359

Watson EB (1985) Henry's law behavior in simple systems and in magmas: Criteria for discerning concentration-dependent partition coefficients in nature. Geochim Comsochim Acta 49:917-923

Watson EB, Harrison TM (1983) Zircon saturation revisited: temperature and composition effects in a variety of crustal magma types. Earth Planet Sci Lett 64:295-304

Watson EB, Ben Othman D, Luck JM, Hofmann AW (1987) Partitioning of U, Pb, Cs, Yb, Hf, Re, and Os between chromian diopsidic pyroxene and haplobasaltic liquid. Chem Geol 62:191-208

Wood BJ, Blundy JD (1997) A predictive model for rare earth element partitioning between clinopyroxene and anhydrous silicate melt. Contrib Mineral Petrol 129:166-181

Wood BJ, Blundy JD, Robinson JAC (1999) The role of clinopyroxene in generating U-series disequilibrium during mantle melting. Geochim Cosmochim Acta 63:1613-1620

Wood BJ, Trigilia R (2001) Experimental determination of aluminous clinopyroxene-melt partition coefficients for potassic liquids with applications to the evolution of the Roman province potassic magmas. Chem Geol 172:213-223

Wood BJ, Blundy JD (2001) The effect of cation charge on crystal-melt partitioning of trace elements. Earth Planet Sci Lett 188:59-71

Wood BJ, Blundy JD (2002) The effect of H_2O on crystal-melt partitioning of trace elements. Geochim Comsochim Acta 66:3647-3656

Wood BJ, Blundy JD (2003) Trace element partitioning. *In:* Treatise on Geochemistry. Carlson R (ed) Elsevier (in press)

4

Timescales of Magma Chamber Processes and Dating of Young Volcanic Rocks

Michel Condomines

Laboratoire Dynamique de la Lithosphère (LDL, UMR CNRS 5573) et ISTEEM
Université Montpellier 2
place Eugène Bataillon
34095 Montpellier cedex 5, France

Pierre-Jean Gauthier and Olgeir Sigmarsson

Laboratoire Magmas et Volcans (LMV, UMR CNRS 6524)
5 rue Kessler
63038 Clermont-Ferrand Cedex, France

1. INTRODUCTION

One of the initial goals of measuring U-series disequilibria in young volcanic rocks was the development of new dating methods in an age range (about 0-300 ka), where few other methods were applicable at that time (Cerrai et al. 1965; Cherdyntsev et al. 1967; Kigoshi 1967; Taddeucci et al. 1967; Allègre 1968). In the following years, several studies showed that the ^{238}U-^{230}Th method, the first to be applied, was indeed able to give reliable ages on rocks from various volcanoes (e.g., Allègre and Condomines 1976; Condomines et al. 1978). But it was soon realized (e.g., Oversby and Gast 1968; Allègre and Condomines 1976; Capaldi et al. 1976) that U-series disequilibria could not only be used to date volcanic rocks, but also to infer the timescales of magma transfer and evolution in the crust, and that these methods could give rather unique information on such important parameters. During the last twenty years, besides continuing efforts to develop and discuss the U-series dating methods, many studies have addressed the problem of the timescales of magmatic processes, from melting in the mantle, to crystallization and differentiation of the magmas in their reservoir(s), and magma degassing near the surface. Moreover, as far as studies on volcanic minerals are concerned, determining the eruption age and the timescales of crystallization are intimately linked. It is the purpose of this chapter to review some of the recent advances in the fields of magma chamber processes and dating (the timescales of partial melting in the mantle are treated in other chapters of this book: Lundstrom 2003; Bourdon et al. 2003; Turner et al. 2003). General reviews of U-series disequilibria in volcanic rocks have been published by Condomines et al. (1988), Gill et al. (1992), Gill and Condomines (1992), Macdougall (1995), Chabaux (1996), and most of the references of earlier U-series articles can be found in these review articles. Section 2 will recall a few general principles, and useful diagrams which are present in most papers on U-series disequilibria in volcanic rocks. In section 3, a few simple models of magma evolution will be considered (closed vs. open systems...) with the corresponding terminology, and several approaches to infer the timescales of magma chamber processes will be described in the rest of this section (disequilibria in single samples, comparison of disequilibria in comagmatic rocks, evolution of disequilibria during the eruptive history of a volcano, studies on minerals, timescales of magma degassing). Section 4 will address more specifically the problem of determining the eruption ages of young volcanic rocks, through mineral isochrons, or whole rock dating. Because dating eruption ages through mineral isochrons relies on the basic assumption that the time interval between crystallization and eruption is short, we have chosen to first discuss the timescales of crystallization before presenting examples of successful datings of eruptions and the conditions required in any attempt to use U-series methods for such determinations (section 4.1).

2. GENERAL PRINCIPLES AND USEFUL DIAGRAMS

2.1. Radioactive disequilibria as dating tools

The principles of radioactive disequilibria in the U or Th series and their applications in geochronology have been described in detail in the book "Uranium-series disequilibrium: Applications to Earth, Marine and Environmental Sciences" edited by Ivanovich and Harmon (1992), and they are recalled in the first chapter of the present book (Bourdon et al. 2003). The existence of a radioactive disequilibrium in a given "parent" (1) – "daughter" (2) pair (i.e., an activity ratio $(N_2/N_1) \neq 1$ in the usual case where the half-life T_1 of the "parent" is much longer than the half-life T_2 of the "daughter") is a necessary condition to be able to date the geochemical fractionation that first created the disequilibrium. Note that the words parent and daughter are written here in quotation marks because, contrary to classical isotope systems like Rb-Sr, both parent and daughter can be radioactive and radiogenic if they belong to intermediate nuclides of the decay series. Figure 1a illustrates this principle of dating the geochemical fractionation, and the well known Figure 1b, first proposed by Williams (1987), further emphasizes that, in closed systems, radioactive equilibrium is nearly reestablished after about 5 half-lives of the daughter nuclide. The behavior of several of the most useful parent-daughter pairs is summarized in this figure (the different shape of the curves in Fig. 1b compared to the curve in Fig. 1a is simply a consequence of the logarithmic scale on the abscissa). With the above condition that $T_1 \gg T_2$, the equation describing the evolution of the daughter activity (N_2) as a function of time in a closed system is:

$$(N_2) = (N_1)(1 - e^{-\lambda_2 t}) + (N_2)_0 e^{-\lambda_2 t}, \qquad (1)$$

which can also be written:

$$[(N_2/N_1) - 1] = [(N_2/N_1)_0 - 1] e^{-\lambda_2 t}. \qquad (2)$$

$(N_2)_0$ and $(N_2/N_1)_0$ represent the initial activity and activity ratio, respectively, just after the fractionation between "parent" and "daughter" nuclides.

2.2. Isochron diagrams

In magmatic processes, both parent and daughter nuclides are usually present in the solid sources, magmas and crystallizing minerals, so that $(N_2)_0$, which is a priori unknown, cannot be neglected. In order to solve Equation (1) for t, the age of fractionation, both terms of this equation are divided by the concentration of a stable isotope (or the activity of a long-lived isotope) of the daughter element. Such a normalization, similar to those used in other classical radiometric methods (Rb-Sr, Sm-Nd…) allows the use of isochron diagrams. In the case of the ^{238}U-^{230}Th method, this type of isochron diagram was first independently proposed by Kigoshi (1967) and Allègre (1968). Its principle is summarized in Figure 2. Isochron diagrams are not only useful to date volcanic minerals as in the example of Figure 2, but they also offer a convenient way to present disequilibria results, in diagrams where chemical fractionation between parent and daughter nuclides (horizontal vectors) are easily distinguished from the effects of radioactive decay (vertical vectors, showing the movement towards radioactive equilibrium and the equiline). The most widely used diagram, in an age range from 0 to ~350 ka, is the (^{230}Th/^{232}Th) vs. (^{238}U/^{232}Th) diagram illustrated in Figure 2. In

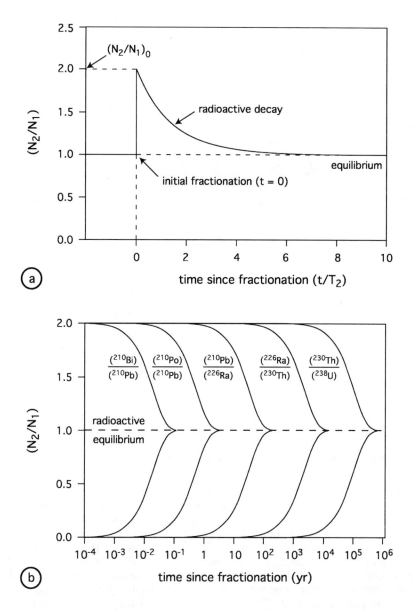

Figure 1. (a) Schematic representation of the evolution by radioactive decay of the daughter-parent (N_2/N_1) activity ratio as a function of time t after an initial fractionation at time 0. The initial $(N_2/N_1)_0$ activity ratio is arbitrarily set at 2. Time t is reported as t/T_2, where T_2 is the half-life of the daughter nuclide. Radioactive equilibrium is nearly reached after about 5 T_2. (b) Evolution of (N_2/N_1) activity ratios for various parent-daughter pairs as a function of time since fractionation (after Williams 1987). Note that the different shape of the curves in (a) and (b) is a consequence of the logarithmic scale on the x axis in (b).

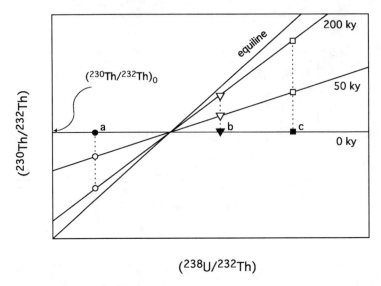

Figure 2. (^{230}Th/^{232}Th) vs. (^{238}U/^{232}Th) isochron diagram (parentheses denote activity ratios). In this example, three minerals a, b and c crystallize at time t = 0 with the same (^{230}Th/^{232}Th)$_0$ initial ratio. They will then evolve as closed systems and will define isochrons whose slope is $1-e^{-\lambda t}$. The isochrons rotate around their intercept with the equiline, which represents radioactive equilibrium between ^{238}U and ^{230}Th. The ordinate of this intercept gives the (^{230}Th/^{232}Th)$_0$ initial ratio.

the age range 0 to 8 ka, the (^{226}Ra)/Ba vs. (^{230}Th)/Ba diagram is also useful: note that, in this case, no long-lived or stable Ra isotope is available, and thus Ba is used as an analog to Ra because of its similar chemical behavior. On a still shorter timescale (0-100 a), (^{210}Pb)/Pb vs. (^{226}Ra)/Pb is applicable. Other isochron diagrams like (^{228}Ra/^{226}Ra) vs. (^{232}Th/^{226}Ra), in an age range 0 to 30 a, could be envisaged although their use would be limited to the study of very recent crystallization history of some minerals, because of the absence of ^{228}Ra-^{232}Th disequilibria in silicate magmas.

2.3. Diagrams showing the evolution of disequilibria vs. time

In studies concerned with the evolution of a magma through time, as represented by the composition of volcanics erupted during the history of a single volcano, it is particularly useful to follow the evolution of radioactive disequilibria vs. time. This can be done in diagrams where the daughter/parent activity ratios of the volcanics, e.g., (^{230}Th/^{238}U) ratios (or isotope ratios like the (^{230}Th/^{232}Th) ratios) at the time of eruption are reported versus the age of eruption θ, often expressed as $e^{\lambda_2 \theta}$ (Fig. 3). This diagram was first proposed and explained in an early study of ^{238}U-^{230}Th disequilibria in Irazu volcano (Allègre and Condomines 1976). It is based on Equation (1) or (2) applied to a magma which stays in the crust long enough to have its disequilibrium (N_2/N_1) ratio modified by radioactive decay. In this simple model, a magma is formed by partial melting at a time T in the past (T is the age before present day and corresponds to the last equilibration of the melt with its solid source), with an initial disequilibrium $(N_2/N_1)_T$. This magma is assumed to be quickly isolated in a magma chamber where it evolves in a closed system without significant parent/daughter fractionation, and is periodically tapped by successive eruptions. Assuming a negligible transfer time of the magma from the chamber to the surface, the $(N_2/N_1)_\theta$ ratios of the lavas erupted at time θ (from

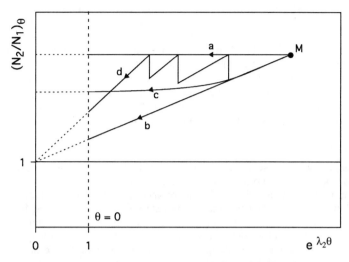

Figure 3. Diagram showing the evolution of the $(N_2/N_1)_\theta$ activity ratio measured in several volcanic rocks of ages θ, erupted from the same volcano, as a function of their eruption ages (expressed as $e^{\lambda_2 \theta}$). Four cases are illustrated. In (a), the $(N_2/N_1)_\theta$ ratios remain constant because of a short transfer time of the magmas towards the surface. In (b), the storage time in a closed-system reservoir is long enough to allow a linear decrease of $(N_2/N_1)_\theta$ ratios since the time of isolation of the magma M (in the case where the initial (N_2/N_1) ratio is >1). In (c), the reservoir is continuously replenished, and the $(N_2/N_1)_\theta$ ratios might eventually reach a constant, steady-state value. Case (d) corresponds to successive periods of closed-system evolution separated by short episodes of reinjection of the same basic magma.

present day) represent the (N_2/N_1) ratio of the magma chamber, and their evolution follow Equation (2):

$$\left[(N_2/N_1)_\theta - 1\right] = \left[(N_2/N_1)_T - 1\right] e^{-\lambda_2(T-\theta)} \quad (3)$$

This equation shows that on a plot of $(N_2/N_1)_\theta$ versus $e^{\lambda_2 \theta}$, the volcanics erupted from this reservoir should define a straight line whose y intercept (at $e^{\lambda_2 \theta} = 0$, i.e., in the future at $\theta = -\infty$) is 1 (radioactive equilibrium, cf. Fig. 3). A linear relationship is also obtained if isotope ratios, like $(^{230}Th/^{232}Th)_\theta$, are reported versus $e^{\lambda_2 \theta}$ (see Fig. 11 in Condomines et al. 1988). This latter diagram is similar to the well known isotope evolution diagrams used in Sr or Nd isotope geochemistry, where $^{87}Sr/^{86}Sr$ or $^{143}Nd/^{144}Nd$ ratios are reported versus time.

3. TIMESCALES OF MAGMA TRANSFER AND MAGMA CHAMBER PROCESSES

3.1. Preliminary comments and definitions

This section will illustrate how the study of U-series disequilibria in volcanic rocks or minerals can give critical information on the timescales of magmatic processes operating during magma transfer towards the surface. The mechanisms by which these disequilibria are created and sustained during partial melting of the mantle and the timescales associated with melting processes are treated in other chapters in this volume (Bourdon et al. 2003; Lundstrom 2003; Turner et al. 2003). We will thus consider the evolution of a magma from the time of its last equilibration within the mantle to its eruption at the surface. During this transfer the magma can stay in one or several reservoirs, and its composition usually evolves from a primitive towards a more differentiated composition by crystal fractionation, sometimes accompanied by crustal contamination, and it loses its volatiles by degassing. The shape of what is usually termed the "plumbing system" of the volcano is rarely well known, and the models of magma conduits and reservoirs are necessarily very schematic. In the best cases, the depth and size of the reservoir(s) can be roughly constrained by petrological and geophysical data. Despite these uncertainties, U-series disequilibria results can be fruitfully discussed in the framework of simple models of magma chamber evolution. Each of the above magmatic processes may have its own characteristic timescale, which can be estimated using various nuclides of appropriate half-lives.

As far as magma reservoirs are concerned, it is important to distinguish between closed and open-system evolution. In the case of "closed-system" evolution of a magma chamber, a magma becomes isolated in its reservoir, with no replenishment, and evolves by crystal fractionation. The magma in the reservoir might be tapped by one or more eruptions, whose volumes represent variable proportions of the total volume of the reservoir. This kind of closed-system behavior is more likely for volcanoes which have long repose times between eruptions and do not show persistent eruptive activity. In open-system magma chambers, the chamber is more or less continuously replenished by a new basic magma, which might mix with the differentiating magma in the reservoir, and is sampled by successive eruptions. Volcanoes in persistent eruptive activity clearly have an open-system behavior. In fact, many volcanoes are probably characterized by alternating episodes of closed and open-system evolution, although the link between the eruptive activity and the behavior of the magma reservoir is not always easy to establish. In open-system magma chambers, depending on the relative values of input rate, crystallization rate, and output rate, the volume of liquid magma in the chamber may increase, decrease, or remain constant. Pyle (1992) developed a model describing the evolution of U-series nuclides in the case of a constant volume crystallizing magma chamber. Albarède (1993) showed how the geochemical fluctuations in volcanic series can be used to infer residence times of the magmas in their reservoirs. Other models describing the evolution of trace elements in inflating, deflating, or steady state (constant volume of liquid magma) magma chambers have been developed for example by Caroff (1997). The influence of magma reinjections on the evolution of ^{238}U-^{230}Th disequilibria has been modeled in detail by Hughes and Hawkesworth (1999). The effects of closed vs. open-system evolution on the variations of $(N_2/N_1)_\theta$ ratios are illustrated in Figure 3.

In a steady state magma chamber, the residence time of the magma, τ, can be defined in the same way as residence times in other earth's reservoirs (ocean, crust...), as the ratio M/Φ, where M is the mass of liquid magma and Φ the input rate (mass of magma per unit time) of magma in the reservoir. As magma injection in open-system magma chambers of most active volcanoes is probably an episodic and not a continuous phenomenon, it could appear that few magma chambers could strictly be at physical steady state. However, if the output rate of lavas is nearly constant when integrated over an eruptive period much longer than the repose time between eruptions, the steady state model may be applied. Similarly, if the activity of a given U-series nuclide remains constant in the erupted lavas, this could be explained by an open-system magma chamber having reached a steady state activity, only if the repose time between eruptions is much shorter than the half-life of this nuclide. If the repose time is similar to the half-life, then the evolution of activity would be better explained by a succession of closed and open-system models. Note that, for a nuclide having a decay constant λ, the time t_S necessary to reach a steady state activity in a magma chamber with a residence time τ is given by $t_S \gg \tau/(1+\lambda\tau)$ (Pyle 1992; Condomines 1994).

Timescales of Magma Chamber Processes & Dating Volcanic Rocks 131

Various terms are currently used when researchers discuss the timescales of magma transfer: transfer time, residence time, storage time, differentiation time... These are sometimes confusing, and their meaning should be clarified by reference to the above models. For example, residence time as defined above should be applied only to open-system magma chambers. Storage time or differentiation time could apply to closed-system magma chambers and designates the time spent by a magma in the reservoir, from the time of its isolation to a given state of differentiation, as sampled by a particular eruption. Transfer time is a general term, simply referring to the time interval between an assumed instantaneous parent-daughter fractionation (either due to partial melting or to some other process higher in the crust) and eruption of a given magma batch. It is implicitly assumed that this magma batch has evolved as a closed system. For example, when stating that the existence of ^{226}Ra-^{230}Th disequilibria in lavas imply a transfer time t of the magma (between the Ra-Th fractionation event and eruption) of less than 8 ka, this time limit is deduced from the time needed to reach radioactive equilibrium in a closed system. In an open system, where a magma chamber is constantly renewed by "fresh" magma having a high and constant (^{226}Ra/^{230}Th) ratio, the residence time in the chamber could be much longer than 8 ka. This is illustrated in Figure 4, where (N_2/N_1) ratios measured in a lava at the surface are plotted against t/T_2 or τ/T_2, where T_2 is the half life of the daughter nuclide.

It may be worth mentioning the analogy between the laws of radioactive equilibrium in decay chains ($N_1/T_1 = N_2/T_2 = ... N_i/T_i$) and those of successive magma reservoirs at steady state (i.e., with the same input and output rates Φ, where $M_1/\tau_1 = M_2/\tau_2 = ... M_i/\tau_i$), as illustrated in Figure 5.

3.2. Disequilibria in single volcanic samples

One of the advantages of U-series disequilibria is that a single measurement on a volcanic rock can place some constraint on the transfer time between the parent-daughter fractionation event and the eruption. Indeed, if the (N_2/N_1) activity ratio is different from 1, the transfer time is shorter than the time needed to reach equilibrium, i.e., about 5 half-lives of the daughter nuclide for a closed-system evolution. This simple reasoning can be illustrated by considering the two Ra-Th pairs: ^{226}Ra-^{230}Th and ^{228}Ra-^{232}Th, which reach equilibrium in about 8 ka and 30 a respectively. The coexistence of (^{226}Ra/^{230}Th) $\neq$ 1 and (^{228}Ra/^{232}Th) = 1, as observed in most active volcanoes erupting silicate lavas, indicates a transfer time between Ra/Th fractionation and eruption in the range 30 a – 8 ka. If the last Ra-Th fractionation is due to partial melting, then this transfer time applies to the transfer from the mantle to the surface.

The case where (^{226}Ra/^{230}Th) $\neq$ 1 and (^{228}Ra/^{232}Th) $\neq$ 1 has been only observed in the active carbonatite volcano Oldoinyo Lengai (Tanzania), first in the lavas erupted during the 1963-67 eruption by Williams and Gill (1986), a result which was later confirmed by Pyle et al. (1991) in their study of the 1988 eruption. In their remarkable paper, Williams and Gill (1986) were able to calculate the age of Ra-Th fractionation, by assuming a similar Ra-Th initial fractionation for both pairs: they found an age of 7 years before eruption for the exsolution of the carbonatite magma (very Ra enriched, with the highest (^{226}Ra/^{230}Th) ratios (~63) ever measured in a volcanic rock) from a nephelinite magma (Fig. 6). This result probably represents the best example of the kind of information that can be obtained from the measurement of a single sample.

Although precise ages generally cannot be deduced from U-series data on single silicate volcanic samples (because of the unknown initial value of the (N_2/N_1) activity ratio) it was nevertheless very significant to discover the ubiquity of ^{230}Th-^{238}U disequilibria (constraining the transfer time of the magmas to less than 350 ka), and later on

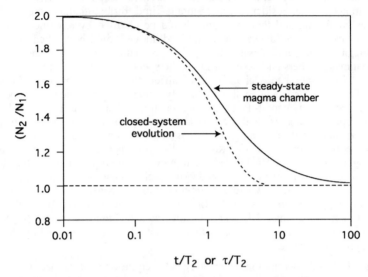

Figure 4. Evolution of the (N_2/N_1) ratio in a reservoir in the two cases of closed system evolution (as a function of t/T_2, where t is the time since fractionation), or in an open-system, steady-state reservoir (the steady-state (N_2/N_1) ratio is plotted as a function of τ/T_2, where τ is the residence time of the magma in the reservoir). Initial fractionation results in an arbitrarily chosen ratio of 2, which is kept constant for the influent magma in the continuously replenished reservoir. The diagram shows that radioactive equilibrium is reached sooner in a closed system evolution. It also illustrates the fact that the radioactive parent-daughter pair should be chosen such as T_2 is commensurate with the residence time of the magma in the reservoir (e.g., τ/T_2 between 0.1 and 10). If T_2 is much longer than the residence time τ, then the (N_2/N_1) ratio will remain close to the initial value (here 2). If T_2 is much shorter than τ, equilibrium will be nearly established in the reservoir.

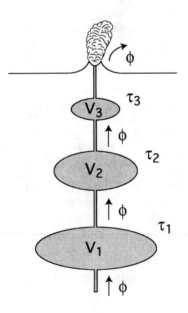

Figure 5. A schematic representation of superposed steady-state reservoirs of constant volumes V_i (fractional crystallization is omitted in this schema). At steady-state, $V_1/\tau_1 = V_2/\tau_2 = ...$, where τ is the residence time. This is analogous to the law of radioactive equilibrium between nuclides 1 and 2: $N_1/T_1 = N_2/T_2 = ...$ A further interest of this simple model is to show that residence times by definition depend on the volume of the reservoirs.

Timescales of Magma Chamber Processes & Dating Volcanic Rocks

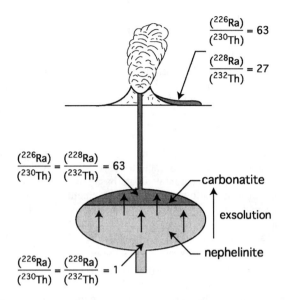

Figure 6. Schematic representation of the model used by Williams et al. (1986) to calculate the age of the Oldoinyo Lengai (Tanzania) carbonatite magma. The model assumes an instantaneous Ra-Th fractionation produced by the exsolution of a carbonatite melt from a nephelinite parental magma in radioactive equilibrium for both Ra-Th pairs. The existence of ^{228}Ra-^{232}Th disequilibria indicates that the fractionation occurred shortly before eruption, and thus the (^{226}Ra/^{230}Th) ratios have not significantly changed since the exsolution. By assuming the same Ra-Th fractionation for both pairs, the (^{226}Ra/^{230}Th) in the carbonatite gives the (^{228}Ra/^{232}Th) ratio just after the exsolution, and its age can then be calculated from the equation:

$$[(^{228}Ra/^{232}Th) - 1] = [(^{226}Ra/^{230}Th) - 1]e^{-\lambda t}$$

where λ is the ^{228}Ra decay constant. With the measured values of these ratios (27 and 63 respectively), an age of 7.2 a is obtained.

of ^{226}Ra-^{230}Th disequilibria (reducing this transfer time to less than 8 ka). For example, the discovery of large ^{226}Ra-^{230}Th (Rubin and MacDougall 1988), or ^{231}Pa-^{235}U disequilibria (Pickett et al. 1997) in MORB samples played an important role, not only to address the problem of magma transfer times, but also in the development of dynamic melting models able to sustain such disequilibria over long periods. In subduction zone volcanoes, the combined study of ^{230}Th-^{238}U and ^{226}Ra-^{230}Th disequilibria place unique constraints on the timescales of mantle metasomatism (U and Ra enrichments by fluids derived from dehydration of the subducted slab), melting and transfer of the magmas towards the surface. These aspects are developed in another chapter (Turner et al. 2003) and in many recent articles (e.g., Gill and Williams 1990; Sigmarsson et al. 1990; Reagan et al. 1994; Hawkesworth et al. 1997; Turner et al. 2000a,b, 2001; Sigmarsson et al. 2002).

3.3. Comparison of disequilibria in comagmatic lavas: timescales of differentiation

Comparison of disequilibria in basic and differentiated magmas. Measurements on a single sample can only give limited information (with the notable exception of Oldoinyo Lengai), and do not allow us to go much beyond this general and rather imprecise notion of transfer time (see section 3.1). It is sometimes possible to evaluate the differentiation time necessary to evolve from a basic magma to a more evolved one, by comparing the disequilibria measured in basic and differentiated rocks emitted in a single volcano or volcanic area. For example, if a basic magma enriched in ^{230}Th has a higher $(^{230}Th/^{232}Th)_0$ ratio (at the time of eruption) than a differentiated magma with a similar U/Th ratio, one can calculate the time needed, in a closed system evolution, to decrease the $(^{230}Th/^{232}Th)$ of the magma by radioactive decay: this time t is calculated from the slope on the isochron diagram: $1-e^{-\lambda t} = [(^{230}Th/^{232}Th)_{0,B} - (^{230}Th/^{232}Th)_{0,D}] / [(^{230}Th/^{232}Th)_{0,B} - (^{238}U/^{232}Th)]$, where the subscripts 0,B and 0,D indicate the ratios at the time of eruption of the basic and differentiated magma respectively, and $(^{238}U/^{232}Th)$ is the ratio of both magmas. The basic hypothesis in this model is that the basic magma which gave rise to the more evolved one by crystal fractionation had the same Th isotope ratio as the basic magma directly erupted at the surface, a reasonable but not necessarily true assumption. Such an approach has been used in several studies. For example, Widom et al. (1992), in their study of the Fogo trachytes (Sao Miguel, Azores), estimated at 90 ka the differentiation time from a parental alkali basalt magma to the trachytic magma, which further evolved in a shallow, zoned magma chamber for a maximum of 4.6 ka. However Claude-Ivanaj et al. (2001) measured an alkali basalt erupted just after the 1563 AD trachytic eruption on Sao Miguel, and found the same $(^{230}Th/^{232}Th)$ ratio as those measured in the 1563 AD trachytes by Widom et al. (1992), which suggests a short differentiation time if this basalt represents the parental magma of the trachytes.

In their work on the Laacher See eruption (Eifel, Germany), Bourdon et al. (1994) inferred a differentiation time for the phonolitic magma of about 100 ka, from the difference in Th isotope ratio between a possible parental basanite and the Laacher See phonolite. Bohrson and Reid (1998) found that the differentiation from trachytes to rhyolites in Socorro Island, Mexico took at most 40 to 50 ka. Another recent study by Thomas (Hawkesworth et al. 2000) of the youngest magmatic cycle in Tenerife (Canary islands) showed a general decrease of the initial $(^{230}Th/^{238}U)_0$ ratios of the lavas with increasing differentiation from basanite to phonolite (both magmas have similar U/Th ratios). This trend is interpreted to reflect a long differentiation time of about 200 ka to produce the more evolved phonolite, although a plot of $(^{230}Th/^{238}U)_0$ ratios vs. Zr (an incompatible element taken as a differentiation index) suggests that the first part of crystal fractionation in the deep crust was much longer than the ultimate differentiation at shallower levels (Hawkesworth et al. 2000). An alternative explanation for the lower $(^{230}Th/^{232}Th)$ and $(^{230}Th/^{238}U)$ ratios in the evolved tephrites and phonolites is that they derive from an assimilation-fractional crystallization process (AFC) involving assimilation of old syenitic rocks in radioactive equilibrium, a model supported by variations in $^{87}Sr/^{86}Sr$ ratios (Sigmarsson et al. 1992a).

Partial melting of old volcanic rocks in radioactive equilibrium as an alternative to long storage and differentiation times can also be illustrated by the study of Sigmarsson et al. (1992b) on Hekla volcano in Iceland. $(^{230}Th/^{232}Th)$ ratios have been measured in nearly all known historic and prehistoric eruptions, which makes this volcano the most extensively studied for its ^{230}Th-^{238}U disequilibria. However, the activity of Hekla is limited to the Holocene, which is a too short period to expect significant variations of $(^{230}Th/^{232}Th)$ ratios by decay in a magma chamber. The lower $(^{230}Th/^{232}Th)$ ratios found in silicic rocks (dacites and rhyolites) compared to the basalts and basaltic andesites are explained by a model of partial melting of the hydrothermally altered basaltic crust (in radioactive equilibrium and thus with lower $(^{230}Th/^{232}Th)$ ratios), which produces a silicic (dacitic) liquid. Mixing of this melt with the basaltic andesite magma, and crystal fractionation of the dacitic melt giving rise to rhyolites at the top of the reservoir, result in a zoned magma chamber. The time necessary to obtain a full zonation of the magma chamber is estimated from the repose times between eruptions, which varies from 400 to 2600 years. In this model, crustal melting, mixing, and crystal fractionation all operate on a short timescale.

On a still shorter timescale, Sigmarsson (1996) used ^{210}Pb-^{226}Ra disequilibria in Surtsey and Heimaey (Iceland) lavas to infer the differentiation time from the Surtsey alkali basalt erupted in 1963-1967 to the Heimaey hawaiites and mugearites (1973 eruption). The (^{210}Pb/^{226}Ra) ratios increase from the basalts to the evolved lavas, which can be explained by a closed system differentiation of about 10 years, a value similar to the time elapsed between the two eruptions. Thus, the Heimaey eruption is interpreted as a consequence of the injection in the crust of a small volume of basaltic magma, 10 years before eruption, a duration long enough to allow its differentiation towards hawaiite and mugearite compositions.

Whole rock isochrons. Differentiation processes can lead to a chemical zonation of a magma chamber. If these magma batches with different compositions are erupted during a single eruption or during a restricted period of time, then it may be possible to date the differentiation process by using whole-rock isochrons, provided that the parent-daughter fractionations are large enough and the differentiation time is short compared to the half-life of the daughter nuclide. Examples of whole-rock isochrons are rare. Crystal fractionation involving major minerals do not fractionate significantly U and Th. Possible causes of variations of the U/Th ratios include fluid-related processes which can mobilize U and not Th, or the fractionation of accessory phases like zircon, sphene or apatite in silicic magmas. The two examples below illustrate these processes.

One example of whole-rock isochrons is provided by the data of Villemant and Fléhoc (1989) on K-rich volcanic rocks from Vico (Central Italy). A large range in U/Th ratios of lavas and ignimbrite units is thought to result from U mobilization by fluids at several depths in the magmatic system (U depletion in basic magma due to the percolation of hydrothermal fluids, and U enrichment of highly differentiated magmas in the shallow reservoir). Three whole-rock isochrons gave ages at 93, 188 and 259 ka, in agreement with previous K-Ar ages (Fig. 7a). It means that the U-mobilization process must have occurred a few millennia at most before eruption.

In their study of the Laacher See eruption, Bourdon et al. (1994) obtained an isochron from their data on phonolitic pumice glasses (probably with U/Th ratios similar to those of the whole rocks) and glasses from cumulate nodules. The fractionation of U/Th ratios is attributed in that case to the crystallization of accessory U and Th-enriched phases such as sphene and apatite. The age of 14.3 ± 6.5 ka is similar to the ages deduced from mineral isochrons (see section 3.5) and to the eruption age of 12.5 ± 0.5 ka (Fig. 7b). Thus differentiation within the phonolitic magma occurred shortly before eruption.

Ra and Th are more easily fractionated when feldspars are involved in the crystallization (plagioclases and mainly alkali feldspars, because Ra and Ba have much higher partition coefficients than Th in these minerals). One whole-rock isochron has been obtained by Evans, in a (^{226}Ra)/Ba-(^{230}Th)/Ba diagram, for seven trachyte samples from Longonot volcano, Kenya (Fig. 4 in Hawkesworth et al. 2000). The large Ba/Th fractionation between the various trachytes is attributed to sanidine fractionation, which also result in (^{226}Ra/^{230}Th) ratios less than 1. The well defined isochron gives an age of 4.3 ka (+1.4, −0.9 ka) for this fractionation, which is indistinguishable from the eruption ages (between 3280 and 5650 years BP). In this case also, differentiation within the trachytic magma body has occurred very quickly, and at most a few centuries before eruptions.

3.4. Evolution of disequilibria during the eruptive history of a volcano: magma storage or residence times

The principle of such studies has been explained in section 2.3 (see also Condomines et al. 1988). Despite the interest of this approach, which was soon realized in early works

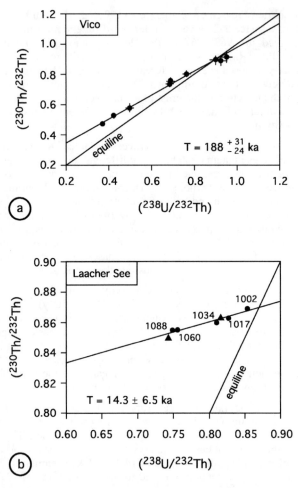

Figure 7. Two examples of whole-rock isochrons. (a) Isochron obtained on a lava unit of the Vico volcano (Central Italy) (Villemant and Fléhoc 1989). (b) Isochron defined by glass and groundmass analyses of the Laacher See phonolite (Bourdon et al. 1994). In both examples, differentiation occurred shortly before eruption since calculated ages are indistinguishable from eruption ages.

of the 1970-1990 period (e.g., Allègre and Condomines 1976; Condomines and Allègre 1980; Condomines et al. 1982; Newman et al. 1984) relatively few studies have presented a detailed evolution of radioactive disequilibria through time in a given volcano or volcanic area. This is certainly in part due to the difficulty to gather a suite of well dated samples from a single volcano, whose history and magmatic evolution is also well known. On the timescale resolvable by ^{230}Th-^{238}U disequilibria (0-350 ka), one may use U-Th mineral isochrons to date ancient volcanic rocks and thus deduce their $(^{230}$Th/^{232}Th$)_0$ ratios at the time of eruption (if isochron ages are close to eruption ages, see section 3.5 below), or alternatively rely on samples dated by another method (e.g., K-Ar, ^{39}Ar-^{40}Ar) to recalculate their initial Th isotope ratios. On the shorter timescale covered by ^{226}Ra-^{230}Th disequilibria (0-8 ka), it can be even more difficult to find a suite of well dated samples, because eruption ages should be obtained from historical accounts (the number of volcanoes for which the historical record spans more than a few centuries is very small) or archaeological or ^{14}C dates.

^{230}Th-^{238}U disequilibria. The long-term (~10^5 years) evolution of ($^{230}Th/^{232}Th$)$_0$ ratios has been studied in a few basaltic volcanoes from ocean islands. Newman et al. (1984) reported a few data on Mauna Kea (Hawaii) and Marion island, which showed no significant variation of ($^{230}Th/^{232}Th$)$_0$ ratios with time. On a larger set of samples, Condomines et al. (1988) reached a similar conclusion for the Piton de la Fournaise in Reunion island, where the small variations in ($^{230}Th/^{232}Th$)$_0$ ratios most likely result from source heterogeneities rather than radioactive decay in an evolving magma chamber. This suggests that the transfer time (or residence time in an open system magma chamber) is short compared to the ^{230}Th half-life. In their detailed study of Mt. Etna, Condomines et al. (1982) interpreted the variation of ($^{230}Th/^{232}Th$)$_0$ ratios to result from closed system evolution in a long-lived (~200 ka) magma chamber filled by an alkali basaltic magma, with episodic injections and mixing of magmas with tholeiitic affinities. To this long term evolution were superimposed multiple and short episodes of crystal fractionation in shallower reservoirs. However this model might not be applicable to the recent volcanic evolution, because the large ^{226}Ra excesses [($^{226}Ra/^{230}Th$) between 1.4 and 2.6] measured in historic lavas cannot easily be maintained in a long lived magma reservoir.

^{226}Ra-^{230}Th disequilibria. The existence of ^{226}Ra-^{230}Th disequilibria in many basaltic volcanoes in permanent eruptive activity suggests that the magma residence time in their open system magma chambers is of the order of the ^{226}Ra half-life (a few centuries to a few ka). The most complete study so far of the evolution of ^{226}Ra-^{230}Th disequilibria in a single volcano is that of Condomines et al. (1995) on the historic lavas of Mt. Etna, where the historic record spans almost two millennia. This study shows that ($^{226}Ra/^{230}Th$) ratios (recalculated at the time of eruption) decrease with differentiation in hawaiites and mugearites, as do the Ba/Th ratios, an effect which is attributed to plagioclase fractionation in the shallow plumbing system of Mt. Etna and confirms the similar behavior of Ra and Ba in this case. The constant (^{226}Ra)$_0$/Ba ratio measured in historic lavas of the last two millennia (Fig. 8) is interpreted to reflect the existence of a steady state open-system deep magma chamber at a depth of about 15 km, filled with an hawaiitic magma: the (^{226}Ra)$_0$/Ba ratio reaches a steady state, because of the opposite effects of ^{226}Ra radioactive decay and reinjection of basaltic magma with a higher (^{226}Ra)$_0$/Ba ratio. According to the models proposed by Pyle (1992) and Condomines (1994), the residence time of the magma in the reservoir can be calculated if the ($^{226}Ra/^{230}Th$) (or (^{226}Ra)/Ba) ratio) of the influent basaltic magma is known. Assuming an upper limit for this ratio is represented by the value of the 1974 basaltic and eccentric eruption, Condomines et al. (1995) were able to calculate a maximum residence time of about 1500 years and a maximum volume of the deep reservoir of 150-300 km^3, much smaller than the volume inferred from seismic studies (this latter probably includes the volume of the outer part of the magma reservoir composed of cumulates and interstitial melt). The present period of activity at Mt. Etna, since 1970 AD, is characterized by the eruption of more basic lavas with anomalous enrichments in Ra and alkaline elements (K, Rb, Cs), and higher $^{87}Sr/^{86}Sr$ ratios, attributed to a selective contamination in the sedimentary basement (see Condomines et al. 1995, and references therein). This basic magma progressively mixes with the previous, more differentiated magma, in the upper plumbing system. From the analysis of geochemical variations during this period, Condomines et al. (1995) infer a residence time of a few tens of years, and a volume of about 0.5 km^3 for the shallow plumbing system. This residence time is in good agreement with the values deduced by Albarède (1993). The principle of calculation of the residence time from measurements of radioactive disequilibria is schematically illustrated in Figure 9, in the simplified case of a magma chamber without fractional crystallization (Pyle 1992).

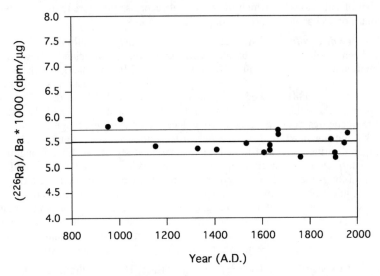

Figure 8. $(^{226}Ra)_0$/Ba ratios (recalculated at the time of eruption) vs. eruption ages of the pre-1970 lavas from Mt. Etna (Condomines et al. 1995). The constant $(^{226}Ra)_0$/Ba ratios are interpreted by a model of steady-state reservoir (see text for explanation). The upper and lower horizontal lines correspond to ± 5% of the mean (the maximum estimated error on the $(^{226}Ra)_0$/Ba ratios).

A similar model is proposed for the Piton de la Fournaise, where there are significant differences between the constant $(^{226}Ra/^{230}Th)$ ratios of the lavas inside the Enclos caldera and the higher ratios of the lavas erupted outside this caldera, which are assumed to have bypassed the shallow magma chamber. Sigmarsson et al. (in preparation) explain these data by the existence of a steady state shallow reservoir and calculate a magma residence time of about 400 years.

Another recent study on historic, basaltic lava flows (1848 to 1977 AD) from Karthala volcano (Comores, Indian ocean) by Claude-Ivanaj et al. (1998) reveals significant variations of the $(^{226}Ra)_0$/Ba ratios, correlated with $^{87}Sr/^{86}Sr$ ratios, which are explained by mantle source heterogeneities preserved by short transfer times of the magmas (a few centuries at most).

^{226}Ra-^{230}Th disequilibria have also been measured in several prehistoric and historic lava flows from Hawaiian volcanoes (e.g., Cohen and O'Nions 1993; Hémond et al. 1994; Sims et al. 1999; Pietruszka et al. 2001). All these authors assume that the $(^{226}Ra/^{230}Th)$ ratios have not been affected by radioactive decay in the shallow open-system reservoirs of tholeiitic volcanoes like Kilauea, and this is supported by independent evidence for short magma residence times (e.g., ≤ 200 years; Pietruszka and Garcia 1999).

$(^{226}Ra/^{230}Th)$ ratios may also vary during a single eruption. Sigmarsson et al. (1998) analyzed the lavas of the long 1730-36 eruption on Lanzarote (Canary Islands), whose nearly primitive compositions evolved from basanites to alkali basalts and tholeiites. The $(^{226}Ra/^{230}Th)$ variations are explained by differences in the mantle sources (with variable proportions of garnet pyroxenites and lherzolites) and melting degrees, or alternatively by the addition of Ra rich fluids to the mantle sources. $(^{226}Ra/^{230}Th)$ ratios are preserved by very short transfer times of the magmas (a few decades) in the absence of any magma

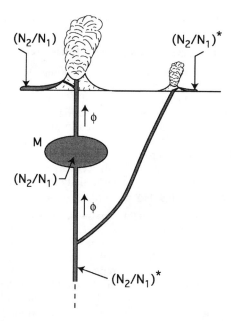

Figure 9. A schematic and ideal model showing how the residence time of the magma in a steady-state reservoir of constant mass M, replenished with an influx Φ of magma and thoroughly mixed, can be calculated from disequilibrium data, in the simplifying case where crystal fractionation is neglected (Pyle 1992). The mass balance equation describing the evolution through time of the concentration $[N_2]$ (number of atoms of the daughter nuclide per unit mass of magma) in the reservoir is:

$$M\,d[N_2]/dt = \lambda_1 M[N_1] - \lambda_2 M[N_2] + \Phi[N_2]^* - \Phi[N_2].$$

At steady state, $d[N_2]/dt=0$, and the residence time $\tau = M/\Phi$ can be written as

$$\tau = [(N_2)^* - (N_2)] / \lambda_2 [(N_2) - (N_1)] \text{ or } \tau = [(N_2/N_1)^* - (N_2/N_1)] / \lambda_2 [(N_2/N_1) - 1],$$

where parentheses denote activities or activity ratios (note that $(N_1) = (N_1)^*$ because of the long half-life of the parent nuclide and the absence of crystal fractionation). If the $(N_2/N_1)^*$ ratio is known, then the residence time can be calculated from the measurement of the (N_2/N_1) ratio in lavas erupted from the central conduit. An eccentric eruption, whose magma has bypassed the reservoir, may provide a value for the $(N_2/N_1)^*$ ratio.

reservoir in the crust. Thomas et al. (1999) also found variable ($^{226}Ra/^{230}Th$) ratios in two samples of the 1730-36 eruption and one sample of the 1824 eruption. They consider that the lower ($^{226}Ra/^{230}Th$) ratios in basanites derived from low-degree melts reflect a significant transfer time of these magmas. However, if transfer times were long enough to affect ($^{226}Ra/^{230}Th$) ratios, the good mixing relationship observed in the ($^{226}Ra/^{230}Th$)-$1/(^{230}Th)$ diagram (Sigmarsson et al. 1998) would not be preserved.

Other variations in ($^{226}Ra/^{230}Th$) ratios observed in the alkali basalts of the 1718 AD eruption on Pico (Azores) by Claude-Ivanaj et al. (2001) are also interpreted as primary features acquired in the mantle and not affected by decay during transfer to the surface.

Variations in ($^{226}Ra/^{230}Th$) ratios are also apparent in the Ardoukoba (Asal Rift) 1978 basaltic eruption studied by Vigier et al. (1999). ($^{226}Ra/^{230}Th$) ratios decrease with increasing Th content, as do the (^{226}Ra)/Ba ratios. These variations cannot be entirely explained by plagioclase fractionation, which led the authors to propose a model of a laterally zoned fissural reservoir, with several injections of basaltic magmas. In this case, the oldest magma would have stayed in the reservoir for about 1.9 ka. However small Sr isotope variations suggest that source heterogeneities might also have contributed to the (^{226}Ra/^{230}Th) variations.

3.5. Studies on minerals and crystallization ages: timescales of crystallization

Another way to place constraints on the magma storage time in its reservoir comes from the analysis of minerals present as phenocrysts in a volcanic rock. Analyses of minerals were undertaken from the beginning of ^{230}Th-^{238}U disequilibria studies in volcanic rocks (around 1965), with the aim of dating these rocks (i.e., to find their eruption ages), and soon mineral data were reported in the (^{230}Th/^{232}Th) vs. (^{238}U/^{232}Th) isochron diagram (Fig. 2). The principle of this isochron diagram has been recalled in section 2.2. A "perfect" mineral isochron, whose slope allows calculation of a meaningful crystallization age, requires that all minerals crystallized in a short time interval compared to the ^{230}Th half-life, that they were initially in Th isotope equilibrium (with the same (^{230}Th/^{232}Th) ratio) and that they evolved as closed systems after crystallization.

In the first studies reporting mineral analyses, it was implicitly assumed that crystallization occurred shortly before eruption (at least in an interval that was within the age limits due to analytical uncertainties). The number of works including analyses of minerals has greatly increased in the last 10 years, and this is largely due to the development of mass spectrometric measurements. Indeed mass spectrometry allow analyses of much smaller amounts of mineral phases, with a generally better precision (Goldstein and Stirling 2003). While in several cases, the ages obtained from mineral isochrons were in agreement with eruption ages derived from other dating methods (mainly ^{14}C, K-Ar or ^{39}Ar-^{40}Ar methods), in many other cases ^{230}Th-^{238}U ages were significantly older than inferred eruption ages. In some instances however, ^{230}Th-^{238}U isochrons give younger ages than ^{39}Ar-^{40}Ar ages (e.g., three volcanic rocks from the Alban Hills, Italy, dated between 33 and 11 ka by Voltaggio et al. 1994). Several examples are detailed below. The difference between crystallization age and eruption age is usually interpreted as a minimum storage time of the magma in its reservoir. We will see, however, that the interpretation of mineral isochrons giving ages older than eruption ages is not straightforward.

Data on mineral separates in present day volcanic rocks. Since every dating method (including the K-Ar or ^{14}C systems) can be affected by several geochemical perturbations which may lead to erroneous ages, the best test for the ^{230}Th-^{238}U mineral isochrons consists in the analysis of presently erupted lavas or historic lavas of well known eruption dates. Rather surprisingly the data obtained on such samples are not so numerous (some examples are illustrated in Fig. 10). Early data showed that, in some cases, there were inter-laboratory analytical discrepancies, especially in Th isotope ratios measured on mineral separates extracted from the same lava flows (this was the case for the 1971 lava from Mt. Etna and 1944 lava from Mt. Vesuvius: Capaldi and Pece 1981; Hémond and Condomines 1985; Capaldi et al. 1985). This emphasizes the fact that ^{230}Th-^{238}U mineral analyses

---

Figure 10 (*on facing page*). Several examples of mineral isochrons obtained on near zero age volcanic rocks. The figure compares the isochron diagrams obtained from both ^{238}U-^{230}Th and ^{226}Ra-^{230}Th data. In (a) and (b), the U-Th isochrons give near zero crystallization ages, in agreement with eruption ages (associated with sample numbers or volcano names). In (c) and (d), U-Th crystallization ages are older than eruption ages, and yet Ra-Th isochrons suggest young crystallization ages (see discussion in the text). (a1-a3) Results for andesites and dacite from Mt. St. Helens (MSH; Volpe et al. 1991). (b) Data for the Nevado del Ruiz dacitic pumice (Schaefer et al. 1993). (c) Mt. St. Helens basalt (Volpe et al. 1991). (d) Mt. Vesuvius phonotephrite (Black et al. 1998). WR: whole-rock; Gl: glass (Nevado del Ruiz) or groundmass (Mt. St. Helens and Vesuvius); Pl: plagioclase; Px: pyroxene; Mt: magnetite or magnetic separate; Ol: olivine; Le: leucite. (^{226}Ra)/Ba and (^{230}Th)/Ba ratios are expressed in dpm/µg Ba.

Timescales of Magma Chamber Processes & Dating Volcanic Rocks 141

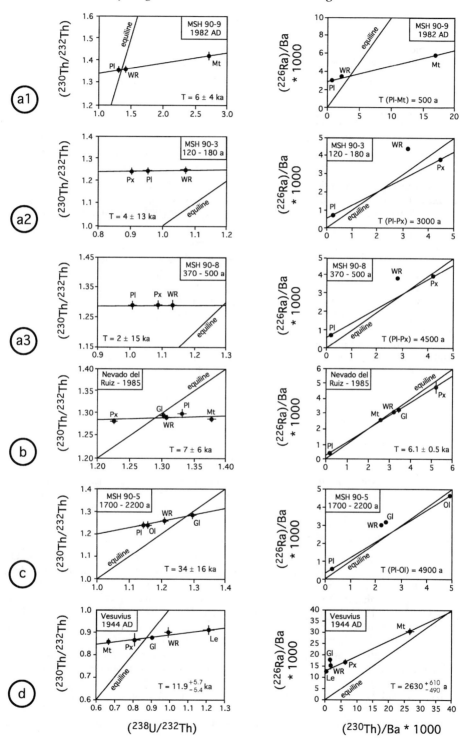

Figure 10. (*caption on facing page*)

remain difficult, and need very careful chemistry, in order to avoid problems linked to inclusions of insoluble accessory phases, insoluble precipitates, etc. Leaching procedures, sometimes used to purify mineral separates, should also be undertaken with great caution as they may induce differential U and Th leaching from the minerals themselves.

Volpe et al. (1991) analyzed minerals (plagioclase, pyroxene, olivine, and magnetite) in the 1982 dacitic dome of Mt. St Helens and in five other samples of dacite, andesite and basalt less than 2200 years old from this volcano: they found four near-zero age U-Th isochrons (Fig. 10 a1-a3), among them the 1982 dacite, giving an age of 6 ± 4 ka (2σ), the magnetite having a slightly higher ratio than the whole rock and plagioclase. But an andesite and a basalt define mineral isochrons of 27 ± 12 and 34 ± 16 ka (2σ) respectively, in excess of the eruption age, between 1700 and 2200 years before 1980 (Fig. 10c).

Schaeffer et al. (1993) analyzed mineral phases extracted from the 1985 dacitic pumice erupted by Nevado del Ruiz (Colombia): pyroxene, plagioclase, magnetite, glass and whole rock define an U-Th isochron with a near-zero age (7 ± 6 ka, 2σ; Fig. 10b).

In their study of Soufrière (St Vincent, Antilles), Heath et al. (1998) analyzed four volcanic rocks erupted less than 4 ka ago. The 1979 pyroclastic flow (basaltic andesite) provides a mineral U-Th isochron giving an age of 46 ± 27 ka (2σ). The three other samples give ages of 56 to 77 ka (cf. Fig. 12c below). All the mineral and groundmass data from the four rocks define an average age of 58 ± 7 ka (2σ), interpreted as reflecting a long storage time of the magma (minerals and melt), perhaps near the walls of the reservoir, in a layer which was maintained at a near-liquidus temperature because of the heat flux produced in the main part of the reservoir by injections of fresh magma. However, this interpretation remains to be reconciled with the study of Zellmer et al. (1999), who obtained an age of a few hundred years for a Soufriere plagioclase crystal from a study of Sr diffusion profile in this mineral. Similar timescales of a few hundred years were also obtained for plagioclase crystals from El Chichon volcano (Mexico), on the basis of $^{87}Sr/^{86}Sr$ diffusion profiles (Davidson et al. 2001).

The products of the 1944 eruption (phonotephrites) of Mt. Vesuvius were studied in detail by Black et al. (1998). Minerals extracted from the lava give an U-Th isochron with an age of 11.9 ± 5.6 ka (2σ; Fig. 10d). A cumulate ejected during the same eruption reveals an age of 39.4 ± 9.1 ka (2σ).

In a recent study, Zellmer et al. (2000) found a zero-age U-Th isochron for the 1940 Kameni dacite in Santorini (Aegean volcanic arc). The existence of $(^{226}Ra/^{230}Th)$ ratios lower than 1 in Kameni dacites suggests that plagioclase fractionation took place less than 1 ka before eruption, in agreement with estimates based on Sr diffusion profiles in plagioclases (Zellmer et al. 1999).

Summarizing these recent U-Th results on near-zero age volcanic rocks (including the results of Hémond and Condomines (1985) on the 1971 lava from Mt. Etna and excluding the Vesuvius cumulate) shows that 7 out of 14 samples also give near-zero crystallization ages (or at least that the time difference between crystallization and eruption is less than a few ka). The other half of samples give crystallization ages significantly different from eruption ages (>10 ka). Interestingly, in many of the above examples, $^{226}Ra-^{230}Th$ have also been measured in mineral phases and/or whole rocks and reported in the $(^{226}Ra)/Ba$-$(^{230}Th)/Ba$ isochron diagram (see Fig. 10). In 6 samples giving U-Th crystallization ages within a few ka of the eruption, $^{226}Ra-^{230}Th$ "isochrons" (or a reasoning based on whole rock results in the case of Santorini) suggest crystallization ages between 0.5 and 6 ka, in agreement with U-Th ages. It should be noted however that most "isochrons" from Mt. St. Helens rocks are two-points isochrons based on two minerals, and that the whole rocks do not plot on these isochrons (Fig. 10a2, a3). As discussed by Volpe et al. (1991), this might be explained in several ways, including a different behavior of Ra and Ba, an open system evolution of the melt after crystallization of the minerals, a mixing of old and young crystals (either as separate crystals or as old crystal cores surrounded by younger rims), a prolonged and/or non simultaneous crystallization of the various minerals. In view of the rather short half-life of ^{226}Ra, this last hypothesis would not seem unlikely. It could even appear surprising to find a very good ^{226}Ra-^{230}Th isochron as that defined by the Nevado del Ruiz andesite minerals (6.1 ± 0.5 ka (2σ); Schaefer et al. 1993; Fig. 10b). In that case, the minerals should have crystallized in a very short time interval and have not reequilibrated through diffusive exchange with the melt. The possibility of reequilibration can be evaluated through a calculation of characteristic length scales of Ra diffusion. These can be estimated in plagioclase using published values for Ba diffusion coefficients in this mineral (Cherniak 2002). A Ra diffusion coefficient $\mathcal{D}$ of around $1.6.10^{-19}$ m^{2}s^{-1} is estimated for an An45 plagioclase at a temperature of 1080°C (the average plagioclase composition and andesitic magma temperature are taken from Schaefer et al. 1993). The characteristic length scale of diffusion (l = $\sqrt{\mathcal{D}t}$) over a timescale t of 6000 years is thus about 0.17 mm, which is significant compared to the size of the smallest crystals, but negligible for the largest ones (the actual sizes of the crystals range from < 20 µm to 5 mm). This suggests that Ra and Ba diffusive reequilibration between minerals and melt at magmatic temperatures on timescales of a few thousand years can only be significant in small crystals (< 1 mm).

The possibility that Ra and Ba have different partitioning behavior is suggested by the work of Blundy and Wood (1994). These authors experimentally determined partition coefficients for several alkaline-earth elements (Mg, Ca, Sr, Ba) in plagioclase and clinopyroxene, and showed that partition coefficients could be predicted by a model based on the size and elasticity of crystal lattice sites. Because of its larger ionic radius, Ra should be more incompatible than Ba in these minerals. In that case, minerals in equilibrium with the melt would have different (^{226}Ra)/Ba ratios at the time of crystallization and would not plot on an horizontal line in the (^{226}Ra)/Ba-(^{230}Th)/Ba isochron diagram. Such an example is illustrated in Figure 11. As developed in the caption of this figure, the age of crystallization of a given mineral can be derived if the (D_{Ra}/D_{Ba}) ratio for this mineral is known. This approach has recently been used by Cooper et al. (2001) as discussed below. It should be noted that all the ages usually calculated from plagioclase and/or clinopyroxene-groundmass data in the (^{226}Ra)/Ba-(^{230}Th)/Ba isochron diagram are older than the true crystallization ages if Ra is more incompatible than Ba in these minerals.

In the three samples from Mt. St Helens and Vesuvius giving crystallization ages more than 10 ka older than eruptions, ^{226}Ra-^{230}Th disequilibria are nevertheless present in mineral phases, groundmass or whole rocks. For example, the minerals of the 1944 Vesuvius lava giving a crystallization age of 11.9 ± 5.6 ka (2σ), which should have been in ^{226}Ra-^{230}Th equilibrium had they evolved as closed systems, define a rather good ^{226}Ra-^{230}Th isochron with an age of 2.63 ± 0.55 ka (2σ) (Fig. 10d). Even the minerals of the Vesuvius cumulate (39.4 ± 9.1 ka crystallization age) indicate a ^{226}Ra-^{230}Th age of 3.27 ± 0.62 ka (2σ) (Black et al. 1998). These data strongly suggest that some Ra-Th fractionation occurred shortly before eruption, either because crystallization of mineral rims or as a consequence of chemical exchanges between already crystallized minerals and melt. The situation is further complicated in the case of Vesuvius by a probable late Ra enrichment by a fluid phase (Black et al. 1998) and the existence of cumulative floating minerals like the leucites which show huge Ra excesses [(^{226}Ra/^{230}Th) > 50].

Another example of ^{226}Ra-^{230}Th isochron dating is provided by the study of Reagan et al. (1992) on phonolites erupted in 1984 and 1988 at Mt. Erebus (Antarctica). Glass-

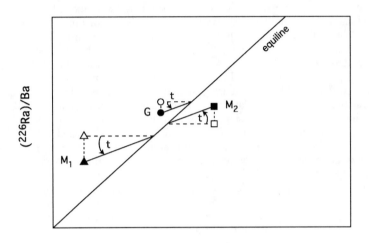

(^{230}Th)/Ba

Figure 11. Theoretical evolution, in a (^{226}Ra)/Ba-(^{230}Th)/Ba isochron diagram, of groundmass (G) and minerals (M_1 and M_2) with different partition coefficients for Ra and Ba. Minerals are assumed to crystallize simultaneously and instantaneously (e.g., in less than 200 years). Open symbols represent the position of minerals and melt in chemical equilibrium at the time of crystallization. Ra is assumed to be more incompatible than Ba in both minerals, with $(D_{Ra}/D_{Ba})_{M1} < (D_{Ra}/D_{Ba})_{M2} < 1$. After a time t of closed system evolution, minerals and groundmass plot as indicated in the diagram by filled symbols. The true age of crystallization cannot be directly obtained from any line drawn through the filled symbols: the M_1G and M_2G lines give ages respectively older and younger than the true age (the respective slopes might even be >1 or < 0), and the best fit line through M_1, M_2, G can give an older, younger or similar age compared to the true crystallization age, depending on the initial position of minerals and groundmass.

Note that, if the $(D_{Ra}/D_{Ba})_M$ of one mineral is known, then the crystallization age t can be calculated by solving a system of two equations with two unknown parameters: t and $[(^{226}Ra)/Ba]_{G,0}$, the initial ratio of the melt. Indeed if minerals and melt evolve as closed systems, their $(^{226}Ra)/Ba$ ratios are given by the following equations, where y and x represent $(^{226}Ra)/Ba$ and $(^{230}Th)/Ba$ ratios respectively:

$$y_G = x_G \times (1 - e^{-\lambda t}) + y_{G,0} \times e^{-\lambda t},$$

and:

$$y_M = x_M \times (1 - e^{-\lambda t}) + \left(\frac{D_{Ra}}{D_{Ba}}\right)_M \times y_{G,0} \times e^{-\lambda t}$$

The $[(^{226}Ra)/Ba]_{G,0}$ ratio can thus be derived from the above equations:

$$y_{G,0} = \frac{(x_G \times y_M) - (x_M \times y_G)}{y_M - x_M - \left(\frac{D_{Ra}}{D_{Ba}}\right)_M \times (y_G - x_G)}$$

and the crystallization age t is calculated as:

$$t = -\frac{T}{\ln 2} \times \ln\{(y_G - x_G)/(y_{G,0} - x_G)\},$$

where T is the ^{226}Ra half-life.

anorthoclase pairs for both rocks define similar ages of about 2.4 ka. But the anorthoclase megacrysts continued to grow until the eruption, as demonstrated by an excess of ^{228}Th over ^{232}Th, a consequence of an enrichment of ^{228}Ra over ^{232}Th in the outer rim of the mineral.

Cooper et al. (2001) recently analyzed ^{226}Ra-^{230}Th disequilibria in the phenocrysts of the 1955 basaltic lava flow from Kilauea. This rather evolved lava erupted in the Kilauea rift zone contains plagioclase and clinopyroxene phenocrysts. The authors discuss their mineral data, by taking into account the possible differences in Ra and Ba partition coefficients as explained above (Fig. 11), and the presence of glass inclusions in the minerals. They conclude that the interval between phenocryst crystallization and eruption was at least 550 years, unless a high proportion (> 30%) of crystals are xenocrysts from an earlier batch of evolved magma. They consider that this age represents the average duration of storage of the 1955 evolved magma batch in the rift zone, and suggest that primitive lavas erupted by the main conduits at the summit crater may have much shorter storage times.

Data on mineral separates in older volcanic rocks. Recent studies comparing U-Th ages and eruption ages obtained by other methods have concerned mainly subduction related volcanoes, but also volcanoes from the East African Rift or intraplate volcanism in the Eifel area (Germany). These data are summarized below and some of the examples showing a large difference between crystallization ages and eruption ages are illustrated in Figure 12.

One of the first detailed evidence of the complexity of mineral data comes from the work of Pyle et al. (1988) on an andesite from Santorini. The minerals extracted from this rock fall on two isochrons: olivine, magnetite and plagioclase define a 92 ka isochron, while whole rock, pyroxene and zircon plot on a 79 ka isochron. These data are thought to reflect mixing of two crystal populations from two different magmas: a crystal poor, dacitic magma and cumulate minerals from an earlier basic magma (Fig. 12a). In a more recent study, Zellmer et al. (1999) obtained a whole rock-plagiocase-magnetite U-Th isochron, with an age of 85 ± 40 ka (2σ), indistinguishable from the 67 ± 9 ka ^{39}Ar-^{40}Ar eruption age for a dacite from Santorini.

Volpe et al. (1992) obtained U-Th isochron ages for andesites from Mt. Shasta (California) which were 27 to 28 ka older than eruption ages (≤ 4.5 ka).

Three U-Th isochrons are available from the Nevados de Payachata volcano (Central Volcanic Zone of Chile) (Bourdon et al. 2000): one rhyolite sample gives an age similar to the eruption age, while two other give older ages, in particular a dacite whose crystallization age is 100 ka older than the eruption age.

In the Taupo volcanic zone of New Zealand, the 26.5 ka Oruanui eruption was studied by Charlier and Zellmer (2000). Three fractions of zircons (sub 63 µm; 63-125 µm; 125-250 µm) were extracted from the rhyolitic pumice, which together with the whole rock respectively define three ages from 5.5 to 12.3 ka before eruption (Fig. 12b). Microscopic observation of the zircons showed that they are composed of a core surrounded by euhedral rims, and the preferred explanation of the authors is that zircons represent mixtures in variable proportions of old crystal cores crystallized 27 ka before eruption and crystal rims crystallized just before eruption.

Several volcanic centers from the Kenya Rift Valley in East Africa were studied by Black et al. (1997, 1998). Four U-Th isochrons were obtained from trachytes from Emuruangogolak and Paka volcanoes (Black et al. 1998): their crystallization ages range from 9.3 to 39.5 ka and are in full agreement with *in situ* ^{39}Ar-^{40}Ar laser dating of

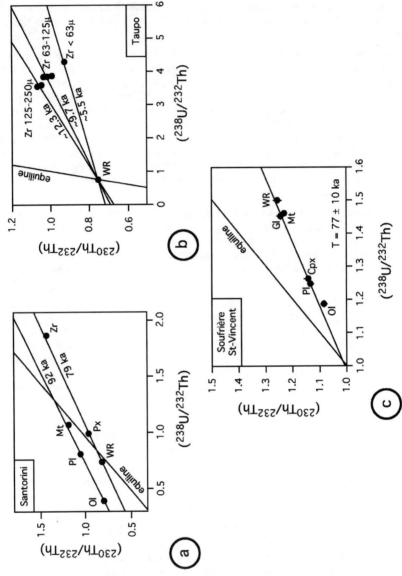

Figure 12. Several examples of complex mineral isochrons, or isochrons giving crystallization ages older than eruption ages (see text for a detailed discussion). (a) data from a Santorini andesite (Pyle et al. 1988) interpreted as evidence of incorporation in a dacitic magma of an older mineral cumulate (Ol, Pl, Mt isochron). (b) Data obtained on zircons (Zr) from a rhyolitic pumice in the Taupo volcanic zone (New Zealand) (Charlier and Zellmer 2000). Note that the three ages obtained are ages before the 26.5 ka eruption because the reported (^{230}Th/^{232}Th) ratios have been recalculated at the time of eruption. (c) Data from a Soufrière andesite (St Vincent, after Heath et al. 1998) interpreted as an old magma (minerals + melt).
(*continued on facing page*)

Figure 12 continued. (d)-(g) Mineral isochrons for rhyolites from the Olkaria volcanic center (Kenya). (d) and (f): alpha spectrometry results from Black et al. (1997). (e) and (g) TIMS results from Heumann and Davies (2002). All the rhyolites have eruption ages between 3.3 and 9.2 ka. Note that the same sample (570) analyzed in both studies gives rather different ages (f and g). Same abbreviations of mineral names as in Figure 10 + Qz: quartz; KF: alkali feldspar; Amph: amphibole; Bt: biotite.

sanidine crystals (11 to 38 ka), which suggests a short storage time of trachytic magma before eruption. Black et al. (1997) also analyzed peralkaline rhyolites from the Olkaria volcanic center in Kenya. Their eruption ages are between 3.3 and 9.2 ka BP. U-Th isochrons indicate crystallization ages between 9.8 and 50.5 ka: three samples give ages within a few millennia of the eruption ages (Fig. 12f), three samples reveal older ages (from 23 to 47 ka before eruption, Fig. 12d). In a more recent work, other peralkaline rhyolites from the same Olkaria volcanic center were analyzed by Heumann and Davies (2002). Both Rb-Sr and U-Th isochrons give crystallization ages of about 25 ka, which are ~16 ka older than the eruption age. However some of the major minerals have anomalously high U and/or Th contents, which are attributed to the presence of inclusions of U and/or Th rich accessory minerals (zircon in fayalite, and chevkinite, a Ti-REE rich accessory phase, in other major minerals, like biotite, amphibole, quartz...). The mineral isochrons (Fig. 12e) might thus represent mixing lines due to variable amounts of chevkinite inclusions in major minerals, and the age of 25 ± 10 ka (2σ) given by the slope would be the crystallization age of chevkinite. The glass and fayalite data give an older age of 47 ± 0.2 ka (2σ) which could reveal the crystallization age of zircon (Heumann and Davies 2002). It may be worth mentioning that internal isochrons were obtained on the same rhyolite sample (#570) by both Black et al. (1997) and Heumann and Davies (2002) (Fig. 12f,g), with significantly different ages of 15 ± 2 ka and 50 ± 25 ka (2σ) respectively. There are also large differences in Th/U ratios of the glasses. The discrepancy between these results might in part be due to analytical problems linked to the presence of accessory phases, which are very difficult to fully dissolve using conventional acid digestion.

The importance of accessory phases is also emphasized in the recent study by Heuman et al. (2002), who analyzed ^{230}Th-^{238}U disequilibria in Long Valley postcaldera rhyolites erupted between 100 and 150 ka ago. Glass analyses yield a Rb-Sr isochron with an age of 257 ± 39 ka interpreted by a feldspar fractionation event 150 ka before eruption. Among other results, one U-Th mineral isochron in a rhyolite from the Deer Mountain Dome defines an age of 236 ± 1 ka (2σ), with amphibole, biotite, sanidine and zircon plotting on the isochron, glass and whole rock lying below the isochron. These ages are in agreement with those previously obtained by Reid et al. (1997) through *in situ* dating of zircons (see next subsection). As all major minerals contain inclusions of accessory phases (zircon and probably allanite), the mineral isochron is interpreted by the authors as a mixing array defined by variable proportions of accessory phases in major minerals (cf. Heumann and Davies 2002): they conclude that zircon and allanite crystallized during two distinct and short (~1 ka) episodes at 250 ± 3 and 187 ± 9 ka (2σ) respectively (Fig. 13). To explain the long storage time of the rhyolitic magma, Heumann et al. (2002) consider that small batches of rhyolitic magma differentiated by a filter press mechanism could have remained isolated near the roof of the magma chamber until the eruption, their complete crystallization being prevented by the heat flux produced by the large volume of magma stored in the main magma chamber.

Two studies have been published on volcanic rocks from Eifel (Germany). Bourdon et al. (1994) analyzed the Laacher See phonolites erupted 12.5 ka ago. Three pumice samples provided mineral isochrons with a mean age (13 ± 3 ka, 2σ) identical to the eruption age. Another pumice sample suggested an older age (~30 ka) while mineral data from various cumulate nodules could be interpreted as evidence of crystallization of some minerals 10 to 20 ka before eruption. In the East Eifel volcanic field (Fornicher Kopf), Peate et al. (2001) tried to date a basanite lava flow but the data failed to give a meaningful mineral isochron, which is attributed by the authors to the presence of old xenocrysts together with an open system bebaviour of phlogopite that experienced recent U exchange.

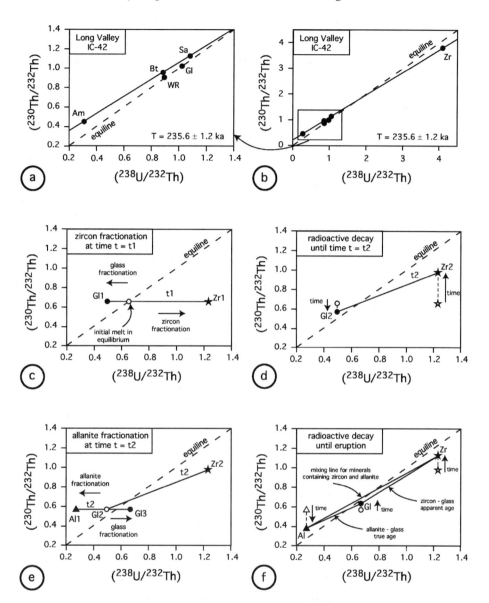

Figure 13. (a) and (b) Mineral U-Th isochron obtained by Heumann et al. (2002) on a post-caldera rhyolite from Long Valley (a is an enlarged part of b). (c)-(f) Model of crystallization given by Heumann et al. (2002) to explain the above data. Zircon (Zr) crystallizes in the melt at time t_1, then allanite (Al) crystallizes at t_2. (f) shows the present situation after radioactive decay. The major minerals plot on the zircon-allanite line because most of their U and Th contents are assumed to come from inclusions in variable proportions of these U and Th-rich phases. The only exact age is the age defined by the allanite-glass pair. Note that the glass does not plot on the mineral isochron and that, in this model, the mean age defined by the mineral isochron corresponds to the mean crystallization age of the accessory phases only, and does not give any indication on the crystallization of major minerals. Am: amphibole; Bt: biotite; Gl: glass; Sa: sanidine; WR: whole-rock.

***In situ* dating of U-Th rich accessory minerals.** Recent instrumental developments have allowed the first *in situ* ^{230}Th-^{238}U dating by ion microprobe of U rich accessory minerals like zircons. The first study using this method was published by Reid et al. (1997). They analyzed several zircons extracted from two postcaldera rhyolites belonging to the Long Valley magmatic system. Model ages are calculated from the slopes of the whole rock-zircon two-points isochrons, the U-Th characteristics of the whole rock being independently measured by thermal ionization mass spectrometry. In spite of a rather large dispersion of the model ages, most of the zircons from both rhyolites are within error of a weighted mean age around 230 ka, much older than the eruption ages of 0.6 and 115 ka (Fig. 14). If these ages represent crystallization ages of zircon in a same upper crustal rhyolitic magma chamber, they imply that the zircons remained more than 100 ka in this reservoir, which could have been maintained in a molten state by influx and differentiation of a large volume of basic magma.

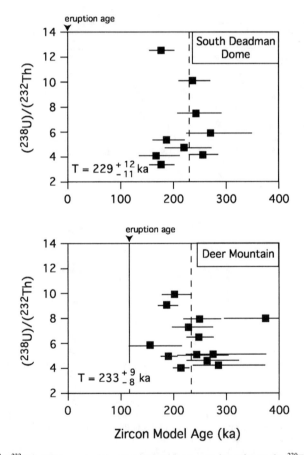

Figure 14. (^{238}U/^{232}Th) ratios vs. model ages obtained by *in situ* ion microprobe ^{230}Th-^{238}U dating of zircons extracted from two rhyolites associated with the Long Valley magmatic system (California, USA) (modified from Reid et al. 1997). The eruption ages of the two rhyolites are indicated on the figure. Only the results used for calculating a weighted mean age (dashed line) for each sample are shown here (with 1σ error bars). Note the absence of correlation between zircon ages and their highly variable (^{238}U/^{232}Th) ratios (compare to Fig. 12b).

Using a similar ion microprobe technique, Bacon et al. (2000) analyzed zircons extracted from a granodiorite block ejected during the climactic eruption of Mount Mazama which formed the Crater Lake caldera (~ 7700 years ago). Several zircon grains plot on an U-Th isochron giving an age of 112 ± 24 ka, interpreted as the crystallization age of the shallow granodiorite pluton, in agreement with a less precise *in situ* ^{238}U-^{206}Pb age on the same zircons.

Lowenstern et al. (2000) used the same approach to date several zircons found in granitoid xenoliths ejected by Medicine Lake volcano (California) during two eruptions, 1065 and 2000 years ago. The ages obtained at 29 and 90 ka are thought to represent previous intrusive episodes in this magmatic system.

One of the most studied examples of a large silicic magma chamber is the Long Valley caldera, formed after the emission of more than 750 km^3 of high silica rhyolite (Hildreth and Mahood 1986). Unfortunately, this eruption is too old (~760 ka) to be subjected to a ^{230}Th-^{238}U analysis. Several ^{39}Ar-^{40}Ar and Rb-Sr studies (e.g., Van den Bogaard and Schirnick 1995; Halliday et al. 1989; Christensen and De Paolo 1993; Christensen and Halliday 1996) have suggested that this rhyolitic magma has been stored in the reservoir for 0.5 to 1.7 Ma before eruption. However Reid and Coath (2000) recently dated some zircons by *in situ* U-Pb ion microprobe analysis. Taking into account the corrections for initial ^{230}Th-^{238}U disequilibrium in zircons, they conclude that zircons could not have crystallized more than 100 ka before eruption, a significantly shorter interval between crystallization and eruption than those indicated by previous studies.

Discussion. From the above review, it appears that interpretation of mineral data is rather complex. In several cases, Th isotope homogeneity has been demonstrated in zero age rocks, which allows the use of mineral isochrons to determine eruption ages (see section 4.1). In other cases (and sometimes in the very same volcanoes showing agreement between crystallization age and eruption age) crystallization ages are much older (10^4 to 10^5 a) than the eruption ages. The significance of these mineral isochrons is still a matter of debate: do they really represent a crystallization age? If so what do they date: a mean crystallization age of major minerals, the age of crystallization of some accessory phases? What is the meaning of this age for the timescale of magma storage, differentiation or residence time?

One possible cause of Th isotope heterogeneity between minerals could be that they crystallized in a heterogeneous magma formed by incomplete mixing of two (or more) magmas with different (^{230}Th/^{232}Th) ratios (see Gill et al. 1992): however, in that case it would be very fortuitous that the minerals define a good isochron. While some data might be explained in this way, this process seems rather unlikely in most cases.

Another often advocated explanation is the incorporation in the magma of old cumulate minerals (or xenocrysts crystallized from a magma batch of a different composition). In some cases, minerals in equilibrium with the magma and the groundmass could define an isochron, the cumulate minerals falling onto another steeper isochron (e.g., Pyle et al. 1988, Fig. 12a). In general, however, if cumulate crystals belong to the same mineral species as phenocrysts, mineral separates will be mixtures in variable proportions of xenocrysts and phenocrysts, and they will not define good isochrons. When it is possible to calculate the crystallization age of the cumulate minerals, this gives an indication on the lifetime of the magma chamber or magmatic system but does not allow us to infer the magma storage or residence time.

The suggestion recently made by Heumann et al. (2002) that U-Th crystallization ages in Long Valley rhyolites can reflect crystallization of accessory minerals like zircon or allanite and not of major minerals, needs to be seriously considered in other studies as well, because the behavior of these accessory phases (and their growth rates) will be different from those of the major minerals. Indeed, crystallization of accessory minerals like zircon depends on the trace-element (Zr) solubility in the melt, whereas crystallization of major minerals are governed by phase diagram relationships. The times spent in the magma chamber by major and accessory minerals might thus be very different. Moreover, the presence of such refractory phases, like zircons, as inclusions in major minerals emphasizes the necessity of a complete dissolution of mineral separates, which requires acid digestion in sealed pressure bombs or alkaline fusion procedures. Incomplete dissolution might result in selective U and/or Th leaching from the accessory phases, affecting the position of the data in the isochron diagram. This is especially true for silicic volcanic rocks, but might be of importance also for more basic rocks, where the magnetic separates sometimes contain zircon inclusions. To what extent some results might be affected by these problems is difficult to evaluate, but this analytical difficulty must be kept in mind.

In fact, it is probably easier to explain dispersed mineral and groundmass data, than some nearly perfect isochrons defined by all mineral and groundmass data and giving ages older than the eruption ages (e.g., Volpe et al. 1991, 1992; Black et al. 1997, 1998; Heath et al. 1998; Fig. 12c,d,e,g). This implies that not only the minerals are old but that the melt has the same old age when the magma is erupted. The isochron can only be preserved during a storage time of 10^4 to 10^5 years in the magma chamber if Th isotopes do not reequilibrate between melt and minerals. This condition is probably met for major minerals larger than 1 mm, in view of the very low Th diffusion coefficients determined for diopside ($\mathcal{D} = 4.10^{-21}$ m^2s^{-1} at 1200°C, Van Orman et al. 1998). The characteristic length scale of Th diffusion in diopside would only be about 0.1 mm in 100 ky. Th appears to diffuse even more slowly in zircon ($\mathcal{D} = 1.2.10^{-24}$ m^2s^{-1} at 1100°C, Lee et al. 1997, or even $\mathcal{D} \sim 10^{-26}$ m^2s^{-1} at 1200°C from an extrapolation of the data obtained by Cherniak et al. 1997).

The fact that the isochron is well defined suggests that all the minerals grew during a short episode (a few ka at most) and not during extended and successive episodes. The explanation proposed by several authors (Heath et al. 1998; Heumann et al. 2002) is that this crystal-rich melt might remain stored near the walls or the roof of the magma chamber at a near constant temperature between liquidus and solidus curves, because of the heat flux produced by the main part of the large magma chamber (and possible influxes of new magma). In some cases, especially for silicic rocks, it could even be envisaged that the erupted magmas are produced by partial melting, just before eruption, of an already consolidated and older border zone.

While the persistence of a crystal-rich melt in the magma chamber for 10^4 to several 10^5 years can rather easily be envisaged for large silicic magma chambers with long repose times between eruptions, it is very difficult to explain in other, smaller central volcanoes (e.g., Mt. St Helens, Soufrière, Vesuvius) which have had numerous eruptions between the inferred age of the old magma and the eruption age of the studied rock. How this old magma could have remained in the chamber without being drained by subsequent eruptions is one of the most puzzling questions raised by these old mineral-groundmass isochrons. A systematic study of mineral isochrons in successive eruptions could help to clarify this problem.

It should be noted that, even when good U-Th mineral isochrons indicate old crystallization ages, ^{226}Ra-^{230}Th disequilibria can be found in minerals, groundmass and whole rocks. It is the case for Mt. St Helens and Mt. Shasta (Volpe et al. 1991, 1992) and Vesuvius (Black et al. 1998) (see Fig. 10c,d): this requires either a selective Ra enrichment in the magma shortly before eruption accompanied by further crystallization and/or Ra exchange between mineral rims and melt, or some sort of mixing of the "old" magma with a newly injected one. In any case, combined ^{230}Th-^{238}U and ^{226}Ra-^{230}Th studies on recently erupted volcanic rocks suggest that the complex processes occurring in the magma chamber are far from being completely understood.

It is worth mentioning, as recently reviewed by Hawkesworth et al. (2000), that most data showing crystallization ages much older than eruption ages come from the more silicic rocks and often from alkaline or peralkaline provinces. Hawkesworth et al. (2000) remarked that these silica rich magmas have a higher viscosity and are thus more likely to keep old minerals in suspension near the walls or roof of the magma chamber. This is especially true for isolated accessory minerals like zircons which usually form very small crystals (in spite of its high density, zircon has a lower settling velocity compared to larger major phenocryst phases, as shown by Reid et al. 1997). Moreover, most of these volcanoes have a very episodic volcanic activity with sometimes very long repose times between eruptions, and their behavior is probably dominated by alternating episodes of closed and open-system evolution, which makes them more susceptible of revealing complex magmatic processes such as magma mixing, magma-cumulate mixing, aging of small batches of magma... When comparing the results provided by U-series studies on minerals to the other methods of evaluating the residence time of crystals in the magma (crystal size distribution or trace element and Sr isotope diffusion profiles in crystals), Hawkesworth et al. (2000) emphasize the fact that these latter suggest short crystal residence times (a few tens to hundreds of years). These estimates are in agreement with results from ^{226}Ra-^{230}Th disequilibria, or ^{230}Th-^{238}U disequilibria in rocks for which crystallization ages and eruption ages are indistinguishable.

Further studies would be necessary to assess the significance of old (10^4 to 10^5 a) magma (crystals and groundmass) ages. For example, it could be particularly interesting to analyze minerals in both U-Th and Ra-Th systems in the case of a complex eruption thought to have tapped a zoned magma chamber. Another useful approach would be to study, on a single volcano whose eruptive history is well known, both the evolution of ^{230}Th-^{238}U and/or ^{226}Ra-^{230}Th disequilibria in whole rocks, and crystallization ages deduced from mineral data. As shown in Figure 15, these two approaches are necessarily linked.

3.6. Timescales of magma degassing

Radionuclides of interest for the purpose of studying magma degassing are ^{226}Ra and its main daughters, namely ^{222}Rn, ^{210}Pb, ^{210}Bi, and ^{210}Po. All the isotopes located between ^{222}Rn and ^{210}Pb (i.e., ^{218}Po to ^{214}Po) have extremely short half-lives, ranging from about 0.16 ms up to about half an hour. They will not be considered any further in this discussion since, in a closed system, they return to radioactive equilibrium with their parent ^{222}Rn within a few hours after fractionation. Thus, it can be assumed that ^{210}Pb is directly produced by decay of ^{222}Rn. ^{226}Ra daughters have half-lives that range from a few days (T ^{222}Rn = 3.82 days; T ^{210}Bi = 5.01 days; T ^{210}Po = 138.4 days) up to 22.3 years for ^{210}Pb. Disequilibria among these nuclides are thus preserved for time periods of the order of a few weeks up to a century. Hence, they can be used to study recent fractionation events, most likely to be associated with shallow magmatic processes. Furthermore, these isotopes have contrasted gas-liquid partition coefficients. While radium forms no volatile compound at magmatic temperatures and mostly remains in the molten lava (Lambert et al. 1985-86), the noble gas radon is continuously and thoroughly lost from the degassing magma as long as major gas species, acting as carriers of trace gaseous species, are efficiently released (Gill et al. 1985; Gauthier et al. 1999). Lead, bismuth, and polonium are known to form chemical compounds, mostly halogenides and sulfides, that are more or less volatile at magmatic temperatures (e.g., Pennisi et al. 1988; Symonds et al. 1994). It is generally accepted that in hot basaltic systems, lead is slightly

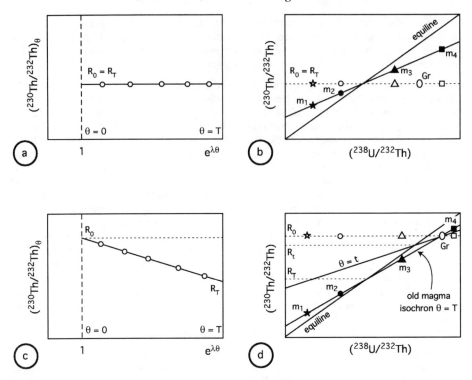

Figure 15. Diagrams showing two contrasted models of magma evolution and their consequences on the location of minerals (m_1, m_2...) and groundmass (Gr) in the U-Th isochron diagram. In (a), the (^{230}Th/^{232}Th) ratio of the magma (as deduced from the (^{230}Th/^{232}Th)$_\theta$ ratio of the lavas of eruption age θ) remains constant (e.g., because of a short transfer time). If the present-day melt incorporates an old mineral cumulate, it will not plot on the mineral isochron (Gr in b). Newly formed minerals (open symbols) would have the same (^{230}Th/^{232}Th) ratio as the groundmass. In (c), the magma is stored in a closed-system reservoir, and its (^{230}Th/^{232}Th) ratio increases through time, if it is located to the right of the equiline, with an (^{238}U/^{230}Th) ratio >1 (see Fig. 3, showing the reverse case). In that case, the magma itself (minerals and melt) might get old, and, in (d), the groundmass will plot on the isochron along with the minerals, if they have not reequilibrated with the melt. If the U/Th ratio of the melt is not significantly affected by fractional crystallization, all the groundmass (melt) data will plot at the same point in the isochron diagram, whatever their eruption ages. If successive but short crystallization episodes occurred in this reservoir and if these magma batches were somehow preserved in the reservoir (zoned magma chamber?), then a variety of mineral isochrons may be obtained but the range in (^{230}Th/^{232}Th) initial ratios given by the intercept of these isochrons with the equiline should increase during the evolution of the reservoir. That means that the oldest eruptions will have (^{230}Th/^{232}Th) initial ratios close to R_T, the present-day eruptions could show a range of (^{230}Th/^{232}Th) initial ratios between R_T and R_0 (such as R_t at θ = t; R_0 is the ratio of newly formed minerals shown as open symbols). Interpretation of mineral isochrons should benefit from coupled mineral and whole-rock studies over a long period of magmatic activity.

volatile (with a gas-melt partition coefficient D_{Pb} < 1) while bismuth is moderately volatile (1 < D_{Bi} < 10) (e.g., Gauthier et al. 2000). On the other hand, polonium usually is thoroughly degassed (D_{Po} >> 1000) from freshly erupted basalts (Gill et al. 1985). Because of these contrasted partition coefficients, ^{226}Ra daughters are strongly fractionated during magmatic degassing processes. As a consequence, ^{210}Po-^{210}Bi-^{210}Pb radioactive disequilibria are often observed in volcanic gases, with (^{210}Po/^{210}Pb) > (^{210}Bi/^{210}Pb) > 1, (Polian and Lambert 1979; Lambert et al. 1985-86; Le Cloarec et al.

1986, 1992, 1994; Gauthier et al. 2000; Le Cloarec and Pennisi 2001). In the same way, both ^{210}Po-^{210}Pb and ^{210}Pb-^{226}Ra disequilibria are common in freshly erupted lavas, with both (^{210}Po/^{210}Pb) and (^{210}Pb/^{226}Ra) activity ratios usually lower than one (Krishnaswami et al. 1984; Le Cloarec et al. 1984; Gill et al. 1985; Rubin and MacDougall 1989; Rubin et al. 1994; Sigmarsson 1996; Gauthier and Condomines 1999 and references therein).

Short-lived disequilibria between ^{226}Ra daughters in lavas and in volcanic gases at open conduit volcanoes can be accounted for by two simple dynamic degassing models (Gauthier and Condomines 1999; Gauthier et al. 2000). These models have the same conceptual framework as the pioneering model developed by Lambert et al. (1985-86) and rely on some basic assumptions. Deep undegassed magma is assumed to be in radioactive equilibrium for ^{226}Ra daughters which all have short half-lives compared to the timescales of magma transfer. Since degassing activity is persistent on many active open-conduit volcanoes, it is reasonable to assume that deep undegassed magma is continuously brought by convection into the shallow degassing reservoir, where efficient radionuclide degassing takes place through water exsolution (Gauthier et al. 1999; 2000). When this reservoir contains a constant mass of magma M (physical steady-state), the influx ϕ_0 of deep magma entering the reservoir is exactly balanced by a flux of gas ϕ_G and a flux of degassed lava ϕ_L leaving the reservoir (Fig. 16). The output of degassed lava is sustained either by lava eruption at the surface (ϕ_E), or by recycling at depth (ϕ_R), or else by sill-like intrusion (ϕ_I). For low-viscosity magmas such as basalts or basaltic andesites, it may be assumed that the magma reservoir is quickly mixed and homogenized, so that these three components (erupted, recycled and intruded) have the same chemical composition, which is also the chemical composition of the magma within the reservoir. Accordingly, the output of degassed lava may be considered as a single component (ϕ_L), no matter the way the degassed magma leaves the reservoir. Thus, the variation through time of the activity of any radionuclide I_k in the degassing magma is merely given by a mass balance equation that includes a term of radioactive ingrowth from its parent I_{k-1}, a term of radioactive decay, and three terms of mass transfer:

$$\frac{d(I_k)_L}{dt} = \lambda_k (I_{k-1}) - \lambda_k (I_k) + \frac{\phi_0}{M}(I_k)_0 - (1-\alpha)\frac{\phi_0}{M}(I_k)_L - \alpha\frac{\phi_0}{M}(I_k)_G \quad (4)$$

where λ_n is the decay constant of nuclide I_k; α is defined as the fraction of volatiles initially dissolved in the deep magma and ultimately released in the gas phase, such as fluxes can be written as: $\phi_G = \alpha\,\phi_0$ and $\phi_L = (1-\alpha)\,\phi_0$; parentheses denote activities in the deep undegassed magma (0), the degassed lava (L), and the gas phase (G). Also note that ϕ_0/M corresponds to the renewal rate of the degassing magma chamber (per unit of time), which is equal to $1/\tau$, where τ is the magma residence time in the degassing reservoir.

Model of radon degassing. Because the half-life of ^{226}Ra is considerably longer than the timescale of magma degassing (Lambert et al. 1985-86; Gauthier et al. 2000), it is assumed that its activity remains constant during degassing. By applying successively Equation (4) to ^{222}Rn and ^{210}Pb, it is possible to link (^{210}Pb/^{226}Ra)$_L$ activity ratios in the degassed lava to the degassing of radon (Gauthier and Condomines 1999). It can be then demonstrated that (^{210}Pb/^{226}Ra)$_L$ activity ratios in the degassed lava depend on a) two unknown parameters: the renewal rate of the degassing reservoir (ϕ_0/M), and the duration of the continuous degassing process (t); and b) two parameters that can be determined from independent studies: the fractional loss of radon (f) which can be estimated by monitoring ^{222}Rn ingrowth in freshly erupted lavas (e.g., Gill et al. 1985), and the initial volatile content (α) which can be determined through the study of volatile contents in melt inclusions (e.g., Johnson et al. 1994).

Time evolution of (^{210}Pb/^{226}Ra)$_L$ activity ratios in the degassed lava can be drawn for

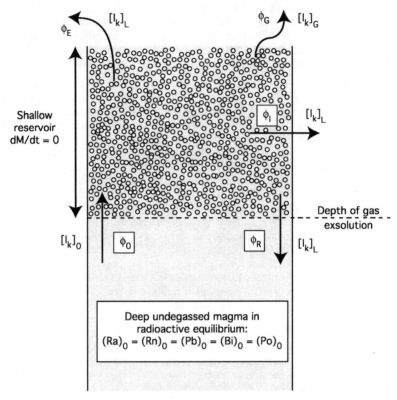

Figure 16. Schematic representation of a degassing magma reservoir in a physical steady-state (mass M of magma constant). ϕ and $[I_k]$ denote fluxes and radionuclide I_k concentrations, respectively. Indices 0, L, G, E, I, R, refer to deep undegassed magma (in radioactive equilibrium), degassed lava, gas phase, and erupted, intruded, or recycled degassed magma, respectively (after Gauthier and Condomines 1999).

different renewal rates ϕ_0/M, using realistic values for both α and f (Fig. 17). Curves are drawn for a value of α fixed at 0.03, but it is worth noting that this parameter has only little influence on the evolution of $(^{210}Pb/^{226}Ra)_L$ activity ratios. Also, curves in Figure 17 are drawn for a complete radon degassing (f = 1). Incomplete degassing would simply shift these curves upwards and, hence, would yield activity ratios closer to unity. When the degassing reservoir is not replenished with deep undegassed magma ($\phi_0/M = 0$), ^{210}Pb (whose radioactive ingrowth is not sustained any more in the degassed magma that is ^{222}Rn-depleted) becomes rapidly depleted with respect to ^{226}Ra. For instance, after 9 years of continuous radon degassing, significant ^{210}Pb-^{226}Ra disequilibria may be observed in lavas, with activity ratios as low as 0.75. This effect of radon degassing on Ra-Pb fractionation still exists when the magma chamber is continuously replenished with deep undegassed magma in radioactive equilibrium, but it is lowered for increasing values of ϕ_0/M. Finally, when ϕ_0/M is considerably higher than λ_{Pb} (i.e., when ϕ_0/M is higher than 3 yr^{-1}, or else when the magma residence time τ is shorter than 4 months), ^{226}Ra-^{210}Pb equilibrium is maintained in erupted lavas in spite of sustained radon degassing. Accordingly, provided that α and f can be determined from independent studies, time-series analyses of ^{226}Ra-^{210}Pb disequilibria in freshly erupted products from an active volcano allow determination of the renewal rate of the shallow degassing reservoir.

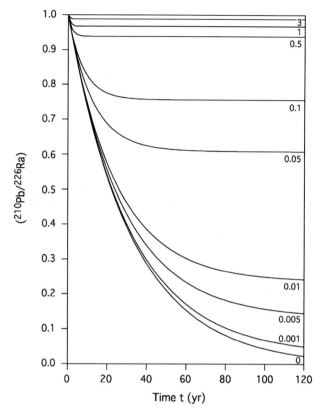

Figure 17. Time-evolution of $(^{210}Pb/^{226}Ra)_L$ activity ratios in the degassed magma for different values of the magma chamber renewal rate ϕ_0/M (figures on curves). Curves are drawn from the equation (Gauthier and Condomines 1999):

$$\left(\frac{^{210}Pb}{^{226}Ra}\right)_L = 1 - \frac{\lambda_{Pb} f}{(1-\alpha)\frac{\phi_0}{M} + \lambda_{Pb}} \times \left[1 - \exp\left(-\left((1-\alpha)\frac{\phi_0}{M} + \lambda_{Pb}\right)t\right)\right]$$

with $\alpha = 0.03$ and $f = 1$ (complete radon degassing) (after Gauthier and Condomines 1999).

This model has been applied to the case of Stromboli (Italy) and Merapi (Indonesia) volcanoes (Gauthier and Condomines 1999). At Stromboli, ^{226}Ra-^{210}Pb disequilibria in unaltered lava samples from the XXth century have been monitored by gamma-ray spectrometry, yielding a constant steady-state $(^{210}Pb/^{226}Ra)_L^0$ activity ratio (recalculated at the time of the eruption) of 0.975 ± 0.015. For a chemical steady-state, $(^{210}Pb/^{226}Ra)_L$ activity ratios do not depend any more on the duration of the degassing process (at steady-state, Eqn. 4 writes $d(I_k)_L/dt = 0$). The steady-state $(^{210}Pb/^{226}Ra)_L$ activity ratio at Stromboli allows direct determination of the renewal rate ϕ_0/M in the range 0.7-3.2 yr^{-1}, corresponding to a magma residence time τ in the shallow degassing reservoir between 110 and 520 days. Measurements of $(^{210}Pb/^{226}Ra)_L^0$ activity ratios in lava domes erupted at Merapi between 1984 and 1995 reveal some significant Ra-Pb disequilibria in erupted products. $(^{210}Pb/^{226}Ra)_L^0$ activity ratios decrease from 1 in 1984 down to 0.75 in 1992, before they increase back to the equilibrium value of 1 which is reached in 1995 (Fig. 18). This evolution can be explained by the succession of a 9-year period of complete radon degassing (f = 1) from a poorly renewed magma batch ($\phi_0/M <$ 0.045 yr^{-1}, that is $\tau > 22$ yr), followed by the injection of deep undegassed magma in radioactive equilibrium which progressively mixes with the degassed magma. By applying a model of progressive injection and instantaneous mixing in a magma reservoir of constant mass M (Condomines et al. 1982b), the injection rate ϕ_0/M between 1992 and 1995 can be estimated at about 0.5 yr^{-1}, yielding a residence time τ of 2 yr. It is worth noting that the sudden reinjection of magma in 1992 coincides with the beginning of a new eruptive cycle characterized by both more numerous dome collapse events and higher lava production rates at the surface (Fig. 18).

Model of ^{210}Pb, ^{210}Bi, and ^{210}Po degassing. For a purpose of clarity, it is considered here that the degassing reservoir has reached a chemical steady-state (i.e., radionuclide activities in the degassing reservoir are constant, that is $d(I_k)_L/dt = 0$ in Eqn. 4). This assumption usually is valid for very active basaltic systems like Stromboli, where erupted products display an almost constant chemical composition as shown above, and where the degassing reservoir is quickly and continuously replenished with deep undegassed magma. By applying Equation (4) successively to ^{210}Pb, ^{210}Bi, and ^{210}Po, $(^{210}$Bi$/^{210}$Pb$)_G$ and $(^{210}$Po$/^{210}$Pb$)_G$ activity ratios in the gas phase at the time of gas exsolution can be related to both the magma dynamics in the degassing reservoir (i.e., the renewal rate ϕ_0/M, or else the magma residence time τ) and the respective volatilities of the three nuclides, which are

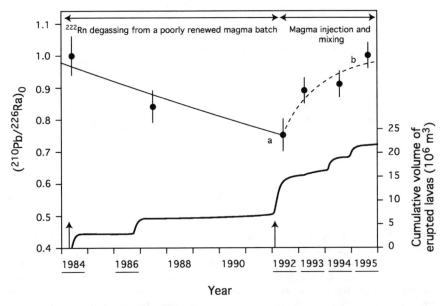

Figure 18. Time-evolution of $(^{210}$Pb$/^{226}$Ra$)_L^0$ activity ratios in Merapi lavas erupted between 1984 and 1995. Curve (a) is drawn from the equation given in the caption of Figure 17 with $\alpha = 0.03$, f = 1, and $\phi_0/M \leq 0.045$ yr^{-1}. Curve (b) is drawn from an equation of progressive magma injection and mixing (modified from Condomines et al. 1982), such as: $(R - R_0) = (R_i - R_0) \times \exp(-(\lambda_{Pb} + \phi_0/M) t)$, where R represents $(^{210}$Pb$/^{226}$Ra$)_L$ activity ratios, R_i is the $(^{210}$Pb$/^{226}$Ra$)_L$ activity ratio of the magma initially present in the reservoir ($R_i = 0.75$), and R_0 is the $(^{210}$Pb$/^{226}$Ra$)_L$ activity ratio of the magma injected into the reservoir ($R_0 = 1$). Vertical arrows indicate the beginning of a new eruptive cycle. Years during which pyroclastic eruptions have occurred are underlined. The thick curve represents the cumulative volume of erupted lavas between 1984 and 1995 (modified from Gauthier and Condomines 1999).

expressed by using a gas-liquid partition coefficient, defined as: $D_{Ik} = (I_k)_G/(I_k)_L$ (Gauthier et al. 2000). It is worth noting that if a significant time θ elapses between the degassing of the radionuclides and their emission into the atmosphere, the two shortest-lived isotopes (^{210}Bi and, to a lesser extent, ^{210}Po) may undergo significant radioactive decay within gas bubbles en route to the surface. Activity ratios in the gas phase, at the time θ after exsolution, are then obtained using the basic radioactive decay equations.

Provided both the initial pre-eruptive volatile content α and the gas-liquid partition coefficients D's can be determined (from melt inclusion studies and from concentration measurements in both lavas and gases, respectively; see for instance Gauthier et al. 2000 and references therein), activity ratios in the gas phase only depend on two unknown parameters: the magma residence time τ (inversely proportional to the renewal rate of the magma chamber ϕ_0/M), and the escape time of gases θ. The variations of both (^{210}Bi/^{210}Pb)$_G$ and (^{210}Po/^{210}Pb)$_G$ activity ratios in the gas phase as a function of the magma residence time inside the degassing reservoir are shown in Figure 19. Both activity ratios increase for increasing values of the residence time τ, which is explained by the contribution to the gas phase of ingrowth atoms of ^{210}Bi and ^{210}Po that are produced by radioactive decay of ^{210}Pb in the degassed magma sojourning in the degassing reservoir. This effect is more pronounced for ^{210}Po due to its higher volatility.

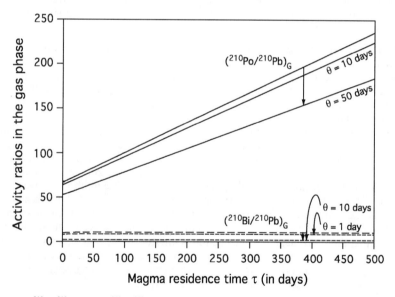

Figure 19. (^{210}Po/^{210}Pb)$_G$ and (^{210}Bi/^{210}Pb)$_G$ activity ratios in the gas phase as a function of the magma residence time τ in the degassing reservoir. Curves are drawn from the equations:

$$\left(\frac{^{210}Bi}{^{210}Pb}\right)_G = \frac{\frac{\lambda_{Bi} \times \tau}{D_{Pb}} + \alpha + \frac{1-\alpha}{D_{Pb}}}{\frac{\lambda_{Bi} \times \tau}{D_{Bi}} + \alpha + \frac{1-\alpha}{D_{Bi}}} \quad \text{and} \quad \left(\frac{^{210}Po}{^{210}Pb}\right)_G = \frac{\lambda_{Po} \times \tau}{D_{Bi}} \times \left(\frac{^{210}Bi}{^{210}Pb}\right)_G + \alpha + \frac{1-\alpha}{D_{Pb}} }{\frac{\lambda_{Po} \times \tau}{D_{Po}} + \alpha + \frac{1-\alpha}{D_{Po}}}$$

written at the time θ (see Gauthier et al. 2000), with α = 0.03, D_{Pb} = 0.5, D_{Bi} = 5, and D_{Po} >> 1000. Solid and dashed curves refer to (^{210}Po/^{210}Pb)$_G$ and (^{210}Bi/^{210}Pb)$_G$ activity ratios at θ = 0 (thick curves) and θ > 0 (thin curves), respectively. Vertical arrows indicate radioactive decay within gas bubbles for different values of the escape time of gases θ (figures on curves) (after Gauthier et al. 2000).

On the other hand, both activity ratios decrease for increasing values of the escape time θ, (^{210}Bi/^{210}Pb)$_G$ ratios being more affected than (^{210}Po/^{210}Pb)$_G$ ratios because of the very short half-life of ^{210}Bi (T = 5.01 d).

This dynamic degassing model has been applied for the first time to the case of Stromboli volcano in Italy (Gauthier et al. 2000). Gross α- and β-countings of ^{210}Po, ^{210}Bi, and ^{210}Pb in gas emissions sampled over 12 years yield (^{210}Po/^{210}Pb)$_G$ activity ratios in the range 70-210, and (^{210}Bi/^{210}Pb)$_G$ activity ratios between 10 and 32. At Stromboli the pre-eruptive volatile content α is estimated at about 3% and Pb, Bi, Po gas-liquid partition coefficients are in the range 0.3-0.5, 5-10, and >> 1000, respectively. Using these parameters, modeling shows that the escape of gases at Stromboli is always very fast, and probably never exceeds a few hours, no matter how intense the volcanic activity is, suggesting that degassing takes place at very shallow depth beneath the summit craters (a few hundred meters at most). This is in good agreement with previous determinations based on acoustic and seismic studies (e.g., Vergniolle et al. 1996; Chouet et al. 1999). Magma residence times are very short, ranging from 10 to 20 days up to about 200 days. They correspond to very fast renewal rates ϕ_0/M, in the range 1.7-36.5 yr^{-1}, which produce the sustained and permanent degassing activity observed at Stromboli. Furthermore, magma residence times seem to be in qualitative agreement with the intensity of volcanic activity observed at the surface, the shortest residence times corresponding to episodes of lava fountaining and sustained strombolian explosions, while the highest ones are found during periods of quiescent degassing and before effusive activity (Gauthier et al. 2000).

In summary, degassing of a magma batch stored at shallow depth with little, if any, turnover may last for tens of years, and probably even longer. On the other hand, the extraction of gas bubbles and gas emission into the atmosphere is a very short process occurring over a few hours at most, when degassing takes place at a depth of a few hundred meters. Magma residence times in shallow degassing reservoirs range from tens of days (Stromboli) to tens of years (Merapi). Since the mass of the shallow degassing reservoir is given by the simple relation M = $\phi \tau$, these estimates of magma residence times may also be used to calculate the mass and thus volume of degassing magma. Calculated volumes range from 10^6 m^3 (Stromboli) to 10^7 m^3 (Merapi). It is important to emphasize that the estimates of magma residence times derived from the two independent and complementary degassing models (one for ^{226}Ra-^{210}Pb disequilibria in lavas and the other for ^{210}Pb-^{210}Bi-^{210}Po disequilibria in gases) are in very good agreement. This strengthens the validity of this approach and demonstrates the potential of short-lived nuclides to characterize the dynamics of shallow degassing reservoirs. It should also be mentioned that the short magma residence times inferred above only apply to these small and shallow degassing "reservoirs" (or simply the upper part of the plumbing systems) fed by undegassed magma rising from a deeper magma chamber, characterized by longer magma residence times (see sections 3.3 to 3.5; Fig. 5).

4. DATING YOUNG VOLCANIC ROCKS

4.1. Mineral isochrons

The principle for a successful dating of eruption ages from mineral isochrons have been recalled in section 3.5. From the previous discussion, it is clear that U-Th mineral isochrons sometimes give crystallization ages in excess of eruption ages. If the difference between these ages is less than 10^4 a, this will introduce a maximum 10% error on a 100 ka old volcanic rock, which is still usually within analytical uncertainties on the age. Dating younger rocks will of course increase the relative error. The discrepancy will also increase if the interval between crystallization and eruption is larger than 10^4 a. However in many examples, U-Th mineral isochrons give ages in agreement with eruption ages. Some examples have already been described in section 3.5. Many successful dates were published during the 1965-1990 period (see the review by Gill et al. 1992). Most of these studies were applied to basaltic volcanoes like Stromboli (Eolian islands), Etna, Chaîne des Puys in the French Massif Central, Piton de la Fournaise in Réunion Island (Condomines et Allègre, 1980; Condomines et al. 1982a, 1982b, 1988), but also to some silicic volcanics from Japan (Fukuoka 1974). Other recent examples include an U-Th isochron obtained by Peate et al. (1996) on a basalt from the Albuquerque volcanic field (New Mexico) recording a geomagnetic excursion, which gives an age of 156 ± 29 ka in agreement with K-Ar ages, three isochrons reported by Bourdon et al. (1994) on the Laacher See (Eifel, Germany) phonolites (see section 3.5), and the isochron obtained on the Puy de Dôme trachyte in the Chaîne des Puys (Condomines 1995) with an age of 12.1 ± 1 ka, identical to ^{14}C and thermoluminescence ages. Figure 20 illustrates four examples of U-Th successful datings.

The main difficulty lies in the selection of samples that will eventually give a crystallization age close to the eruption age. However, careful mineralogical and petrological studies, including electron microprobe analyses to check for chemical equilibrium or disequilibrium between phenocrysts and groundmass, could help to eliminate samples containing xenocrysts or showing evidence of a complex magmatic history. Another approach consists in the selection of volcanic rocks nearly aphyric, or at least poor in phenocrysts. If the microlites display U-Th fractionations large enough to provide an isochron, then the age will certainly give the eruption age.

It should also be mentioned that most successful datings have been obtained either on basaltic or only slightly differentiated basic lavas, or on silicic lavas of small volume eruptions. Basaltic volcanoes in permanent activity probably have open-system magma chambers with rather short residence times of the magmas (a few 10^3 years) as suggested by ^{226}Ra-^{230}Th disequilibria in Etna (see section 3.4). They might represent favorable examples to apply U-Th mineral isochron dating. Other basaltic and nearly primitive lavas from simple monogenic volcanoes or fissure eruptions could also be dated, because such magmas are unlikely to have been stored for a long duration in the crust. For silicic rocks, successful dating can certainly be achieved for small volume eruptions, for which volcanological, petrological and geochemical data suggest that they are derived from the differentiation of a small and shallow magma chamber with a short storage time. Conversely, determining the eruption ages of rocks emitted from large volume silicic magma chambers (e.g., Long Valley) is certainly a very risky task. In spite of the above criteria taking into account the volcanological context, it is difficult to ascertain that the data will not be disturbed by such problems as magma-cumulate mixing. A good, though not absolute, test is the analysis of a zero age (or well dated) volcanic rock from the volcano to be studied, and whenever possible, it is always profitable to compare U-series results with the ages obtained from independent dating methods.

The use of $(^{226}Ra)/Ba$-$(^{230}Th)/Ba$ mineral isochrons to date eruption ages of very young volcanic rocks in the range 0 to 10^4 a is necessarily limited because the available data described in section 3.5 show that crystallization ages might be a few 10^2 to 10^3 a older than eruption. However, there is one example where ^{226}Ra-^{230}Th data on minerals have been used to infer eruption ages, although the method does not involve true mineral isochrons. Voltaggio et al. (1995) measured ^{226}Ra-^{230}Th disequilibria in four shoshonitic lavas from Vulcano (Eolian Islands) using a leaching method. For a given rock, they analyzed by gamma spectrometry the residues obtained from successive leaching steps of the rock powder, and reported their results in a $(^{226}Ra/^{228}Ra)$-$K/(^{228}Ra)$ diagram (Fig. 21).

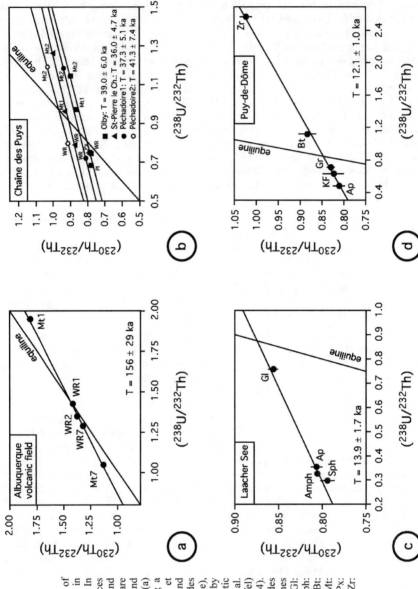

Figure 20. Several examples of U-Th mineral isochrons used in dating young volcanic rocks. In these cases, the differences between crystallization ages and the known eruption ages are small (a few 10^2 to 10^3 a) and within age uncertainties. (a) Albuquerque basalt, recording a geomagnetic excursion (Peate et al. 1996). (b) Basalts and hawaiites from the Chaîne des Puys (Massif Central, France), including the 39 ± 6 ka old Olby flow with a reversed magnetic direction (Condomines et al. 1982a). (c) Laacher See (Eifel) phonolite (Bourdon et al. 1994). d) Puy de Dôme (Chaîne des Puys) trachyte (Condomines 1997). WR: whole-rock; Gl: glass; Gr: groundmass; Amph: amphibole; Ap: apatite; Bt: biotite; KF: alkali feldspar; Mt: magnetite; Pl: plagioclase; Px: pyroxene; Sph: sphene; Zr: zircon.

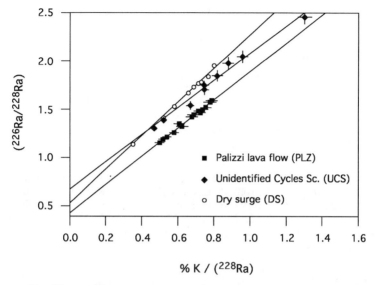

Figure 21. (^{226}Ra/^{228}Ra)-K/(^{228}Ra) diagram proposed by Voltaggio et al. (1995). In this diagram, equivalent to a (^{226}Ra/^{230}Th)-K/(^{230}Th) diagram, are reported the data resulting from successive leachings of K-rich volcanic rocks from Vulcano (Eolian Islands). For each rock, the data define a straight line, whose intercept on the y-axis gives the (^{226}Ra/^{230}Th) ratio of the Th-enriched accessory phase and thus the age of the rock (UCS: 2.9 ± 0.4 ka; DS: 2.1 ± 0.3 ka; PLZ: 1.5 ± 0.2 ka). K/(^{228}Ra) ratios are expressed in weight % K/dpm.g^{-1} (results from Voltaggio et al. 1995; see text for a detailed explanation).

Instead of Ba, K is taken as an analog for Ra. This diagram is equivalent to a (^{226}Ra/^{230}Th)-K/(^{230}Th) diagram, because (^{228}Ra)=(^{232}Th) and the (^{230}Th/^{232}Th) ratio is constant in minerals of these very recent rocks. The analyzed residues and the whole rock plot on a straight line in such a diagram. The interpretation of the authors is that this straight line represents a mixing line between major minerals and groundmass, and a Th-enriched accessory phase contained in the groundmass, which is assumed to be perrierite, a REE-Ti bearing silicate found in several volcanic rocks from Central Italy. Successive leachings progressively dissolve this accessory phase and produce the observed linear array. The diagram is similar in its principle to the well known (^{230}Th/^{234}U) vs. (^{232}Th/^{234}U) Osmond-type mixing diagram used for dating impure carbonates. In this latter, the y-intercept of the straight line defines the (^{230}Th/^{234}U) ratio of the pure carbonate component ([Th]~0). Similarly, the (^{226}Ra/^{230}Th) ratio of the Th-enriched accessory phase is deduced from the y-intercept, and the age of this phase is then calculated, assuming a negligible initial ^{226}Ra content, from the simple equation: (^{226}Ra/^{230}Th) = $1 - e^{-\lambda t}$. Voltaggio et al. (1995) obtained ages from 1.5 to 2.9 ka, which seem in good agreement with K-Ar estimates, archaeological and historical data.

4.2. Dating of whole-rocks

Dating of whole rocks by any U-series method is possible only if the initial $(N_2/N_1)_0$ ratio is known (see Eqn. 1). In several cases, this condition can be met as will be shown in the examples below.

^{230}Th-^{238}U and ^{231}Pa-^{235}U dating of MORB samples. Several recent attempts to date eruption ages of recent mid-ocean ridge basalts through ^{230}Th-^{238}U and ^{231}Pa-^{235}U

disequilibria measured in whole rocks have proved successful. This is a very important advance because of the difficulty to date these rocks by other methods (the low K content does not allow K-Ar dating of such young rocks, and the low U-Th contents make U-series analysis of minerals very difficult).

Goldstein et al. (1992), in their study of MORB from the Juan de Fuca and Gorda ridges in the Pacific, assumed that the $(^{230}Th/^{232}Th)_0$ initial ratio (at the time of eruption) could have remained constant for several 10^5 a, for a given ridge segment. This ratio is estimated from the present day ratio of very young axial MORB, and the ages of older, off-axis MORB can then be calculated. The authors obtained ages from a few ka to 220 (±37) ka, which were in good agreement with the spreading rates and MORB geographical locations relative to the ridge.

Goldstein et al. (1993, 1994) applied the same method as well as the ^{231}Pa-^{235}U method to date young MORB from the East Pacific Rise (EPR) at 9-10°N and the Juan de Fuca-Gorda ridges. Again, very young axial MORB were used to infer the initial $(^{231}Pa/^{235}U)_0$ ratio. The results show a remarkable agreement between the ages determined by these two independent methods in the range 0 to 130 ka (Fig. 22), suggesting that the assumption of constant initial $(^{230}Th/^{232}Th)_0$ and $(^{231}Pa/^{235}U)_0$ ratios is valid.

A similar approach led Sturm et al. (2000) to propose some ^{230}Th-^{238}U and ^{231}Pa-^{235}U ages for the Mark area, South of the Kane Fracture Zone of the Mid-Atlantic Ridge. ^{226}Ra-^{230}Th disequilibria were also measured, and helped in the selection of the youngest samples to infer the $(^{230}Th/^{232}Th)_0$ and $(^{231}Pa/^{235}U)_0$ initial ratios.

^{230}Th-^{238}U dating of samples from subaerial volcanoes. When several volcanic rocks covering a significant period of the eruptive activity of a volcano can be dated (either by mineral isochrons or by other dating methods), the evolution through time of the $(^{230}Th/^{232}Th)_0$ or $(^{230}Th/^{238}U)_0$ initial ratios will be revealed (see section 3.4). If these ratios remain nearly constant, then they may be used to calculate the ages of other lavas

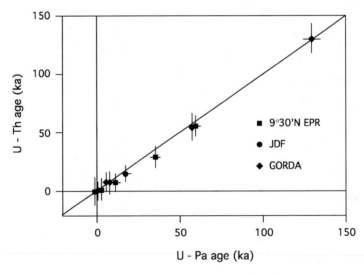

Figure 22. Comparison of the ages obtained by the ^{238}U-^{230}Th and ^{235}U-^{231}Pa dating methods on MORB samples from the East Pacific, Juan de Fuca and Gorda ridges. [Reprinted from Goldstein et al. 1993, with permission from Elsevier Science].

from this volcano, from their measured (^{230}Th/^{232}Th) ratios. One such example can be illustrated in the case of Piton de la Fournaise (Réunion Island) for which (^{230}Th/^{232}Th)$_0$ or (^{230}Th/^{238}U)$_0$ ratios display very limited variations (Condomines et al. 1988). Figure 23 shows how an approximate age can be obtained from the measured (^{230}Th/^{232}Th) ratio in a lava of unknown age from this volcano.

^{226}Ra-^{230}Th (and (^{226}Ra)/Ba) dating of MORB. The (^{226}Ra)/Ba method for dating very young MORB (< 8 ka) was originally proposed by Rubin and Macdougall (1990). Its principle is based on the assumption that Ra and Ba have the same behavior during partial melting and potential fractional crystallization of MORB magmas. Thus a same mantle source undergoing various melting degrees will give magmas with the same (^{226}Ra)$_0$/Ba, provided that the transfer time of these magmas is short compared to the ^{226}Ra half-life (e.g., ≤ 200 a). If this (^{226}Ra)$_0$/Ba ratio can be estimated, then the age of young MORB can be calculated from their measured (^{226}Ra)/Ba ratios. Rubin and Macdougall (1990) actually use an empirical correlation between (^{226}Ra/^{230}Th) and Ba/Th ratios in the samples having the largest ^{226}Ra excesses (and thus assumed to be the youngest) to estimate the (^{226}Ra/^{230}Th)$_0$ initial ratio of each sample and deduce its age.

The same method was also applied to a few MORB from the Juan de Fuca – Gorda ridges on the EPR (Volpe and Goldstein 1993).

^{226}Ra-^{230}Th (and (^{226}Ra)/Ba) dating of subaerial volcanics. Systematic analyses of ^{226}Ra-^{230}Th disequilibria in recent and well dated rocks from active volcanoes allow studies of magmatic evolution as explained in section 3.4. If the variations through time of (^{226}Ra/^{230}Th)$_0$ or (^{226}Ra)$_0$/Ba ratios are sufficiently well constrained, one may use the

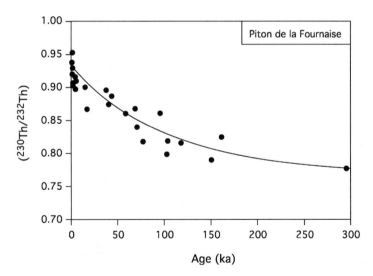

Figure 23. Measured (^{230}Th/^{232}Th) ratios in basalts from Piton de la Fournaise (Réunion Island) as a function of their eruption ages deduced from mineral isochrons. These ratios decrease with increasing eruption ages as a result of post-eruptive radioactive decay. The curve shows the theoretical evolution by radioactive decay for a rock with a Th/U ratio of 3.95 and a (^{230}Th/^{232}Th) ratio of 0.93, similar to the values measured in presently erupted lavas. An approximate age can thus be obtained from the measured (^{230}Th/^{232}Th) ratio of an old sample. Part of the dispersion around the theoretical curve are due to small source heterogeneities (slightly variable (^{230}Th/^{232}Th) and Th/U ratios), also evidenced by ^{87}Sr/^{86}Sr ratios (Condomines et al. 1988, and unpublished results).

curves representing these variations to infer the initial ratios for a rock of unknown age (in the range 0 to 8 ka), and thus calculate its age from the measured $(^{226}Ra/^{230}Th)$ or $(^{226}Ra)/Ba$ ratios. Of course, the simplest case occurs when these ratios have monotonous variations or even are constant. The results obtained on historical lava flows erupted during the last two millennia (and before 1970) at Mt. Etna (Condomines et al. 1995), already described in section 3.4, show a good correlation between $(^{226}Ra)_0$ and Th contents (which also means that $(^{226}Ra/^{230}Th)_0$ ratios decrease with increasing differentiation and Th content, the $(^{230}Th/^{232}Th)$ ratio being constant), and a constant $(^{226}Ra)_0/Ba$ ratio (see Fig. 8). From the $(^{226}Ra)_0$-Th correlation or the constant $(^{226}Ra)_0/Ba$ ratio, ages of lava flows of a given Th or Ba content can be determined. If the eruption age falls in the age range used to define the above correlations, one can be confident about the validity of the calculated age. If it is outside this range, then the calculated age is based on the assumption that the ^{226}Ra-^{230}Th disequilibria followed the same evolution during the past millennia. Examples of such dates on prehistoric lava flows from Etna are given in Condomines et al. (1995). New samples from Etna and Merapi (Indonesia) are now being dated: Mt. Merapi, a permanently active andesitic volcano, indeed seems to present ^{226}Ra-^{230}Th systematics very similar to that of Mt. Etna (Condomines et al., in preparation). The ages of lava flows can be conveniently estimated by comparison with the isochron curves drawn in the diagram of Figure 24 for the case of Mt. Etna.

^{210}Po-^{210}Pb dating method. An original application of this method based on very short lived nuclides (equilibrium between ^{210}Po and ^{210}Pb is reached after only 2 years) was published by Rubin et al. (1994). These authors collected and analyzed volcanic glasses from freshly erupted lava flows on the East Pacific Rise. They found that $(^{210}Po/^{210}Pb)$ ratios were lower than unity. By assuming that ^{210}Po was completely degassed on eruption (a reasonable assumption owing to the high volatility of Po, see section 3.6), they could calculate the eruption ages of the lava flows from the equation:

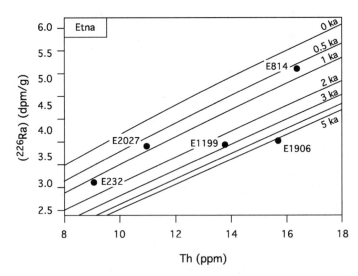

Figure 24. Measured (^{226}Ra) vs. Th diagram used to infer the ages of lava flows from Mt. Etna in the age range 0 to 8 ka. The curves labeled 0, 0.5, 1... ka correspond to isochron curves. The isochron 0 is defined by the $(^{226}Ra)_0$-Th correlation observed in historical lava flows of the last two millennia (Condomines et al. 1995). Five samples of previously unknown ages are reported in this diagram (Condomines et al. 1995, and unpublished results).

$(^{210}\text{Po}/^{210}\text{Pb}) = 1 - e^{-\lambda t}$. Rubin et al. (1994) were able to show the existence of two eruptive episodes separated by about 7 months and to precisely determine their ages.

5. SUMMARY AND CONCLUSIONS

The development of U-series studies in volcanic rocks over the past twenty years has led to significant advances in our understanding of magmatic processes and their timescales as well as in their use as dating tools for young volcanic rocks. They have also raised a number of important questions which hopefully will promote more work in this field.

Because U- or Th-series disequilibria date chemical fractionations between various nuclides of the radioactive decay chains, it is fundamental to understand their behavior during the main magmatic processes (partial melting, fractional crystallization, assimilation, fluid circulation...). For example, ^{230}Th-^{238}U disequilibria are mainly created during partial melting, and Th/U ratios are little modified by crystal fractionation involving major minerals (but U and Th can be fractionated by the crystallization of U,Th-rich accessory phases in silicic magmas). Fluid transfer can also affect U, either during partial melting in the mantle (e.g., in subduction zones whose magmas are often enriched in U through slab-derived fluids), or, in some cases, during differentiation (e.g., in ultrapotassic magmas, Villemant et al. 1989). ^{226}Ra-^{230}Th disequilibria are influenced by partial melting, crystal fractionation involving feldspars, and fluid transfer affecting Ra. When a fractionation process has been identified, then U-series disequilibria offer the possibility to constrain the time interval between this event and eruption of a volcanic rock. This is an unique advantage of U-series methods based on the fact that the half-lives of several nuclides are of the same magnitude as the timescales of the main magmatic processes.

These timescales may be constrained in several ways. Measurements on single samples gives a rough indication of the transfer time of a magma from the time of the last fractionation until the eruption. For example, if most of the ^{226}Ra excesses are produced during mantle melting, the absence of disequilibrium between ^{228}Ra and ^{232}Th constrain the transfer time from mantle to surface of all presently erupted silicate magmas to be in the range 30 a to 8 ka. Comparison between disequilibria in basic and more differentiated lavas from the same volcanic complex can give important information on the timescale of differentiation. Comparison of $(^{230}\text{Th}/^{232}\text{Th})$ ratios suggest that differentiation from basalt to an intermediate magma in deep magma chambers might take several 10^4 a. In other cases, ^{210}Pb-^{226}Ra disequilibria suggest a very short differentiation time in small and/or shallow magma chambers (~10 a in Surtsey and Heimaey, Iceland).

Whole rock U-Th isochrons can also provide an age for the differentiation process, but they are rare because of the difficulty of fractionating the U/Th ratio through crystallization of major minerals. In the two reported examples (Laacher See phonolites, and Vico ultrapotassic volcanics), whole-rock isochrons show that differentiation occurred shortly before eruption. A similar conclusion was reached in the case of the Longonot volcano (Kenya), whose trachytes define the only reported $(^{226}\text{Ra})/\text{Ba}$-$(^{230}\text{Th})/\text{Ba}$ whole-rock isochron.

Further constraints can be obtained by systematic studies of the evolution of U-series disequilibria through time in a given volcano. The number of works considering the evolution of $(^{230}\text{Th}/^{238}\text{U})$ or $(^{230}\text{Th}/^{232}\text{Th})$ ratios is very small, but it appears that, in the studied basaltic volcanoes from oceanic islands (Marion, Mauna Kea, Piton de la Fournaise), these ratios remain nearly constant with time. This suggests that the residence time of the magmas in these continuously replenished magma chambers is short compared to the ^{230}Th half-life and might be better approached by ^{226}Ra-^{230}Th disequilibria. Such a study on Etna indicates a residence time ≤ 1500 a. Other studies on basaltic eruptions (Lanzarote, Canary Islands; Karthala, Comores; Ardoukoba, Asal Rift) show negligible transfer times of the magma, except for Ardoukoba where the magma might have been stored at depth for about 2 ka.

The interpretation of U-Th and Ra-Th mineral isochrons can often be very complicated. In some cases, U-Th isochrons indicate crystallization ages in agreement with eruption ages, and thus a short residence of the minerals in the magma chamber, whilst, in other cases, they give crystallization ages much older than the known eruption ages (several 10^4 to 10^5 a older). In some instances, these ages can be explained by incorporation of old cumulate minerals or xenocrysts. Old ages are also found for zircons extracted from silicic rocks, which can now be subjected to *in situ* SIMS analysis. Accessory minerals apparently can be stored for long periods (≥ 10^5 a) in silicic magma chambers, and it has been suggested that some major mineral isochrons do in fact date the crystallization of accessory minerals in inclusions within the major phases. Whether major minerals that often define good isochrons with the groundmass could also be stored for such long periods in the magma chamber is still not clear, although several studies suggest that a crystal-rich layer of magma might be maintained at a nearly constant temperature near the walls of the magma chamber by the heat flux generated by the main part of this chamber.

When measured in very young volcanic rocks, ^{226}Ra-^{230}Th disequilibria suggest crystallization ages of a few 10^2 to 10^3 a older than eruption, even for rocks displaying much older U-Th isochron ages. This suggests that the mineral rims continued to grow shortly before eruption or that Ra was subjected to chemical exchange or mobilization whereas Th isotopes were not reequilibrated.

The fact that most of the rocks that show crystallization ages older than eruption ages come from andesitic volcanoes or silicic volcanoes suggest that the higher viscosity of these silica–rich magmas might play a role in preventing old crystals to settle in the magma chamber. The long repose time between eruptions in large silicic volcanoes could also explain the preservation of "old" magma in the reservoir. But it is difficult to explain the persistence of old magma batches in central volcanoes with frequent eruptions (i.e., when the inferred age of the old magma is much larger than the average repose time between eruptions).

Further advances in the interpretation of mineral data could come from *in situ* analyses (unfortunately only possible presently on U-Th rich accessory minerals like zircon), or from detailed studies of particularly large phenocrysts, whose successive growth zones could be sampled (e.g., through microdrillings) and analyzed. ^{226}Ra measurements in such crystals could allow a direct determination of their growth rates. A systematic study of successive, well-dated eruptions of a given volcano, combining U-series measurements in both whole rocks and minerals, should also help with the interpretation of mineral data.

From the available studies, it seems that permanently active basaltic or andesitic volcanoes might have magma residence times in their open-system reservoirs of a few 10^2 to 10^3 a. Longer storage times are possible for other volcanoes from subduction zones (10^4 to 10^5 a) and certainly also for large silicic magma chambers (10^5 to 10^6 a). But it is probably too early to draw a general conclusion on the magma storage time, differentiation time or residence time (in open-system magma chambers), and large differences might exist between volcanoes. As rightly emphasized by Hawkesworth et al. (2000), the time that a magma can spend in its reservoir before it consolidates depends on the size and depth (i.e., the thermal contrast with the surrounding rocks) of this reservoir. A large and deep magma chamber will have a longer life span than a small and shallow one. Two (or more) magma chambers at different depths are often advocated to explain the whole compositional variation of erupted lavas. In the deep reservoir, the basic magma could undergo limited differentiation leading to an intermediate composition, and further differentiation in (a) small and shallower magma chamber(s) could produce highly differentiated magmas. The timescales of these two differentiation processes are certainly different. In the very schematic representation of open-system steady state magma chambers of Figure 5, the ratio of residence times in the two reservoirs is simply the ratio of their volumes. As it may be expected from this analogy, the half-lives of the U-series nuclides define their domain of application: the longest-lived (^{230}Th, ^{231}Pa, ^{226}Ra), as discussed above, are appropriate to study partial melting in the mantle or deep crustal reservoirs, the shortest-lived give information on processes occurring in shallow magma chambers.

Indeed, the timescales of shallow magmatic processes like degassing can be constrained by using very short-lived isotopes, like ^{210}Pb, ^{210}Po, and ^{210}Bi. Measurements of ^{210}Pb-^{226}Ra disequilibria in lavas of active volcanoes, like Stromboli and Merapi, suggest that the residence time of the magma in the shallow plumbing system, where most of the degassing takes place, may vary from a few days or tens of days in a vigorously renewed magma chamber (Stromboli) to a few years in a less active magma chamber (Merapi). Similar results are obtained from the measurements of ^{210}Pb-^{210}Bi-^{210}Po disequilibria in volcanic gases and aerosols.

Dating the eruption age of young volcanic rocks (up to about 400 ka) can be successfully achieved from U-Th mineral isochrons, with the possible pitfalls due to incorporation of old cumulate minerals or other processes detailed in section 3.5, which could give crystallization ages much older than eruption ages. Detailed petrological-geochemical analyses and proper consideration of the volcanological context could help to select the most appropriate samples. Working on microlites extracted from phenocryst-poor volcanic rocks might also be a fruitful approach to obtain eruption ages. The best examples of successful datings come from permanently active basaltic volcanoes (e.g., Hawaii, Fournaise, Etna) or single basaltic fissural eruptions, or from simple volcanoes having produced small amounts of differentiated lavas, likely to originate from differentiation in a small and shallow magma chamber.

Dating of whole rocks from the measurements of their ^{230}Th-^{238}U or ^{231}Pa-^{235}U disequilibria has been successful in the case of MORB, and can also be applied to volcanoes for which the evolution of these disequilibria through time has been studied in detail (e.g., Piton de la Fournaise). Similarly ^{226}Ra-^{230}Th disequilibria may be useful dating tools in the age range 0 to 8 ka, for MORB or continental volcanoes where a detailed knowledge of their variations in well dated samples is available.

ACKNOWLEDGMENTS

We are grateful to Chris Hawkesworth and Simon Turner for their general comments on the manuscript and to Mary Reid and Georg Zellmer for their thoughtful and constructive reviews, which helped to improve the first version of this chapter.

REFERENCES

Albarède F (1993) Residence time analysis of geochemical fluctuations in volcanic series. Geochim Cosmochim Acta 57:615-621
Allègre CJ (1968) ^{230}Th dating of volcanic rocks: a comment. Earth Planet Sci Lett 5:209-210

Allègre CJ, Condomines M (1976) Fine chronology of volcanic processes using ^{238}U-^{230}Th systematics. Earth Planet Sci Lett 28:395-406

Bacon CR, Persing HM, Wooden JL, Ireland TR (2000) Late pleistocene granodiorite beneath Crater Lake caldera, Oregon, dated by ion microprobe. Geology 28:467-470

Black S, Macdonald R, Kelly MR (1997) Crustal origin for peralkaline rhyolites from Kenya: Evidence from U-series disequilibria and Th-isotopes. J Petrol 38:277-297

Black S, Macdonald R, Barreiro BA, Dunkley PN, Smith M (1998) Open system alkaline magmatism in northern Kenya: Evidence from U-series disequilibria and radiogenic isotopes. Contrib Mineral Petrol 131:364-378

Black S, Macdonald R, De Vivo B, Kilburn CRJ, Rolandi G (1998) U-series disequilibria in young (A.D. 1944) Vesuvius rocks: Preliminary implications for magma residence times and volatile addition. J Volcanol Geotherm Res 82:97-111

Blundy JD, Wood BJ (1994) Prediction of crystal-melt partition coefficients from elastic moduli. Nature 372:452-454

Bohrson WA, Reid MR (1998) Genesis of evolved ocean island magmas by deep- and shallow-level basement recycling, Socorro Island, Mexico: Constraints from Th and other isotope signatures. J Petrol 39:995-1008

Bourdon B, Zindler A, Wörner G (1994) Evolution of the Laacher See magma chamber: Evidence from SIMS and TIMS measurements of U-Th disequilibria in minerals and glasses. Earth Planet Sci Lett 126:75-90

Bourdon B, Joron JL, Claude Ivanaj C, Allègre CJ (1998) U-Th-Pa-Ra systematics for the Grande Comore volcanics: Melting processes in an upwelling plume. Earth Planet Sci Lett 164:119-133

Bourdon B, Woerner G, Zindler A (2000) U-series evidence for crustal involvement and magma residence times in the petrogenesis of Parinacota Volcano, Chile. Contrib Mineral Petrol 139:458-469

Bourdon B, Turner S, Henderson GM, Lundstrom CC (2003) Introduction to U-series geochemistry. Rev Mineral Geochem 52:1-21

Capaldi G, Cortini M, Gasparini P, Pece R (1976) Short lived radioactive disequilibria in freshly erupted volcanic rocks and their implications for the pre-eruption history of a magma. J Geophys Res 81:350-358

Capaldi G, Pece R (1981) On the reliability of the ^{230}Th-^{238}U dating method applied to young volcanic rocks. J Volcanol Geotherm Res 11:367-372

Capaldi G, Cortini M, Pece R (1985) On the reliability of the ^{230}Th-^{238}U dating method applied to young volcanic rocks – Reply. J Volcanol Geotherm Res 26:369-376

Caroff M, Lagabrielle Y, Spadea P, Auzende JM (1997) Geochemical modelling of nonsteady state magma chambers: a case study from an ultrafast spreading ridge (East Pacific Rise, 17° to 19° S). Geochim Cosmochim Acta 61:4367-4374

Cerrai E, Dugnani Lonati R, Gazzarini F, Tongiorgi E (1965) Il metodo iono/uranio per la determinazione dell'età di minerali vulcanici recenti. Rendiconti Soc Mineral Ital, anno XXI:47-62

Chabaux F (1996) Déséquilibres radioactifs ^{238}U-^{230}Th-^{226}Ra, ^{235}U-^{231}Pa et phénomènes magmatiques. CR Acad Sci 323 (II a):897-910

Charlier B, Zellmer G (2000) Some remarks on U-Th mineral ages from igneous rocks with prolonged crystallisation histories. Earth Planet Sci Lett 183:457-469

Cherniak DJ, Hanchar JM, Watson EB (1997) Diffusion of tetravalent cations in zircon. Contrib Mineral Petrol 127:383-390

Cherniak DJ (2002) Ba diffusion in feldspar. Geochim Cosmochim Acta 66:1641-1650

Cherdyntsev VV, Kislitsina GI, Kuptsov VM, Kuzmina YA, Zverev VL (1967) Radioactivity and absolute age of young volcanic rocks. Geokhimiya 7:755-762

Chouet B, Saccorotti G, Dawson P, Martini M, Scarpa R, De Luca G, Milana G, Cattaneo M (1999) Broadband measurements of the sources of explosions at Stromboli volcano, Italy. Geophys Res Lett 26:1937-1940

Christensen JN, DePaolo DJ (1993) Time scales of large volume silicic magma systems: Sr isotopic systematics of phenocrysts and glass from the Bishop Tuff, Long Valley, California. Contrib Mineral Petrol 113:100-114

Christensen JN, Halliday AN (1996) Rb-Sr ages and Nd isotopic compositions of melt inclusions from the Bishop Tuff and the generation of silicic magma. Earth Planet Sci Lett 144:547-561

Claude-Ivanaj C, Bourdon B, Allègre CJ (1998) Ra-Th-Sr isotope systematics in Grande Comore Island: A case study of plume-lithosphere interaction. Earth Planet Sci Lett 164:99-117

Claude-Ivanaj C, Joron JL, Allègre CJ (2001) ^{238}U-^{230}Th-^{226}Ra fractionation in historical lavas from the Azores: long-lived source heterogeneity vs. metasomatism fingerprints. Chem Geol 176:295-310

Cohen AS, O'Nions RK (1993) Melting rates beneath Hawaii: Evidence from uranium series isotopes in recent lavas. Earth Planet Sci Lett 120:169-175

Condomines M (1978) Age of the Olby-Lascamp geomagnetic polarity event. Nature 276:257
Condomines M, Allègre CJ (1980) Age and magmatic evolution of Stromboli volcano from ^{230}Th-^{238}U data. Nature288:354-357
Condomines M, Morand P, Camus G, Duthou JL (1982a) Chronological and Geochemical study of lavas from the Chaîne des Puys, Massif Central, France: evidence for crustal contamination. Contrib Mineral Petrol 81:296-303
Condomines M, Tanguy JC, Kieffer G, Allègre CJ (1982b) Magmatic evolution of a volcano studied by ^{230}Th-^{238}U disequilibrium and trace elements systematics: The Etna case. Geochim Cosmochim Acta 46:1397-1416
Condomines M, Hémond C, Allègre CJ (1988) U-Th-Ra radioactive disequilibria and magmatic processes. Earth Planet Sci Lett 90:243-262
Condomines M (1994) Comment on: "The volume and residence time of magma beneath active volcanoes determined by decay series disequilibria methods." Earth Planet Sci Lett 122:251-255
Condomines M, Tanguy JC, Michaud V (1995) Magma dynamics at Mt. Etna: Constraints from U-Th-Ra-Pb radioactive disequilibria and Sr isotopes in historical lavas. Earth Planet Sci Lett 132:25-41
Condomines M (1997) Dating recent volcanic rocks through ^{230}Th-^{238}U disequilibrium in accessory minerals: Example of the Puy de Dôme (French Massif Central). Geology 25:375-378
Cooper KM, Reid MR, Murrell MT, Clague DA (2001) Crystal and magma residence at Kilauea Volcano, Hawaii: ^{230}Th-^{226}Ra dating of the 1955 East Rift eruption. Earth Planet Sci Lett 184:703-718
Davidson J, Tepley F, III, Palacz Z, Meffan-Main S (2001) Magma recharge, contamination and residence times revealed by *in situ* laser ablation isotopic analysis of feldspar in volcanic rocks. Earth Planet Sci Lett 184:427-442
Davies GR, Halliday AN (1998) Development of the Long Valley rhyolitic magma system: Strontium and neodymium isotope evidence from glasses and individual phenocrysts. Geochim Cosmochim Acta 62:3561-3574
Fukuoka T (1974) Ionium dating of acidic volcanic rocks. Geochem J 8:109-116
Gauthier P-J, Condomines M (1999) ^{210}Pb-^{226}Ra radioactive disequilibria in recent lavas and radon degassing: Inferences on the magma chamber dynamics at Stromboli and Merapi volcanoes. Earth Planet Sci Lett 172:111-126
Gauthier P-J, Condomines M, Hammouda T (1999) An experimental investigation of radon diffusion in an anhydrous andesitic melt at atmospheric pressure: Implications for radon degassing from erupting magmas. Geochim Cosmochim Acta 63:645-656
Gauthier PJ, Le Cloarec MF, Condomines M (2000) Degassing processes at Stromboli Volcano inferred from short-lived disequilibria (^{210}Pb-^{210}Bi-^{210}Po) in volcanic gases. J Volcanol Geotherm Res 102:1-19
Gill J, Williams R, Bruland K (1985) Eruption of basalt and andesite lava degasses ^{222}Rn and ^{210}Po. Geophys Res Lett 12:17-20
Gill JB and Williams RW (1990) Th isotope and U-series studies of subduction-related volcanic rocks. Geochim Cosmochim Acta 54:1427-1442
Gill JB, Condomines M (1992) Short-lived radioactivity and magma genesis. Science 257:1368-1376
Gill JB, Pyle D, Williams RW (1992) Igneous rocks. *In:* Uranium series disequilibrium. Applications to Earth, Marine, and Environmental Sciences (2nd ed.). M Ivanovich and RS Harmon (eds.), Oxford Science Publications, Oxford University Press, New York
Goldstein SJ, Murrell MT, Janecky DR, Delaney JR, Clague DA (1992) Geochronology and petrogenesis of MORB from the Juan de Fuca and Gorda ridges by ^{238}U-^{230}Th disequilibrium. Earth Planet Sci Lett 109:255-272
Goldstein SJ, Murrell MT, Williams RW (1993) ^{231}Pa and ^{230}Th chronology of mid-ocean ridge basalts. Earth Planet Sci Lett 115:151-159
Goldstein SJ, Perfit MR, Batiza R, Fornari DJ, Murrell MT. (1994) Off-axis volcanism at the East Pacific Rise detected by uranium-series dating of basalts. Nature 367:157-159
Goldstein SJ, Stirling CH (2003) Techniques for measuring uranium-series nuclides: 1992-2002. Rev Mineral Geochem 52:23-57
Halliday AN, Mahood GA, Holden P, Metz JM, Dempster TJ, Davidson JP (1989) Evidence for long residence times of rhyolitic magma in the Long Valley magmatic system — The isotopic record in precaldera lavas of Glass Mountain. Earth Planet Sci Lett 94:274-290
Hawkesworth CJ, Turner SP, McDermott F, Peate DW, van Calsteren P (1997) U-Th isotopes in arc magmas: Implications for element transfer from the subducted crust. Science 276:551-555
Hawkesworth CJ, Blake S, Evans P, Hughes R, MacDonald R, Thomas LE, Turner SP, Zellmer G (2000) Time scales of crystal fractionation in magma chambers — Integrating physical, isotopic and geochemical perspectives. J. Petrol. 41:991-1006

Heath E, Turner SP, Macdonald R, Hawkesworth CJ, van Calsteren P (1998) Long magma residence times at an island arc volcano (Soufriere, St. Vincent) in the Lesser Antilles: Evidence from ^{238}U-^{230}Th isochron dating. Earth Planet Sci Lett 160:49-63

Hémond Ch, Condomines M (1985) On the reliability of the ^{230}Th-^{238}U dating method applied to young volcanic rocks – Discussion. J Volcanol Geotherm Res 26:365-369

Hémond Ch, Hofmann AW, Heusser G, Condomines M, Raczek I, Rhodes JM (1994) U-Th-Ra systematics in Kilauea and Mauna Loa basalts, Hawaii. Chem Geol 116:163-180

Heumann A, Davies GR (2002) U-Th disequilibrium and Rb-Sr age constraints on the magmatic evolution of peralkaline rhyolites from Kenya. J Petrol 43:557-577

Heumann A, Davies GR, Elliott T (2002) Crystallization history of rhyolites at Long Valley, California, inferred from combined U-series and Rb-Sr isotope systematics. Geochim Cosmochim Acta 66:1821-1837

Hildreth W and Mahood GA (1986) Ring-fracture eruption of the Bishop Tuff. Geol Soc Amer Bull 97:396-403

Hughes RD, Hawkesworth CJ (1999) The effects of magma replenishment processes on ^{238}U-^{230}Th disequilibrium. Geochim Cosmochim Acta 63:4101-4110

Ivanovich M and Harmon RS, Editors (1992) Uranium-series disequilibrium: Applications to Earth, Marine and Environmental Sciences (2nd edition). Oxford Science Publications, Oxford University Press, New York

Johnson MC, Anderson AT, Rutherford MJ (1994) Pre-eruptive volatile contents of magmas. Rev Min 30:281-330

Kigoshi K (1967) Ionium dating of igneous rocks. Science 156:932-934

Krishnaswami S, Turekian KK, Bennett JT (1984) The behavior of ^{232}Th and the ^{238}U decay chain nuclides during magma formation and volcanism. Geochim Cosmochim Acta 48:505-511

Lambert G, Le Cloarec MF, Ardouin B, Le Roulley JC (1985-86) Volcanic emission of radionuclides and magma dynamics. Earth Planet Sci Lett 76:185-192

Le Cloarec M-F, Pennisi M (2001) Radionuclides and sulfur content in Mount Etna plume in 1983-1995: New constraints on the magma feeding system. J Volcanol Geotherm Res 108:141-155

Le Cloarec MF, Lambert G, Le Guern F, Ardouin B (1984) Echanges de matériaux volatils entre phases solide, liquide et gazeuse au cours de l'éruption de l'Etna de 1983. CR Acad Sci Paris 298-II:805-808

Le Cloarec MF, Lambert G, Le Roulley JC, Ardouin B (1986) Long-lived decay products in Mount St. Helens emissions: An estimation of the magma reservoir volume. J Volcanol Geotherm Res 28:85-89

Le Cloarec M-F, Allard P, Ardouin B, Giggenbach WF, Sheppard DS (1992) Radioactive isotopes and trace elements in gaseous emissions from White Island, New Zealand. Earth Planet Sci Lett 108:19-28

Le Cloarec M-F, Pennisi M, Corazza E, Lambert G (1994) Origin of fumarolic fluids emitted from a nonerupting volcano: Radionuclide constraints at Vulcano (Aeolian Islands, Italy). Geochim Cosmochim Acta 58:4401-4410

Lee JKW, Williams IS, Ellis DJ (1997) Pb, U and Th diffusion in natural zircon. Nature 390:159-162

Lowenstern JB, Persing HM, Wooden JL, Lanphere M, Donnelly Nolan J, Grove TL (2000) U-Th dating of single zircons from young granitoid xenoliths: New tools for understanding volcanic processes. Earth Planet Sci Lett 183:291-302

Lundstrom CC (2003) Uranium-series disequilibria in mid-ocean ridge basalts: observations and models of basalt genesis. Rev Mineral Geochem 52:175-214

Macdougall JD (1995) Using short-lived U and Th series isotopes to investigate volcanic processes. Annu Rev Earth Planet Sci 23:143-167

Newman S, Finkel RC, MacDougall JD (1984) Comparison of ^{230}Th-^{238}U disequilibrium systematics in lavas from three hot spot regions: Hawaii, Prince Edward and Samoa. Geochim Cosmochim Acta 48:315-324

Oversby VM, Gast PW (1968) Lead isotope composition and uranium decay series disequilibrium in recent volcanic rocks. Earth Planet Sci Lett 5:199-206

Peate DW, Chen JH, Wasserburg GJ, Papanastassiou DA (1996) ^{238}U-^{230}Th dating of a geomagnetic excursion in Quaternary basalts of the Albuquerque volcanoes field, New Mexico (USA). Geophys Res Lett 23:2271-2274

Peate DW, Mangini A, Leyk HJ, van Calsteren P (2001) Pitfalls in ^{230}Th-^{238}U dating of young Quaternary volcanic rocks: A case study from Fornicher Kopf (East Eifel volcanic field, Germany). Quaternary Sci Rev 20:1927-1933

Pennisi M, Le Cloarec MF, Lambert G, Le Roulley JC (1988) Fractionation of metals in volcanic emissions. Earth Planet Sci Lett 88:284-288

Pietruszka AJ, Garcia MO (1999) The size and shape of Kilauea Volcano's summit magma storage reservoir: a geochemical probe. Earth Planet Sci Lett 167:311-320

Pietruszka AJ, Rubin KH, Garcia MO (2001) ^{226}Ra-^{230}Th-^{238}U disequilibria of historical Kilauea lavas (1790-1982) and the dynamics of mantle melting within the Hawaiian plume. Earth Planet Sci Lett 186:15-31

Pickett DA, Murrell MT. (1997) Observations of ^{231}Pa/^{235}U disequilibrium in volcanic rocks. Earth Planet Sci Lett 148:259-271

Polian G, Lambert G (1979) Radon daughters and sulfur output from Erebus volcano, Antartica. J Volcanol Geotherm Res 6:125-137

Pyle D, Ivanovich M, Sparks RSJ (1988) Magma-cumulate mixing identified by U-Th disequilibrium dating. Nature 331:157-159

Pyle DM (1992) The volume and residence time of magma beneath active volcanoes determined by decay-series disequilibria methods. Earth Planet Sci Lett 112:61-73

Pyle DM, Dawson JB, Ivanovich M (1991) Short-lived decay series disequilibria in the natrocarbonatite lavas of Oldoinyo Lengai, Tanzania: Constraints on the timing of magma genesis. Earth Planet Sci Lett 105:378-396

Reagan MK, Volpe AM, Cashman KV (1992) ^{238}U and ^{232}Th-series chronology of phonolite fractionation at Mount Erebus, Antartica. Geochim Cosmochim Acta 56:1401-1407

Reagan MK, Morris JD, Herrstrom EA, Murrell MT. (1994) Uranium series and beryllium isotope evidence for an extended history of subduction modification of the mantle below Nicaragua. Geochim Cosmochim Acta 58:4199-4212

Reid MR, Coath CD, Harrison TM, McKeegan KD (1997) Prolonged residence times for the youngest rhyolites associated with Long Valley Caldera: ^{230}Th-^{238}U ion microprobe dating of young zircons. Earth Planet Sci Lett 150:27-39

Reid MR, Coath CD (2000) *In situ* U-Pb ages of zircons from the Bishop Tuff: No evidence for long crystal residence times. Geology 28:443-446

Rubin KH, Macdougall JD (1988) ^{226}Ra excesses in mid-ocean-ridge basalts and mantle melting. Nature 335:158-161

Rubin KH, Macdougall JD (1989) Submarine magma degassing and explosive magmatism at Macdonald (Tamarii) seamount. Nature 341:50-52

Rubin KH, Macdougall JD (1990) Dating of neovolcanic MORB using (^{226}Ra/^{230}Th) disequilibrium. Earth Planet Sci Lett 101:313-322

Rubin KH, Macdougall JD, Perfit MR (1994) ^{210}Po-^{210}Pb dating of recent volcanic eruptions on the sea floor. Nature 368:841-844

Schaefer SJ, Sturchio NC, Murrell MT, Williams SN (1993) Internal ^{238}U-series systematics of pumice from the November 13, 1985, eruption of Nevado del Ruiz, Colombia. Geochim Cosmochim Acta 57:1215-1219

Sigmarsson O, Condomines M, Morris JD, Harmon RS (1990) Uranium and ^{10}Be enrichments by fluids in Andean arc magmas. Nature 391:883-886

Sigmarsson O (1996) Short magma chamber residence time at an Icelandic volcano inferred from U-series disequilibria. Nature 382:440-442

Sigmarsson O, Condomines M, Ibarrola E (1992a) ^{238}U-^{230}Th radioactive disequilibria in historic lavas from the Canary Islands and genetic implications. J Volcanol Geotherm Res 54:145-156

Sigmarsson O, Condomines M, Fourcade S (1992b) A detailed Th, Sr and O isotope study of Hekla: differentiation processes in an Icelandic volcano. Contrib Mineral Petrol 112:20-34

Sigmarsson O, Carn S, Carracedo JC (1998) Systematics of U-series nuclides in primitive lavas from the 1730-36 eruption on Lanzarote, Canary Islands, and implications for the role of garnet pyroxenites during oceanic basalt formation Earth Planet Sci Lett 162:137-151

Sigmarsson O, Chmeleff J, Morris J, Lopez-Escobar L (2002) Origin of ^{226}Ra-^{230}Th disequilibria in arc lavas from Southern Chile and implications for magma transfer time. Earth Planet Sci Lett 196:189-196

Sims KWW, DePaolo DJ, Murrell MT, Baldridge WS, Goldstein S, Clague D, Jull M (1999) Porosity of the melting zone and variations in the solid mantle upwelling rate beneath Hawaii: Inferences from ^{238}U-^{230}Th-^{226}Ra and ^{231}Pa-^{235}U disequilibria. Geochim Cosmochim Acta 63:4119-4138

Sturm ME, Goldstein SJ, Klein EM, Karson JA, Murrell MT. (2000) Uranium-series age constraints on lavas from the axial valley of the Mid-Atlantic Ridge, MARK area. Earth Planet Sci Lett 181:61-70

Symonds RB, Rose WI, Bluth GJS, Gerlach TM (1994) Volcanic-gas studies: Methods, results, and applications. Rev Min 30:1-66

Taddeucci A, Broecker WS, Thurber D (1967) ^{230}Th dating of volcanic rocks. Earth Planet Sci Lett 3:338-342

Thomas LE, Hawkesworth CJ, Van Calsteren P, Turner SP, Rogers NW (1999) Melt generation beneath ocean islands: A U-Th-Ra isotope study from Lanzarote in the Canary Islands. Geochim Cosmochim Acta 63:4081-4099

Turner SP, Bourdon B, Hawkesworth C, Evans PJ (2000a) ^{226}Ra-^{230}Th evidence for multiple dehydration events, rapid melt ascent and the time scales of differentiation beneath the Tonga-Kermadec island arc. Earth Planet Sci Lett 179:581-593

Turner SP, George RMM, Evans PJ, Hawkesworth CJ, Zellmer GF (2000b) Time-scales of magma formation, ascent and storage beneath subduction-zone volcanoes. Phil Trans R Soc 358:1443-1464

Turner SP, Evans P, Hawkesworth CJ (2001) Ultrafast source-to-surface movement of melt at island arcs from ^{226}Ra-^{230}Th systematics. Science 292:1363-1366

Turner S, Bourdon B, Gill J (2003) Insights into magma genesis at convergent margins from U-series isotopes. Rev Mineral Geochem 52:255-315

Van den Bogaard P, Schirnick C (1995) ^{40}Ar/^{39}Ar laser probe age of Bishop Tuff quartz phenocrysts substantiate long-lived silicic magma chamber at Long Valley, United States. Geology 23:759-762

Van Orman J.A., Grove T.L., Shimizu N (1998) Uranium and thorium diffusion in diopside. Earth Planet Sci Lett 160:505-519

Vergniolle S, Brandeis G, Mareschal J-C (1996) Strombolian explosions 2. Eruption dynamics determined from acoustic measurements. J Geophys Res 101:20449-20466

Vigier N, Bourdon B, Joron JL, Allègre CJ (1999) U-decay series and trace element systematics in the 1978 eruption of Ardoukoba, Asal Rift: Timescale of magma crystallization. Earth Planet Sci Lett 174:81-97

Villemant B, Flehoc C (1989) U-Th fractionation by fluids in K-rich magma genesis: the Vico volcano, Central Italy. Earth Planet Sci Lett 91:312-326

Volpe A (1992) ^{238}U-^{230}Th-^{226}Ra disequilibrium in young Mt. Shasta andesites and dacites. J Volcanol Geotherm Res 53:227-238

Volpe AM, Hammond PE (1991) ^{238}U-^{230}Th-^{226}Ra disequilibria in young Mount St. Helens rocks: Time constraint for magma formation and crystallization. Earth Planet Sci Lett 107:475-486

Volpe AM, Goldstein SJ (1993) ^{226}Ra-^{230}Th disequilibrium in axial and off-axis mid-ocean ridge basalts. Geochim Cosmochim Acta 57:1233-1241

Voltaggio M, Branca M, Tuccimei P, Tecce F (1995) Leaching procedure used in dating young potassic volcanic rocks by the ^{226}Ra/^{230}Th method. Earth Planet Sci Lett 136:123-131

Voltaggio M, Andretta D, Taddeucci A (1994) ^{230}Th-^{238}U data in conflict with ^{40}Ar/^{39}Ar leucite ages for quaternary volcanic rocks of the Alban Hills, Italy. Eur J Mineral 6:209-216

Widom E, Schmincke HU, Gill JB (1992) Processes and timescales in the evolution of chemically zoned trachyte: Fogo A, Sao Miguel, Azores. Contrib Mineral Petrol 111:311-328

Williams RW (1987) Igneous geochemistry – Reading radioactive rock clocks. Nature 325:573-574

Williams RW, Gill JB, Bruland KW (1986) Ra-Th disequilibria systematics: Timescale of carbonatite magma formation of Oldoinyo Lengai volcano, Tanzania. Geochim Cosmochim Acta 50:1249-1259

Zellmer GF, Blake S, Vance D, Hawkesworth C, Turner S (1999) Plagioclase residence times at two island arc volcanoes (Kameni Islands, Santorini, and Soufriere, St. Vincent) determined by Sr diffusion systematics. Contrib Mineral Petrol 136:345-357

Zellmer G, Turner S, Hawkesworth C (2000) Timescales of destructive plate margin magmatism: New insights from Santorini, Aegean volcanic arc. Earth Planet Sci Lett 174:265-281

5

Uranium-series Disequilibria in Mid-ocean Ridge Basalts: Observations and Models of Basalt Genesis

Craig C. Lundstrom

Department of Geology
University of Illinois at Urbana Champaign
245 NHB, 1301 W. Green St.
Urbana, Illinois, 61801, U.S.A.

1. INTRODUCTION

Mid-ocean ridges account for more than 75% of the annual magmatic output of planet earth and are a critical piece of the plate tectonic model. Understanding the mantle melting process beneath ridges is essential for discerning basic physical and chemical processes of the earth. For instance, ridges play an intrinsic role in the geochemical flux balance for many elements on the earth and provide a tectonic framework for investigating the physical process of solid mantle flow. Ridges are also the simplest tectonic regime to study the mantle melting process because interactions with the lithosphere are thought to be minimal. Yet major questions remain about the rates and timescales of melt transport and solid mantle flow. Because uranium series (U-series) nuclides are sensitive to processes occurring at timescales similar to their half-lives (e.g., Allègre and Condomines 1976; McKenzie 1985), U-series disequilibria can provide constraint on the velocities of ascending melts and upwelling solid mantle as well as constraint on the depth of melting beneath mid-ocean ridges.

This chapter begins with a brief introduction to the most important decay series nuclides for studying melting processes. Next, I discuss analytical issues involved with measuring U-series disequilibria in mid-ocean ridge basalts (MORB), including sample preparation and instrumental techniques. I summarize the observations of U-series disequilibria in MORB to date and applications of U-series data to geochronology of MORB. Geochemical data bearing on the issue of secondary contamination are used to argue that the majority of U-series disequilibria observed in MORB reflect the melting process and do not reflect alteration processes. Given the observation of disequilibria, geochemical models for MORB genesis are described and assessed for their ability to realistically explain the disequilibria observations. Finally, I discuss the interpretations of the combined U-series models and observations and relate these to constraints on MORB genesis from other geochemical and petrological tracers.

1.1. A primer on U-series disequilibria

Chapter 1 (Bourdon et al. 2003) provides an introduction to the workings of decay chains including the concept of secular equilibrium. The short-lived nuclides most commonly used for studying magma genesis are ^{230}Th ($t_{1/2}$ = 75.4 kyrs) and ^{226}Ra ($t_{1/2}$ =1600 yrs) in the ^{238}U decay chain and ^{231}Pa ($t_{1/2}$ =32.8 kyrs) in the ^{235}U decay chain. The relatively long half-lives make these nuclides particularly suited to investigating processes that occur over timescales similar to their decay. For instance, these nuclides will be sensitive to the process of solid upwelling at cm/yr rates or melt ascending at m/yr rates over kilometer scale distances. Although this discussion will focus on studies using these longer-lived nuclides, shorter-lived nuclides have also been used extensively in igneous petrology (e.g., ^{210}Pb, ^{210}Po, ^{222}Rn, and ^{228}Ra with half-lives < 25 yrs). Melting processes may create disequilibria between these shorter-lived nuclides and their parents.

However, the short half-lives of these nuclides make them more useful for constraining the timescales of processes occurring closer to the eruption, such as degassing or crystal fractionation at crustal depths (Condomines et al. 2003).

U-series disequilibria are powerful for studying mantle melting because secular equilibrium between all members of the chain should exist for any parcel of solid mantle upwelling from depth. Prior to formation of any fluid phase (early formed melt or volatile rich fluid), upwelling material will have been solid for times that are much longer than the half-life of any short-lived nuclide (all but ^{238}U, ^{235}U and ^{232}Th). Secular equilibrium should therefore exist for the entire chain because few processes could fractionate a parent-daughter prior to melting and none could fractionate nuclides over length scales outside of those affected by melting.

Because of this, U-series methods have significant advantages over standard geochemical techniques. It has become increasingly clear over the past 15 years that the mantle source for MORB is heterogeneous on a relatively small spatial scale (Allègre and Turcotte 1986). In contrast to incompatible trace element ratios that vary widely with mantle source composition, ratios of U-series nuclides are set by secular equilibrium providing the ability to examine process despite the occurrence of source heterogeneity. However, parent-daughter fractionation during melting depends on source mineralogy meaning that lithologic heterogeneity can greatly affect the amount of U-series disequilibria generated during melting.

2. OBSERVED U-SERIES DISEQUILIBRIA IN MORB

2.1. Analysis of U-series nuclides

Measurement of U-series disequilibria in MORB presents a considerable analytical challenge. Typical concentrations of normal MORB (NMORB) are variable but are generally in the 50-150 ppb U range and 100-400 ppb Th range. Some depleted MORB have concentrations as low as 8-20 ppb U and Th. The concentrations of ^{230}Th, ^{231}Pa, and ^{226}Ra in secular equilibrium with these U contents are exceedingly low. For instance, the atomic ratio of ^{238}U to ^{226}Ra in secular equilibrium is ~2.5 × 10^6 with a quick rule of thumb being that 50 ng of U corresponds to ~20 fg of ^{226}Ra and ~15 fg of ^{231}Pa. Thus, dissolution of a gram of MORB still requires measurement of fg quantities of these nuclides by any mass spectrometric techniques.

The exclusive sample material for U-series analysis of MORB is the glassy rind of erupted basalt for two reasons: 1) glass represents the composition of the original magma (before significant crystallization occurs); and 2) glass alters less readily than crystalline basalt (because it is less porous) and alteration can be more easily recognized. Because of the low concentrations of U and Th in MORB, relatively large amounts of glass (0.3 to 5 grams) can be required for analysis; unfortunately, such quantities are not always available for these precious samples. The glass is usually hand picked and leached prior to dissolution. Preparation techniques for mass spectrometric studies have included leaching with weak inorganic acids (HCl), oxalic acid, hydrogen peroxide, and citrate buffered dithionate. These leaching techniques are aimed at removing common alteration products found on the seafloor such as ferro-manganese oxides and palagonite. Reinitz and Turekian (1989) examined techniques for eliminating surface contamination on MORB glasses prior to U-series analysis and compared two methods of sample preparation: 1) simple inorganic acid leaching and 2) a citrate-dithionate treatment. Both methods were shown to remove significant amounts of surface contamination and were considered effective within the precision of this alpha counting study.

Analysis techniques for U, Th, Ra and Pa have evolved a great deal over the past 15 years. A more detailed discussion is given in Goldstein and Stirling (2003). Early U-series studies of MORB used alpha spectrometry to measure activities directly. Typically, upwards of 10g of glass were handpicked and cleaned for analysis, spiked with a yield tracer and purified by ion chromatography. The low activities of MORB samples often required counting for weeks to months and resulted in counting statistics based precision of 5-10% (2σ). Recently, more sensitive mass spectrometric techniques including thermal ionization mass spectrometry (TIMS), secondary ion mass spectrometry (SIMS) and inductively coupled plasma mass spectrometry (ICP-MS) have been developed for U-series analysis. In these types of analysis, concentrations of nuclides are measured by isotope dilution and then converted to activities by multiplying by the decay constant of that nuclide.

Mass spectrometric techniques for analysis of ^{230}Th-^{238}U disequilibria were first developed to date corals for paleoclimate research (Edwards et al. 1987). Soon thereafter, workers at Los Alamos National Laboratory (LANL) developed methods for silicate analysis by TIMS (Goldstein et al. 1989). Typical TIMS analysis of MORB requires 0.5 to 1 gram of material in order have an analyzable load of 100 ng of Th. TIMS analyses of U and Th last ~2-3 hrs and produce a precision of 0.5-2% (2σ). SIMS techniques for measuring Th isotopes have also been developed (England et al. 1992; Layne and Sims 2000). Analysis of Ra and Pa isotopes by TIMS was developed in the early 1990's significantly increasing the sensitivity over decay counting analysis (Volpe et al. 1993; Cohen and Onions 1993; Pickett et al. 1994; Chabaux et al. 1994).

Thermal ionization is an inefficient way of producing Th ions, and smaller samples can be analyzed using SIMS, and more recently, ICP techniques. Multi-collector ICP-MS is poised to make further improvements by reducing sample size, increasing precision and increasing sample throughput (Luo et al. 1997). Currently, sensitivities >6×10^8 cps for a 10 ppb solution aspirating at ~90 µl/min allow samples of less than 5 ng Th to be analyzed with precisions in the per mil range. Thus, this technique could further revolutionize the study of U-series disequilibria in MORB by reducing sample size and analysis time as it is already doing for carbonate dating. Decreasing sample size requirements should increase the number of MORB samples available for analysis. The requirement of gram-sized samples eliminates many important samples from current U-series analysis simply because of the lack of material, limiting the questions that can be addressed by U-series methods.

2.2. Observed U-series disequilibria in MORB relative to other tectonic settings

Since the earliest measurements of U-series disequilibria in young volcanic rocks (e.g., Oversby and Gast 1968), disequilibrium between parent and daughter nuclides has been observed in samples from every tectonic setting on the planet from mid-ocean ridges to intra-plate/ocean island settings to convergent margins. The disequilibria produced in each setting are systematically distinct (Fig. 1).

Of the three parent-daughter pairs most relevant to mantle melting, (^{230}Th)/(^{238}U) data are by far the most numerous. Figure 1A shows an equiline or U-Th isochron diagram, which plots (^{238}U)/(^{232}Th) (which is U/Th, an incompatible element ratio, converted to an activity ratio (activity denoted by parentheses)) versus the Th isotope ratio, also converted to an activity ratio. The amount of deviation from the equiline indicates the amount of ^{230}Th-^{238}U disequilibria (lines of equal % excess are shown). Ridges have higher U/Th than intra-plate basalts (Fig. 1A), indicating that U was more compatible than Th during upper mantle depletion and extraction of the crust (Sun and McDonough 1989). MORB and ocean island basalts (OIB) have a similar range in (^{230}Th)/(^{238}U) with almost all samples in both settings having ^{230}Th excess (general range 0-40% ^{230}Th excess). In contrast, convergent margins have a wider range in U/Th with

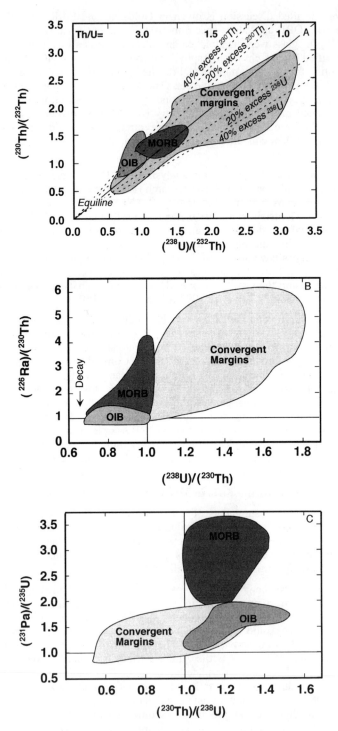

Figure 1. (caption on facing page)

several arcs having much higher U/Th. Some convergent margin samples, especially those with high U/Th, have ^{238}U excess (sometimes >50% ^{238}U excess). Both of these features are interpreted to reflect the transfer of U but not Th from the slab to the mantle wedge by fluids (Turner et al. 2003).

In contrast to the modest (^{230}Th)/(^{238}U) observed in MORB, (^{226}Ra)/(^{230}Th) and (^{231}Pa)/(^{235}U) can exceed 3.5 at ridges. Although many fewer (^{226}Ra)/(^{230}Th) and (^{231}Pa)/(^{235}U) data exist, systematic differences between tectonic settings are still apparent. Age uncertainties on (^{226}Ra)/(^{230}Th) may greatly affect the fields shown in Figure 1B resulting in samples with significant age since eruption decaying back toward (^{226}Ra)/(^{230}Th) = 1. Nevertheless, (^{226}Ra)/(^{230}Th) in several MORB (whether age constrained or not) range from 2 up to 4 whereas the highest measured (^{226}Ra)/(^{230}Th) in age constrained OIB is ~2 (Bourdon and Sims 2003). Convergent margins have ^{226}Ra excesses as large or larger than MORB (Turner et al. 2001); however these excesses are exclusively found in samples with ^{238}U excess and are interpreted to reflect fluid transfer from the slab to the mantle wedge (Turner et al. 2003).

MORB systematically have higher ^{231}Pa excesses than either convergent margins or intra-plate settings. Whereas a few convergent margin settings have ^{231}Pa deficits, most have (^{231}Pa)/(^{235}U) of 1.0-1.8 (Turner et al. 2003). (^{231}Pa)/(^{235}U) in intra-plate settings generally positively correlates with (^{230}Th)/(^{238}U) and varies from small (1.1) to maximum values of ~2. In contrast, when age constrained by the presence of ^{226}Ra excess, MORB melts have (^{231}Pa)/(^{235}U) ranging between 2 and 3.6. Thus, there is almost complete separation between the fields of (^{231}Pa)/(^{235}U) for MORB and the fields of (^{231}Pa)/(^{235}U) for intra-plate and convergent margin settings.

2.3. Observed U-series disequilibria at specific ridge locations

Table 1 lists MORB studies to date and Figures 2 and 3 present data from different areas of ridge. The first study of U-series disequilibria in MORB was the pioneering work of Condomines et al. (1981) (Fig. 2A). These workers analyzed samples having a relatively wide range in composition (Mg# 72 to 57) from the FAMOUS region of the Mid-Atlantic Ridge (MAR 37°N) by combined alpha spectrometry (for U and Th isotopic ratios) and mass spectrometry (isotope dilution measurements for U and Th

---

Figure 1 (*on facing page*). Fields of U-series disequilibria data within the major tectonic settings. (A) An equiline or isochron diagram showing the fields for MORB, OIB and convergent margins. (^{238}U)/(^{232}Th) is essentially the U/Th ratio converted into activities by multiplying the ratio by the decay constants of ^{238}U and ^{232}Th while (^{230}Th)/(^{232}Th) is the Th isotope ratio converted to activities. The one to one line, referred to as the equiline, denotes equal activities of ^{230}Th and ^{238}U. Sample lying to the left of the equiline have excess ^{230}Th while those lying to the right have ^{238}U excess. All OIB and almost all MORB have ^{230}Th excess while convergent margins often have ^{238}U excess (shown more clearly in B). The presence of ^{230}Th excess in MORB and OIB is generally attributed to melting in the presence of garnet or aluminous clinopyroxene (Blundy and Wood 2003). MORB and OIB have a restricted range in U/Th compared to convergent margins which can have much higher U/Th. Both the higher U/Th and ^{238}U excess in convergent margins are attributed to the role of U addition by fluids from the subducting slab (Turner et al. 2003). (B) (^{226}Ra)/(^{230}Th) vs. (^{238}U)/(^{230}Th). The lack of age constraint on many samples, particularly MORB, leads to some uncertainty on the vertical dimension of the fields; MORB and OIB fields extend down to (^{226}Ra)/(^{230}Th) = 1 due to unknown ages since eruption (note that a few OIB samples have small ^{226}Ra deficits). MORB are limited to only small ^{238}U excess but can have relatively large ^{226}Ra excesses. OIB have the smallest ^{226}Ra excesses, possibly due to long transit times through the lithosphere. Convergent margin samples have the highest ^{226}Ra excesses but these occur only in samples with large ^{238}U excess. (C) (^{231}Pa)/(^{235}U) vs. (^{230}Th)/(^{238}U). OIB and convergent margins have systematically lower ^{231}Pa excess than MORB. The larger ^{231}Pa excesses in MORB may reflect a longer melting column beneath ridges relative to OIB where a lithospheric lid limits the melting column length.

Table 1. Summary of published U-Th-Ra-Pa data for mid-ocean ridge basalts.

Area of Ridge	$\frac{(^{230}Th)}{(^{238}U)}$	$\frac{(^{226}Ra)}{(^{230}Th)}$	$\frac{(^{231}Pa)}{(^{235}U)}$	Reference(s)
Juan de Fuca	TIMS	TIMS	TIMS	Goldstein et al. 1989, 1991; Lundstrom et al. 1995; Volpe and Goldstein 1993
Gorda Ridge	TIMS	TIMS	TIMS	Goldstein et al. 1989, 1991; Volpe and Goldstein 1993; Rubin et al. 1998
8-10°N EPR (including Siquieros Transform)	TIMS	TIMS	TIMS	Goldstein et al. 1993, 1994; Lundstrom et al. 1999; Rubin et al. 1994; Zou et al. 2002; Sims et al. 2002
20-27°S EPR	α	α	---	Rubin and Macdougall 1988
36-37°N MAR	α		---	Condomines et al. 1981
37-40°N MAR	TIMS	---	---	Bourdon et al. 1996a
11-13°N EPR	α	α	---	Rubin and Macdougall 1988; Reinitz and Turekian 1989; Ben Othman and Allegre 1990
21°N EPR	α	---	---	Newmann et al. 1983
29°N MAR, AAD, Tamayo	TIMS	---	---	Bourdon et al. 1996b
33°S MAR	TIMS	TIMS	TIMS	Lundstrom et al. 1998
Reykjanes	TIMS	TIMS	---	Peate et al. 2000
22-23°N MAR	TIMS	TIMS	TIMS	Sturm et al. 2000

concentrations). Based on variations in U-series data as well as $^{87}Sr/^{86}Sr$, these authors suggested that roughly half of their samples reflected contamination of U or Th or Sr (or all 3) by a seawater or sediment-derived component. Specifically, they observed a large range in both U-Th concentrations and the amount of ^{230}Th-^{238}U disequilibrium. One sample also had $(^{234}U)/(^{238}U)$ elevated by 10% above secular equilibrium, consistent with a seawater derived contaminant. Although the measured Th/U varied greatly within the study (from 1.7 to 3.1), the authors attributed most of this variation to contamination and instead inferred a relatively constant Th/U of ~3 for the region based on the uniformity of Th/U within samples categorized as "Olivine basalts." As it turns out, such variability in Th/U appears to be normal for MORB (particularly for the MAR) so it is not clear to what extent these data do reflect contamination.

Other early alpha spectrometry studies examined U-series disequilibria in East Pacific Rise (EPR) samples. Newman et al. (1983) measured samples from the RISE study area (21°N EPR) finding significant variations in Th/U and $(^{230}Th)/(^{238}U)$ (Fig. 2B). Rubin and Macdougall (1988) measured $(^{230}Th)/(^{238}U)$ and $(^{226}Ra)/(^{230}Th)$ for samples from both the northern (~10-12°N) and southern (~20-25°) EPR (Fig. 2C,D). This study was first to document large ^{226}Ra-^{230}Th disequilibria in young MORB and interpreted the ^{226}Ra excesses to reflect kinetic effects such as diffusion controlled partitioning rather than equilibrium melting. Reinitz and Turekian (1989) found little ^{230}Th-^{238}U

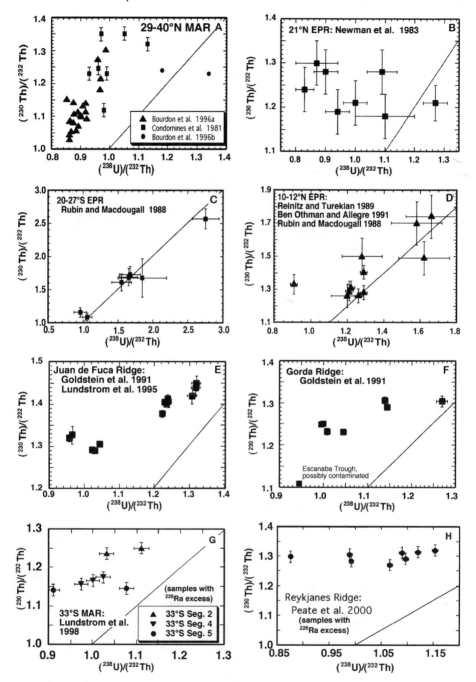

Figure 2. (A)-(H) Equiline diagrams for areas of ridge from the global ridge system. Note that data from individual areas of ridge often form trends indicating a relationship between Th/U and $(^{230}Th)/(^{238}U)$. The equiline diagram is an x-y plot with both axes normalized to (^{232}Th). Therefore a likely explanation for the trends is that they reflect mixing of melts having distinct Th/U.

disequilibria but large ^{226}Ra-^{230}Th disequilibria in 6 samples from the 10-11°N area of the EPR just north of the Clipperton Transform (Fig. 2D). Ben Othman and Allègre (1990) also studied ^{238}U-^{230}Th disequilibria in MORB from the northern EPR between the Clipperton and Orozco Transforms (10-12°N EPR) (Fig. 2D). These authors analyzed samples known to be derived from a heterogeneous source based on their ^{87}Sr/^{86}Sr signatures and found a strong relationship between (^{230}Th)/(^{232}Th) and Th (and U) concentrations of their samples. This study was the first to suggest that small-scale mafic heterogeneities in the mantle source control much of the U-series systematics of MORB.

For the most part, the alpha spectrometry data has held up well with time despite the difficulties with hand picking and analyzing large quantities of low concentration samples. However, a few alpha spectrometry data appear to be biased toward higher values (^{230}Th)/(^{232}Th) and (^{238}U)/(^{232}Th) than those determined by mass spectrometry. Rubin and Macdougall (1988) and Rubin and Macdougall (1992) report samples with (^{230}Th)/(^{232}Th) > 1.5 (up to 2.5) that are significantly higher than mass spectrometry data from the same geographical locations. Given the lack of observation of high (^{230}Th)/(^{232}Th) by mass spectrometry, these data and the interpretations resulting from them should be treated with caution.

The first mass spectrometric work on MORB analyzed both on-axis and off-axis dredge samples from the Juan de Fuca and Gorda Ridges for ^{230}Th-^{238}U (Goldstein et al. 1989; 1992) (Fig. 2E,F). Volpe and Goldstein (1993) analyzed ^{226}Ra in a sub-set of these samples by TIMS, confirming the conclusions from alpha spectrometry that high ^{226}Ra excesses were present in fresh MORB. Importantly, most on-axis basalts had ^{226}Ra excesses while off-axis basalts were in secular equilibrium for ^{226}Ra-^{230}Th, consistent with the disequilibria being solely produced by the melting process. Because ^{226}Ra excesses decay over an 8000 yr timescale, any ^{226}Ra-^{230}Th disequilibria will constrain (^{230}Th)/(^{238}U) (and ^{231}Pa excess) to essentially no decay since eruption. A clear observation from the samples bearing ^{226}Ra excess was the linear relationship between the Th/U of a sample and the amount of ^{230}Th-^{238}U disequilibria. The Juan de Fuca and Gorda Ridges defined two separate but parallel trends with the Juan de Fuca trend offset to higher (^{230}Th)/(^{232}Th).

U-series work continues in the NE Pacific. Goldstein et al. (1993) reported combined ^{230}Th-^{238}U and ^{231}Pa-^{235}U for 7 samples from the Juan de Fuca and Gorda Ridges and used these data to develop a technique for dating off-axis basalts (see Section 2.4). The concordance of the two dating methods argued that the melting process along both of these ridges was relatively uniform and predictable through time. Mass spectrometric analyses of ^{230}Th, ^{226}Ra and ^{231}Pa excesses in age constrained samples (known to have erupted in the 1980s) from the New Mounds area of the Juan de Fuca ridge further reinforced the concept of uniformity in the magma generation process (Lundstrom et al. 1995) (Fig. 2E). Combined with the LANL data (Goldstein et al. 1992; Goldstein et al. 1993; Volpe and Goldstein 1993), all samples from the Juan de Fuca ridge form good linear correlations for three parent-daughter pairs as a function of Th/U, strongly implying they resulted from binary mixing of melts. Lundstrom et al. (1995) interpreted this systematic behavior to reflect mixing of melts derived from heterogeneous sources which melted at different depths. Several recent eruptions documented by seismology and repeat bathymetric surveys have occurred on the Juan de Fuca and Gorda ridges in the last 10 years. Rubin et al. (1998) analyzed several of these by ^{210}Po-^{210}Pb disequilibria determining two-month eruption windows centered on early 1996.

The area between the Siquieros and Clipperton Transforms on the East Pacific Rise at 8-10°N (often referred to as "9°N") has been a focus of intense study over the past 10 years. The first U-series work in this area involved combined ^{230}Th and ^{231}Pa analyses of samples collected by submersible, rock coring and dredging along the southern portion of 9°N (Goldstein et al. 1993) (Fig. 3). Volpe and Goldstein (1993) analyzed three of these

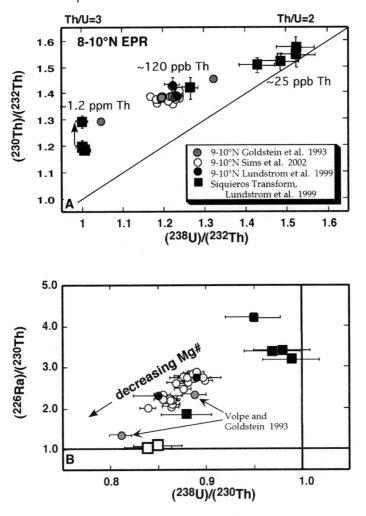

Figure 3. (A) Equiline diagram for 8-10°N EPR, a well studied ridge in terms of U-series disequilibria and other geochemical tracers. The presence of ^{226}Ra excess in all but the two open square samples constrains ages since eruption to be <8 kyr. Thus negligible decay of ^{231}Pa and ^{230}Th excess has occurred since eruption. One of the open square samples has been dated by combined ^{231}Pa and ^{230}Th excesses; the black arrow points to the age corrected position of this sample (see Section 2.4; Lundstrom et al. 1999). The data for both axial ridge samples and samples from the Siquieros Transform form a well defined trend that is interpreted by to reflect mixing of melts derived from different depths in the melting column. Whether these melts reflect derivation from a heterogeneous source or reflect progressive depletion of an originally homogeneous source is currently debated. Th concentrations vary by a factor of ~50 between the high Th/U and low Th/U end members. The open symbol samples have no age constraint (^{226}Ra excess) and are enriched MORB derived from a source with significantly higher ^{87}Sr/^{86}Sr. All other samples have identical ^{87}Sr/^{86}Sr. (B) (^{226}Ra)/(^{230}Th) vs. (^{238}U)/(^{230}Th) for the same samples are positively correlated. Since several of these samples have known eruption dates, the correlation does not reflect the effect of post-eruption aging but rather reflects differences in the disequilibria in the erupted magma. Notably, this trend also corresponds to a progressive decrease in Mg# (Sims et al. 2002) indicating a probable relationship between the extent of crustal level differentiation (Mg#) and the mantle melting process as indicated by (^{238}U)/(^{230}Th). Note that all samples from 8-10°N EPR having ^{226}Ra excess have (^{231}Pa)/(^{235}U) > 2.5.

samples for $(^{226}\text{Ra})/(^{230}\text{Th})$ which ranged from 1.3-2.2. The presence of ^{226}Ra excess constrained $(^{231}\text{Pa})/(^{235}\text{U})$ to 2.3-2.76 and $(^{230}\text{Th})/(^{238}\text{U})$ to ~1.1-1.23 at the time of eruption. The 9°N area is known as a robust, magma rich portion of ridge with observed MORB having a high degree of chemical homogeneity. Consistent with this, Th and U concentrations, Th/U (2.4-2.55) and $(^{230}\text{Th})/(^{232}\text{Th})$ (1.34-1.39) for all but 3 of the Goldstein et al. (1993) samples show limited variation compared with most MORB suites. Whether the chemical homogeneity of the basalts from this area reflects the mantle source or reflects the efficiency of mixing due to the continuous presence of a magma lens in the crust (Sinton and Detrick 1992) is currently actively debated.

In order to examine the U-series disequilibria in basalts having greater chemical diversity than those from the 9°N ridge axis, Lundstrom et al. (1999) analyzed samples from the Siquieros Transform and Lamont Seamounts (located at ~9°50' N) (Fig. 3). Samples ranged from enriched MORB (EMORB) having distinctly higher $^{87}\text{Sr}/^{86}\text{Sr}$ and Th concentrations (1.2 ppm) to primitive, high MgO basalts having highly depleted incompatible elements contents (Th ~25 ppb). Sims et al. (2002) analyzed the high MgO basalts for $^{87}\text{Sr}/^{86}\text{Sr}$ finding that they were identical in composition to samples from the 9°N ridge axis. Notably, the disequilibria data for all samples (from EMORB to NMORB from the 9°N axis to the depleted MORB) lay along common trends as a function of their Th/U similar to the behavior on the Juan de Fuca Ridge (Fig. 3). Based on concentrations progressively increasing with Th/U, Lundstrom et al. (1999) inferred that typical NMORB on the ridge axis reflected mixing of melts generated from distinct enriched and depleted mantle sources with a 20:1 volumetric ratio of depleted MORB mixed with EMORB accounting for observed NMORB. The most depleted basalts had $(^{226}\text{Ra})/(^{230}\text{Th})$ greater than 4 resulting in an inverse correlation between $(^{226}\text{Ra})/(^{230}\text{Th})$ and $(^{230}\text{Th})/(^{238}\text{U})$ or Th/U. Unfortunately, none of the Siqueiros samples had any age constraint such that definitively attributing this inverse correlation to melting was impossible.

The discovery of volcanic eruptions in 1991 and 1992 generated great interest in U-series studies at 9-10°N because these eruptions produced samples with known eruption ages. These age constrained axial samples indicate the inverse correlation between $(^{226}\text{Ra})/(^{230}\text{Th})$ and $(^{230}\text{Th})/(^{238}\text{U})$ exists upon eruption and results from melting and differentiation processes (Sims et al. 2002) (Fig. 3). Age constraint for these samples comes from both submersible observation (Haymon et al. 1993) and $^{210}\text{Po}-^{210}\text{Pb}$ disequilibria (Rubin et al. 1994). Notably, the inverse relationship between ^{230}Th excess and ^{226}Ra excess exists not only for the entire 8-10°N area but also solely for samples from the 9°N ridge axis (Sims et al. 2002). ^{226}Ra excesses decrease with Mg# which is interpreted to reflect either open system magma chamber residence times or transport times of hundreds of years. There is also a good inverse correlation between Mg# and ^{230}Th excess that is quite surprising because neither magma chamber residence times nor fractional crystallization should affect the amount of $(^{230}\text{Th})/(^{238}\text{U})$. Sims et al. (2002) interpret this observation to reflect distinct initial Mg# and fractionation histories for melts derived from different depths in the melting column.

A recent study of basalts at 9°30'N examined U-series disequilibria in samples ranging from on-axis locations to 33 km off-axis in both directions (Zou et al. 2002). The data from this study do not conform to the trends of the data observed from the earlier 8-10°N EPR studies (Fig. 3). Although there is no age constraint on these samples, significant $^{230}\text{Th}-^{238}\text{U}$ disequilibria exist in samples located far off-axis implying that eruptions of new magmas occur up to ~30 km off axis. This is significantly farther off-axis than inferred previously (Goldstein et al. 1994). One puzzling observation of these data is that the most chemically enriched samples (broadly characterized as EMORB) have ^{238}U excess; enriched MORB previously analyzed have exclusively had ^{230}Th

excess. Although this behavior has yet to be observed anywhere else in the global ridge system, it could relate to the off-axis nature of this volcanism.

Bourdon et al. (1996a) reported the first mass spectrometric analyses of MORB from the Mid-Atlantic Ridge investigating the role of hotspot interactions on the generation of U-series disequilibria (Fig. 2A). This study examined MORB along the progressively shallowing ridge leading up to the Azores platform at 38-40°N MAR (FAZAR area), finding that the amount of ^{230}Th excess was inversely related to the axial depth. These authors also suggested that a reduction in the rate of solid upwelling could be observed near the ends of spreading segments relative to the middle of spreading segments (a transform fault effect). The FAZAR data alone, along with a global compilation of existing (^{230}Th)/(^{238}U) data, produce correlations between the amount of ^{230}Th excess and the axial depth of a ridge (Bourdon et al. 1996a; Bourdon et al. 1996b). This correlation was interpreted to reflect greater amounts of melting in the garnet stability field occurring beneath more shallow ridges (see Section 4). In addition, Bourdon et al. (1996b) presented data for a few samples from the Tamayo Fracture zone (22-23°N EPR), the Australian-Antarctic Discordance and the northern MAR. Of particular interest is a depleted basalt from 29-30°N MAR that had a small but analytically reproducible excess of ^{238}U, the first such measurement by mass spectrometry, which was attributed to melting within the spinel peridotite stability field (Fig. 2A).

Lundstrom et al. (1998a) examined (^{230}Th)/(^{238}U), (^{231}Pa)/(^{235}U) and (^{226}Ra)/(^{230}Th) disequilibria in samples from 33°S MAR (Fig. 2G). This area consists of several geochemically distinct segments, one of which has the largest magnitude mantle Bouguer gravity anomaly observed on any ridge and a corresponding spike in enriched geochemical characteristics (Michael et al. 1994). The gravity anomaly was interpreted to reflect increased solid upwelling rates beneath the enriched segment. Although this study examined dredge samples from the slow spreading MAR, the presence of (^{226}Ra)/(^{230}Th) disequilibria constrained (^{230}Th)/(^{238}U) and (^{231}Pa)/(^{235}U) to essentially their eruptive values for 7 out of 15 samples. ^{231}Pa excesses in the gravity anomaly segment were significantly lower than those from the adjacent segments or any other zero-age MORB thus far measured (Table 2).

Other recent studies from the MAR include an examination of the chronology of samples from the MARK area (~23°N MAR) and analyses from the ridges leading up to Iceland. Like the southern MAR, samples from 23°N MAR had (^{231}Pa)/(^{235}U) > 3.0 but had low (^{230}Th)/(^{238}U) (1.03 to 1.05) (Sturm et al. 2000). Peate et al. (2001) examined the (^{230}Th)/(^{238}U) systematics along the Reykjanes ridge south of Iceland with age constraints provided by ^{226}Ra excess (7 out of 12 samples had ^{226}Ra excess) (Fig. 2H). No relationship between ^{230}Th excess and axial depth was observed for these samples that ranged from bathymetric depths of 300 to 1825 m. Age constrained samples formed a trend on an equiline diagram which had a near horizontal slope. Finally, a study of (^{230}Th)/(^{238}U) from the Kolbeinsey ridge north of Iceland (Sims et al. 2001) found disequilibria ranging from moderate ^{230}Th excesses to small but significant ^{238}U excesses in the most depleted samples with the data also defining a near-horizontal slope on an equiline diagram.

Overall, U-series disequilibria in MORB show several systematic behaviors that provide insight into the melting process occurring beneath ridges. Shown by Figures 2 and 3, (^{230}Th)/(^{238}U) repeatedly correlates with Th/U for sample suites from both the Pacific and Atlantic ridges. Age constrained samples from 9°N EPR and the Juan de Fuca Ridge indicate an inverse correlation between ^{226}Ra excess and ^{230}Th excess. To make this observation more robust, however, more ^{226}Ra analyses of age-constrained samples, particularly those representing the high Th/U end of the trends, are needed. Lastly,

Table 2. Observed disequilibria and models for 33°S MAR (Lundstrom et al. 1998)

Depth (km)	gt/cpx mode	W_s (cm/yr)	ϕ_{max}	$\dfrac{(^{230}Th)}{(^{238}U)}$	$\dfrac{(^{231}Pa)}{(^{235}U)}$	F (%)
observed at segment 2, 33°S MAR:				1.13-1.20	2.85-2.91	
70	0.12/0.08	1.5	0.15%	1.07	2.47	18
80	0.14/0.09	1.5	0.15%	1.19	2.76	22
observed at segment 4, 33°S MAR (gravity anomaly segment):				1.16-1.19	1.77-2.08	
80	0.14/0.09	5	0.15%	1.13	1.88	22
110	0.20/0.12	8	0.15%	1.17	1.94	34
80	0.14/0.09	2	0.35%	1.14	1.91	22
DM	0.20/0.12	1.5	0.15%	1.21	1.95	
DM	0.12/0.08	1.5	0.15%	1.17	1.65	
observed at segment 5, 33°S MAR:				1.06-1.25	3.16-3.57	
70	0.12/0.08	1.5	0.05%	1.10	3.19	18
80	0.14/0.09	1.5	0.05%	1.24	3.63	22

Model results reflect calculations of ingrowth melting using either the Spiegelman and Elliott (1993) equilibrium transport model (models with numerical values for the depth of melt initiation) or the Richardson and McKenzie (1994) dynamic melting model (DM). Melting terminates in all models at 25 km depth. W_s is the assumed initial solid upwelling rate and ϕ_{max} is the maximum porosity at the top of the melting column. Calculations use the garnet and clinopyroxene modes given along with the following partition coefficients: $^{gt/l}D_U = 0.015$; $^{gt/l}D_{Th} = 0.0015$; $^{cpx/l}D_U = 0.01$; $^{cpx/l}D_{Th} = 0.015$; $^{cpx\,or\,gt/l}D_{Pa} = 0.00067$. The productivity in all models is assumed to be 0.4%/km. F is the degree of melting and is equal to (depth − 25)*0.4%.

MORB samples having ^{226}Ra excess and thus age constraint always have $(^{231}Pa)/(^{235}U) > 2$ and sometimes exceeding 3.5.

Several studies have emphasized that the trends in disequilibria are consistent with mixing of melts (Ben Othman et al. 1990; Lundstrom et al. 1995; Bourdon et al. 1996b; Lundstrom et al. 1999; Sims et al. 2002). One strong argument for mixing comes from plotting the disequilibria such that both axes have a common denominator. The linearity of the data, in particular for 9°N EPR (Fig. 3) and the Juan de Fuca, is striking and fully consistent with mixing of melts. This is in agreement with many previous studies of MORB which have emphasized that mixing of melts is a ubiquitous process occurring beneath ridges (Dungan and Rhodes 1978).

Based on the observations at 9°N EPR, the end member melts forming the mix have been postulated to reflect derivation from different depths in the melting column. Because it contains ^{230}Th excess, has higher incompatible element contents and sometimes has slight enrichment of middle rare earth elements (REE) over heavy REE, the high Th/U end member is interpreted to reflect melting at greater depths in the melting column (in the presence of either garnet or aluminous clinopyroxene) (Lundstrom et al. 1999; Sims et al. 2002). The low Th/U end member has low ^{230}Th excess (or may even have ^{238}U excess in some cases), lower incompatible element concentrations and higher ^{226}Ra excesses. This has led to the interpretation that the low Th/U end member reflects melting of a depleted source at shallow depths: either this source begins to melt at more shallow depths because of its refractory nature (previous source depletion) or it simply reflects source depletion that occurred during progressive upwelling and melting of the solid. Evidence for shallow melting forming this end member includes the short transport times required by the large ^{226}Ra excesses and the lack of any ^{230}Th excess as expected for melts derived from depths greater than ~60 km.

The observation of ^{238}U-^{230}Th equilibrium or ^{238}U excess in some depleted samples further reinforces the hypothesis that the low Th/U end member results from melting at shallow depths in the presence of diopsidic clinopyroxene. For instance, the Siquieros depleted MORB have ^{238}U-^{230}Th equilibrium despite their youth based on the presence of ^{226}Ra excess (Lundstrom et al. 1999); these samples also have primitive major element compositions (Mg# = 0.71) that closely resemble multiply saturated melts of spinel lherzolite at 1 GPa pressure. ^{238}U excesses have also been observed in several depleted, primitive basalts from the Garrett Transform (Tepley et al. 2001) and one basalt from the MAR (Condomines et al. 1981), all of which also resemble experimental melts of peridotite at 1 GPa. The major element composition for the ^{238}U excess bearing sample of Bourdon et al. (1996b) is not known; however, it comes from an area of the MAR known to have erupted primitive high Al_2O_3 MORB nearly identical in composition to the Siquieros basalts (Schilling et al. 1983). The ^{238}U excess samples from Kolbeinsey also are highly depleted in incompatible trace elements (Sims et al. 2001). Thus, MORB which lack any signature of melting in the garnet stability field (lack of ^{230}Th excess) consistently have primitive major element compositions resembling primary melts of spinel lherzolite at 1 GPa (Lundstrom et al. 2000).

The origin of observed variations in Th/U in MORB is currently debated. Some advocate that the range in Th/U primarily reflects variations in source composition (original compositional heterogeneity before melting; Lundstrom et al. 1999). Others argue that all Th/U variation is attributable to changes in Th/U during the melting process itself (depletion of Th relative to U in the source during progressive melting; Sims et al. 2002). Since the disequilibria correlate with Th/U, these two interpretations fundamentally differ in whether or not the disequilibria reflect a signature of source heterogeneity on the melting process. The observations regarding this debate are somewhat ambiguous. On the one hand, EMORB with distinct $^{87}Sr/^{86}Sr$ lie along the same mixing trend as all other basalts at 9°N EPR, consistent with these basalts mixing with the depleted, primitive basalts from Siquieros to form typical NMORB (Lundstrom et al. 1999). Wide variations in Th concentration occur over small spatial scales within the Siquieros Transform (50 fold difference occurs within samples only 35 km apart), possibly indicating a marble cake mantle source. On the other hand, both the NMORB and depleted MORB from 9°N have identical long-lived isotope signatures (for Sr, Nd, Pb and Hf isotopes) arguing that these samples reflect melting of a source that is homogeneous over the length scale of melting (Sims et al. 2002). Progressive source depletion during melting should result in the residual solid having lower Th/U at more shallow depths so melting could explain the variations. The question remains whether the variations in Th/U observed dominantly reflect the signature of source or depletion from melting.

Further discoveries about U-series disequilibria in MORB are certain to come. Large portions of ridge now characterized for major and trace elements and radiogenic isotopes are yet to have even cursory studies of U-series disequilibria. For instance, only two data from the entire Indian Ocean spreading system have been published to date (Bourdon et al. 1996b). Two challenges with unknown outcomes will impact the future of U-series disequilibria in MORB: 1) improvements in technique will increase analytical sensitivity (allowing for smaller samples) and throughput; and 2) development of new age dating techniques could provide needed age constraint on samples making a greater proportion of the disequilibria measurements in MORB meaningful for constraining the timescales of melting and differentiation.

2.4. Applications: dating of MORB using U-series disequilibria

Because parent-daughter disequilibrium is ubiquitously generated by mantle melting, the decay of ^{230}Th, ^{231}Pa, ^{226}Ra, ^{210}Po, and ^{210}Pb back to equilibrium with their parents can be used for dating or at least constraining the age of basalts since eruption. Unfortunately, accurate age dating requires knowledge of the amount of initial disequilibria at the time of eruption, which is not straightforwardly known. At the very least, the presence of any parent-daughter disequilibria can provide important age limits on samples by constraining a sample's maximum age to less than 5-6 times the half-life of the daughter nuclide.

The first attempts to develop a dating method followed the approach of constructing an isochron diagram normalizing a radioactive nuclide to a stable element analog. Reinitz and Turekian (1989) used K as a stable analog for ^{226}Ra on northern EPR samples to calculate ages for two samples which had lower (^{226}Ra)/(^{230}Th) than expected based on their K/U ratio. Similarly, Rubin and Macdougall (1990) combined (^{226}Ra)/(^{230}Th) measured by alpha spectrometry with Ba/Th (Ba measured by ICPMS) using Ba as an analog for Ra. For this dating method (or the K analog method) to be valid, two assumptions must be met: 1) there must be a constant initial (^{226}Ra)/(^{230}Th) and (Ba or K)/Th in the mantle source; and 2) Ra and Ba (or K) must behave identically during the melting process.

There is little support for either of these assumptions. It is unlikely that the Ba/Th ratio in the source would remain constant given the variability in incompatible element ratios observed in the mantle. If Ba/Th reflects mixing between melts having distinct Ba/Th (as appears to be the case with Th/U), then the calculated date will be meaningless. Second, Ba is unlikely to be an analog for the behavior of ^{226}Ra in the melting column. D_{Ba} is not identical to D_{Ra} meaning that batch melting or fractional crystallization will fractionate Ba and Ra (Cooper et al. 2001; Blundy and Wood 2003). Even if these partition coefficients were identical, the support of ^{226}Ra by ^{230}Th during dynamic processes means that its concentration cannot be estimated by any stable element analog such as Ba (see the Section 3 discussion of ingrowth melting processes). If such a process does account for the observed disequilibria, Ba/Th will not provide information about changes in (^{226}Ra)/(^{230}Th).

A method for dating basalts over longer timescales uses measured ^{230}Th and ^{231}Pa excesses in off-axis basalts and calculates an age assuming that the disequilibria in basalts erupted at the ridge axis now are equal to the initial disequilibria in these systems in the past (Goldstein et al. 1992). Initially, these workers dated off-axis basalts from the Juan de Fuca and Gorda ridges using solely the decay of ^{230}Th excess. Although the progressive decrease in ^{230}Th excess with distance off-axis was qualitatively consistent with the hypothesis, the ages obtained had no quantitative validation other than general agreement with asymmetric spreading rates based on paleomagnetism. Subsequently, Goldstein et al. (1993) used both ^{231}Pa and ^{230}Th excesses to calculate ages for off-axis samples from the Juan de Fuca and Gorda ridges as well as 9°N EPR. The observed concordance of the two systems for samples ranging from 0 to 130 kyrs in age provides validation to both dating methods because the ^{231}Pa and ^{230}Th systems are independent of one another in terms of the rates of decay. Although the use of present day ^{231}Pa and ^{230}Th excesses to indicate the initial disequilibria in the past is clearly an assumption, the fact that both systems give ages within error of each other suggests that the method works for these areas of ridge. There are, however, important melting dynamics that are overlooked by this assumption and the method should be used with caution at ridges where melting and mixing processes are less uniform. With this chronometer in hand, Goldstein et al. (1994) showed that some samples from the 9°N EPR up to 4 km off the axis were recently erupted indicating the width of the neovolcanic zone at the EPR extends outside the axial summit trough.

Other studies have used the techniques of Goldstein et al. (1993) and Rubin and Macdougall (1990) to date MORB samples. Lundstrom et al. (1998a) applied the combined ^{231}Pa-^{230}Th excess dating method to basalts from the southern MAR. For two segments of ridge, the technique produced concordant ages between the two systems for 3 samples; however, large errors on the ages prohibited clear validation of the dating technique. Lundstrom et al. (1999) used measured ^{231}Pa and ^{230}Th excesses in an EMORB from the 9°N axis as the initial disequilibria for an EMORB from the Siquieros Transform resulting in concordant ^{231}Pa and ^{230}Th ages. Sturm et al. (2000) used both the ^{231}Pa-^{230}Th method and the (^{226}Ra)/(^{230}Th)-Ba/Th method to date 4 samples from the MARK area of the MAR. The ages for all 3 chronometers were within error although the propagated errors on these relatively young ages (~10 kyr) did not provide a robust validation of the age dating techniques.

One other highly useful chronometer is measurement of ^{210}Po-^{210}Pb disequilibria. ^{210}Po has a half-life of 138.4 days making the chronometer active for ~2 yrs. ^{210}Po-^{210}Pb fractionation is based on Po but not Pb partitioning into volatiles during degassing (Gill et al. 1985). By repeat analysis of ^{210}Po, Rubin et al. (1994) constrained the time of eruption of several samples from 9°N EPR to windows of ~100 days. These dates are consistent with eruption windows based on submersible observation. Thus, this technique can provide critical age constraints for other U-series parent-daughter pairs but requires that samples be collected and analyzed as soon as possible after eruption.

2.5. Assessing secondary contamination in creating U-series disequilibria

A great deal of work has gone into assessing the role of secondary processes in creating the observed U-series disequilibria in MORB. Contamination of U-series nuclides in the ridge environment is highly possible given the low U and Th concentrations in MORB and the relatively higher concentrations of ocean sediments and seawater. A key issue is that Mn oxide coatings that commonly form on the seafloor preferentially sequester daughter nuclides such as ^{230}Th from U rich seawater. Thus sample preparation must be rigorous. In addition, the ridge environment contains hydrothermally derived materials, often with high Ba concentrations forming the mineral barite and high Ra activities (Von Damm et al. 1985) that are known to interact with and contaminate newly formed crust in certain localities. Because of this, the large ^{226}Ra excesses in MORB have often been speculated to reflect secondary contamination. For instance, consideration of the experimental constraints on mantle permeability and the inability to explain the low melt porosities needed by U-series models of melt flow led Faul (2001) to argue that ^{226}Ra excess resulted from an unspecified contamination process.

Bourdon et al. (2000) performed the most systematic evaluation of the role of contamination thus far. These workers examined the effectiveness of two methods, (1) handpicking and acid leaching together and (2) acid leaching alone, in removing surface contamination by measuring ^{10}Be in several MORB samples having ^{231}Pa and ^{230}Th excesses. Since ^{10}Be is only created at the earth's surface through cosmogenic processes, it is not present above detection limits in the MORB source or in pristine MORB but is present in oceanic sediments and Mn oxides (^{10}Be is scavenged from seawater similar to ^{230}Th). The results showed that the samples that were acid leached alone had small but elevated ^{10}Be. However, there was no detectable ^{10}Be in the handpicked, acid leached glasses. Bourdon et al. (2000) calculated the maximum weight fraction of contaminant in the handpicked glass to be <10^{-4} to 10^{-5}, meaning that less than 15% and 1% of the observed ^{231}Pa and ^{230}Th excesses in the samples, respectively, could be explained by incorporation of sedimentary material. Thus, this study eliminates incorporation of sediments as the sole cause for the creation of ^{231}Pa-^{235}U and ^{230}Th-^{238}U disequilibria.

In addition to ^{10}Be, a number of other geochemical tracers such as $(^{234}$U$)/(^{238}$U$)$, B, Cl and Ba consistently argue that the U-series disequilibria in MORB do not reflect secondary contamination. One important check that all U-series studies of MORB routinely perform is analysis of $(^{234}$U$)/(^{238}$U$)$. Since $(^{234}$U$)/(^{238}$U$)$ of seawater is well established as 1.15 (Thurber 1962; Henderson and Anderson 2003), incorporation of seawater or seawater derived components will result in positive deviations from secular equilibrium, a condition found for pristine young volcanic rocks.

Boron concentrations and isotopes are also useful geochemical tracers of contamination in MORB. Boron concentrations are low (<2 ppm) in unaltered ocean floor basalt but high in altered basalts (>8 ppm B) (Spivack and Edmond 1987; Ryan and Langmuir 1993). Goldstein et al. (1989) measured B concentrations in their samples and found them to be less than 1.6 ppm, inconsistent with contamination. More recently, B isotopes have been used to assess contamination since large differences in δ^{11}B are known to exist between seawater, sediments, and unaltered MORB. Sims et al. (2002) reported that δ^{11}B for their 9°N EPR samples were inconsistent with incorporation of any seawater or seawater-derived material.

Because the concentration difference between MORB and hydrothermal chimney materials for both Ba and Ra is greater than 10^6, incorporation of small amounts of barite from hydrothermal systems could create the high $(^{226}$Ra$)/(^{230}$Th$)$ observed in MORB. However, if ^{226}Ra excess results from this process, then Ba concentrations (and Ba/Th) should be equally raised by such contamination. However, Ba/Th in highly depleted glasses from the Siquieros Transform that have $(^{226}$Ra$)/(^{230}$Th$)$ greater than 3 are typical for MORB and do not have anomalously high Ba contents (Lundstrom et al. 1999).

Measurement of Cl concentrations provides another technique for assessing contamination (Michael and Cornell 1998). Hydrothermally altered crust contains high Cl concentrations and incorporation of such material into a magma will raise its Cl/K. Michael and Cornell (1998) have shown that contamination results in Cl/K elevated from a baseline value for a given area of ridge. Sims et al. (2002) reported Cl/K for a number of young MORB from the Siquieros/9°N EPR area having large ^{226}Ra excesses. No correlation was found between Cl/K and ^{226}Ra excesses nor did any of the samples with ^{226}Ra excess have anomalously high Cl/K.

A final argument against contamination creating the observed U-series disequilibria is the systematic behavior of the excesses for a given area of ridge (e.g., the behavior at 9°N EPR and the Juan de Fuca/Gorda ridges). The coherent behavior correlates with changes in incompatible trace element concentrations and therefore is unlikely to reflect a contamination process that would be expected to be more random in nature. Thus, although contamination must always be addressed on an individual sample basis, there is overwhelming evidence that the ^{230}Th-^{238}U, ^{231}Pa-^{235}U and ^{226}Ra-^{230}Th disequilibria in some if not most MORB do not reflect contamination but rather reflect creation by the melting process.

3. MODELS OF U-SERIES DISEQUILIBRIA GENERATION DURING MELTING

Although U-series disequilibria have been observed in MORB for over 20 years, consensus about how the disequilibria are generated by the melting process does not yet exist. The inability to explain the disequilibria by simple melting models has fostered new ideas about how mantle melting occurs. Here, I summarize the different models to date and argue that most of the disequilibria in MORB reflect generation by the process whereby differences in element residence times create the disequilibria ("ingrowth models"—see Section 3.3). Ultimately, the verdict of whether ingrowth melting models satisfactorily explain both U-series and other geochemical observations in MORB remains a test for the future where larger amounts of U-series data are combined with full geochemical characterization.

Melting models for creating disequilibria can be broadly distinguished into two types based on whether time plays a role in the model. In one category are more traditional melting models like batch or fractional melting where all fractionation between elements is due solely to chemical partitioning of elements between melt and residual solid. I refer to these types of models as time-independent models because time does not play a role in creating the disequilibria. In such models, U-series nuclides are treated identically to other incompatible elements. In the second type of model, referred to here as ingrowth models (also known as residence time models), the unique trait of short-lived radioactivity and support of a nuclide's concentration by a parent nuclide leads to generation of disequilibria by differences in element residence times. These models are based on both time and mass conservation during dynamic processes. Both types of models critically depend on the specifics of element partitioning.

3.1. Constraints on element partitioning

Over the past 10 years a great deal of effort has gone into constraining the values of the partition coefficients for U and Th between mantle minerals and basalt. Chapter 3 (Blundy and Wood 2003) reviews this work in more detail; however, a basic review relevant to observations in MORB and modeling is given here.

Uranium and thorium are large, highly charged cations that are not readily incorporated into mantle minerals. While Th has a constant valence of +4, U can vary depending on oxygen fugacity between the +4, +5 and +6 valence states (Calas 1979). Maximum partition coefficients for U and Th into clinopyroxene and garnet are $\sim 10^{-2}$ making bulk partition coefficient in peridotite assemblages of the order 10^{-3} (Blundy and Wood 2003). Because of increased garnet and pyroxene modes, pyroxenite sources will have higher bulk partition coefficients and therefore could generate significant disequilibria during melting (Hirschmann and Stolper 1996).

The relative sense of U and Th partitioning (D_U/D_{Th}) shifts from greater than one at greater mantle depths to less than one at shallow depths. Measurements of garnet-melt partitioning consistently show $D_U > D_{Th}$ (Beattie 1993b; LaTourrette et al. 1993; Hauri et al. 1994; Salters and Longhi 1999). D_U/D_{Th} for clinopyroxene-melt varies with composition. Aluminous clinopyroxene like those on the spinel lherzolite solidus at 1.5 GPa have $D_U > D_{Th}$ and therefore can produce ^{230}Th excesses by melting spinel lherzolite (Wood et al. 1999; Landwehr et al. 2001). However, Ca-rich, Al-poor clinopyroxenes consistently show $D_U/D_{Th} < 1$ indicating that melting at shallow depths cannot produce ^{230}Th excess but could produce ^{238}U excess (LaTourrette and Burnett 1992; Beattie 1993a; Lundstrom et al. 1994; Hauri et al. 1994). Since most MORB have ^{230}Th excess, at least some portion of the magma was likely derived from depths greater than ~60 km. However, if the interpretation that NMORB reflects a 20:1 mix of low Th/U melt with high Th/U melt is correct (based on the Siquieros-9°N data; Lundstrom et al. 1999), then only 5% of the volume of melt beneath a ridge actually reflects derivation from this depth.

Although no measurements of D_{Ra} or D_{Pa} have yet been made, both of these elements are likely to be highly incompatible. For instance, D_{Ba} for clinopyroxene-melt is

~2 × 10^{-4} (Beattie 1993a; Lundstrom et al. 1994) and D_{Ra} should be less than D_{Ba}. This is because Ra^{2+} is slightly larger than Ba^{2+} (Shannon 1976) and therefore its substitution for Ca will create a greater lattice strain than Ba (Blundy and Wood 2003). The absolute values of D_{Ra} and D_{Pa} actually make little difference to melting model calculations as long as Ra and Pa are more than 10 times more incompatible than their parents (Lundstrom et al. 2000). This is because the retention of the more compatible element controls the amount of disequilibria produced. Therefore, understanding the absolute values of D_U or D_{Th} is much more important for constraining models of ^{231}Pa and ^{226}Ra excess generation than knowledge of either D_{Ra} or D_{Pa}.

3.2. Time-independent melting models

Disequilibria between members of the U-series decay chain indicate that the melting process chemically fractionates U from Th, Th from Ra, and U from Pa. The low values of D_U and D_{Th} require extremely small degrees of melting in time-independent melting models to produce the magnitude of the observed fractionations. For instance, Figure 4 uses a batch melting calculation with relatively high values for D_U and D_{Th}. This model will maximize fractionation produced by any time-independent model (fractional melting or models using more realistic partition coefficients will produce less fractionation). Nevertheless, to match the observed Ra-Th and Pa-U fractionations at 9°N EPR requires F < 0.2% (Fig. 4). This is because the important parameter ratio in any time-independent model is D/F (where F is the degree of melting) which must be near or greater than one for one of the partition coefficients in order to generate any disequilibria. Thus the small values for D_U and D_{Th} require F to be a few per mil or less in time independent models.

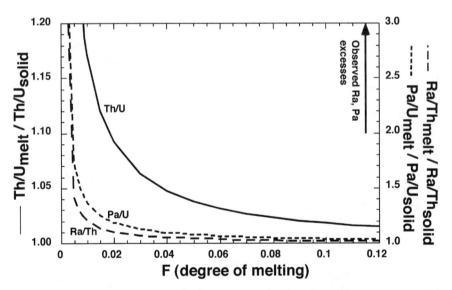

Figure 4. Fractionation of Th-U, Pa-U and Ra-Th relative to source as a function of degree of melting for batch melting. The bulk partition coefficients used for U (0.0042) and Th (0.0021) are near upper limits for garnet or spinel lherzolite based on experimental measurements. D_{Ra} (4 × 10^{-5}) and D_{Pa} (4 × 10^{-4}) are more than an order of magnitude less compatible than U and Th and the amount of fractionation is much less dependent on these values. The extent of fractionation is maximized both by using batch rather than fractional melting and by using high values for D_U and D_{Th}. Nevertheless, explaining the observed 10-40% excess of ^{230}Th observed in MORB requires F < 0.02 while ^{226}Ra or ^{231}Pa excesses of 100-250% require F < 0.2%. Maximum fractionations (at F = 0) using these partition coefficients correspond to Th/U = 2, Pa/U = 10, Ra/Th = 53.

Four observations suggest the disequilibria in MORB do not simply reflect time-independent melting. First, the degree of melting inferred from the U-series data do not match with those predicted based on the major element characteristics of MORB (estimated to be 5-30% melting). Second, if U and Th behave like the relatively less incompatible light rare earth elements (Johnson et al. 1990), the solid residue after MORB extraction (abyssal peridotites) is unlikely to retain enough U or Th to be a complementary residue to the disequilibria in MORB. Third, time-independent models of disequilibria are inconsistent with the observed relationship between different parent-daughter pairs. For instance, time-independent models predict simultaneous enrichment in Ra over Th and Th over U while age constrained MORB from 9°N EPR show an inverse correlation between these two systems. Fourth, because D_U/D_{Th} changes from >1 deep in the mantle to <1 at shallow depths, most MORB might be expected to be ^{238}U enriched relative to ^{230}Th which they are not. Finally, global observations show that $(^{230}Th)/(^{238}U)$ correlates with axial depth (Bourdon et al. 1996b) which is not easily explained by time-independent models.

Mixing of a small degree volatile-rich melt from the initiation of melting with a high degree tholeiitic melt could provide an alternative explanation for the disequilibria. However, many clear inconsistencies exist with the observed data. For instance, if small degree melts from the beginning of melting create the disequilibria, then the U-series excesses should always correlate with enrichments in other incompatible elements. However, ^{226}Ra excesses are highest in basalts that are highly depleted in incompatible elements (Lundstrom et al. 1999). Furthermore, if small degree melts create the ^{230}Th and ^{231}Pa excesses in MORB, then melts of the residue after this melting must never occur; only a few MORB have ^{238}U excess over ^{230}Th and no ^{235}U excess over ^{231}Pa has ever been observed. A caveat to this argument is that if re-melting of the residue does occur after some time has elapsed, then this argument would no longer hold as the daughter ^{231}Pa would grow into the ^{235}U rich residue. As explained in Section 3.3, this is essentially a variant of ingrowth melting; in other words, the generation of disequilibria by continuous melting of a solid will reflect the timescale of melting versus that of support by decay of the nuclide's parent (McKenzie 2000).

Time-independent melting models create U-series disequilibria by preferentially retaining one nuclide over another in the mantle residue after melting. Thus, assuming that Pa is infinitely incompatible, $(^{231}Pa)/(^{235}U) = 2.5$ in MORB (a conservative low average for age constrained MORB) requires that 60% of the U originally in the source is retained in the mantle after melting. If true, such a model would have important implications for the extraction and fractionation of elements between the crust and mantle through time. Elliott (1997) evaluated and discussed the relative importance of time-independent models versus ingrowth models and concluded that while ocean island basalts could reflect a combination of both types of melting, the observations for MORB support the disequilibria being generated solely by ingrowth melting. For the reasons listed above, time-independent melting does not appear to represent an adequate explanation for the observed U-series disequilibria.

3.3. Ingrowth melting models

Because of the difficulty in explaining the observed U-series excesses by time-independent models, interpretations of how disequilibria are created have evolved into models based on residence times. In these models, a melt phase coexists with the solid mantle but moves relative to it due to a driving force, most typically buoyancy. The physical situation under ridges can be referred to as two-phase flow because both the solid and the liquid flow. McKenzie (1984) and Scott and Stevenson (1984, 1986) derived the equations describing flow in a viscously deforming porous media. McKenzie

(1985) was the first to show the effects of two-phase flow on U-series disequilibria although incorporation of the melt transport equations into ingrowth models did not occur until the work of Spiegelman and Elliott (1993).

The physical process of melt ascent during two-phase flow models is typically based on the separation of melt and solid described by Darcy's Law modified for a buoyancy driving force. The melt velocity depends on the permeability and pressure gradients but the actual microscopic distribution of the melt (on grain boundaries or in veins) is left unspecified. The creation of disequilibria only requires movement of the fluid relative to the solid.

It is important to clearly define some of the terminology related to ingrowth models, particularly since some misuse of terms occurs due to confusion with the terms of time-independent models. Melt porosity (ϕ) is the fraction of the total volume of rock being filled by melt. In two-phase flow models, melt moves relative to the solid and so porosity at any location in the column reflects the melt production beneath that location and the permeability of the partially molten matrix. Porosity is distinct from the degree of melting, F, which is the total amount of melt produced from an initial amount of unmelted solid. Thus, a parcel of solid in a melting column can undergo a total degree of melting realistic to the major element constraints on MORB (F = 0.05-0.3) while the porosity at a given depth reflects the physical ability of the melt to move relative to the solid and can be much smaller than F. Note that the term "melt fraction" is ambiguous and should be avoided as it is sometimes used to mean porosity and sometimes used to mean degree of melting (in time-independent melting models). The term threshold porosity is used in "dynamic melting" models to signify a porosity below which a partially molten matrix is not permeable but above which melt can escape. In an open system where melt can move relative to the solid, a melting column can have low melt porosity if the partially molten matrix is highly permeable; low melt porosity, however, does not require the degree of melting to be small.

When melt moves relative to solid and chemically exchanges with the solid, elements will move at different effective velocities. Consider a situation where a fluid moves interstitially through a solid and elements exchange between the melt and solid (for simplicity we will ignore the issue of melting in this example). The effective velocity (w_{eff}) of an element in one dimension can be approximately expressed as

$$w_{eff} \approx W + \frac{(w-W)}{(1+D/\phi)} \qquad (1)$$

where D is the partition coefficient of the element, and W and w are the solid and fluid velocities, respectively (Eqn. 12 of Spiegelman and Elliott 1993). The effective velocity of any element therefore can be no faster than the fluid and no slower than the solid. A perfectly incompatible element will travel at the melt velocity while a strongly compatible element will travel at the solid velocity. As D/ϕ increases, the effective velocity of an element decreases. Thus, two elements having different partition coefficients will have different effective velocities and therefore different transit times over a given transport distance. It should be clear from Equation (1) that the ratio of D to ϕ is particularly critical for changing elemental residence times.

Differences in residence time can lead to increased support of a daughter nuclide if the parent residence time in the melting column exceeds that of the daughter (see Bourdon et al. 2003). We can quantify this effect using the ^{231}Pa-^{235}U system as an example (Fig. 5). First, assume a melt velocity of 1 m/yr, a solid upwelling rate of 1 cm/yr (inconsequential to this calculation), D_U = 0.002, D_{Pa} = 1 × 10^{-4}, a melt porosity of 0.002 and a 30 km length of melting column. Using these values within Equation (1),

Disequilibria in Mid-ocean Basalts: Basalt Genesis 195

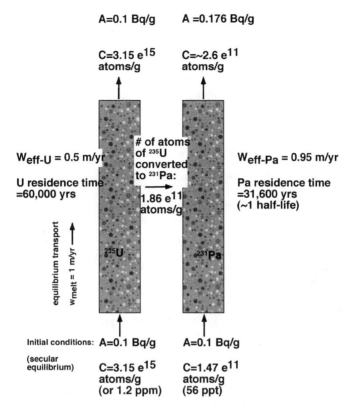

Figure 5. A semi-quantitative illustration of how differences in residence time can produce ^{231}Pa-^{235}U disequilibria. "A" stands for activity while "C" stands for concentration. Due to differences in partitioning, U has an effective velocity that is half that of Pa and has a residence time of 60,000 years in the 30 km long melt column (32,000 year residence time for Pa). The two nuclides enter the bottom of the column in secular equilibrium with 0.1 Bq/g activity corresponding to the concentrations of ^{235}U and ^{231}Pa given. Over its 32,000 year residence in the column, a little more than half of the original ^{231}Pa decays. However the decay of ^{235}U during its 60,000 year residence in the column resulting in 70% more ^{231}Pa than existed initially at the base of the column. Note that this calculation does not account for the decay of some of the ^{231}Pa produced in the melting column and so the estimated (^{231}Pa)/(^{235}U) of ~1.7 will be higher than a fully quantitative numerical calculation. Nonetheless, it readily shows how differences in partitioning affect residence time and thus disequilibria.

^{235}U will move towards the surface at an effective velocity of 0.5 m/yr resulting in a 60,000 yr residence time over the 30 km path. Pa will move only slightly slower than the melt, taking ~32,000 years (1 half-life) to go through the column. Thus, half of the original ^{231}Pa entering at the bottom arrives at the top of the column. The longer residence time of U in the column results in production of more ^{231}Pa by decay of ^{235}U than is lost by the decay of ^{231}Pa. Assuming secular equilibrium exists initially at the column bottom with all nuclides having activities of 0.1 Bq/g, then the longer residence time of U dynamically supports the Pa in the column at a higher activity level than secular equilibrium. During a 60 kyr residence time, much less than 1% of the ^{235}U will decay producing 1.9×10^{11} atoms/g of ^{231}Pa in the column. Added to the half of the original ^{231}Pa which had not yet decayed, the total ^{231}Pa concentration at the top of the column would be 1.7× higher than started at the bottom (such that (^{231}Pa)/(^{235}U)=1.7);

this calculation is not exact as some of the ^{231}Pa produced within the column will also decay lessening the (^{231}Pa)/(^{235}U) slightly. Nevertheless, this simple example using parameters typical of those used in mantle melting models shows how easily ^{231}Pa excess can be generated when elements have different effective velocities during magma ascent.

In the above example, ^{231}Pa-^{235}U disequilibrium reflects different residence times created solely by solid-fluid partitioning during fluid movement (a chromatographic effect). Addition of melting to the calculation will produce even greater differences in residence time and therefore greater potential for creating disequilibria. When melting first occurs, the more incompatible nuclide partitions into the fast moving melt while the less incompatible nuclide remains in the slowly ascending solid. This can lead to the generation of large residence time differences and large amounts of disequilibria. Thus, a large residence time difference between two nuclides is created during the initial stages of melting and it is solely this process that creates disequilibria in dynamic melting models (Section 3.4).

A final distinction between the time independent models and the ingrowth models is that ingrowth models do not require that any phase be residual to hold back one nuclide in the mantle preferentially to its daughter or parent in order to generate disequilibria. Thus, trace phases present at the initiation of melting could influence the generation of disequilibria yet not leave any other signature on the melt because they subsequently melt out. This principle also applies to ^{230}Th excesses generated by melting in the garnet peridotite stability field but preserved at the surface despite continuous melt-solid equilibration within the spinel peridotite stability field (Spiegelman and Elliott 1993).

3.4. A review of U-series melting models

Since the first development of ingrowth models (McKenzie 1985), several variations on how residence time differences can create disequilibria during melting have been proposed. These range from models based on diffusion controlled partitioning to those involving melting columns having two scales of melt porosity. The different models boil down to one factor controlling the magnitude of the disequilibria generated: time. The time of a nuclide in the melting column will reflect several parameters including the partition coefficients, the solid and melt velocities, the productivity (the amount of melting per reduction in pressure), the length of the melting column and the solid diffusion coefficients.

The relationship between the various types of models is depicted in Figure 6. Ingrowth models can be broadly divided into two types: 1) models where melt is assumed to be in equilibrium with the entirety of all mineral grains during melting (equilibrium partitioning models); and 2) models where solid-state diffusion within minerals controls the transfer of the element from the solid to the melt (diffusion control models). For equilibrium partitioning models (#1), there are two further variations: a) models where melt transport during ascent occurs in chemical isolation with the surrounding solid (dynamic melting or disequilibrium transport models) and b) models where melt continuously re-equilibrates with the solid during transport (equilibrium transport models). Ingrowth models can also be separated based on whether melt transport is explicitly included in the calculation. Models in which the transit times of different elements reflect the physics of two-phase flow are similar in number (Spiegelman and Elliott 1993; Iwamori 1994; Lundstrom 2000; Jull et al. 2002) to models where transport is simply treated as instantaneous (McKenzie 1985; Williams and Gill 1989; Qin 1993; Richardson and McKenzie 1994). Neglecting transport allows an analytical approximation for calculating the disequilibria generated (see below). An analytical approximation for solving transport based melting models is given in Bourdon et al. (2003) while the full equations for melting with transport times accounted for is given in the Appendix.

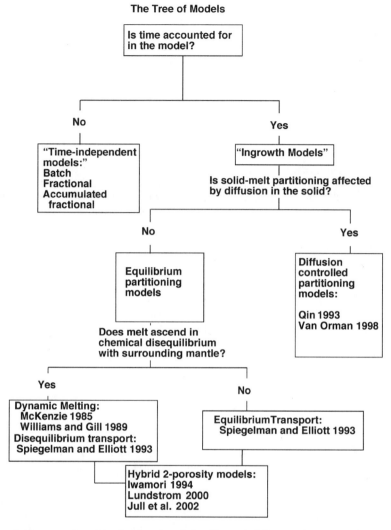

Figure 6. The tree of models showing the relationships and differences between various U-series melting models.

McKenzie (1985) proposed "dynamic melting" for the process of creating disequilibria. In this model, melt is extracted from the solid after it reaches the threshold porosity. If the threshold porosity is similar in magnitude to the partition coefficient of one nuclide of a parent-daughter pair, the two nuclides will be fractionated as melt is extracted, resulting in a residence time difference between the two nuclides at the top of the melting column. If the threshold porosity is too large, then essentially no fractionation occurs before melt removal and no disequilibria is generated. Williams and Gill (1989) used a similar dynamic melting model to examine the combined ^{226}Ra-^{230}Th and ^{230}Th-^{238}U systems and emphasized that the $(^{230}Th)/(^{232}Th)$ of a melt does not necessarily reflect the $(^{238}U)/(^{232}Th)$ of the source. Richardson and McKenzie (1994) extended the dynamic model to two dimensions by pooling melts over a 2D region where upwelling rates varied based on a solid mantle flow model. These authors gave a useful analytic approximation to dynamic melting:

$$(\text{parent}/\text{daughter}) = \frac{\lambda_d(D_p + \phi_o) + \Gamma}{\lambda_d(D_d + \phi_o) + \Gamma} \qquad (2)$$

where Γ is the melting rate in units of inverse time, ϕ_o is the threshold porosity, and subscripts p and d refer to parent and daughter, respectively. All of these models assume chemically isolated transport and instantaneous melt ascent to the surface. For ^{226}Ra excesses to be explained by dynamic melting models, the disequilibria must be created at the base of the melting column and melt ascent rates must be on the order of 50 m/yr or greater for a melt to ascend from the garnet stability field within one half-life of ^{226}Ra. Because incompatible elements are removed from the solid and never further interact with the solid, dynamic melting models are similar to near-fractional melting in that the solid residue becomes severely depleted with progressive melting.

Spiegelman and Elliott (1993) were the first to develop models that specifically accounted for the physics of melt transport. Although this work emphasized a model in which melt continuously re-equilibrated with the solid as it ascended (equilibrium transport), it also presented a model with disequilibrium transport (a model similar to dynamic melting except that it accounted for ascent times). Spiegelman and Elliott (1993) showed that given equal parameters, equilibrium transport models created greater parent-daughter disequilibria than disequilibrium transport models. This result is fully understandable in terms of residence time differences; in essence, the continuous interaction of equilibrium transport models results in residence time differences being generated throughout the melting column. In contrast, residence time differences are only generated at the initiation of melting in dynamic or disequilibrium transport melting models. Lundstrom et al. (1994) applied the equilibrium transport model to the 9°N EPR data and suggested that the differences in disequilibria between N-MORB and E-MORB might be explained solely as differences in the length of the melting column. Bourdon et al. (1996b) used a similar model to explain the observed ^{230}Th excess-axial depth relationship by varying the depth of melting initiation. Spiegelman (2000) has developed a web site (*www.ldeo.columbia.edu/~mspieg/UserCalc/*) that allows users to vary input parameters and examine the controls on parent-daughter disequilibria using an equilibrium transport model.

Both dynamic melting and equilibrium transport melting require that the porosity when two nuclides are fractionated from one another is similar to the size of the larger of the partition coefficients for the two nuclides. Given the low values of the experimental determinations of D_U and D_{Th}, the porosities required to explain the observational data in these models are generally less than 0.5% and often times closer to 0.1%. Such low porosity estimates have been criticized based on physical grounds given the low estimated mantle permeability derived from the extent of melt connection observed in experiments (Faul 2001).

One possibility for increasing the minimum porosity needed to generate disequilibria involves control of element extraction by solid-state diffusion (diffusion control models). If solid diffusion slows the rate that an incompatible element is transported to the melt-mineral interface, then the element will behave as if it has a higher partition coefficient than its equilibrium partition coefficient. This in turn would allow higher melt porosities to achieve the same amount of disequilibria as in pure equilibrium models. Iwamori (1992, 1993) presented a model of this process applicable to all elements that suggested that diffusion control would be important for all elements having diffusivities less than

10^{-19} m^2/s in the solid. Qin (1992, 1993) developed a similar model to explain the creation of U-series disequilibria and calculated optimum melting rates for melting beneath ridges based on the observed disequilibria in MORB.

More recent work has suggested diffusion control models are unlikely to explain the disequilibria observations. Iwamori (1994) concluded that local chemical disequilibrium probably had an insignificant effect on the activity ratios given realistic melting conditions except that it might limit the amount of ^{226}Ra excess generated. Note that all of these diffusion-based models were premised on estimated U and Th diffusion coefficients since no experimental determination existed. Van Orman et al. (1998) recently addressed this deficiency by presenting U and Th diffusion coefficients in diopside. As expected for large, high charges ions, both elements had low diffusion coefficients (on order of 10^{-21} m^2/s at 1200°C) with U diffusing slightly faster than Th. Thus although the slow diffusion rates might keep U and Th in the solid longer if diffusion controlled their effective partitioning, such a model cannot explain the observed ^{230}Th excess in MORB if melting occurs within the spinel lherzolite stability field (Van Orman et al. 1998).

Of the three types of ingrowth models discussed (dynamic melting, equilibrium transport and diffusion controlled partitioning), equilibrium transport models have been most successful at explaining the observable U-series data including the large excesses of ^{226}Ra and ^{231}Pa and the relationship between axial depth and ^{230}Th excess (Bourdon et al. 1996a). However, continuous chemical equilibration between ascending melt and the surrounding solid is not consistent with other geochemical observations that mandate that MORB cannot maintain equilibrium with mantle at shallowest depths. For instance, the observation that MORB is undersaturated in orthopyroxene at low pressure (O'Hara 1968) as well as the strong depletion in trace elements in abyssal peridotites (Johnson et al. 1990) argues that some of the melt beneath ridges is channelized during ascent at shallow depths. Spiegelman and Kenyon (1992) examined how melt channelization affects equilibration and concluded that a vein network with veins 10 cm apart would cause extensive chemical disequilibrium between solid and melt. Thus, creating a physical scenario for chemical disequilibrium during ascent is not difficult.

This has lead to the development of hybrid "two porosity" models that combine elements of equilibrium transport with chemically isolated transport in channels. These models have sprung forth from the hypothesis of Kelemen et al. (1995) that discordant dunites within exposed ophiolite mantle are former melt channels formed during the solid's adiabatic rise beneath a ridge axis. Based on this, Kelemen et al. (1997) proposed that the MORB melting column might consist of melt ascending by porous flow distributed at two scales of porosity: a network of high porosity melt channels which pass through and drain the ambient low porosity mantle that supplies melt. Indeed, Spiegelman and Elliott (1993) discussed the idea that two scales of melt porosity could be important to U-series melting models.

Iwamori (1994) was the first to develop such a model describing the rate of extraction of melt from a peridotite melt source into channels by a suction parameter, "S." By varying the suction term, the melting model could vary between a pure equilibrium transport model and a pure dynamic melting model. However, because this model assumed instantaneous transport within the channels, the dynamic melting end member always produced larger disequilibria than the equilibrium transport end member. Lundstrom (2000) developed a similar model having channels that served to drain the melting peridotite region. Instead of assuming instantaneous transport, melt porosities and velocities in the channel were calculated based on the flux balance from a cylindrical melting region surrounding the channel. This two porosity model showed that large (^{226}Ra)/(^{230}Th) and (^{231}Pa)/(^{235}U) disequilibria could be generated when S = 0.5 (half of the melt going to channels and half flowing porously in the lherzolite) and that depletions of trace elements in the solid similar to those observed in abyssal peridotites could also be produced. Because the sense of U-Th partitioning can change dramatically depending on depth (see Section 3.1), this study did not attempt to model the $(^{230}Th)/(^{238}U)$ generated but rather focused on explaining the large ^{231}Pa and ^{226}Ra excesses observed in melts interpreted to reflect shallow melting. Jull et al. (2002) recently reported a two porosity model that generated large ^{226}Ra excesses at shallow depths by ~60% of the total melt flux within a melting column flowing through low porosity peridotite at shallow depths. Such a model could explain the inverse relationship between ^{230}Th excess and ^{226}Ra excess observed at 9°N EPR as well as correlations between La/Yb, Th/U and U-series disequilibria.

A number of issues remain for future modeling work. One clear deficiency in ingrowth models to date is treating melting beneath ridges as a steady state process for U-series nuclides. Steady state assumes that the solid mantle entering the melting zone is homogeneous in composition for all elements including U and Th. However, it is clear that mantle heterogeneity strongly affects the distribution of U and Th in the mantle. For instance, Th concentrations, which vary by a factor of 100 between the most enriched and depleted basalts from EPR seamounts, correlate with changes in $^{87}Sr/^{86}Sr$ implying they reflect source variations (Niu and Batiza 1997; Niu et al. 2002). More complex non-steady state models that follow a concentration pulse derived from melting a blob of enriched mantle remains a goal for future work. In addition, two-dimensional models of stable trace elements show that dramatic changes in trace element patterns occur relative to expectations of batch or fractional melting result when melt and solid flow independently of one another (Spiegelman 1996). Such models have yet to be explored for U-series elements. Physical models for the development of channels (Spiegelman et al. 2001) are now being exploited to examine the trace element variability due to the presence of channels. Similar models for U-series disequilibria could provide a completely new view of interpreting the excesses observed in MORB.

4. RELATING MODELS TO OBSERVATIONS

Because of the large number of parameters that affect the generation of disequilibria in ingrowth melting models, it is not possible to invert the observable data for "best fits" to melting beneath ridges. Instead, varying parameters within forward models and relating the results to observations are typically used to gain insight into melting processes beneath ridges. Such forward models are best constrained by the simultaneous use of three parent-daughter pairs. For instance, a major control on the production of both ^{230}Th and ^{231}Pa excesses in ingrowth models is D_U and a successful model should be able to match all of the observed disequilibria. Likewise, D_{Th} plays a critical role in determining the amount of both ^{230}Th excess and ^{226}Ra excess.

The disequilibria systematics at 33°S MAR provide one example of how the disequilibria, combined with modeling, provide insight into the dynamics of melting (Lundstrom et al. 1998). Of note at 33°S MAR was the observation of significantly lower ^{231}Pa excess in samples from the segment having the large mantle Bouguer gravity anomaly. Because the gravity low was interpreted to reflect increased solid upwelling velocity beneath this segment (Michael et al. 1993), models of upwelling rate variation were explored and it was found that increasing the upwelling velocity by a factor of 2 could explain the difference in ^{231}Pa excess between the anomalous segment and surrounding segments. However, the data were equally well explained by variations in other parameters such as increased melt porosity or melting by a dynamic melting process rather than an equilibrium transport process (Table 2).

Comparison of different models also plays a role in distinguishing the style of the melting process beneath ridges. For instance, the inverse relationship between (^{226}Ra)/(^{230}Th) and (^{230}Th)/(^{238}U) for 9°N EPR samples with clear age constraint (Sims et al. 2002) indicates that disequilibria in these two parent-daughter pairs must be generated at different locations in the melting column. Time-independent melting models or dynamic melting models generally will predict positive correlation between these two parent-daughter pairs. Equilibrium transport models on the other hand have the ability to generate ^{226}Ra excess shallow in the melting column by a chromatographic process. Jull et al. (2002) have reproduced the observed inverse relationship using a two porosity model where the ^{230}Th excesses are created deep in the melting column while ^{226}Ra excesses reflect melt-rock interactions at relatively shallow depths.

Two global observations of U-series disequilibria, combined with ingrowth models, have further constrained the ridge melting process. Bourdon et al. (1996b) found that the amount of ^{230}Th excess in a ridge basalt correlated with the axial depth of the sample (Fig. 7). These authors attributed this global correlation to higher mantle temperatures causing melting to initiate at greater depth beneath more shallow ridges and quantitatively explained the relationship by varying the depth of melt initiation within the garnet stability field using an equilibrium transport model. Bourdon et al. (1996b) assessed the role of mantle heterogeneity in creating the observed trend and argued it was hard to clearly distinguish source (mineralogical) heterogeneity from temperature variations; both were possibly associated with shallower ridges.

Lundstrom et al. (1998b) observed a different systematic behavior in the global MORB data. By examining the trends of ^{230}Th-^{238}U data from different areas of ridge on an equiline diagram, these authors found that the value of the slope of the data varied

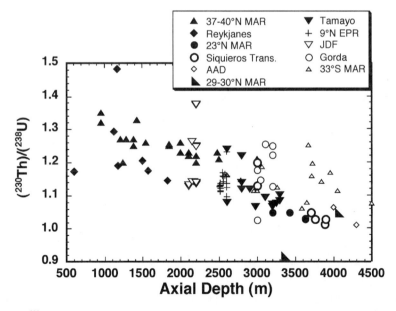

Figure 7. ^{230}Th excess as a function of bathymetric depth following Bourdon et al. (1996b) with addition of data published since 1996. The negative correlation is consistent with shallower ridges having higher mantle temperatures and therefore intersecting the mantle solidus at greater depths. Bourdon et al. (1996b) quantitatively modeled the increase in ^{230}Th excess with depth of melting using an equilibrium transport melting column in which the melting column length increased.

with the half-spreading rate of a ridge. Lundstrom et al. (1998b) interpreted the trends on an equiline diagram to reflect mixing of end member melts derived from enriched and depleted sources with each end member producing disequilibria by an equilibrium transport process. The variation in slope was then modeled by assuming that the enriched and depleted sources had distinct mineralogy and that the solid mantle upwelled at a rate equal to the half-spreading rate. The good correspondence between the observable data and the models suggests a near one to one relationship between the half spreading rate and the solid upwelling rate, consistent with passive corner flow describing solid mantle flow beneath ridges. Since this study, the few observational data added to the global database (2 studies) appear to further substantiate this observation (Fig. 8). With age constraint from ^{226}Ra excess, the Reykjanes ridge forms a near horizontal array, falling in its predicted position on the slope spreading rate trend (Peate et al. 2001) (Fig. 2H). The data from Kolbeinsey ridge also appear consistent with this model having a wide range in Th/U at near constant (^{230}Th)/(^{232}Th). However, long-lived isotopes do not indicate that source heterogeneity exists to explain the variation in Th/U (Sims et al. 2001).

Although these two observations and models give different interpretations about the important factors governing the creation of (^{230}Th)/(^{238}U) disequilibria beneath ridges, it is important to note that the two models are not in contradiction of one another. Indeed, examination of the NE Pacific ridges are consistent with both models: both ridges have similar slopes consistent with the spreading rate hypothesis while the Juan de Fuca has

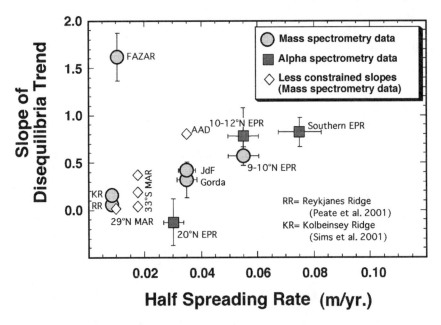

Figure 8. Slope of regressed ^{238}U-^{230}Th disequilibria trends from equiline diagrams versus the half-spreading rates of different ridges. See Lundstrom et al. (1998b) for details of sample selection and regression procedure. Note that the FAZAR data (Bourdon et al. 1996b) form a good mixing trend with samples from Azores indicating that this area of ridge is strongly influenced by hot spot melting. New data added since Lundstrom et al. (1998b) include those from the Reykjanes Ridge (Peate et al. 2001) and Kolbeinsey Ridge (Sims et al. 2001). These ridges from the slow spreading northern MAR appear to be consistent with the hypothesis that the slope of the ^{238}U-^{230}Th trend correlates with the half spreading rate. The observed relationship can be reproduced by an equilibrium transport model where the solid upwelling rate is linearly related to the half spreading rate (see Lundstrom et al. 1998b).

higher average disequilibria, consistent with its more shallow average axial depth. Both of the global $(^{230}\text{Th})/(^{238}\text{U})$ models (Bourdon et al. 1996b; Lundstrom et al. 1998b) provide clear testable hypotheses that remain incompletely assessed at present. The development of ICP-MS techniques should help in increasing the number of data available to test these global models. Because constraining the variability in $(^{230}\text{Th})/(^{238}\text{U})$ at single areas of ridge is critical to both models, effort should be placed in obtaining complete data sets from single areas of ridge. These should include samples spanning as wide a compositional range as possible and include age constraint on samples (e.g., measurement of ^{226}Ra excess).

5. INTEGRATING U-SERIES DISEQUILIBRIA MODELS WITH CONSTRAINTS FROM OTHER GEOCHEMICAL TRACERS

5.1. Comparison of melting at ridges to other tectonic settings

Comparison of the disequilibria in MORB with that of other tectonic settings, in combination with models, allows distinction of the important aspects of the melting process at each setting. By combining ^{230}Th-^{238}U disequilibria with Sm-Nd modeling, Sims et al. (1995) pointed out important differences between melting beneath mid-ocean ridges and Hawaii. Whereas the range in both $(^{230}\text{Th})/(^{238}\text{U})$ and α_{Sm-Nd} (where α_{Sm-Nd} is the calculated Sm-Nd fractionation from a 1.7 Ga mantle isochron model of Nd isotopes; see Sims et al. 1995) in Hawaiian basalts could be explained by similar degrees of batch melting of a homogeneous solid, these two geochemical probes produced very different degree of melting estimates within MORB. Indeed, although the inferred Sm-Nd fractionation for Hawaii is larger than MORB as expected, the amount of ^{230}Th-^{238}U fractionation for Hawaii and MORB is the same. This observation is consistent with the interpretation that $(^{230}\text{Th})/(^{238}\text{U})$ in MORB dominantly reflects an ingrowth melting process. In addition, this study showed that $(^{230}\text{Th})/(^{238}\text{U})$ and α_{Sm-Nd} for MORB samples from the Juan de Fuca ridge were linearly correlated, reinforcing the interpretation that mixing of diverse magmas is critically important to interpreting mid-ocean ridge melting processes.

MORB have significantly higher ^{231}Pa excesses than either OIB or convergent margins (Fig. 1). Although the generation of ^{231}Pa-^{235}U disequilibria at convergent margins is not well understood, the greater ^{231}Pa excess in MORB relative to OIB is consistent with the basic understanding of melting in these two environments. For instance, the cold lithospheric lid beneath ocean islands plays a clear role in limiting the minimum depth of melting and keeping melting within the garnet peridotite stability field. In contrast MORB can reflect melting to more shallow depths within the spinel peridotite stability field and the longer melting column beneath ridges allows greater amounts of ingrowth of ^{231}Pa (Lundstrom et al. submitted). Similarly, Bourdon et al. (1998) suggested that larger clinopyroxene/garnet ratios in the MORB source relative to OIB sources resulted in higher ^{231}Pa excess in MORB. The observation that large ^{231}Pa excesses are found in both the end member melts observed at 9°N EPR indicates that all MOR magmas probably reflect some amount of equilibrium transport melting over distances of tens of kilometers (Lundstrom et al. 2000).

5.2. The relationship between Th isotopes and long-lived isotope systems

Because the $(^{230}\text{Th})/(^{232}\text{Th})$ of any material after 350,000 years will depend on the material's Th/U, the behavior of Th isotopes in the mantle is intrinsically different from that of Sr, Nd and Pb isotopes. In principle, Th isotopes could provide information about discrete recent depletion and enrichment events. Unfortunately, this interpretation is not straightforward if Th isotopes are significantly modified from the original mantle source value by ingrowth during the melting process.

The mantle array of $(^{230}\text{Th})/(^{232}\text{Th})$ versus $^{87}\text{Sr}/^{86}\text{Sr}$ has been used repeatedly for examining Th geochemistry relative to Sr with particular emphasis on identifying recent changes in mantle sources by deviations from a linear array (Condomines et al. 1981; Ben Othman and Allègre 1990; Rubin and Macdougall 1992). Bourdon and Sims (2003) present an up-to-date $(^{230}\text{Th})/(^{232}\text{Th})$-$^{87}\text{Sr}/^{86}\text{Sr}$ array based on mass spectrometric data alone. The data form a well-defined trend that includes a distinct curvature toward higher $(^{230}\text{Th})/(^{232}\text{Th})$ for MORB samples at low, near constant $^{87}\text{Sr}/^{86}\text{Sr}$. It is this deviation of some MORB above the mantle array that has been interpreted to reflect repeated melting of a source to form high $(^{230}\text{Th})/(^{232}\text{Th})$ MORB (Rubin and Macdougall 1992). Bourdon and Sims (2003) argue that this curvature indicates the importance of ingrowth in MORB relative to OIB.

However, it is not clear that this plot can clearly separate the issue of source history (long-term depletion of the mantle) from recent melting processes (ingrowth) for the following reasons. First, it is impossible to know the relative importance of ingrowth effects versus fractionation of Th-U for any MORB or OIB sample. Ingrowth melting can significantly change $(^{230}\text{Th})/(^{232}\text{Th})$ but distinguishing the effect of ingrowth in MORB relative to OIB is not straightforward. O'Nions and McKenzie (1993) argued that the measured Th/U of a basalt is likely a better indicator of the Th/U in the mantle source than $(^{230}\text{Th})/(^{232}\text{Th})$ given the low values of D_{Th} and D_U. Second, based on the observation that variations in Sr, U and Th concentrations are clearly tied to source variations (e.g., Niu et al. 2002; Niu and Batiza 1997), mixing of melts from heterogeneous sources must play a role in the observed Th-Sr array. Given the relative compatibilities of Sr and Th in the mantle, the curvature observed in the Th-Sr array is as likely to reflect the mixing systematics as it is to distinguish recent melting events.

Indeed, a plot of Th/U versus $^{87}\text{Sr}/^{86}\text{Sr}$ for oceanic mantle basalts produces a strongly hyperbolic array (Fig. 9). Since Th/U cannot reflect ingrowth effects, the curvature must result from the mixing process. The shape of the hyperbola is well matched by a model of binary mixing of melts which have U, Th and Sr concentrations equal to the most depleted and most enriched basalts found in Pacific seamounts (Lundstrom et al. 2000). Note that all three types of basalts shown (Atlantic MORB, Pacific MORB, Pacific seamount basalts) produce similar curvature with $^{87}\text{Sr}/^{86}\text{Sr}$ only increasing once Th/U > 2.5. If mixing is the explanation for the hyperbolic shape, then Th/U will be a more sensitive indicator of addition of small amounts of enriched component than $^{87}\text{Sr}/^{86}\text{Sr}$. If so, the lack of variation in $^{87}\text{Sr}/^{86}\text{Sr}$ within a set of ridge samples cannot be construed to mean a homogeneous source, particularly if variations in Th/U exist.

Because it simultaneously examines Th/U in the mantle source and any $(^{230}\text{Th})/(^{238}\text{U})$ disequilibria generated by ingrowth processes during melting, the $(^{230}\text{Th})/(^{232}\text{Th})$ versus $^{87}\text{Sr}/^{86}\text{Sr}$ diagram confuses two distinct aspects of Th geochemistry and therefore should be avoided. Instead, assessment of Th isotopes relative to Sr, Nd, or Pb isotopes should be broken into two distinct plots: 1) Th/U versus $^{87}\text{Sr}/^{86}\text{Sr}$ to assess the role of mixing in affecting Th-Sr systematics and 2) $(^{230}\text{Th})/(^{238}\text{U})$ versus $^{87}\text{Sr}/^{86}\text{Sr}$ to assess the role of mantle heterogeneity in influencing the generation of ^{230}Th excess.

5.3. Reconciling U-series interpretations with other geochemical observations

Over the past 15 years, great strides have been made in understanding the processes of melting and melt extraction beneath ridges. Critical constraints on ridge melting have come from a variety of geochemical and geophysical observations, numerical modeling and laboratory experiments. The U-series constraints discussed here are a small piece of the overall puzzle. Although there remain discrepancies between different techniques, there appears to be convergence of ideas toward a more unified model of MORB genesis. For instance, regardless of the interpretation about whether ingrowth or time-independent

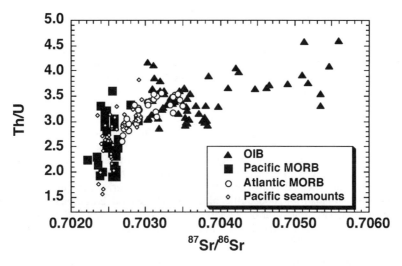

Figure 9. Th/U versus $^{87}Sr/^{86}Sr$ for oceanic mantle basalts. The data form a distinct hyperbola consistent with binary mixing of melts. Models of binary mixing using observed enriched and depleted basalts from seamounts off the EPR successfully reproduce the curvature of the array (see Lundstrom et al. 2000). Because of the strong curvature of mixing, lack of variation in $^{87}Sr/^{86}Sr$ in MORB does not necessarily indicate a homogeneous source whereas variation in Th/U can. Note that Pacific seamounts (Niu et al. 2002) span beyond the range of Pacific MORB extending to both more enriched and more depleted compositions. This is consistent with melts being less homogenized beneath seamount settings relative to ridge-axis settings where passage through crustal magma lens efficiently mixes magmas. Data from Rubin and Macdougall (1988), Macdougall and Lugmair (1986), Newman et al. (1983), Ben Othman and Allègre (1990), Goldstein et al. (1989, 1991), Eaby et al. (1984), Davis and Clague (1987), Condomines et al. (1981), Dupre et al. (1981), Hemond et al. (1988), Sigmarsson et al. (1992a,b), Newman et al. (1984), Williams and Gill (1992), Widom et al. (1997), Hilton et al. (1990), Lundstrom et al. (1995, 1998b, 1999), Lundstrom (unpublished), Perfit (pers comm., 1997), Fornari et al. (1988), Sims et al. (1995), Bourdon et al. (1996), Dosso et al. (1999), Claude-Ivanaj et al. (2001), Niu et al. (2002) and Hanan (unpublished).

melting best explains the U-series data, all U-series models agree that melt must be extractable at quite small melt porosity. This interpretation is consistent with the results of the MELT experiment which used seismic and magnetotelluric techniques to image the melt zone beneath the southern EPR. The general results of this experiment showed that melt porosity beneath the entire ridge was less than 1-2% indicating efficient extraction of melt (Toomey et al. 1998). The 1-2% estimate actually results from a conversion between the observed seismic wave velocity reductions and the assumed effects of partial melt on seismic wave propagation velocity. Since this relationship is not well known at present, the <1-2% porosity should be considered a maximum value and is therefore consistent with the porosity needed by the U-series models.

Geochemical interpretations of MORB genesis have also converged over the past 15 years from a starting point that seemed remarkably different. For instance, recent U-series observations have been modeled using an equilibrium transport melting process (Lundstrom et al. 1995; Bourdon et al. 1996b) which is equivalent to batch melting for non-decay series elements. This interpretation contradicts the conclusions of other geochemical studies that infer that melting beneath ridges is a near-fractional process. The major observations supporting near-fractional melting include melt inclusion studies (Sobolev and Shimizu 1993), Lu-Hf systematics (Salters and Hart 1989), studies of trace elements in abyssal peridotites (Johnson et al. 1990), and the global systematics of Fe in MORB (Klein and Langmuir 1987).

The convergence has been catalyzed by important gains in understanding of the physical process of melt transport based on observations in ophiolites. Kelemen and co-workers (Kelemen et al. 1995, 1997) have shown that mantle dunites (discordant dunites) represent pathways of channelized melt ascent in the upwelling mantle beneath ridges. Dunites within the Oman ophiolite have rare earth element patterns consistent with integrated MORB melts whereas surrounding harzburgites have highly fractionated rare earth element patterns similar to abyssal peridotites (Kelemen et al. 1995). Dunites are suggested to reflect channeling caused by porously flowing magma that dissolves pyroxene and precipitates olivine during ascent (Kelemen et al. 1997; Spiegelman et al. 2001). Based on these works, the idea of porous flow melt organizing into channels has lead to more sophisticated "two porosity" models of U-series disequilibria generation (Iwamori 1994; Lundstrom 2000; Jull et al. 2003). These models combine the elements of equilibrium transport argued for by the large excesses of ^{226}Ra and ^{231}Pa with the channel system needed to produce the highly fractionated trace element patterns of abyssal peridotites.

The critical relationship between the U-series data and dunite channel models is the observation of two end member melts having distinct disequilibria generated by the melting process (e.g., Fig. 3). The melting column in the dunite channel model of Kelemen et al. (1997) consists of two solid materials, a lherzolite/harzburgite mantle source which produces melt during upwelling and a dunite channel network for efficiently draining this melt by chemically isolated magma ascent. The dunite channel model would thus predict two end member melts with signatures of distinct depth and this is what is indeed observed from the U-series data. Regardless of the interpretation about whether the U-series end member melts reflect derivation from heterogeneous or homogeneous sources, as long as the high Th/U melt ascends through dunite channels, there are no contradictions between the U-series data and the Lu/Hf, abyssal peridotite or melt inclusion data (Lundstrom et al. 2000). The melt from depth moving up through the dunite channel network contains the ^{230}Th excess and the "garnet signature" of the Lu/Hf system. The low Th/U end member, on the other hand, has a signature of shallow porous flow melting based on its U-series characteristics (large ^{226}Ra excess, sometimes ^{238}U excess) and is interpreted to reflect melt that enters the dunite conduit only at shallow depth (or perhaps never at all). The transfer of melt from the harzburgite into the dunite channel at shallow depths will produce the near-fractional melting signature of the abyssal peridotites and can also produce the observed melt inclusion population. Finally, the global trends of FeO, which also require preservation of signatures of melting at depth, are consistent with models that combine batch and fractional melting, similar to the two porosity U-series models (Asimow 1999).

There is clearly a great more to be learned about ridge melting processes and it may well turn out that the two porosity model will be eliminated as a possible model for MORB genesis by future data. The important point is that the development of two porosity models has provided a framework for testing future data. For instance, by examining the systematics of Cr and La/Yb of MORB and ultra-depleted melt inclusions, Jull et al. (2002) have shown that MORB glasses with high magnesium number are consistent with mixing of melts produced at distinct depths in the melting column rather than progressive depletion by incremental polybaric melting. Many further tests of the two porosity model are possible and certain to come with addition of new data.

6. CONCLUDING REMARKS

U-series disequilibria have become a powerful technique for examining the melting process beneath mid-ocean ridges. All available evidence points to the disequilibria reflecting a primary signature of the mantle melting process. The generation of ^{230}Th-^{238}U disequilibria is sensitive to mineralogy allowing U-series disequilibria to provide information about the depth of melting and indicating that at least some portion of MORB magma reflects melting near the garnet-spinel transition. The disequilibria appear to be best explained as resulting from an "ingrowth" melting process in the melting column reflecting the process of two-phase flow at low melt porosity. As such, information about the time scales of melting, solid upwelling and melt ascent may be retrieved from U-series data.

However, U-series work on mid-ocean ridges remains an expanding field of research and new discoveries and understanding is certain to come. A large portion of the mid-ocean ridge system has yet to be explored for U-series studies. With rapid analytical advances, the stage is set for many new discoveries as well as robust testing of hypotheses thus far developed. Comprehensive data sets in which U-series data are combined with a full set of other geochemical tracers will critically advance the field. Refinements in constraints on partition coefficients, diffusion coefficients and melt-mineral reactions will provide better control on forward models of U-series disequilibria generation. Finally, the integration of geochemical and geophysical data and models will allow comparison of estimates of solid mantle upwelling and melt porosity beneath mid-ocean ridges and testing of numerical models of solid and melt flow.

ACKNOWLEDGMENTS

I would like to thank Paul Asimow, Ken Sims and Bernard Bourdon for their reviews which have significantly improved the manuscript. Comments by Marc Spiegelman, Tim Elliott, Jeff Standish, Frank Tepley and Scott Clark were also greatly appreciated. Support for this work was provided in part by NSF OCE-9910921.

REFERENCES

Allègre CJ, Condomines M (1976) Fine chronology of volcanic processes using ^{238}U-^{230}Th systematics. Earth Planet Sci Lett 28:395-406
Allègre CJ, Turcotte DL (1986) Implications of a two-component marble-cake mantle. Nature 323:123-127
Asimow PD, Hirschmann MM, Stolper EM (1997). An analysis of variations in isentropic melt productivity. Phil Trans Roy Soc 355:255-281
Asimow PD (1999) A model that reconciles major- and trace-element data from abyssal peridotites. Earth Planet Sci Lett 169:303-319
Beattie P (1993a) The generation of uranium series disequilibria by partial melting of spinel peridotite; constraints from partitioning studies. Earth Planet Sci Lett 117:379-391
Beattie P (1993b) Uranium-thorium disequilibria and partitioning on melting of garnet peridotite. Nature 363:63-65
Blundy J, Wood B (2003) Mineral-melt partitioning of uranium, thorium and their daughters. Rev Mineral Geochem 52:59-123
Bourdon B, Langmuir CH, Zindler A (1996a) Ridge-hotspot interaction along the Mid-Atlantic Ridge between 37°30' and 40°30'N; the U-Th disequilibrium evidence. Earth Planet Sci Lett 142:175-189
Bourdon B, Zindler A, Elliott T, Langmuir CH (1996b) Constraints on mantle melting at mid-ocean ridges from global ^{238}U-^{230}Th disequilibrium data. Nature 384:231-235
Bourdon B, Joron J-L, Claude-Ivanaj C, Allègre CJ (1998) U-Th-Pa-Ra systematics for the Grande Comore volcanics: melting processes in an upwelling plume. Earth Planet Sci Lett 164:119-133
Bourdon B, Goldstein SJ, Bourles D, Murrell MT, Langmuir CH (2000). Evidence from ^{10}Be and U-series disequilibria on the possible contamination of mid-ocean ridge basalt glasses by sedimentary material. Geochem Geophys Geosyst 2000GC000047

Bourdon B, Sims KWW (2003)U-series constraints on intraplate basaltic magmatism. Rev Mineral Geochem 52:215-254
Bourdon B, Turner S, Henderson GM, Lundstrom CC (2003) Introduction to U-series geochemistry. Rev Mineral Geochem 52:1-21
Calas G (1979) Etude experimentale du comportement de l'uranium dans les magmas: états d'oxydation et coordinance. Geochim Cosmochim Acta 43:1521-1531
Chabaux F, Ben Othman D, Birck JL (1994) A new Ra-Ba chromatographic separation and its application to Ra mass-spectrometric measurement in volcanic rocks. Chem Geol 114:191-197
Claude-Ivanaj C, Joron J-L, Allègre CJ (2001) 238U-230T-226Ra fractionation in historical lavas from the Azores: long-lived source heterogeneity versus metasomatism fingerprints. Chem Geol 176:295-310
Cohen AS, O'Nions RK (1991) Precise determination of femtogram quantities of radium by thermal ionization mass spectrometry. Anal Chem 63:2705-2708
Cooper KM, Reid MR, Murrell MT, Clague DA (2001) Crystal and magma residence at Kilauea Volcano, Hawaii; (^{230}Th-^{226}Ra) dating of the 1955 East Rift eruption. Earth Planet Sci Lett 184:703-718
Condomines M, Morand P, Allègre CJ (1981) ^{230}Th-^{238}U radioactive disequilibria in tholeiites from the FAMOUS zone (Mid-Atlantic Ridge, 36°50'N); Th and Sr isotopic geochemistry. Earth Planet Sci Lett 55:247-256
Condomines M, Gauthier P-J, Sigmarsson O (2003) Timescales of magma chamber processes and dating of young volcanic rocks. Rev Mineral Geochem 52:125-174
Davis AS, Clague D (1987) Geochemistry, mineralogy, and petrogenesis of basalt from the Gorda Ridge. J Geophys Res 92:10,467-10,483
Dosso L, Bougault H, Langmuir C, Bollinger C, Bonnier O, Etoubleau J (1999) The age and distribution of mantle heterogeneity along the Mid-Atlantic Ridge (31-41°N). Earth Planet Sci Lett 170:269-286
Dungan MA, Rhodes JM (1978) Residual glasses and melt inclusions in basalts from DSDP Legs 45 and 46; evidence for magma mixing. Contrib Mineral Petrol 67:417-431
Dupre B, Lambert D, Rousseau D, Allègre CJ (1981) Limitations on the scale of mantle heterogeneities under oceanic ridges. Nature 294:552-554
Eaby JS, Clague DA, Delaney JR (1984) Sr isotopic variations along the Juan de Fuca Ridge. J Geophys Res 89:7883-7890
Edwards RL, Chen JH, Wasserburg GJ (1987) ^{238}U-^{234}U-^{230}Th-^{232}Th systematics and the precise measurement of time over the past 500,000 years. Earth Planet Sci Lett 81:175-192
Elliott T (1997). Fractionation of U and Th during mantle melting; a reprise. Chem Geol 139:165-183.
England J, Zindler A, Reisberg L, Rubenstone JL, Salters V, Marcantonio F, Bourdon B, Brueckner H, Turner HJ, Weaver S, Read P (1992) The Lamont-Doherty Geological Observatory Isolab 54 isotope ratio mass spectrometer. Int J Mass Spec Ion Proc 121:201-240
Faul UH (2001) Melt retention and segregation beneath mid-ocean ridges. Nature 410:920-923
Fornari DJ, Perfit MR, Allan JF, Batiza R, Haymon R, Barone A, Ryan WBF, Smith T, Simkin T, Luckman MA (1988) Geochemical and structural studies of the Lamont Seamounts; seamounts as indicators of mantle processes Earth Planet Sci Lett 89:63-83
Gill J, Williams R, Bruland K (1985) Eruption of basalt and andesite lava degasses ^{222}Rn and ^{210}Po. Geophys Res Lett 12:17-20
Goldstein SJ, Murrell MT, Janecky DR (1989) Th and U isotopic systematics of basalts from the Juan de Fuca and Gorda ridges by mass spectrometry. Earth Planet Sci Lett 96:134-146
Goldstein SJ, Murrell MT, Janecky DR, Delaney JR, Clague DA (1992) Erratum; Geochronology and petrogenesis of MORB from the Juan de Fuca and Gorda ridges by ^{238}U- ^{230}Th disequilibrium. Earth Planet Sci Lett 109:255-272
Goldstein SJ, Murrell MT, Williams RW (1993) ^{231}Pa and ^{230}Th chronology of mid-ocean ridge basalts. Earth Planet Sci Lett 115:151-159
Goldstein SJ, Perfit MR, Batiza R, Fornari DJ, Murrell MT (1994) Off-axis volcanism at the East Pacific Rise detected by uranium-series dating of basalts. Nature 367:157-159
Goldstein SJ, Stirling CH (2003) Techniques for measuring uranium-series nuclides: 1992-2002. Rev Mineral Geochem 52:23-57
Hauri EH, Wagner TP, Grove TL (1994). Experimental and natural partitioning of Th, U, Pb and other trace elements between garnet, clinopyroxene and basaltic melts. Chem Geol 117:149-166.
Haymon RM, Fornari DJ et al. (1993) Volcanic eruption of the mid-ocean ridge along the East Pacific Rise crest at 9°45-52'N: Direct submersible observations of seafloor phenomena associated with an eruption event in April, 1991. Earth Planet Sci Lett 119:85-101
Hemond C, Condomines M, Fourcade S, Allègre CJ, Oskarsson N, Javoy M (1988) Thorium, strontium and oxygen isotopic geochemistry in recent tholeiites from Iceland; crustal influence on mantle-derived magmas. Earth Planet Sci Lett 87:273-285
Henderson GM, Anderson RF (2003) The U-series toolbox for paleoceanography. Rev Mineral Geochem 52:493-531

Hilton DR, Barling J, Wheiler GE (1990) The effect of shallow-level contamination on the helium isotope systematics of ocean island lavas. Nature 348:59-62

Hirschmann MM, Stolper EM (1996) A possible role for pyroxenite in the origin of the garnet signature in MORB. Contrib Mineral Petrol 124:185-208

Iwamori H (1992) Melt-solid flow with diffusion-controlled chemical reaction. Geophys Res Lett 19:309-312

Iwamori H (1993) Dynamic disequilibrium melting model with porous flow and diffusion-controlled chemical equilibration. Earth Planet Sci Lett 114:301-313

Iwamori H (1994) ^{238}U-^{230}Th-^{226}Ra and ^{235}U-^{231}Pa disequilibria produced by mantle melting with porous and channel flows. Earth Planet Sci Lett 125:1-16

Johnson KTM, Dick HJB, Shimizu N (1990) Melting in the oceanic upper mantle; an ion microprobe study of diopsides in abyssal peridotites. J Geophys Res 95:2661-2678

Jull M, Kelemen PB, Sims KW (2002) Consequences of diffuse and channeled porous melt migration on U-series disequilibria. Geochim Cosmochim Acta 66:4133-4148

Kelemen PB, Shimizu N, Salters VJM (1995) Extraction of mid-ocean-ridge basalt from the upwelling mantle by focused flow of melt in dunite channels. Nature 375:747-753

Kelemen, PB, Hirth G, Shimizu N, Spiegelman M, Dick HJB (1997) Review of melt migration processes in the adiabatically upwelling mantle beneath oceanic spreading ridges. Phil Trans Royal Soc 355:283-318

Klein EM, Langmuir CH (1987) Global correlations of ocean ridge basalt chemistry with axial depth and crustal thickness. J Geophys Res 92:8089-8115

Landwehr D, Blundy J, Chamorro PEM, Hill E, Wood B (2001) U-series disequilibria generated by partial melting of spinel lherzolite. Earth Planet Sci Lett 188:329-348

LaTourrette TZ, Burnett DS (1992) Experimental determination of U and Th partitioning between clinopyroxene and natural and synthetic basaltic liquid Earth Planet Sci Lett 110:227-244

LaTourrette TZ, Kennedy AK, Wasserburg GJ (1993) Thorium-uranium fractionation by garnet; evidence for a deep source and rapid rise of oceanic basalts. Science 261:739-742

Layne G, Sims KW (2000) Secondary ion mass spectrometry for the measurement of Th^{232}/Th^{230} in volcanic rocks. I J Mass Spect Ion Process 203:187-198

Lundstrom CC, Shaw HF, Ryerson FJ, Phinney DL, Gill JB, Williams Q (1994) Compositional controls on the partitioning of U, Th, Ba, Pb, Sr and Zr between clinopyroxene and haplobasaltic melts; implications for uranium series disequilibria in basalts. Earth Planet Sci Lett 128:407-423

Lundstrom CC, Gill J, Williams Q, Perfit MR (1995) Mantle melting and basalt extraction by equilibrium porous flow. Science 270:1958-1961

Lundstrom CC, Gill J, Williams Q, Hanan BB (1998a) Investigating solid mantle upwelling beneath mid-ocean ridges using U-series disequilibria. II. A local study at 33°S Mid-Atlantic Ridge. Earth Planet Sci Lett 157:167-181

Lundstrom CC, Williams Q, Gill JB (1998b) Investigating solid mantle upwelling rates beneath mid-ocean ridges using U-series disequilibria. 1. A global approach. Earth Planet Sci Lett 157:151-165

Lundstrom CC, Sampson DE, Perfit MR, Gill J, Williams Q (1999) Insights into mid-ocean ridge basalt petrogenesis; U-series disequilibria from the Siqueiros Transform, Lamont Seamounts, and East Pacific Rise. J Geophys Res 104:13,035-13,048

Lundstrom C (2000) Models of U-series disequilibria generation in MORB; the effects of two scales of melt porosity. Phys Earth Planet Inter 121:189-204

Lundstrom CC, Gill J, Williams Q (2000) A geochemically consistent hypothesis for MORB generation. Chem Geol 162:105-126

Lundstrom CC, Hoernle K, Gill J (submitted) U-series disequilibria in volcanic rocks from the Canary Islands: asthenospheric versus lithospheric melting. submitted to Geochim Cosmochim Acta

Luo XZ Rehkamper M Lee DC, Halliday AN (1997) High precision Th^{230}/Th^{232} and U^{234}/U^{238} measurements using energy-filtered ICP magnetic sector multiple collector mass spectrometry. I J Mass Spect Ion Process 171:105-117

Macdougall JD, Lugmair G (1986) Sr and Nd isotopes in basalts from the East Pacific Rise; significance for mantle heterogeneity, Earth Planet Sci Lett 77:273-284

McKenzie D (1984) The generation and compaction of partially molten rock. J Petrol 25:713-765

McKenzie D (1985) ^{230}Th-^{238}U disequilibrium and the melting processes beneath ridge axes. Earth Planet Sci Lett 72:149-157

McKenzie D (2000) Constraints on melt generation and transport from U-series activity ratios. Chem Geol 162:81-94

Michael PJ, Forsyth DW, Blackman DK, Fox PJ, Hanan BB, Harding AJ, Macdonald KC, Neumann GA, Orcutt JA, Tolstoy M, Weiland CM (1994) Mantle control of a dynamically evolving spreading center; Mid-Atlantic Ridge 31-34°S. Earth Planet Sci Lett 121:451-468

Michael PJ, Cornell WC (1998) Influence of spreading rate and magma supply on crystallization and assimilation beneath mid-ocean ridges; evidence from chlorine and major element chemistry of mid-ocean ridge basalts. J Geophys Res 103:18,325-18,356

Newman S, Finkel RC, Macdougall JD (1984) Comparison of ^{230}Th-^{238}U disequilibrium systematics in lavas from three hot spot regions; Hawaii, Prince Edward and Samoa. Geochim Cosmochim Acta 48:315-324

Newman S, Finkel RC, Macdougall JD (1983) ^{230}Th-^{238}U disequilibrium systematics in oceanic tholeiites from 21°N on the East Pacific Rise. Earth Planet Sci Lett 65:17-33

Niu YL, Batiza R (1997) Trace element evidence from seamounts for recycled oceanic crust in the Eastern Pacific mantle. Earth Planet Sci Lett 148:471-483

Niu Y, Regelous M, Wendt IJ, Batiza R, O'Hara MJ (2002) Geochemistry of near-EPR seamounts: importance of source versus process and the origin of enriched mantle component. Earth Planet Sci Lett 199:327-345

O'Hara MJ (1968) The bearing of phase equilibria studies in synthetic and natural systems on the origin and evolution of basic and ultrabasic rocks. Earth Sci Rev 4:69-133

O'Nions RK, McKenzie D (1993) Estimates of mantle thorium/uranium ratios from Th, U and Pb isotope abundances in basaltic melts. Phil Trans Royal Soc 342:65-77

Oversby V, Gast PW (1968) Lead isotope compositions and uranium decay series disequilibrium in recent volcanic rocks. Earth Planet Sci Lett 5:199-206

Peate DW, Hawkesworth CJ, van Calsteron PW, Taylor RN, Murton BJ (2001) ^{238}U- ^{230}Th constraints on mantle upwelling and plume-ridge interaction along the Reykjanes Ridge. Earth Planet Sci Lett 187:259-272

Pickett DA Murrell MT, Williams RW (1994) Determination of femtogram quantities of protactinium in geologic samples by thermal ionization mass spectrometry. Anal Chem 66:1044-1049

Qin Z (1992) Disequilibrium partial melting model and its implications for trace element fractionations during mantle melting. Earth Planet Sci Lett 112:75-90

Qin Z (1993) Dynamics of melt generation beneath mid-ocean ridge axes; theoretical analysis based on ^{238}U-^{230}Th-^{226}Ra and ^{235}U-^{231}Pa disequilibria. Geochim Cosmochim Acta 57:1629-1634

Reinitz I, Turekian KK (1989) ^{230}Th/^{238}U and ^{226}Ra/^{230}Th fractionation in young basaltic glasses from the East Pacific Rise. Earth Planet Sci Lett 94:199-207

Richardson C, McKenzie D (1994) Radioactive disequilibria from 2D models of melt generation by plumes and ridges. Earth Planet Sci Lett 128:425-437

Rubin KH, Macdougall JD (1988) ^{226}Ra excesses in mid-ocean-ridge basalts and mantle melting. Nature 335:158-161

Rubin KH, Macdougall JD (1990) Dating of neovolcanic MORB using ^{226}Ra/ ^{230}Th disequilibrium. Earth Planet Sci Lett 101:313-322

Rubin KH, Macdougall JD (1992) Th-Sr isotopic relationships in MORB. Earth Planet Sci Lett 114:149-157

Rubin KH, Macdougall JD, Perfit MR (1994) ^{210}Po-^{210}Pb dating of Recent volcanic eruptions on the sea floor. Nature 368:841-844

Rubin KH, Smith MC, Perfit MR, Christie DM, Sacks LF (1998). Geochronology and geochemistry of lavas from the 1996 North Gorda Ridge eruption. Deep Sea Res 45:2571-2597

Ryan JG, Langmuir CH (1993) The systematics of boron abundances in young volcanic rocks. Geochim Cosmochim Acta 57:1489-1498

Salters VJM, Hart SR (1989) The hafnium paradox and the role of garnet in the source of mid-ocean-ridge basalts. Nature 342:420-422

Salters VJM, Longhi J (1999) Trace element partitioning during the initial stages of melting beneath mid-ocean ridges. Earth Planet Sci Lett 166:15-30

Schilling JG, Zajac M, Evans R, Johnston T, White W, Devine JD, Kingsley R (1983) Petrologic and geochemical variations along the Mid-Atlantic Ridge from 29°N to 73°N. Am J Sci 238:510-586

Scott DR, Stevenson DJ (1986) Magma ascent by porous flow. J Geophys Res 91:9283-9296

Scott DR, Stevenson DJ (1984) Magma solitons. Geophys Res Lett 11:1161-1164

Sigmarsson O, Condomines M, Fourcade S (1992) A detailed Th, Sr and O isotope study of Hekla; differentiation processes in an Icelandic volcano. Contrib Mineral Petrol 112:20-34

Sigmarsson O, Condomines M, Fourcade S (1992) Mantle and crustal contribution in the genesis of recent basalts from off-rift zones in Iceland; constraints from Th, Sr and O isotopes. Earth Planet Sci Lett 110:149-162

Sims KWW, DePaolo DJ, Murrell MT, Baldridge WS, Goldstein SJ, Clague DA (1995) Mechanisms of magma generation beneath Hawaii and mid-ocean ridges; uranium/ thorium and samarium/ neodymium isotopic evidence. Science 267:508-512

Sims K, Mattielli N, Elliott T, Kelemen PB, DePaolo DJ, Mertz DF, Devey C, Murrell MT (2001) ^{238}U and ^{230}Th excesses in Kolbeinsey Ridge Basalts. EOS Trans. AGU 82 (47):Fall meeting suppl. Abstract V12A-0952

Sims K, Goldstein SJ, Blichert-Toft J, Perfit M, Kelemen PB, Fornari D, Michael PJ, Murrell MT, Hart SR, DePaolo DJ, Layne G, Ball L, Jull M, Bender J (2002) Chemical and isotopic constraints on the genesis and transport of magmas beneath the East Pacific Rise. Geochim Cosmochim Acta 66:3481-3505

Sinton JM, Detrick RS (1992) Mid-ocean ridge magma chambers. J Geophys Res 97:197-216

Sobolev AV, Shimizu N (1993) Ultra-depleted primary melt included in an olivine from the Mid-Atlantic Ridge. Nature 363:151-154

Spiegelman M, Kenyon PM (1992) The requirements for chemical disequilibrium during magma migration. Earth Planet Sci Lett 109:611-620

Spiegelman M, Elliott T. (1993) Consequences of melt transport for uranium series disequilibrium. Earth Planet Sci Lett 118:1-20

Spiegelman M, (1996) Geochemical consequences of melt transport in 2-D: the sensitivity of trace elements to mantle dynamics. Earth Planet Sci Lett 139:115-132

Spiegelman M, Kelemen PB, Aharonov E (2001) Causes and consequences of flow organization during melt transport: The reaction infiltration instability in compactible media. J Geophys Res 106:2061-2077

Spivack AJ, Edmond JM (1987) Boron isotope exchange between seawater and the oceanic crust. Geochim Cosmochim Acta 51:1033-1043

Sturm ME, Goldstein SJ, Klein EM, Karson JA, Murrell MT (2000) Uranium-series age constraints on lavas from the axial valley of the Mid-Atlantic Ridge, MARK area. Earth Planet Sci Lett 181:61-70

Sun S, McDonough WF (1989) Chemical and isotopic systematics of ocean basalts: implications for mantle composition and processes. *In:* Magmatism in the Ocean Basins. Saunders AD, Norry MJ (eds) Blackwell Scientific Publ. Oxford, p 313-345

Tepley FJT, Lundstrom CC, Sims K (2001) U-series Disequilibria in MORB from Transforms and Implications for Mantle Melting. EOS Trans. AGU 82 (47):Fall meeting suppl. Abstract V51E-07

Thurber DL (1962) Anomalous U^{234}/U^{238} in nature. J Geophys Res 67:4518-4520

Toomey DR, Wilcock WSD, Solomon SC, Hammond WC, Orcutt JA (1998) Mantle seismic structure beneath the MELT region of the East Pacific Rise from P and S wave tomography. Science 280:1224-1227

Turner S, Bourdon B, Gill J (2003) Insights into magma genesis at convergent margins from U-series isotopes. Rev Mineral Geochem 52:255-315

Turner S, Evans P, Hawkesworth C (2001) Ultra-fast source-to-surface movement of melt at island arcs from ^{226}Ra-^{230}Th systematics. Science 292:1363-1366

Van Orman JA, Grove TL, Shimizu N (1998) Uranium and thorium diffusion in diopside. Earth Planet Sci Lett 160:505-519

Volpe AM, Olivares JA, Murrell MT (1991) Determination of Radium isotope ratios and abundances in geologica samples by thermal ionization mass spectrometry. Anal Chem 63:916-919

Volpe AM, Goldstein SJ (1993) ^{226}Ra-^{230}Th disequilibrium in axial and off-axis mid-ocean ridge basalts. Geochim Cosmochim Acta 57:1233-1241

Von Damm KL, Edmond JM, Grant B, Measures CI, Walden B, Weiss RF (1985) Chemistry of submarine hydrothermal solutions at 21°N, East Pacific Rise. Geochim Cosmochim Acta 49:2197-2220

Widom E, Carlson RW, Gill JB, Schmincke HU (1997) Th-Sr-Nd-Pb isotope and trace element evidence for the origin of the Sao Miguel enriched mantle source. Chem Geol 140:49-68

Williams RW, Gill JB (1992) Th isotope and U-series disequilibria in some alkali basalts. Geophys Res Lett 19:139-142

Williams RW, Gill JB (1989) Effects of partial melting on the uranium decay series. Geochim Cosmochim Acta 53:1607-1619

Wood BJ, Blundy JD, Robinson JAC (1999) The role of clinopyroxene in generating U-series disequilibrium during mantle melting. Geochim Cosmochim Acta 63:1613-1620

Zou H, Zindler A, Niu Y (2002) Constraints on melt movement beneath the East Pacific Rise from ^{230}Th-^{238}U disequilibrium. Science 295:107-110

APPENDIX:
TRANSPORT-BASED MODELS FOR
CREATING U-SERIES DISEQUILIBRIA

Here, I review a one-dimensional model for melt moving relative to solid following the work of Spiegelman and Elliott (1993). The physical model described provides the parameters used in the equations tracking residence times differences for decay chain nuclides and thus generating disequilibria. Assuming steady state, the transfer of mass between the solid and melt is described by:

$$\frac{d}{dz}(\phi \rho_f w) = \dot{\Gamma} \tag{A1}$$

$$\frac{d}{dz}((1-\phi)\rho_s W) = -\dot{\Gamma} \tag{A2}$$

where $\dot{\Gamma}$ is the melting rate, ρ_f is the melt density and ρ_s is the solid density with porosity and solid and fluid velocities as noted above. Equation (A1) states that the change in the mass of melt within a system depends on the production from melting ($\dot{\Gamma}$). Equation (A2) states that the change in solid mass depends on that removed by melting. The simplest treatment of melting rate assumes that it linearly depends on the solid upwelling rate:

$$\dot{\Gamma} = \rho_s W_o \frac{dF}{dz} \tag{A3}$$

In reality, the melt productivity, dF/dz (degree of melting per km decompression), is likely to be non-linear with depth (Asimow et al. 1997). Darcy's law can then be used to describe the flow of melt given a permeable matrix and a driving force:

$$\phi(w - W) = \frac{k}{\eta}(1-\phi)\Delta\rho g \tag{A4}$$

In this formulation buoyancy ($\Delta\rho g$) is the driving force while k is the permeability and η is the melt viscosity. The permeability is often taken as

$$k = \frac{\phi^n a^2}{C} \tag{A5}$$

where a is the grain size, C is a geometrical constant and n, the exponent on the porosity, is 1-3. Given these equations a numerical model can be constructed to determine the porosity distribution as a function of depth. Typically, one chooses ϕ_{max}, the porosity at the top of the melting column, then solves for the porosity distribution as a function of depth using numerical techniques. To simplify this step, one could choose n = 1 so that the porosity simply linearly increases from 0 at the initiation of melting to ϕ_{max} at the top of the column (as long as the melting rate is constant). Varying n from 1 to 3 does change the amount of disequilibria generated but only slightly relative to the size of disequilibria that is typically being modeled. Once the porosity as a function of depth is known, the melt velocity at any depth can be calculated based on the amount of solid melted below that depth in the column. In other words, using Equations (2)-(4) from Spiegelman and Elliott (1993), one can substitute terms for w and W into the U-series transport equations (Eqns. A8-A10) below.

Given the above description of the physics of the melt and solid flow process, a conservation of mass equation can be written to describe the transport of a decay chain nuclide:

$$\frac{d}{dt}\left[\rho_f \phi c_f^d + \rho_s(1-\phi)c_s^d\right] + \frac{d}{dz}\left[\rho_f \phi w c_f^d + \rho_s(1-\phi)W c_s^d\right] = \\ \lambda^p\left[c_f^p \rho_f \phi + c_s^p \rho_s (1-\phi)\right] - \lambda^d\left[c_f^d \rho_f \phi + c_s^d \rho_s (1-\phi)\right] \quad (A6)$$

where C_s is the concentration in the solid, C_f is the concentration in the fluid and λ refers to the decay constants while superscripts d and p denote daughter and parent, respectively. Equation (A6) states that the total change in the mass of a daughter nuclide (the left hand side of the equation which includes time dependent change and the change due to advection) reflects daughter produced by decay of parent (1st term on the right hand side) minus the amount of daughter that has decayed (2nd term). The intimate link between parent and daughter nuclides in ingrowth models comes from the incorporation of the parent nuclide decay term in the equation for the daughter nuclide concentration. Each of the four major terms making up Equation (A6) contains a set of brackets ([]), each with two sub-terms. These sub-terms correspond to the amount of a nuclide in the fluid and solid phases, respectively, with the amount of fluid and solid at any point designated by ϕ and $(1-\phi)$.

If chemical equilibrium between the melt and the solid is assumed throughout the melting column, the definition of the partition coefficient (D):

$$D = \frac{C_s}{C_f} \quad (A7)$$

can be substituted into Equation (A6) to eliminate C_s. Assuming steady state conditions leads to the basic formulation of the equilibrium transport model (Spiegelman and Elliott 1993):

$$\frac{d}{dz}\left(\rho_f \phi w + \rho_s(1-\phi)WD^d\right)c_f^d = \\ \lambda^p c_f^p\left[\rho_f \phi + \rho_s(1-\phi)D^p\right] - \lambda^d c_f^d\left[\rho_f \phi + \rho_s(1-\phi)D^d\right] \quad (A8)$$

Equation (A8) states that the change in the amount of a daughter nuclide must equal the amount that decays in from the parent minus the amount of the daughter nuclide decaying away. By coupled accounting of the different decay series nuclides, the coupled differential equations can be solved numerically. For instance, a simple, not elegant, method of solution is to numerically integrate dc_f/dz over the melting column length with each nuclide concentration being updated with each increment dz.

In contrast to the full equilibrium transport model, melt could be incrementally removed from the melting solid and isolated into channels for melt ascent. This model is the disequilibrium transport model of Spiegelman and Elliott (1993). Instead of substituting Equation (A7) in for c_s, the problem becomes one of separately keeping track of the concentrations of parent and daughter nuclides in the solid and the fluid. In this case, assuming steady state, two equations are used to account for the daughter nuclide:

$$\frac{d}{dz}\left(\rho_s(1-\phi)W\right)c_s^d = -\frac{\dot{\Gamma}c_s^d}{D^d} + \left[\lambda^p c_s^p - \lambda^d c_s^d\right]\rho_s(1-\phi) \quad (A9)$$

$$\frac{d}{dz}\left(\rho_f \phi w\right)c_f^d = \frac{\dot{\Gamma}c_s^d}{D^d} + \left[\lambda^p c_f^p - \lambda^d c_f^d\right]\rho_f \phi \quad (A10)$$

Equation (A9) accounts for the change in the amount of daughter within the solid while Equation (A10) accounts for that in the fluid. The transfer of the nuclide from the solid to the fluid phase is governed by the first term on the right side of both equations. Note that in this formulation, as given in Spiegelman and Elliott (1993), the time of melt transport is accounted for and depends on the physically based transport velocity. Like Equation (A8), both equations can be solved simultaneously by numerical integration or more sophisticated numerical method.

6 U-series Constraints on Intraplate Basaltic Magmatism

Bernard Bourdon

Laboratoire de Géochimie et Cosmochimie IPGP-CNRS UMR7579
4, Place Jussieu
75252 Paris cedex 05 France

Kenneth W. W. Sims

Department of Geology and Geophysics
Woods Hole Oceanographic Institution
Woods Hole, Massachusetts, 02543, USA

1. INTRODUCTION

Intraplate magmatism represents approximately one tenth of the flux of magma to the Earth's surface (Sleep 1990). This type of magmatism has received considerable attention from petrologists and geochemists as it generally exhibits a wider range of chemical compositions than the more uniform mid-ocean ridge basalts. Hence, it is rather paradoxical that our understanding of intraplate magmatism is rather poor. In this chapter, we review the insights that have been gained from using U-series measurements (combined with other chemical and isotopic constraints) to better understand the sources and processes related to intraplate volcanism.

Several unique constraints can be obtained from measurement of U-series disequilibria in basalts. First, U-series fractionation can tell us about the residual phases present during melting as small differences in partitioning behavior between the nuclides will induce distinct signatures. Second, as has been shown by the earlier work of Allègre and Condomines (1982), Th isotope ratios can be used to infer the Th/U ratio of the mantle source providing another useful probe for mapping mantle heterogeneities. Lastly, as detailed below, the time-dependence of U-series fractionation during melting and melt migration can place constraints on several rate-dependent parameters such as the melt production rate, and melt velocities.

An important feature of hotspot magmatism is that in many cases, the timing of hotspot activity seems to be decoupled from the motion of the lithospheric plate. This observation, which has been the basis for proposing the existence of mantle plumes, suggests that magmas erupted at hotspots should reveal something about the nature of the deeper mantle. Understanding the processes of hotspot magmatism should also tell us about the nature of convective motion responsible for hotspots.

In the following section, we first review some of the outstanding issues that need to be resolved to better understand intraplate magmatism. We then discuss some of the specific advances that were possible using U-series. Specific features related to magma differentiation and crustal processes are dealt with in the chapter in this volume by Condomines et al. (2003).

2. DIFFICULTIES IN CONSTRAINING HOTSPOT MELTING PROCESSES

2.1. Source composition and source heterogeneities

It has long been known that the source of hotspots is different from the source of mid-ocean ridge magmatism (Gast 1968), both on the basis of trace element systematics (e.g., rare earth elements) and long-lived isotope systems (Rb-Sr, Sm-Nd, U-Pb, Lu-Hf, Re-Os). Both types of tracers indicate an enrichment of hotspot material relative to the depleted mantle sampled at mid-ocean ridges. What is less clear is whether this source material is also distinct in terms of its major elements (Langmuir and Hanson 1980). There are two main reasons why this debate is still ongoing: firstly, as hotspot magmas often represent smaller extents of melting (Gast 1968), the experimental constraints on the composition of these lavas are difficult to obtain because it has been more difficult to perform experiments with low extents of melting until recently (Baker et al. 1995). Secondly, the amount of volatiles (CO_2 and H_2O) is generally greater in hotspot lavas than in mid-ocean ridge basalts (MORB). These volatiles significantly depress the solidus temperature at a given pressure (Olfasson and Eggler 1983; Falloon and Green 1989, see Fig. 1). Melt productivity will also be affected and the composition of melts will be greatly different. As shown by the early work of Kushiro (1968), a greater pressure in CO_2 will result in more silica-undersaturated compositions while water will have the opposite effect. Recovering the initial source composition is difficult, such that a full parameterization of all these effects is not a trivial task and often results in non-unique solutions.

It has also been recognized that some minor phases, such as hydrous phases (amphibole, phlogopite), or accessory phases (rutile, apatite, etc.) are sometimes present in the source of hotspots (Class et al. 1997, 1998; Späth et al. 1996; Hoernle et al. 1993). These phases have a strong potential for fractionating trace elements that can be used to trace both the nature of the source material and the melting process (LaTourrette et al. 1995; Sigmarsson et al. 1998). For trace element geochemists, this has been a major challenge to identify the presence of these phases and in some cases, invert the trace element data to calculate the bulk partition coefficients and degree of melting (Minster et al. 1978; Sims and DePaolo 1996). The presence of hydrous phases makes it more difficult to parameterize melting reactions and predict melt compositions for a given pressure and temperature because there are fewer experimental constraints (Wendlandt and Eggler 1980; Olfasson and Eggler 1983).

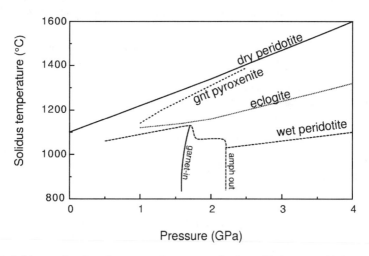

Figure 1. Solidus as a function of pressure and temperature for dry peridotite, wet peridotite, pyroxenite and eclogite. This diagram illustrates the range in solidus temperatures potentially found in the intraplate tectonic setting where compositional differences and differences in volatile contents are expected compared with the mid-ocean ridge (dry peridotite).

An added complexity stems from the generally large compositional range found in a single locality (e.g., Dupré et al. 1982). This is often interpreted as reflecting heterogeneities within the source of the hotspot (e.g., Sims et al. 1995; 1999; Lassiter and Hauri 1998). Up to five distinct components have been identified on a global basis (e.g., Zindler and Hart 1986). Some of the components can be clearly fingerprinted such as the N-MORB mantle source or the lithospheric mantle. It is far more difficult to establish what the composition of recycled components should look like (White and Hoffmann 1982; Staudigel et al. 1995; Hart 1989). One of the debates has been to decide whether a pyroxenite component is present in the source of ocean island lavas in addition to a peridotitic component (Allègre and Turcotte 1985; Hirschmann et al. 1995; Sigmarsson et al. 1998). Melting of a pyroxenite should yield compositions and melting temperature (Fig. 1) that are markedly different from a peridotite (Petermann and Hirschmann 1999). Unfortunately, there are very few constraints on the major element composition of pyroxenite melts. Furthermore, the modal abundances of minerals in pyroxenite differ greatly from peridotite, which should also affect the relative elemental fractionation during melting.

All of these complexities need to be considered when dealing with U-series disequilibrium data. Mineral modes, role of volatile contents or melting rate for a given lithology will potentially affect the fractionation of U-series nuclides during melting. Although much progress is being made in this regard, to date, the experimental data necessary to assess the effect of all these parameters is not always available. These issues are further discussed in Section 3.

2.2. Role of the lithosphere

In most cases, the lithosphere acts as an upper boundary in the melting regime of hotspots except when they are located on the axis of a spreading center (Iceland, Galapagos). Thus, this cold lithospheric cap adds yet another complexity in understanding hotspot magmatism. First, the depth of the base of the lithosphere is not always well known and as a result the final pressure of melting is not always well constrained. Second, the lithosphere will act as a cold boundary that will perturb the temperature field created by upwelling of hot material (Watson and MacKenzie 1991). It is generally believed that the lithosphere will only be able melt once the lithosphere has been heated by the impinging plume (Chase and Liu 1991; Ribe 1988). This process should take at least/approximately 10-20 Ma considering the rate of heat diffusion in solids and the temperature-dependence of viscosity (Olson et al. 1988). Even if the lithosphere does not melt extensively, the interaction of plume melts with the lithosphere can potentially affect the trace element and isotope signatures of hotspot magmas (Chase and Liu 1991; Class et al. 1997, 1998).

While we do have constraints on the composition of the lithosphere from peridotite nodules carried by hotspot magmas, the information they carry is not always relevant to understanding the origin and composition of hotspot magmas. For example, the pressure determined on the basis of fluid and melt inclusions can indicate an intermediate pressure of 10 kbar (Schiano et al. 1998). This would indicate that these nodules come from the middle of the lithosphere and not from its base where magmas are likely to interact. As the lithosphere has been stable over a timescale of up to several Ga, it may have had a complex history of depletion followed by metasomatism (Pearson et al. 1995; Hawkesworth et al. 1989). The addition of minute amounts of metasomatic melt can wildly fractionate trace elements (Galer and O'Nions 1986; MacKenzie 1989) and ultimately result in long-lived isotope signatures (Sr, Nd, Hf and Pb) that are significantly different from the asthenospheric mantle (Hawkesworth et al. 1990).

2.3. Complexities in the melting region

The presence of a lithosphere with a thickness up to 100 km above the plume head obscures observations that could be made in terms of heat flow, gravity field or seismic structure. Establishing the temperature and flow fields beneath a hotspot thus becomes a difficult exercise. Several key parameters (Fig. 2) are poorly constrained and mostly result from theoretical fluid dynamics model, which underlines their large uncertainty. The temperature anomaly within the hotspot region is generally estimated to be approximately 200 ± 100°C with large uncertainties (Shilling 1991; Sleep 1990). These temperature anomalies will induce smaller densities in the plume and the flux of the density anomalies is called buoyancy flux as defined in (Sleep 1990):

$$B = \Delta\rho Q = \rho\alpha\Delta T Q \tag{1}$$

where Q is the volume flux, ρ the density, ΔT the temperature and α the coefficient of thermal expansion. Sleep's (1990) estimate of buoyancy fluxes was based on a fixed

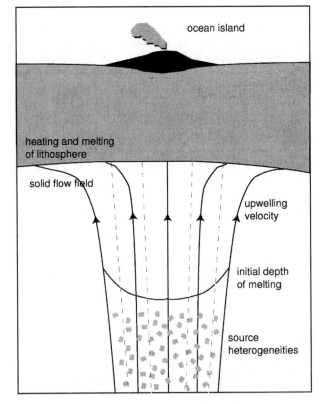

Figure 2. Schematic diagram of the top section of a mantle plume emphasizing the parameters involved in melting at hotspots. Solid lines indicate the flow lines for the solid mantle while the dashed lines indicates melt trajectories (adapted from Ribe and Smooke 1987). The size of the melting region is a function of the mantle temperature relative to ambient mantle, the amount of cooling of the overlying lithosphere and the width of the upwelling. Melting can also be influenced by heterogeneities that can modify solidus pressures and melting rates. The amount of melting produced per unit of time depends on the mantle upwelling velocity and is thus expected to be smaller on the edges of the plume. Some melting of the lithosphere can take place once the plume has heated the lithosphere. See text for more details.

point approach, since this time there have been several refinements modelling the buoyancy fluxes beneath hotspots (Ribe 1996). For example it has been argued that a rigorous estimate of buoyancy flux should take into account the excess buoyancy gained from melt depletion (Ribe and Christensen 1999). It is nevertheless widely acknowledged that there are large uncertainties in the fluxes, which ultimately relate to the size of the melting region (Ribe 1996).

The mantle viscosity which controls much of the dynamics of mantle plumes is not well known (Watson and Mackenzie 1991; Hauri et al. 1994). There are also additional effects such as when the mantle contains even small amounts of volatiles, as these volatiles will significantly lower the viscosity of the mantle (Hirth and Kohlstedt 1996). Once the volatiles are extracted, the increased viscosity will slow down the upwelling mantle (Ito et al. 1999).

At mid-ocean ridges, mass conservation dictates that there should be some scaling between the surface plate velocities and the vertical motions if one assumes simple corner flow. In the case of hotspots, there is no direct geophysical observation that can be used to quantify the upwelling velocity of the plume (although buoyancy flux can provide a qualitative scale). It could be argued however, that the fixity of hotspots relative to the plate motion suggests that the upwelling velocity of hotspot has to be significantly greater than the plate velocities (Rabinowicz et al. 1990). It has also been suggested that plume heads can be deflected by the plate motion, which would suggest that their upwelling velocity is affected by the slow plate motion (Griffiths and Campbell 1991). Thus, there are considerable uncertainties on the absolute upwelling velocities beneath hotspots or active intraplate magmatism.

The location of islands in the hotspot region, which must related to the shape of the melting region, appears to be very complex. Hotspot tracks do not define a single line of islands. There is usually more than one active volcano at a time (e.g., Hawaii), there can be two lineaments of volcanoes rather than one (Hawaii), or the islands can be arranged in a horse-shoe shape (Marquesas, Cape Verde, Galapagos). These latter observations have been explained by several theories: (1) the plume head could spread as a gravity current that cools and slows down as its own viscosity increases (Bercovici and Lin 1996), (2) ambient mantle can be entrained and give rise to a ring or horseshoe shape (Griffiths and Campbell 1991), or (3) the head of the plume can be shaped as a torus (Griffiths 1986).

The melt migration pattern beneath a hotspot is also expected to be very different from what is observed beneath ridges or subduction zones, where focusing of the melt leads to a narrow region of magmatism. Here, there is no lateral pressure gradient in the matrix that can force melt to move faster than the matrix horizontally (Ribe and Smooke 1987). Thus, based on this model, melt is expected to move almost vertically, which should: (1) prevent thorough mixing of melts and (2) result in a relatively wide zone of magmatism.

This section underlines some important features of hotspot magmatism that are relevant for interpreting U-series data in intraplate settings. It is important to note that in general, not all of these complexities have been incorporated in melting models for U-series as most of these models were originally designed to explain U-series data in mid-ocean ridge lavas.

3. THE ROLE OF SOURCE HETEROGENEITIES ON U-SERIES FRACTIONATION IN HOTSPOT MAGMATISM

3.1. Identifying residual mineral phases

An important property of the U-series decay chains is that the daughter and parent isotopes attain a steady state of radioactive equilibrium if not disturbed (i.e., there is no chemical fractionation) in approximately 0.5 Ma. This equilibrium implies that all the nuclides have similar activities (see Introduction, this volume). If the mantle has remained in a solid-state, there is no reason for the U-series chain to be perturbed prior to melting. Upon melting, it is expected that there will be some fractionation between the nuclides that have chemically distinct properties, namely U, Th, Ra and Pa. Any fractionation that can be measured in intraplate lavas must therefore be recent. This is a key feature of U-series disequilibria compared with other trace element systems. By comparison, when rare earth elements are used to study melting processes, one has to make assumptions about the source composition. Additionally, in the case of basaltic lavas which mainly crystallize olivine, pyroxene, plagioclase and Fe-Ti oxides, fractional crystallization is not expected to significantly fractionate the U-series nuclides from one another and thus the measured disequilibria is generally attributed to melting. However, there is some potential for plagioclase and Fe-Ti oxides to fractionate Th-Ra and U-Pa, respectively, in more evolved basaltic lavas (Blundy and Wood 2003).

There are now several experimental measurements of D_U and D_{Th} values for the relevant mantle phases (Blundy and Wood 2003): garnet, clinopyroxene, orthopyroxene, and olivine. While Ra and Pa are thought to be highly incompatible during mantle melting, their absolute D values have never been measured and are only inferred. Based on the fact that Ra and Ba have similar ionic radii (1.40 Å and 1.35 Å) and charge (+2 for both), it is generally assumed that $D_{Ra} \approx D_{Ba}$, which has been measured. However, recent theoretical work (Blundy and Wood 1994) and measurements of Ra and Ba in glass and plagioclase separates from both Hawaiian lavas and MORB (Cooper et al. 2000; Cooper et al. submitted) indicates that most likely $D_{Ra} < D_{Ba}$. For Pa, no experimental measurements have been made and no analogs have been found, therefore it has been generally assumed that the bulk partitioning behavior of Pa is similar to that of U^{5+} (Lundstrom et al. 1994). Another constraint for D_{Pa} comes from measurements of clinopyroxene and coexisting glass, which indicate that D_U/D_{Pa} for clinopyroxene is greater than two (Pickett and Murrell 1997). Inversion of U-Pa data in Hawaiian lavas, using the relationship between Sm/Nd fractionation and U-Pa disequilibria (Sims et al. 1999) also indicates that Pa is considerably more incompatible than U and Th. In this discussion, we will assume that Ra is incompatible and behaves like Ba (an alkali-earth like Ra with a smaller ionic radius) and that Pa is a highly incompatible element whose absolute D is very small. It is worth pointing out that as shown by Blundy and Wood (2003), it is possible to provide theoretical estimates for D_{Pa} and D_{Ra}.

As explained below, most melting models are highly sensitive to the D values chosen for the parent elements U and Th. For example, in the case of simple batch melting, it can be shown that the ^{235}U-^{231}Pa fractionation will only depend on the partition coefficients for U as long as D_{Pa} is essentially zero:

$$\left(\frac{^{231}Pa}{^{235}U}\right) = \frac{D_U + F(1-D_U)}{F} \qquad (2)$$

Thus, the absolute value of D_{Pa} and D_{Ra} is not critical to the results, as long as they are ten times smaller than D_U and D_{Th} respectively. In order to use U-series disequilibria for deriving useful information about melting conditions, it is important to assess the expected fractionation from various melting conditions and mantle sources. The diagram on Figure 3 illustrates some of the potential fractionation for (^{230}Th/^{238}U), (^{231}Pa/^{235}U) and (^{226}Ra/^{230}Th) activity ratios, which is discussed below in detail. Most melting regimes will yield excesses of ^{230}Th, ^{231}Pa and ^{226}Ra relative to their respective parents.

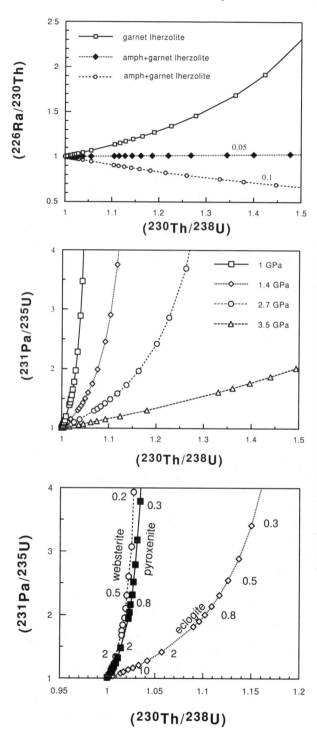

Figure 3. Potential fractionation in U-series due to batch melting of: (a) Garnet peridotite (12% garnet, 8% clinopyroxene) compared with amphibole-bearing garnet peridotite with the mode of amphibole in the garnet-lherzolite labeled on the melting curves (b) Lherzolite at a range of pressures of melting. The model curves are labeled with pressures in GPa (c) Mafic composition: garnet pyroxenite (70% cpx and 30% garnet), eclogite (50% cpx and 50% garnet), and websterite (20% garnet, 50% clinopyroxene and 30% orthopyroxene). Curves are labeled with degree of melting in percent. The model curves were calculated using the partition coefficients of Landwehr et al. (2001) and Hauri et al. (1994) for pyroxenites and LaTourrette et al. (1995) for amphibole. The model of Blundy and Wood (1994) was then used to estimate partition coefficients for Ra. As had been argued by Stracke et al. (2000), the mafic compositions do not produce large ($^{230}Th/^{238}U$) fractionation because of the dominance of clinopyroxene.

Melting of a mixed peridotite-pyroxenite source. The observation that ocean island and intraplate basalts have (^{230}Th/^{238}U) >1 indicates that garnet is required as a residual phase in their mantle source (Cohen et al. 1994; Hémond et al. 1994; Sims et al. 1995). There are recent debates about the possibility that high-pressure clinopyroxene might also produce (^{230}Th/^{238}U) greater than one (Landwehr et al. 2001). However, as pointed out in Landwehr to produce the large ^{230}Th excesses seen in most oceanic basalts (both MORB and OIB) garnet is required as a residual component during melting. It is also important to note that the REE patterns of intraplate volcanics generally display clear evidence for a garnet-bearing residue. Additionally, in many cases, the thickness of the lithosphere generally precludes significant melting at shallow depths suggesting once again that melting occurred in the presence of garnet. Although a pyroxenitic component has often been suggested as a potential source of chemically and isotopically enriched signatures in oceanic basalts (Zindler et al. 1984; Allègre, and Turcotte 1986; Hirschmann and Stolper 1996, Lassiter and Hauri 1999), its role in causing these variations remains controversial.

If pyroxenites are present in the source of ocean island lavas, then the fractionation in ^{230}Th-^{238}U and ^{231}Pa-^{235}U is predictably distinct (Hirschmann and Stolper 1996; Stracke et al. 2000) from what would be seen in peridotites. Using the partition coefficients for garnet-pyroxenite (Hauri et al. 1994; La Tourrette et al. 1993) and garnet-peridotite (Salters and Longhi 1999) allows one to compare trace element fractionations during melting of these source components (Stracke et al. 2000). In order to take into account the source variation, a commonly used parameter is δ(Lu/Hf), which is defined with the following equation:

$$\delta(\text{Lu/Hf}) = \frac{(\text{Lu/Hf})_{2Ga} - (\text{Lu/Hf})_m}{(\text{Lu/Hf})_{2Ga}} \qquad (3)$$

This parameter is a direct measure of the Lu/Hf fractionation, which is greater for garnet-bearing lithologies such as pyroxenite. Melting of garnet-peridotite results in melts having smaller Lu/Hf fractionation {defined by δ(Lu/Hf)} and significantly larger ^{230}Th excesses compared to melts from a garnet-pyroxenite with similar degrees of melting (Figs. 3b,c and 4). Sm/Nd fractionation (defined by δ(Sm/Nd) shows large variations with the degree of melting, but the variation is larger in garnet-peridotite melts than in garnet-pyroxenite melts. In the case of melting of a pyroxenite-peridotite mixture, it is expected that the pyroxenite will melt to a large degree while the peridotite will melt to a smaller extent. For increasing extents of melting, the melt will become richer in a peridotite component which is characterized by larger (^{230}Th/^{238}U) and smaller δ(Lu/Hf). As shown on Figure 4, the Hawaiian data trend is exactly the reverse of the predicted trend for a mixed pyroxenite-peridotite lithology.

In addition, garnet-peridotite melts are expected to have a more depleted isotopic signature than garnet-pyroxenite melts (see e.g., Allègre and Turcotte 1986; Hirschmann et al. 1995; Lundstrom et al. 1998; Sigmarsson et al. 1998). Thus, garnet-pyroxenite melts are clearly distinguishable from garnet-peridotite melts in terms of combined δ(Lu/Hf), δ(Sm/Nd), (^{230}Th/^{238}U), ^{176}Hf/^{177}Hf, and ^{143}Nd/^{144}Nd. Based on all these arguments and measurements of U-Th disequilibria and Nd and Hf isotope for young Hawaiian basalts, Stracke et al. (2000) suggested that the combined Hf-Nd-Th isotope and trace element data can distinguish between melts derived from peridotitic and pyroxenitic sources, and exclude the existence of garnet-pyroxenite in the source of Hawaiian basalts (Stracke et al. 2000).

An alternative view presented by Reiners (2002), suggests that in general pyroxenites do not have more enriched signatures than peridotites in their trace element

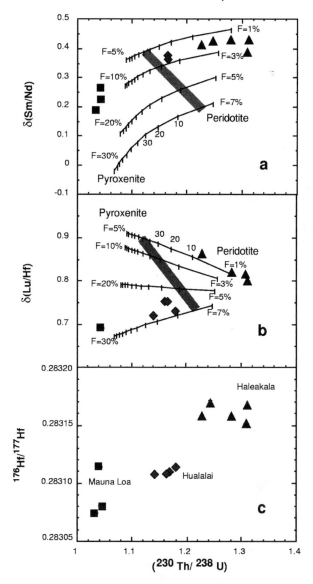

Figure 4. (^{230}Th/^{238}U) vs. $\delta_{(Sm/Nd)}$, $\delta_{(Lu/Hf)}$, and ^{176}Hf/^{177}Hf. $\delta_{(Lu/Hf)}$ is an indicator of the amount of residual garnet, whereas $\delta_{(Sm/Nd)}$ is an indicator of the degree of melting. For the Hawaiian basalts, the highest degree melts (low $\delta_{(Sm/Nd)}$) with the least prominent garnet signature (low $\delta_{(Lu/Hf)}$) are the most enriched melts (low ^{176}Hf/^{177}Hf) with the lowest ^{230}Th-excesses. Curves are mixing lines between high degree garnet-pyroxenite and low degree garnet-peridotite derived melts. Degree of melting (F) for each endmember is indicated at the end of the curve, tick marks are for 10% mixing intervals. Solid circles are the Hawaiian data from this study. Shaded areas are the expected melting trends for a garnet-pyroxenite bearing source. All melting calculations are done by incremental batch melting with small melt increments (0.1%) and similar porosities (0.1%), i.e., dynamic melting. The trace element composition for the garnet-pyroxenite component is similar to depleted mid ocean ridge basalts (N-MORB), and the trace element composition of the garnet-peridotite is similar to MORB–source mantle (according to the geochemical earth reference model, GERM). Melt transport times are equal for both melting of garnet-pyroxenite and garnet-peridotite (mantle matrix ascent rates are 1 cm/yr and melt ascent rates are 400 cm/yr). Further explanations and sources of partition coefficients are given in the text.

compositions (see compilation of Sm/Nd in Hirschmann and Stolper 1996). The model proposed by Reiners also considers different process for aggregating melts of pyroxenites and peridotites. Rather than considering progressive melting of a mixed lithology source, Reiners proposes that large degree-melt of pyroxenite mix with low degree melts of peridotites. This result is exactly the opposite trends observed in trace element diagrams. Further studies of heterogeneous sources will thus be necessary in the future.

Another evaluation of the pyroxenite hypothesis can be found in the evaluation of a suite of data from the Canaries islands (Thomas et al. 1999). These authors estimated the maximum mode of residual garnet for the 1730-1736 Lanzarote eruption modelling

REE's and found that less than 5% residual garnet is required. This clearly precludes a garnet pyroxenite mineralogy, which would have a much greater garnet mode.

Another argument against a pyroxenite component comes from U and Th measurements in pyroxenites and associated peridotites plotted as a histogram (Fig. 5). These data show that pyroxenites from the lithospheric mantle (which, in this respect, might be different from pyroxenite from the bulk mantle) do not have a greater Th/U than peridotite as had been argued by Sigmarsson et al. (1998) for the case of the Canaries and Hawaii. While (^{230}Th/^{238}U) and Th/U data form an inclined array that could represent a mixing line, the endmembers are unlikely to be pyroxenite and peridotite. If one is to interpret these arrays with mixing, they could represent mixtures between depleted and enriched peridotites.

Melting of the lithospheric mantle. There are several signatures that can be used to trace lithospheric melting with U-series disequilibria. First, lithospheric melting should be shallower and therefore have a weaker garnet signature than asthenospheric melting. Second, the lithosphere can be hydrated and contain phases that are only stable at intermediate pressure such as amphibole. Third, the lithosphere can be characterized by a distinct Th/U ratio that can be traced using (^{230}Th/^{232}Th) activity ratios. This last point will be discussed in Section 3.3.

At shallow pressure (less than 1 GPa), Th becomes slightly more compatible than U in clinopyroxene (Landwehr et al. 2001) and garnet is not stable even in pyroxenite or eclogite. On the other hand, it is believed (mostly based on observations) that Pa remains more incompatible than U through the entire length of the melt column. Thus, melting at intermediate pressure should yield small ^{230}Th or ^{238}U excesses, but significant ^{231}Pa excess. This has been observed in the lithospheric mantle in the Colorado plateau by

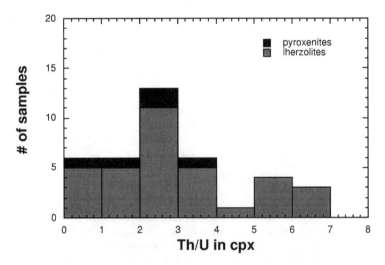

Figure 5. Histogram Th/U for clinopyroxenes in peridotites and pyroxenites from the Ronda peridotite massif. Concentrations were measured by isotope dilution mass spectrometry in acid-leached clinopyroxenes. This histogram shows that pyroxenites do not have larger Th/U ratios than peridotites. Thus, the correlation found between (^{230}Th/^{238}U) and Th/U cannot be explained by mixing of peridotite and pyroxenite melts as advocated in Sigmarsson et al. (1998). Data from Hauri et al. (1994) and Bourdon and Zindler (unpublished). It can be shown with a simple Student t-test that the two populations are indistinguishable.

Asmerom et al. (2000). The depth of melting in this region is approximately 2 GPa while the prediction for equal D_{Th} and D_U is 0.8-1.3 GPa (Landwehr et al. 2001). This apparent discrepancy might reflect what Asmerom et al. (2000) suggested is a reequilibration of the melts produced at the base of the static lithosphere at shallower pressure. If there is such a reequilibration or some form of assimilation, then the initial excess ^{230}Th would decrease while the ^{231}Pa excess could be maintained as advocated in the Comores (Bourdon et al. 1998) and the East Pacific Rise (Sims et al. 2002).

Another index of a shallow production or equilibration of melt are the ^{226}Ra deficits found in some ocean island basalts. Several hydrous minerals such as amphibole and phlogopite are thought to incorporate more Ba than Th. As Ba is used is an analogue for Ra, it can be inferred that the amphibole partition coefficient for Ra will also be greater than for Th. Using the approach of Blundy and Wood (1994), the partition coefficient of Ra in amphibole can be estimated to be 0.03, which is much greater than the value for Th in this same mineral (LaTourrette et al. 1995). Thus, it can be predicted that melting of an amphibole or phlogopite-bearing peridotite should yield ^{226}Ra deficits or smaller (^{226}Ra/^{230}Th) (Claude-Ivanaj et al. 1998, 2001) if the mode of amphibole is greater than a few percent. Such deficits were first reported in the Azores by Widom et al. (1992) for a differentiated trachyte and interpreted as due to alteration. Claude-Ivanaj et al. (1998) also found lavas from the La Grille volcano, Comores that had an amphibole signature on the basis of their trace element compositions (Class and Goldstein 1997) and smaller ^{226}Ra excess than the nearby Karthala lavas that has no apparent amphibole signature. Deficits for Samoan lavas have also been reported for lavas from Samoa and once again the deficits are interpreted to be a result of melting in the presence of amphibole or phlogopite (Sims and Hart, in prep).

Melting with accessory phases. It has been advocated that in some cases, accessory phases such as apatite, monazite, rutile or zircon could play a role in fractionating U-series during melting. These phases are potentially important as they have large partition coefficients for U and Th. This was discussed theoretically by Beattie (1993) who showed that based on the mantle abundances of P_2O_5 and ZrO_2, it was unlikely that there were significant amounts of apatite and zircon in the mantle. For these phases to significantly affect the relative compatibility of Th and U, their abundances has to be such that:

$$X_a \approx \frac{\sum_{i \neq a} X_i D_i}{D_a} \quad (4)$$

Beattie (1993) argued rather convincingly, that only zircon could have a significant effect, but zirconium is also present in garnets and clinopyroxenes such that zircon is extremely unlikely in mantle assemblages.

There are several trace-element ratios that should be diagnostic indicators of many of these accessory phases. For example, if one assumes that rutile is a residual phase during melting of pyroxenite (Sigmarsson et al. 1998), then considering the large Nb partition coefficient in rutile, large Nb deficits are expected in the melt relative to U. In the case of Lanzarote (Sigmarsson et al. 1998), it is found that the samples with the smallest Nb/U ratios do not have the largest (^{230}Th/^{238}U) activity ratios (Fig. 6). Similarly, the most radiogenic ^{87}Sr/^{86}Sr ratios, which should indicate the presence of pyroxenite, do not correspond to the lowest Nb/U ratios that should be found in the rutile-bearing pyroxenite. Thus, it is unlikely, that rutile is a residual phase during melting (Sims and DePaolo 1997). While the presence of these phases could explain some of the fractionation found in OIB, both its presence and its influence on U-Th disequilibria are yet to be demonstrated.

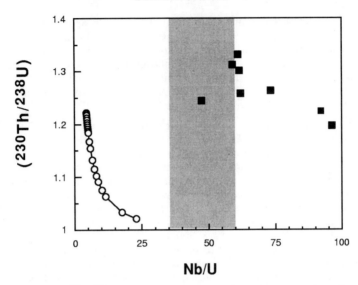

Figure 6. Nb/U versus (^{230}Th/^{238}U) diagram to illustrate the effect of accessory phases (e.g. rutile) on trace element fractionation in OIB. It is generally found that Nb/U ratio show a relatively restricted range in ocean island basalts (Hofmann 1986) except when there is evidence for sediment recycling in their source (Sims and DePaolo 1997). The data for Lanzarote are shown here for reference (Sigmarsson et al. 1998; Carracedo et al. 1990). The curve is for batch melting of a rutile-bearing pyroxenite with the mode of rutile estimated such that (^{230}Th/^{238}U) matches observed values ($X^{rutile} = 0.01$). The partition coefficients used in this calculation are as follows: $D_U^{garnet} = 0.04$, $D_U^{cpx} = 0.01$, $D_{Nb}^{rutile} = 29$. Degree of melting ranges between 0.1 and 20%. If rutile were present as a residual phase in the source ocean island basalts, it would have a dramatic effect on their Nb/U ratios, in contrast with what is observed.

3.2. Role of source heterogeneities on melting processes

In this section, we discuss how source heterogeneities could affect the rate of magma production per unit of time, which is an essential parameter in predicting U-series fractionation. It can be shown simply that if one considers that the rate of melt migration is fast relative to the half life of the daughter isotope, then the magnitude of the fractionation between a daughter isotope and its parent depends directly on the rate of magma production (see discussion below). It has been argued that the melting rate, (generally given in kg m^{-3} a^{-1}) could vary depending on the source composition (Hirschmann et al. 1999; Turner et al. 1997), but also on phase transitions as the mantle upwells (Asimow et al. 1995). These effects will also depend on the process of magma extraction. This can be shown simply in the case of wet melting. If the melt is rapidly extracted from the source region and no longer interacts with the matrix, once the volatiles are completely partitioned into the melt, further melting will be equivalent to melting of a dry peridotite. It has also been shown that more fertile peridotite sources have slower melting rates, at least at the inception of melting (Hirschmann et al. 1999), while pyroxenite sources should melt at a faster melting rate (Petermann and Hirschmann 1999).

Although these considerations are theoretically reasonable, there are very few case studies where these complexities have been explicitly demonstrated. However, it is important to note that several studies have shown that source heterogeneities could play a role in producing the observed U-Th disequilibria (Hirschmann et al. 1999, Turner et al. 1997). It has been argued in the case of the Azores region that the volatile content in the source is larger than in N-MORB mantle. As argued by Turner et al. (1997), the lower (^{230}Th/^{238}U) activity ratios found for the Sao Miguel lavas in the Azores could be due to a faster melting rate of an enriched source.

3.3. Tracing mantle sources

Global systematics. An another important aspect of U-series disequilibrium measurements is that they provide another probe for mantle sources. This application of U-series disequilibria was initially pioneered by Condomines et al. (1988) and Allègre and Condomines (1982) and subsequently discussed in great detail by Gill et al. (1992). If one assumes that the mantle melts according to a batch melting process, then measured (^{230}Th/^{232}Th) ratios in lavas should reflect the Th/U ratio in the mantle source of the lavas, which can be calculated with the following equation:

$$\left(\frac{^{232}Th}{^{238}U}\right)_s = \frac{\lambda_{238}}{\lambda_{232}\left(^{230}Th/^{232}Th\right)} \qquad (5)$$

where λ_{232} and λ_{238} are decay constants for ^{232}Th and ^{238}U respectively and parentheses denote activity ratios. This provides a very important tool for understanding mantle processes.

However, it is important to note that this argument requires rapid melt production [e.g., as is observed for Hawaii (Sims et al. 1995; 1999; Elliott 1997)]. On the other hand, if the melting rate (melt production per unit of time) is slow, then ingrowth of ^{230}Th during melting becomes important and the resulting the ^{230}Th/^{232}Th in the melt can be greater than that of the source (Williams and Gill 1989).

While other radiogenic isotope tracers (Sr, Nd, Hf and Pb) respond slowly to change in parent/daughter ratio (e.g., Rb/Sr), in the case of Th isotopes, after approximately 300 ka, the (^{230}Th/^{232}Th) ratios directly reflects the mantle source composition. Thus, the global correlation found between Th/U$_s$ and ^{87}Sr/^{86}Sr for MORB and OIB indicate that the Th/U ratio in the mantle source is long-lived and consistent with the indication given by ^{87}Sr/^{86}Sr (Fig. 7). This negative correlation supports the idea that the depleted mantle is complementary to the more enriched continental crust. This correlation is also consistent with both the trends observed in Th isochron diagrams (Fig. 11a) and the correlation observed between Sr and Nd isotope ratios for MORB and OIB.

This correlation has been further extended with recent measurement of basalts from Samoa, which have extremely high ^{87}Sr/^{86}Sr (0.708) and low (^{230}Th/^{232}Th) (Sims and Hart, in prep), which clearly shows that the functional form of this relationship is hyperbolic rather than linear, as has been previously suggested (Fig. 7a). While this plot emphasizes the influence which source compositions can have on the (^{230}Th/^{232}Th) of OIBs like Samoa, MORB, on the other hand, show little variability in ^{87}Sr/^{86}Sr, but significant variability in (^{230}Th/^{232}Th), indicating that there is a significant component of ^{230}Th which has been ingrown as a result of slow melting (see below). In this context it is important to note that if ingrowth was the dominant effect for all basalts (OIB and MORB), then the Th/U ratio measured in the basalt would be more representative of its Th/U source ratio. However, as has been previously pointed out by other studies (Condomines and Sigmarsson 1999), and this studies compilation of data, the correlation between (^{230}Th/^{232}Th) and ^{87}Sr/^{86}Sr is better defined than the correlation between Th/U and ^{87}Sr/^{86}Sr (Fig. 7b), suggesting that for at least OIBs, (^{230}Th/^{232}Th) is a better estimate of the Th/U source ratio.

As advocated by Allègre et al. (1986), the Th-Sr can be used to determine the Th/U ratio of the bulk silicate Earth based on an estimated value of ^{87}Sr/^{86}Sr of 0.7045-0.705.

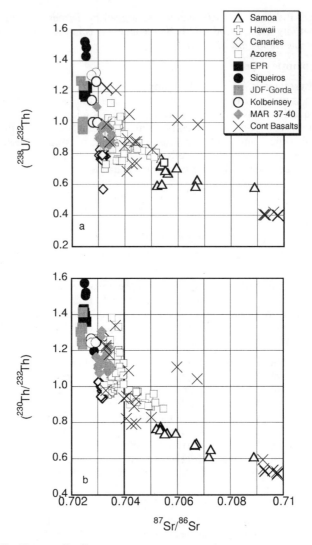

Figure 7. (a) $(^{238}U/^{232}Th)$ vs. $^{87}Sr/^{86}Sr$ diagram for mid-ocean ridge, ocean island and continental basalts compiled from the literature (Turner et al. 1997; Bourdon et al. 1996; Dosso et al. 1999; Claude-Ivanaj et al. 2001; Sims et al. 1995, 1999, 2002; Reid and Ramos 1996; Sigmarsson et al. 1998; Williams et al. 1992; Asmerom and Edwards 1995; Asmerom 1999; Thomas et al. 1999 and unpublished data from Sims and Hart, Sims and Elliott, and Bourdon and Turner). (b) $(^{230}Th/^{232}Th)$ vs. $^{87}Sr/^{86}Sr$ diagram for mid-ocean ridge, ocean island and continental basalts compiled from the literature (Turner et al. 1997; Bourdon et al. 1996; Dosso et al. 1999; Claude-Ivanaj et al. 1998, 2001; Bourdon et al. 1998; Sims et al. 1995, 1999, 2002; Reid and Ramos 1996; Sigmarsson et al. 1998; Williams et al. 1992; Asmerom and Edwards 1995; Asmerom 1999; Thomas et al. 1999 and unpublished data from Sims and Hart, Sims and Elliott and Bourdon and Turner). This new data set based on mass spectrometry data shows a hyperbolic rather than a linear trend. If this is interpreted as reflecting a mixing line, this indicates that the Th/Sr ratios in the enriched components (low $(^{230}Th/^{232}Th)$ ratios) is greater than the Th/Sr ratio in the depleted component. The lavas (in particular continental basalts from the Great Rift which are the two continental basalts plotting well above the trend) must have had a distinct evolution and can be explained by a recent metasomatism of the lithosphere (see text).

The corresponding Th/U is ~ 4.2. A similar approach can be used with $^{208}Pb^*/^{206}Pb^*$ where as there $^{208}Pb^*/^{206}Pb^*$ is defined as follows:

$$\frac{^{208}Pb^*}{^{206}Pb^*} = \frac{\frac{^{208}Pb}{^{204}Pb} - \left(\frac{^{208}Pb}{^{204}Pb}\right)_{CD}}{\frac{^{206}Pb}{^{204}Pb} - \left(\frac{^{206}Pb}{^{204}Pb}\right)_{CD}} \tag{6}$$

where the subscript CD refers to the Canyon Diablo troilite.

There is a positive trend between $^{208}Pb^*/^{206}Pb^*$ and Th/U_s (Eqn. 5) which is better defined using a compilation of mass spectrometry data (Fig. 8) than in the original paper by Allègre et al. (1986). Another line that can be defined in this diagram is the $^{208}Pb^*/^{206}Pb^*$ ratio calculated assuming that the bulk silicate Earth (BSE) has evolved as a closed system since 4.55 Ga for a range of Th/U ratios. The correlation defined by OIB and MORB yields an intercept of 4.2, which has been interpreted as defining the Th/U ratio of the BSE (Allègre et al. 1986). For this argument to be complete, data for the continental crust should plot on the other side of the closed-system line on this diagram since this reservoir represents roughly 50% of the Th-U budget of the Earth.

Tracing recent metasomatism. As pointed out above, the global systematics in Th isotope geochemistry generally show a consistency between long-lived isotope systems and Th isotopes, which respond rapidly to mantle processes. Deviations from these systematics make is possible to put some constraints on processes that may have modified the mantle source composition prior to melting. This is particularly useful in tracing metasomatism. Metasomatic melts should have fractionated U from Th. It these melts get frozen in the lithospheric mantle, they will, over a period of time greater than

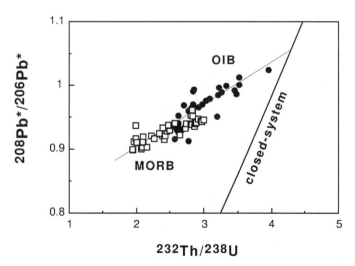

Figure 8. $^{208}Pb^*/^{206}Pb^*$ vs. Th/U_s (derived using Eqn. 5 in the text) diagram for mid-ocean ridge and ocean island basalt based on a recent data set with mostly mass spectrometry measurements (Turner et al. 1997; Bourdon et al. 1996; Dosso et al. 1999; Claude-Ivanaj et al. 1998, 2001; Sims et al. 2002). The data show a relatively well defined array that intersect a closed-system line for the bulk Earth starting with an initial lead isotope composition equal to Canyon Diablo (T = 4.55 Ga). This intersect was used by Allègre et al. (1986) to define the Th/U ratio of the Earth.

the half-life of ^{230}Th, impart a specific (^{230}Th/^{232}Th) signature. A recent metasomatic enrichment will not necessarily modify the Sr or Pb isotope signature (because of the long half-lives of ^{87}Rb and U and Th isotopes) but the Th isotopes will be affected significantly (because of the short half-life of ^{230}Th). Thus, a decoupling between (^{230}Th/^{232}Th) and ^{87}Sr/^{86}Sr or ^{208}Pb*/^{206}Pb* should be detected. Elevated (^{230}Th/^{232}Th) would for example correspond to an enrichment of U relative to Th (Reid 1995, Reid and Ramos 1996). This is particularly obvious in the case of the Great Rift lavas (So. Idaho), which are well above the Th-Sr trend defined for continental and oceanic basalts (Fig. 7). It is hard to put a precise time scale on these effects because for a given end-point several paths are possible. However, it is possible to estimate/establish an upper limit for the timescale of enrichment. In the case of the Crater of the Moon lavas, for ^{238}U/^{204}Pb ranging between 8 and 10, it would take 100 to 200 Ma to significantly modify the ^{208}Pb*/^{206}Pb* ratios. This relatively recent enrichment has been interpreted by Reid (1995) as reflecting metasomatism by U-rich fluid related to the Yellowstone mantle plume, which would have lowered the Th/U ratios. An alternative explanation would be that the lithosphere has been metasomatized by U-rich subduction related fluids. A similar signature of metasomatized lithospheric mantle was found by Reid and Ramos (1996) in central Basin and Range lavas in the Southwestern United States. The estimated Th/U ratio from the (^{230}Th/^{232}Th) (Eqn. 5) of these lavas correlates negatively with their ^{143}Nd/^{144}Nd ratios. These authors (Reid and Ramos 1996) argue that this observed negative correlation cannot reflect contamination by lower crust with high Th/U ratios. Rather, they favor a model where the variation in Th/U$_s$ is due to contamination of a source that has been progressively enriched in U since the Proterozoic.

In contrast to the Great Rift lavas (Reid 1995), continental lavas from the Gaussberg volcano are characterized by high Th/U ratios (low ^{230}Th/^{232}Th) (Williams et al. 1992). These high Th/U (up to 7.5) ratios are attributed to repeated metasomatic events in the subcontinental lithosphere. Recent experiments by LaTourrette et al. (1995) have shown that the partition coefficients of U and Th in phlogopite are small (10^{-3}) and close to each other. Thus, the large Th/U found in the Gaussberg lavas can no longer be explained by the presence of phlogopite. Similarly, residual phlogopite does not preclude the large ^{230}Th excesses found in the Gaussberg lavas. It is likely that the large ^{230}Th excess (1.44-1.58) can be attributed to small degree melts with a garnet residue and does not require disequilibrium melting as advocated in Williams et al. (1992).

4. MELTING PROCESSES AND RELATION TO CONVECTIVE STRUCTURE OF PLUME

4.1. Time dependent melting models for hotspot magmatism

There is considerable evidence indicating that the dominant mode of melt production in intraplate volcanism is adiabatic decompression of solid upwelling mantle material in the form of mantle plumes. In this scenario, the formation and transport of basaltic magma during adiabatic decompression of mantle rock can be characterized in terms of several rate dependent parameters: the velocity of the upwelling (solid) mantle, the rate of melting, the melt velocity and the storage time of magma before eruption (e.g., Stolper et al. 1981; Turcotte 1982; McKenzie 1984, 1985; Richter and McKenzie 1984; Ribe 1985, 1987; DePaolo 1996). While numerous physical models have now been developed which describe the generation and transport of basaltic magma from the Earth's mantle, they suffer in application because of an incomplete knowledge of the physical properties of solid and liquid rock material, which control the time and length scales. Although it would be desirable to measure these physical parameters in order to improve this situation, and some progress is being made in this regard, the critical properties- the effective viscosities of solid and liquid rock material and the permeabilities of partially molten materials- are hard to measure or even estimate.

For this reason it has been desirable to pursue other approaches to constrain the timescales of magma genesis; and in this regard, U-series disequilibria have played an important role. The half-lives of ^{230}Th (75,300 yr) and ^{226}Ra (1,600 yr) of the ^{238}U decay series bracket the time scales over which melting and melt extraction are thought to occur, and large differences in the solid/liquid partitioning behavior of U, Th, Pa and Ra produce large parent-daughter fractionations during the early stages of melting. Therefore, measurement of ^{238}U-^{230}Th and ^{230}Th-^{226}Ra disequilibria in young mantle-derived basalts should, in principle, provide information on the rate of melting, the melt migration velocity and the extent of mantle melting.

Several models relating the isotopic effects of U-series disequilibria to the timescales of the melting process have now been proposed (e.g., McKenzie 1985; Williams and Gill 1989; Spiegelman and Elliott 1993; Qin 1992; Iwamori 1994; Richardson and McKenzie 1994). While these models differ mainly in their treatment of the melt extraction process (i.e., reactive porous flow vs near fractional melting), because they incorporate the effect of radioactive ingrowth and decay of the daughter isotopes (e.g., ^{230}Th, ^{226}Ra, ^{231}Pa) during melt production they are able to predict mantle melting rates and melt transport rates, based upon U-Th-Ra and U-Pa disequilibria. These mantle melting rates derived from U-series can, in turn, be combined with theoretical predictions of melt productivity (e.g., McKenzie and Bickle 1988; Asimow et al. 1997) to calculate a model solid mantle model upwelling rate.

U-Th-Ra and U-Pa have now been measured in several OIB suites, by high precision mass spectrometric methods (Sims et al. 1995, 1999; Cohen and O'Nions 1993; Turner et al. 1997; Bourdon et al. 1998; Thomas et al. 1999; Claude-Ivanaj et al. 1998, 2001; Pietruszka et al. 2001; Sims and Hart, in prep.). Though a number of factors such as source heterogeneity and the degree of partial melting can complicate the interpretation of the U-series data in terms of the timescales of the melting processes, the influence of mantle upwelling on the measured disequilibria is clearly distinguishable. For example, the influence of mantle upwelling is seen on the scale of individual plumes like Hawaii (Fig. 13), where the solid mantle upwelling velocity determined from U-Th and U-Pa disequilibria, decreases as a function of radial distance from the plume center (Sims et al. 1999). In the following section, we briefly review the two different end-member melting models, which incorporate the timescales of the melting process. We then show, through examples, how the U-series data can then be compared with these models to place unique temporal constraints on many of the magmatic processes associated with intraplate volcanism.

Chromatographic porous flow. The steady state chromatographic porous flow melting model of Spiegelman and Elliot (1993) accounts for the duration of melt generation (i.e., the time it takes the upwelling mantle to traverse the melt column), the relative velocity of the upward-percolating melt and the residual solid, and the effects of continuous melt-solid interaction during melt transport. The model assumes that chemical equilibrium is maintained between migrating liquid and the solid matrix. This corresponds to an infinite Damköhler number, which represents the ratio of the time scale necessary for melt migration divided by the time scale for equilibration by diffusion:

$$D_a = \frac{3DL\rho_s(1-\phi)}{w_0 a \rho_m \phi} \quad (7)$$

where L denotes the length of the melting column, a is the grain size and D the diffusion coefficient. Other parameters are given in Table 1. Continuous melt-solid interaction produces chromatography; different elements travel through the melt column at different velocities according to their relative melt/solid partition coefficients (i.e., D values). For non-radioactive elements, the steady-state chromatographic porous flow melting model produces trace-element enrichments that are identical to those of equilibrium batch melting (Spiegelman and Elliott 1993). However, for the U-decay-series isotopes, which have half-lives that are comparable to the melt migration time scales, the chromatographic effect has a significant influence on the resulting concentrations and activity ratios. If the daughter isotope is more incompatible than the parent isotope, which is true for all of the systems considered here, the residence time of the daughter isotope in the melting column is shorter relative to the parent nuclide as the daughter nuclide travels preferentially with the melt which moves faster (this is because the daughters are more incompatible than their parent nuclides, see also Appendix A.3). This implies that the production of daughter nuclides is enhanced relative to its decay in the melting column and this produces large excesses in daughter isotopes in the resultant basaltic liquids. The extent of this daughter nuclide enhancement depends on the half life of the daughter nuclide, and consequently it is very large for Ra, and much smaller for ^{230}Th and ^{231}Pa. The extent of ingrowth, is also dependent upon the length of the melting column, with longer melt columns leading to greater enhancements of the daughter nuclides (see appendix for a more quantitative discussion of this).

The 1-D chromatographic porous flow model of Spiegelman and Elliott (1993) has seven variable parameters (which are not all independent): the solid velocity (W), the liquid velocity (w), the porosity (ϕ), the height of the melting column (Δz), the melt fraction (F_{max}), the melting rate (Γ) and the mineral melt partition coefficients for the trace-elements being considered. The simplest way to use the U-series data and the chromatographic porous flow model to obtain estimates on time dependent parameters of melting is to use experimental constraints to estimate the height of the melt column, the melt production rate and the mineral partition coefficients and then to compare the measured U-series data with a forward model which calculates porosity and melting rate. Both numerical and analytical solutions can be used to calculate the U-series activity ratios for the 1-D chromatographic porous flow of Spiegelman and Elliott (1993). These solutions show that (^{226}Ra/^{230}Th) disequilibrium is controlled mostly by the porosity of the melt region (which controls the velocity of the melt relative to the solid) for melting

Table 1. Main parameters used in melting models.

Parameters	Notation	Units	Range
Degree of melting	F	---	0-0.2
Upwelling rate	W	m a^{-1}	0-1
Melting rate	Γ	kg m^{-3} a^{-1}	10^{-3}-10^{-5}
Solidus temperature	T	K	1000-1400
Length of melting column	Z	km	0-50
Plate velocity	V	m a^{-1}	0.01-0.1
Matrix porosity	ϕ	---	0-0.05
Solid density	ρ_s	kg m^{-3}	3350
Melt density	ρ_m	kg m^{-3}	2700
Lithosphere thickness	h	km	0-100

rates ranging less than 2×10^{-2} kg m^{-3} a^{-1}. For faster melting rates, the short melting timescale does not allow enough in-growth of ^{226}Ra and becomes a limiting factor. In contrast, (^{230}Th/^{238}U) and (^{231}Pa/^{235}U) disequilibria are controlled more by the melting rate (which is related to the solid mantle upwelling rate) (Figs. 9 and 10). However, because the numerical solution takes into account a depth dependent porosity structure, whereas the analytical solutions assume a constant porosity over the length of the melting column, the absolute values of these determined parameters are significantly different depending on the initial assumptions of the porosity structure of the mantle (see Appendix or Sims et al. 1999, which explicitly discusses the contrast between the analytical and numerical solution for the model by Spiegelman and Elliott 1993). Also, as will be discussed below, the trace element partition coefficients used for the modeling can have significant influence over the resulting parameters.

Finally, with regard to the chromatographic melting model, it is also important to note that because this model takes into account the time-scale of melt migration, the velocity of the melt can be determined explicitly. In the case of the analytical solutions however we have assumed that porosity and melt velocity are also constant throughout the melting column. Whereas with the numerical solution, melt velocity varies as a function of the porosity distribution of the melt column, and therefore in this case, it is more useful to calculate melt transport times by integrating the melt velocity over the length of the melt column.

Dynamic melting. In the "dynamic melting" model, as formulated by McKenzie (1985), melts produced in the upwelling mantle remain trapped, and in equilibrium with the solid residue until a critical threshold porosity is reached, after which any produced melt in excess of the threshold value escapes instantaneously such that the porosity remains constant. Dynamic melting takes into consideration the time-scale of the melting process and consequently, like the Spiegelman and Elliott model, produces enhanced excesses of short-lived daughter nuclides like ^{226}Ra. Dynamic melting differs from chromatographic melting in that melts move instantly to the surface as they form and do not react with the solid on the way. Because most of the uranium is extracted near the base of the melting column, disequilibria are created only at the bottom of the melt column instead of throughout the entire length of the melt column as is observed with chromatographic melting. Unlike chromatographic porous flow where melt velocities are determined explicitly, with dynamic melting, melt velocities are constrained by the shortest half-life of the daughter nuclides not in radioactive equilibrium with its immediate parent (e.g., ^{226}Ra). Above the point where the threshold porosity is reached, the porosity of the melt zone is constant. Whereas the Spiegelman and Elliott model is an "infinite Damköhler number" model (melt-solid reaction is very fast relative to melt migration velocity), the dynamic melting model is a "zero-Damköhler number" model. Of course, it is likely that reality by necessity somewhere in between (e.g., see Hart 1993; Sims et al. 2002). The dynamic melting model also differs from the Spiegelman and Elliott model in that the trace element enrichments in the melts are intermediate between batch melting and accumulated fractional melting, depending on the value of the threshold porosity relative to the distribution coefficients of the element(s) considered (Williams and Gill 1989; Zou and Zindler 1996). This porosity plays an important role in slowing down the extraction of the incompatible element from the solid. With a small porosity, dynamic melting is close to fractional melting where incompatible elements are efficiently stripped from the residue, while with a large porosity, it is more akin to batch melting where incompatible elements reside in the solid longer.

Dynamic melting models provide results that are similar to those obtained from the chromatographic porous flow melting model in that the extent of (^{226}Ra/^{230}Th)

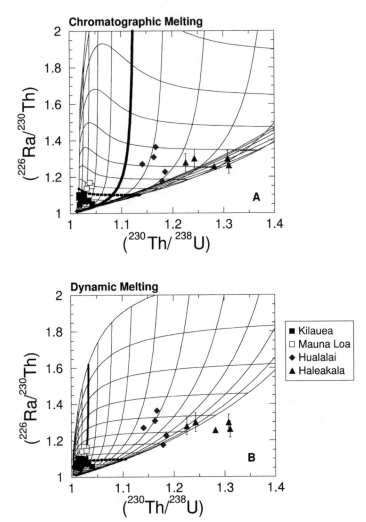

Figure 9. Comparison of the measured (^{226}Ra/^{230}Th) and (^{230}Th/^{238}U) for the Hawaiian basalts (symbols) to the calculated values (curves) for (A) chromatographic melting (Spiegelman and Elliott 1993) and (B) dynamic melting (McKenzie 1985) through a 1D melt column using the partition coefficients of LaTourrette et al. (1992, 1993). Data are a compilation of mass spectrometric measurements from Cohen et al. (1994); Sims et al. (1995, 1999); Pietruszka et al. (2001). For chromatographic melting, the results are calculated for a 50 km melt column and the horizontally trending curves show activity ratios for constant maximum porosity (ϕ_{max}) in percent, while the vertically trending contours show the activity ratios for a constant upwelling rate (W_S) in cm/yr. For dynamic melting the horizontally trending curves show activity ratios for constant threshold porosity (ϕ_{thres}) in percent, while the vertically trending contours show the activity ratios for a constant upwelling rate (W_S) in cm/yr. In both plots, the bold solid contour shows a reference upwelling rate of 10 cm/a, while the bold dashed curve shows a reference porosity of about 1% (phi max for chromatographic melting and phi critical for dynamic melting). Spacing between curves is calculated in log units: increments between curves of solid upwelling rate are $10^{0.2}$ cm/a, and between curves of constant maximum porosity $10^{0.13}$%. Note that with chromatographic melting the decrease of (^{230}Th/^{238}U) at very low upwelling rates is a result of the melting rate approaching the time-scale of the half-life of ^{230}Th. Error bars for (^{226}Ra/^{230}Th) reflect the uncertainties in ^{14}C ages (see Sims et al. 1999).

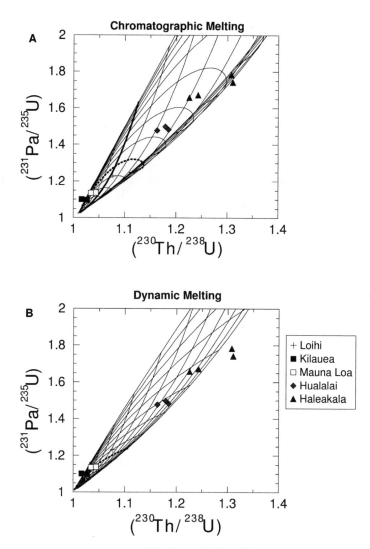

Figure 10. Comparison of the measured $(^{231}Pa/^{235}U)$ and $(^{230}Th/^{238}U)$ for Hawaiian basalts (Sims et al. 1999) to the calculated values (curves) for (A) chromatographic melting (Spiegelman and Elliott 1993) and (B) dynamic melting (McKenzie 1985) through a 1D melt column using the partition coefficients of LaTourrette et al. (1992, 1993). For chromatographic melting- the results are calculated for a 50 km melt column and the horizontally trending curves show activity ratios for constant maximum porosity (ϕ_{max}) in percent, while the vertically trending contours show the activity ratios for a constant upwelling rate (W_S) in cm/yr. For dynamic melting the horizontally trending curves show activity ratios for constant threshold porosity (ϕ_{thres}) in percent, while the vertically trending contours show the activity ratios for a constant upwelling rate (W_S) in cm/yr. In both plots, the bold solid contour shows a reference upwelling rate of 10cm/yr, while the bold dashed curve shows a reference porosity of about 1% (phi max for chromatographic melting and phi critical for dynamic melting). Spacing between curves is calculated in log units: increments between curves of solid upwelling rate are $10^{0.2}$ cm/a, and between curves of constant maximum porosity $10^{0.13}$%. The decrease of $(^{231}Pa/^{238}U)$ and $(^{230}Th/^{238}U)$ at very low upwelling rates is a result of the melting rate approaching the time-scale of the half-lives of ^{231}Pa and ^{230}Th.

disequilibrium is controlled mainly by the porosity of the melt region, and (^{230}Th/^{238}U) and (^{231}Pa/^{235}U) disequilibrium mainly by the melting rate (Figs. 9 and 10). However, with dynamic melting the time-scale of melt migration is not considered. Therefore, for a given set of partition coefficients, the dynamic melting model requires threshold or escape porosities, which are lower (about a factor of 1.5) than the maximum porosity at the top of the melt zone predicted by a chromatographic melting model. With dynamic melting, the inferred melting rates are also slightly lower. This stems from the shorter residence time for the parent nuclide in the melting column obtained with dynamic melting for a given melting rate. In order to get the same residence time for the parent nuclide (i.e., same time for in-growth of the daughter nuclide), the melting rate has to be slower.

4.2. Sources of uncertainty in these models

Uncertainties in sample ages and magma storage times. Because of the relatively short half-lives of the U-series daughter isotopes, as samples age either on the surface or during magma storage, the parent/daughter activity ratios will decay toward their equilibrium value. If this sample age is old and unknown, or if the magma storage is significant and not corrected for, one would infer too small an amount of parent/daughter fractionation for the melting process. Therefore it is important to explicitly know the sample age and to have some constraints on magma storage time when evaluating melting processes using U-series disequilibria. This issue is most significant for ^{226}Ra; and is probably not a significant source of uncertainty for the longer-lived ^{230}Th and ^{231}Pa. For example, it was shown by Vigier et al. (1999) that tholeiitic lavas from the 1978 Ardoukoba eruption in the Afars had (^{226}Ra/^{230}Th) ranging from 1.93 and 1.35. Since the (^{226}Ra/^{230}Th) ratios correlated with indices of magma differentiation, these results were interpreted as reflecting variable residence times in a magma chamber. Since ^{226}Ra-^{230}Th is largely controlled by the porosity of the melt zone, uncertainties in sample ages and magma storage times translate to large uncertainties in the calculated porosity.

Uncertainties in measured U, Th and Ra partition coefficients. Calculated porosities and melting rates from both the chromatographic and dynamic melting models are highly dependent upon the absolute and relative values chosen for the U, Th, Pa and Ra partition coefficients (Table 2). The length of time that the parent element spends in the melt column is critical to the ingrowth models, therefore, these models are most sensitive to the D values chosen for the parent elements U and Th. For the combined U-Th-Ra disequilibria, the value chosen for D_{Th} is particularly important, as Th is both a parent and daughter isotope and therefore its D value affects both (^{230}Th/^{238}U) and (^{226}Ra/^{230}Th) disequilibria. For example, with the chromatographic porous flow model, in order to produce a given degree of (^{230}Th/^{238}U) and (^{226}Ra/^{230}Th) disequilibrium, decreasing the value of D_{Th} by a factor of 3 (while holding D_U and D_{Ra} constant) changes the inferred porosity by an order of magnitude for large melting rates and by up to a factor of 3 for low melting rates. In the above modeling, it is also assumed that the U-Th-Ra partitioning is controlled by major phases (olivine, pyroxene and garnet) and that partitioning of U, Th, Pa and Ra between the melt and solid is constant throughout the melting column. While it can be reasoned that the partitioning of U, Th, (and Ra and Pa?) is controlled by the major mantle phases (Beattie 1993a,b; La Tourrette et al. 1993) it has been recently shown that the values of mineral/melt partition coefficients can be highly dependent on the chemical compositions of the minerals as well as the temperature and pressure of the experiments. In fact, recent high-pressure experiments show that D_U and D_{Th} values can vary by as much as a factor of 10 due to variations in mineral composition and pressure (Beattie 1993a,b; La Tourrette 1993; Lundstrom et al. 1994; Blundy and Wood, 2003; Salters and Longhi, in review). The large variations between the different experimentally determined D_U and D_{Th} values for clinopyroxene and garnet (see Table 2)

Table 2. Experimentally determined mineral/melt partition coefficients ($\times 10^3$) for U, Th, and Ba in clinopyroxene and garnet and calculated bulk partition coefficients ($\times 10^3$) for a garnet peridotite source.

Clinopyroxene			Garnet			Garnet Peridotite				Ref.
U	Th	Ba	U	Th	Ba	U	Th	Ra	Pa	
0.9	1.3	0.5	9.6	1.5	0.01	1.2	0.29	0.04		(1)
3.5	3.5	**	41	1.5	**	5.3	2.6		**	(2)
4.5	10	**	15	19	**	2.2	1.0		**	(3)
10	15	**	**	1.7	**	2.6	1.4	(0.01)	(0.13)	(4)

References: (1) Beattie 1993a,b, (2) Salters and Longhi, in review, (3) LaTourette et al. 1992, 1993, (4) Lundstrom et al. 1994.

Note: Bulk partition coefficients for the garnet peridotite source were calculated using Ol = 59%, Opx = 21%, Cpx = 8%, Gt = 12%. Lundstrom et al. (1994) bulk D values use D^U and D^{Th} for garnet from LaTourette et al. (1993) and assume that $D^{Pa} = D^{U\beta^+}$ and $D^{Ra} = 1.0 \times 10^{-5}$. Beattie bulk D^{Ra} value assumes that $D^{Ra} \approx D^{Ba}$. The inverted D^U-D^{Th} and D^U and D^{Pa} values are from correlations of (^{230}Th/^{238}U) and (^{231}Pa/^{235}U) with Sm/Nd fractionation (Fig. 3) (see Sims et al. 1995 for details) and are in the range of bulk D values calculated from experimental measurements. This inversion is based upon the equations for batch melting and does not account for the contribution of ingrown ^{230}Th and ^{231}Pa in the measured (^{230}Th/^{238}U) and (^{231}Pa/^{235}U) disequilibria; therefore these estimates of the bulk D^U, D^{Th} and D^{Pa} maximize U/Th and U/Pa fractionation.

result in large variations in the calculated porosities and melting rates for both chromatographic and dynamic melting models (Figs. 9 and 10). Until these issues are resolved and the effect of chemical composition, temperature and pressure, and melt structure on the measured partitioning values for U, Th, (and Pa and Ra) are established, calculated solid mantle upwelling rates and porosities based upon (^{230}Th/^{238}U) (^{226}Ra/^{230}Th) and (^{231}Pa/^{235}U) disequilibria must be considered as preliminary estimates only.

4.3. Observational constraints

U-series systematics for ocean island basalts. On a global basis, ocean island basalts are characterized by (^{230}Th/^{238}U) activity ratios greater than one, which suggests that they were generated deep by melting of garnet or high-pressure clinopyroxene. This is particularly clear when looking at the compilation shown on Figure 11 where all the OIB plot to the left of the equiline. Most earlier studies, in particularly in Iceland, had focused on crustal and differentiation processes and are dealt with in another chapter (Condomines et al. 2003). It is worth noting however, that as pointed out by Sigmarsson et al. (1992), that the tholeiites that are thought to come from the center of the plume where the maximum of the gravity anomaly is located are characterized by larger (^{230}Th/^{238}U) activity ratios (up to 1.2) than the alkali basalts that are at the periphery. This observation bears some similarities with the observations made in Hawaii as detailed below. This observation was interpreted as reflecting larger degrees of melting beneath the center of the plume but in light of more recent observations could reflect variations in mantle upwelling rates as well (see below). Chabaux and Allègre (1994) had shown that there is a general positive correlation between (^{230}Th/^{238}U) and (^{226}Ra/^{230}Th), which they then interpreted as due to variable degree of melting. This correlation is not so clear as new mass spectrometry data become available for the Comores, Hawaii, Canaries, Azores (Fig. 11b). There is at this stage no clear systematics in a (^{226}Ra/^{230}Th) versus (^{230}Th/^{238}U) diagram as shown in Figure 11b. More importantly, there is no simple picture based on local systematics. The Azores show no simple relationship, the Canaries show a negative correlation, while Hawaii is positively correlated in the (^{226}Ra/^{230}Th) versus (^{230}Th/^{238}U) diagram. Possible explanations for the generally lower (^{226}Ra/^{230}Th) found in intraplate

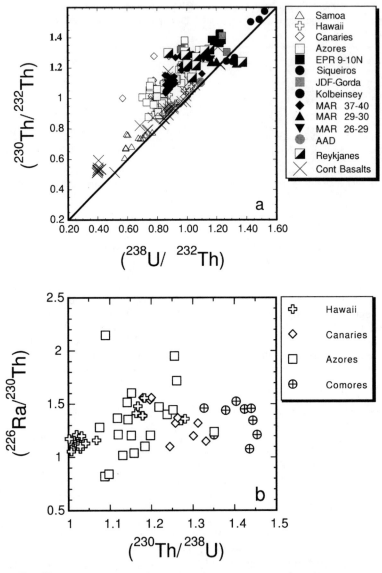

Figure 11. ^{230}Th-^{238}U isochron diagram for mid-ocean ridge, ocean island and continental basalts compiled from the literature [with the exception of the Reykjanes ridge data from Peate et al. (2000)—for which no Sr data exists—references are given in the caption of Fig. 7, and unpublished data from Sims and Hart, Sims and Elliott and Bourdon and Turner]. All of the OIB lavas plot to the left of the equiline where the activity of ^{230}Th is equal to the activity of ^{238}U. This indicates that the OIB lavas are generated by melting of a source containing high-pressure clinopyroxene or garnet (see text). (^{226}Ra/^{230}Th) plotted against (^{230}Th/^{238}U) for mid-ocean ridge, ocean island and continental basalts compiled from the literature (references indicated in the caption of Fig. 7 and unpublished data from Sims and Hart, and Bourdon and Turner). See text for discussion.

magmas could be due to decay of excess ^{226}Ra during transfer through the lithospheric plate or storage in a magma chamber. Alternatively, the lower ^{226}Ra excess could be controlled in some cases by melting in the presence of hydrous phases, which retain Ba (and therefore Ra) preferentially to Th.

More information can also be gained by considering ^{231}Pa-^{235}U systematics (Pickett and Murrell 1997; Bourdon et al. 1998; Sims et al. 1999; Bourdon and Turner, in prep). As pointed out by Pickett and Murrell (1997), the ocean island basalts define a linear trend that is clearly distinct from the MORB trend. This observation still holds based on a larger database (Fig. 12). Bourdon et al. (1998) had shown that these two trends could be explained by variable cpx:garnet ratio in the mantle source. In the case of shallow melting (MORBs), with a high cpx:garnet ratio, there will be less ^{230}Th-^{238}U fractionation while there will be still significant ^{231}Pa-^{235}U fractionation (Sims et al. 2002). In contrast, with a low cpx:garnet ratio (OIB), the ^{230}Th-^{238}U fractionation will be much greater for a given (^{231}Pa/^{235}U) ratio (see Fig. 12). This can be shown in the context of a simple batch melting model. The recent partitioning data of Landwehr et al. (2001) as a function of pressure produces melting trends that are consistent with melting at a pressure deeper than the garnet-spinel transition (Fig. 12). As argued by Bourdon et al. (1998), the range in (^{230}Th/^{238}U) in OIB can also be produced by variation in the upwelling rates (see below). An alternative is that the range in (^{230}Th/^{238}U) for OIB is controlled by the amount of residual garnet in the source as argued by Thomas et al. (1999) who found a

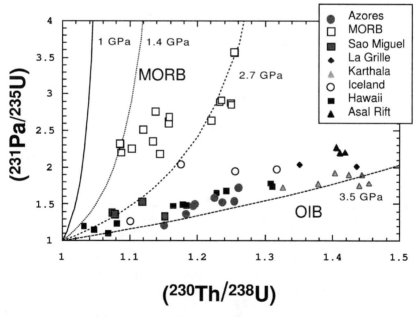

Figure 12. Compilation of ^{230}Th-^{238}U and ^{231}Pa-^{235}U activity ratios for ocean island basalts and mid-ocean ridge basalts (Pickett et al. 1997; Lundstrom et al. 1998; Bourdon et al 1998; Sims et al. 1999; 2002; Vigier et al. 1999; Bourdon and Turner, in prep). The data show two clearly distinct linear trends. The model curves for melting of lherzolite at various pressures (taken from Fig. 4) are also shown for reference and show that the OIB trend can be generated by melting at an initial pressure of melting greater than for MORB. This is a clear indication of the influence of garnet on the ^{230}Th-^{238}U fractionation in OIB. As advocated by Pickett and Murrell (1997), ^{231}Pa-^{235}U fractionation can be produced by clinopyroxene and does not require garnet as a residual phase.

weak correlation between (^{230}Th/^{238}U) and La/Yb ratios, an index of garnet involvement. Further studies including more comprehensive data sets will be necessary to sort out these issues.

Constraints on mantle upwelling velocities from ^{230}Th disequilibrium. Chabaux and Allègre (1994) showed that there is a broad correlation between averaged (^{230}Th/^{238}U) over hotspot localities and the buoyancy flux as estimated by Sleep (1990). The original interpretation of Chabaux and Allègre (1994) was that the range in (^{230}Th/^{238}U) that roughly correlated with (^{226}Ra/^{230}Th) was due to variation in the degree of melting. In fact, this correlation can also be interpreted as reflecting variation in the mantle upwelling rates. It can be shown that there is a direct link between the buoyancy flux and the mantle upwelling velocity (and hence the melting rate). In the case of a pipe flow, the buoyancy flux Q as defined by Sleep (1990) should scale with the square of the mantle upwelling velocity W (Bourdon et al. 1998). The volumetric flux within a cylinder is equal to:

$$Q = W\pi r^2 \tag{8}$$

The average velocity in a pipe with a circular section is given by:

$$W = \frac{\Delta\rho g r^2}{8\mu} \tag{9}$$

Where μ is the viscosity. It follows that the buoyancy flux $B(=\Delta\rho Q)$ is proportional to the square of the average upwelling velocity in the pipe:

$$B = \left(\frac{8\pi\mu}{g}\right) W^2 \tag{10}$$

For the range of buoyancy fluxes found in mantle plumes (0.5-11 M g s^{-1}), the mantle upwelling velocities should vary by a factor of 3-4. This was shown by Bourdon et al. (1998) to be consistent with the range in ^{230}Th-^{238}U and ^{231}Pa-^{235}U disequilibrium found for ocean island basalts. Based on more recent data, this correlation still seems to hold but there is clearly a large variability within a single locality (Sims et al. 1999; Thomas et al. 1999; Sigmarsson et al. 1998). This variability could be explained by variation in source composition or local variation in the mantle upwelling across the plume head (see below). These observations will need to be further documented by more comprehensive data set including several islands/volcanoes from a single hotspot.

Hawaiian plume dynamics: evidence for a fast upwelling plume. One of the most widely studied (and perhaps the most well understood) examples of ocean island volcanism is the Hawaiian Archipelago. The islands of Hawaii are at the southeastern end of a linear chain of volcanic islands, which form part of the 6000 km long Hawaiian-Emperor volcanic ridge. These islands are generally thought to be the surface expression of a stationary mantle plume that has been sitting beneath the westward moving Pacific plate for about 75 Ma. Based on the volume and composition of these basalts, as well as geophysical observations such as the geoid height and bathymetric swell, it is generally thought that the northwest motion of the Pacific plate is moving the hotspot center off the main island of Hawaii, and that the new submarine volcano of Loihi to the southeast represents the initial stages of melting produced at the leading edge of the plume.

Numerous studies on lavas from the Hawaiian Islands have shown that over the life-span of a volcano, the major- and trace-element chemistry of the erupted basalts vary significantly and systematically (MacDonald and Katsura 1964; Chen and Frey 1985). This variation is believed to be the result of movement of the Pacific plate, over the stationary mantle plume, and is typified into four distinct phases of volcano evolution: (1) a pre-shield stage, during which both tholeiitic and alkaline lavas are erupted; (2) a tholeiitic shield-building stage, which represents about 98% of the volcanoes total volume; (3) a post-caldera or post-shield stage during which the lavas become more alkalic in composition and erupt as small parasitic cones, forming a thin veneer over the shield stage tholeiites; and (4) a late or rejuvenated alkalic volcanism, which occurs after an ill-defined period of quiescence. While the time required for the entire evolution of one volcano (ca. 2-5 million years) is much longer than the time-span addressable by U-Th disequilibria, it is possible to obtain from several different volcanoes a suite of historic or young radiocarbon dated lavas which range in composition from primitive tholeiite to primitive alkali basalt and collectively encompass all four stages of Hawaiian volcano evolution.

Past U-series work on Hawaii, has included both early alpha counting studies (Krishnaswami 1984; Newman et al. 1984; Reinitz and Turekian 1991; Hémond et al. 1994) and more recent mass spectrometric studies using TIMS, SIMS and PIMMS (Cohen and O'Nions 1993; Sims et al. 1995, 1999; Layne and Sims 2000; Pietruszka et al. 2001). Most U-series studies have focused on the active shield-stage tholeiitic volcanoes- Kilauea and Mauna Loa (Krishnaswami 1984; Newman et al. 1984; Reinitz and Turekian 1991; Cohen et al. 1993; Hémond et al. 1994; Pietruszka et al. 2001). However, U-Th-Ra and U-Pa disequilibria have also been measured in young tholeiitic and alkaline lavas from the preshield-stage submarine volcano Loihi (Sims et al. 1995; 1999) and alkaline lavas from the post-shield stage volcanoes- Hualalai, and Mauna Kea and rejuvenated stage volcano- Haleakala (Sims et al. 1995, 1999). U-Th disequilibria have also been measured in young (or not quite so young) alkaline submarine flows from the South Arch (Sims et al. 1995).

The early U- and Th-series measurements, utilizing radioactive counting techniques, demonstrated the existence of significant radioactive disequilibria in lavas from Kilauea, suggesting that chemical fractionation between U-Th and Th-Ra occurred during the formation of the Kilauean magmas, on a time scale of less than 350 ka and 8 ka, respectively. However, the origin of this fractionation- melting or secondary process, such as assimilation of altered oceanic crust- was highly uncertain (Krishnaswami 1984; Hémond et al. 1994). In a subsequent alpha counting study, Reinitz and Turekian (1991) measured the activities of several U-series nuclides in a suite of lavas erupted during the first two years of the ongoing Pu' O'o eruption (from 1983 to present). Their results confirmed the excess of ^{230}Th over ^{238}U, but more importantly demonstrated that the activity of ^{230}Th was correlated with the lavas incompatible trace element ratios, suggesting that partial melting was responsible for the fractionation of U from Th.

Recent mass spectrometric studies have consistently demonstrated that the ^{230}Th excesses in tholeiitic lavas from Loihi, Kilauea, and Mauna Loa are small (2 to 6% excesses) compared to the large ^{230}Th excesses seen in MORB (up to 25%), which are also mostly tholeiites. The ^{230}Th excesses in the Hawaiian tholeiites are relatively small (Cohen et al. 1993; Hémond et al. 1994; Sims et al. 1995, 1999; Pietruszka et al. 2001) and constant (Pietruszka et al. 2001), whereas the young alkalic lavas from the post-shield stage volcanoes- Hualalai, and Mauna Kea and rejuvenated stage volcano-Haleakala have much larger ^{230}Th excesses, from 15 to 30% (Sims et al. 1995, 1999). Furthermore, these ^{230}Th excesses vary systematically across the range of compositions observed for Hawaiian lavas, with the smallest excess observed in the tholeiitic lavas and the largest excesses seen in the basanitic lavas from Haleakala (Sims et al. 1995, 1999). The observation that (^{230}Th/^{238}U) disequilibria in Hawaiian lavas are correlated with the lavas extent of melting, as inferred by its trace-and major-element chemistry, provides further evidence that U-Th fractionation is related to the melting process (Sims et al. 1995, 1999). In contrast to (^{230}Th/^{238}U), (^{226}Ra/^{230}Th) disequilibria in the shield stage tholeiitic lava samples are much larger, with excesses ranging from 7 to 15% (Cohen et al. 1993; Hémond et al. 1994; Sims et al. 1999; Pietruszka et al. 2001; Cooper et al. 2001). For the alkalic lavas from Hualalai and Haleakala ^{226}Ra excesses are even larger, up to 30% (Sims et al. 1999). (^{231}Pa/^{235}U) in Hawaiian lavas (Pickett and Murrell 1997; Sims et al. 1999) shows the largest extents of parent-daughter fractionation with ^{231}Pa excesses ranging from 10 to 15% for the tholeiitic lavas from Kilauea, Mauna Loa and Loihi, up to 78% for the basanitic lavas from Halekala. Like (^{230}Th/^{238}U) the magnitude of the (^{231}Pa/^{235}U) disequilibria, is related to the lavas major-element and trace element chemistry, and by inference melt fraction. The relative extent of U-Th-Ra and U-Pa disequilibria observed in the Hawaiian lavas suggest that during melting of $D_{Ra} < D_{Th} < D_U$ and $D_{Pa} < D_U$. The observation that (^{230}Th/^{238}U) is greater than one in all Hawaiian lavas requires melting to be deep and most likely in the presence of garnet (which is logical considering the thickness of the lithosphere beneath Hawaii approximately 80 km).

While it can be shown that the ^{238}U-^{230}Th and ^{235}U-^{231}Pa disequilibria are mainly sensitive to melt fraction because of the long half-lives of ^{230}Th and ^{231}Pa (Sims et al. 1995, 1999; Elliott 1997) interpretation of ^{226}Ra data, because of its much shorter half-life, requires models that treat explicitly the time-scales of melt generation and melt extraction (e.g., McKenzie 1985; Spiegelman and Elliott 1993). Using a forward modeling approach to compare their data with the dynamic melting formalism of McKenzie (1984), Cohen et al. (1993) and Hémond et al. (1994) showed that measurements of ^{238}U-^{230}Th-^{226}Ra disequilibria in tholeiitic lavas from Kilauea and Mauna Loa can provide previously unattainable estimates of porosity (< 1%), solid mantle melting rate (~ 10^{-3} kg m^{-3} a^{-1}) and melt transport times (< 1.7 ka) beneath these volcanoes. Using a similar forward modeling approach, Sims et al. (1999) showed that both chromatographic porous flow and dynamic melting of a garnet peridotite source can adequately explain the ^{238}U-^{230}Th-^{226}Ra and ^{235}U-^{231}Pa disequilibria in their suite of tholeiitic-to-basanitic lavas, but that in both cases the range of allowable melting conditions was highly dependent on the U-Th and (to a lesser extent) Ra and Pa partition coefficients. Using the observed range of experimental partition coefficients for U and Th (Beattie 1993a,b; La Tourrette 1993; Lundstrom et al. 1994; Salters and Longhi 1999), they found that for chromatographic porous flow, the calculated maximum porosity in the melting zone ranges from 0.3-3% for tholeiites and 0.1-1% for alkali basalts and basanites, and solid mantle upwelling rates range from 40 to 100 cm yr^{-1} for tholeiites and from 1 to 3 cm yr^{-1} for basanites. For dynamic melting, the escape or threshold porosity is 0.5-2% for tholeiites and 0.1-0.8% for alkali basalts and basanites, and solid mantle upwelling rates range from 10 to 30 cm yr^{-1} for tholeiites and from 0.1 to 1 cm yr^{-1} for basanites. While it is clear that the Hawaiian U-series data do not distinguish the mode of melting beneath Hawaii- dynamic or chromatographic- both models show that there is a significant range in the predicted solid mantle upwelling rates in the different volcanoes that extend from the leading edge to the trailing edge of the Hawaiian plume (Fig. 13).

Despite considerable evidence for the existence of a deep-rooted mantle plume beneath Hawaii and numerous geochemical studies, including several measurements of U-series disequilibria, there have been very few studies constraining the nature of the melting region and the dynamics of the plume by combining geophysical and geochemical observations. Watson and McKenzie (1991) used an axisymmetric plume model to calculate the melt production rate and bulk major and trace element chemistry of magmas. However, they did not try to account for the spatial and compositional variability seen in the chemistry of the erupted lavas, nor, did they try to take into account

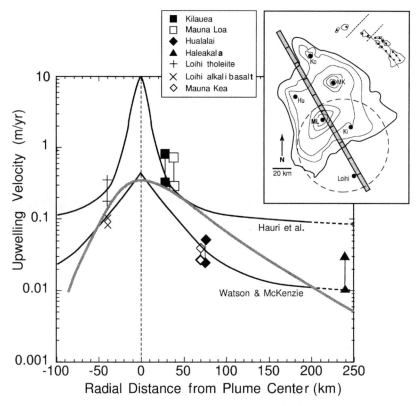

Figure 13. Solid mantle upwelling rate calculated from ^{238}U-^{230}Th-^{226}Ra and ^{235}U-^{231}Pa disequilibria versus the radial position of the volcanoes relative to the center of the Hawaiian plume. The center of the plume (see inset) is located so that the active volcanoes Loihi, Kilauea and Mauna Loa are positioned appropriately for their model ages and eruption rates (DePaolo and Stolper 1996). Also shown are the model curves for solid mantle upwelling from Watson and McKenzie (1991), Hauri et al. (1994) and Jull and Ribe (2002). The model of Watson and McKenzie (1991) assumes a constant viscosity across the plume whereas the Hauri et al. model incorporates a temperature-dependent viscosity. The upwelling rates from the plume model of Jull and Ribe (2002) take into account the effect of motion of the Pacific plate and depletion buoyancy on the dynamics of the plume, whereas the calculations of Hauri et al. (1994) and Watson and McKenzie (1991) do not. The ^{230}Th and ^{231}Pa data show a clear asymmetry in upwelling velocity which is evident in the results of Jull and Ribe (2002). These estimates will need to be refined with more accurate values of distribution coefficients and better models for the melt productivity, but they illustrate that the U-series disequilibrium isotopic data may provide a much-needed constraint on the dynamics of the plume.

the effect of the overriding Pacific plate on the Hawaiian plume dynamics. Richardson and Mckenzie (1994) used the plume models developed by Watson and McKenzie (1991) to calculate the effects of an upwelling plume on U-series disequilibria, and to use these results to determine the melting rate (related to solid mantle upwelling rate) and porosity of the melting region. However, because there was very little U-series data from volcanoes other than Kilauea and Mauna Loa, they made no attempt to correlate spatial variability in the U-series activity ratios along the plume track with the dynamics of the plume. More recent work of Sims et al. (1999) which looked at U-series in young lavas extending from the center of the inferred Hawaiian plume out to its fringes, has shown that that solid mantle upwelling rates are highest in the plume center and lowest at its leading and trailing edges (Fig. 13). This observation has been further confirmed by the study of Pietruszka et al. (2001) which also indicates large upwelling rates for Kilauean lavas. As shown in Figure 13, upwelling rates calculated from U-series measurements are comparable to upwelling velocities calculated from the axis symmetric plume models of Watson and McKenzie (1991) and Hauri (1994), however, these U-series upwelling rates also show an asymmetry not predicted by these plume models. The observation that the solid mantle upwelling is greater for Loihi, than Hualalai and Haleakala appears to be more consistent with the recent plume model of Ribe and Christensen (1999) and Jull and Ribe (2002) which takes into account the influence of the overriding Pacific plate on the plumes upwelling dynamics.

5. SUMMARY AND PERSPECTIVES

This chapter has attempted to show the potential of U-series for unraveling the complexities of intraplate magmatism. While there are still many uncertainties, which limit the inferences that can be obtained from U-series measurements, there are several results that appear to be more robust. First, the nature and magnitude of U-Th and U-Pa fractionation suggests that melting beneath hotspots is in general deeper than beneath ridges. Second, the presence of ^{226}Ra excess suggests that melt migration must be relatively fast inside the lithosphere. Third, the range in U-Th and U-Pa in hotspot lavas (particularly in Hawaii) strongly indicates that the melting rate can be constrained with U-series disequilibria. Forth, there are a few locations where lithospheric melting can be demonstrated using U-series disequilibria and this is an important constraint for thermal models of hotspots.

While over the past ten years, our ability to measure U-series disequilibria and interpret this data has improved significantly it is important to note that many questions still remain. In particular, because of uncertainties in the partition coefficients, fully quantitative constraints can only be obtained when more experimental data, as a function of P and T as well as source composition, become available. Furthermore, the robustness of the various melting models that are used to interpret the data needs to be established and 2D and 3D models need to be developed. However, full testing of these models will only be possible when more comprehensive data sets including all the geochemical parameters are available for more locations and settings.

ACKNOWLEDGMENTS

We would like to thank Thomas Kokfelt, Aaron Pietruszka and Craig Lundstrom for their reviews that have significantly contributed to improving the manuscript. Discussions with Neil Ribe, Matthew Jull, Stan Hart, Peter Kelemen, Marc Spiegelman, Tim Elliott, Alberto Saal, Kari Cooper, Pierre Gauthier and Christophe Hémond have helped in shaping some of the ideas presented in this manuscript. Support for this work was provided in part by the National Science Foundation (EAR-9909473 to KWWS).

REFERENCES

Allègre CJ, Condomines M (1982) Basalt genesis and mantle structure studied through Th-isotopic geochemistry. Nature 299:21-24
Allègre CJ, Dupré B, Lewin E (1986) Thorium/uranium ratio of the Earth. Chem Geol 56:217-227
Allègre CJ, Turcotte D (1986) Implications of a two-component marble-cake mantle. Nature 323:123-127
Asimow PD, Hirschmann MM, Ghiorso MS, O'Hara MJ Stolper EM (1995) The effect of pressure-induced solid-solid phase transitions on decompression melting of the mantle. Geochim Cosmochim Acta 59:4489-4506

Asmerom Y(1999) Th-U fractionation and mantle structure. Earth Planet Sci Lett 166:163-175
Asmerom Y, Edwards R L (1995) U-series isotope evidence for the origin of continental basalts. Earth Planet Sci Lett 134:1-7
Asmerom Y, Cheng H, Thomas R, Hirschmann M, Edwards RL (2000) Melting of the Earth's lithospheric mantle inferred from protactinium-thorium-uranium isotopic data. Nature 406:293-296
Baker MB, Hirschmann MM, Ghiorso MS, Stolper EM (1995) Compositions of near-solidus peridotite melts from experiments and thermodynamic calculations. Nature 375:308-311
Beattie PD (1993) Uranium-thorium disequilibria and partitioning on melting of garnet peridotite. Nature 363:63-65
Bercovici D, Lin J (1996) A gravity current model of cooling mantle plume heads with temperature dependent buoyancy and viscosity. J Geophys Res 101:3291-3309
Blundy J, Wood B (1994) Prediction of crystal-melt partition coefficients from elastic moduli. Nature 372:452-454
Bourdon B, Langmuir C, Zindler A (1996) Rigde-hotspot interaction along the Mid-Atlantic ridge between 37°30 and 40°30 N: the U-Th disequilibrium evidence. Earth Planet Sci Lett 142:175-196
Bourdon B, Joron J-L, Claude-Ivanaj C, Allègre CJ (1998) U-Th-Pa-Ra systematics for the Grande Comore volcanics: melting processes in an upwelling plume. Earth Planet Sci Lett 164:119-133
Bourdon B and Turner SP (in prep) Upwelling rates beneath hotspots: evidence from U-series in basalts from the Azores region.
Carracedo JC, Badiola ER, Soler V (1990) Aspectos volcanologicos y estructurales, evolucion opetrologica e implicaciones en riesgo volcanico de la erupcion de 1730 en Lanzarote, Islas Canarias. Estud Geol 46:25-55
Chabaux F, Allègre CJ (1994) ^{238}U-^{230}Th-^{226}Ra disequilibria in volcanics- a new insight into melting conditions. Earth Planet Sci Lett 126:61-74
Chen C-Y Frey, F.A (1985) Trace element and isotopic geochemistry of lavas from Haleakala volcano, East Maui, Hawaii. implications for the origins of Hawaiian basalts. J. Geophys Res 90:8743-8768
Class C, Goldstein SL, Altherr R, Bachèlery P (1998) The process of plume-lithosphere interactions in the ocean basins- the case of Grande Comore. J Pet 39:881-903
Class C, Goldstein SL (1997) Plume-lithosphere interaction in the ocean basins: constraints from the source mineralogy. Earth Planet Sci Lett 150:245-260
Claude-Ivanaj C, Bourdon B, Allègre CJ (1998) Ra-Th-Sr isotope systematics in Grande Comore Island: a case study of plume-lithosphere interaction. Earth Planet Sci Lett 164:99-117
Claude-Ivanaj C, Joron J-L, Allègre CJ (2001) ^{238}U-^{230}Th-^{226}Ra fractionation in historical lavas from the Azores: long-lived source heterogeneity vs. Metasomatism fingerprints. Chem Geol 176:295-310
Cohen AS, O'Nions RK (1993) Melting rates beneath Hawaii: evidence from uranium series isotopes in recent lavas. Earth Planet Sci Lett 120:169-175
Cohen AS, O'Nions RK, Kurz MD (1996) Chemical and isotopic variations in Mauna Loa tholeiites. Earth Planet Sci Lett 143:111-124
Cooper KM, Reid, Mary R, Murrell, MT, Clague, DA (2001) Crystal and magma residence at Kilauea Volcano, Hawaii. ^{230}Th- ^{226}Ra dating of the 1955 East Rift eruption. Earth Planet Sci Lett 184:703-718
Condomines M, Hémond C, Allègre CJ (1988) U-Th-Ra radioactive disequilibria and magmatic processes. Earth Planet Sci Lett 90:243-262
Condomines M, Sigmarsson O (2000) ^{238}U-^{230}Th disequilibria and mantle processes : a discussion. Chem Geol 162:95-104
Dupré B, Lambret B, Allègre CJ (1982) Isotopic variations within a single oceanic island: the Terceira case. Nature 299:620-622
Falloon TJ, Green DH (1989) The solidus of carbonated, fertile peridotite. Earth Planet Sci Lett 94:364-370
Galer SJG, O'Nions RK (1986) Magmagenesis and the mapping of chemical and isotopic variations in the mantle. Chem Geol 56:45-61
Gast PW (1968) Trace element fractionation and the origin of tholeiitic and alkaline magma type. Geochim Cosmochim Acta 32:1057-1086
Griffiths RW (1986) The differing effects of compositional and thermal buoyancies on the evolution of mantle diapirs. Phys Earth Planet Inter 43:261-273
Griffiths RW, Campbell IH (1991) On the dynamics of long-lived plume conduits in the convective mantle. Earth Planet Sci Lett 103:214-222
Hart SR (1988) Heterogeneous mantle domains: signatures, genesis and mixing chronologies. Earth Planet Sci Lett 90:273-296
Hauri EH, Whitehead JA, Hart SR, (1994) Fluid dynamic and geochemical aspects of entrainment in mantle plumes. J Geophys Res 99:24275-24300

Hawkesworth CJ, Kempton PD, Rogers NW, Ellam RM, van Calsteren PW (1990) Continental mantle lithosphere, and shallow level enrichment processes in the Earth's mantle. Earth Planet Sci Lett 96:256-268

Hémond C, Condomines M, Fourcade S, Allègre CJ, Oskarsson N, Javoy M (1988) Thorium, strontium and oxygen isotopic geochemistry in recent tholeiites from Iceland: crustal influence on mantle-derived magmas. Earth Planet Sci Lett 87:273-285

Hémond C, Hofmann AW, Heusser G, Condomines M, Rhodes JM, Garcia MO (1994) U-Th-Ra systematics in Kilauea and Mauna Loa basalts. Contrib Mineral Petrol 116:163-180

Hirose K (1997) Partial melt compositions of carbonated peridotite at 3 GPa and role of CO_2 in alkali basalt magma generation. Geophys Res Lett 24:2837-2840

Hirschmann MM, Stolper EM (1996) A possible role for pyroxenite in the origin of the garnet signature in MORB. Contrib Mineral Petrol 124:185-208

Hirschmann MM, Asimow PD, Ghiorso MS, Stolper EM (199) Calculation of peridotite partial melting form thermodynamic model of minerals and melts III. Controls on isobaric melt production and the effect of water on melt production. J Petrol. 40:831-851

Hirth G Kohlstedt DL (1996) Water in the oceanic upper mantle: implications for rheology, melt extraction and the evolution of the lithosphere. Earth Planet Sci Lett 144:93-108

Hoernle K, Gill J, Schmincke HU (1993) Extreme fractionation of (Nb, K, Th)/U ratios during metasomatism of Jurassic oceanic lithospheric mantle. Eos Trans Am Geophys U 74:633

Huang Y, Hawkesworth C, Carlsteren Pv, Smith I, Black P (1997) Melt generation models for the Auckland volcanic field, New Zealand: constraints from U-Th isotopes. Earth Planet Sci Lett 149:67-84

Ito G, Shen Y, Hirth G, Wolfe CJ (1999) Mantle flow, melting, and dehydration of the Iceland mantle plume. Earth Planet Sci Lett 165:81-96

Jull M, Ribe N (2002) The geochemistry of hawaian plume dynamics. Geochim Cosmochim Acta 66:A375 (abstr.)

Kushiro I (1968) Compositions of magmas formed by partial zone melting of the Earth's upper mantle. J Geophys Res 73:619-634

Landwehr D, Blundy J, Chamorro-Perez E, Hill E, Wood B (2001) U-series disequilibria generated by partial melting of spinel lherzolite. Earth Planet Sci Lett 188:329-348

Langmuir CH, Hanson GN (1980) An evaluation of major element heterogeneity in the mantle sources of basalts. Phil Trans R Soc London A 297:383-407

Lassiter JC, Hauri EH (1998) Osmium-isotope variations in Hawaiian lavas: evidence for recycled oceanic lithosphere in the Hawaiian plume. Earth Planet Sci Lett 164:483-496

LaTourrette TZ, Kennedy AK, Wasserburg GJ (1993) Thorium-uranium fractionation by garnet: evidence for a deep source and rapid rise of oceanic basalts. Science 261:739-742

Liu M, Chase CG (1991) Evolution of Hawaiian basalts: a hotspot melting model. Earth Planet Sci Lett 104:151-165

Loper DE, Stacey FD (1983) The dynamical and thermal structure of deep mantle plume. Phys Earth Planet Int 33:304-317

Lundstrom CC, Shaw HF, Ryerson FJ, Phinney DL, Gill JB, Williams Q (1994) Compositional controls on the partitioning of U, TH, Ba, Pb, Sr and Zr between clinopyroxene and haplobasaltic melts: implications for uranium series disequilibria in basalts. Earth Planet Sci Lett 128:407-423

MacDonald GA, Katsura T (1964) Chemical composition of Hawaiian lavas. J Petrol 5:82-133

McKenzie D (1985) ^{230}Th-^{238}U disequilibrium and the melting processes beneath ridge. Earth Planet Sci Lett 72:149-157

McKenzie (1989) Some remarks on the movement of small melt fractions in the mantle. Earth Planet Sci Lett 95:53-72

Minster JF, Allègre CJ (1978) Systematic use of trace elements in igneous processes. Part III: Inverse problem of batch melting in volcanic suites. Contrib Mineral Petrol 68:37-52

Morgan JW (1971) Convection plumes in the lower mantle. Nature 230:42-43

Newman S., Finkel R.C. Macdougall J.D. (1984) Comparison of ^{230}Th-^{238}U disequilibrium systematics in lavas from three hot spot regions: Hawaii, Prince Edward and Samoa. Geochim Cosmochim Acta 48:315-324

Olafsson M, Eggler DH (1983) Phase relations of amphibole, amphibole-carbonate, and phlogopite-carbonate peridotite: petrologic constraints on the asthenosphere. Earth Planet Sci Lett 64:305-315

Olson P, Schubert G, Anderson C, Goldman P (1988) Plume formation and lithosphere erosion: a comparison of laboratory and numerical experiments. J Geophys Res 93:15065-15084

Pearson DG, Shirey SB, Carlson RW, Boyd FR, Nixon PH (1995) Stabilisation of Archean lithospheric mantle: A Re-Os isotope isotope study of peridotite xenoliths. Earth Planet Sci Lett 134:341-357

Pickett DA, MT Murrell (1997) Observations of ^{231}Pa/^{235}U disequilibrium in volcanic rocks. Earth Planet Sci Lett 148:259-271

Pietruszka AJ, Rubin KH, Garcia MO (2001) ^{226}Ra-^{230}Th-^{238}U disequilibria of historical Kilauea lavas (1790-1982) and the dynamics of mantle melting within the Hawaiian plume. Earth Planet Sci Lett 186:15-31

Rabinowicz M, Ceuleneer G, Monnereau M, Rosemberg C (1990) Three-dimensional models of mantle flow across a low-viscosity zone: implications for hotspot dynamics. Earth Planet Sci Lett 99:170-184

Reid MR(1995) Processes of mantle enrichment and magmatic differentiation in the eastern Snake River Plain: Th isotope evidence. Earth Planet Sci Lett, 131:239-254

Reid MR, Ramos FC (1996) Chemical dynamics of enriched mantle in the southwestern United States: Thorium isotope evidence. Earth Planet Sci Lett, 138:67-81.

Ribe NM (1988) Dynamical geochemistry of the Hawaiian plume. Earth Planet Sci Lett 88:37-46

Ribe NM (1996) The dynamics of plume-ridge interaction 2. Off-ridge plumes. J Geophys Res 101:16195-16204

Ribe NM Smooke MD (1987) A stagnation point flow model for melt extraction from a mantle plume. J Geophys Res 92:6437-6443

Ribe NM, Christensen UR (1999) The dynamical origin of Hawaiian volcanism. Earth Planet Sci Lett 171:517-531

Schiano P, Bourdon B, Clochiatti R, Massare D, Varela ME, Bottinga Y (1998) Low-degree partial melting trends recorded in upper mantle minerals. Earth Planet Sci Lett 160:537-550

Schilling JG (1973) Iceland mantle plume geochemical evidence along Reykjanes Ridge. Nature 242:663-706

Schilling JG (1991) Fluxes and excess temperatures of mantle plumes inferred from their interaction with migrating mid-ocean ridges. Nature 352:397-403

Sigmarsson O, Condomines M, Fourcade S (1992) A detailed Th, Sr and O isotope study of Hekla : differentiation processes in an Icelandic Volcano. Contrib Mineral Petrol 112:20-34

Sigmarsson O, Condomines M, Fourcade S (1992) Mantle and crustal contribution in the genesis of recent basalts from off-rift zones in Iceland: constraints from Th, Sr and O isotopes, Earth Planet Sci Lett 110:149-162

Sigmarsson O, Condomines M, Ibarrola E (1992) ^{238}U-^{230}Th radioactive disequilibria in historic lavas from the Canary Islands and genetic implications. J Volcanol Geotherm Res 54:145-156

Sigmarsson O, Carn S, Carracedo JC (1998) Systematics of U-series nuclides in primitive lavas from the 1730-36 eruption in Lanzarote, Canary Islands and the implications for the role of garnet pyroxenite during oceanic basalt formation. Earth Planet Sci Lett 162:137-151

Sims KWW, DePaolo DJ, Murrell MT, Baldridge WS, Goldstein SJ, Clague DA (1995) Mechanisms of magma generation beneath Hawaii and mid-ocean ridges: uranium/thorium and samarium/neodymium isotopic evidence. Science 267:508-512

Sims KWW, DePaolo DJ, Murrell MT, Baldridge WS, Goldstein S, Clague D, Jull M (1999) Porosity of the melting zone and variations in the solid mantle upwelling rate beneath Hawaii: inferences from ^{238}U-^{230}Th-^{226}Ra and ^{235}U-^{231}Pa disequilibria. Geochim Cosmochim Acta 63:4119-4138

Sleep NH (1990) Hotspots and mantle plumes: some phenomenology. J Geophys Res 95:6715-6736

Späth A, Roex APL, Duncan RA (1996) The geochemistry of lavas from the Comores archipelago, Western Indian Ocean: petrogenesis and mantle source region characteristics. J Petrol 37:961-991

Spiegelman M Elliott T (1993) Consequences of melt transport for uranium series disequilibrium in young lavas. Earth Planet Sci Lett 118:1-20

Stracke A, Salters VJM, Sims KWW (1999) Assessing the presence of garnet-pyroxenite in the mantle sources of basalts through combined hafnium-neodymium-thorium isotope systematics. Geochem Geophys Geosyst 1:1999GC000013

Thomas LE, Hawkesworth CJ, Van Calsteren P, Turner SP, Rogers NW (1999) Melt generation beneath ocean islands: a U-Th-Ra isotope study from Lanzarote in the Canary Islands. Geochim Cosmochim Acta 63:4081-4099

Turner S, Hawkesworth C, Rodgers N, King P (1997) U-Th disequilibria and ocean island basalt generation in the Azores. Chem Geol 139:145-164

Vigier N, Bourdon B, Joron J-L, Allègre CJ (1999) U-Th-Ra disequilibria in Ardoukoba tholeiitic basalts (Asal rift): timescales of crystallization. Earth Planet Sci Lett 174:81-98

Watson S, McKenzie D (1991) Melt generation by plumes: a study of Hawaiian volcanism. J Petrol 32:501-537

Wendlandt RF, Eggler DH (1980) The origins of potassic magmas: 2. Stability of phlogopite in natural spinel lherzolite and in the system $KalSiO_4$-MgO-H_2O-CO_2 at high pressures and high temperatures. Am J Sci 280:421-458

Widom E, Schmincke H-U, Gill J (1992) Processes and timescales in the evolution of a chemically zoned trachyte: Fogo A, Sao Miguel, Azores. Contrib Mineral Petrol 111:311-328

Widom RE, Carlson RW, Gill JB, Schmincke H-U (1997) Th-Sr-Nd-Pb isotope and trace element evidence for the origin of the Saõ Miguel, Azores enriched mantle source. Chem Geol 140:49-68

Williams RW, Gill JB (1989) Effects of partial melting on the uranium decay series. Geochim Cosmochim Acta 53:1607-1619

Wilson TJ (1963) A possible origin of the Hawaian islands. Canad J Phys 41:863-870

Zimmerman ME, Zhang S, Kohlstedt DL, Karato S (1999) Melt distribution in mantle rocks deformed in shear. Geophys Res Lett 26:1505-1508

Zindler A, Hart SR (1986) Chemical geodynamics. Ann Rev Earth Planet Sci 14:493-571

Zindler A, Staudigel H, Batiza R (1984) Isotope and trace element geochemistry of young Pacific seamounts: implications for the scale of upper mantle heterogeneity. Earth Planet Sci Lett:70:175-190

APPENDIX:
ANALYTICAL SOLUTIONS FOR TIME DEPENDENT MELTING MODELS

A.1. Dynamic melt transport

Analytical solutions for ^{238}U-^{230}Th-^{236}Ra disequilibria during dynamic melting (near fractional) were first derived by McKenzie (1985) and have subsequently been presented in several papers pertaining to the production of U-series disequilibria in basaltic melts (Williams and Gill 1989; Beattie 1993a; Chabaux and Allègre 1994; Sims et al. 1999).

Assuming a constant melt rate (Γ), by mass, and porosity (ϕ), the (^{230}Th/^{238}U) and ^{226}Ra/^{230}Th) can be expressed as a function of ϕ and Γ:

$$\left(\frac{^{230}Th}{^{238}U}\right) = \frac{F_{Th}(K_U + \lambda_{Th})}{F_U(K_{Th} + \lambda_{Th})} \tag{A1}$$

$$\left(\frac{^{226}Ra}{^{230}Th}\right) = \frac{F_{Ra}(K_{Th} + \lambda_{Ra})}{F_{Th}(K_{Ra} + \lambda_{Ra})}\left\{1 + \frac{\lambda_{Th}(K_U - K_{Th})}{(K_{Th} + \lambda_{Ra})(K_U + \lambda_{Th})}\right\} \tag{A2}$$

where

$$F_i = \frac{\phi \rho_f}{D_i \rho_s(1-\phi) + \phi \rho_f} \tag{A3}$$

and

$$K_i = \frac{F_i(1-D_i)}{\phi \rho_f}\Gamma_0 \tag{A4}$$

D_i is the bulk mineral/melt partition coefficient and ρ_f and ρ_s are the densities of the melt and peridotite (2800 and 3300 kg m^{-3}). A similar expression to Equation (A1) can be derived for (^{231}Pa/^{235}U).

A.2. Chromatographic melt transport

Numeric solutions for ^{238}U-^{230}Th-^{236}Ra disequilibria during chromatographic melting were first derived by Spiegelman and Elliott (1993). While Spiegelman and Elliott pointed out that the system of equations for the production of U-series disequilibria during chromatographic melting could be solved analytically, these expressions were never presented. Subsequent efforts by Sims (1995) showed that there was a simple analytical approximation for the chromatographic melting expressions. These two solutions, which were originally presented in Sims et al. (1999) are given below along with a comparison between results obtained from analytical solutions and those obtained numerically.

Analytical solutions. Following the approach of Spiegelman and Elliott (1993), the effects of transport of melt through a melting column on the chemistry of radiogenic isotopes can be separated from melting by expressing the concentration of an element as:

$$c_i = \alpha_i c_{bi} \tag{5}$$

Where c_i is the concentration of an element measured at the surface, c_{bi} is the concentration of a stable element due to batch melting and α_i is the enrichment factor due to transport. The concentrations in the melt are given by the batch melting equation, despite the fact that the porosity is small and the melt moves relative to the solid, because of the assumption of continuous re-equilibration of solid and liquid. Fractional melts are produced only if the solid does not re-equilibrate with the melt that passes through (see Appendix of Spiegelman and Elliott 1993).

For the decay chain ^{238}U-^{230}Th-^{226}Ra, the change in α_i with position in a 1D melt column for each isotope can be expressed by the following [Note that in Eqn. 15 of Spiegelman and Elliott (1993), there is a typographical error]

$$\frac{d\alpha_0}{d\zeta} = -\lambda_o \tau_0 \alpha_0 \tag{A6}$$

$$\frac{d\alpha_1}{d\zeta} = \lambda_1 (\tau_0 \alpha_0 - \tau_1 \alpha_1) \tag{A7}$$

$$\frac{d\alpha_2}{d\zeta} = \lambda_2 (\tau_1 \alpha_1 - \tau_2 \alpha_2) \tag{A8}$$

Where ζ is the dimensionless distance (z/d) along a column of length d, λ_i are the decay constants (Ln(2)/$t_{1/2}$), and $\tau_i = d/w'_{eff}$ is the effective velocity is given by

$$w^i_{eff} = \frac{\rho_f \phi w + \rho_s (1-\phi) D_i W}{\rho_f \phi + \rho_s (1-\phi) D_i} \tag{A9}$$

Where ρ_f and ρ_s are the melt and solid densities, respectively, ϕ is the porosity, w is the melt velocity. W is the solid upwelling velocity and D_i is the bulk partition coefficient. Note that if the partition coefficient D_i is $\ll 1$, the effective velocity approaches the melt velocity, and also that the difference in effective velocity between elements with different D_i decreases at larger porosities. In the following text, the subscripts 0, 1, and 2 are taken to refer to ^{238}U, ^{230}Th, and ^{226}Ra, respectively.

Equations (A6-A8) form a coupled system of differential equations, whereby the decay of each parent can increase the concentration of the daughter element as melt migrates through the melting region. To evaluate the solutions of these equations for the enrichment factor α_i, we can make the problem analytically tractable by assuming that both the melt velocity (w) and porosity (ϕ) remain constant over the melt column using only their average values. (It should be emphasized that the assumption of constant porosity is made with reference to the migration of melt only, and that for the purposes of melting of the solid matrix, the degree of melting at the solidus is zero and increases linearly upwards.) For a 1D steady-state melting column, the average melt velocity is given by

$$w = \frac{\Gamma d}{\rho_f \phi 2} \tag{A10}$$

Where Γ is the melting rate, which we assume to be constant is given by

$$\Gamma = \frac{W \rho_s F_{max}}{d} \tag{A11}$$

Solving Equations (A6-A8), we obtain

$$\alpha_0(\zeta) = \alpha_0^o e^{-\lambda_0 \tau_0 \zeta} \tag{A12}$$

$$\alpha_1(\zeta) = \frac{\lambda_1 \tau_0 \alpha_0^o}{\lambda_1 \tau - \lambda_0 \tau_0} e^{-\lambda_0 \tau_0 \zeta} + \left[\alpha_1^o - \frac{\lambda_1 \tau 0 \alpha_0^o}{\lambda_1 \tau_1 - \lambda_0 \tau_0} \right] e^{-\lambda_1 \tau_1 \zeta} \quad \text{(A13)}$$

$$\alpha_2(\zeta) = \frac{\lambda_2 \lambda_1 \tau_1 \tau_0 \alpha_0^o}{(\lambda_1 \tau_1 - \lambda_0 \tau_0)(\lambda_2 \tau_2 - \lambda_0 \tau_0)} e^{-\lambda_0 \tau_0 \zeta} + \frac{\lambda_2 \tau_1}{\lambda_2 \tau_2 - \lambda_1 \tau_1} \left[\alpha_1^o - \frac{\lambda_1 \tau_0 \alpha_1^o}{\lambda_1 \tau_1 - \lambda_0 \tau_0} \right] e^{-\lambda_1 \tau_1 \zeta}$$
$$+ \left[\alpha_2^o - \frac{\lambda_2 \lambda_1 \tau_1 \tau_0 \alpha_0^o}{(\lambda_1 \tau_1 - \lambda_0 \tau_0)(\lambda_2 \tau_2 - \lambda_0 \tau_0)} - \frac{\lambda_2 \tau_1}{\lambda_2 \tau_2 - \lambda_1 \tau_1} \left(\alpha_1^o - \frac{\lambda_1 \tau 0 \alpha_0^o}{\lambda_1 \tau_1 - \lambda_0 \tau_0} \right) \right] e^{-\lambda_2 \tau_2 \zeta} \quad \text{(A14)}$$

Where α_0^o, α_1^o and α_2^o are constants of integration. For secular equilibrium at the base of the melting column, we set $\alpha_i^o = 1$.

Figure A1(a) shows the constant value of porosity used in the analytic model (dashed curve), compared to the porosity distribution for a 1D melt column in which the upward flux of melt is required to remain constant (see Spiegelman and Elliott 1993). The solid curves in Figure A1(b) show values of α_i calculated from equations (A12-A14) along the (dimensionless) length of the melting column for the ^{238}U decay chain with a constant porosity of 0.1% and solid upwelling velocity of 1 cm/yr.

As has been shown by Sims et al. (1999) an analytical approximation to chromatographic melting can also be obtained by solving Equations (A7) and (A8) while holding the values of α_0 and α_1 held constant, respectively.

$$\alpha_0(\zeta) = \alpha_0^o e^{-\lambda_0 \tau_0 \zeta} \quad \text{(A15)}$$

$$\alpha_1(\zeta) = \alpha_1^o e^{-\lambda_1 \tau_1 \zeta} + \frac{\tau_0}{\tau_1} \alpha_0 \left[1 - e^{-\lambda_1 \tau_1 \zeta} \right] \quad \text{(A16)}$$

$$\alpha_2(\zeta) = \alpha_2^o e^{-\lambda_2 \tau_2 \zeta} + \frac{\tau_1}{\tau_2} \alpha_1 \left[1 - e^{-\lambda_2 \tau_2 \zeta} \right] \quad \text{(A17)}$$

Holding the values of α_0 and α_1 constant is equivalent to saying that the change in α for a given daughter isotope over the melt column is large compared to that of the parent element. It is evident in Figure A1(b) shows that this approximate analytical solution (dashed curves) is nearly identical to the full analytic solution. The reason for this can be seen by considering the terms $\lambda_0 \tau_0$, $\lambda_1 \tau_1$, and $\lambda_2 \tau_2$, in the full analytic solution (Equations A12-A14), which represent the ratio of the effective transport time with the time taken for concentration of the isotope to decrease by half. For ^{238}U, $e^{-\lambda_0 \tau_0 \zeta} \approx 1$, while for ^{226}Ra, $e^{-\lambda_2 \tau_2 \zeta} \approx 0$. With these approximations, α_1 in Equations (A13) and (A16) are identical, and this is evident in Figure A1(b). For α_2, Equation (A14) reduces to (A17) if $\lambda_1 \tau_1 \ll \lambda_2 \tau_2$, which is nearly true for ^{230}Th and ^{226}Ra. Therefore, for the ^{238}U decay chain, the system of ordinary differential Equations (A6-A8) can essentially be treated as if they are decoupled.

Comparison of analytical solutions with the full numerical solution. Figure A1(c) shows the full numerical solution for α of Spiegelman and Elliot (1993). It is evident that in comparison of Figure A1(c) with A1(b), the numerical and analytic solutions differ significantly, with the analytic solution overestimating (^{226}Ra/^{230}Th). In Figure (A2), comparison of the activity ratios calculated from the full analytic solution (light solid curves), the approximate analytic solution (light dotted curves), and the full numerical solution (dark solid curves) is shown for (^{226}Ra/^{230}Th) and ^{230}Th/^{238}U) using a melt column of 50 km and total melt fraction of 15% and constant partition coefficients from Lundstrom et al. (1994) for garnet peridotite. The maximum porosity for the melt column

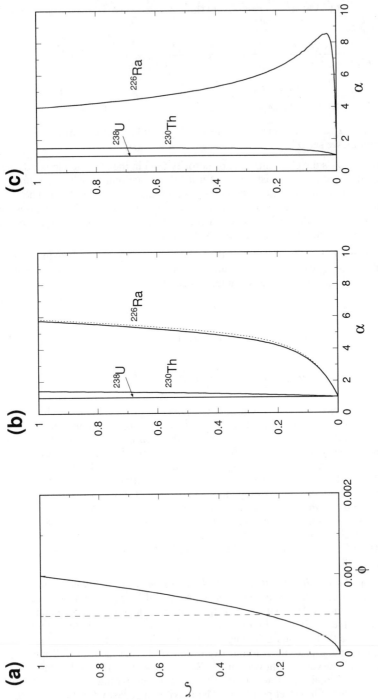

Figure A1. a) Porosity distribution (ϕ) for a 1D melt column (solid curve) assuming constant melt flux (see Spiegelman and Elliott 1993). Average porosity is shown as the dashed line. b) Enrichment factors (α) calculated from the analytical solution (solid curves) and approximate analytical solution (dotted curves) for ^{238}U, ^{230}Th and ^{226}Ra. c) Enrichment factors (α) calculated from the numerical solution of Spiegelman and Elliott (1993) for ^{238}U, ^{230}Th and ^{226}Ra. In these plots, depth (z) is non-dimensionalized. See text for explanation.

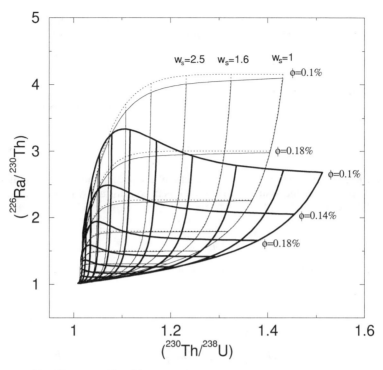

Figure A2. (^{226}Ra/^{230}Th) and (^{230}Th/^{238}U) calculated from the analytical solution (solid light curves), approximate analytical solution (dotted light curves) and full numerical solution (solid dark curves). Horizontal curves represent constant maximum porosity (ϕ_{max}), while vertical curves represent constant upwelling rates (W_s) in cm/yr. Selected contours are labeled. Contours range from 1-100 cm/a and 0.1-10% for upwelling velocity and maximum porosity, respectively. See text for explanation.

is varied from 0.1-10% and the solid upwelling velocity is varied from 1-100 cm/yr. It is clear that there is a major difference between the analytic and numerical solutions. Considering that (^{226}Ra/^{230}Th) is controlled mainly by the porosity, while (^{230}Th/^{238}U) is controlled mainly by the upwelling rate, it can be concluded that the analytic solution overestimates the porosity and under-estimates the upwelling velocity.

From the results in Figure A2, it is not clear that the analytic solutions using the average porosity and melt velocity do a very good job at approximating the numerical solution. Both the (^{226}Ra/^{230}Th) and (^{230}Th/^{238}U) values are significantly different from the numerical solution, indicating the importance of taking into consideration the details of the porosity distribution and melt velocity along the melt column. However, because these analytical solutions, particularly the analytical approximation, can be easily calculated with MATLAB scripts or in Excel™ type spreadsheets, these solutions can be broken into discrete increments of the melt column with the total results summed over the whole melt column. This allows the analytical approximation to be used in models which have discrete multiple porosities.

A.3. Box-model for equilibrium melting

In what follows, we present a simplified box-model that capture the essential of the chromatographic porous flow model with the only assumption that there is equilibration between the melt and the matrix. If one assumes that the melting column is in steady-state, then the following mass balance equations for U and Th can be written:

$$M_c \frac{dU_c}{dt} = \dot{M}_i U_i - \dot{M}_m U_m - \dot{M}_s U_s \tag{A18}$$

where m denotes the melt, s the solid and i the material input at the base of the melting column. The dotted parameters are mass fluxes. The corresponding equation for Th (or Pa) would be:

$$M_c \frac{dTh_c}{dt} = \dot{M}_i Th_i - \dot{M}_m Th_m - \dot{M}_s Th_s - \lambda_{Th} M_c Th_c + \lambda_U M_c U_c \tag{A19}$$

At steady-state, the left-hand side of the equation disappears and the equations can be simplified if one makes the assumption that melt is in equilibrium with the solid at the top of the melting column:

$$D_{Th} = \frac{Th_s}{Th_m} \text{ and } D_U = \frac{U_s}{U_m}$$

Then, the degree of melting for the solid at the top of the melt column is:

$$F = \frac{\dot{M}_m}{\dot{M}_i} \tag{A20}$$

Equations (A1) and (A2) can be combined:

$$\frac{Th_i}{U_i} = \frac{Th_m}{U_m} \left[F \frac{U_m}{U_i} + (1-F) \frac{U_m}{U_i} \frac{D_{Th}}{D_U} \right] - \frac{M_c}{\dot{M}_i} \frac{U_m}{U_i} \left[\lambda_U - \lambda_{Th} \frac{Th_m}{U_m} \right] \tag{A21}$$

If we assume that the material coming in the box is in secular equilibrium, this equation can be further simplified and multiplied by λ_{Th}/λ_U to yield activity ratios:

$$1 = \left(\frac{Th}{U} \right)_m \left[\frac{F + D_{Th}(1-F)}{F + D_U(1-F)} \right] - \lambda_{Th} \frac{M_c}{\dot{M}_i} \left[\frac{U_c}{U_i} - \frac{Th_c}{Th_i} \right] \tag{A22}$$

If one expresses the Th/U activity ratio in the melt:

$$\left(\frac{Th}{U} \right)_m = \left[\frac{F + D_U(1-F)}{F + D_{Th}(1-F)} \right] (1 + \lambda_{Th} [\tau_U - \tau_{Th}]) \tag{A23}$$

where τ_U and τ_{Th} are the residence times of U and Th, respectively, in the melting column. The residence times can be expressed as follows:

$$\tau_U = \frac{M_c U_c}{\dot{M}_i U_i} \tag{A24}$$

In Equation (A23), the first term of the right-hand side corresponds to a batch melting model and the second term corresponds to an in-growth term due to a longer residence time of U relative to Th. This equation is identical to Equation (31) in Spiegelman and Elliott in their Appendix A.4. It is obtained with a much simpler derivation and no assumption about melt transport. Our approach highlights that in the case of equilibrium melting, the fractionation between short-lived isotopes is simply dependent on the relative residence time of the elements in the melting column that comes out of a simple box model and the half-life of the daughter nuclide.

7 Insights into Magma Genesis at Convergent Margins from U-series Isotopes

Simon Turner
Department of Earth Sciences
Wills Memorial Building
University of Bristol
Bristol, BS8 1RJ, United Kingdom

Bernard Bourdon
Laboratoire de Géochimie et Cosmochimie
IPGP-CNRS UMR7579
4, Place Jussieu, 75252
Paris cedex 05 France

Jim Gill
Department of Earth Sciences
University of California
Santa Cruz, California, 95064, USA

1. INTRODUCTION

Convergent margins (oceanic and continental arcs) form one of the Earth's key mass transfer locations, being sites where melting and transfer of new material to the Earth's crust occurs and also where crustal materials, including water, are recycled back into the mantle. Volcanism in this tectonic setting constitutes ~15% (0.4-0.6 km^3/yr) of the total global output (Crisp 1984) and the composition of the erupted magmas is, on average, similar to that of the continental crust (Taylor and McLennan 1981). Moreover, many arc volcanoes have been responsible for the most hazardous, historic volcanic eruptions. Yet, despite their importance, many fundamental aspects of convergent margin magmatism remain poorly understood. Key among these are the rates of processes of fluid addition from the subducting plate. Furthermore, in stark contrast to the ocean ridges, where adiabatic decompression provides a simple and robust physical model for partial melting, no consensus has yet been reached about the physics of the partial melting process and the mechanism of melt extraction beneath arcs.

Preceding chapters concerned with partial melting in this volume (Lundstrom 2003; Bourdon and Sims 2003) have discussed how the differing half-lives and distribution coefficients of the various U-series nuclides result in disequilibria through in-growth. This provides important information on the nature and timing of mantle partial melting processes. In convergent margin settings the differential fluid mobility of U and Ra relative to Th and Pa provides an additional source of fractionation leading to in-growth and this is crucial to understanding the timing and mechanisms of fluid addition. Here we review the role that the proliferation of high quality U-series isotope data, over the last decade, have had in obtaining precise information on time scales and the development of quantitative physical models for convergent margin magmatism. Our approach is to use trace element and isotope data to determine the nature of the contributing source components and then to use the U-series isotopes to constrain processes and time scales. We begin with a brief review of the evidence for multiple source component contributions to most arc magmas and then summarize the empirical and experimental evidence for the behavior of the U-series nuclides of interest in aqueous fluids. Subsequent sections look in turn at the U-series evidence for the timing and mechanisms of sediment and fluid addition and the important new constraints that U-series disequilibria are beginning to place on partial melting and melt extraction processes beneath arc volcanoes. These are followed by brief discussions of the few data available from rear arc lavas, the processes which might modify source-derived U-series signals and finally some of the implications of U-Th isotopes as long term tracers of the evolution of the terrestrial reservoirs.

2. CONVERGENT MARGIN MAGMATISM

Convergent margin magmatism is widely regarded to reflect partial melting that is in some way linked to the lowering of the peridotite solidus through addition of fluids released by dehydration reactions in the subducting plate (e.g., Tatsumi et al. 1986; Davies and Bickle 1991). For example, the relatively constant ~110 km depth to Benioff zone was, for a long time, taken as evidence that an isobaric dehydration reaction in the subducting plate is responsible for melting (Gill 1981; Tatsumi et al. 1986). Stolper and Newman (1994) noted a strong positive correlation between H_2O content and inferred degree of melting in lavas from back arc spreading centers behind the Marianas. A strong dependence on fluid addition is also consistent with the apparent positive correlation between volcanic output and rate of subduction from east to west along the Aleutian arc (Marsh 1987; George et al. 2003). Conversely, there is an absence of volcanism in South America above the area of subduction of the Chile ridge. This will be the hottest part of the wedge but also the driest, again suggesting that partial melting is linked to fluid addition. Exceptions can occur when the subducting plate is sufficiently young to be hot enough for the altered oceanic crust to undergo hydrous partial melting (Peacock et al. 1994). However, these distinctive, adakitic lavas (Defant and Drummond 1990) appear to be volumetrically minor in the present day convergent margins (but see also Kelemen et al. 2003) and their U-series systematics will be considered separately under the section on partial melting. Finally, amphibole-rich sub arc mantle xenoliths provide direct evidence for hydration of the mantle wedge (e.g., Maury et al. 1992; Blatter and Carmichael 1998).

2.1. Geochemical signatures of source components

A distinctive geochemical signature of island arc lavas is that, compared to mid-ocean ridge and ocean island basalts (MORB and OIB), they are enriched in large ion lithophile elements (LILE), such as Rb, Ba, K, Sr and Pb, and depleted in high field strength elements (HFSE), such as Ta, Nb and Ti, relative to the rare earth elements (REE) (e.g., Gill 1981; Hawkesworth et al. 1997a). Similar trace element signatures are observed in sub arc mantle xenoliths (Maury et al. 1992). Furthermore, Nb and often Zr, Hf, and Ti abundances in arc lavas are usually depleted with respect to MORB, at a given MgO content. Consequently, the arc mantle source is often thought to be even more depleted than the MORB source. Note that with an ionic charge of +4 and ionic radius of ~1 Å, Th has an ionic potential (or field strength) <2 and is classified as a HFSE and whilst U has a similar ionic radius, the fluid mobility of U^{6+} (see Section 4) has lead to U being considered as similar to a LILE element. If the mantle wedge peridotite is similar to that of the MORB source, then the incompatible trace element data from arc lavas appear to require contributions from two additional components. Figure 1a emphasizes this by showing that the compositions of arc lavas extend well beyond those of MORB and OIB and demonstrates the need for both an enriched component with elevated La/Sm and a component with high Ba/Th.

2.2. The enriched component—sediment or OIB?

The upper portions of the subducting plate are comprised of hydrothermally altered oceanic crust and an overlying layer of sediments, sometime jointly referred to as "the

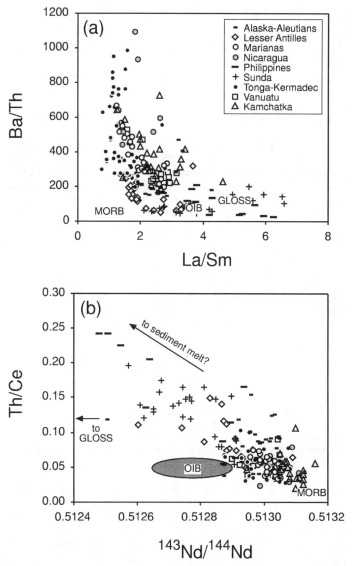

Figure 1. (a) Ba/Th versus La/Sm for all published along-arc suites of arc rocks for which both mass spectrometric U-Th isotope data and a comprehensive range of accompanying trace element and radiogenic isotope data are available (see Table 1 for data sources). For clarity on this and most subsequent diagrams only samples with $SiO_2 < 60$ wt. % are plotted. Also plotted are typical compositions for MORB and OIB (Sun and McDonough 1989) and GLOSS (Plank and Langmuir 1998). The array demonstrates the need for contributions from two components in addition to MORB-source type mantle wedge—an enriched component with elevated La/Sm (sediment) and a component with high Ba/Th (fluid). (b) Th/Ce versus $^{143}Nd/^{144}Nd$ illustrating that the low $^{143}Nd/^{144}Nd$ component in arc lavas is subducted sediment not OIB-type mantle. The trajectory toward higher Th/Ce ratios than GLOSS provides one line of evidence that the sediment component is added as a partial melt.

slab." The presence of ^{10}Be and negative Ce anomalies in many arc lavas can be taken as unambiguous evidence for a contribution from the subducted sediments (e.g., Hole et al. 1984; Morris et al. 1990), although neither are universal features of arc lavas. Nevertheless, at first glance, OIB-type mantle, which also has elevated La/Sm and thus low time-integrated $^{143}Nd/^{144}Nd$ relative to MORB, might be considered as a candidate for the enriched component (e.g., Morris and Hart 1983). However, OIB are also characterized by Nb ≥ La and low Th/Ce ratios whereas the enriched component in arc lavas has low Nb/La and high Th/Ce at low $^{143}Nd/^{144}Nd$ (Fig. 1b) and so the enriched component is unlikely to reflect OIB material trapped within the mantle wedge (Hawkesworth et al. 1997a). Sediments have high Th/Ce and Figure 1b shows that the arc lavas form an array toward high Th/Ce and low $^{143}Nd/^{144}Nd$ which is broadly characteristic of average global subducted sediment GLOSS (Plank and Langmuir 1998). The significance of the array projecting towards higher Th/Ce than GLOSS will be discussed later. Many oceanic arc lavas are congested in the lower right corner of Figure 1b where there is a range in $^{143}Nd/^{144}Nd$ and low Th/Ce and broad overlap with MORB and OIB. For these lavas, it is primarily the observation of low Nb/La (and the presence of ^{10}Be and negative Ce anomalies in some instances) which argues for the presence of a sediment component and an absence of an OIB component (e.g., Elliott et al. 1997).Thus, on a plot of Th/Nb versus $^{143}Nd/^{144}Nd$ (not shown) the different arcs form steep arrays rather than shallow trends toward a low Th/Nb, low $^{143}Nd/^{144}Nd$ OIB component. An exception to this is the northernmost islands of the Tonga arc where decreases in $^{143}Nd/^{144}Nd$ are accompanied by increases in Ta/Nd (Regelous et al. 1997; Turner et al. 1997; Wendt et al. 1997). However, Turner and Hawkesworth (1998) have argued that this OIB signal was introduced into the sub arc mantle in the form of volcaniclastic sediments carried on the subducting plate rather than requiring the presence of OIB mantle in the wedge.

2.3. The fluid component

Undegassed, primitive arc lavas are uncommon but available analyses generally show elevated volatile contents relative to MORB (e.g., Harris and Anderson 1984; Sisson and Layne 1993), although some exceptions do occur (Sisson and Bronto 1998). Phase relations indicate that the amount of H_2O available in the subducting plate at ~ 100 km depth beneath arc front volcanoes will be in the range 1-5% (Poli and Schmidt 1995). Experimental data suggest that the high LIL/HFSE ratios in arc lavas reflect the relative mobility and immobility (respectively) of these elements in aqueous fluids (Section 4.3) and empirical evidence supports this. For example elements like Cl, Ba and U all correlate positively with H_2O content in the Mariana back arc lavas (Stolper and Newman 1994). Consequently, the high Ba/Th component in Figures 1 and 2 is inferred to reflect fluid addition. Importantly, the highest LIL/HFSE ratios (e.g., Ba/Th) are usually found in those rocks with the highest $^{143}Nd/^{144}Nd$ (and lowest $^{87}Sr/^{86}Sr$ ± $^{206}Pb/^{204}Pb$) ratios (Fig. 2a) from which it has been inferred that the fluid end-member is most typically derived more from the subducting altered basaltic crust rather than the overlying sediments (Miller et al. 1994; Turner et al. 1996; Turner and Hawkesworth 1997). This is consistent with flux calculations by Staudigel et al. (1996) which showed that the altered oceanic crust potentially can contribute ~ 70% of the U recycled into the arc crust. However, we also note that several studies have provided evidence for transport of some of the sediment components in a fluid (Morris et al. 1990; Class et al. 2000; Hochstaedter et al., 2001; Sigmarsson et al. 2002) and so the relative role of altered oceanic crust versus sediment as source of the fluid apparently varies substantially between and even within arcs.

In summary, many recent studies have argued for three components in most arc lavas: depleted mantle, basalt-derived fluid, and sediment (in bulk, via fluid, or via melt)

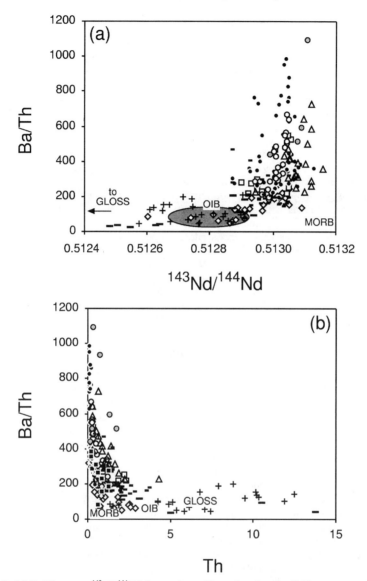

Figure 2. (a) Ba/Th versus ^{143}Nd/^{144}Nd for arc lavas illustrating that the fluid component (elevated Ba/Th) has high ^{143}Nd/^{144}Nd and must therefore generally be derived from the subducting altered oceanic crust. (b) Ba/Th ratio versus Th content from which is can be inferred that Th is less mobilized in the fluid component and that the relative effect of the fluid component is more pronounced in the more depleted lavas with the lowest Th contents. Symbols and data sources as in Figure 1.

(e.g., Kay 1980; Ellam and Hawkesworth 1988; Miller et al. 1994; Turner et al. 1996, 1997; Elliott et al. 1997; Hawkesworth et al. 1997a). Debate continues about whether and where enriched mantle (e.g., Morris and Hart 1983) and basalt melts (adakites) are involved (e.g., Kelemen et al. 2003; George et al. 2003).

2.4. Relative depletion of the mantle wedge

A final and most important point to emphasize before turning to the implications of U-series isotope data is that the nature of the arc lava array in Figure 2a requires that the relative contributions of the fluid and sediment components vary both within individual arcs as well as from arc to arc. Figure 2b shows that Ba/Th ratio varies inversely with Th content and thus the largest fluid signals are found in those lavas with the lowest Th contents (Condomines and Sigmarsson 1993; McDermott and Hawkesworth 1991). This provides one line of evidence that Th is not strongly mobilized in the fluid (see Section 4) and it is argued that much of the Th abundances in arc lavas are controlled by the competing effects of prior melt depletion of the mantle wedge (Ewart and Hawkesworth 1987; Woodhead et al. 1993; Stolper and Newman 1994), which acts to lower the Th contents, and addition of sediment which has a high Th/U ratio relative to MORB (Fig. 3) and thus has the opposite effect of enriching Th (see Section 5.1). For this reason, sediment-rich arc lavas with high Th contents are often referred to as "enriched" and the nature of the arrays in Figures 2a and b are consistent with a mass balance model in which a broadly similar fluid flux is added to sources containing a variable sediment component (Condomines and Sigmarsson 1993; Turner et al. 1996, 1997; Elliott et al. 1997; Hawkesworth et al. 1997a, b). Note that flux melting models (e.g., Stolper and Newman 1994; Eiler et al. 2000; Thomas et al. 2002) in which the decreases in Th content on Fig. 2b are thought to reflect increasing degrees of partial melting due to a higher fluid flux (i.e., and thus higher Ba/Th) do not predict the broad associated change in $^{143}Nd/^{144}Nd$ observed in Fig. 2a.

3. U-SERIES ISOTOPES IN ARC LAVAS

Since publication of the second edition of Ivanovich and Harmon (1992) there has been a proliferation of high quality U-series isotope data, the majority of which have been analyzed by thermal- and, more recently, plasma-source mass spectrometry. Table 1 represents an updated version of Table 7.1 from Gill et al. (1992) and summarizes (to the best of our knowledge) specific arc studies which have been divided into a three-tier hierarchy of global surveys, along-arc studies and studies of individual volcanoes. For the purposes of much of the discussion in this chapter and for most of the figures we concentrate on the data sets from along-arc suites of arc rocks for which both mass spectrometric U-series isotope data and a comprehensive range of accompanying trace element and radiogenic isotope data are available. For clarity we have restricted the dataset to only samples with $SiO_2 < 60$ wt. % unless otherwise specified. The data from numerous studies of individual volcanoes have also been omitted since the variations in recent lavas from individual volcanoes are likely to dominantly reflect the effects of magma chamber processes which are less relevant to this chapter and are discussed separately by Condomines et al. (2003).

In Figure 3 we have plotted histograms of U/Th, $(^{230}Th/^{238}U)$, $(^{231}Pa/^{235}U)$ and $(^{226}Ra/^{230}Th)$ in arc lavas and compared these with MORB (data sources from Lundstrom 2003). This shows that the majority of arc lavas have higher U/Th ratios and the opposite sense of fractionation of $(^{230}Th/^{238}U)$ to MORB with the majority being characterized by excesses of ^{238}U over ^{230}Th, though many are close to ^{230}Th-^{238}U equilibrium and some have ^{230}Th-excesses (e.g., Reagan and Gill 1989; Reagan et al. 1994; George et al. 2003). This observation has not changed significantly since the early global surveys of Gill and Williams (1990), McDermott and Hawkesworth (1991) and Condomines and Sigmarsson (1993) excepting that the advances in analytical techniques have allowed more highly depleted arc rocks to be analyzed and this has increased the total number analyzed that have large ^{238}U-excesses. Note that the U/Th ratio of GLOSS is lower than most MORB

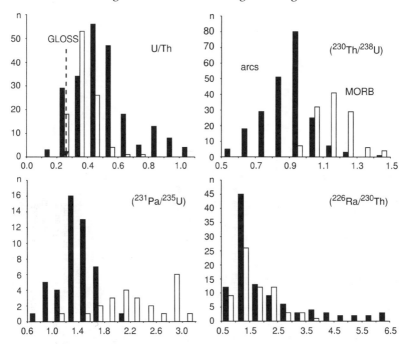

Figure 3. Histograms of U/Th, $(^{230}Th/^{238}U)$, $(^{231}Pa/^{235}U)$ and $(^{226}Ra/^{230}Th)$ for arc lavas (data sources as in Figure 1) compared with MORB (data sources from Lundstrom 2003). Note that the U/Th ratio of GLOSS (Plank and Langmuir 1998) is lower than that in most MORB or arc lavas.

or arc lavas (Fig. 3). Due to analytical difficulties ^{231}Pa-^{235}U data are much scarcer but the histogram shows that arc lavas have the same sense of ^{231}Pa-^{235}U disequilibria to MORB but that the absolute size of the disequilibria is typically less, and similar to OIB. Exceptions to this are the lavas from Tonga which are characterized by excesses of ^{235}U over ^{231}Pa, or $(^{231}Pa/^{235}U) < 1$ (Bourdon et al. 1999). The third U-series system for which there are appreciable mass spectrometric data is the ^{226}Ra-^{230}Th system and the histogram shows that whilst the sense and median size of fractionation is similar to that observed in MORB, the arc data extend to $(^{226}Ra/^{230}Th)$ ratios > 6 that are far larger than those observed in MORB.

The most important observations about U-series isotopes in arc lavas for this chapter are: (1) the widespread excess of ^{238}U over ^{230}Th but deficit of ^{235}U with respect to ^{231}Pa; and (2) the extreme ^{226}Ra enrichments in some arc lavas. We will explore the profound implications of these for magma genesis and transport at subduction zones. The conclusions apply most convincingly to the oceanic arcs where the observations are most extreme (the volcanic fronts of Tonga, Marianas, and eastern Sunda, and one or two volcanoes in some other arcs). Whether the conclusions apply elsewhere is harder to verify but there is no convincing reason with respect to U-series data to believe that they do not.

4. BEHAVIOUR OF THE U-SERIES NUCLIDES IN AQUEOUS FLUIDS

Most of the differences between arc magmas and those from other tectonic settings result from oxidation (Wood et al. 1990; Blatter and Carmichael 1998; Parkinson and Arculus 1999) and hydration (Blatter and Carmichael 1998) of arc magma sources by

Table 1. Summary of published U-Th-Pa-Ra data from detailed studies of convergent margin rocks.

Global	Arcs	Volcanoes	$^{230}Th/^{238}U$	$^{235}U/^{231}Pa$	$^{226}Ra/^{230}Th$	Reference(s)
Island arcs						
	Aleutians		α, TIMS	TIMS	α, TIMS	Gill and Williams (1990); McDermott and Hawkesworth (1991); Pickett and Murrell (1997); Hawkesworth et al. (1997a); Turner et al. (2001)
	Eolian		α, T-, P-IMS	---	TIMS	Newman et al. (1984); Turner et al. (1998); George et al. (2003)
		Stromboli	α	---	γ	Capaldi et al. (1983)
	Lesser Antilles		α, TIMS	---	α, TIMS	Capaldi et al. (1978); Gauthier and Condomines (1999)
		Soufriere	α, TIMS	---	γ, TIMS	Turner et al. (1996); Chabaux et al. (1999); Turner et al. (2001)
	Marianas		TIMS	---	TIMS	Heath et al. (1998); Turner et al. (2001)
			α, TIMS	---	TIMS	Newman et al. (1984); Elliott et al. (1997); Turner et al. (2001)
	New Britain		α	---	α	Gill et al. (1993)
	Philippines		TIMS	---	TIMS	McDermott et al. (1993); Turner et al. (2001)
	Sunda		TIMS	---	TIMS	Condomines and Sigmarsson (1993); Turner and Foden (2001)
		Merapi	α, TIMS	---	α, TIMS	Gauthier and Condomines (1999); Turner and Foden (2001)
	Banda		α	---	α	Hoogewerff et al. (1997)
		Batur	α	---	α	Rubin et al. (1989)
		Galunggung	α, TIMS	---	α, TIMS	Gerbe et al. (1992); Turner and Foden (2001)
		Sangeang Api	TIMS	---	TIMS	Turner et al. (2003)
	Tonga-Kermadec		TIMS	TIMS	TIMS	Regelous et al. (1997); Turner et al. (1997, 2000); Bourdon et al. (1999)
			α, TIMS	---	TIMS	Condomines and Sigmarsson (1993); Turner et al. (1999); Turner et al. (2001)
	Vanuatu		α, TIMS	TIMS	α, TIMS	Gill and Williams (1990); McDermott and Hawkesworth (1991); Pickett and Murrell (1997); Hawkesworth et al. (1997a); Turner et al. (2001)
Continental arcs						
	Aegean		α, TIMS	---	γ, TIMS	Pyle et al. (1988); Druitt et al. (1999); Zellmer et al. (2000)
		Santorini	TIMS	---	TIMS	George et al. (2002)
	Alaska					
	Central Andes					
		Nevado del Ruiz	TIMS	---	α	Schaefer et al. (1993)
		Parinacota	TIMS	---	TIMS	Bourdon et al. (2000)
	Southern Andes		α	---	TIMS	Sigmarsson et al. (1992, 2002)
	Austral Andes		α	---	TIMS	Sigmarsson et al. (1998)
	Cascades		α	---	---	Newman et al. (1986)
		St Helens	α, TIMS	---	α, TIMS	Bennett et al. (1982); Volpe and Hammond (1991)
		Shasta	TIMS	---	TIMS	Newman et al. (1986); Volpe (1992)
	Kamchatka		TIMS	TIMS	TIMS	Turner et al. (1998); Dosseto et al. (2003)
	Nicaragua		α, TIMS	TIMS	TIMS	Condomines and Sigmarsson (1993); Reagan et al. (1994); Herrstrom et al. (1995); Thomas et al. (2002)
	Costa Rica		α, TIMS	TIMS	TIMS	Clark et al. (1998); Thomas et al. (2002)
		Turrialba	α, TIMS	---	α, TIMS	Reagan and Gill (1989); Thomas et al. (2002)
	Italy					
	Vesuvius		α	---	---	Capaldi et al. (1982); Villemant and Flehoc (1989); Black et al. (1998)

α = alpha counting; γ = gamma counting; TIMS = thermal ionization mass spectrometry; PIMS = plasma ionization mass spectrometry aqueous fluids. The single-most striking U-series observation, that is largely unique to the arc environment, is the tendency towards excesses of ^{238}U over ^{230}Th. This has been attributed for over two decades to the addition of U by fluids from the subducting plate (e.g., Gill, 1981; Allègre and Condomines 1982). The inference that aqueous fluids have a particularly strong effect on partitioning between U and its daughter nuclides dictates that any interpretation requires constraints on their solid/fluid partitioning behavior. As we will see later (Section 8), whether or not small amounts of Th and Pa are mobilized in subduction zone fluids is critical to many of the U-series time scale interpretations. We will now look at the evidence for U-series nuclide solid/fluid partitioning in closer detail.

4.1. General empirical evidence

The behavior of the U-series nuclides at the Earth's surface, in the oceans and in ground waters is discussed at length in chapters 8-14 of this volume and the reader is referred to these for details of the following summary. U is characterized by two main oxidation states (+4 and +6) with the hexavalent state occurring under oxidizing conditions in which the uranyl ion (UO_2^{2+}) is highly fluid mobile. This is manifest at the Earth's surface by elevated U concentrations in surficial and ground waters (Porcelli 2003) and a long residence time in the oceans (Cochran and Masque 2003). Ra, whose closest chemical analogue is Ba, is an alkaline earth element which forms the +2 oxidation state and is also commonly mobile in fluids. Th on the other hand only forms the +4 oxidation state and is generally immobile in aqueous fluids at the Earths surface as evidenced by its insolubility in the laboratory (Goldstein and Stirling 2003), surface and ground waters and its short residence time in the oceans (Cochran and Masque 2003). Less information is available for Pa which adopts the +5 oxidation state and for which the best chemical analogue may be the HFSE Nb or Ta (Lundstrom et al. 1994). Various lines of evidence suggest that Pa is highly fluid immobile (Guillaumont et al. 1968), and certainly Pa is even harder to keep in solution than Th in the laboratory (Goldstein and Stirling 2003). At depths within the crust, the formation of barite and uranium deposits but absence of thorium deposits suggests that U and Ra are still fluid mobile whilst Th is much less so and we infer the same for Pa. More difficult is inferring the behavior of these nuclides at the temperatures and pressures appropriate to the dehydration of subducting oceanic plates beneath arcs and in the likely presence of additional solute species like Na and Cl. For information on relative fluid mobility under these conditions we are restricted to inferences from empirical observations and the few available results from very difficult fluid-mineral partitioning experiments.

4.2. Empirical observations from arc lavas

The fact that the that highest U/Th ratios are found in the arc lavas with the lowest Th contents (Gill and Williams 1990; McDermott and Hawkesworth 1991; Condomines and Sigmarsson 1993; Hawkesworth et al. 1997a,b), that excesses of ^{238}U over ^{230}Th are rare apart from arc lavas, and that ^{238}U-excesses often correlate positively with fluid-mobile elements within arc lavas, provides strong empirical evidence that U is significantly more fluid mobile than Th. Secondly, Stolper and Newman (1994) found a good positive correlation between U and H_2O in the Marianas back arc lavas. Thirdly, Hawkesworth et al. (1997a) showed that Th/Nd ratios in arc lavas converge towards those of MORB at high ^{143}Nd/^{144}Nd ratios rather than the opposite correlation for Ba/Th in Figure 2a which would be expected if Th was more mobilized than Ba in the fluid. Finally, Plank and Langmuir (1993) investigated the correlations between sedimentary influx and volcanic output for a range of elements enriched in arc lavas from a number of arcs worldwide. Plots of those elements extrapolated to 6% MgO (to correct for magma differentiation) and normalized to Na at 6% MgO (to correct for relative dilution due to variation in the degree of melting) versus sediment flux yield strong positive correlations implicating the sediments as the major source of those elements. Importantly, the correlation for Th passes through the origin implying zero Th flux at zero sediment input whereas those for U and Ba intercept at elevated Ba_6/Na_6 and U_6/Na_6 (Fig. 4) which requires a semi-constant flux of U and Ba in addition to that supplied by the subducting sediment and that is inferred to be the fluid contribution. Thus, this approach also suggests that U and Ba are added in the fluid component much more than Th.

An important aspect in the preceding discussion is the need to separate the fluid and sediment components spatially and (as we will see) also temporally. Quantitative mass balance estimates (e.g., McCulloch and Gamble 1991; Stolper and Newman 1994; Ayers 1998) often conclude that there is as much, or even more, Th and U in the bulk "slab component" (i.e., sediment plus fluid from the altered oceanic crust). However, if the sediment component added is in U-Th isotope equilibrium (or returns to this state prior to fluid addition; see Section 5.3), then addition of only 0.02 ppm U in the fluid will result in significant ^{238}U-excesss in the composite source (e.g., Condomines and Sigmarsson 1993; Turner et al. 1997).

4.3. Experimental constraints on mineral/fluid partitioning

Experimental verification of the observational inferences is complicated by the difficulty of performing mineral/fluid partitioning experiments at high temperatures and pressures and their dependence on fO_2 and complexing agents in the fluid (especially CO_3^{2-} and Cl^-). Nevertheless some attempts have been made (e.g., Tatsumi et al. 1986; Brenan et al. 1994, 1995; Keppler 1996; Ayers et al. 1998; Johnson and Plank 1999).

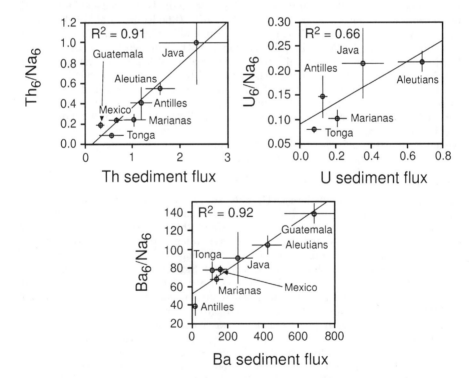

Figure 4. Plots of Th_6/Na_6, U_6/Na_6 and Ba_6/Na_6 versus sediment flux for a number of arcs. Modified from Plank (1993) and Plank and Langmuir (1993).

Tatsumi et al. (1986) noted that the mobility of elements in fluids released during dehydration of serpentinite increased with ionic radius. This correctly predicts the greater mobility of the LILE relative to the HFSE and REE but does not account for the observed U/Th fractionation in arc lavas.

Brenan et al. (1994, 1995) conducted experiments to directly determine mineral/fluid partition coefficients for a range of mantle minerals with incompatible element concentrations similar to those that might occur naturally. In evidence of the difficulty of such experiments, they expressed concern about the attainment of equilibrium and adherence to Henry's law as well as apparatus design and analysis of very low concentration run products (Brenan et al. 1995). These authors also noted the strong influence of oxygen fugacity on U partitioning which decreases by over an order of magnitude on going from log fO_2 = QFM – 4 (Fig. 5a), which might be expected in the MORB source, to QFM + 2 which is likely to be more typical in the mantle wedge beneath arcs (Blatter and Carmichael 1998; Parkinson and Arculus 1999). In order to circumvent some of the experimental difficulties, Keppler (1996) used an indirect approach whereby fluid/andesite-melt partition coefficients were determined and $D^{fluid/clinopyroxene}$ was calculated from $D^{fluid/melt} \div D^{clinopyroxene/melt}$. The results of the Brenan et al. (1995) and Keppler (1996) experiments are summarized in Table 2 along with calculated U/Th and inferred U/Pa and Ra/Th mineral/fluid partition coefficients.

First we compare the results for clinopyroxene which was the only phase common to both of these studies. The results from the pure H_2O experiments are somewhat inconsistent. Keppler (1996) found that Th was more mobile than U, contrary to all observations, whereas Brenan et al. (1995) found the reverse. However, subduction zone fluids are almost certain to contain solutes like Na and Cl derived from seawater and Keppler and Wyllie (1990) showed that the solubility of U, but not Th, is enhanced by the presence of Cl, although the salinities used by Keppler (1996) were very high. Both Brenan et al. (1995) and Keppler (1996) found that U was an order of magnitude more fluid mobile than Th when NaCl was present, although, in the Brenan et al. (1995) experiments, the absolute $D^{clinopyroxene/fluid}$ for Th was lower in the presence of NaCl than in H_2O alone (Table 2), which is inconsistent with the findings of Keppler and Wyllie (1990). However, this may reflect the effects of oxygen fugacity noted above. The results for Nb, taken as an analogue for Pa (Lundstrom et al. 1994), are even more variable. Keppler (1996) found Nb to be fluid immobile whereas Brenan et al. (1995) found it to be quite strongly partitioned into fluids in equilibrium with a range of mantle phases (Table 2) which goes against most observational evidence. It seems likely that most of these inconsistencies reflect the difficulty of the experiments. In contrast, the results for Ba, taken as an analogue for Ra, are entirely consistent with this element being highly fluid mobile with respect to anhydrous phases.

Of the other phases analyzed by Brenan et al. (1994, 1995), neither olivine nor orthopyroxene will contain appreciable incompatible element concentrations. Garnet will be an important residual phase during either dehydration or melting of eclogite and also during mantle melting beneath some rear arc volcanoes (there is little evidence that garnet is a residual phase during mantle melting for most arc front lavas). However, is it important to realize that, like clinopyroxene (Fig. 5a), an order of magnitude decrease in D_U is also expected for garnet with increasing oxygen fugacity (Brenan pers. comm.), so that U is significantly more fluid mobile than Th with respect to garnet at fO_2 = QFM +2. Finally, although rutile will not be stable in the mantle wedge, it may be present as a trace phase in eclogite (Ryerson and Watson 1987).

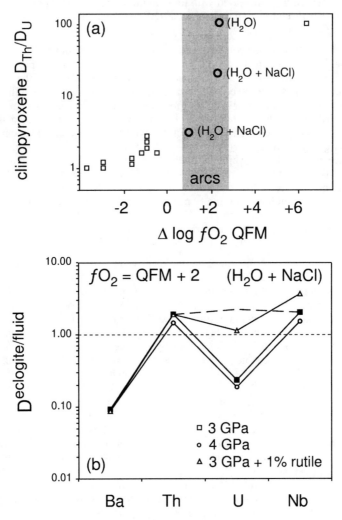

Figure 5. (a) Ratio of D_{Th}/D_U for clinopyroxene/fluid (large circles; Brenan et al. 1995) or clinopyroxene/melt (small squares; Lundstrom et al. 1994) partitioning as a function of oxygen fugacity relative to the QFM buffer. Shaded bar indicated range of oxygen fugacities determined for arcs (Blatter and Carmichael 1998; Parkinson and Arculus 1999). (b) Log-plot of eclogite/fluid (H_2O + NaCl) partition coefficients from Table 2 for Ba, U, Th and Nb based on partition coefficients from Brenan et al. (1994, 1995) and Keppler (1996) and assuming that fO_2 = QFM + 2 from (a). Ba and Nb are assumed to represent reasonable analogies for Ra and Pa. Eclogite mineralogy at 3 and 4 GPa is taken from Schmidt and Poli (1998) with amphibole used to represent the hydrous phases present because mineral-fluid partition coefficients are unavailable for other hydrous minerals (this approximation is likely to have little influence on the results which are strongly controlled by the garnet/clinopyroxene ratio ± rutile). Dashed line is for 3 GPa at fO_2 = QFM − 2 and indicates that there would be little, or even the reverse sense of U/Th and U/Nb (and by implication U/Pa) fractionation at low oxygen fugacity conditions.

Table 2. Experimental mineral/fluid partition coefficients[a]

Mineral	Fluid	Ba	Th	U	Nb	$D_{U/Th}$	$D_{U/Pa}$[b]	$D_{Ra/Th}$[c]
Clinopyroxene	H_2O[d]	0.02	0.13	3.33	1.75	25.67	1.90	0.17
	H_2O + NaCl[d]	7.88E-04	0.25	0.01	3.33	0.04	2.86E-03	3.15E-03
	H_2O[e]	7.95E-04	3.60	0.24	0.17	0.07	1.41	2.21E-04
	H_2O + NaCl[e]	4.8E-04	3.13	0.26	0.20	0.08	1.33	1.53E-04
Cpx @ fO_2=QFM+2[f]	H_2O + NaCl	4.8E-04	3.13	0.31	3.33	0.10	0.09	1.53E-04
Garnet	H_2O[e]	3.3	0.08	0.88	0.05	10.60	17.25	39.16
	H_2O + NaCl[e]	4.8E-05	0.13	0.72	0.06	5.54	12.63	3.69E-04
Gt @ fO_2=QFM+2[g]	H_2O + NaCl	4.8E-05	0.13	0.01	0.06	0.10	0.23	3.69E-04
Amphibole	H_2O + NaCl[e]	0.51	0.93	0.35	1.10	0.38	0.32	0.55
Olivine	H_2O + NaCl[e]	0.00	0.01	0.01	0.25	1.07	0.03	3.19E-03
Rutile	H_2O[h]	n/a	0.10	89	164	890	0.54	n/a
Eclogite @ 3GPa[i]	H_2O + NaCl	0.092	1.895	0.236	2.014	0.12	0.12	0.049
Eclogite @ 4GPa[j]	H_2O + NaCl	0.087	1.437	0.187	1.512	0.13	0.12	0.060
Eclogite @ 3GPa+ 1% rutile	H_2O + NaCl	0.087	1.887	1.122	3.643	0.59	0.31	0.046

[a] An average is quoted when more than one determination was available
[b] Inferred from $D_{U/Nb}$
[c] Inferred from $D_{Ba/Th}$
[d] Keppler (1996)
[e] Brenan et al. (1995)
[f] Using clinopyroxene-fluid (H_2O + NaCl) D's from Brenan et al. (1995), excepting D_{Nb} from Keppler (1996) and assuming D_U = $D_{Th/10}$
[g] Using garnet-fluid (H_2O + NaCl) D's from Brenan et al. (1995), excepting D_{Nb} from Keppler (1996) and assuming D_U = $D_{Th/10}$
[h] Brenan et al. (1994); n/a = not analyzed
[i] Assuming garnet:clinopyroxene:amphibole[k] = 28:54:18, mineral-fluid D's from Brenan et al. (1995) and fO_2 = QFM + 2
[j] Assuming garnet:clinopyroxene:amphibole[k] = 44:39:17, mineral-fluid D's from Brenan et al. (1995) and fO_2 = QFM + 2
[k] Based on modal proportions given by Schmidt and Poli (1998)

4.4. Composition of fluids released from the subducting plate

During subduction, both the hydrated oceanic crust and over-riding sediments undergo a progressive sequence of up-grade metamorphic reactions, many of which involve dehydration (e.g., Poli and Schmidt 1995; Schmidt and Poli 1998; Johnson and Plank 2000). However, as discussed above, it would appear that the altered oceanic crust is the major source of fluid for most arc lavas. The altered oceanic crust becomes progressively converted to eclogite and the ratio of garnet to clinopyroxene increases rapidly beneath the arc front volcanoes from 34:66 at 3 GPa to 53:47 at 4 GPa (Schmidt and Poli 1998). In Table 2, we give the calculated eclogite/fluid partition coefficients for fO_2 = QFM +2 at both 3 GPa and 4 GPa and for eclogite + 1% rutile at 3 GPa and these values are plotted on Figure 5b. For this oxygen fugacity D_{Th}/D_U for both clinopyroxene and garnet was assumed to be 10 on the basis of the median value on Figure 5a. We have used amphibole as a proxy for the additional phases (phengite and lawsonite) in the eclogite because there are no available mineral/fluid partition coefficients for these minerals. However, phengite is likely to have a very low D_U (Becker et al. 1999) and because clinopyroxene and garnet dominate the mode (Schmidt and Poli 1998) and thus the bulk D's (Table 2) this is unlikely to significantly alter the conclusions. The importance of oxygen fugacity is re-emphasized by the dashed line on Figure 5b which indicates that there will be minimal U/Th and U/Pa fractionation at fO_2 = QFM – 2. Thus, if oxygen fugacity varies significantly between or within subduction zones this could have a major effect on U-series fractionation.

Notwithstanding the complexities discussed above, results are encouragingly consistent with empirical observations in that the relative order of fluid compatibility is Ba (and by inference Ra) > U > Th > Nb (and by inference Pa). In detail, Ba is highly fluid mobile which is consistent with all observations (D < 0.1) and the eclogite/fluid partition coefficient for U averages around 0.2. The eclogite/fluid partition coefficients for Th and Nb range from 1.9 to 1.4 and 1.5 to 2, respectively, from 3 to 4 GPa. Again, this is consistent with the inferred relatively immobile nature of these elements. However, the presence of 1% residual rutile has a significant effect in increasing the eclogite/fluid partition coefficient for U to 1.1 and Nb to 3.6 at 3 GPa. So, although the experimental data predict more U than Th in the fluid under all conditions, neither Th nor Nb are predicted to be completely fluid immobile and the relative fractionation of U/Th decreases in the presence of residual rutile. Thus, it seems likely that the thermal conditions of the subducting plate and the presence or absence of rutile may also play an important role in determining the relative fractionation of the U-series nuclides in subduction zone fluids.

In Table 3 we have calculated the composition of fluids produced by dehydration of subducting altered oceanic crust that has a composition based on that of Staudigel et al. (1996). The model assumes that the altered oceanic crust is in U-series equilibrium, bulk eclogite/fluid D's for a rutile-free mineralogy at 3 GPa from Table 2 in which $D_{Ra} = D_{Ba}$ and $D_{Pa} = D_{Nb}$, and a Rayleigh distillation process with 2, 3 and 5% H_2O release. Taking 3% H_2O as the likely amount of fluid added to the wedge beneath arc front volcanoes (Davies and Bickle 1991; Poli and Schmidt 1995), we calculate that this fluid will have the following composition: U/Th = 35.9, Ba/Th = 6470, ($^{238}U/^{232}Th$) = 113, ($^{235}U/^{231}Pa$) = 7.6 and ($^{226}Ra/^{230}Th$) = 15 and this will be used in the subsequent discussion and figures with the acknowledgement that this composition will vary somewhat if different mineral/fluid partition coefficients and/or a different altered oceanic crust composition was used.

4.5. Chromatographic interaction with the mantle wedge

Several authors have suggested that the composition of subduction zone fluids is likely to change by chromatographic interaction during their passage through the mantle

Table 3. Calculated compositions of fluids produced by
dehydration of the altered oceanic crust[a]

% fluid	AOC[b]	2	3	5
Ba (ppm)	31	276	249	203
Th (ppm)	0.072	0.038	0.039	0.039
U (ppm)	0.36	1.430	1.383	1.293
Nb (ppm)	1.18	0.592	0.595	0.601
U/Th	5.0	37.3	35.9	33.2
U/Nb	0.31	2.42	2.32	2.15
Ba/Th	431	7193	6470	5217
(^{238}U/^{232}Th)	15.3	117	113	104
(^{238}U/^{230}Th)	1.0[c]	7.9	7.6	7.0
(^{235}U/^{231}Pa)	1.0[c]	7.9	7.6	7.0
(^{226}Ra/^{230}Th)	1.0[c]	16.7	15.0	12.1
apparent U-Th age (kyr)[d]	-	14	15	16
apparent U-Pa/Nb age (kyr)[e]	-	0.040	0.042	0.045

[a] Assuming a Rayleigh distillation process and bulk eclogite/fluid D's for 3 GPa without rutile from Table 2
[b] Altered oceanic crust composition based on Staudigel et al. (1996)
[c] Assuming U-series equilibrium in altered oceanic crust
[d] Two point isochron tied to composition in U-series equilibrium with (^{238}U/^{232}Th) = 1
[e] Two point isochron tied to composition in U-series equilibrium with 0.047 ppm U, 2.33 ppm Nb

wedge (Navon and Stolper 1987; Kelemen et al. 1990; Hawkesworth et al. 1993; Stolper and Newman 1994). In the mantle wedge, clinopyroxene is likely to be the main host of incompatible trace elements and the mineral/fluid partition coefficients for clinopyroxene under oxidizing conditions in Table 2 suggest that the effect of such interaction would be to increase the U/Th, U/Nb and Ba/Th (and by analogy U/Pa and Ra/Th) ratios of the fluids. The magnitude of such effects is unknown and will be lessened if the mantle wedge becomes increasingly reducing with distance away from the subducting plate or if the fluid transport velocities are too fast to allow significant equilibration en route (see sections 6 and 7.2).

In summary, a key aspect to the utility of U-series isotopes in the study of arc lavas is that whereas Th and Pa are observed and predicted to behave as relatively immobile high field strength elements (HFSE), Ra and (under oxidizing conditions) U behave like large ion lithophile elements (LILE) and are significantly mobilized in aqueous fluids. Fluid-wedge interaction will only serve to increase these fractionations. Just how robust the experimental partition coefficients are remains to be established by future experiments.

5. SEDIMENT ADDITION, MASS BALANCE FOR Th AND TIME SCALES

The presence of ^{10}Be or negative Ce anomalies, or both, in some arc lavas provide unambiguous evidence for a contribution from subducted sediments to the source of arc lavas (Morris et al. 1990; Hole et al. 1984). In addition, Pb isotopes commonly are elevated above the northern hemisphere reference line (NHRL) and arc lavas form arrays pointed toward the isotopic composition of locally subducting sediment (Kay 1980; Gill 1981). Consequently, we look next at the impact of sediment addition on the abundances of U-series nuclides and the resulting implications for the time scale and process of mass transfer of the sediment component beneath arc volcanoes.

5.1. Mass balance for Th content and ^{230}Th/^{232}Th ratios

The implication of Figure 1 is that the Th content of arc lavas is dominated by the sediment component and this can be verified by simple mass balance. In N-MORB reflect ~10% partial melting then MORB-source mantle is likely to have ~ 0.01 ppm Th (Sun and McDonough 1989) or lower, if there has been back-arc melt extraction, whereas GLOSS has ~6.9 ppm (Plank and Langmuir 1998). If a primitive arc lava contains 0.5 ppm Th and represents a 15% partial melt then its source had a Th content of ~0.075 ppm, assuming for the purposes of illustration that Th is completely incompatible during melting. By simple mass balance, the sediment contribution required to produce a source with this Th content can be estimated to 100 × (0.075-0.011)/(6.9-0.011) or ~1%. Lower Th in primitive melts or smaller percent melting require less sediment; lower Th in sediments requires more. Regardless, even 1% sediment accounts for 80-98% of the Th in the source of most arc magmas.

This conclusion is supported by the calculations of Plank and Langmuir (1993) who showed that the Th output of arc volcanoes is directly proportional to the sediment flux of Th (Fig. 4). Using an arc growth rate of 1.1 km^3/yr (Reymer and Schubert 1984), a density of 2.8 g/cm^3 and a Th content of 1 ppm, and assuming that the arc mantle source is similar to that of N-MORB, the annual flux of Th from subducted sediment is estimated at ~2.7 × 10^9 g (Hawkesworth et al. 1997a). From these estimates and an annual flux of Th from subducted sediment of 9 × 10^9 g/yr (Plank and Langmuir 1998), Hawkesworth et al. (1997a) calculated that ~30% of the Th in subducted sediments is returned to the crust in arc magmas, leaving 70% to be recycled beyond the arc into the deep mantle. Using the fluxes provided by Plank and Langmuir (1998) this equates to ~ 2.7 × 19^9 g/yr entering the arc crust and 6.3 × 10^9 g/yr continuing on into the mantle. Because 0.5 ppm Th is a relatively high value for some primitive arc melts (see Fig. 2b), the percent and amount of recycled Th may be even higher in places.

An important consequence of the Th abundance mass balance is that the Th isotopes of the mantle wedge will usually be dictated by that of the subducting sediment. Nicaragua provides a particularly clear example where the distinctive signature of subducted, carbonate-rich sediment with high U/Th and (^{230}Th/^{232}Th) also characterizes the adjacent arcs lavas (McDermott and Hawkesworth 1991; Reagan et al., 1994; Thomas et al., 2002). More generally the subducting sediment will be dominantly pelagic, in which case the (^{230}Th/^{232}Th) ratio of the sediment modified mantle wedge will typically be significantly lower than MORB mantle. This is illustrated by a MORB-source – sediment mixing curve in Figure 6 which shows that the Th isotopic composition of a sediment-wedge mixture becomes effectively that of the sediment after 1% sediment addition. Thus, so long as the subducting sediments can be assumed to be in secular equilibrium (see Section 5.2) the (^{230}Th/^{232}Th) ratio of the sediment-wedge mixture can be estimated directly from the U/Th ratio of the subducting sediments. Moreover, if the composition of the subducting sediments is relatively constant along a given arc and if the sediment contribution is ≥ 1%, the (^{230}Th/^{232}Th) ratio of the sediment-wedge mixture will be invariant. As we shall see this not only has fundamental implications for how the time scales of fluid addition are determined (see Section 6), but it also provides one means of determining the rate of transfer, or recycling, of the sediment component back into the crust via the arc lavas.

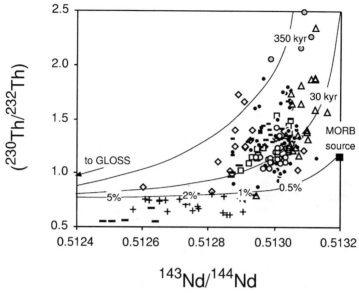

Figure 6. (^{230}Th/^{232}Th) versus ^{143}Nd/^{144}Nd for arc lavas. The mixing curve between a nominal MORB mantle and subducted sediment illustrates that the Th isotope signature of the mantle wedge will be dictated by as little as 1% subducted sediment (assuming no ^{230}Th in-growth). The model uses a MORB-source with 0.004 ppm U, 0.011 ppm Th, 0.67 ppm Nd and ^{143}Nd/^{144}Nd = 0.51321 and GLOSS (Plank and Langmuir 1998) and assumes that both are initially in U-Th equilibrium. The upper curves represent the Th isotope ratios which will be reached 30 and 350 kyr after addition of a constant fluid flux of 0.021 ppm U which results in variable ^{238}U-excesses whose magnitude depends on the mass balance between this U addition and the Th content resulting from the sediment- mantle wedge mix. The degree of melting used is 15% and U and Th are assumed to be completely incompatible. Symbols and data sources as in Figure 1.

5.2. Mechanism of sediment transfer—implications for the temperature of the wedge?

Sediment may be added by bulk mixing via imbricate thrusting (Bebout and Barton 2002), dehydration (Class et al. 2000), or melting (Johnson and Plank 1999). The latter two may differ in their P-T conditions and, therefore, residual mineralogy as well as relevant partition coefficients. In general, fluids are less effective transport agents than melts (i.e., trace elements are more soluble in melt than in pure water or even brine), but fluid/solid partitioning can fractionate some elements, notably Ba-Th and U-Th, more than melt/solid. However, as pressure increases, the distinction between "fluid" and "melt" decreases as their mutual solubility increases and they approach a critical end-point.

It is widely thought that melts can transport Th, light REE, and Zr-Hf much more than fluids can (e.g., Johnson and Plank 1999), although this is a relative statement in which the relevant partition coefficients are poorly known and variable. However, lavas from several arcs have Nd/Ta and Th/Ce ratios that are significantly higher than those in their input bulk sediments (see Fig. 1b) and so it has been suggested that the sediment component is transferred as a partial melt formed in the presence of residual accessory phases (Nichols et al. 1994; Johnson and Plank, 1999) which retain HFSE (Elliott et al. 1997; Turner and Hawkesworth 1997; Johnson and Plank, 1999). At the Marianas volcanic front, the lavas with the highest Th contents (highest sediment contribution) lie on or near the U-Th equiline with (^{230}Th/^{232}Th) ~1.04 (Elliott et al. 1997). However, the bulk subducting sediment (Plank and Langmuir 1998) has a much lower (^{230}Th/^{232}Th) ratio of 0.58 (see Fig. 7). Because mass balance calculations show that the Th isotope composition of the wedge-sediment mix is controlled by the sediment, the U/Th ratio of the sediment component has been modified. Elliott et al. (1997) concluded that the sediment component was added as a partial melt which had a higher U/Th ratio than that of the bulk sediments. Note that this cannot be due to melting in the presence of residual rutile, as originally suggested by Elliott et al. (1997), because U is much more compatible than Th in rutile (Blundy and Wood 2003). Some other residual phase, such as apatite, may be responsible.

This discussion leads us to perhaps one of the most serious apparent discrepancies between current geochemical and geophysical models. Experimental data demonstrate that partial melting of sediments at shallow levels (3 GPa) requires temperatures ≥ 670 °C (Nichols et al. 1994; Johnson and Plank 1999) and thus a thermal structure in the mantle wedge several hundred degrees hotter than that predicted by many numerical thermal models (e.g., Davies and Stevenson 1992; Peacock 1996). However, the viscosity of the mantle wedge may be lower than often assumed (Billen and Gurnis 2001) and numerical models that incorporate a temperature-dependant viscosity predict a component of upward flow beneath arcs and thus a hotter temperature structure (e.g., Furukawa 1993a, b; Kincaid and Sacks 1997). Such models may also help to reconcile the high eruption and equilibration temperatures inferred for some arc lavas (e.g., Sisson and Bronto 1998; Elkins Tanton et al. 2001).

5.3. Time scale of sediment transfer

Three lines of evidence suggest that the sediment component is added to the arc mantle before the principal fluid component that appears in arc magmas. First, the mechanical mixing of sediment and peridotite observed in outcrops of subduction

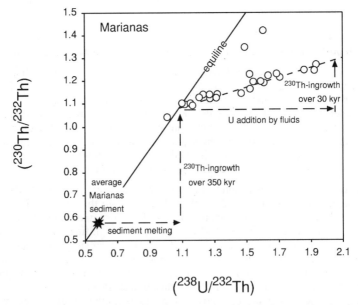

Figure 7. U-Th equiline diagram for lavas from the Marianas with dashed arrows showing the preferred interpretation of the sediment melt and fluid transfer times (after Elliott et al. 1997).

complexes occurs beneath the forearc (Bebout and Barton 2002). Secondly, geochemical observations of arc lavas also suggest addition of sediment prior to events that most affect U-series disequilibria. Returning to the case of the Marianas, the increased U/Th ratio inferred to have been produced from the sediment component by partial melting would result in the sediment melt having ^{238}U-excess. Because the Marianas lava array projects back to the equiline, Elliott et al. (1997) argued that the sediment melt was added sufficiently long ago for it to have decayed back up to the equiline (see Fig. 7) which would require that the transfer time of this component was at least 350 kyr. Of course, this is a minimum estimate because the U/Th isotope system cannot resolve beyond 350 kyr. The Marianas argument applies only if the subducting sediment has a high Th/U ratio (e.g., is dominated by clay more than carbonate or organic-rich components) and it also loses force if some lavas along the $(^{230}\text{Th}/^{232}\text{Th})$-$(^{238}\text{U}/^{232}\text{Th})$ correlation line are Th-enriched instead of in equilibrium (e.g., Reagan et al. 1994). Thirdly, in addition to these same arguments applying to Tonga, a Pb isotopic signature there, that occurs only in the northernmost lavas, has been attributed to Louisville volcaniclastic sediments which are presently being subducted some 1000 km further south (Regelous et al. 1997; Turner and Hawkesworth 1997). Because the line of volcaniclastics is being subducted obliquely beneath the arc, the locus of their subduction has migrated southwards over time and from this Turner and Hawkesworth (1997) inferred a sediment transfer time of 2-4 Myr. In the simplest model, such long transfer times require the transfer of sediment into the mantle wedge at shallow levels and decoupling of convection in the wedge from the subducting plate in order to slow the rate of transfer of the sediment component to the site of partial melting (Turner and Hawkesworth 1997). We note in passing that there is some evidence that flow within the mantle wedge may be oriented along the arc parallel to the trench rather than being directly coupled to the subducting plate (Turner and Hawkesworth 1998; Smith et al. 2001).

In summary, there is permissive evidence that the sedimentary component is added to the mantle wedge before the fluid addition and melting that leads to arc magmatism.

6. FLUID ADDITION TIME SCALES

In the following sections we look at the evidence for the timing of fluid addition to the sources of arc magmas from U-series disequilibria data and then discuss the extent to which some apparently conflicting time scale information might be reconciled.

6.1. U addition time scales

Fluids produced by dehydration reactions in the subducting, altered oceanic crust preferentially add U (along with other fluid mobile elements) to the mantle wedge. So long as this is the principal cause of U/Th fractionation (but see sections 7 and 8), U-Th isotopes can be used to estimate the time elapsed since fluid release into the mantle wedge. Firstly, U and Th have similar and small distribution coefficients in most mantle minerals at depths < 80 km (Blundy and Wood 2003) and so neither partial melting or gabbroic crystal fractionation are likely to have much effect on elemental U/Th ratios at the degrees of melting inferred for arcs (>5%). Secondly, continental upper crustal materials generally have low U/Th and so bulk crustal contamination would act to decrease U/Th. Therefore, since the vast majority of arc lavas have U/Th ratios exceeding anything found in MORB or OIB and frequently preserve ^{238}U-excesses (Fig. 3), the high U/Th ratios are generally accepted to reflect U addition to their mantle source by fluids from the subducting plate.

In practice, information on the timing of fluid release can either be obtained from along-arc suites of lavas which form inclined arrays on U-Th equiline diagrams, where the age = $-1/\lambda^{230} \times \ln(1 - \text{slope})$, or if the initial ($^{230}\text{Th}/^{232}\text{Th}$) ratio is constrained for an individual sample. For example, Sigmarsson et al. (1990) obtained an inclined U-Th isotope array for lavas from the southern Andes and suggested that the slope of this array reflected the time (20 kyr) since U addition by fluids. Similarly, Elliott et al. (1997) showed that lavas from the Marianas form a 30 kyr U-Th isotope array (Fig. 7). In total about 15 arcs have now been studied for U-Th disequilibria indicating that the time since U addition by fluids from the subducting oceanic crust varies from 10 to 200 kyr prior to eruption (Fig. 8a), if the correlations have strict time significance (see Section 8). All of these arrays show variable degrees of scatter and any chronological interpretation should be viewed as the time-integrated effect of U addition rather than to imply that U addition occurred at a discrete and identical time along the length of an arc. Indeed, variations can occur within the single arc due to the effects of tectonic collisions on the rate of subduction, such as occurs in New Britain (Gill et al. 1993) and Vanuatu (Turner et al. 1999), or possibly even through the life of a single volcano such as Santorini (Zellmer et al. 2000).

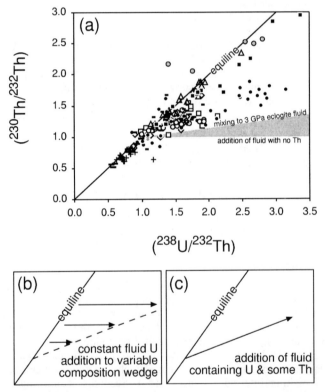

Figure 8. (a) U-Th equiline diagram for arc lavas. Symbols and data sources as in Figure 1. (b) Schematic U-Th equiline diagram showing how addition of a constant U fluid flux to a wedge with variable composition could in principle produce an inclined array. (c) Schematic U-Th equiline diagram showing how addition of a fluid containing some Th could in principle produce an inclined array. Shaded envelope in (a) indicates the range of compositions which could be produced by adding either a fluid containing no Th or the calculated 3 GPa eclogite fluid from Table 3 to a mantle wedge in equilibrium with an arbitrary ($^{238}\text{U}/^{232}\text{Th}$) ratio of 1. The majority of arc lavas lie above this envelope (even though their arrays often tie back to the equiline around ($^{238}\text{U}/^{232}\text{Th}$) = 1). The slope of a hypothetical array produced by addition of the latter fluid corresponds to an age of 15 kyr.

An underlying assumption in these interpretations is that U addition by fluids is the only cause of U/Th fractionation (see Fig. 8b,c) and if the same were true of U/Pa ratios then U addition should likewise produce U-Pa arrays which record a similar time to the U-Th arrays. So far the only arc where this may be true is Tonga, where Turner et al. (1997) showed that the combined Tonga and Kermadec lava array scatters around a ~ 50 kyr U-Th isochron (Fig. 9a). Subsequently, Bourdon et al. (1999) showed that the Tonga lavas are characterized by excesses of ^{235}U over ^{231}Pa, and by normalizing to Nb, obtained a U addition age of 60 kyr (Fig. 9b). This result supports the interpretation that the U-Th arrays have time significance (see Section 8). In the original age calculation made by Bourdon et al. (1999) the most depleted sample (26837) was omitted because of its very high (^{235}U)/Nb ratio which was most likely due to imprecision on its very low Nb

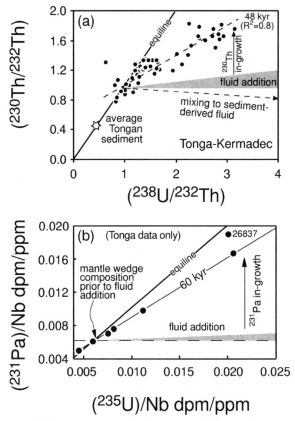

Figure 9. (a) U-Th isotope data from the Tonga-Kermadec island arc (Turner et al. 1997; Regelous et al. 1997). The dashed line indicates that an regression line through the full Tonga-Kermadec data set has an age of 48 kyr. Shaded envelope encompasses likely fluid compositions ranging from Th-free to an fluid in equilibrium with eclogite at 3 GPa (as in Table 3 and Fig. 8). (b) U-Pa isotopes (Bourdon et al. 1999) normalized to Nb on the basis that it behaves as a stable analogue of Pa (see text). Note the addition of the Tofua sample 26837 using the revised Nb concentration (T. Plank pers. comm.). The arrows show that if the fluid contains U but no Pa then the lavas from Tonga record similar U addition times in both the U-Th and U-Pa systems. As in (a) the shaded envelope encompasses the likely range of fluid compositions and illustrates that the Pa content of the fluid is predicted to be so low that the age determined from the isochron requires virtually no correction for the fluid composition (i.e., the assumption of addition of U only is justifiable).

concentration. Using a revised Nb measurement of 0.23 ppm (T. Plank pers. comm.), this point lies closer to, but still slightly above the array defined by the other 6 samples (Fig. 9b). When the 7 Tongan samples analyzed for U-Th-Pa are considered, their U-Th and U-Pa arrays yield apparent ages of 35 +24/−15 kyr and 60 +20/−25 kyr, respectively. Thus, they agree within 1σ analytical error of each other and with a regression through the complete Tonga-Kermadec U-Th dataset which yields 48 kyr ($r^2 = 0.8$), although the errors are large. Finally, Bourdon et al. (2003) have used a Th/U versus Pa/U concordia diagram to show that the Tonga samples are consistent with addition of U 60 kyr ago. Consequently, it may be that a single U-addition event is recorded in the Tongan lavas, but more such studies are needed to determine where else this may apply.

More typically, arc lavas preserve ^{231}Pa-excesses (Fig. 3) which is inconsistent with *only* fluid addition of U and subsequent in-growth of ^{231}Pa and, instead, suggests additional in-growth of ^{231}Pa during partial melting as well. The balance between addition and in-growth of all U-series nuclides is a substantive issue and we will return to it below.

6.2. Ra addition time scales

^{226}Ra has a much shorter half life (1600 yr) than its parent ^{230}Th (75 kyr) and so provides the opportunity to look at very recent fractionation between Ra and Th. Global surveys by Gill and Williams (1990) and Turner et al. (2001) have shown that most arc lavas are characterized by ^{226}Ra-excesses. Reagan et al. (1994), Chabaux et al. (1999) and Sigmarsson et al. (2002) noted positive correlations between (^{226}Ra/^{230}Th) and (^{238}U/^{230}Th) and, like Gill and Williams (1990), inferred that the ^{226}Ra-excesses may reflect very recent metasomatism of the wedge. Turner et al. (2001) have subsequently shown that ^{226}Ra-excesses is arcs can greatly exceed those in other tectonic settings that the largest ^{226}Ra-excesses are found in those arc lavas with the highest Ba/Th ratios. The latter are usually taken as a good index of fluid addition (Fig. 10a) and these two traits are greatest in Th-poor oceanic arcs but it is important to verify that this is indeed a fluid signal.

Most arc lavas are plagioclase-phyric and because Ba and Ra are mildly compatible in plagioclase (Bundy and Wood 2003) it might be thought that the elevated Ba/Th and ^{226}Ra in arc lavas results from plagioclase accumulation. However, because D_{Plag} for Ra is predicted to be < 0.1 for mafic to intermediate magmatic systems (Bundy and Wood 2003), the Ra concentration in a rock will always be dominated by the groundmass component and the ^{226}Ra-excesses cannot have resulted from plagioclase accumulation. The elevated Ba/Th ratios are also unlikely to reflect gabbroic fractionation or crustal contamination because Ba is more compatible than Th during in gabbroic mineral assemblage and crustal materials have low Ba/Th relative to most arc lavas. In any case, the highest Ba/Th ratios occur in those arc rocks with the lowest SiO_2 and $^{87}Sr/^{86}Sr$ ratios and so the observed ^{226}Ra-excesses are inferred to be a mantle signature. Because Ba is more incompatible than Th during mantle melting (Blundy and Wood 2003), the ^{226}Ra-excesses in arc lavas might have been developed during partial melting (e.g., Turner and Hawkesworth 1997). However, both elements are highly incompatible, making them hard to fractionate by melting processes and Bourdon et al. (2003) noted that the largest ^{226}Ra-excesses also tend to occur in those arc lavas with the highest Sr/Th ratios (Fig. 10b). This latter correlation cannot be a melting signature because Sr is more compatible than Th during mantle melting.

Therefore, since both Ba and Sr are fluid mobile, the large ^{226}Ra-excesses at high Ba/Th and Sr/Th ratios are inferred to result from fluid addition to the mantle wedge. At the other end of the arrays in Figure 10 it is notable that the intercept for any arc is with the Ba/Th, not (^{226}Ra/^{230}Th) axis. This contrasts strikingly with MORB, for example,

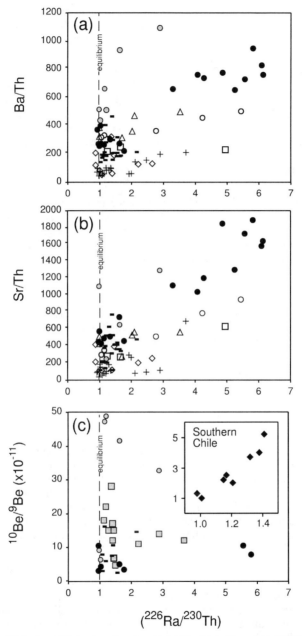

Figure 10. ^{226}Ra-^{230}Th disequilibria plotted against (a) Ba/Th, (b) Sr/Th and (c) ^{10}Be/^{9}Be. The positive correlations with Ba/Th and Sr/Th which are sensitive to fluid additions (cf. Figs. 1 and 2), suggest that the observed ^{226}Ra-excesses result from Ra addition by fluids from the subducting plate. The positive correlation in southern Chile (Sigmarsson et al. 2002) shown in the insert to (c) is also consistent with derivation of the ^{226}Ra-excesses from the subducting plate (see text). Symbols and data sources as in Figure 1. In (c) grey squares are data from New Britain (Gill et al. 1993); Tonga Be isotopes are unpublished data from Turner, George and Morris (in prep.).

where $(^{226}Ra/^{230}Th)$ is high even when Ba/Th is low. This means that the most recent addition of Ba and Ra is superimposed on a mantle previously enriched in Ba, and that whatever excess ^{226}Ra is produced during low-porosity melting in the mantle wedge is overwhelmed by the slab contribution.

Recently, Sigmarsson et al. (2002) showed that there is a good positive correlation between $(^{226}Ra/^{230}Th)$ and $^{10}Be/^{9}Be$ in lavas from southern Chile. More commonly, $(^{226}Ra/^{230}Th)$ is not well correlated with $^{10}Be/^{9}Be$ (Fig. 10c) consistent with decoupling between the sediment and fluid signatures. Nevertheless, since ^{10}Be is unambiguously derived from the subducted sediments, the Chile data provides independent evidence that ^{226}Ra-excesses are derived from the subducted plate. It also indicates that, in Chile, fluids from the altered oceanic crust may scavenge ^{10}Be from the overlying sediments as they pass through them from the underlying altered oceanic crust and out into the mantle wedge.

In summary, there is now convincing evidence that ^{226}Ra is added to the mantle wedge source of arcs by fluids and that the most recent of these additions must have occurred less than 8 kyr ago. The evidence depends on the magnitude of $(^{226}Ra/^{230}Th)$ and the quality of its correlation with subduction parameters. The evidence is most robust for oceanic arcs and it usually hinges on data for a few volcanoes, usually the ones with the lowest Th concentrations.

6.3. Reconciling the U and Ra time scales

At face value, the ^{226}Ra evidence for fluid addition in the last few 1000 years appears inconsistent with the interpretation that U/Th disequilibria resulted from fluid addition 10-200 kyr ago and we now discuss some possible models for reconciling these disparate time scales.

6.4. Single-stage fluid addition

Chabaux et al. (1999) noted a good correlation between $(^{226}Ra/^{230}Th)$ and $(^{238}U/^{230}Th)$ in lavas from the Lesser Antilles and, although this correlation is weaker with new mass spectrometric data for Kick 'em Jenny (Turner et al. 2001), $(^{226}Ra/^{230}Th)$ and $(^{238}U/^{230}Th)$ are broadly correlated globally in arc lavas of known eruption age (Fig. 11a). Section A1 of the Appendix provides a method for estimating the range of U-Th-Ra disequilibria in the wedge prior to melting. Sigmarsson et al (2002) also observed a strong correlation between $(^{226}Ra/^{230}Th)$ and $(^{238}U/^{230}Th)$ in lavas from southern Chile. On the basis of this, Chabaux et al. (1999) and Sigmarsson et al (2002) suggested that both U and Ra were added together in a single very recent fluid addition. Such an interpretation requires that the U-Th (and U-Pa in the case of Tonga) arrays do not simply reflect the time elapsed since U addition and, in principle, this could arise in several ways as discussed by Elliott et al. (1997).

Firstly, the slope of the U-Th arrays might reflect addition of a constant U flux to a mantle wedge with variable U-Th concentrations and isotope ratios (Fig. 8b). However, the mass balance calculations show that the Th isotope composition of the mantle wedge will *usually* be dictated by the composition of the subducted sediment so this is generally only likely if the sediment composition varies significantly and U-Th correlates positively with other sediment indicators within the arc. Vanuatu provides an interesting exception to this because in this arc the composition of the wedge (especially as seen with Pb isotope data) does vary, not due to sediment addition, but from to a change from Pacific to Indian mantle in the area of the D'Entrecastreaux collision. Significantly, here the lavas erupted above the Indian and Pacific portions of the wedge define two separate, inclined U-Th arrays which intersect the equiline at the compositions predicted for Indian

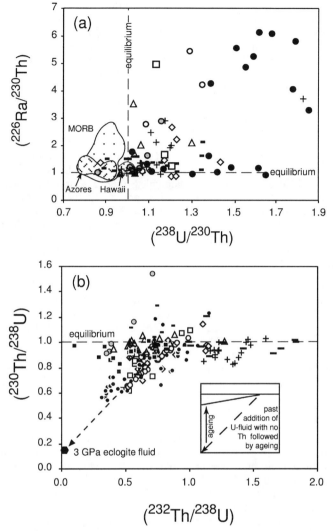

Figure 11. (a) Ra-Th versus U-Th isotopes for arc lavas. (b) $(^{230}Th/^{238}U)$ versus $(^{232}Th/^{238}U)$ after Bourdon et al. (2003) with composition of 3 GPa eclogite fluid from Table 3 also plotted. Inset shows interpretation for lava arrays not directed towards the fluid (see text for explanation and discussion). Symbols and data sources as in Figure 1.

and Pacific mantle, suggesting that the slope of those arrays does faithfully reflect the time since U addition (see discussion in Turner et al. 1999).

A second possibility is that the inclined U-Th arrays reflect addition of fluids containing some Th (Fig. 8c) and Pa from a source with higher U/Th than the mantle wedge (e.g., Yokoyama et al. 2002). As discussed earlier, U is significantly more fluid-mobile than Pa or Th, but the absolute magnitude remains to be confirmed. Bourdon et al. (2003) used a plot of $(^{231}Pa/^{235}U)$ versus Nb/U ratios for the Tonga lavas to show that a

Nb-free fluid (at Nb/U=0) would have a $(^{231}Pa/^{235}U)$ ratio of 0.7 if the Pa-U data were to reflect mixing with a fluid containing Pa. This contrasts with our estimate of a $(^{231}Pa/^{235}U)$ ratio in the fluid of 0.13 (Table 3) and would require that Pa is almost as mobile as U, which is inconsistent with inferences based on the mobility of Nb (Table 2). Therefore, the high $(^{231}Pa/^{235}U)$ intercept is more likely due to time than fluid-mobility of Pa, but the relative roles remain to be determined. Bourdon et al. (2003) also used an analogous plot of $(^{230}Th/^{238}U)$ versus $(^{232}Th/^{238}U)$ to explore the same relationship for the U-Th isotope system (Fig. 11b). Addition of U only results in mixing lines through the origin which are not observed. The high $(^{230}Th/^{238}U)$ intercept must result from either addition of Th as well as U, or in-growth over time subsequent to U addition. The observed trajectories for some arc data can be explained with our estimate of a $(^{230}Th/^{238}U)$ ratio in the fluid of 0.14, but others can only be explained if the $(^{230}Th/^{238}U)$ ratio of the putative fluid is ~0.5 and thus D_{Th}/D_U is <2 and this seems unlikely on the basis of the experimental data unless there are large fractions of rutile in the subducting plate (Tables 2,3). For other arc data, such as Kamchatka, the discrepancy with experimental data is even larger. Moreover, the largest U-excesses occur above the coldest subducting plates where full transformation to eclogite beneath the arc volcanoes is least likely to have occurred. Therefore, it is highly likely that a reasonable proportion of the ^{230}Th in arc lavas reflects in-growth from excess ^{238}U and, therefore, time.

In conclusion, the inclined U-Pa and U-Th arrays appear to have some time significance because their interpretation as simply the result of recent mixing with a fluid containing Th and Pa as well as U requires fluid partition coefficients for Th and Pa well in excess of those observed experimentally. The corollary is that that there must be a decoupling between Ra-Th and Th-U disequilibria. A further possibility is a combination of the two end-member models discussed above into one in which some Th and Pa addition by fluids is followed by some in-growth due to ageing. In this case (discussed further below) the age inferred from the U-Th and U-Pa arrays is necessarily less straight forward to interpret.

6.5. Two-stage fluid addition

Models in which the fluid does not contain appreciable Th or Pa, require a decoupling between Ra-Th and U-Th-Pa disequilibria in order to reconcile the different implied ages of fluid addition. However, unlike U, ^{226}Ra lost to the mantle wedge during initial dehydration continues to be replenished in the subducting altered oceanic crust by decay from residual ^{230}Th (Fig. 12a) on timescales of 10's kyr until all of the residual ^{230}Th has decayed away (350 kyr after removal of the last U). Thus, if dehydration reactions and fluid addition occur step-wise or as a continuum (Schmidt and Poli 1998), the ^{226}Ra-excesses will largely reflect the last increments of fluid addition whereas U-Th (and U-Pa in the case of Tonga) isotopes reflect the integrated time elapsed since the onset of U addition. As a simple approximation to this, two-stage fluid addition was modeled quantitatively by Turner et al. (2000a) and these results are reproduced in Figure 12b (see Section A2 of the Appendix for details) which shows the evolution of $(^{226}Ra/^{230}Th)$ and $(^{238}U/^{230}Th)$ over time in the mantle wedge. ^{226}Ra-excesses resulting from the initial fluid addition decay away, but new ^{226}Ra-excesses result from the second fluid addition. If little U remains to be added by the second fluid addition, U/Th isotopes can remain effectively undisturbed by this event and record the integrated time elapsed since the onset of fluid addition. The results of this modeling provided a reasonable approximation of the Tonga-Kermadec ^{238}U-^{230}Th and ^{226}Ra-^{230}Th versus Ba/Th data (see Turner et al. 2000a).

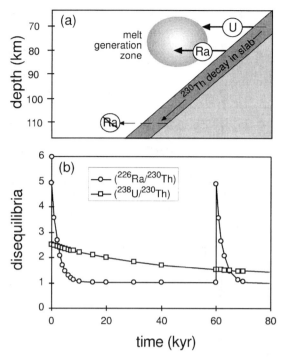

Figure 12. (a) Illustration of the spatial distribution of progressive distillation of fluid mobile elements from the subducting slab. Labeled arrows indicate the locations of U addition, Ra addition, and the point after which no further ^{226}Ra is produced in the plate. (b) Numerical simulation of two-stage fluid addition for the Tonga-Kermadec arc assuming that the degree of previous depletion was 5%(see Section A2 in the Appendix for details). Modified from Turner et al. (2000a).

6.6. Continuous fluid addition

The succession of dehydration reactions in the subducting plate could result in a nearly continuous release of aqueous fluids into the mantle wedge (Schmidt and Poli 1998) and an intuitively more satisfying solution to the two-stage fluid addition approximation, involves continuous fluid addition. This has been modeled by Dosseto et al. (2003) and Yokoyama et al. (2002) and the results of Dosseto et al. (2003) are summarized in Figure 13. In this approach it is assumed that the fluid dehydrates continuously from the slab and simple first-order differential equations are used to describe the evolution of the concentrations of relevant nuclides in both the slab and the mantle wedge. A Rayleigh distillation law is used to describe the dehydration process. Dosseto et al. (2003) have attempted to reproduce data from the most mafic lavas of the Central Kamchatka Depression with this model using published experimental data and also by inverting for mineral/fluid partition coefficients with a Monte-Carlo simulation. Thomas et al. (2002) attempted something similar for data from Nicaragua and Costa Rica. The surprising result is that both groups showed that in all cases the results were unsatisfactory unless significant quantities of Th and Pa are transported in the fluid along with U. However, both efforts sought to explain an atypically large range of high (^{230}Th/^{232}Th) ratios and doing the same for Marianas or Tonga may be less difficult. Nevertheless, continuous fluid addition models will certainly require further reassessment before they can be thought to have any general application.

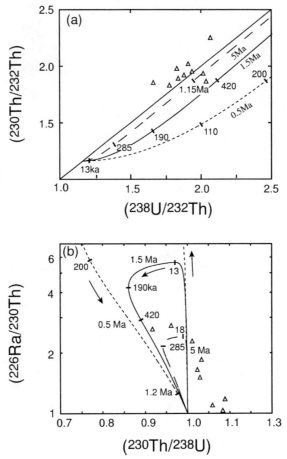

Figure 13. Model curves for continuous fluid addition to a MORB-type mantle source in (a) a ^{230}Th-^{238}U isochron diagram, and (b) a (^{226}Ra/^{230}Th) versus (^{230}Th/^{238}U) diagram. The time constant for dehydration is indicated next to the curve in Ma while the tick marks give the time since the inception of dehydration. These curves are calculated using the model described in Dosseto et al. (2003) where a fuller description is given. In this model, the fluid is released following a Rayleigh distillation process with a time constant τ such that $f = f_{max}(1-e^{-t/\tau})$ where is the total amount of fluid released (f_{max}= 0.03). The fluid is added to the source continuous and the source then undergoes batch melting with the following partition coefficients D_U = 0.0026, D_{Th} = 0.0014, D_{Pa} = 1.3 × 10^{-4} and D_{Ra} = 10^{-5}. The slab/fluid partition coefficients used in this calculation were determined by Montecarlo simulation (D_U= 0.16, D_{Th} = 74, D_{Ra}= 0.006) so as to fit the data for the Kamchatka arc shown as solid circles (Dosseto et al. 2003). The partition coefficient for U indicates that uranium would have to be less mobile than suggested by the experimental data in Table 2.

6.7. Mechanisms of fluid addition

One might anticipate that there should be some link between the time scales of fluid addition and some physical parameters of subduction zones (Jarrard 1986). In general such correlations have not yet been found, although there may be a weak positive correlation between the time since the inception of U addition and the rate of subduction (Turner et al. 2000b).

It has been suggested that fluid transfer could occur horizontally across the wedge by a series of hydration-dehydration reactions (Davies and Stevenson 1992). However, horizontal migration across the wedge can only occur at a velocity controlled by the rate and angle of descent of the subducting plate. This predicts fluid transfer time scales >> 350 kyr so that any ^{238}U-excess (let alone ^{226}Ra) originating from the subducting plate would have decayed away (Turner and Hawkesworth 1997; Schmidt and Poli 1998). Unlike the plethora of hydrous phases in the subducting plate, amphibole will be the main hydrous phase in the mantle wedge and the horizontal migration model could be reconciled if both the observed U- and Ra-excesses were generated by a final amphibole dehydration reaction in the wedge prior to melting (Regelous et al. 1997; Sigmarsson et al. 2002). However, whilst wedge amphibole could inherit high Ba/Th from the slab fluid and pass this trait along each time the wedge amphibole breaks down, it is unlikely that any ^{226}Ra-excess would survive. The mineral/fluid partition coefficients for Ba and Th appear to be similar in amphibole (Table 2) and so fluids produced in the presence of residual amphibole would not be predicted to have high Ba/Th yet those lavas with the largest ^{238}U- and ^{226}Ra-excesses also have the highest Ba/Th ratios (Fig. 10). Moreover, it seems more likely that the amphibole would be melt instead of dehydrate at the final stage and, at the degrees of melting inferred for arc front lavas (10-20%), it is likely to be consumed during melting. If amphibole were to be residual during melting, the high amphibole/melt partition coefficient Ba relative to Th (La Tourette et al. 1995) would predict low Ba/Th ratios. Similarly, Bourdon et al. (2003) showed that a single dehydration-reaction could not explain the presence of both ^{231}Pa-excesses and -deficits as observed in the Tonga-Kermadec arc. By implication, the major U/Th and Th/Ra fractionation is inferred to occur during fluid release from the subducting plate where redox conditions are the most strongly oxidizing (Blatter and Carmichael 1998; Parkinson and Arculus 1999). Thus, the combined U-Th-Ra isotope data would seem to require that fluid transfer occurs via a rapid mechanism. Such mechanisms may include hydraulic fracturing (Davies 1999) and experiments by Mysen et al. (1978) have indicated that H$_2$O will infiltrate lherzolite at 12 mm per hour at 800 °C and 3.5 GPa. This would allow fluids from the subducting plate to traverse 20 km of mantle wedge to reach the melt generation zone in only 200 years carrying a (^{226}Ra/^{230}Th) signal largely unchanged from that of the fluid which left the subducting plate.

The U-Th time scales imply that fluid addition commences prior to the initiation of partial melting within the mantle wedge which is inferred to be more or less synchronous with ^{226}Ra addition. This is consistent with recent geophysical analysis which suggests that the onset of melting is controlled by an isotherm and thus the thermal structure within the wedge (England 2001). This contrasts with the existing paradigm that an isobaric dehydration reaction in the subducting plate or in the mantle wedge is responsible for the relatively constant depth to Benioff zone (Gill 1981; Tatsumi et al. 1986). Rapid fluid transfer also implies that fluids escape from the slab as a free phase and are not instantaneously locked in a hydrous phase such as serpentinite or Mg-rich chlorite. This may place strong constraints on the thermal structure of the wedge beneath arc volcanoes.

7. PARTIAL MELTING AND MELT ASCENT RATES

The occurrence of relatively anhydrous arc lavas, modeling of trace element behavior, tomographic imaging of the mantle wedge, and correlations between lava major element composition and thickness of the overlying lithosphere have lead several recent studies to argue for a component of decompression melting in the production of arc lavas (Plank and Langmuir 1988; Pearce and Parkinson 1993; Zhao and Hasegawa, 1993;

Sisson and Bronto 1998; Parkinson and Arculus 1999; Hochstaedter et al. 2001). We now investigate the available U-series constraints on partial melting beneath arcs.

7.1. ^{231}Pa- and ^{230}Th-excess evidence for a partial melting signature in the wedge

In MORB and OIB, ^{230}Th- and ^{231}Pa-excesses are interpreted to reflect the effects of dynamic melting or chromatographic melting during which daughter isotope in-growth is facilitated by different residence times of U and Th in the upwelling matrix and melt (see Lundstrom 2003; Bourdon and Sims 2003). Consequently, ^{230}Th-excesses are considered a hallmark of mantle partial melting and although they are uncommon in arc lavas they are not absent (see Fig. 3). Thus, in the arc setting, it is likely that ^{230}Th is controlled by in-growth resulting *both* from U added by the fluid component *and* differences in the residence times of U and Th in the melt zone during partial melting.

It is particularly notable that ^{230}Th-excesses tend to be found in those arc lavas erupted through thicker crust which, combined with the angle of subduction, will control the thickness and thermal structure of the mantle wedge. ^{230}Th-excesses are found behind the fronts of the Marianas (Gill and Williams 1990), New Britain (Gill et al. 1994) and eastern Sunda (Gill and Williams 1990; Hoogewerff et al. 1997; Turner et al. 2003) arcs and in Nicaragua (Reagan et al. 1994), Costa Rica (Reagan and Gill 1989; Clark et al. 1998; Thomas et al. 2002) and Kamchatka (Turner et al. 1998; Dosseto et al. 2003). There is also a progression from ^{238}U-excesses to ^{230}Th-excesses on passing from oceanic to continental basement in both the Kermadec-New Zealand (Turner et al. 1997) and Aleutian-Alaska arcs (George et al. 2003). These observations lend support to models in which the thickness of the overlying lithosphere influences partial melting beneath arc volcanoes either because the melting region in the wedge is upwelling, or because the thickness of the overlying lithosphere controls the thermal structure of the mantle wedge (Plank and Langmuir 1988; Kincaid and Sacks 1997; George et al. 2003).

As discussed above, addition of U by fluids will produce excesses of ^{235}U over ^{231}Pa. Therefore, an extremely important observation, shown on Figure 14, is that the great majority of arc lavas are characterized by the reverse sense of fractionation (Pickett and Murrell 1997; Bourdon et al. 1999; Thomas et al. 2002). Subsequently, Bourdon et al. (1999) showed that only lavas from the Tonga-Kermadec arc preserve both ^{231}Pa excesses and deficits (Fig. 14a). Their interpretation was that, in Tonga, fluid addition resulted in $(^{231}Pa/^{235}U) < 1$ and this is preserved because partial melting occurred in the absence of significant amounts of residual clinopyroxene (Fig. 14b) which minimized any subsequent fractionation of Pa/U. However, the lavas from Kermadec, like most arc lavas, have $(^{231}Pa/^{235}U)$ ratios > 1, and this appears to provide unequivocal evidence for U/Pa fractionation during the partial melting process (Pickett and Murrell 1997; Bourdon et al. 1999; Thomas et al. 2002). Note that the displacement of the Tonga field relative to that for the Kermadec lavas on Figure 14b argues that the low HFSE concentrations in these lavas reflect prior depletion of the mantle wedge rather than just larger degrees of partial melting.

In summary, an important and exciting recent development in the application of U-series isotopes to arc lavas is the recognition that it is possible to distinguish elemental fractionation due to fluid addition from those due to partial melting, in particular using the ^{235}U-^{231}Pa system. This is a crucial point because the melting signature of arc magmas has been rather elusive in other geochemical tracers.

7.2. Ra evidence for melt ascent rates

Theoretical calculations suggest that the segregation and ascent time scales for basaltic magmas from a partially molten matrix are likely to be short (McKenzie 1985).

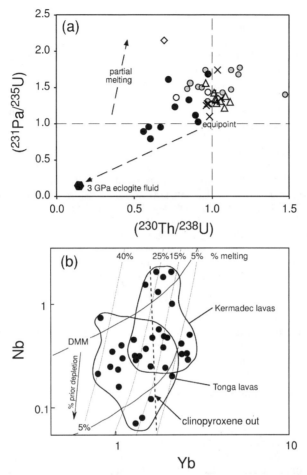

Figure 14. (a) Plot of U-Pa versus U-Th disequilibria for arc lavas. The Tonga samples are those $(^{231}Pa/^{235}U)$ and $(^{230}Th/^{238}U) < 1$ as expected from U addition by subduction zone fluids. However, all other arc lavas have $(^{231}Pa/^{235}U) > 1$ suggesting that there was a subsequent increase in $(^{231}Pa/^{235}U)$ due to the partial melting process, schematically illustrated by the dashed arrow, which is inferred to have a much greater effect on $(^{231}Pa/^{235}U)$ ratios than $(^{230}Th/^{238}U)$ ratios. Symbols and data sources as for Figure 1. Additional data for Mexico, Stromboli and the Andes from Pickett and Murrell (1997) are plotted as crosses. (b) Plot of Nb_8 versus Yb_8 (where subscripts refer to compositions regressed to 8% MgO) for Tonga-Kermadec lavas showing that the Tonga lavas are derived from a more depleted source in which clinopyroxene is likely to have been exhausted during partial melting explaining the lack of fractionation of $(^{231}Pa/^{235}U)$ ratios during melting.

One of the seemingly inescapable conclusions from the ^{226}Ra-^{230}Th disequilibrium data, at least for Tonga and the Mariana volcanic fronts, is that significantly less than 8000 years and arguably only a few half lives (ca. 1000-3000 yrs) can have elapsed since the generation of the ^{226}Ra-excesses observed in the arc lavas plotted on Figure 10.

Porous melt flow is likely to be unstable over large distances in the mantle resulting in a transition to channeled magma flow (Aharonov et al. 1995) and this may be swift if melting rates are high and the threshold porosity is quickly exceeded. The arc ^{226}Ra-^{230}Th

disequilibria data seem to indicate that the segregation of the melt from its matrix into channeled ascent does indeed occur on a very rapid time scale. If a single-stage fluid addition model is assumed then the initial (^{226}Ra/^{230}Th) predicted in Table 3 is 15 for 3% fluid release (see also Yokoyama et al. 2002). In that case the largest observed (^{226}Ra/^{230}Th) ratios in Tonga (~ 6.1, Turner et al. 2000a) could have decayed by ~ 60% or one half life and the calculated ascent rates would be ~ 60 meters per year. However, the two-stage fluid addition model of Turner et al. (2000a) predicts maximum initial (^{226}Ra/^{230}Th) ratios of 6-7 and so, if partial melting occurs at 80-100 km depth beneath arc volcanoes, then the required magma ascent rates are arguably of the order of 100's to 1000's meters per year (Turner et al. 2000a, 2001). Either way, these ascent rates rule out models of melt migration in diapirs (Hall and Kincaid 2001) and even the kind of equilibrium porous flow typically invoked in other tectonic environments (cf. Lundstrom 2003; Bourdon and Sims 2003). In principle, ascent rates could be significantly faster (McKenzie 2000) and values of 1.8 km per day were estimated from seismic data in Vanuatu (Blot 1972) and 26 km per day from xenoliths in Mexico (Blatter and Carmichael 1998). As argued by Bourdon et al. (2003), the fast magma ascent rates indicated by Ra-Th disequilibria appear to be independently required in order to preserve slab-derived isotopic signatures of highly compatible elements such as Os (Woodland et al. 2002; Alves et al. 2002) which would otherwise be expected to be lost during equilibration with the wedge peridotite during melt ascent (Hauri 1997).

7.3. Models to reconcile the Pa, Th and Ra data

It is important to note that the ^{226}Ra-^{230}Th disequilibria do not preclude a partial melting origin for the ^{231}Pa-excesses. However, they do require that the residence time of Ra in the melting column was short enough, and thus that the melt velocity was fast enough, to prevent ^{226}Ra from decaying back to values solely attributable to partial melting of the mantle wedge.

7.4. Batch and equilibrium porous flow melting models

A simple batch melting model with instantaneous extraction of the melts predicts that, for degrees of melting greater than 5%, there will be little U-Th-Pa fractionation. Thus, the only models that can produce significant disequilibria involve in-growth during the melting process (see Lundstrom 2003; Bourdon and Sims 2003). Whilst equilibrium porous flow melting models involving chromatographic re-equilibration of the melts on their way to the surface (Spiegelman and Elliott 1993; Bourdon et al. 1999) can produce the observed ^{230}Th- and ^{231}Pa-excesses, they fail to preserve large ^{226}Ra/^{230}Th (e.g., Bourdon et al. 2003) as illustrated in Figure 15. Therefore, the most likely models to reconcile the arc Ra and Pa data will involve some variation on the theme of dynamic melting involving protracted melt production but fast melt extraction (McKenzie, 1985; Williams and Gill 1989). In these models slab components reach the surface quickly but the mantle undergoes partial melting over a time interval comparable to the half lives of ^{230}Th and ^{231}Pa. Initial attempts include dynamic melting subsequent to fluid addition (Elliott 2001; Bourdon et al. 2003; George et al. 2003) and fluxed in-growth melting (Thomas et al. 2002; Bourdon et al. 2003) as schematically illustrated on Figure 16.

7.5. Dynamic melting

For a given melting rate, dynamic melting of mantle wedge material bearing a range of ^{238}U-excesses leads to greater in-growth of ^{230}Th in those sources beginning with the largest ^{238}U-excesses. In other words, the more fluid enriched mantle undergoes the greatest ^{230}Th increase during partial melting with the net effect of rotating U-Th arrays anticlockwise (Fig. 17a). Thus, Elliott (2001) simulated the sloped U-Th array from the Marianas by dynamic melting following fluid addition of U. The model does not require,

Magma Genesis at Convergent Margins

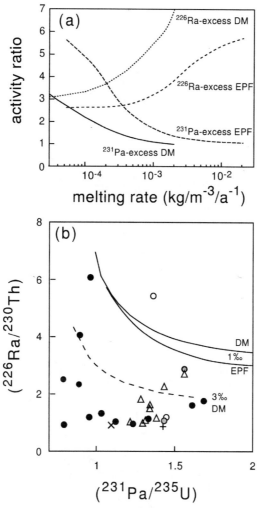

Figure 15. (a) Comparison of equilibrium porous flow (EPF; Spiegelman and Elliott 1993) and dynamic melting (DM; Williams and Gill 1989) models for ^{226}Ra-^{230}Th and ^{231}Pa-^{235}U activity ratios in the melt as a function of melting rate. Both models assume a matrix porosity of 0.001, a total extent of melting of 10% and initial (^{226}Ra/^{230}Th) = 7 and (^{231}Pa/^{235}U) = 0.8. If the melting rate is too large there is not enough time to built in-grow ^{231}Pa, whereas if it is too small, the initial ^{226}Ra-excess in the source decays rapidly and the maximum ^{226}Ra-excess is 3. In the case of the EPF model, the melting rate is scaled to an upwelling velocity assuming 1% melt produced per kbar of decompression. Total degree of melting is 10%. In order to explain the range of (^{231}Pa/^{235}U) and (^{226}Ra/^{230}Th) found in arc lavas, the melting rate needs to be of the order of 10^{-3}-10^{-4} kg.m^{-3}.a^{-1}, assuming dynamic melting and the partition coefficients used here (D_{Pa}= 0.0001, D_U= 0.003 D_{Th}= 0.002 D_{Ra}= 10^{-5}). For equilibrium porous flow to explain the data, the melting rate (and by implication the upwelling velocity) has to be extremely fast (up to 200 cm.a^{-1}) in order to preserve the large (^{226}Ra/^{230}Th) produced at the base of the melting column and the corresponding (^{231}Pa/^{235}U) ratios are low. (b) (^{226}Ra/^{230}Th) versus (^{231}Pa/^{235}U) calculated for various melting models. Melting rates were varied from 0.002-0.003 (DM) and 0.006-0.03 (EPF). The model curves are estimated based on initial (^{226}Ra/^{230}Th) =7 prior to melting. Matrix porosity in per mil is indicated next to the curves. The solid curves assume an initial (^{231}Pa/^{235}U) = 1 in the mantle wedge source whilst the dashed curve is calculated with an initial (^{231}Pa/^{235}U) = 0.8, simulating previous addition of U by fluid to the mantle wedge. The curves are compared with the published data from arcs (symbols and data sources as in Figure 1).

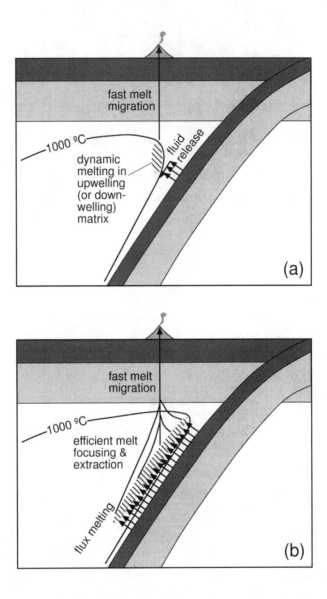

Figure 16. Cartoon cross-sections illustrating two current arc melting models. (a) Fluid release followed by dynamic melting with rapid melt extraction once the hydrated peridotite crosses its solidus. Dynamic melting can occur in either upwelling or downwelling matrix. (b) Flux in-growth melting where each fluid addition promotes a proportional amount of melting of peridotite in which in-growth is occurring in response to the preceding melt extraction increment. All melt fractions are efficiently focused and rapidly extracted.

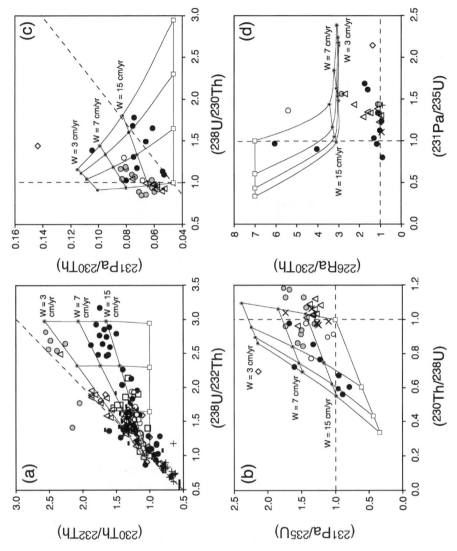

Figure 17. U-series diagrams showing the effects of in-growth produced during dynamic partial melting of an arc source region containing variable ^{238}U-excesses (cf. Figs. 8a, 14a, 15b and 19). Note in (a) that dynamic melting results in inclined arrays that resemble the effects of U addition and ^{230}Th in-growth during ageing or addition of a fluid containing some Th (George et al. 2003). The model assumes a source containing 0.15 ppm U and 0.47 ppm Th, which is initially in secular equilibrium, to which three different amounts of U (0.1, 0.2, 0.3 ppm) have been added. Dynamic melting was calculated following Williams and Gill (1989) assuming a residual porosity of 0.004 and a source mineralogy composed of 56% olivine, 22% clinopyroxene and 22% orthopyroxene. Bulk distribution coefficients for U, Th, Ra and Pa of 0.0081, 0.0066, 1.07 × 10^{-6} and 1.75 × 10^{-6}, respectively, were based on Blundy and Wood (2003). The total extent of melting was 15% and the model grid lines are labeled with the rate of matrix flow through the melting region (W) which could reflect upwelling, in a decompression model, or downwelling due to induced convection in the mantle wedge (values encompass the range of observed subduction rates).

nor does it preclude, the presence of Th or Pa in the fluid component. In these models in-growth can occur either because the matrix is mildly upwelling or because its is being progressively dragged downwards through the melting region by induced convection in the mantle wedge. Figure 17 shows that dynamic melting can replicate much of the range of the U-series observations in arc lavas but few well defined trends have been identified from the few data available for Figures 17b-d. The dynamic melting effects are inferred to occur after fluid addition when the hydrated peridotite crosses its solidus (Fig. 16a). Thus, the onset of melting in these models is assumed to be controlled by the position of an isotherm in the wedge as suggested by Tatsumi et al. (1986) and England (2001) and there is no *requirement* for a link between degree of partial melting and amount of fluid addition.

George et al. (2003) proposed model of dynamic melting in an upwelling matrix may be a viable explanation for the range of (^{230}Th/^{232}Th) ratios observed in the Alaska-Aleutian arc where ^{230}Th-excesses dominate in the continental sector of the arc and ^{226}Ra-excesses are sufficiently small to be attributable to the effects of partial melting alone. However, in this arc the horizontal base to the U-Th isotope array requires either that the fluid did not contain Th, or else that it had the same (^{230}Th/^{232}Th) ratio as the subducting sediments (George et al. 2003). In contrast, Bourdon et al. (2003) took the agreement between the U-Th and U-Pa ages in the depleted Tonga arc to indicate that U-Th systematics may be only weakly fractionated during melting beneath arcs. They used this and the pressure dependence of U/Th mineral/melt partitioning (see Blundy and Wood, 2003) to constrain the depth of the melting zone, and then developed a model of dynamic melting with fast melt extraction in a mildly upwelling wedge to reconcile the ^{226}Ra- and ^{231}Pa-excesses found in the Kermadec and other arc lavas (Fig. 15).

At first sight, upwelling may seem to be precluded by the confines of the overriding plate but, as noted earlier, upwelling of the sources of arc magmas has been inferred for a variety of reasons quite independent of the U-series disequilibria requirements (e.g., Plank and Langmuir 1988; Pearce and Parkinson, 1993; Parkinson and Arculus 1999). As noted earlier, the viscosity of the mantle wedge may be significantly lowered by the presence of volatiles (Billen and Gurnis 2001) and numerical models that incorporate a temperature-dependant viscosity also predict a component of upward flow beneath arcs (e.g., Furukawa 1993a,b; Kincaid and Sacks 1997). Small density and viscosity contrasts arising from the addition of volatiles and the presence of partial melt may be sufficient to cause gravitational instabilities leading to localized, upwelling that controls the surficial distribution and life-span of arc volcanoes (Brémond d'Ars et al. 1995). Such models might predict a positive correlation between crustal thickness and extent of melting induced disequilibria (e.g., ^{231}Pa-excess) if crustal thickness in some way relates to the overall wedge thickness (e.g., Plank and Langmuir 1988). Figure 18 shows that such a correlation may exist within the available data for oceanic arcs but that the behavior of continental arcs is more complex. Alternatively, those arcs with thicker crust may have a hotter wedge (Kincaid and Sacks 1997) and thus be characterized by larger melting rates (Bourdon et al. 2003).

7.6. Flux melting

Initial attempts at flux melting models have been made by Thomas et al. (2002) and Bourdon et al. (2003). Both assess variations on the theme of flux melting in which slab-derived fluids bearing various (including zero) amounts of U, Th, and Pa cause melting of mantle sources that have just been depleted by melt extraction in the previous melting increment. The recent depletion results in preferential in-growth of ^{231}Pa ± ^{230}Th in the matrix which is transferred to subsequent melts. In this type of model melting occurs along a lengthy zone of the wedge parallel to the subducting plate and all the melt

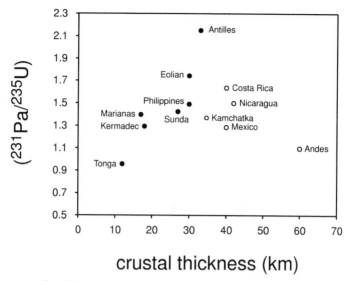

Figure 18. Plot of (^{231}Pa/^{235}U) versus crustal thickness (from Gill 1981) for each arc studied (averages are plotted where more than one analysis was available—data sources in Table 1) showing that there is a reasonable positive correlation within the oceanic arcs (filled circles).

fractions need to be efficiently extracted and focused into a zone beneath the arc front volcanoes (Fig. 16b). A link between amount of fluid addition and the degree of melting is an intrinsic and testable aspect of this type of model (e.g., Davies and Bickle 1991; Stolper and Newman 1994; Hirschmann et al. 1999; Eiler et al. 2000; Thomas et al. 2002). The chief variables (besides mantle/melt partition coefficients, porosity and permeability, total degree of melting, melting rate, and duration of chemical equilibrium that are common to all U-series melting models) are the relationship between the amount and rate of fluid addition and the degree of mantle melting and the relative solubilities of U, Th, and Pa in the fluid (see Section A3 of the Appendix for details).

Thomas et al. (2002) evaluated ^{231}Pa in-growth in downwelling mantle accompanied by continuous flux melting and instantaneous melt extraction. They note that the widespread positive correlation between ^{238}U-excess and (^{231}Pa/^{230}Th) is inconsistent with simple radioactive decay alone (Fig. 19), and implies a link between recent addition of U (via fluid) and recent in-growth of ^{231}Pa (via melting). They attribute the linkage to flux melting but Figure 17c shows that similar relationships can result from dynamic melting. The flux melting model developed is able to explain many differences between ^{230}Th-enriched lavas from Costa Rica and ^{238}U-enriched lavas from Nicaragua by greater flux melting in Nicaragua, but only when fluid addition is continuous, Th and Pa are somewhat fluid-mobile, and melt ascent is too fast to sustain chemical equilibrium. Their model also requires that (a) melts are continuously produced during continuing fluid release from the subducting plate and (b) that those fluids are derived from carbonate-rich sediments with high (^{230}Th/^{232}Th).

One aspect of the flux melting models which led Bourdon et al. (2003) to prefer a dynamic melting model is that the time scale required to in-grow ^{231}Pa (assuming that Pa is relatively fluid immobile) requires the down-dip length of the melting region to be on the order of 100 km. This is similar to the depth range over which fluid is released from

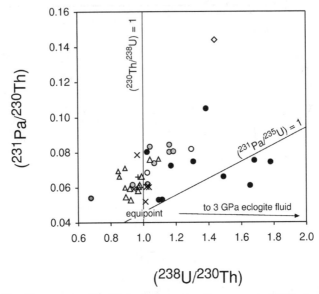

Figure 19. (^{231}Pa/^{230}Th) versus (^{238}U/^{230}Th) diagram after Thomas et al. (2002). Symbols and data sources as in Figure 1 excepting Costa Rica samples which are plotted here as open triangles so as to distinguish them from lavas from Nicaragua and samples from Mexico, Stromboli and the Andes (Pickett and Murrell 1997) which are plotted as crosses. As noted by Thomas et al. (2002) several of the arcs form trends in which (^{231}Pa/^{230}Th) increases with increasing (^{238}U/^{230}Th). Addition of the 3 GPa fluid from Table 3 forms a shallow negative trend from the equipoint. The arc lavas which form trends steeper than and lie above the (^{231}Pa/^{235}U) = 1 line arguably require a combination of fluid addition, ageing and ^{231}Pa in-growth due to partial melting in proportion to the amount of fluid added (Thomas et al. 2002).

the downgoing plate (Schmidt and Poli 1998; Thomas et al. 2002). However, assuming that wedge temperatures near the surface of the subducting plate are unlikely to be sufficient for partial melting much shallower than 80 km in any thermal model (Davies and Stevenson, 1992; Furukawa 1993a,b; Peacock, 1996), this requires that the melting zone extends up to 200 km behind the arc front and that the melts are very efficiently focused back to a predominant region beneath the arc front (e.g., Davies and Stevenson, 1992; Spiegelman and McKenzie 1987). However, this appears to significantly exceed the distance inferred from seismic data (Wyss et al. 2001).

The important conclusion from these studies is that ^{226}Ra and ^{231}Pa data add apparently robust and important, but different, insights into arc magma genesis which are less ambiguous than the ^{230}Th-^{238}U results alone. The ^{226}Ra-^{230}Th results require that melting and melt ascent are faster in subduction zones (at least for the volcanic front of some oceanic arcs) than usually inferred for mid-ocean ridges, and that at least the final fluid addition is effectively coincident with melting and rapid ascent. The entire process of slab dehydration, melt ascent, and differentiation can take less than a few thousand years. This indicates a whole new class of melt ascent mechanisms, probably made possible by the depleted nature of the mantle, the deformation style and thermal structure in the wedge corner, and the presence of water, but possibly also applicable to MORB and OIB (Turner et al. 2001). In contrast, the ^{231}Pa-^{235}U results require that melts are derived from a mantle with a recent melting history, either during upwelling or downwelling lasting scores to hundreds of thousands of years. The challenge is to reconcile these U-series time frames with one another and to other observations. For example, some arc magmas have major element compositions that imply continual re-equilibration to shallow depths (e.g., Grove et al. 2002) whereas the Ra-Th systematics seem to preclude that.

It is also important to note that none of these models are yet without their difficulties. For example, in any model it is difficult to explain the very large range in (^{230}Th/^{232}Th) found in the Kamchatka arc (Turner et al. 1998). If the flux melting models are adopted then, whilst the linear arrays on U-Th isotope diagrams may broadly reflect the duration of fluid fluxing, they may not provide a direct indication of the time since U addition (see further discussion in Section 8). Tonga may be an exception: if the fluid contains Th and Pa then none of the models predict the deficits of ^{231}Pa relative to ^{235}U in the Tonga arc or the similarity between the time scales derived from U-Th and U-Pa systematics in that arc. Fully reconciling the constraints from the Ra-Th and Pa-U data remains an important challenge to be further addressed by future arc studies and clearly holds great promise for unraveling further the physics of the partial melting process in this tectonic environment.

7.7. Partial melting of the subducting oceanic crust

The subducting basaltic oceanic crust is generally thought not hot enough to undergo partial melting except when the plate is very young (Peacock et al. 1994). Under atypical circumstances, however, it may be hot enough for the subducting basaltic crust to melt sufficiently that the resulting melts can reach the surface with some of the elemental and isotopic features of their MORB source intact. Partial melting versus dehydration of subducted basaltic crust produces distinctive, adakitic lavas due to the already fractionated nature of the protolith and the presence of significant amounts of residual garnet (Defant and Drummond 1990). In the Austral Andean zone of southern Chile, south of the point of subduction of the Chili ridge, very young (< 20 Ma) oceanic crust is presently being subducted. Lavas erupted above this zone were analyzed by Sigmarsson et al. (1998) who showed that they are characterized by ^{230}Th-excesses and high La/Yb and Sr/Y ratios which they ascribed to melting in the presence of residual garnet. A slab-melt signature has also been recognized in the Central Kamchatka Depression located above a zone of plate-tearing. Here Dosseto et al. (2003) have shown that the lavas with higher La/Yb and ^{230}Th-excesses (inferred to have a slab melt component) also have lower (^{226}Ra/^{230}Th) and (^{231}Pa/^{235}U) which they ascribed to residual phengite and rutile which should retain Ra and Pa respectively. However, it would be erroneous to make the assumption that all arc lavas that have ^{230}Th-excesses were formed in this way (Bourdon et al. 2000a; George et al. 2003). Even other "adakitic" arc lavas with ^{230}Th- and ^{226}Ra-excesses (e.g., Shasta in the Cascades: Volpe 1992) can be interpreted as mostly mantle-derived melts through processes similar to those described in the section above (Grove et al. 2002).

8. DISCUSSION OF U-SERIES TIME SCALE IMPLICATIONS FOR ARC LAVAS

The preceding sections have established that the U-series information from arc lavas is a complex end-product of several different processes and this is especially true for U-Th isotopes. The two observations that seem to be most robust are (1) the need to transfer the ^{226}Ra-excess signal from the subducting plate into the melting zone and to the surface in less than a few 1000 years, and (2) the need for a partial melting process which allows for ^{231}Pa in-growth. There are a number of different models which can take account of these two requirements but which have different implications for the U-Th isotope arrays.

In an extreme end-member case where both Th and Pa are considered very mobile, the U-Th arrays (and U-Pa in the case of Tonga) would simply point to the U-Th-Pa composition of the fluid (e.g., Chabaux et al. 1999; Sigmarsson et al. 2002; Yokoyama et al. 2002). However, our calculations of likely fluid compositions (Table 3) make this seem unlikely (note the position of the calculated fluid on Figs. 8, 9, 11, 14, 19). Note that such models also generally *require* that the fluid is derived from the altered oceanic crust. If the fluid was derived from pelagic sediment it would have a much lower (^{230}Th/^{232}Th) ratio (~ 0.5 in the case of Tonga) such that fluid addition would result in a negative array on the U-Th equiline diagram and the U-Th array would underestimate the integrated time of U addition (see Fig. 9). This is not true for those margins where the subducting sediment is rich in hemi-pelagic components characterized by high U/Th (and therefore high (^{230}Th/^{232}Th) ratios), such as central America (Herrstrom et al. 1995; Thomas et al. 2002) but they are exceptional. Conversely, this model would fail to explain the high (^{230}Th/^{232}Th) ratios of the Kamchatka lavas (Turner et al. 1998; Dosseto et al. 2003) since there is no carbonate sediment input beneath this arc.

Perhaps the simplest model, most applicable to Tonga-Kermadec, involves a two-stage addition of a fluid containing no Th or Pa followed by dynamic melting in an upwelling matrix (e.g., Bourdon et al. 2003). In this model, the slope of the U-Th arrays reflects the integrated time of U addition and, in Tonga, the absence of significant residual clinopyroxene during partial melting leads to minimal ^{235}U-^{231}Pa fractionation (thus U-Pa isotopes record a similar U addition time scale). In Kermadec, partial melting results in ^{231}Pa-excesses which can be used to constrain the rate of matrix flow through the melting region. In this case it is assumed that the U-Th systematics are not affected by partial melting due to the similarity of the U and Th mineral/melt partition coefficients at the pressure of the melt generation zone (Bourdon et al. 2003).

An important question in this model is to what extent are the derived time scales compromised if there is a small amount of Th and Pa in the fluid as predicted by the data in Table 2. In Table 3 we calculate the apparent U-Th and Nb-normalized U-Pa ages of the fluid composition derived earlier. The U-Th apparent age is 14-16 kyr, so that the 50-60 kyr time scale inferred in Tonga would overestimate the true integrated U addition time scale by 14-16 kyr (Fig. 9a). However, the U-Pa apparent age is only 40-45 years due to the minimal mobilization of Pa in the fluid. This implies that the U-Pa age is the more robust estimate (Fig. 9b). These results contrast with those of Yokoyama et al. (2002) who suggested that subduction fluids could have a very large effect on the U-Th equiline diagram. Firstly, their results were based on calculations using a high eclogite/fluid partition coefficient for U largely because they did not take into account the effects of fO_2 on U partitioning into garnet. Secondly, they allowed ^{230}Th to in-grow in the accumulating fluid in the subducting plate prior to its release and during it's transport through the wedge to the melting zone. However, all ^{226}Ra-excess would have decayed in such fluids. In reality, this model is effectively indistinguishable from the Tonga-Kermadec two-stage fluid addition interpretation; in both cases, it is the time elapsed since U-Th fractionation that is constrained. The relative amounts of time spent during fluid accumulation, passage through the wedge or in the melting zone remain unknown. In this respect it is interesting that Bourdon et al. (2003) speculated that the time required for fluid build-up prior to hydrofracturing may be on the order of 10's kyr.

In the model advocated by Elliott et al. (2001) and George et al. (2003) the U-Th systematics *are* affected by partial melting leading to increases in (^{230}Th/^{232}Th). If the fluid does not contain appreciable Th (or Pa), then the slope of the U-Th array is simply a function of the matrix flow rate through the melting zone (Fig. 17a), rather than the time since U addition, and this bears similarities with the approach used to derive upwelling (and melting) rates in MORB and OIB (see Lundstrom 2003; Bourdon and Sims 2003). In principle the U-Th arrays in this model could be produced by ^{230}Th in-growth in a downwelling matrix and the rate of matrix flow controls the extent of in-growth rather than any link to the amount of fluid addition and extent of melting (i.e., the amount of in-growth should be constant for a given subduction rate).

Thomas et al. (2002) have explored the flux melting models both with and without the presence of significant Th and Pa in the fluid. If the fluid contains only U, then the slopes of the U-Th and U-Pa arrays reflect the composite effects of ageing since U addition and ^{230}Th and ^{231}Pa in-growth during the partial melting process. Since the mineral/melt distribution coefficients are very similar for U and Th but very different for Pa (Blundy and Wood 2003), U-Th and U-Pa isotopes will contain information about the melting rate (which is a function of the amount of fluid addition in these models) whereas U-Pa isotopes will be more sensitive to the residual porosity in the melt region due to the very small partition coefficients for Pa. If the fluid also contains appreciable Th and Pa, then extracting information on a specific process (composition of the fluid, time since U addition, rate of melting) will necessarily be more difficult. However, there is no reason this should be impossible, especially if our knowledge of mineral-fluid partition coefficients and melting processes can be improved.

In conclusion, the significance and interpretation of the information derived from U-series isotopes in arc lavas is critically dependant on the fluid addition and melting model assumed. However, we stress that useful information is always contained in the U-series isotopes which ever model is ultimately adopted. It is incorrect (and misleading) to say, for instance, that the slopes of U-Th arrays have no time significance if there is Th in the fluid or if ^{230}Th in-growth occurs during the melting process. Rather, they contain important, albeit integrated, information about the composition of the fluid, the timing of its addition and the nature of the melting process which careful modeling should allow us to de-convolve.

9. REAR ARC LAVAS

Rear lavas provide a unique opportunity to investigate models for the progressive distillation of elements from the subducting plate and changes in magma composition in response to changes in melt fraction and residual mineralogy with increasing depth. Ryan et al. (1995), Woodhead et al. (1998), Hochsteadter et al. (2001) and others have documented that fluid-sensitive indexes like U/Th, Ba/La and Sr/Nd often decrease behind the volcanic front of most arcs suggesting a decrease in the relative fluid contribution and/or progressive depletion of these elements in the subducting plate which results in changes in the composition of the fluids released. Stolper and Newman (1994) showed that H_2O played an important role in the genesis of the Marianas back arc lavas but also suggested that the aqueous fluids became progressively stripped of all but the most fluid mobile elements with increasing passage through the mantle.

U-series data from across arc traverses are available from the New Britain (Gill et al. 1993), Kamchatka (Turner et al. 1998) and eastern Sunda (Hoogewerff et al. 1997; Turner and Foden 2001) arcs as well as from individual rear-arc volcanoes such as Bogoslof in the Aleutians (Newman et al. 1984; Turner et al. 1998; George et al. 2003) and Merelava in Vanuatu (Turner et al. 1999). Additionally, lavas from back arc spreading centres in the Lau Basin behind the Tonga arc have been analyzed by Gill and Williams (1990) and Peate et al. (2001). These data are plotted on diagrams of (^{230}Th/^{238}U) and (^{226}Ra/^{230}Th) against depth to Benioff zone in Figure 20. Although there is much scatter, Figure 20a shows that whilst in general arc front lavas are characterized

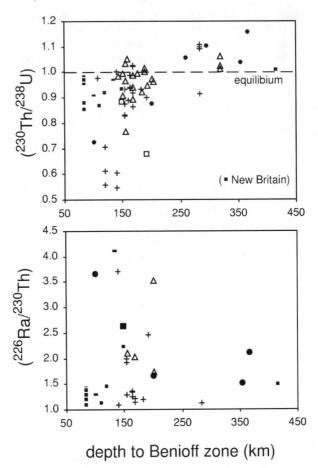

Figure 20. (a) (^{230}Th/^{238}U) and (b) (^{226}Ra/^{230}Th) versus depth to Benioff zone. Note that the age of the Lau Basin glasses (plotted as filled circles) is unknown and so the (^{226}Ra/^{230}Th) ratios should be treated as minimum values. Data are from New Britain (Gill et al. 1993), Kamchatka (Turner et al. 1998) and Sunda (Hoogewerff et al. 1997; Turner et al. 2001) arcs. Also plotted are average values for the volcanic front and several rear-arc volcanoes from the following arcs: Bogoslof - Aleutians (Turner et al. 1998, 2001; George et al. 2003), Lau Basin and Niuafo'ou - Tonga (Regelous et al. 1997; Turner et al. 1997, 2000a; Peate et al. 2001), Merelava in Vanuatu (Turner et al. 1999). Symbols and data sources as in Figure 1 unless specified differently.

by ^{238}U-excesses, these tend to decrease with increasing distance behind the arc. In principle, this could be consistent with decreasing U in the fluids, decreasing amount of fluid added to the wedge behind the volcanic front, increasing depth of melting, or any combination thereof. In contrast, there is some evidence that ^{226}Ra-excesses reach a maximum slightly behind the arc front (Fig. 20b, Gill et al. 1993; George et al. 2003) which may lend support to the model in which ^{226}Ra continues to in-grow in the subducted plate and be released beyond the point of maximum U flux. ^{226}Ra-excesses are also found far beyond the arc but these are associated with ^{230}Th-excesses suggesting that both disequilibria reflect partial melting effects more than fluid flux (Regelous et al. 1997; Turner et al. 1997, 1998, 2000a; Peate et al. 2001).

Magma Genesis at Convergent Margins 297

In detail, some of the trends within individual across-arc traverses are complicated. For example, within Kamchatka U/Th ratio increases with depth to the Benioff zone (Turner et al. 1998), although there is no simple correlation with (^{230}Th/^{238}U) (Fig. 20a). In Tonga, there is a relatively abrupt switch from ^{238}U-excesses to ^{230}Th-excesses some 250 km behind the arc front (Peate et al. 2001). H$_2$O/Ce ratios are still higher than MORB in the central and eastern Lau Basin and degree of melting is correlated with H$_2$O content, so H$_2$O demonstrably still plays an important role in back arc magma genesis (Stolper and Newman 1994; Peate et al. 2001). However, there is evidence that the composition of the fluids changes with distance from the arc in Tonga (Pearce et al. 1995; Peate et al. 2001) and, amongst the eastern Lau Basin lavas, the more H$_2$O-rich lavas have the larger ^{230}Th-excesses rather than ^{238}U-excesses (Peate et al. 2001). Finally, neither the dynamic or flux melting models readily explain rear-arc volcanism.

10. MODIFYING PROCESSES

This chapter is primarily concerned with the interpretation of U-series isotope disequilibria originating from arc lava sources and those created during partial melting. However, it is important to summarize the types of processes that could modify this signal and therefore potentially lead to erroneous interpretations.

10.1. Time since eruption

Sample age is clearly important, especially for ^{226}Ra-^{230}Th disequilibria. Ideally, only samples of known age should be analyzed or else U-series investigations should be combined with a geochronological study. Unfortunately, this is not always possible and samples ages have often only been estimated to be "< 10 ka," so as to imply that an age correction is unnecessary for Th isotopes, or as "this century" or, worst of all, "historic" for the purposes of ^{226}Ra. The appropriate decay equations can be used to investigate the magnitude of error introduced by a given age uncertainty. This issue is particularly important for the interpretation of ^{226}Ra-^{230}Th disequilibria in pre-historical samples. Note that the Ra data in Figures 10, 11a and 21 have been corrected to the date of eruption. In some cases, especially for Tonga, this involves assuming that the uppermost unvegetated lavas are from one of the most recent magmatic eruptions of the volcano, and correcting to the mean age of possible eruption ages. Assuming too old an eruption age increases apparent (^{226}Ra/^{230}Th) at the time of eruption. However, most samples in those figures are from eruptions this century with no major ambiguity about sample age.

10.2. Alteration and seawater interaction

Alteration is always a cause for concern in geochemical investigations and the best approach will always be to avoid samples with visual or chemical evidence for alteration. The differential fluid mobility of U, Th, Pa and Ra undoubtedly provides the potential for weathering or hydrothermal circulation to disturb the U-series signatures of arc lavas. In a study of lavas from Mt. Pelée on Martinique, Villemant et al. (1996) found that dome-forming lavas were in U-Th equilibrium whereas plinian deposits from the same eruptions had small ^{238}U-excesses which they interpreted to reflect hydrothermal alteration. However, whilst the addition of U could be due to hydrothermal alteration, the plinian deposits were also displaced to lower ^{230}Th/^{232}Th ratios which cannot. Instead, the two rock types may just be from separate magma batches.

Seawater interaction is a potential source of several U-series nuclides, and is strongly enriched in B, Cl and Sr. Thus, abundances of these elements and/or their isotopes can often be used to test for seawater contamination (e.g., Turner et al. 2000a). Because ^{234}U will be located in mineral sites that have been damaged by recoil effects, ^{234}U is likely to

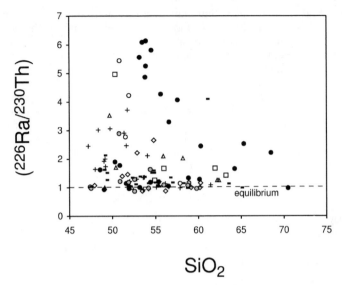

Figure 21. (^{226}Ra/^{230}Th) disequilibria versus SiO$_2$ as an index of differentiation (on this diagram lavas with SiO$_2$ > 60% been plotted). The general decrease in (^{226}Ra/^{230}Th) with increasing SiO$_2$ within most individual arc datasets suggests that basalts evolve to andesites and dacites within millennia.

be preferentially mobilized relative to ^{238}U (Chabaux et al. 2003). Seawater is enriched in ^{234}U and so ^{234}U-excesses may indicate recent seawater interaction in arc lavas. As shown by Yokoyama et al. (2002), seawater contamination can have a dramatic effect: a (^{234}U/^{238}U) only 3% higher than equilibrium (1.03) equates to 20% seawater contamination, which will shift (^{232}Th/^{238}U) from 0.6 to 0.45! Note that the revised decay constant for ^{234}U (Cheng et al. 2000) results in some published data giving equilibrium (^{234}U/^{238}U) more faithfully. Similarly, Bourdon et al. (2000b) showed that a (^{234}U/^{238}U) ratio only 1.5% higher than equilibrium can shift (^{231}Pa/^{235}U) from 2.5 to 3.4. Careful selection and leaching of submarine samples eliminates contamination by Mn-oxides which are enriched in ^{231}Pa and ^{230}Th (Bourdon et al. 2000b).

10.3. Magma chambers processes

Plutons provide undeniable evidence that rising magmas can stall in magma chambers and many magmatic processes can fractionate U-series nuclides and also potentially last for times long relative to their half-lives allowing for nuclide in-growth or decay. The encouraging thing is that the correlations described above are as good as they are because it means that such modification has been minor. Nonetheless, we now review several processes that seem to have demonstrable effects.

10.4. Radioactive decay

Radioactive decay during residence in magma chambers is critical for interpretation of the shorter lived isotopes such as ^{226}Ra. Several studies have shown negative correlations between (^{226}Ra/^{230}Th) and indices of fractionation (e.g., Gill and Williams 1990; Vigier et al. 2000; Turner et al. 2000a, 2001). In oceanic arcs, there is a fairly reproducible negative correlation between (^{226}Ra/^{230}Th) and SiO$_2$ (Fig. 21) suggesting that andesites and dacites may take a few millennia to evolve from basaltic parental magmas. Less commonly, some studies have suggested that magma residence times in oceanic arcs can be significant even relative to the half-life of ^{230}Th (Heath et al. 1998; Hawkesworth et al. 2000). In continental arcs, residence times also appear to be longer and fractionation combined with assimilation of pre-existing crystal mushes has been argued to produce rhyolites on time scales of 10^4-10^5 years (Reagan et al. 2003). However, since the subsurface residence times of magmas is specifically addressed elsewhere by Condomines et al. (2003), here we will restrict ourselves to briefly considering some of the processes which could occur during this residence to modify the deep level U-series isotope signals.

10.5. Crystal fractionation or accumulation

Crystal fractionation is the most obvious but since U, Th, Pa and Ra are all highly incompatible elements in gabbroic phases (Blundy and Wood 2003), U-series nuclides in arc basalts and andesites will be remarkably robust to the effects of crystal fractionation until the saturation of K-feldspar or accessory phases is attained. The same does not hold for evolved dacites and rhyolites because zircon, apatite and sphene all have the ability to strongly fractionate U-Th-Pa-Ra (Blundy and Wood 2003). Conversely, the bulk entrainment of gabbroic phases is also likely to have minimal effect upon U-series isotopes because of the orders of magnitude higher concentrations in the magma. As noted in Section 6.2, the very low D_{Plag} for Ra (Bundy and Wood 2003), means that Ra concentrations in plagioclase are 10's fg/g compared with 100's fg/g in the groundmass and so ^{226}Ra-excesses cannot reflect plagioclase accumulation. An exception is the behavior of the short-lived nuclide ^{222}Rn which is efficiently extracted by degassing in shallow-level magma chambers. Its longer lived daughter product ^{210}Pb may also be sensitive to extraction in a volatile phase (Condomines et al. 2003). Moreover, Pb can be moderately compatible in plagioclase (Blundy and Wood 2003). However, these properties have the exciting potential to be exploited to investigate the time scales of magma degassing (Gauthier and Condomines 1999).

10.6. Magma recharge

Recharge of a magma chamber with fresh inputs may trigger eruption (Sparks et al. 1977) and can significantly alter the U-series disequilibria of the resident magma either because the new input has not undergone decay or because the new input has a different source signal to the resident magma. The former case has been investigated numerically for a fixed volume system by Hughes and Hawkesworth (1999) who showed that magmas can be maintained out of secular equilibrium over residence times that are significant relative to the half-lives of ^{226}Ra and ^{230}Th by periodic mixing with influxes of new magma. This buffering effect is sensitive to the periodicity of the replenishment events, relative to the half-lives of ^{226}Ra and ^{230}Th, and to the ratio of input to chamber volume. These two parameters also control the size of oscillations in activity ratios and the time scale over which steady state conditions can be reached. If replenishment occurs on a time scale that is short relative to the half-life of ^{226}Ra then ^{230}Th-^{238}U disequilibria will be maintained close to its starting value and ^{226}Ra-^{230}Th will approach equilibrium as the replenishment time scale increases (Hughes and Hawkesworth 1999). In the case of a chamber with variable size, similar results were obtained by Vigier et al. (1999), who showed that the actual residence time of the magma could be longer than that calculated assuming simple closed-system decay.

10.7. Crustal contamination

The passage of magma through the crust and any storage time in magma chambers provides the opportunity for crustal interaction. As with any geochemical study of igneous rocks, the possible effects of crustal contamination have to be considered when attempting to interpret U-series disequilibria in terms of source processes. However, U-

Th isotopes have the advantage of being sensitive to contamination even by very young materials and thus can be usefully employed to investigate both the time scales of assimilation and the nature of the contaminant. We illustrate this latter potential with reference to selected case studies and the U-Th equiline diagram in Figure 22 recalling that mixing lines are linear on this diagram.

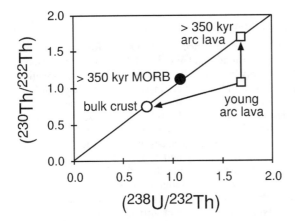

Figure 22. U-Th equiline diagram showing U-Th isotopes may be sensitive to the effects of assimilation of pre-existing (>350 kyr) arc basalts or continental crust. This is much more complicated using long-lived isotopic systems (e.g., $^{87}Sr/^{86}Sr$ or $^{143}Nd/^{144}Nd$) because in these systems pre-existing arc basalts will be indistinguishable from new magmas and crustal assimilation will be hard to distinguish from subducted sediment addition unless assimilation is coupled to differentiation.

In oceanic arcs, likely contaminants include the oceanic crust upon which the arc is built and the older arc lavas and sediments that comprise the upper arc crust. The long-lived radiogenic isotopes Sr, Nd and Pb are very poor tracers of the first two materials because they are too young to have evolved diagnostic isotope ratios and older arc lavas are likely to have similar long-lived isotope and trace element ratios to the younger ones of interest to U-series investigations. Equally, distinguishing contamination by ancient crustal materials from subducted sediment input is difficult. These features mean that contamination in oceanic arc lavas is usually a subtle feature that is hard to detect. However, the same is not necessarily true of Th isotopes because these change and reach equilibrium on a time scale of 350 kyr. Sediments and ancient continental crust have low U/Th and thus low ($^{230}Th/^{232}Th$), whereas since most arc lavas begin with ^{238}U-excesses, their $^{230}Th/^{232}Th$ evolution is conveniently in the opposite direction towards high ($^{230}Th/^{232}Th$). As illustrated on Figure 22, this may facilitate detection of contamination by such materials. Similarly, pre-existing arc lavas that are > 350 kyr old will have decayed up to the equiline and will have higher ($^{230}Th/^{232}Th$) ratios than the incoming arc lavas, although contamination with this material may be hard to distinguish from ageing of the incoming arc lavas during residence in magma chambers.

A further means by which U-series isotopes can be used to identify mixing or contamination processes is when the disequilibria in two parent-daughter systems with very different half-lives both decrease with increasing fractionation. For example, lavas from Galunggung in the Sunda arc trend towards both ($^{238}U/^{230}Th$) and ($^{226}Ra/^{230}Th$) = 1 with increasing SiO_2 (Turner and Foden 2001). Given the difference between the half-lives of these two isotope systems, these arrays cannot record the time taken for differentiation and the simplest explanation is that they reflect mixing between mantle derived magmas and silicic crustal melts or more evolved magmas in which U-Th-Ra isotopes have returned to secular equilibrium. These lavas show no change in $^{143}Nd/^{144}Nd$ with increasing SiO_2 (Turner and Foden 2001), and so the most likely candidate for the contaminant is small degree partial melts of pre-existing basalts. An important implication is that these crustal melts were not significantly out of ^{238}U-^{230}Th or ^{226}Ra-^{230}Th equilibrium and therefore that this type of crustal contamination does not mask the mantle source component signatures.

Continental arcs provide maximum potential for crustal assimilation especially where the crustal column is thick. Ancient continental crust has high time-integrated Th/U ratios and consequently much lower (^{230}Th/^{232}Th) ratios (~0.4-0.7) than the MORB source (~1.2). Therefore, Th isotopes are very sensitive to even small additions of ancient crustal material which will lead to a mixing trajectory toward low (^{230}Th/^{232}Th) ratios. If the wedge-sediment mix has a higher (^{230}Th/^{232}Th) ratio than the local crustal basement then contamination will result in displacement below this value. Thus, George et al. (2003) found in the Alaska-Aleutian arc that lavas erupted in the oceanic sector had (^{230}Th/^{232}Th) ratios equal to or higher than the subducting sediment whereas some lavas erupted through the continent on the northeastern Alaskan peninsula were significantly displaced to lower (^{230}Th/^{232}Th).

The Andes is the area for which Hildreth and Moorbath (1988) originally developed their MASH model (Melting, Assimilation, Storage and Homogenization) for crustal interaction. Here, lavas from the southern volcanic zone of Chile lie on continental crust of average thickness (35-40 km) and their O and Sr isotope signatures can generally be explained by sediment addition consistent with the presence of a ^{10}Be signal (Sigmarsson et al. 1990). The U-Th-Ra systematics of these lavas appear to be unaffected by crustal contamination (Sigmarsson et al. 1990, 2002). The preservation of ^{226}Ra-excesses in most of the southern volcanic zone lavas requires sufficiently rapid transit through the crust that there may have been no time for the formation of large magma chambers in which major contamination could take place. One exception to this is a lava from San Jose which lies on significantly thicker crust (55-60 km) and has elevated ^{87}Sr/^{86}Sr and ^{18}O isotopes indicative of crustal contamination. This lava is also characterized by ^{226}Ra-^{230}Th equilibrium and lies slightly above the U-Th isotope array for the remainder of the lavas (Sigmarsson et al. 1990, 2002) suggesting that it may have had a greater crustal residence time during which contamination occurred.

In contrast to the southern volcanic zone, Parinacota volcano lies on very thick continental crust (> 70 km) in the central volcanic zone of Chile. Bourdon et al. (2000a) showed that young Parinacota lavas encompass a wide range of U-Th disequilibria. ^{238}U-excesses were attributed to fluid addition to the mantle wedge but ^{230}Th-excesses in lavas from the same volcano are more difficult to explain. The lavas with ^{230}Th-excesses also have low (^{230}Th/^{232}Th) (< 0.6) characteristic of lower continental crust characterized by low Th/U and in their preferred model, Bourdon et al. (2000a) attributed the ^{230}Th-excesses to contamination by partial melts, formed in the presence of residual garnet, of old lower crustal materials.

11. TIME-INTEGRATED U-Th-Pb EVOLUTION OF THE CRUST-MANTLE

Convergent margins are generally considered to be the principle present-day tectonic setting where new continental crust is formed (~ 1.1 km^3/yr, Reymer and Schubert 1984). As illustrated on Figure 23, this new crustal material is characterized by Th/U ratios that are even lower than the Th/U ratio of the MORB mantle (2.6, Sun and McDonough 1989) yet the Th/U ratio of the bulk continental crust (3.9, Rudnick and Fountain 1995) is close to the Th/U ratio of the bulk silicate earth (see Bourdon and Sims 2003). There are several possible explanations for this paradox. Firstly, it is possible that the processes that formed the continental crust in the past were different to those in operation today. Since

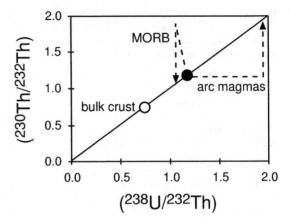

Figure 23. U-Th equiline diagram illustrating trajectories for present day MORB production (^{230}Th in-growth due to dynamic melting in the presence of aluminous clinopyroxene or garnet followed by ^{230}Th decay back to the equiline) and arc magma production (U addition followed by ^{230}Th in-growth back to the equiline due to ageing ± melting processes). Neither process seems to be able to account for the present day composition of bulk crust.

the mantle was some 200°C hotter, melting of the subducting oceanic curst is likely to have been more prevalent (Peacock et al. 1994) and this is the process that is thought to have produced TTG (tondjhemite-tonalite-granite) series in the Archaean crust (Defant and Drummond 1990). As discussed above, during slab melting, residual garnet retains U relative to Th and this is consistent both with the observation of ^{230}Th-excess in putative slab melts from the Austral Andean zone in southern Chile (Sigmarsson et al. 1998), and the need for residual garnet to explain the Sm/Nd ratio of the crust (O'Nions and McKenzie 1988). Secondly, it is likely that the surface of the Earth was less oxidizing in the Archaean such that U was less preferentially transported to oceans and recycled (Hawkesworth et al. 1997a,b).

Alternatively, it can be argued that U and Th are both highly incompatible during melting and that there is only minor fractionation between U and Th during the formation of *most* arc lavas which lie close to U-Th equilibrium (see Fig. 3). In that case, the higher Th/U ratio of the bulk continental crust would reflect a higher Th/U ratio of the mantle in the past when the bulk of the continental crust was formed. In this model, the present day low Th/U ratio of the upper mantle would be the result of more recent evolution of the mantle. The long-term evolution of Th/U ratios in mantle rocks can be tracked with both ^{208}Pb-^{206}Pb and ^{230}Th/^{232}Th systematics (Bourdon and Sims 2003). Pb is also preferentially removed to the crust at subduction zones leading to long term secular depletion of the mantle in Pb (Miller et al. 1994). Thus, Galer and O'Nions (1985) showed that MORB and OIB are characterized by present-day Th/U ratios that are lower than their time-integrated Th/U ratios inferred from Pb isotopes (the so-called kappa conundrum) and called upon a steady-state model for exchange between the lower and upper mantle to explain this observation. However, the operation of the mantle at steady-state for long-lived isotope systems has been questioned by Albarède (2001) and alternative explanations have been suggested by Zartmann and Haines (1988), McCulloch (1993) and Elliott et al. (1999). As shown by Elliott et al. (1999), an alternative explanation to explain the lower Th/U ratio of the upper mantle relative to the time-integrated Th/U ratio from Pb isotopes, is to recycle preferentially U relative to Th in subduction zones. Estimates of recycling fluxes based on Th-U disequilibrium data for the Marianas (Elliott et al. 1997, 1999) have shown that U is preferentially recycled in the mantle compared with Th. However, this will only happen when U is in its fluid-mobile U^{6+} form and this would have commenced only when the atmosphere and hydrosphere became sufficiently oxidizing some time in the Post-Archaean Earth (Holland 1984). The actual reasons for build up of oxygen in the atmosphere are still a hotly debated issue (e.g., Kasting 2001) but it is important to note how the chemistry of the deep-earth might be fundamentally affected by the chemistry of the atmosphere and how the fractionation of elements such as U and Th is only enhanced in the hydrosphere and shallow Earth where they have largely different behavior compared with the reducing conditions of the deep mantle.

12. FUTURE WORK

The use of U-series disequilibria in unraveling the physical processes of fluid transfer, partial melting, melt migration and modification at convergent margins is still a new and rapidly expanding field of research. Indeed, some of the full implications of the available data are only just beginning to be appreciated and many have barely been explored. In tandem with this, advances in analytical techniques are allowing for more rapid and precise analysis of ever smaller amounts of materials. These will inevitably lead to new data sets which can only help improve our understanding of convergent margin processes. It must be stressed that these need to be undertaken upon fully characterized and well dated, ideally primitive lavas. Studies of fore-arc and back-arc seamounts would help to map fluid release and melting effects along the slab. However, these data will need to be combined with better numerical models if their full significance is to be realized and an outstanding need in this regard is for a testable melting model for arcs. This will need to be evolved with geophysical constraints and independent information about the P, T and H_2O conditions of melting beneath arcs. For example, the flux melting models imply a testable link between melting rate and subduction velocity. While the current 1-D models seem to explain many of the U-series observations in arcs, they certainly do not capture some of the complexity that should arise in 2- or 3-D models. Thus, there are many potential rewards available from future melt and fluid inclusion studies on primitive arc lavas. Finally, additional experiments on fluid-mineral partition coefficients at the appropriate conditions and mantle melting under hydrous conditions also need to be conducted to allow for better forward modeling of the U-series data.

ACKNOWLEDGMENTS

Our present understanding of U-series disequilibria in arc lavas has benefited from interaction with too many people to acknowledge them all individually. We are especially grateful to Claude Allègre, Jon Davidson, Tony Dosseto, Tim Elliott, Chris Hawkesworth, Terry Plank, Mark Reagan, and the participants of the State Of The Arc meetings for many lively discussions over the years. For Simon Turner this represents the culmination of an 8 year research fellowship from the Royal Society which he gratefully acknowledges. Simon Turner also thanks IPGP for funding a one month visit during which the initial version of this chapter was written. Mark Reagan, Marc Hirschmann and Olgeir Sigmarsson provided careful reviews which helped to improve the final version.

REFERENCES

Albarède F (2001) Radiogenic ingrowth in systems with multiple reservoirs: applications to the differentiation of the mantle-crust system. Earth Planet Sci Lett 189:59-73
Aharonov E, Whitehead JA, Kelemen PB, Spiegelman M (1995) Channeling instability of upwelling melt in the mantle. J Geophys Res 100:20433-20450
Allègre CJ, Condomines M (1982) Basalt genesis and mantle structure studied through Th-isotopic geochemistry. Nature 299:21-24
Allègre CJ, Dupré B, Lewin E (1986) Thorium/uranium ratio of the Earth. Chem Geol 56:219-227

Alves S, Schiano P, Capmas F, Allègre CJ (2002) Osmium isotope binary mixing arrays in arc volcanism. Earth Planet Sci Lett 198:355-369

Ayers JC, Dittmer SK, Layne GD (1997) Partitioning of elements between peridotite and H_2O at 2.0-3.0 GPa and 900-1000°C, and application to models of subduction zone processes. Earth Planet Sci Lett 150:381-398

Bebout GE, Barton MD (2002) Tectonic and metasomatic mixing in a high-T subduction zone melange - insights into the geochemical evolution of the slab-mantle interface. Chem Geol 187:79-106

Becker H, Jochum KP, Carlson RW (1999) Constraints from high-pressure veins in eclogites on the composition of hydrous fluid in subduction zones. Chem Geol 160:291-308

Bennett JT, Krishnaswami S, Turekian KK, Melson WG, Hopson CA (1982) The uranium and thorium decay series nuclides in Mt. St. Helens effusives. Earth Planet Sci Lett 60:61-69

Billen MI, Gurnis M (2001) A low viscosity wedge in subduction zones. Earth Planet Sci Lett 193:227-236

Black S, Macdonald R, DeViro B, Kilburn CRJ, Rolandi G (1998) U-series disequilibria in young (A.D. 1944) Vesuvius rocks: Preliminary implications for magma residence times and volatile addition. J Volcanol Geotherm Res 82:97-111

Blatter DL, Carmichael ISE (1998) Hornblende peridotite xenoliths from central Mexico reveal the highly oxidized nature of subarc upper mantle. Geology 26:1035-1038

Blundy J, Wood B (2003) Mineral-melt partitioning of uranium, thorium and their daughters. Rev Mineral Geochem 52:59-123

Bourdon B, Turner S, Allègre C (1999) Melting dynamics beneath the Tonga-Kermadec island arc inferred from ^{231}Pa-^{235}U systematics. Science 286:2491-2493

Bourdon B, Wörner G, Zindler A (2000a) U-series evidence for crustal involvement and magma residence times in the petrogenesis of Parinacota volcano, Chile. Contrib Mineral Petrol 139:458-469

Bourdon B, Bourlès D, Goldstein SJ, Murrell MT, Langmuir CH, Allègre CJ (2000b) Evidence from ^{10}Be and U-series disequilibria on the possible contamination of MORB glasses by sedimentary material. Geochem Geophys Geosyst 2000GC000047

Bourdon B, Turner S, Dosseto A (2003) Dehydration and partial melting in subduction zones: constraints from U-series disequilibria. J Geophys Res (in press).

Bourdon B, Sims KWW (2003) U-series constraints on intraplate basaltic magmatism. Rev Mineral Geochem 52:215-254

Blot C (1972) Volcanisme et séismes du manteau supérieur dans l'Archipel des Nouvelles-Hebrides. Bull Volcanol 36:446-461

Brémond d'Ars J, Jaupart C, Sparks RSJ (1995) Distribution of volcanoes in active margins. J Geophys Res 100:20421-20432

Brenan JM, Shaw HF, Phinney DL, Ryerson FJ, (1994) Rutile-aqueous fluid partitioning of Nb, Ta, Hf, Zr, U and Th: implications for high field strength element depletions in island-arc basalts. Earth Planet Sci Lett 128:327-339

Brenan JM, Shaw HF, Ryerson FJ, Phinney DL (1995) Mineral-aqueous fluid partitioning of trace elements at 900 °C and 2.0 GPa: constraints on the trace element chemistry of mantle and deep crustal fluids. Geochim Cosmochim Acta 59:3331-3350

Brophy JG, Whittington CS, Park Y-R (1999) Sector-zoned augite megacrysts in Aleutian high alumina basalts: implications for the conditions of basalt crystallization and the generation of calc-alkaline series magmas. Contrib Mineral Petrol 135:277-290

Capaldi G, Del Pezzo E, Ghiara MR, Guerra I, La Volpe L, Lirer L, Lo Basico A, Luongo G, Martini M, Munno R, Pece R, Rapolla A, Scarpa R (1978) Stromboli and its 1977 eruption. Bull Volcanol 41:259-285

Capaldi G, Cortini M, Pece R (1982) Th isotopes at Vesuvius: evidence for open-system behavior of magma-forming processes. J Volcanol Geotherm Res 14:247-260

Capaldi G, Cortini M, Pece R (1983) U and Th decay-series disequilibria in historical lavas from the Eolian islands, Tyrrhenian Sea. Isot Geosci 1:39-55

Chabaux F, Hémond C, Allègre CJ (1999) ^{238}U-^{230}Th-^{226}Ra disequilibria in the Lesser Antilles arc: implications for mantle metasomatism. Chem Geol 153:171-185

Cheng H, Edwards RL, Hoff J, Gallup CD, Richards DA, Asmerom Y (2000) The half-lives of uranium-234 and thorium-230. Chem Geol 169:17-33

Clark SK, Reagan MK, Plank T (1998) Trace element and U-series systematics for 1963-1965 tephras from Irazú volcano, Costa Rica: implications for magma generation processes and transit times. Geochim Cosmochim Acta 62:2689-2699

Class C, Miller DM, Goldstein SL, Langmuir CH (2000) Distinguishing melt and fluid subduction components in Umnak volcanics. Aleutian arc. Geochem Geophys Geosys 1: paper number 1999GC000010

Condomines M, Sigmarsson O (1993) Why are so many arc magmas close to ^{238}U-^{230}Th radioactive equilibrium? Geochim Cosmochim Acta 57:4491-4497

Crisp JA (1984) Rates of magma emplacement and volcanic output. J Volcanol Geotherm Res 20:177-211

Davies JH (1999) The role of hydraulic fractures and intermediate-depth earthquakes in generating subduction-zone magmatism. Nature 398:142-145

Davies JH, Bickle MJ (1991) A physical model for the volume and composition of melt produced by hydrous fluxing above subduction zones. Phil Trans R Soc Lond 335:355-364

Davies JH, Stevenson DJ (1992) Physical model of source region of subduction zone volcanics. J Geophys Res 97:2037-2070

Defant MJ, Drummond MS (1990) Derivation of some modern arc magmas by melting of young subducted lithosphere. Nature 347:662-665

Dosseto A, Bourdon B, Joron JL, Dupré B (2003) U-Th-Pa-Ra study of the Kamchatka arc: new constraints on the genesis of arc lavas. Geochim Cosmochim Acta (submitted)

Druitt TH, Edwards L, Mellors RM, Pyle DM, Sparks RSJ, Lanphere M, Davies M, Barriero B (1999) Santorini Volcano. Geol Soc Lond Mem 19

Eiler, JM, Crawford A, Elliott T, Farley KA, Valley JW, Stolper EM (2000) Oxygen isotope geochemistry of oceanic-arc lavas. J Petrol 41:229-256

Ellam RM, Hawkesworth CJ (1988) Elemental and isotopic variations in subduction related basalts: evidence for a three component model. Contrib Mineral Petrol 98:72-80

Elkins Tanton LT, Grove TL, Donnelly-Nolan J (2001) Hot, shallow mantle melting under the Cascades volcanoc arc. Geology 29:631-634

Elliott T, Plank T, Zindler A, White W, Bourdon B (1997) Element transport from slab to volcanic front at the Mariana arc. J Geophys Res 102:14991-15019

Elliott T, Zindler A, Bourdon B (1999) Exploring the Kappa conundrum: the role of recycling in the lead isotope evolution of the mantle. Earth Planet Sci Lett 169:129-145

Elliott T, Heumann A, Koetsier G (2001) U-series constraints on melting beneath the Marianas. In: Intra-oceanic subduction systems: Tectonic and Magmatic Processes. Geol Soc Lond Abstr

England PC (2001) Why are the arc volcanoes where they are? EOS Trans, Amer Geophys Union 82:F1156

Ewart A, Hawkesworth CJ (1987) The Pleistocene-Recent Tonga-Kermadec arc lavas: Interpretation of new isotopic and rare earth data in terms of a depleted mantle source model. J Petrol 28:495-530

Furukawa F (1993a) Magmatic processes under arcs and formation of the volcanic front. J Geophys Res 98:8309-8319

Furukawa F (1993b) Depth of the decoupling plate interface and thermal structure under arcs. J Geophys Res 98:20,005-20,013

Galer SJG, O'Nions RK (1985) Residence time of thorium, uranium and lead in the mantle with implications for mantle convection. Nature 316:778-782.

Gauthier P-J, Condomines M (1999) ^{210}Pa-^{226}Ra radioactive disequilibria in recent lavas and radon degassing: inferences on the magma chamber dynamics at Stromboli and Merapi volcanoes. Earth Planet Sci Lett 172:111-126

George R, Turner S, Hawkesworth C, Morris J, Nye C, Ryan J, Zheng S-H (2003) Melting processes and fluid and sediment transport rates along the Alaska-Aleutian arc from an integrated U-Th-Ra-Be isotope study. J Geophys Res (in press).

Gill JB (1981) Orogenic Andesites and Plate Tectonics. Springer-Verlag, New York

Gill JB, Williams RW (1990) Th isotope and U-series studies of subduction-related volcanic rocks. Geochim Cosmochim Acta 54:1427-1442

Gill JB, Pyle DM, Williams RW (1992) Igneous rocks. In: Uranium-series disequilibrium. Ivanovich M, Harmon RS (eds) Oxford University Press, Oxford, p 207-258.

Gill JB, Morris JD, Johnson RW (1993) Timescale for producing the geochemical signature of island arc magmas: U-Th-Po and Be-B systematics in Recent Papua New Guinea lavas. Geochim Cosmochim Acta 57:4269-4283

Goldstein SJ, Stirling CH (2003) Techniques for measuring uranium-series nuclides: 1992-2002. Rev Mineral Geochem 52:23-57

Grove TL, Parman SW, Bowring SA, Price RC, Baker MB (2002) The role of an H_2O-rich fluid component in the generation of primitive basaltic andesites and andesites from the Mt. Shasta region, N. California. Contrib Mineral Petrol 142:375-396

Guillaumont R, Bouissières G, Muxart Y (1968) Protactinium chemistry. 1. Aqueous solutions for penta and tetravalent protactinium. Actinid Rev 1:135

Hall PS, Kincaid C (2001) Diapiric Flow at Subduction Zones: A Recipe for Rapid Transport. Science 292:2472-2475

Harris DM, Anderson AT (1984) Volatiles H_2O, CO_2, and Cl in a subduction related basalt. Contrib Mineral Petrol 87:120-128
Hauri EH (1997) Melt migration and mantle chromatography, 1:simplified theory and conditions for chemical and isotopic decoupling. Earth Planet Sci Lett 153:1-19
Hawkesworth CJ, Gallagher K, Hergt JM, McDermott F (1993) Mantle and slab contributions in arc magmas. Ann Rev Earth Planet Sci 21:175-204
Hawkesworth C, Turner S, Peate D, McDermott F, van Calsteren P (1997a) Elemental U and Th variations in island arc rocks: implications for U-series isotopes. Chem Geol 139:207-222
Hawkesworth CJ, Turner SP, McDermott F, Peate DW, van Calsteren P (1997b) U-Th isotopes in arc magmas: implications for element transfer from the subducted crust. Science 276:551-555
Hawkesworth C, Blake S, Evans P, Hughes R, Macdonald R, Thomas L, Turner S, Zellmer G (2000) The time scales of crystal fractionation in magma chambers - integrating physical, isotopic and geochemical perspectives. J Petrol 41:991-1006
Heath E, Turner SP, Macdonald R, Hawkesworth CJ, van Calsteren P (1997) Long magma residence times at an island arc volcano (Soufriere, St. Vincent) in the Lesser Antilles: evidence from ^{238}U-^{230}Th isochron dating. Earth Planet Sci Lett 160:49-63
Herrstrom EA, Reagan MK, Morris JD (1995) Variations in lava composition associated with flow of asthenosphere beneath southern Central America. Geology 23:617-620
Hildreth W, Moorbath S (1988) Crustal contributions to arc magmatism in the Andes of Central Chile. Contrib Mineral Petrol 98:455-489
Hirschmann MM, Asimow PD, Ghiorso MS, Stolper EM (1999) Calculation of peridotite partial melting from thermodynamic models of minerals and melts. III Controls on isobaric melt production and the effect of water on melt production. J Petrol 40:831-851
Hochstaedter A, Gill J, Peters R, Broughton P, Holden P, Taylor B (2001) Across-arc geochemical trends in the Izu-Bonin arc: contributions from the subducting slab. Geochem Geophys Geosys 2:paper number 2000GC000105
Hole MJ, Saunders AD, Marriner GF, Tarney J (1984) Subduction of pelagic sediments: implications for the origin of Ce-anomalous basalts from the Mariana islands. J Geol Soc London 141:453-472
Holland HD (1984) The Chemical Evolution of the Atmosphere and Oceans. Princeton University Press, Princeton
Hoogewerff JA, van Bergen MJ, Vroon PZ, Hertogen J, Wordel R, Sneyers A, Nasution A, Varekamp JC, Moens HLE, Mouchel D (1997) U-series, Sr-Nd-Pb isotope and trace-element systematics across an active island arc-continent collision zone: implications for element transfer at the slab-wedge interface. Geochim Cosmochim Acta 61:1057-1072
Hughes RD, Hawkesworth CJ (1999) The effects of magma replenishment processes on ^{238}U-^{230}Th disequilibrium. Geochim Cosmochim Acta 63:4101-4110
Ivanovich M, Harmon RS (1992) Uranium-series disequilibrium. Oxford University Press, Oxford
Jarrard RD (1986) Relations among subduction parameters. Rev Geophys 24:217-284
Johnson MC, Plank T (1999) Dehydration and melting experiments constrain the fate of subducted sediments. Geochem Geophys Geosys 1:paper number 1999GC000014
Kasting JF (2001) The rise of atmospheric oxygen. Science 293:819-820
Kay RW (1980) Volcanic arc magmas: implications of a melting-mixing model for element recycling in the crust-upper mantle system. J Geol 88:497-522
Kelemen PB, Johnson KTM, Kinzler RJ, Irving AI (1990) High-field strength element depletions in arc basalts due to mantle-magma interaction. Nature 345:521-524
Kelemen PB, Yogodinsky GM, Scholl DW (2003) Along strike variation in lavas of the Aleutian island arc: implications for the genesis of high Mg# andesite and the continental crust. *In*: The Subduction Factory. Eiler JM (ed) AGU Geophys Monogr Ser (in press)
Keppler H, Wyllie P (1990) Role of fluids in transport and fractionation of uranium and thorium in magmatic processes. Nature 348:531-533
Keppler H (1996) Constraints from partitioning experiments on the composition of subduction-zone fluids. Nature 380:237-240
Kincaid C, Sacks IS (1997) Thermal and dynamic evolution of the upper mantle in subduction zones. J Geophys Res 102:12,295-12,315
Kirby S, Engdahl ER, Denlinger R (1996) Intermediate-depth intraslab earthquakes and arc volcanism as physical expressions of crustal and uppermost mantle metamorphism in subducting slabs. *In*: Subduction Top to Bottom. Bebout GE et al. (eds) AGU Geophys Monogr Ser 96:195-214
La Tourette T, Hervig RL, Holloway JR (1995) Trace element partitioning between amphibole, phlogopite and basanite melt. Earth Planet Sci Lett 135:13-30

Lundstrom CC, Shaw H, Ryerson F, Phinney D, Gill J, Williams Q (1994) Compositional controls on on the partitioning of U, Th, Ba, Pb, Sr and Zr between clinopyroxene and haplobasaltic melts; implications for uranium series disequilibria in basalts. Earth Planet Sci Lett 128:407-423

Lundstrom CC (2003) Uranium-series disequilibria in mid-ocean ridge basalts: observations and models of basalt genesis. Rev Mineral Geochem 52:175-214

Marsh BD (1987) Petrology and evolution of the N.E. Pacific including the Aleutians. Pacific Rim Congress 87:309-315

Maury RE, Defant MJ, Joron J-L (1992) Metasomatism of the sub-arc mantle inferred from trace elements in Philippine xenoliths. Nature 360:661-663

McCulloch MT (1993) The role of subducted slabs in an evolving earth. Earth Planet Sci Lett 115:89-100

McCulloch MT, Gamble JA (1991) Geochemical and geodynamical constraints on subduction zone magmatism. Earth Planet Sci Lett 102:358-374

McDermott F, Hawkesworth C (1991) Th, Pb, and Sr isotope variations in young island arc volcanics and oceanic sediments. Earth Planet Sci Lett 104:1-15

McDermott F, Defant MJ, Hawkesworth CJ, Maury RC, Joron JL (1993) Isotope and trace element evidence for three component mixing in the genesis of the North Luzon arc lavas (Philippines). Contrib Mineral Petrol 113:9-23

McKenzie D (1985) The extraction of magma from the crust and mantle. Earth Planet Sci Lett 74:81-91

McKenzie D (1985) ^{230}Th-^{238}U disequilibrium and the melting process beneath ridge axes. Earth Planet Sci Lett 72:149-157

McKenzie D (2000) Constraints on melt generation and transport from U-series activity ratios. Chem Geol 162:81-94

Miller DM, Goldstein SL, Langmuir CH (1994) Cerium/lead and lead isotope ratios in arc magmas and the enrichment of lead in the continents. Nature 368:514-520

Morris JD, Hart SR (1983) Isotopic and incompatible element constraints on the genesis of island arc volcanics from Cold Bay and Amak Island, Aleutians, and implications for mantle structure. Geochim Cosmochim Acta 47:2015-2030

Morris JD, Leeman BW, Tera F (1990) The subducted component in island arc lavas: constraints from Be isotopes and B-Be systematics. Nature 344:31-36

Mysen BO, Kushiro I, Fujii T, (1978) Preliminary experimental data bearing on the mobility of H_2O in crystalline upper mantle. Carnegie Inst Washington Yearbook 77:793-797

Navon O, Stolper E (1987) Geochemical consequences of melt percolation: the upper mantle as a chromatographic column. J Geol 95:285-307

Newman S, Macdougall JD, Finkel RC (1984) ^{230}Th-^{238}U disequilibrium in island arc lavas: evidence from the Aleutians and the Marianas. Nature 308:266-270

Newman S, Macdougall JD, Finkel RC (1986) Petrogenesis and ^{230}Th-^{238}U disequilibrium at Mt. Shasta, California, and in the Cascades. Contrib Mineral Petrol 93:195-206

Nichols GT, Wyllie PJ Stern CR (1994) Subduction zone melting of pelagic sediments constrained by melting experiments. Nature 371:785-788

O'Nions RK, McKenzie DP (1988) Melting and continent generation. Earth Planet Sci Lett 90:449-456

Parkinson IJ, Arculus RJ (1999) The redox state of subduction zones: insights from arc-peridotites. Chem Geol 160:409-423

Peacock SM (1996) Thermal and petrologic structure of subduction zones. *In*: Subduction Top to Bottom. Bebout GE et al. (eds) AGU Geophys Monogr Ser 96:119-133

Peacock SM, Rushmer T, Thompson AB (1994) Partial melting of subducting oceanic crust. Earth Planet Sci Lett 121:227-244

Pearce JA, Parkinson IJ (1993) Trace element models for mantle melting: application to volcanic arc petrogenesis. J Geol Soc Lond 76:373-403

Pearce JA, Ernewein M, Bloomer SH, Parson LM, Murton BJ, Johnson LE (1995) Geochemistry of Lau Basin volcanic rocks: influence of ridge segmentation and arc proximity. *In*: Volcanism associated with extension at consuming plate margins. Smellie JL (ed) Geol Soc Lond Spec Publ 81:53-75

Peate DW, Kokfelt TF, Hawkesworth CJ, van Calsteren PW, Hergt JM, Pearce JA (2001) U-series isotope data on Lau Basin glasses: the role of subduction-related fluids during melt generation in back-arc basins. J Petrol 42:1449-1470

Pickett DA, Murrell MT (1997) Observations of ^{231}Pa/^{235}U disequilibrium in volcanic rocks. Earth Planet Sci Lett 148:259-271

Plank T (1993) Mantle melting and crustal recycling in subduction zones. PhD Dissertation, Columbia University, New York City, New York

Plank T, Langmuir CH (1988) An evaluation of the global variations in the major element chemistry of arc basalts. Earth Planet Sci Lett 90:349-370

Plank T, Langmuir CH (1993) Tracing trace elements from sediment input to volcanic output at subduction zones. Nature 362:739-743

Plank T, Langmuir CH (1998) The chemical composition of subducting sediment and its consequences for the crust and mantle. Chem Geol 145:325-394

Poli S, Schmidt MW (1995) H_2O transport and release in subduction zones: experimental constraints on basaltic and andesitic systems. J Geophys Res 100:22,299-22,314

Pyle DM, Ivanovich M, Sparks RSJ (1988) Magma-cumulate mixing identified by U-Th disequilibrium dating. Nature 331:157-159

Reagan MK, Gill JB (1989) Coexisting calkalkaline and high-Nb basalts from Turrialba volcano, Costa Rica: implications for residual titanites in arc magma sources. J Geophys Res 94:4619-4633

Reagan MK, Morris JD, Herrstrom EA, Murrell MT (1994) Uranium series and beryllium isotope evidence for an extended history of subduction modification of the mantle below Nicaragua. Geochim Cosmochim Acta 58:4199-4212

Reagan MK, Sims KW, Erich J, Thomas RB, Cheng H, Edwards RL, Layne G, Ball L (2003) Timescales of differentiation from mafic parents to rhyolite in North American continental arcs. J Petrol (in press)

Regelous M, Collerson KD, Ewart A, Wendt JI (1997) Trace element transport rates in subduction zones: evidence from Th, Sr and Pb isotope data for Tonga-Kermadec arc lavas. Earth Planet Sci Lett 150:291-302

Reymer A, Schubert G (1984) Phanerozoic addition rates to the continental crust and crustal growth. Tectonics 3:63-77

Rubin KH, Wheller GE, Tanzer MO, MacDougall JD, Varne R, Finkel R (1989) ^{238}U decay series systematics of young lavas from Batur volcano, Sunda arc. J Volcanol Geotherm Res 38:215-226

Rudnick RL, Fountain DM (1995) Nature and composition of the continental crust: a lower crustal perspective. Rev Geophys 33:267-309

Ryan JG, Morris JD, Tera F, Leeman WP, Tsvetkov A (1995) Cross-arc geochemical variations in the Kurile arc as a function of slab depth. Science 270:625-627

Ryerson FJ, Watson EB (1987) Rutile saturation in magmas: implications for Ti-Nb-Ta depletion in island-arc basalts. Earth Planet Sci Lett 86:225-239

Schaefer SJ, Sturchio NC, Murrell MT, Williams SN (1993) Internal ^{238}U-series systematics of pumice from the November 13, 1985, eruption of Nevado del Ruiz, Columbia. Geochim Cosmochim Acta 57:1215-1219

Schmidt MW, Poli S (1998) Experimentally based water budgets for dehydrating slabs and consequences for arc magma generation. Earth Planet Sci Lett 163:361-379

Sigmarsson O, Condomines M, Morris JD, Harmon RS (1990) Uranium and ^{10}Be enrichments by fluids in Andean arc magmas. Nature 346:163-165

Sigmarsson O, Martin H, Knowles J (1998) Melting of a subducting oceanic crust from U-Th disequilibria in austral Andean lavas. Nature 394:566-569

Sigmarsson O, Chmeleff J, Morris J, Lopez-Escobar L (2002) Origin of ^{226}Ra-^{230}Th disequilibria in arc lavas from southern Chile and magma transfer time. Earth Planet Sci Lett 196:189-196

Sisson TW, Layne GD (1993) H_2O in basalt and basaltic andesite inclusions from four subduction-related volcanoes. Earth Planet Sci Lett 117:619-635

Sisson TW, Bronto S (1998) Evidence for pressure-release melting beneath magmatic arcs from basalt at Galunggung, Indonesia. Nature 391:883-886

Smith GP, Weins DA, Fischer KM, Dorman LM, Webb SC, Hildebrand JA (2001) A complex pattern of mantle flow in the Lau backarc. Science 292:713-716

Sparks RSJ, Sigurdsson H, Wilson L (1977) Magma mixing: a mechanism of triggering acid explosive eruptions. Nature 267:315-318

Spiegelman M, McKenzie D (1987) Simple 2-D models for melt extraction at mid-ocean ridges and island arcs. Earth Planet Sci Lett 83:137-152

Staudigel H, Plank T, White W, Schminke H-U (1996) Geochemical fluxes driving seafloor alteration of the basaltic upper oceanic crust: DSDP sites 417 and 418. In: Subduction: top to bottom. Bebout GE, Scholl DW, Kirby SH, Platt JP (eds) Am Geophys Union Geophys Monogr 96:19-38

Stolper E, Newman S (1994) The role of water in the petrogenesis of Mariana trough magmas. Earth Planet Sci Lett 121:293-325

Sun SS, McDonough WF (1989) Chemical and isotopic systematics of oceanic basalts: implications for mantle composition and processes. In: Magmatism in ocean basins. Saunders AD, Norry MJ (eds) Geol Soc Lond Spec Publ 42:313-345

Tatsumi Y, Hamilton DL, Nesbitt RW, (1986) Chemical characteristics of fluid phase released from a subducted lithosphere and origin of arc magmas: evidence from high-pressure experiments and natural rocks. J Volcanol Geotherm Res 29:293-309

Taylor RS, McLennan SM (1981) The composition and evolution of the continental crust: rare earth element evidence from sedimentary rocks. Phil Trans R Soc Lond 301:381-399

Thomas RB, Hirschmann MM, Cheng H, Reagan MK, Edwards RL (2002) (^{231}Pa/^{235}U)-(^{230}Th/^{238}U) of young mafic volcanic rocks from Nicaragua and Costa Rica and the influence of flux melting on U-series systematics of arc lavas. Geochim Cosmochim Acta 66:4287-4309

Turner S, Hawkesworth C, van Calsteren P, Heath E, Macdonald R, Black S (1996) U-series isotopes and destructive plate margin magma genesis in the Lesser Antilles. Earth Planet Sci Lett 142:191-207

Turner S, Hawkesworth C (1997) Constraints on flux rates and mantle dynamics beneath island arcs from Tonga-Kermadec. Nature 389:568-573

Turner S, Hawkesworth C, Rogers N, Bartlett J, Worthington T, Hergt J, Pearce J, Smith I (1997) ^{238}U-^{230}Th disequilibria, magma petrogenesis and flux rates beneath the depleted Tonga-Kermadec island arc. Geochim Cosmochim Acta 61:4855-4884

Turner S, Hawkesworth C (1998) Using geochemistry to map mantle flow beneath the Lau Basin. Geology 26:1019-1022

Turner S, McDermott F, Hawkesworth C, Kepezhinskas P (1998) A U-series study of lavas from Kamchatka and the Aleutians: constraints on source composition and melting processes. Contrib Mineral Petrol 133:217-234

Turner SP, Peate DW, Hawkesworth CJ, Eggins SM, Crawford AJ (1999) Two mantle domains and the time scales of fluid transfer beneath the Vanuatu arc. Geology 27:963-966

Turner S, Bourdon B, Hawkesworth C, Evans P, (2000a) ^{226}Ra-^{230}Th evidence for multiple dehydration events, rapid melt ascent and the time scales of differentiation beneath the Tonga-Kermadec island arc. Earth Planet Sci Lett 179:581-593

Turner SP, George RMM, Evans PE, Hawkesworth CJ, Zellmer GF (2000b) Time-scales of magma formation, ascent and storage beneath subduction-zone volcanoes. Phil Trans R Soc Lond 358:1443-1464

Turner S, Foden J (2001) U, Th and Ra disequilibria, Sr, Nd and Pb isotope and trace element variations in Sunda arc lavas: predominance of a subducted sediment component. Contrib Mineral Petrol 142:43-57

Turner S, Evans P, Hawkesworth C, (2001) Ultra-fast source-to-surface movement of melt at island arcs from ^{226}Ra-^{230}Th systematics. Science 292:1363-1366

Turner S, Foden J, George R, Evans P, Varne R, Elburg M, Jenner G (2003) Rates and processes of potassic magma evolution beneath Sangeang Api volcano, east Sunda arc, Indonesia. J Petrol (in press)

Vigier N, Bourdon B, Joron JL, Allègre CJ (1999) U-decay series and trace element systematics in the 1978 eruption of Ardoukoba, Asal rift: timescale of magma crystallisation. Earth Planet Sci Lett 174:81-97

Villemant B, Flehoc C (1989) U-Th fractionation by fluids in K-rich magma genesis: the Vico volcano, central Italy. Earth Planet Sci Lett 91:312-326

Villemant B, Boudon G, Komorowski JC (1996) U-series disequilibrium in arc magmas induced by water-magma interaction. Earth Planet Sci Lett 140:259-267

Volpe AM, Hammond PE (1991) ^{238}U-^{230}Th-^{226}Ra disequilibrium in young Mt. St. Helens rocks: time constraint for magma formation and crystallization. Earth Planet Sci Lett 107:475-486

Volpe A M (1992) ^{238}U-^{230}Th-^{226}Ra disequilibrium in young Mt. Shasta andesites and dacites. J Volcanol Geotherm Res 53:227-238

Wendt JI, Regelous M, Collerson KD, Ewart A (1997) Evidence for a contribution from two mantle plumes to island arc lavas from northern Tonga. Geology 25:611-614

Williams RW, Gill JB (1989) Effects of partial melting on the uranium decay series. Geochim Cosmochim Acta 53:1607-1619

Woodhead J, Eggins S, Gamble J (1993) High field strength and transition element systematics in island arc and back-arc basin basalts: evidence for multi-phase melt extraction and a depleted mantle wedge. Earth Planet Sci Lett 114:491-504

Woodhead JD, Eggins SM, Johnson RW (1998) Magma genesis in the New Britain arc: further insights into melting and mass transfer processes. J Petrol 39:1641-1668

Woodland SJ, Pearson DG, Thirlwall MF (2002) A Platinum Group Element and Re–Os Isotope Investigation of Siderophile Element Recycling in Subduction Zones: Comparison of Grenada, Lesser Antilles Arc, and the Izu–Bonin Arc. J Petrol 43:171-198

Wood BJ, Bryndzia LT, Robinson JAC (1990) Mantle oxidation state and its relationship to tectonic environment and fluid speciation. Science 248:337-345

Wood BJ, Blundy JD, Robinson JAC (1999) The role of clinopyroxene in generating U-series disequilibrium during mantle melting. Geochim Cosmochim Acta 63:1613-1620

Wyss M, Hasegawa A, Nakajima J (2001) Source and path of magma for volcanoes in the subduction zone of northeastern Japan. Geophys Res Lett 28:1819-1822

Yokoyama T, Kobayashi K, Kuritani T, Nakamura E (2002) Mantle metasomatism and rapid ascent of slab components beneath island arcs: evidence from ^{238}U-^{230}Th-^{226}Ra disequilibria of Miyakejima volcano, Izu arc, Japan. J Geophys Res (in press)

Zartman RE, Haines SM (1988) The plumbotectonic model for Pb isotopic systematics among major terrestrial reservoirs – a case for bi-directional transport. Geochim. Cosmochim Acta 52:1327-1339

Zellmer G, Turner S, Hawkesworth C (2000) Timescales of destructive plate margin magmatism: new insights from Santorini, Aegean volcanic arc. Earth Planet Sci Lett 174:265-281

Zhao D, Hasegawa A. (1993) P-wave tomographic imaging of the crust and upper mantle beneath the Japan Islands. J Geophys Res 98:4333-4353

APPENDIX:
EQUATIONS FOR SIMULATING MELTING AND DEHYDRATION MODELS IN ARCS

A1. Single-stage model

The fluid compositions given in Table 3 and plotted on various figures were calculated assuming Rayleigh distillation of altered oceanic crust using the partition coefficients for rutile-free eclogite at 3 Gpa from Table 2. However, the composition of the mantle wedge prior to melting reflects a mass balance of this fluid plus wedge peridotite and sediment. Here we present a simple mass balance calculation which indicates how the U-Th or Th-Ra composition can be estimated in the composite mantle wedge source prior to melting. In this model, the fluid release by dehydration occurs as a single event releasing both U and Ra (and small amounts of Th or Pa based on the partition coefficients estimated for Th or Pa in Table 2). This dehydration can be thought of as either a "batch" or a "Rayleigh" process, the latter being more efficient in stripping the elements from the altered oceanic crust. The mineral fluid partition coefficients compiled in Table 2 and/or A1 can be used to calculate the fluid concentration. If C_f is the concentration in the fluid, then the composition of the metasomatized mantle wedge prior to melting is:

$$C_w = C_{dw} X_{dw} + X_{sed} C_{sed} + (1 - X_{sed} - X_{dw}) C_f$$

where the sum of all the fractions (X_i) from the depleted wedge (dw), sediment (sed) and slab fluid (f) equals 1. The concentration of a given element in the depleted mantle wedge can be estimated, as follows, assuming that the wedge has become variably depleted during a back-arc melt extraction event:

$$C_{dw} = \frac{D C_{MORB}}{D + F(1 - D)}$$

where C_{MORB} represents the composition of the MORB source, D is a partition coefficient during melting and F the degree of melting in the back-arc. Using the partition coefficients given in Table 2, this model produces almost horizontal arrays on the ^{238}U-^{230}Th isochron diagram for a constant amount of fluid addition. As argued in the text, inclined arrays can then be produced by dynamic melting (George et al. 2003; Lundstrom 2003; Bourdon and Sims 2003). If we allow for greater amounts of Th in the slab fluid, then inclined arrays can also be produced with a single stage fluid addition without the need for ^{230}Th in-growth during melting or ageing.

A2. Two stage-model for fluid addition

This model assumes that fluid dehydration takes place in two distinct stages (Fig. 12). The first stage produces most of the ^{238}U-^{230}Th signature while the second stage adds mainly ^{226}Ra. The model assumes that Th is not fluid mobile and that no Th is lost from the altered oceanic crust (slab) during either dehydration stage. Dehydration transport of U from the altered oceanic crust during the first stage is assumed to occur via a Rayleigh distillation process according to the following equation describing the activity of residual uranium in the slab:

$$^{238}U_1 = {}^{238}U_0 (1 - F_1)^{\frac{1}{D_U} - 1}$$

where F_1 is the fraction of fluid released during the first stage, U_0 is the activity of U in the slab, U_1 is the activity of residual U in the slab and D_U is the fluid/solid partition coefficient. Note that here the elemental symbols U, Th and Ra represent activities of the nuclides considered (^{238}U, ^{230}Th and ^{226}Ra) rather than concentration. In this calculation, we assume that the initial U concentration in the slab is 0.1-0.3 ppm and F_1 is 0.02-0.03. This first step efficiently removes U from the slab. Had we chosen batch dehydration, the amount of U left in the slab would be greater and there could be some U in the second stage fluid addition.

The time between the two dehydration stages is Δt and is taken to match the slope of array on the U-Th diagram. The activity of ^{230}Th in the wedge after this time is:

$$^{230}Th_2 = {^{230}Th_1} e^{-\lambda \Delta t} + {^{238}U_1}(1-e^{-\lambda \Delta t})$$

where

$$^{230}Th_1 = \frac{^{230}Th_0}{1-F_1}$$

where λ is the decay constant of ^{230}Th. After Δt, ^{226}Ra added to the mantle wedge during the first dehydration stage will have decayed back to secular equilibrium with the ^{230}Th in the wedge (^{230}Th$_2$) such that:

$$^{226}Ra_2 = {^{230}Th_2}$$

The activity of ^{226}Ra available to be added by the second stage dehydration is that which has in-grown from ^{230}Th left behind in the slab. Thus, we can calculate the ^{226}Ra and U activities in the second step dehydration fluid:

$$Ra_2^f = \frac{Ra_2}{F_1}\left(1-(1-F_2)^{1/D_{Ra}}\right)$$

$$U_2^f = \frac{U_1}{F_2}\left(1-(1-F_2)^{1/D_U}\right)$$

where Ra_2^f and U_2^f are the integrated Ra and U activities during fractional dehydration of the slab, F_2 is the fraction of fluid released in the second stage (up to 0.02) and U_1 is the residual U content in the slab before the second fractional dehydration (first equation). An outcome of the Rayleigh distillation process is that the amount of U in the second-stage fluid is negligible compared with the first stage.

In the second step, we calculate the mass balance for all the components present in the mantle wedge: (1) depleted peridotite (dw), (2) sediment (sed), (3) first fluid (mass fraction F_1) and (4) second fluid (mass fraction F_2). The depleted peridotite is assumed to be a depleted peridotite (8 ppb) that is further depleted during batch extraction of melt in a back-arc (F = 0-10%). In the case of Tonga-Kermadec the mantle wedge appears to be variably depleted (e.g., Turner et al. 1997) and so it is appropriate to use a range of previous depletion in the modeling discussed here. The fraction of sediment can be estimated from Pb or Nd isotopes (e.g., 5‰ for Tonga-Kermadec). For ^{230}Th, there is no input from the second fluid because Th is again assumed to be fluid-immobile and the only contribution of ^{230}Th comes from decay from U_1^f. To a first approximation, we assume that the ratio of slab mass to mantle mass is equal to 1. This parameter could of course be tuned and incorporated in the mass balance equation given below. For U, the first dehydration stage removed all the U from the slab. Thus the budget of ^{230}Th in the mantle wedge is given by:

Magma Genesis at Convergent Margins 313

$$^{230}Th_{mix} = {}^{230}Th_w X_w + (1-X_w)^{230}Th_1^f$$

$$^{230}Th_1^f = \overline{^{238}U_1^f}\left(1-e^{-\lambda \Delta t}\right)$$

$$\overline{^{238}U_1^f} = {}^{238}U_0\left(1-(1-F_1)\right)^{\frac{1}{D_U}}$$

Finally, the ^{226}Ra content in the mixture wedge + fluid 1+ fluid 2 is:

$$^{226}Ra_{mix} = {}^{226}Ra_w(1-F_1-F_2) + F_1\,{}^{230}Th_1^f + F_2\,{}^{226}Ra_2^f$$

$$^{226}Ra_w = {}^{230}Th_w$$

$$^{230}Th_w = {}^{230}Th_{dw}X_{dw} + (1-X_{dw})^{230}Th_{sed}$$

Where w denotes the mantle wedge (w). The total fraction of fluid added is inferred to be close to 0.03, a reasonable estimate for the fraction of fluid contained in the oceanic crust; for example, $F_1 = 0.01$ and $F_2 = 0.02$. A sediment component (sed) is also added to the variably depleted mantle wedge (dw). The sediment composition can be based on values for the composition of the sediment entering the trench using, for example, the compilation of Plank and Langmuir (1998). Figure 12 illustrates an example of the two-stage model used to simulate the Tonga-Kermadec arc whilst matching the 30-50 ka timescale from the U-Th array.

A3. Continuous dehydration and melting

This section describes the continuous flux melting model used in Bourdon et al. (2003) and has many similarities with the model of Thomas et al. (2002). A significant difference is that the model described here keeps track of the composition of the slab as it dehydrates. This model is based on mass balance equations for both the mantle wedge and the slab. We assume secular equilibrium in the U-series decay chain initially:

$$\left(^{238}U_{oc}^o\right) = \left(^{230}Th_{oc}^o\right) = \left(^{226}Ra_{oc}^o\right)$$

In this equation, the parentheses denote activities. The model operates with a series of time-steps. Between two time steps, a fraction of fluid (df) is added to the mantle wedge. The activity of Ra in the fluid is calculated from the composition of the slab at the preceding step (similar equations for U, Th, Ra, Pa):

$$\left(^{226}Ra_i^f\right) = \frac{\left(^{226}Ra_{oc}^{i-1}\right)}{D^{s/f} + f(1-D^{s/f})}$$

At each step, a fraction of fluid f is added to the mantle wedge from the slab. The bulk partition coefficients used for fluid dehydration can be derived from published mineral/fluid partition coefficients (see Tables A1 and A2). The composition of the residual slab is estimated as follows after Δt which is the time step between two melt extractions (similar equation for Th and Pa):

$$\left(^{230}Th_{oc}^i\right) = D_{Th}^{s/f}\left(^{230}Th_f^{i-1}\right)e^{-\lambda_{230}\Delta t} + D_U^{s/f}\left(^{238}U_f^{i-1}\right)\left(1-e^{-\lambda_{230}\Delta t}\right)$$

A similar but more complex equation is derived for Ra:

$$\left(^{226}Ra_{oc}^i\right) = D_{s/f}^{Ra}\left(^{226}Ra_f^i\right)e^{-\lambda_{226}t} + \frac{\lambda_{226}D_{s/f}^{Th}}{\lambda_{226}-\lambda_{230}}\left(^{230}Th_f^i\right)\left(e^{-\lambda_{230}t} - e^{-\lambda_{226}t}\right) +$$

$$\lambda_{226}\lambda_{230}D_U^{s/f}\left(^{238}U_f^{i-1}\right)\left[\alpha e^{-\lambda_{226}t} + \beta e^{-\lambda_{230}t} + \gamma e^{-\lambda_{238}t}\right]$$

Table A1. Partition coefficients used in dehydration and melting models

	U	Th	Ra	Pa
$D^{solid/fluid}$	0.02	0.2	0.002	0-1
$D^{solid/melt}$	0.003	0.0014	1×10^{-5}	1×10^{-5}

D's were calculated assuming 60% cpx, 40% garnet and no rutile using values from Keppler (1996) and garnet/melt cpx/melt partition coefficients to estimate garnet/fluid partitioning.

Table A2. Parameters used in melting and dehydration models

Parameters	Notation	Units	Range
Degree of melting	F	none	0-0.2
Upwelling rate	W	m a^{-1}	0-1
Melting rate	Γ	kg m^{-3} a^{-1}	10^{-3}-10^{-5}
Solidus temperature	T	K	1000-1400
Length of melting column	Z	km	0-50
Matrix porosity	φ	none	0-0.005
Solid density	ρ_s	kg m^{-3}	3350
Melt density	ρ_m	kg m^{-3}	2700

where

$$\alpha = \frac{1}{(\lambda_{238} - \lambda_{226})(\lambda_{230} - \lambda_{226})}$$

$$\beta = \frac{1}{(\lambda_{238} - \lambda_{230})(\lambda_{226} - \lambda_{230})}$$

$$\gamma = \frac{1}{(\lambda_{230} - \lambda_{230})(\lambda_{226} - \lambda_{238})}$$

The initial U activity in the mantle wedge (U_w) is set to an arbitrary value of 1 and all the other nuclides are scaled relative to U_w. The initial U activity in the oceanic crust is twice the activity in the mantle wedge. The Th/U ratios of the mantle wedge and the slab are both equal to 2.5. This value is relevant for modeling the higher (^{230}Th/^{232}Th) observed in some arc lavas. Fluid is added to a portion of mantle wedge, and the mass fraction of fluid (f) and the composition of the mixture at time step i is given by (same equation for all the nuclides):

$$(^{230}Th_w^i) = (^{230}Th_w^{i-1})(1-f) + (^{230}Th_f^i)f$$

The composition of the melt is then estimated with a batch melting model as follows assuming that there is a total porosity of dfm+f_r, where f_r is the mass fraction of melt prior to this new melting event:

$$\left(^{230}\text{Th}_m^i\right) = \frac{\left(^{230}\text{Th}_w^i\right)}{D_{\text{Th}}^{s/m} + (\text{dfm} + f_r)\left(1 - D_{\text{Th}}^{s/m}\right)}$$

We assume that a constant mass fraction f_r remains in the mantle wedge after melt extraction. As in Section A2 of the Appendix, the ratio of slab mass to wedge mass is assumed to be equal to 1 but more complex models are also possible. The bulk composition of the mantle wedge after melt extraction is calculated with the following equation after each extraction increment:

$$C_R^T = \left(D(1-f_r) + f_r\right)C_m$$

where C_R^T and C_m represent the activity in the matrix after melt extraction and in the melt respectively. Before the subsequent dehydration/melting event, the metasomatized mantle wedge behaves as closed system and the decay equations (which are similar to the equations given for the slab) are used to track the abundances of U-series nuclides.

8   The Behavior of U- and Th-series Nuclides in Groundwater

Donald Porcelli
Department of Earth Sciences
University of Oxford
Parks Rd.
Oxford, OX1 3PR, United Kingdom
don.porcelli@earth.ox.ac.uk

Peter W. Swarzenski
Coastal Marine Geology Program
US Geological Survey
St. Petersburg, Florida, 33701, USA
pswarzen@usgs.gov

1. INTRODUCTION

Groundwater has long been an active area of research driven by its importance both as a societal resource and as a component in the global hydrological cycle. Key issues in groundwater research include inferring rates of transport of chemical constituents, determining the ages of groundwater, and tracing water masses using chemical fingerprints. While information on the trace elements pertinent to these topics can be obtained from aquifer tests using experimentally introduced tracers, and from laboratory experiments on aquifer materials, these studies are necessarily limited in time and space. Regional studies of aquifers can focus on greater scales and time periods, but must contend with greater complexities and variations. In this regard, the isotopic systematics of the naturally occurring radionuclides in the U- and Th- decay series have been invaluable in investigating aquifer behavior of U, Th, and Ra. These nuclides are present in all groundwaters and are each represented by several isotopes with very different half-lives, so that processes occurring over a range of time-scales can be studied (Table 1). Within the host aquifer minerals, the radionuclides in each decay series are generally expected to be in secular equilibrium and so have equal activities (see Bourdon et al. 2003). In contrast, these nuclides exhibit strong relative fractionations within the surrounding groundwaters that reflect contrasting behavior during release into the water and during interaction with the surrounding host aquifer rocks. Radionuclide data can be used, within the framework of models of the processes involved, to obtain quantitative assessments of radionuclide release from aquifer rocks and groundwater migration rates. The isotopic variations that are generated also have the potential for providing fingerprints for groundwaters from specific aquifer environments, and have even been explored as a means for calculating groundwater ages.

The highly fractionated nature of the ^{238}U and ^{232}Th series nuclides is illustrated by the measured activities in some representative waters in Figure 1. The highest activities are typically observed for ^{222}Rn, reflecting the lack of reactivity of this noble gas. Groundwater ^{222}Rn activities are controlled only by rapid *in situ* decay (Table 1) and supply from host rocks, without the complications of removal by adsorption or precipitation. The actinide U, which is soluble in oxidizing waters, is present in intermediate activities that are moderated by removal onto aquifer rocks. The long-lived parent of a decay series, ^{238}U, does not have a radioactive supplier, while ^{234}U is a radiogenic nuclide; both of these nuclides have half-lives that are long compared to

Table 1. Radionuclides important in groundwater studies.

Nuclide	Half-life[1]	Atoms/dpm[2]	Factors Controlling Groundwater Concentrations
^{238}U	4.47×10^9 a	2.35×10^{15}	weathering, adsorption
^{234}Th	24.1d	3.47×10^4	recoil, strong adsorption, decay
^{234}U	2.45×10^5 a	1.29×10^{11}	weathering, adsorption
^{230}Th	7.57×10^4 a	3.98×10^{10}	recoil, weathering, strong adsorption
^{226}Ra	1.60×10^3 a	8.42×10^8	recoil, strong adsorption, decay, surface production
^{222}Rn	3.83d	5.52×10^3	recoil, decay, surface production
^{210}Pb	22a	1.2×10^7	recoil, strong adsorption, decay
^{232}Th	1.39×10^{10} a	7.31×10^{15}	weathering, strong adsorption
^{228}Ra	5.75a	3.02×10^6	recoil, strong adsorption, decay, surface production
^{224}Th	1.91a	1.00×10^6	recoil, strong adsorption, decay, surface production
^{224}Ra	3.64d	5.24×10^3	strong adsorption, decay, surface production
^{235}U	7.13×10^8 a	3.75×10^{14}	weathering, adsorption
^{223}Ra	11.7d	1.68×10^4	recoil, strong adsorption, decay, surface production

1. From compilation by Bourdon et al. (2003).
2. dpm is decays per minute, and so is 60 Bq.

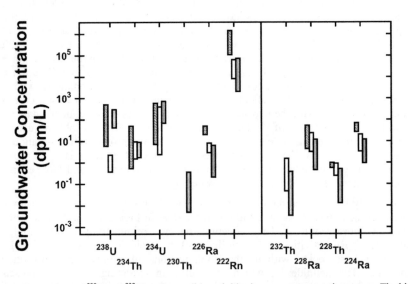

Figure 1. Comparison of ^{238}U and ^{232}Th series nuclide activities in some representative waters. The highest concentrations are typically observed for ^{222}Rn. The isotopes of Th and Ra are strongly depleted due to interaction with the host rock, while somewhat higher activities of U are generally maintained. Data from Krishnaswami et al. (1982) on the left, Tricca et al. (2001) in the center, and Luo et al. (2000) on the right.

groundwater ages and so are generally not substantially affected by decay within aquifer systems. The alkaline earth Ra, and to a greater extent the less soluble actinide Th, are readily removed from groundwater by water-rock interactions, and so are strongly depleted. Both of these elements have very short-lived as well as longer-lived isotopes, and so their isotope compositions reflect processes over a range of time scales. Overall, since the set of nuclides in each decay series are tied by decay systematics, with daughter nuclide productions and distributions dependent upon parent distributions, combined studies of these elements can generate considerable information regarding radionuclide water-rock interactions and weathering release rates.

Many studies have evaluated the behavior of these elements in the hydrosphere, and have been reviewed earlier (Gascoyne 1992; Osmond and Cowart 1992). A large body of data has been gathered using widely practiced counting techniques, although this cannot provide precise data on the longer lived nuclides. The development of high precision mass spectrometric techniques (Chen et al. 1986; Goldstein and Stirling 2003) have made it possible for more subtle variations in U isotopes to be explored, and also provide the potential for doing so with long-lived Th isotopes. However, these techniques have yet to be fully exploited in this field and data remain limited.

Research into the behavior of radionuclides continues to be active, greatly motivated by the necessities of understanding and predicting migration of contaminant actinides and related species. The focus of this review is on the use of the isotope systematics of the U- and Th-decay series in understanding natural radionuclide migration. In general, quantifying aquifer parameters that control trace element behavior are difficult at many sites by direct measurements, since the host rocks typically cannot be as readily sampled as groundwater, and average values for host rocks recovered from boreholes are difficult to obtain. However, decay series systematics of nuclides in groundwater provide the potential for quantifying some chemical parameters of water-rock interaction from direct measurements of waters alone. Various studies have attempted to model the transport of the range of U, Th, Ra, Rn, and Pb radioisotopes in groundwaters by combining the systematics of the nuclides within the decay chains, and so providing a framework for calculating such critical parameters as retardation factors. These particular elements have the additional appeal of providing natural analogues for radioactive wastes, so that evaluations of the transport behavior of U and Th series nuclides can have predictive power for assessing the possible migration of low level anthropogenic waste nuclides in any environment. A key difference is in the supply functions; where the effects of the supply rate of naturally occurring nuclides within the aquifer can be distinguished, the controls on nuclides introduced only at the aquifer recharge boundary can be quantified. Many of the earlier advances in modeling naturally occurring nuclides have been summarized by Osmond and Cowart (1992), Ku et al. (1992), and Ivanovich et al. (1992a). While there are many different approaches that have been used to infer radionuclide behavior, these are all based on the fundamental connections between nuclides by radioactive decay, and simple modeling of known processes of input by weathering and recoil into groundwater as well as interaction with aquifer host rock surfaces by sorption and precipitation.

In this review, the general principles incorporated in most modeling approaches will be outlined. While the principles of the models are relatively straightforward, the derivations of the equations involved are involved. The intention here is not to derive the specific equations used by the various studies to obtain quantitative results. These are considered in detail by Ku et al. (1992) and the studies referenced here. Rather, the emphasis is on the qualitative understanding of how the important conclusions are derived as well as identifying and evaluating the underlying assumptions. The detailed discussion of individual processes and parameters are intended to demonstrate where assumptions that underpin model calculations are fully justified, whether additional work is required to support these assumptions, and when experimental or theoretical considerations indicate that alternative parameter values are possible. Overall, it should become clearer where qualitative conclusions regarding radionuclide behavior are firm and to what extent quantitative evaluations might be questioned. Such a systematic evaluation of the controlling processes has not been compiled recently, and is needed for future progress in this field.

Considerable recent attention also has focused on the discharge of groundwaters into surface water. Earlier water budget studies often regarded these inputs are negligible in the absence of methods for their quantification. However, distinct isotopic characteristics of groundwater, imparted by close interaction with aquifer rocks where daughter nuclides are supported by continuing supply from host minerals, can be used to trace these inputs. This is particularly true of Rn and the short-lived isotopes of Ra, which have concentrations that diminish in the absence of direct supply from host rocks, and so can be used to fingerprint patterns of groundwater inflow into highly depleted surface waters. These studies are also reviewed here.

2. NUCLIDE TRANSPORT IN AQUIFERS

A key parameter in understanding the controls on groundwater concentrations and radionuclide transport is the retardation factor R_I. The rate of transport of a groundwater constituent I is decreased by the factor R_I relative to the groundwater flow rate due to interaction with aquifer host rock surfaces. The retardation factor is $R_I = 1 + K_I$, where K_I is the ratio of the inventory atoms adsorbed on surfaces to the inventory in the surrounding groundwater. Thus if 90% of a species is on surfaces, then it will migrate at a rate of 1/10 that of groundwater, while if there is none on the surfaces so that it is behaving conservatively, it will keep apace of groundwater flow. Therefore, a major goal of groundwater models is to obtain values of K_I. A bulk value is used for large-scale transport modeling, although comparison with laboratory experiments is required to determine the specific mechanisms that define this value and are responsible for any changes across an aquifer. Mathematical treatments of simple aquifer models have been used to calculate retardation factors, as well as the rates of weathering, recoil, and adsorption/desorption, and have been extensively developed by Krisnaswami et al. (1982), Davidson and Dickson (1986), Ku et al. (1992), Luo et al. (2000), and Tricca et al. (2000, 2001). These approach the determination of various parameters using somewhat different perspectives and underlying assumptions. The modeling equations for most models are generally subsets of the same general set of equations, as shown by Ku et al. (1992), with different studies utilizing different assumptions regarding parameter inputs or neglecting as insignificant different processes. The details of specific processes, and alternatives, are discussed according to element in subsequent sections. Note that the ratios and concentrations in parentheses refer to activities; others refer to molar concentrations.

2.1. General modeling considerations

The general modeling approach is described here to provide a context for understanding recent observations and identifying which processes, whether generally explicitly considered or neglected, are important. Individual studies generally incorporate further simplifying assumptions. Ku et al. (1992) provide a discussion of the general equations.

Model components. Models generally consider 3 populations of radionuclides:

- *Groundwater*, where concentrations are obtained by direct sampling. Since transport of radionuclides is being considered without regard for the mobile form these are in, bulk activities per unit mass are used. Speciation does not change the calculation of the bulk transport properties of each nuclide, although this is certainly an underlying control on observed behavior and so becomes important only when calculated transport and water/rock interaction rates are being interpreted. Operationally, most samples are filtered to remove particles less than ~0.5 μm to remove material presumably mobilized by pumping. Of particular concern is the inclusion in samples of colloid-bound species, which may not readily exchange with dissolved nuclides and may have different transport properties. Unfortunately, this generally is not considered in isotope models.

- *Host aquifer rock*, which can include inventories in primary minerals and secondary phases generated prior to establishment of the present environment, or actively precipitating phases. For modeling calculations, this strictly represents all radionuclides not actively exchanging with the groundwater, although the distribution of radionuclides is important for interpreting the pattern of releases to groundwater.

- *Atoms adsorbed on host rock surfaces.* This pool is assumed to readily exchange with atoms in groundwater. Bulk abundances are sought in the models, while specific mechanisms that explain these values must be sought separately in laboratory experiments.

It is sometimes useful to consider the abundances in the groundwater and sorbed together as the "mobile" population, in contrast to those fixed within the host rock.

Processes controlling nuclide distributions. The general equations for one-dimensional advective transport along a groundwater flow path of groundwater constituents, and the incorporation of water/rock interactions, are given in such texts as Freeze and Cherry (1979). The equations can be applied to the distribution in groundwater of each isotope I with a molar concentration I_W and parent with P_W to obtain

$$\frac{\partial I_W}{\partial t} + \frac{v \partial I_W}{\partial x} = bw_I I_R + b\varepsilon_I \lambda_P P_R + I_{ADS} k_{-I} + f_I \lambda_P P_{ADS} + \lambda_P P_W - Q_I - \lambda_I I_W - k_I I_W \quad (1)$$

which includes the groundwater flow velocity (v), weathering rate constant (w_I), recoil release fraction ε_I, absorbed molar concentrations of species I (I_{ADS}) and parent P (P_{ADS}), host rock molar concentrations for the species (I_R) and parent (P_R), desorption (k_{-1}) and adsorption (k_1) rate constants, fraction produced by sorbed parent atoms that are directly released to groundwater (f_I), the nuclide and parent decay constants λ_I and λ_P, and precipitation rate (Q_I). The parameter b converts concentrations of weathered rock to those in water. In porous rocks, $b = (1 - n)\rho_R/n\rho_W$ is the mass ratio of rock to water with densities ρ_R and ρ_W, where n is the porosity. Equation (1) can be cast in terms of the activity $(I)_W$ per unit mass by multiplying all terms by the decay constant for I (λ_I). The convention of using parentheses to denote activities per unit mass or activity ratios will be used here. It is worth bearing in mind that unfortunately notation for many of the parameters is not consistent between published studies; the notation used here is summarized in Table 2.

The terms on the left side represent the change of I_W with time and the change of

Table 2. Parameters used in groundwater models.

I_W, $(I)_W$	Groundwater nuclide I molar and activity concentration
P_W, $(P)_W$	Groundwater parent P molar and activity concentration
I_R, $(I)_R$	Host rock nuclide I molar and activity concentration
P_R, $(P)_R$	Host rock parent P molar and activity concentration
I_{ADS}, $(I)_{ADS}$	Adsorbed nuclide I molar and activity concentration (per mass of rock)
P_{ADS}, $(P)_{ADS}$	Adsorbed parent P molar and activity concentration (per mass of rock)
v	Groundwater flow velocity
x	Distance along a groundwater flow line
w_I	First order constant for weathering release of nuclide I
ε_I	Recoil release fraction for nuclide I
λ_I, λ_P	Decay constants of nuclide I and parent P
k_I	First order bulk adsorption rate constant
k_{-I}	First order bulk desorption rate constant
Q_I	Rate of precipitation of nuclide I
f_I	Fraction of nuclide I produced by adsorbed parents and recoiled into water
b	Mass ratio of rock to water, where $b = (1 - n)\, \rho_R / n\, \rho_W$
ρ_W, ρ_R	Densities of groundwater and host rock
n	Porosity of the host rock
R_I	Retardation factor of nuclide I, where $R_I = 1 + K_I$
K_I	Partition coefficient between dissolved and adsorbed atoms

I_W with distance x along a groundwater flow path. The terms on the right side represent the input and removal terms (see Fig. 2a), which in order are:

- *Input by weathering.* Minerals with molar concentrations I_R release nuclides according to rate constant w_I by simple breakdown of mineral structures. Weathering is depicted here as a first-order rate constant for convenience (Tricca et al. 2000), but is constant for small degrees of weathering, and conversion to more familiar rates per unit area can be done by determining the ratio of the mass to the surface area of the weathering minerals. Note that the weathering constant for U- and Th-bearing minerals may be different from that of the bulk rock, and release of nuclide I by w_I may be different than dissolution of the bulk mineral due to e.g., leaching or incongruent breakdown (Eyal and Olander 1990; Read et al. 2002). However, it is generally assumed that all the nuclides in each series are released together, and that U and Th are similarly sited so nuclides from the different series are not separated due to different release rates.

- *Input by recoil.* The parent nuclide population, with molar concentration P_R, releases a fraction ε_I of the daughter by direct recoil or related losses (e.g., rapid migration along its recoil track). Direct recoil of nuclides into groundwater (Kigoshi 1971; Fleischer and Raabe 1978) occurs when daughter nuclides that are generated with high kinetic energies are propelled across the mineral surface and are stopped within the aqueous phase. The direction of recoil is random for each decay, and the rate of ejection into groundwater is determined only by the fraction of parent nuclides within recoil distance of mineral surfaces or channels

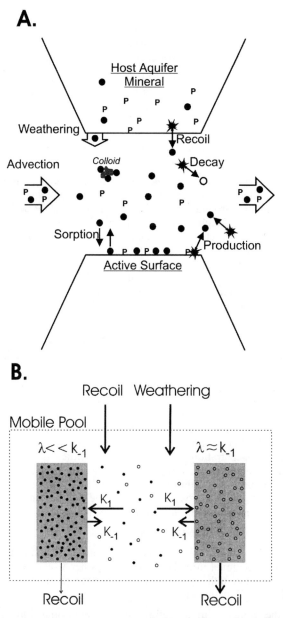

Figure 2. (A.) The radionuclides in an aquifer are divided into three reservoirs; groundwater, the host aquifer minerals, and adsorbed onto active surfaces. Also shown are the processes adding to a daughter nuclide (closed circles) in the groundwater of weathering, advection, recoil from decay of parent atoms ("P") in the aquifer minerals, and production by parent decay, the processes of losses of a radionuclide of advection and decay, and exchange between dissolved and adsorbed atoms.

(B.) In steady state, the mobile pool of nuclides is constant. For a long-lived nuclide (solid circles), the adsorbed abundances are determined by exchange with the groundwater. For a short-lived isotope that has a decay constant that is comparable to the desorption rate constant k_{-1}, decay of sorbed atoms is a significant flux, and so the steady state sorbed abundance is lower (see Eqn. 3).

to the surface, and so can be readily calculated given the distribution of parent nuclides. Recoil of nuclides is the most clearly defined physical supply process, and has been considered to be most amenable to quantification, and so has been the focus of considerable research. This is discussed further in Section 2.2.

- *Input by desorption according to first order rate constant k_{-1}*. Note that this represents the association of atoms on surfaces that are free to exchange with the groundwater. The alternative formulation of specifying a volume and density of a layer containing species with defined migrations velocities (Tricca et al. 2001) is only a mathematical construct and is not consistent with established absorption processes involving association of species directly with surfaces.

- *Input by production from sorbed parent atoms.* Daughter atoms recoiled during α decay are unlikely to come to rest on the surfaces, and are ejected either into the surrounding groundwater or into the mineral. The fraction released, f_1, may be 0.5 if half the daughter atoms are recoiled into the underlying mineral (Tricca et al. 2000), or 1 (Luo et al. 2000) if these atoms are rapidly released by migration back along the recoil track, or a value in between. Unfortunately, there is no experimental evidence to establish which value is appropriate. In contrast, if a daughter atom is produced by low energy β decay, it may remain on the surface, and so the fraction released is zero. However, in this case, this production must be considered when calculating the budget of adsorbed atoms (see below).

- *Input by decay* in solution of parent with molar concentration P_W.

- *Removal by precipitation.* This is represented here by some function Q_I, with the exact form depending upon the controlling parameters. U may precipitate as uraninite where reducing conditions are encountered, and Th may reach the thorianite solubility limit under some conditions (see Tricca et al. 2001). In these cases, precipitation will balance supply by weathering of the major isotope. Alexander and McKinley (1994) noted that precipitation of such minerals is not necessarily a first-order process, and it is related to the difference between concentration and solubility limit rather than to absolute concentrations. Note, however, that during precipitation of low concentration species by partitioning into phases defined by other elements, removal may be proportional to groundwater radionuclide concentrations and so may be governed by a first order rate constant (Ku et al. 1998).

- *Removal by decay* of I in the water.

- *Removal by sorption.* It is generally assumed that adsorption is reversible. Irreversible removal directly from groundwater would be included in the term for precipitation, although slow irreversible incorporation of sorbed atoms is not considered in Equation (1).

Note that the models aim to define general descriptive bulk parameters for an aquifer. Values for bulk adsorption or weathering will reflect the weighted range for a variety of phases present in complex, natural systems, and provide an overall measure of the behavior of the aquifer. In contrast, individual phases can be examined in the laboratory to determine the specific processes involved.

An important consideration is the relative importance of the two processes that supply radionuclides to the dissolved and adsorbed inventories from within the host rock minerals. The recoil term in Equation (1), $b\varepsilon_I\lambda_P P_R$, can be compared to the weathering term, $bw_I I_R$, most clearly when these are expressed using activity concentrations rather than atom concentrations by multiplying each by λ_I to obtain $b\lambda_I \varepsilon_I (P)_R$ and $bw_I(I)_R$. Then noting that in minerals the radionuclides are expected to be approximately in secular equilibrium so that $(P)_R = (I)_R$, it can be seen that weathering is more important when w_I ($\equiv 1/\tau_W$, where τ_W is the mean time required for completely weathering host minerals) is greater than $\lambda_I \varepsilon_I$ ($\equiv \varepsilon_I \ln 2 /^I t_{1/2}$, where $^I t_{1/2}$ is the half life of nuclide I). Values of $\sim 10^{-2}$ have often been inferred for ε_I (see Section 2.2), so that if aquifers were being removed by chemical weathering with a mean time of 10^7 years, any nuclide with a half life of $<10^5$ years will be supplied largely by recoil. This includes all daughter nuclides except ^{234}U. Where ε_I is much smaller, weathering may become important for the supply of ^{230}Th or even ^{226}Ra. However, for the shorter-lived nuclides, it is unlikely that weathering can be important.

Adsorbed nuclides. Obtaining a solution to Equation (1) requires knowledge of the absorbed abundance of the nuclide of interest (I_{ADS}). Assuming that steady state abundances have been achieved at each location, supply by adsorption is balanced by desorption and decay. No supply is expected to occur by decay of adsorbed parent atoms in the cases where α recoil drives daughters instead into solution or into the underlying mineral (although modification may be made in the case of β decay). Then,

$$I_W k_1 = I_{ADS} k_{-1} + \lambda_I I_{ADS}. \tag{2}$$

The dimensionless partition coefficient K_I is defined as the ratio of atoms adsorbed to atoms in solution. Rearranging Equation (2) (Krishnaswami et al. 1982),

$$K_I \equiv \frac{I_{ADS}}{I_W} = \frac{k_1}{k_{-1} + \lambda_I}. \tag{3}$$

It should be emphasized that where this value is calculated from field data, it simply represents the distribution of the radionuclides, and makes no assumptions about whether thermodynamic equilibrium has been achieved, about the precise mechanisms of adsorption, or about the form of the isotherm describing adsorption as a function of concentration. Alexander and McKinley (1994) have objected to this modeling approach by arguing that *in situ* values are valid only if the mechanism responsible can be shown to distribute a species between water and solid so that the equilibrium phase concentration is directly and linearly related to that in the aqueous phase. Also, it was pointed out that where substantial changes in water chemistry are encountered, competition with other species might change the distribution values. However, a linear relationship between the partition coefficient and the water concentration is an issue only where the distribution between water and solid is assumed to remain constant during large changes in total abundances. This is not an issue in describing the present aquifer system, but is a factor when predicting radionuclide behavior under changed conditions. Note that in most cases, the actual number of atoms involved is exceedingly small. Miller and Benson (1983), using numerical models, point out that when species concentrations are sufficiently small relative to the supporting electrolytes in solution and the capacity of the sorbing material, the distribution coefficient is constant. Where the relationship is nonlinear, empirical data must be used to define applicable isotherms. While this may not affect the modeling of naturally occurring nuclides, it complicates extrapolation to high radionuclide concentrations. In that case, the number of sites for sorption on a solid phase may be limited, and beyond this there is no direct relationship between aqueous and dissolved concentrations. Overall, these considerations do not compromise the calculations generally performed in studies of the distribution of naturally occurring radionuclides, but do become critical where conclusions are extrapolated to make predictions involving grossly different concentrations.

The migration rate of a groundwater constituent, relative to the groundwater flow rate, is controlled by the retardation factor, where $R_I \equiv 1 + K_I$. Where $K_I \gg 1$ (e.g., for Th and Ra), $R_I \approx K_I$, and $I_{ADS} + I_W = I_W R_I$. Note that k_1 and k_{-1} are element-specific but not isotope-specific. All isotopes that decay slower than desorption, so that $k_{-1} \gg \lambda_I$, have a value of K_I that is equal to that of a stable isotope (Eqn. 3). The value of K_I may be lower for the shortest-lived nuclides (see Fig. 2b), and so a series of equations derived from Equation (3) applied to different isotopes of the same element may be used to obtain absolute values for the separate rate constants.

Obtaining solutions to the model equations. The general modeling approach has been discussed in considerable detail in Ku et al. (1992). Detailed application of these models to a wide range of nuclides have been performed by Krishnaswami et al. (1982) and Copenhaver et al. (1992, 1993) to single wells, by Tricca et al. (2000, 2001) to data from across a shallow, unconfined, unconsolidated sandy aquifer, and by Luo et al. (2000) to a regional basalt aquifer. Each of these studies, as well as many others, treat radionuclide transport in fundamentally the same way, although the full derivation of the equations is not always presented; rather, subsets of the full equations are often obtained directly by applying simplifying assumptions or implicitly neglecting some processes as inconsequential.

More complete treatment of the derivation of the equations governing radionuclide distributions along a single groundwater flow path are given by Ku et al. (1992) and Tricca et al. (2000). The general strategy has been to combine Equations (1) to (3) for various nuclides. It is possible to fully integrate the equations for each radionuclide and so obtain analytical solutions for the distribution and migration rates. However, the task becomes progressively more difficult along each decay series (see Figs. 3 and 4), since the production rates of each nuclide depend upon the distribution of the parent element (see Tricca et al. 2000). Therefore, all studies have focussed on particular special cases that allow simplifying assumptions. The most common is to assume that the radionuclide distributions are in steady state. In this case, at each location along a flow line the concentrations of all radionuclides in groundwater, surfaces, and host minerals are constant with time. The first term on the left of Equation (1) (the time dependence of the concentration of a nuclide) then can be set to zero. The assumption requires that all controlling processes have been operating steadily over sufficient time to have allowed the accumulation of adsorbed inventories that are in exchange with the groundwater inventories, and that there are no processes such as irreversible removal of adsorbed nuclides that have not allowed this to occur. These conditions, however, do not mean that concentrations are the same along a flow line, and should be carefully distinguished from the assumption that the concentrations in a flowing groundwater do not change. This further assumption is often also made, so that the second term on the left side of Equation (1), which is the effect of advection of nuclides where there are concentration gradients, can also be set to zero.

Where there is an abrupt change in conditions, e.g., where water interacting with the vadose zone enters into the groundwater, strong concentration gradients may occur along groundwater flow lines as concentrations change from initial to steady state values. Groundwater profiles were calculated by Tricca et al. (2000) and it was shown that steady state concentrations are achieved over distances that are inversely proportional to the nuclide decay constant and the partition coefficient. Therefore, while ^{238}U, ^{234}U, and ^{232}Th do not reach constant concentrations over any reasonable distances, ^{226}Ra does over several hundred meters, and all the other radionuclides require less than a few meters. Constant concentrations are achieved when the inventories of a radionuclide in both groundwater and adsorbed on surfaces decay at the same rate at which this radionuclide is

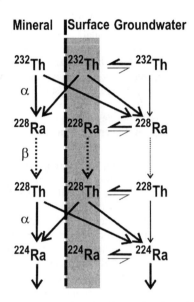

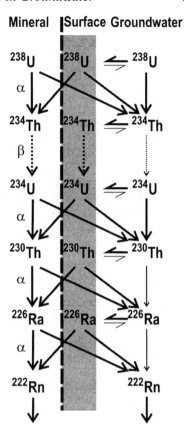

Figure 3. Systematics of radionuclides along the ^{232}Th series. The major and minor fluxes to each nuclide can be readily seen from the arrows shown. The behavior of each nuclide can be evaluated by considering the surface and groundwater populations individually, or together as the "mobile" pool. Nuclides in the decay series within the host rock minerals supply atoms at the surface and in the groundwater by recoil during α decay, so that there are greater abundances in the mobile pool of nuclides progressively along the series. α decay of nuclides at the surface injects atoms back into the minerals as well as into groundwater.

Figure 4. Systematics of radionuclides along the ^{238}U series. The major and minor fluxes to each nuclide can be readily seen from the arrows shown. The behavior of each nuclide can be evaluated by considering the surface and groundwater populations individually, or together as the "mobile" pool. As with the ^{232}Th series (Fig. 3), there is addition to the mobile pool of nuclides produced by α decay along the series. Note, however, that steady state abundances may not be achieved for ^{234}U, which is long-lived and may also be sufficiently mobile in the groundwater for advection to be important.

supplied by recoil as well as adsorbed and dissolved parent inventories. Where recoil is not the dominant supply term, groundwater concentrations will change along with those of the parent nuclide, although if parent concentration gradients are not steep, a quasi-steady state will be reached. Overall, for all but the most long-lived nuclides (^{238}U, ^{234}U, ^{232}Th), constant activities per unit mass are generally reached over short distances (Tricca et al. 2000). Unfortunately, there have not been detailed studies comparing model calculations of the approach to constant groundwater concentrations with field data.

An important assumption in solving Equations (1) to (3) is that the values of all parameters are constant both at each location and along the flow line. Where there are changes within the aquifer, these must be incorporated into the equations. The simplest cases are where all the changes occur abruptly, so that segments of the aquifer can be treated as separate, uniform systems. In many cases, field studies have found that the concentrations and isotopic ratios of short-lived nuclides vary across an aquifer (see sections of specific elements below); this is taken to reflect variations in some of the controlling parameters over a limited area, so that values representing local steady state conditions can be calculated. Such calculations generally ignore the possibility that strong radionuclide concentration gradients are present at the sampling locations.

It is worth emphasizing that these are one-dimensional models and track concentrations in groundwater along a single flow path, neglecting the effects due to such factors as lateral heterogeneities and dispersion. Also, where there is groundwater flow along fractures or preferential pathways, waters moving more slowly may develop distinctive characteristics before joining the main flow paths (e.g., Nitzsche and Merkel 1999), and averages calculated from waters drawn from these flow paths will not adequately represent the relationships of the short-lived nuclides across the aquifer. This may be most pronounced in fractured rock aquifer systems. Therefore, care must be taken in defining the flow regime being modeled.

The model equations can of course be modified to include different processes or more complex descriptions of water/rock interaction within the general approach described here. However, further complexity can only be fruitfully added where independent information can provide additional constraints and limit the range of solutions. An important application of radionuclide transport models is the calculation of the transport of anthropogenic nuclides. In this case, several modifications must be made, including nonsteady state concentrations as the contaminant plume migrates and possible nonlinear adsorption behavior at higher concentrations. Similarly, changes in weathering in the vadose zone due to changes in climate could change the input at the water table and require consideration of time-dependent concentrations where there has been insufficient time to re-establish steady state conditions.

Relative recoil input rates. As discussed above, for all the short-lived nuclides, recoil is the dominant input. Many nuclides are supplied to groundwater by the same physical process of recoil during α decay production, and so their supplies can be related to one another. For simplicity, the recoil rates of the different nuclides are often assumed to be equal within each chain, and proportional to the bulk rock $^{238}U/^{232}Th$ and $^{238}U/^{235}U$ ratios between chains. However, there are several factors that will affect individual recoil release rates.

- *Different locations of parent elements.* ^{238}U and ^{232}Th are generally located in minor phases within host rocks. Due to different U/Th ratios in these phases, recoil from the two chains may be affected by different surrounding matrix characteristics or mineral sizes. Not only might the primary distribution of U and Th be different, but earlier weathering or alteration may also have redistributed U and Th. This is discussed further below.

- *Different recoil lengths.* The distance that a daughter nuclide is recoiled depends upon the recoil energy as well as the host matrix. There are some uncertainties in the precise recoil distances. Kigoshi (1971) gave an average recoil distance of 550 Å for ^{234}Th, although this appears to be too high, since subsequent experiments suggest values of ~200 Å (Fleischer 1980; Sheng and Kuroda 1986). Recoil distances have been calculated for a range of recoil energies and matrices (Hashimoto et al. 1985; Roessler 1989). There are some differences between isotopes where similar matrices were examined, on the order of up to ~10%, but there are greater differences between matrices. For example, while the recoil energies of ^{230}Th (83 keV) and ^{234}Th (72 keV) are similar and so losses are similar from the same matrix, recoil distances for ^{230}Th varies between 160 Å for uraninite, to 413 Å for autunite, to 270 Å for muscovite (Hashimoto et al. 1985). Therefore, different mineral locations between parents or between aquifer rocks can account for over a factor of 2 variation.

- *Different depletions of parents.* Recoil of daughter atoms from a mineral structure results in the depletion near the surface of parent atoms for the next nuclide in the decay chain. This effect advances progressively along each chain. Krishnaswami et al. (1982) calculated that the recoil supply of ^{222}Rn, the result of the fourth α recoil product in the ^{238}U chain, is 48% that of ^{234}Th, the first product. However, if the daughters are adsorbed onto the mineral surfaces, then there is the progressive migration of parents toward the surface either by accumulation on adsorbed sites or by injection from the surface back into the mineral; if all the Ra and Th are adsorbed, then ^{222}Rn supply from ^{226}Ra that is both within the mineral and adsorbed is 30% greater than that of ^{234}Th. Note that selection of the proper value for this process must be consistent with the value of f_I that defines the supply of daughters from adsorbed atoms into the groundwater.

- *Differential losses due to differences in chemistry.* While the physical process of recoil makes no distinctions due to the contrasting chemistries of the nuclides involved, subsequent further release by leaching of daughter nuclides along recoil tracks (Fleisher 1980) may depend upon nuclide chemistry. Such a process is generally considered to be part of the recoil process in modeling treatments, since the other supply term from aquifer solids, weathering, is assumed to be non-fractionating between all parents and daughters. However, in this case differences in behavior must be reflected by differences in recoil release fractions (ε_I). Leaching is considered in more detail below under the treatment of individual elements. However, it is worth noting here that such leaching will only occur when the recoil track crosses the mineral surface or a fast pathway to the surface, even though the daughter has not been directly lost. This can occur if daughters cross mineral boundaries and are implanted into adjacent minerals (e.g., Suksi and Rasilainen 1996), or if a process such as fracturing creates a new mineral face. Nuclides within recoil tracks may also be released if weathering moves the mineral boundary inwards by bulk dissolution of the mineral or by incongruent dissolution that releases those constituents that are incompatible with the secondary phase. However, in many circumstances, the recoil track will not cross a mineral surface unless the daughter is also directly ejected. It has been argued that ^{222}Rn is a special case, and as an unreactive noble gas is preferentially released. This is discussed in detail in Section 2.2.

Distributions in host rock minerals. A key issue in determining the recoil supply of ^{222}Rn and other daughter nuclides, as well as the weathering supply of the long-lived isotopes, is the distribution of ^{238}U and ^{232}Th within host rocks. During the initial formation of igneous rocks, U and Th are incompatible with many major rock-forming minerals and are largely incorporated into various minor phases, but also may be on grain boundaries. For example, analyses of the Götemar Granite in SW Sweden found zircon and monazite to be dominant sources of U and Th, while there were secondary amounts in sphene and magnetite (Smellie and Stuckless 1985). Weathering and secondary processes may redistribute U and Th (see Chabaux et al. 2003). In a granodiorite weathering profile, U in weathered rocks was largely in primary resistates and Th was in fine-grained material such as clays and Fe phases; selective leaching experiments found that U was fractionated from Th by up to 30% (Pliler and Adams 1962). Studies of monazite weathering by Read et al. (2002) found that preferential loss of U occurred, while Th was also leached from primary phases but was incorporated in nearby sites within the host rock in microcrystalline silicate and oxide alteration products. Redistribution between mineral phases may also occur by hydrothermal alteration during initial cooling of igneous rocks, weathering, and during transport of minerals before incorporation into sedimentary deposits. Overall, it is generally difficult to predict the distributions of U and Th in host rocks without direct detailed study of the host aquifer rocks, and any modeling of a particular past process of secondary radionuclide redistribution from groundwater data (e.g., Tricca et al. 2001) has been only hypothetical and essentially unconstrained. Once an aquifer regime has been established, further U and Th redistribution can result from precipitation under presently observed aquifer conditions. In this case, there are often mass balance constraints on the fraction of aquifer U and Th could have been released by *in situ* weathering and the associated removal of major elements. This is particularly true of less soluble elements such as Th, which are not readily mobilized and so often have not been transported far from weathering zones (see Section 2.4).

From the diversity of potential sites for U and Th, it is clear that the weathering release of ^{238}U and ^{232}Th, as well as of the recoil and leaching release of daughter nuclides, must be determined for each site. Nonetheless, most studies assume that ^{238}U and ^{232}Th have similar distributions, so that the values for ε_I (the fraction released by recoil) and w_I (the weathering release constant) is assumed to be approximately equal for all nuclides I.

Colloids. As discussed above, in most models groundwater concentrations are characterized by bulk measurement of all species. One particular component in groundwater that has received particular attention is colloids, which have the potential of increasing the proportion of radionuclides in the groundwater rather than on surfaces, and so in some circumstances may provide an explanation for enhanced migration rates. In general, colloids are a common component of groundwaters. The composition, structure, and size distribution of colloids vary widely and so the population in each aquifer must be characterized separately. Inorganic constituents include weathering products such as Fe or Mn oxyhydroxides and clays as well as precipitates. Organic colloids include humic and fulvic acids, along with microorganisms. Small fragments of primary host rock mineral phases may also be included. Colloids have large surface areas for complexation and ion exchange reactions, and have been shown to effectively transport a variety of constituents. Colloids are often characterized operationally, especially when separation from large volumes is required for trace element analyses, and may contain a mixture of different active components, although bulk characteristics can be used to determine their overall importance to trace element transport. There are now various ultrafiltration separation techniques for separating sufficient quantities of colloids for trace element analyses from large volume samples (Buffle et al. 1992). Unfortunately, different techniques may separate different colloid populations due to differences in exclusion by size, shape, charge, or other characteristics, and a direct comparison between available methods using seawater samples (Buesseler et al. 1996) found a significant range of results that cannot be clearly interpreted. Nonetheless, available data provide qualitative information on the importance of colloid transport in groundwaters.

A general review of actinide transport on colloids, and how this may relate to radionuclide transport studies, is provided by Ivanovich (1991). It has been found that colloids can carry a large fraction of U (Dearlove et al. 1991). Due to the greater reactivity,

Th is likely to be more efficiently carried by colloids, and it has indeed been found on colloids (Short and Lowson 1988). Porcelli et al. (1997, 2001) found that in an organic-rich river a significant fraction of U, and a greater fraction of Th, in colloids that likely included humic acids. Ra is not expected to be so strongly associated with humic acids, but may be readily attached to other colloids due to its strong affinity to surfaces. Overall, these associations of radionuclides with colloids can increase groundwater concentrations above those possible in the presence of smaller, dissolved ligands, or compete with host rock surface adsorption sites. In these cases, a controlling parameter in nuclide transport is the migration rate of the colloids (see review by Ryan and Elimelech 1996).

An important issue is whether there is continuing rapid and complete isotopic equilibration between colloids and other groundwater species, so that no isotopic differences are maintained. If this does not occur, and the colloidal species respond differently than dissolved components to processes such as adsorption, then the net effect is that the nuclides in groundwater cannot be described as a single population, violating a basic assumption of most models. In this case, colloid-bound nuclides must be considered separately and assigned exchange rates with other components. Some studies have found that low fractions of U were on colloids, with no isotopic difference from dissolved U, suggesting isotopic equilibrium occurs (see Osmond and Ivanovich 1992). No U isotopic differences were found in riverine U that was likely associated with humic acids, even as changes occurred along the river (Porcelli et al. 1997, 2001). Laboratory experiments also found that actinides uptake by humic groundwater colloids was reversible with changes in pH (Kim et al. 1992). In contrast, Dearlove et al. (1991) found that U on colloids in organic-rich groundwaters was not in isotopic equilibrium. Substantial differences were found in some samples from a sandy aquifer (Tricca et al. 2001), with higher $^{234}U/^{238}U$ ratios found in the colloids, indicating that the difference was not due to the presence of colloid-sized primary mineral fragments, which do not contain ^{234}U excesses. Since Th and Ra may be more closely associated with colloids than U, it is possible that isotopic differences between groundwater species may be found for these elements, although there is insufficient data available to quantify this.

Models of the systematics of the decay series nuclides generally have not considered the effects of colloid transport. If radionuclides associated with colloids are fully exchangeable with dissolved atoms, then the modeling of groundwater as a single component is valid, and colloids then may provide an explanation for observed increased groundwater concentrations and faster transport. Where rapid isotopic exchange does not occur, a separate component must be considered to account for isotopic variations in the groundwater. However, there is little data quantifying the isotopic composition of colloid-bound radionuclides, and so little basis for modeling isotopic exchange rates between colloids and dissolved species. This is clearly an area for future research.

2.2. Radon and the recoil rate of U-series nuclides

As a noble gas, ^{222}Rn in groundwater does not react with host aquifer surfaces and is present as uncharged single atoms. The radionuclide ^{222}Rn typically has the highest activities in groundwater (Fig. 1). Krishnaswami et al. (1982) argued that ^{222}Rn and all of the other isotopes produced by α decay are supplied at similar rates by recoil, so that the differences in concentrations are related to the more reactive nature of the other nuclides. Therefore, the concentration of ^{222}Rn could be used to calculate the recoil rate for all U-series nuclides produced by α recoil. The only output of ^{222}Rn is by decay, and with a 3.8 day half-life it is expected to readily reach steady state concentrations at each location. Each measured activity (i.e., the decay or removal rate) can therefore be equated with the input rate. In this case, the fraction released, or emanation efficiency, can be calculated from the bulk rock ^{226}Ra activity per unit mass;

$$b\varepsilon_{222}(^{226}Ra)_R = (^{222}Rn)_W \quad (4)$$

where ε_{222} is the fraction of ^{222}Rn produced within the host rock that is released by recoil into groundwater, and b is the conversion factor to account for the ratio between the masses of host rock and groundwater (see Eqn. 1). Since $(^{226}Ra)_R$ generally can be measured or estimated (from an estimated ^{238}U concentration), a measurement of $(^{222}Rn)_W$ leads directly to a value for ε_{222}. The measured ^{222}Rn groundwater activities per unit mass that are obtained can correspond to ^{222}Rn release rates of up to ~10% of the amount being produced in the aquifer rock (Krishnaswami et al. 1982). This requires that ~20% of the ^{226}Ra in the host rock is within recoil distance of the surface. Such surprisingly high rates generally cannot be reached by recoil from typical size aquifer grains with uniform parent Ra concentrations. There are various possible causes for the high release rates, with different implications for the recoil of other nuclides. These include:

- *A dominant proportion of the ^{226}Ra is in small grains.* The size of the grains required to produce emanation rates of up to ~10% are several microns in diameter. This is generally below the typical size of U-bearing minerals in aquifers, although it is possible locally. In this case, the recoil into groundwater of other nuclides in the chain will be similarly high, subject to the scaling parameters discussed in Section 2.1.

- *A dominant proportion of the Ra is in secondary phases.* The concentration of parent ^{226}Ra in surface coatings and secondary phases is often considered the main cause of high emanation rates. For an emanation rate of ~10%, up to 20% of the ^{226}Ra in the rock must be located in such sites. For many environments, there is no evidence for weathering of as much as 20% of a sedimentary aquifer to release the Ra and reprecipitate it onto remaining grains within the present geological environment. The redistribution is likely to have occurred during evolution of the source rocks, weathering, transport of sediment, or final deposition. Clearly, if this occurred at an age much greater than 10^3 years (and so much greater than the 1600 a half life of ^{226}Ra), then the ^{226}Ra must be supported by ^{230}Th, and if much greater than 10^5 years, then the ^{230}Th must be supported by ^{234}U and possibly also ^{238}U. Therefore, the recoil rate of ^{222}Rn is related to the activity of other nuclides further up along the supported chain within the secondary phases. In order to determine how much of the chain is supported in these sites, and so which other radionuclides are similarly subject to the high recoil rates recorded by ^{222}Rn, the time period for the development of these phases must be explicitly considered.

- *^{226}Ra adsorbed on surfaces can also supply ^{222}Rn.* As described above, this would require weathering of a considerable fraction of the host rock. However, there may be circumstances where changing conditions can cause adsorption from Ra-rich waters that have transported ^{226}Ra into the aquifer from elsewhere, for example where brines that readily dissolve Ra mix with fresh groundwaters, resulting in a large increase in Ra partition coefficient and so extensive adsorption of the Ra from the brine (Moise et al. 2000; see below). In this case, the high ^{222}Rn recoil rates do not apply to other nuclides. However, this does not provide a general explanation for high ^{222}Rn recoil rates.

- *Leaching of nuclides implanted into adjacent minerals* has been suggested for the supply of ^{222}Rn into the vadose zone. Where there are intermittent undersaturated conditions, i.e., in soils or rocks where the water table lowers seasonally, the low stopping power of air allows atoms ejected from minerals to be implanted across pore spaces. These atoms will then be available for leaching subsequently (Fleischer and Raabe 1978) and this has consequences for the seasonal supply of ^{222}Rn from soils and at the water table. This surface accumulation can also occur in minerals within low permeability rocks (Suksi and Rasilainen 1996). This raises the possibility that there are available pools of other nuclides in minerals from such rocks that have been eroded and incorporated into aquifer rocks within time periods comparable with the half-life of some nuclides. For example, sediments from the last glaciation may have an implanted near-surface pool of ^{226}Ra, ^{230}Th and ^{234}U that not only can provide high supply rates into surrounding groundwater of recoiled daughters, but can also be leached at a rate unrelated to present ^{222}Rn fluxes. However, this remains to be documented.

- *U and Th may be heterogeneously distributed within the aquifer.* In particular, these may be enriched in fine-grained clay layers or other aquicludes with very low hydraulic conductivities that are not part of the main water-bearing deposits considered in model calculations. If these materials are interspersed within the aquifer rocks, then there is the potential for ^{222}Rn to diffuse out into the main groundwater flow, while other radionuclides are retained by adsorption in the aquiclude. However, such a process may have restricted application due to the short half-life of ^{222}Rn, which limits the distance that may be travelled. Tadolini and Spizzico (1998) found that ^{222}Rn concentrations in an Italian limestone were also much higher than could be supplied by the host rock, and found that it was supplied by ^{226}Ra (presumably supported by ^{238}U) concentrated in nearby layers of detrital material. The migration from such deposits of ^{222}Rn and other radionuclides with high retardation factors but longer half lives has neither been modeled.

- *Diffusive transport in a hypothesized network of nanopores* has been suggested by Rama and Moore (1984) as a mechanism for transporting ^{222}Rn from across a large volume of a mineral to grain boundaries. The result is a high ^{222}Rn release rate that does not apply to other nuclides. High release rates of ^{220}Rn ($t_{1/2} = 1$ minute) have also been reported (Rama and Moore 1984; Howard et al. 1995) that would require even faster diffusive transport, although it is in the ^{232}Th chain, and so may be produced at different sites. Rama and Moore (1990a) found that high emanation rates were observed even from individual crystals that did not exhibit evidence for surface concentrations of parent nuclides. Further, Rama and Moore (1990b) showed that ^{220}Rn diffusion through mineral slabs was uneven, and suggested that this reflected nanopore geometry. Similarly, Andrews and Wood (1972) suggested that ^{222}Rn migrates along dislocation planes and grain boundaries in wall rocks. However, other experiments have not found evidence for such nanopores. Hussain (1995) found in experiments that the activity of ^{212}Pb in water surrounding monazite and zircon grains was not enhanced by rapid escape of its parent ^{220}Rn compared to that of its precursor ^{224}Ra in the ^{232}Th chain. Copenhaver et al. (1993) measured similar recoil rates of ^{222}Rn and ^{224}Ra from core samples. Krishnaswami and Seidemann (1988) found that Ar isotopes produced throughout sample grains during irradiation did not leak out along with ^{222}Rn, suggesting that ^{222}Rn was not released by interconnected pores available to both Ar and Rn. Overall, more direct evidence of the widespread presence of nanopores is required to substantiate this mechanism.

Mathematical treatments of ^{222}Rn release from minerals have been developed (Semkow 1990; Morawska and Phillips 1992), although these are only useful once the issues listed above are resolved. Clearly, additional work is required before a more definitive statement is made about the existence of nanopores. However, it is likely that different features will control the recoil flux of the nuclides for different lithologies and local geological histories, and it will be necessary to investigate the various possibilities at each study site. As discussed above, the ^{222}Rn release rate cannot be simply extrapolated to all other parent nuclides. It should be emphasized that, as discussed further in subsequent sections, ^{222}Rn measurements are often used for obtaining the emanation efficiencies of other radionuclides, which in turn are critical for deriving key parameters for the behavior of these species. Where the ^{222}Rn emanation efficiency obtained from ^{222}Rn measurements can be shown to be compatible with that deduced from the geometry of U- and Th-bearing mineral grains, then a generalized recoil rate obtained from ^{222}Rn may be justified. However, this information necessary to confirm this is often not available. Also, high ^{222}Rn release rates due to preferential siting of ^{226}Ra in secondary sites seems most plausible at present, although this does not always necessarily imply that ^{230}Th, ^{238}U, or ^{232}Th are also in such locations, and the possibility that ^{222}Rn is preferentially released from mineral phases relative to all other radionuclides by diffusion in nanopores or by some other mechanism remains. Where the assumption that all radionuclides are recoiled at the same rate is not correct the model equations are not invalidated, but rather require different input values for ε_1. Some calculated results do not depend upon radionuclide recoil rates relative to that of ^{222}Rn and so are unaffected by these considerations.

2.3. Ra isotopes

There are four naturally occurring isotopes of Ra; ^{228}Ra ($t_{1/2}$ = 5.8 a) and ^{224}Ra (3.7 d) in the ^{232}Th series, ^{226}Ra (1600 a) in the ^{238}U series, and ^{223}Ra (11.7 d) in the ^{235}U series (Table 1). The data for ^{223}Ra are more limited, since it is generally present in low concentrations due to the low abundance of ^{235}U. The differences in half lives and the connections across the different decay series have been used to infer a variety of groundwater and water-rock interaction features. For the short-lived Ra isotopes, the dominant input term to groundwater is recoil, rather than weathering, and steady state concentrations are often achieved (see Section 2.2).

Behavior of Ra in groundwater. The general behavior of Ra has been examined under laboratory conditions and in various environments (see Osmond and Cowart 1992). A major goal of field studies of Ra isotopes have aimed at obtaining bulk, *in situ* values of adsorption rates and so the retardation factors. Note that Ba serves as a very close chemical analogue to Ra but is typically 10^7 times more abundant, and so its behavior is related to that of Ra.

The speciation of Ra is reviewed by Dickson (1990). In low salinity solutions, Ra occurs as uncomplexed Ra^{2+}, while significant complexing as $RaSO_4$, $RaCO_3$, and $RaCl^+$ will only occur in brines with high concentrations of the respective inorganic ligand. Organic complexing has not been considered to be significant in groundwaters (Dickson 1990). It is possible that colloid and particulate transport on clays and iron hydroxides may have a role in Ra transport, although this has not been widely documented. The solubility limit of Ra compounds is generally not reached, but Ra can be precipitated in solid solution within Ca and Ba minerals. For example, in deep brines of the Palo Duro Basin, saturation of sulfate minerals such as barite and gypsum controls dissolved Ra concentrations (Langmuir and Melchior 1985), and Langmuir and Riese (1985) found Ra is incorporated into barite without substantial Ra/Ba fractionation from the groundwater ratio. While Andrews et al. (1989) suggested that calcite precipitation controls dissolved Ra in the Stripa granite based on a correlation between Ra and Ca, Gnanapragasam and Lewis (1991) found that the Ra/Ca ratio in precipitated calcites is ~10^{-2} times the value for the source solution, and so calcite is less likely to be a common control on Ra concentrations.

Adsorption exerts a strong control on Ra in dilute groundwater. Adsorption contstants are strongly dependent upon the type of substrate, solution composition (e.g., Eh, pH, and other cations), and temperature (see Benes 1990). Decreases in adsorption have been observed due to increases in salinity (e.g., Zukin et al. 1987; Krishnaswami et al. 1991; Sturchio et al. 2001). Reasons suggested for this include competition by other, abundant cations for available adsorption sites, increases in mineral surface charge (Mahoney and Langmuir 1991), increases in the stability of inorganic complexes (Hammond et al. 1988), and the presence of strong organic complexes (Langmuir and Riese 1985; Molinari and Snodgrass 1990). Where there are strong changes in groundwater salinity, e.g., by mixing, Ra may be deposited on aquifer surfaces and so may be a local source of ^{222}Rn (Moise et al. 2000).

Initial inputs to groundwater can occur due to weathering in the vadose zone. There are few data for Ra in natural soils (e.g., not impacted by mining wastes) and Frissel and Köster (1990) have reviewed what is known of the mobility in soils of Ra. Partitioning onto soil solids is quite strong, but it is not uniform and is hard to predict due to the effects of precipitation reactions, bioturbation, and varying distributions of organic matter and clays. Groundwater characteristics will reflect vadose zone inputs for some distance below the water table depending upon the half-life of the isotopes (Tricca et al. 2000).

Isotope systematics. There are two Ra isotopes in the ^{232}Th series, which is shown in Figure 3, and determining the behavior of these Ra isotopes requires consideration of the closely related Th isotopes as well. The factors affecting the groundwater concentrations of each relevant nuclide in the series are:

- ^{232}*Th*. The ^{232}Th within the mineral phases produces ^{228}Ra that is ejected by recoil. Any ^{232}Th that has been released by weathering will be largely on the surface, and will produce ^{228}Ra that will be recoiled back into the mineral or into groundwater. Note that there is generally no a priori way of determining the amount of ^{232}Th that is not in the mineral, although this might be calculated from both the amount of ^{232}Th in the water that is exchanging with the surface reservoir, and the distribution factor obtained from Th isotope systematics (Luo et al. 2000; see Section 2.4).

- ^{228}*Ra*. The ^{228}Ra in the water is largely supplied by the ^{232}Th in the minerals and on the surfaces, and since it is strongly adsorbed exchanges with a much larger surface reservoir.

- ^{228}*Th*. ^{228}Ra decays by low energy beta decay to the short-lived ^{228}Ac (with a half life of 6 hours), which rapidly decays to ^{228}Th. It is generally assumed that the behavior of ^{228}Ac, due to its short lifetime, does not affect the subsequent supply of ^{228}Th, and so ^{228}Ac is ignored. Since the decays of both ^{228}Ra and ^{228}Ac are low energy, there is no release of ^{228}Th from the mineral by recoil. Consequently, the ^{228}Th in the surface is supplied only be decay of the parent ^{228}Ra reservoir already there, and the amount in groundwater is determined by interchange with the surface. This therefore directly connects the mobile abundances of ^{228}Th and ^{228}Ra.

- ^{224}*Ra*. The amount in groundwater is largely supplied by recoil from ^{228}Th in the host rock minerals, as well as ^{228}Th on the mineral surfaces (which in turn was supplied by ^{228}Ra that was recoiled from the mineral as well as possibly from ^{232}Th on surfaces).

Note that surface and groundwater nuclides can be considered together as the mobile pool (see Ku et al. 1992; Luo et al. 2000), with the Th and Ra isotopes largely residing at the surface. The supply to this pool is by recoil; weathering for short-lived nuclides can be neglected (see Section 2.2), although it is the source of any ^{232}Th that is outside the host minerals.

The main factors that have been considered in calculating Ra nuclide abundances in the surface and groundwater are as follows:

- Since ^{228}Ra and ^{224}Ra are both produced by recoil from the host mineral, it might be assumed that the production rates are equal. However, the relative recoil rates can be adjusted by considering that the parent nuclides near the mineral surface may not be in secular equilibrium due to ejection losses; i.e., the activity of ^{228}Th may be lower than that of ^{232}Th due to recoil into groundwater of the intermediate nuclide ^{228}Ra. Krisnaswami et al. (1982) calculated that the recoil rate of ^{224}Ra is 70% that of ^{228}Ra if radionuclides are depleted along the decay chain in this way.

- Taking into account the decay of nuclides on the surfaces, the ^{224}Ra on both surfaces and in the groundwater is supplied by two recoil events; directly from the mineral, and indirectly through the supply of ^{228}Ra. Note that not all ^{228}Ra may lead to mobile ^{224}Ra, since some ^{228}Th is recoiled back into the mineral.

- There may be mobile ^{232}Th that supplies ^{228}Ra directly, and so also ^{224}Ra indirectly. Note that there are few ^{232}Th data available, and this flux is often assumed to be insignificant.

The Ra isotopes in the other decay series can be evaluated similarly. ^{226}Ra in the ^{238}U series (Table 1) is the product of the third α decay, and so the effects of near-surface depletion or decay of recoiled precursors must be calculated accordingly. ^{223}Ra in the ^{235}U series is also the product of the third α decay. Further processes that may be considered where circumstances warrant include nonsteady state conditions or removal by precipitation at rates that are fast compared to the decay rate of the Ra nuclides.

^{226}Ra abundances. Activities of ^{226}Ra per unit mass in groundwater vary widely. Fresh groundwaters typically have on the order of 1-2 dpm/L (e.g., King et al. 1982). Saline groundwaters, by contrast, have much higher concentrations that correlate with salinity (e.g., Krishnaswami et al. 1991). This is compatible with the much lower partitioning of Ra onto surfaces in such waters. While the shorter-lived isotopes will generally adjust to local conditions over short distances, ^{226}Ra, with a much longer half-life than the other Ra isotopes, requires a much longer distance along a flowline to achieve a steady state groundwater concentration. This is reached once the recoil rate from host minerals in a volume of aquifer is equal to the activity (i.e., decay rate) of the ^{226}Ra in both the groundwater and on the surfaces (that is, the supply rate to the mobile pool is equal to the decay rate). The distance over which this occurs depends upon the fraction adsorbed, the groundwater flow rate, and the recoil rate (Section 2.1), and may occur within a km (Tricca et al. 2000). Therefore, ^{226}Ra concentration gradients may be found over greater distances tan those of the other Ra isotopes, but still close to abrupt changes in groundwater chemistry.

(^{226}Ra/^{228}Ra) ratios. This ratio in the groundwater is dependent upon the (^{238}U/^{232}Th) ratio of the host rock, and so provides information on the relative recoil rates of nuclides from the two decay series. Supply to the groundwater by recoil produces (^{226}Ra/^{228}Ra) ratios up to 1.75 times that of the rock due to accumulation in the mobile pool of preceding nuclides (Davidson and Dickson 1986). The average upper crust has a

(^{238}U/^{232}Th) activity ratio of ~0.8 (equivalent to a Th/U weight ratio of 3.8), and this is often taken to represent that of the host rocks in the absence of direct measurements, although this can of course be substantially different in rocks such as limestones or other sedimentary deposits. If it is assumed that the groundwater profile is in steady state, that weathering and precipitation are not important for these nuclides, and that the parent nuclides ^{230}Th and ^{232}Th have similar behaviors, then the corresponding terms in Equation (1) can be ignored. Assuming further that the desorption rate is fast compared to the Ra half lives, then the (^{226}Ra/^{228}Ra) ratios in the groundwater and adsorbed on surfaces (and so in the mobile Ra pool) are equal. In this case, the measured groundwater (^{226}Ra/^{228}Ra) ratio reflects the ratio of the supply rates of ^{226}Ra and ^{228}Ra, which is equal to the (^{238}U/^{232}Th) rock ratio, adjusted for any differences in the distributions of ^{238}U and ^{232}Th and the depletion of ^{230}Th relative to ^{232}Th due to recoil along the ^{238}U decay chain. Luo et al (2000) found (^{226}Ra/^{228}Ra) ratios of 0.34 to 1.4 in a basaltic aquifer, compared to a bulk rock (^{238}U/^{232}Th) value of 0.9, and the variation may be due to different distributions of U and Th producing different relative recoil rates, either reflecting different amounts adsorbed onto surfaces (Luo et al. 2000) or within the host rock. The values found by Tricca et al. (2001) in a sandy aquifer of 0.3 to 0.9 are low compared to the likely (but unmeasured) (^{238}U/^{232}Th) ratio of the host rock and may reflect variable losses of Th in the depositing sediments relative to source rocks due for example to sorting of heavy minerals. Local variations are also possible due to changes in supply or adsorption characteristics along a groundwater flow line followed by a much slower return to steady state along the flow line by longer lived ^{226}Ra.

Sturchio et al. (2001) found (^{226}Ra/^{228}Ra) ratios in carbonate aquifers in the central US largely within the range of 1.3-10 (with a complete range of 0.7-17) that generally coincides with the range in the aquifer rocks. There is a strong correlation between concentrations of ^{226}Ra and total dissolved solids, as well as of Ca, Sr, and Ba, and some fluids are saturated with calcite and barite. The highest ^{226}Ra concentrations are up to 6 times that found in the host rock, and it was suggested that this is due to the supply of ^{226}Ra by ^{230}Th enrichments on surfaces. However, concentrating so much ^{230}Th by local weathering is implausible, and this Th is likely derived from layers with higher U concentrations.

(^{224}Ra/^{228}Ra) ratios. As members of the same decay series, these Ra isotopes are the most closely related and differences in groundwater activity ratios cannot be ascribed to differences in aquifer rock distributions of ^{238}U and ^{232}Th. Within host rocks that are in secular equilibrium, (^{224}Ra/^{228}Ra) = 1. Reported values for fresh groundwaters fall in a narrow range, with 0.5-2.1 reported for a sandy aquifer (Tricca et al. 2001), 1.0-2.2 (and one high value of 4.2) for a basaltic aquifer (Luo et al. 2000), and 0.8-1.8 for arkose and glacial drift (Krisnaswami et al. 1982). Much higher values have also been reported (e.g., Davidson and Dickson 1986; Krishnaswami et al. 1991). Various reasons have been discussed for values that are different from the host rock production ratio. Krishnaswami et al. (1982) calculated a ratio of 0.67 due to progressive depletion along the decay series at grain boundaries, and 1.2 if ^{228}Ra and ^{228}Th are adsorbed onto surfaces and continue to supply daughters into the groundwater. Davidson and Dickson (1986) considered the recoiled ^{228}Ra, strongly adsorbed, to be a source of ^{228}Th (with 50% recoiled into solution) and so ^{224}Ra, so that steady state values in groundwater should be (^{224}Ra/^{228}Ra) = 1.5 (see also Luo et al. 2000). Another process is therefore required to explain the highest measured ratios, and some variations may be due to redistribution of ^{228}Th. There are also several effects to be considered due to the much shorter half-life of ^{224}Ra. Along a flowline, ^{224}Ra will reach a steady state concentration more rapidly than ^{226}Ra, and so higher values of (^{224}Ra/^{228}Ra) will be observed in recently recharged waters (Davidson and Dickson 1986; Tricca et al. 2000), or immediately downgradient of Ra

precipitation (Sturchio et al. 1993). Also, ^{224}Ra may decay within the surface layer at a rate comparable to the desorption rate, resulting in a lower effective partitioning value for ^{224}Ra over ^{228}Ra (Eqn. 3; Krishnaswami et al. 1982). In general, it appears that unusually high values are due to circumstances where steady state conditions have not been reached, while smaller variations are due either to the recoil loss and redistribution of ^{228}Th or decay of ^{224}Ra on surfaces where desorption rates are relatively long.

(^{224}Ra/^{222}Rn) ratios. A comparison between Ra and Rn isotopes can be made when the recoil supply rates are related to one another. Notwithstanding the range of possible mechanisms for Rn release (see Section 2.2), if it is assumed that measured ^{222}Rn concentrations provide values for the emanation efficiency of all radionuclides, with the adjustment for the rock (^{238}U/^{232}Th) ratio, then the (^{224}Ra/^{222}Rn) ratio can be used to obtain a value for the partition coefficient of Ra. The concentrations of ^{224}Ra, along with ^{222}Rn, are generally expected to be in steady state due to their short half lives. In this case, the total activity of mobile ^{224}Ra in a volume of aquifer (that is, ^{224}Ra in solution and adsorbed, which is equal to $(1 + K_{226Ra})(^{226}Ra)_w$) and so is equal to the recoil supply rate of ^{224}Ra, and the total activity of ^{222}Rn in solution (with none adsorbed) is equal to the recoil supply rate of ^{222}Rn. While the (^{224}Ra/^{222}Rn) ratio for the mobile pool is then equal to the ratio of the emanation rates, which is then equal to $(^{224}Ra/^{222}Rn)/K_{224Ra}$. Although typical bulk rock (^{238}U/^{232}Th) activity ratios are close to one, so that the recoil rates of ^{224}Ra and ^{222}Rn are similar, the ratio of the recoil supplies of ^{222}Rn and ^{224}Ra might be somewhat different due to different distributions between ^{238}U and ^{232}Th, and this can be confirmed using the (^{226}Ra/^{228}Ra) ratios (see above). Also, ^{222}Rn is further down its decay series and is preceded by 4 α decays, while ^{224}Ra is preceded by 2, and so ^{222}Rn production may be 50% higher due to precursors accumulated on surfaces. Measured ranges of ^{224}Ra/^{222}Rn ratios in fresh groundwaters of $(0.5-2.2) \times 10^{-4}$ (Tricca et al. 2001), $(0.8-1) \times 10^{-4}$ (Luo et al. 2000), and $(0.2-4.4) \times 10^{-4}$ (Krishnaswami et al. 1982) somewhat surprisingly coincide, indicating that the partitioning of Ra onto host surfaces are similar despite the contrasting lithologies. The corresponding partition coefficients are 10^3-10^4. Some variation may be due to desorption rates that are comparable to the decay constant of ^{224}Ra. By comparing (^{224}Ra/^{228}Ra) and (^{224}Ra/^{222}Rn) ratios, Krishnaswami et al. (1982) calculated sorption rates of 3-20 min^{-1} using Equation (3), and clearly similar values can be obtained for each of the other datasets. Higher ^{224}Ra/^{222}Rn ratios have been found in saline groundwaters (Krishnaswami et al. 1991), consistent with the reduced absorption in these waters.

Luo et al. (2000) used a somewhat different method for determining the partitioning of Ra by noting that ^{222}Rn is produced by the total amount of ^{226}Ra both in solution and on surfaces (and so equal to $(1 + K_{226Ra})(^{226}Ra)_W$) as well as by recoil. ^{224}Ra is produced similarly by ^{228}Ra (through the beta decay of ^{228}Th) from within the minerals and from the surface. Combining the respective equations (by assuming that the recoil rates for ^{238}U series nuclides and ^{232}Th series nuclides are proportional to the (^{238}U/^{232}Th) ratio of the bulk rock), values for the recoil of ^{224}Ra that are 0.1-0.2 that of the ^{222}Rn flux were calculated, compared to a value of 0.3 expected by assuming that there was no ^{238}U and ^{232}Th (and associated daughters) on surface sites. This difference does not significantly change the calculated Ra partition coefficients that can then be obtained from ^{226}Ra or ^{224}Ra, but it implies that there is a significant production of radionuclides above that from recoil from ^{238}U in the host mineral alone, and requires a substantial budget of U on mineral surfaces (see Section 2.5).

(^{223}Ra/^{226}Ra) ratios. Since ^{235}U and ^{238}U are present in constant proportions everywhere and so are located in the same locations, there are no fractionations between their decay series due to siting in the host rock. ^{223}Ra and ^{226}Ra are both generated after 3

α decays, and so waters should have a ratio similar to the rock ($^{235}U/^{238}U$) activity ratio of 0.046. Where ^{226}Ra has not reached a steady state concentration in a flowing groundwater after recharge or precipitation due to its longer half-life, higher ratios may be found locally (Davidson and Dickson 1986; Martin and Akber 1999).

Martin and Akber (1999) looked at a confined aquifer of weathered and fresh bedrock below the Ranger U mine in Australia that apparently has not been significantly impacted by mine tailings. ($^{223}Ra/^{226}Ra$) ratios substantially above the rock value of 0.046 were found in samples that were saturated in barite and suggested this was due to precipitation of both Ra isotopes, followed by more rapid return to steady state of the shorter-lived isotope, creating higher ($^{223}Ra/^{226}Ra$) ratios near the sites of precipitation that progressively drop off as ^{226}Ra is added. This data therefore could be used to calculate how much ^{226}Ra was precipitated. An interesting observation at the site was the presence of some low ($^{223}Ra/^{226}Ra$), as well as ($^{224}Ra/^{228}Ra$), ratios that might be explained by the transport to the site of a substantial fraction of the longer lived Ra on colloids and without rapid exchange with dissolved Ra. This allowed a greater concentration of ^{226}Ra and ^{228}Ra to be maintained against more efficient removal of dissolved Ra that contained a greater proportion of the ^{223}Ra and ^{224}Ra.

Summary. Ra isotopes can provide information regarding the supply of ^{238}U series nuclides relative to those of ^{232}Th series, and indicate where significant changes in adsorption or parent element distribution occur along groundwater flow paths. Values for Ra partitioning coefficients, and so retardation factors, can be obtained, but only by assuming that ^{222}Rn provides a reasonable measure of recoil supply of Ra isotopes. In this way, values of 10^3-10^4 for K_{Ra} have been calculated for a range of aquifer lithologies. This is an important result, as it suggests that under natural conditions, low concentrations of Ra will migrate at rates of 10^{-3}-10^{-4} times that of the groundwater, and so is significantly retarded by adsorption. Different studies have used different methods to calculate precise values, based on assumptions regarding the relationship between recoil supplies of the different isotopes. The simplest approach for calculating retardation factors is based on ($^{222}Rn/^{224}Ra$) ratios, although ^{224}Ra data is not always available and more complex calculations can be used with further assumptions. An important issue is how high can Ra concentrations be elevated without changing adsorption rate constants, and how variations in other factors that may accompany releases of anthropogenic nuclides (e.g., pH, or concentrations of other solutes and colloids) affect Ra adsorption. In regions where saline waters occur, mixing and precipitation processes may result in waters that do not have steady state concentrations, and Ra evolutions become more complicated (e.g., Moise et al. 2000). Detailed profile data in these areas can potentially provide more Ra transport information, although care must be taken that changes in values for all input parameters (such as adsorption rates due to changes in adsorbing surface area) is considered. It should be emphasized that while the conclusion that Ra is strongly adsorbed is compatible with laboratory data (e.g., Ames et al. 1983), the values obtained in radionuclide transport models are based on the assumption that measured ^{222}Rn concentrations provide a reasonable measure of the recoil rate of other nuclides in the decay series. While this is plausible (see Section 2.2), the thorough investigation of other possibilities that might lead to significant changes in these values (up to a factor of 10) is still required, although the conclusion that Ra is strongly adsorbed at the freshwater locations studied will likely not be affected.

2.4. Th isotopes

The isotopes of Th certainly have the widest range of half-lives, from the decay series parent ^{232}Th (1.39 × 10^{10}a half life) and the relatively long lived ^{230}Th (7.52 × 10^4 yrs) to a short-lived isotope in each of the decay series: ^{234}Th (24 days), ^{227}Th (18 days), and ^{228}Th (1.9 years). Unfortunately, there are few data for ^{227}Th, which is a part of the less abundant ^{235}U chain, and the long-lived Th isotopes have not been explored as much as the others due to lower activities, and so difficulties in obtaining precise measurements.

Behavior of ^{232}Th in groundwater. ^{232}Th can only be released from aquifer minerals by weathering, and the presence of ^{232}Th in groundwaters throughout aquifers despite strong removal processes indicates that continuous release occurs. Th concentrations in waters are generally very low due to the low solubility of thorianite (ThO$_2$) and strong sorption properties. While very little data has been available, it is clear that Th is highly reactive with mineral surfaces, and is much more strongly sorbed than U. In natural waters, it is present only in the tetravalent form. Langmuir and Herman (1980) have calculated the solubility and speciation of Th in groundwaters. Above a pH of 5, the dominant species is Th(OH)$_4$ in pure water, but is Th(HPO$_4$)$_3$ between 5 and 7 where phosphate is available. Data for EDTA suggests that complexing with organic ligands can increase Th solubility. Solubilities calculated by Langmiur and Herman (1980) for possible groundwater compositions are shown in Figure 5. Available ^{232}Th data include those for saline groundwaters from Missouri carbonates and sandstones of (0.1-9.1) pg/g (Banner et al. 1990) and a similar range for a basaltic Idaho aquifer in the Snake River plain of (0.1-11.5) pg/g (Luo et al. 2000). An unconsolidated unconfined sandy Long Island aquifer in New York has comparable values of (0.01-11.5) pg/g but one sample from the underlying, oxygen-poor (but still oxic) sandy Magothy aquifer has a much higher concentration of 90 pg/g (Tricca et al. 2001). Values from a bedrock spring (7 pg/g) and from within overlying tills (8-27 pg/g) from Sweden are comparable

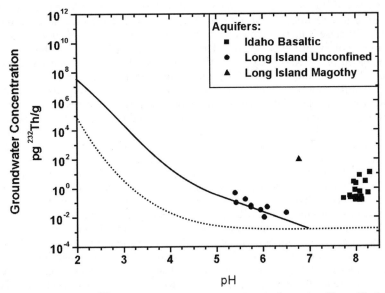

Figure 5. Concentrations of ^{232}Th and pH are shown for groundwaters from a basaltic aquifer (■ Luo et al. 2000), a sandy unconfined (●) and underlying confined (▲) aquifer (Tricca et al. 2001). The dotted line is the solubility of thorianite in pure water, and the solid line is for the presence of typical inorganic ligands (Langmuir and Herman 1980). In the pH range for the samples, phosphates are the dominant inorganic ligands, although there is no phosphate data available for the samples to calculate actual solubility limits. Groundwater concentrations above the solubilities shown can be attained in the presence of organic compounds or colloids.

(Porcelli et al. 2001). All of these data are above the solubility of thorianite in pure water. The unconfined Long Island aquifer data are inversely correlated with pH and fall along the solubility curve for inorganic ligands in a model groundwater composition in Fig. 5 (Langmuir and Herman 1980). However, the phosphate concentrations are not available for these samples, and so actual saturation indices for these samples cannot be calculated. Therefore, it is possible that there are other pH-dependent controlling factors, such as adsorption. Also, a significant proportion of the U in these unconfined Long Island waters is associated with colloids (of unidentified nature), and while there are no colloid data available for Th, it is likely that an even greater proportion of Th is associated with the colloids (Ivanovich et al. 1992a) and so present in the groundwater above inorganic solubility limits. While the high concentrations found in both the Magothy and basaltic aquifers might be ascribed to organic complexing, the samples in all of these studies have been filtered only for particles (~0.5 μm), and so the effect of colloids cannot be evaluated. It should be noted that the Magothy also has elevated Fe concentrations, and where this aquifer has been sampled elsewhere, such high Fe concentrations were present as suspended Fe(III) oxyhydroxides (Langmuir 1969), so that these may be important colloid carriers of actinides. The Th/U ratio of this water is similar to that estimated for the host rock (Tricca et al. 2001), suggesting that any transporting colloids did not substantially fractionate the two actinides. The uptake of Th on colloids has been studied in the laboratory by Lieser and Hill (1992).

Modeling the concentrations of ^{232}Th in groundwater involves considering the influx by weathering and the interactions with the host aquifer by sorption and precipitation. If concentrations are indeed limited by thorianite solubility, then since Th is also strongly adsorbed, sufficient Th has been released by weathering to have also reached steady state sorbed concentrations as well. This implies that the rate of any irreversible incorporation of sorbed Th (e.g., by recrystallization of sorbing Fe phases) is slow compared to the supply by weathering so that the groundwater concentrations are maintained at the solubility limit. However, in regions where changing conditions create lower solubility limits, removal by precipitation can occur. In modeling of ^{232}Th concentrations by considering only weathering and adsorption, steady state concentration profiles are achieved that increase linearly along a flow line, regardless of the distribution coefficient between water and surfaces (Tricca et al. 2001). A consequence of this is that groundwater (^{232}Th/^{238}U) ratios then reflect the relative weathering input rates rather than differences in chemistry; in this case, the ^{232}Th weathering supply is typically orders of magnitude lower than that of ^{238}U. This of course implies very different siting of these elements. However, these profiles presuppose that steady state concentrations have been reached at each location, weathering and adsorption parameter values are constant along the flow path, Th solubility limits have not been reached, Th and U are not irreversibly bound on colloids, and Th and U are not removed by irreversible adsorption (see Section 2.5). Since such linear profiles of either ^{232}Th or ^{238}U are generally not observed, these assumptions commonly are not applicable.

The distribution of ^{232}Th is not controlled by decay processes that can be used to constrain its supply rate, but the ^{232}Th-^{228}Ra pair can be used to provide some constraints on Th adsorption. In an area where ^{228}Ra concentrations are constant, mobile ^{228}Ra is supplied by only recoil and decay of mobile ^{232}Th. The amount of ^{232}Th that is absorbed on surfaces can be obtained if the amount of mobile ^{228}Ra is constrained (i.e., the Ra partition coefficient and the groundwater concentration are known) and can be compared to an estimate of the recoil rate of ^{228}Ra (using the measured ^{222}Rn activity in the groundwater) from the host rock. In this way, Luo et al. (2000) calculated that recoil from the host rock was insufficient to explain the amount of mobile ^{228}Ra, and argued that adsorbed ^{232}Th provided much of the mobile ^{228}Ra; then ^{232}Th partition coefficients can be obtained from the amount of measured groundwater ^{232}Th and the amount adsorbed, and these partition coefficients can be compared with those of other Th isotopes (see below). Note that this assumes not only that ^{222}Rn represents the recoil flux of ^{232}Th series nuclides (adjusted for the rock Th/U ratio), but also that the additional fluxes of ^{228}Ra are due to mobile ^{232}Th that freely exchanges with groundwater ^{232}Th.

^{228}Th and Th retardation factors. ^{228}Th is distinctive in being produced by low energy β decay, from ^{228}Ra. As can be seen in Figure 3, where advection is not important the steady state activity per unit aquifer mass of mobile ^{228}Ra and that of ^{228}Th should be equivalent. In this case, $(^{228}Th)_W R_{228Th} = (^{228}Ra)_W R_{228Ra}$ (where $R_I = K_I + 1$). Therefore, the value of the ratio of the retardation factors R_{228Ra}/R_{228Th} is the inverse of the measured groundwater $(^{228}Th/^{228}Ra)$ ratio (Luo et al. 2000). Groundwater values of $(^{228}Th/^{228}Ra) = (0.6-8) \times 10^{-2}$ (Luo et al. 2000), $(2-5) \times 10^{-2}$ (Ivanovich et al. 1992b), and $(5-20) \times 10^{-2}$ (Tricca et al. 2001) indicate that Th is generally adsorbed 10^1-10^2 times more efficiently than Ra, with the largest difference in the basaltic aquifer of Luo et al (2000). Note that this is based largely on the assumption that ^{228}Th is not supplied from the host minerals, and although this has not been confirmed experimentally it appears to be a reasonable assumption. In this case, these relative adsorption values are reasonably well constrained. Where the value for Ra can be obtained (based on the assumption that ^{222}Rn provides a measure of the recoil input of Ra isotopes), a value for Th can be obtained. If the retardation factor of Ra is 10^4 (see Section 2.3), then that of Th appears to be typically ~10^6.

It should be noted that the ^{232}Th data raises questions about whether Th is saturated or colloid-bound. Saturation will be dictated by the concentration of ^{232}Th, The other, shorter-lived Th isotopes are present in molar concentrations that are typically over 10^5 times lower than that of ^{232}Th, and so separate additions of these nuclides will likely be absorbed through isotopic exchange with the surfaces, rather than by triggering increased precipitation. Therefore, except under the unusual circumstance of large additions of ^{232}Th, the precipitation rate is unlikely to be a dominant flux for the other isotopes. Colloids will only be a factor where nuclides are rapidly irreversibly incorporated and so exhibit different exchange behavior with the host rock than dissolved species. Where colloids are composed of secondary phases from the wall rock that reversibly adsorb nuclides, or include organic compounds that allow continuous exchange with dissolved species, the presence of colloids will not affect modeling calculations. However, as discussed in Section 2.1, colloids may control the value of the calculated partition coefficients.

^{234}Th and sorption rate constants. The concentrations of ^{234}Th are highly variable with respect to those of parent ^{238}U, with $(^{234}Th/^{238}U) = (0.7-7) \times 10^{-2}$ where U contents are high (Luo et al. 2000) to 0.1-14 where U concentrations are generally much lower (Copenhaver et al. 1993; Tricca et al. 2001). This range reflects the very different controls on U and Th chemistry. The ^{234}Th concentrations have been used to calculate Th sorption coefficients. Krisnaswami et al. (1982) calculated adsorption rates by neglecting the desorption rate; while the $(^{234}Th/^{238}U)$ ratio would be greater than 1 in the absence of adsorption due to not only *in situ* decay of ^{238}U but also by recoil, the much lower measured ratios reflect the rapid adsorption of ^{234}Th relative to the ^{234}Th decay constant (which controls how the ^{234}Th activity would reach steady state without adsorption). In this case, adsorption rates of minutes were obtained. With a partition coefficient of 10^6, this corresponds to a desorption rate of ~month. Partition coefficients can be calculated from Equation (3) when the Th desorption coefficient is comparable to at least one half-life. Values of $(^{234}Th/^{228}Th)$, both of which are supplied by one α recoil, should be equal to the $(^{238}U/^{232}Th)$ ratio of the rock and generally close to one if desorption were very fast. Higher values of 6-230 (Luo et al. 2000) and 2-7 (Tricca et al. 2001) are plausibly due to sorption rate constants comparable to the half-life of the shorter-lived ^{234}Th. Note that the alternative explanation, that ^{232}Th is distributed near surfaces preferentially to ^{238}U, is contrary to the (^{226}Ra/^{228}Ra) ratios, which suggest that U series nuclides are more readily supplied (see Section 2.3). Luo et al. (2000) used model equations to calculate progressively smaller partition coefficients for ^{232}Th, ^{228}Th, and ^{234}Th. Note that the ^{232}Th values were calculated from ^{228}Ra groundwater concentrations and using ^{222}Rn from the recoil flux, and assuming that the ^{232}Th in the groundwater was not bound irreversibly in colloids (see above). In this case, ^{232}Th provides the Th partition coefficient when sorption dominates over decay. By combining values for K_1 for the other Th isotopes, adsorption coefficients of 0.1-4.1 min^{-1}, and desorption coefficients of 0.8-2.5 yr^{-1} were obtained (Luo et al. 2000). Copenhaver et al. (1993) found adsorption coefficients of 0.01-4.6 min^{-1}, with the lowest values for sandy deposits. Tricca et al. (2001) assumed that the desorption rate constant was much less than the decay constant of ^{228}Th, and that the total recoil supply of ^{234}Th is known from ^{222}Rn concentrations, and obtained substantially lower values for the partition coefficient of ~10^3. Tricca et al. (2001) also considered that while ^{234}Th is supplied by recoil into groundwater and then sorbed, ^{228}Th is formed largely on the surface and then desorbs. This provides a desorption value of 1 yr^{-1} for the sandy aquifer data.

^{230}Th. Groundwater concentrations of ^{230}Th, with a 75ka half-life, can have significant contributions from both recoil and weathering. Distinguishing different (^{230}Th/^{232}Th) ratios due to weathering of minerals with different Th/U ratios from those due to ^{230}Th additions by recoil is difficult. This is due to the analytical difficulties in obtaining precise counting data for such a low concentration element, and in obtaining mass spectrometric data from such low ^{230}Th/^{232}Th atomic ratios of 10^{-6}-10^{-5} (see Goldstein and Stirling 2003). Luo et al. (2000) found (^{230}Th/^{232}Th) activity ratios of 1.0-1.9, somewhat higher than the value for the host rock. Other studies have found similar ratios of (^{230}Th/^{232}Th) (Ivanovich et al. 1992b). These values may be due to either excess ^{230}Th added by recoil or differences between the rates of weathering of minerals bearing ^{238}U (and so ^{230}Th) and those bearing ^{232}Th. Note that the activity of excess ^{230}Th is (0.3-4) × 10^{-3} that of ^{234}Th, and so there is clearly a deficiency in recoiled ^{230}Th. This may be because adsorbed Th is irreversibly incorporated into the solid phase, perhaps by recrystallization of host Fe phases, on timescales that are long relative to the half life of ^{234}Th but short relative to that of ^{230}Th.

Summary. Further studies of the longer-lived isotopes of Th are essential for understanding the roles of colloids and weathering in the transport of Th. While it is clear that short-lived Th isotopes are strongly absorbed on aquifer materials and have retardation factors that are orders of magnitude greater than Ra, more detailed analysis is more model dependent and so requires further constraints on the processes involved. Modeling the behavior of long-lived Th isotopes appears to require further understanding regarding the irreversible retention of Th onto host rock surfaces. As with Ra, the absolute value for retardation factors remains dependent upon using ^{222}Rn concentrations to obtain rates of recoil supply into the groundwater, and so is subject to the same uncertainties as those discussed in Sections 2.2 and 2.4. While the conclusion that Th is strongly adsorbed will not change, retardation factors could conceivably be substantially lower.

2.5. U isotopes

The ubiquity of groundwater (^{234}U/^{238}U) ratios above the equilibrium value attests to the widespread operation of preferential ^{234}U release processes in aquifers. Since the U isotopes are long-lived, the concentrations cannot be assumed to be in steady state, adding difficulties to unravelling the causes of U isotopic variations. However, U isotopes provide the greatest potential for tracing groundwater flow and studying the migration of the more mobile anthropogenic nuclides.

Behavior of U in groundwater. The general behavior of U was reviewed by Gascoyne (1992). Under oxidizing conditions, U is highly mobile in the hexavalent form, forming soluble complexes primarily with carbonate and phosphate under near-neutral conditions, and with sulphate and fluorides at lower pHs. In saline groundwaters, solubilities are higher, where chloride and sulphate complexes are important. Concentrations are typically close to 1ppb, and values over 1ppm are generally only found in mineralised areas (Osmond and Cowart 1992). Under reducing conditions, U is in the tetravalent state and stable as $U(OH)_4$, and the solubility limit of uraninite, UO_2, sets the maximum U concentration to ~0.06 ppb (Gascoyne 1992). Groundwaters crossing a redox front therefore can precipitate U, which then serves as a source of ^{234}U to the passing groundwaters that can exhibit high ($^{234}U/^{238}U$) ratios (Osmond and Cowart 1992).

^{238}U is released from the host aquifer by weathering. Primary minerals that contain U, as well as Th, are often considered insoluble (e.g., Tole 1985). While enhanced weathering may occur in the vadose zone, the presence of U throughout aquifers, despite removal processes, indicates that U release continues to occur in the saturated zones. Recoil damage may enhance loss of radionuclides (Fleischer 1982, 1988), and the dissolution of actinide-rich minerals also may be enhanced due to accumulated recoil damage (Petit et al. 1985a). The release of radionuclides from secondary minerals and grain boundaries also may be important.

A substantial amount of experimental work has been published on the sorption behavior of U. The ability of clays, carbonates, and surfaces of other minerals to adsorb substantial amounts of U has been well documented. While these data are critical for explaining calculated bulk adsorption parameters and predicting migration behavior under changing conditions, these studies are not directly part of models of decay series isotopic systematics, and so are not discussed here. The goal of the field studies using decay series nuclides is to obtain bulk values for the fraction of radionuclides adsorbed without specifying specific processes and which are likely to be due to reactions involving a mixture of sites. A study by Payne et al. (2001) found that laboratory partitioning data were comparable to the distribution of U in groundwater and adsorbed or incorporated in readily soluble sites in cored materials. Models assume that adsorption is reversible, and Sims et al. (1996) found that most of the U that was passed through a sandstone core was readily recoverable by changing water conditions, although a small fraction was released more slowly, suggesting that adsorbed U becomes progressively more strongly bound within periods of a few months.

Secondary iron minerals are widespread and have highly reactive surface areas, and are important in controlling radionuclide migration. Ferrihydrite was found to sorb greater amounts of U than crystalline forms such as hematite and goethite (Payne et al. 1994). Structural changes in host Fe phases may further bind adsorbed species. U adsorbed by amorphous Fe hydroxides was found to be incorporated into more stable sites during crystallization of Fe minerals (Payne et al. 1994; Ohnuki et al. 1997). Giammar and Hering (2001) found that U sorption onto goethite was initially reversible, but during aging over months a portion of the U was no longer readily exchangeable. The long-term precipitation of Fe minerals therefore may transfer a significant fraction of U and other adsorbed radionuclides more permanently into surface coatings. While these processes may not affect the systematics of very short-lived nuclides, calculations of the evolution of longer lived nuclides over greater distances will be altered, since long-term accumulation on sorption sites will not occur. Rather, there will be a continuing process of irreversible removal. This may make it impossible to achieve a stable profile for longer-lived isotopes, since the exchangeable adsorbed pool is constantly being irreversibly incorporated. This process has not been incorporated into radionuclide models.

Several field and modeling studies have sought to obtain U partitioning information. It is difficult to obtain partitioning data for U from Equation (1). Due to the long half life of ^{234}U, steady state concentrations along a flow line are not expected to be reached, and so the advection term is significant. Further, it is unlikely that the various controlling parameters are constant over the long distances over which U isotope compositions are expected to vary. Steady state concentrations at each location also requires that significant reversibly adsorbed abundances have accumulated, although as discussed above this may not be possible. Tricca et al. (2000) showed that under constant conditions and reversible adsorption behavior, U concentrations are expected to increase linearly with groundwater age, while U isotope compositions will vary due only to mixing between U provided at the water table and that added in the aquifer by weathering and recoil. These characteristics have not been observed. Alternatively, Luo et al. (2000) calculated the distribution of ^{238}U by noting that the mobile ^{234}Th (i.e., $(1 + K_{234Th})(^{234}Th)_W$, representing groundwater and adsorbed species) is the result of recoil from host rock minerals and decay of mobile ^{238}U. Therefore, by using values for the partition coefficient of Th to calculate the total production rate of mobile ^{234}Th, and the recoil rate from within aquifer minerals by assuming the recoil rate of ^{234}Th can be obtained from that for ^{222}Rn, the amount of ^{234}Th supplied by dissolved and adsorbed ^{238}U was obtained; this corresponds to values for the partition coefficient of U (K_U) of $\sim10^3$. Clearly, these calculations are sensitive to the uncertainties in calculating the relevant ^{234}Th partitioning and recoil parameters. A laboratory study of aquifer rocks in this area by Fjeld et al. (2001) found a U retardation factor of 30, and the difference with the field study values reflects either the difficulties of using laboratory data as an average for the larger scales of the aquifer or the limitations of the mathematical model that was applied to the field data. Using similar modeling assumptions, Ivanovich et al. (1992b) deduced U retardation factors of $(0.8-7) \times 10^3$ for a sandstone aquifer. In contrast, as discussed below, some aquifers have U concentrations that appear to fall on conservative mixing trends, suggesting that adsorption is less important there. Overall, since the data does not consistently point to strong retardation of U under all circumstances, it is important to characterize U behavior at each location of interest.

Generation of U isotope variations. Both ^{234}U and ^{238}U are provided in secular equilibrium to the groundwater by simple weathering release. In addition, "excess" ^{234}U is released by recoil from host minerals of ^{234}Th, followed by decay to ^{234}Pa (with a 1 minute half life) and then to ^{234}U. Further release of ^{234}U is often assumed to occur by leaching of ^{234}U along recoil tracks, although in practice this is not readily distinguished from direct recoil and is only plausible where new surfaces are exposed or dry periods occur that allow implantation of recoiled radionuclides into adjacent phases. In sum, the ratio of supply rates of "excess" ^{234}U to ^{238}U in groundwater is equal to the ratio of ^{234}Th (and so ^{234}U) recoil release rate to weathering rate. Where removal processes are not fractionating, this ratio will be retained in groundwater. In this case, the measured ratio of the excess ^{234}U to ^{238}U (that is, $(^{234}U/^{238}U)_W - 1$) is equal to the ratio of recoil supply to weathering (Petit et al. 1985b; Tricca et al. 2001). Note that both recoil and weathering are proportional to the surface area of U-bearing phases; therefore, changes in grain size or the abundances of these phases within the rock will not change the ratio of these supply rates, but changes in water chemistry or in the nature of the U host phases will alter only the weathering rate and so will produce a change in groundwater isotopic composition. Note that preferential losses of ^{234}U will result in a $(^{234}U/^{238}U)$ activity ratio in the weathering mineral that is lower than 1, and so release of this U by weathering will partly balance the recoiled ^{234}U. Therefore, there are limits on the ratio that can be generated by the combined effects of recoil and weathering from a single phase (Petit et al. 1985a,b; Hussain and Lal 1986).

High ratios in groundwaters can be generated where U is concentrated in secondary phases and weathering is limited. For example, U can be precipitated when groundwaters become anoxic since reduced U^{+4} is much more insoluble. Therefore, concentrations are greatly lowered, and ($^{234}U/^{238}U$) ratios increase due to efficient recoil from precipitated phases (Osmond and Cowart 1992). In such cases, the generation of high ($^{234}U/^{238}U$) values in groundwater depends upon limited isotopic exchange with the larger reservoir of precipitated U.

Isotopic variations can also occur due to changes in groundwater chemistry or host rock characteristics. The most obvious change occurs at the water table. Distinctive U characteristics can be generated in waters travelling through more rapidly weathering, organic-rich deposits in the unsaturated zone. Once water with this U reaches the water table, conditions are encountered which often are dramatically different. U will then evolve along a flow line following a mixing trajectory between vadose zone U and that provided by the host aquifer rocks from the combined processes of weathering and recoil (see Tricca et al. 2000). Any adsorbed U will have the same isotopic composition as the groundwater and the concentration will follow that of the groundwater, assuming that the partitioning between surfaces and groundwater remains constant. Where regular groundwater patterns are found, U isotopes might be used to date groundwaters. Using short-lived nuclides to identify the recoil inputs, the accumulation of excess ^{234}U can be tracked. For example, studies of the Milk River aquifer in Canada found that reasonable ages could be determined from changes in $^{234}U/^{238}U$ ratios along a flowline, as long as significant retardation of U in the aquifer was included in the calculations (Ivanovich et al. 1991). Henderson et al. (1999) similarly used U isotope variations to determine pore water flow rates. Overall, this raises the hope that a reliable dating method could be found using U isotopes, but also indicated that independent controls on U behavior are required. Dating of groundwaters using these methods has not yet been applied widely.

Some of the complications involved in multi-scale porosities and evolutionary paths within fractured rock systems have been explored. For example, in the Palmottu site of gneisses in Finland U, as well as Th, was found to be redistributed into fracture coatings of carbonates and Fe oxides (Suksi et al. 1991). Suksi and Rasilainen (1996) have demonstrated how U concentrated in these fracture fillings can implant daughters in the surrounding phases, generating large fractionations between $^{234}U/^{238}U$ and $^{230}Th/^{234}U$. This suggests that phases that may weather at different rates later through changing conditions, new fracturing, or transport to a sedimentary aquifer and can become sources of highly fractionated nuclides into groundwater by leaching or weathering. Different migration patterns could also be discerned in the ($^{234}U/^{238}U$) signatures found in the groundwater-derived U in the coatings (Suksi et al. 2001). Matrix diffusion, where dissolved species can diffuse into rock pores and microfractures away from the main groundwater flow, can greatly affect water-rock interaction processes and migration rates of radionuclides (Neretnieks 1980; Suksi et al. 1992). Variations in ($^{230}Th/^{234}U$) and ($^{234}U/^{238}U$) indicated that leaching of ^{238}U, and to a greater extent of ^{234}U, occurred over a scale of centimeters from fractures (Suksi et al. 2001), so that under these conditions U could be used to map migration paths of different groundwaters.

Preferential leaching of oxidized U. Unlike Th or Ra, U has two oxidation states with very different solubilities. U in minerals is generally present as U^{4+}. However, separation of U from various minerals and rocks by oxidation state has found that there is hexavalent U present with substantially higher $^{234}U/^{238}U$ ratios (Chalov and Merkulova 1968; Kolodny and Kaplan 1970; Suksi et al. 2001). It was suggested that some ^{234}U

atoms are oxidized during α decay due to stripping of electrons (Rosholt et al. 1963), or an increase in positive charge during the two β decays from tetravalent ^{234}Th and pentavalent ^{234}Pa ($t_{1/2}$ = 1 minute) (Kolodny and Kaplan 1970). Alternatively, it has been suggested that since ^{234}U is resident in damaged lattice locations, it is more vulnerable to oxidation by fluids (Kolodny and Kaplan 1970; Cherdyntsev 1971) and so is due to external conditions. A detailed model of ^{234}U oxidation during recoil has been formulated (Ordonez Regil et al. 1989; Adloff and Roessler 1991; Roessler 1983, 1989). Using computer simulations of the recoil process, it was shown that recoiling ^{234}Th atoms push lighter oxygen atoms in front of it, enriching the final resting location in oxidizing species that are responsible for subsequent oxidation of ^{234}U after decay of the ^{234}Th. Note that this does not occur in minerals where U or Th is a major element, since the final resting place is not enriched in oxidizing species, and reactions with surrounding actinide atoms may occur which hinder oxidation. Therefore, in this model, the composition of the phase hosting ^{238}U is a controlling factor for the generation of disequilibrium in the groundwater.

Release of preferentially oxidized ^{234}U by subsequent leaching can only occur where the recoil path crosses the mineral-water interface and exposes the atom to leaching, but has not already led to direct injection of the daughter into solution. This may be applicable under some aquifer-specific conditions, similar to those discussed for preferential leaching. Sediments that have been deposited within the last 10^5 years may include minerals that were within nonporous rocks or under-saturated conditions so that radionuclides have been implanted in adjacent minerals. Recent fracturing may also provide water conduits exposed to such tracks. The development of a hydrated layer on some weathered silicate surfaces may allow enhanced diffusive escape of oxidized U that has been implanted there (Petit et al. 1985b). Also, weathering has the effect of moving the interface inwards to intersect new tracks, although this generally occurs too slowly to account for measured groundwater ^{234}U enrichments. In sum, this mechanism can explain the preferential release of ^{234}U where the necessary conditions for leaching exist, although data is required to confirm that this occurs under natural conditions.

Alternatively, it has been suggested (Petit et al. 1985a,b; Dran et al. 1988) that as U is released by weathering, tetravalent ^{238}U is preferentially precipitated or adsorbed, while hexavalent ^{234}U, oxidized during the recoil process, more readily remains in solution. This will increase the (^{234}U/^{238}U) ratio of the groundwater if isotopic equilibration does not then occur between adsorbed and dissolved U. A consequence of such preferential solution of ^{234}U is that excess ^{234}U in groundwater is not related to the amount of ^{234}Th in solution and on aquifer surfaces, and so the supply rates of the two nuclides cannot be directly linked. This process can provide an explanation for widespread enrichment of ^{234}U that is greater than that produced by direct recoil into groundwater. However, there is no data available to assess whether this generally operates in groundwater systems.

U isotope variations in groundwater. There have been several studies in the last 10 years that have reported (^{234}U/^{238}U) distributions in aquifers; earlier studies have been reviewed by Osmond and Cowart (1992, 2000). Kronfeld et al. (1994) measured values in a phreatic, oxidizing dolomite aquifer in South Africa. Young well waters (with measurable ^{3}H) had (^{234}U/^{238}U) = 2.2-5.8, while values up to 11 were found in springs and speleothems. The host aquifer rock had low U concentrations, and so it was hypothesized that a substantial inventory of ^{238}U resides adsorbed on the surfaces to supply ^{234}U. However, such a reservoir would isotopically exchange with U in the water, and so the high (^{234}U/^{238}U) ratios must be generated from ^{238}U is irreversibly bound in the host rocks. It is interesting to note that if the ^{222}Rn supply rates, calculated from the reported ^{222}Rn concentrations, are equated to that producing excess ^{234}U, then much greater times are required to generate the measured excess ^{234}U concentrations than the possible ages of the waters. These are minimum times, based upon the assumption that there are no losses of U to the aquifer surfaces. This raises the possibility that the ^{234}U excesses were generated under different conditions, such as in the vadose zone. Bonotto and Andrews (1993) suggested that the ratios of ~1.5 in the Carboniferous Limestone of SW England were due to leaching of silicates such as micas that are continually exposed during carbonate dissolution.

U isotope variations in young, shallow groundwaters in a sandy unconfined aquifer were reported by Tricca et al. (2001). A considerable range of ^{234}U excess and ^{238}U concentrations were found in water table wells, with a similar range in wells less than 5km down gradient. Therefore, it is possible that the generation of much of the U characteristics that is seen in young waters occurs outside the aquifer and in the vadose zone. Unfortunately, it is generally difficult to precisely follow groundwater flow paths and account for dispersion effects, so that close to the water table it is often difficult to separate variations in groundwater chemistry due to variable water table inputs from those due to aquifer processes. At greater distances, the changes due to aquifer processes will be clearly greater than the range in water table characteristics.

Several recent studies have used U isotope compositions to trace groundwater flow patterns, and Osmond and Cowart (2000) have discussed the basic principles involved. Roback et al. (2001) combined U and Sr isotope data for the Snake River Plain aquifer to identify the flow of isotopically distinctive recharging waters as well as flow along higher conductivity flow paths. While some regions could be explained by mixing of different water masses, simple mixing trends between U concentrations, $(^{234}U/^{238}U)$, and $^{87}Sr/^{86}Sr$ were not always observed, suggesting that U was not conservative during mixing. Dabous and Osmond (2001) reported that waters from the sandstone Nubian Aquifer in Egypt exhibit broad correlations between $(^{234}U/^{238}U)$ and $1/U$, consistent with mixing between recharging and aquifer waters with distinct $(^{234}U/^{238}U)$ ratios as well as ^{238}U and excess ^{234}U excess concentrations. While the data are not tightly constrained due to secondary processes, the systematic variations indicate that conservative mixing is the dominant process in the area sampled. Similarly, Hodge et al. (1996) found that there is a broad correlation between U and the conservative components Na, Cl, and SO$_4$ in spring waters from California, and argued that this indicated that was behaving conservatively. Overall, it appears that the behavior of U varies between groundwater conditions, although a direct comparison between the different methods for deducing U behavior, including laboratory experiments of U adsorption on aquifer rocks, is necessary.

The unusual occurrence of $(^{234}U/^{238}U)$ ratios less than one were observed in waters from the weathered zone around the Koongarra ore deposit (Yanase et al. 1995). While this could be due to prior leaching and depletion of ^{234}U from the U that is now released by weathering, it was suggested that this reflects implantation of recoil ^{234}U into phases more resistant to weathering during undersaturated conditions in this zone.

Summary. U isotope compositions potentially can provide an important tool for tracing groundwaters from different aquifer conditions. However, clear interpretations of $(^{234}U/^{238}U)$ ratios, and quantification of the responsible processes, remain difficult. Additional work regarding the release rates of ^{234}Th and ^{234}U, and the weathering and retardation rates of U, is still required to complement site evaluations.

2.6. ^{210}Pb

Few studies have addressed the groundwater behavior of Pb. It is clear that Pb is highly surface reactive, since ^{210}Pb that is deposited onto watersheds after production in the atmosphere from ^{222}Rn, has residence times in watersheds on the order of 10^4 years due to retention in soils. ^{222}Rn decays to ^{210}Pb (the four intermediate radionuclides have half lives less than one hour and so are unlikely to significantly affect distribution of ^{210}Pb), and so (^{210}Pb/^{222}Rn) ratios can be used to obtain a retardation factor in the same way that (^{224}Ra/^{222}Rn) ratios are used to determine Ra retardation factors, although in these case the nuclides are in the same decay series chain so that there are fewer issues regarding possible differences in parent nuclide siting. When both have steady state concentrations, the mobile ^{210}Pb is derived from both recoil and ^{222}Rn in solution. If the latter was generated only by recoil, the pool of mobile ^{210}Pb is twice that of ^{222}Rn in groundwater (see Hussain and Krishnaswami 1980), and the groundwater (^{210}Pb/^{222}Rn) ratio is then simply half the inverse of the Pb distribution coefficient. Measured (^{210}Pb/^{222}Rn) values are (0.1-6) × 10^{-4} for sandy aquifers (Copenhaver et al. 1993) and (0.14-1.6) × 10^{-3} for a schist (Yanase et al. 1995). Therefore, paritition coefficients of ~10^3-10^5 are obtained, with the largest assumption being that ^{222}Rn provides a measure of the ^{210}Pb recoil flux into groundwater. Comparison with ^{224}Ra suggests that Pb is generally somewhat more surface-reactive than Ra.

3. GROUNDWATER DISCHARGE INTO ESTUARIES

As discussed in the preceding sections, interactions with host aquifer rocks, including the supply of dissolved constituents as well as removal by adsorption and precipitation, can impart radionuclide signatures to groundwaters that are distinctive from those in surface waters. In particular, the very short-lived nuclides that are present in significant concentrations due to continuous supply from host rocks become strongly depleted in surface waters. All waters flowing into rivers have interacted with rocks or soils to varying degrees. These signals are then modified by residence in the reverie system. Waters in larger surface bodies have even longer residence times with only limited interaction with sediments. Clearly this effect is most pronounced in the ocean basins. One issue that has received considerable attention is the magnitude of groundwater discharge into coastal and estuarine waters, and the relative inputs in these regions from rivers and groundwater. In this environment, radionuclides have the potential for tracing waters from groundwater.

Groundwater discharge to coastal waters and estuaries has been of interest for centuries (Dominica and Schwartz 1990; Seltzer et al. 1973; Seltzer and Logician 1993). Several studies have utilized U- and Th- decay series nuclides to assess groundwater inputs to the coastal ocean (e.g., ^{228}Ra, ^{226}Ra and ^{222}Rn) as well as the short-lived isotopes (^{224}Ra, ^{223}Ra) to assess coastal mixing (Moore 1996; Cable et al. 1996a,b; Toreros et al. 1996; Hussein et al. 1999; Corbett et al. 2000b,c; Moore 2000; Swarzenski et al. 2001; Top et al. 2001; Schwartz and Sharpe 2003). More recently, a renewed interest in submarine groundwater discharge has emerged from studies on coastal eutrophication and contamination (Giblin and Gaines 1990; Reay et al. 1992; Bugna et al. 1996). The ecological importance of subsurface flow to water and nutrient budgets has been repeatedly demonstrated during the last two decades (Bokuniewicz 1980; Johannes 1980; Simmons 1992).

3.1. Background

The complex interactions amongst geological, biological, and geochemical processes at the land-sea margin control the delivery and fate of radionuclides, contaminants, and other natural elements in coastal environments (Swarzenski et al. 2003). For many such constituents, there is at least a fundamental understanding of major source and sink functions and their potential estuarine transformation reactions. For example, rivers can be monitored quite easily for discharge rates into estuaries as well as for elemental concentrations to derive estimates of the integrated fluxes that are introduced to estuaries. Continental constituents can also be delivered into estuaries by groundwaters. However, the importance of the delivery of dissolved constituents by submarine groundwater discharges directly into coastal bottom waters, such as select radionuclides and nutrient species, to coastal bottom waters has often been neglected (Harvey and Odum 1990; Valiela et al. 1990; Simmons 1992). This general omission from coastal mass balance budgets by both hydrologists and oceanographers alike is largely due to the difficulty in accurately identifying and quantifying such submarine groundwater discharges (Burnett et al. 2001b, 2002). However, this flux recently been shown to indeed play a substantial role in the overall delivery of certain radionuclides and other elements to the coastal sea (Moore 1996; Cable et al. 1996a,b; Swarzenski et al. 2001).

Hydrologists and oceanographers do not necessarily share the same vocabulary to define processes, so a few comments on definitions is warranted. Groundwater is most commonly defined as water within the saturated zone of geologic material (Freeze and Cherry 1979). The bottom sediments of an estuary are obviously saturated, so water within submerged sediments (i.e., pore waters or interstitial waters) can be defined as groundwater. Therefore, submarine groundwater discharge includes any upward fluid transfer across the sediment-water interface, regardless of its age, origin and salinity. Exchange across this interface is bi-directional (discharge and recharge), although a net flux is most often upward.

The rates of submarine groundwater discharge within "leaky coastal margins" are controlled by inland recharge rates and the underlying geologic framework. Figure 6 shows the dominant characteristics of a generalized coastal groundwater system influenced by submarine groundwater discharge. Fresh water that flows down gradient from the water table towards the sea may discharge either as diffuse seepage close to

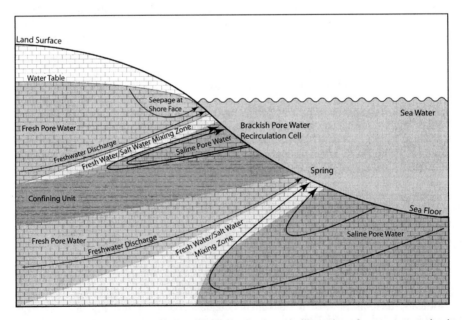

Figure 6. Fluid exchange at the land-sea margins and schematic illustration of processes associated with submarine groundwater discharge.

shore, or directly into the sea either as submarine springs (Swarzenski et al. 2001) or wide scale seepage. The hydraulic gradient that drives freshwater towards the sea can also drive seawater that has intruded into underlying deposits back to the ocean, creating a saltwater circulation cell. Wherever multiple aquifers and confining units co-exist, each aquifer will have its own fresh water/saltwater interface, and deeper aquifers will discharge further offshore (Freeze and Cherry 1979). Submarine groundwater discharges can be spatially as well as temporally variable since there are a variety of both natural and anthropogenic influences (i.e., sea-level, tides, precipitation, dredging, groundwater withdrawals) that can have strong effects (Zektzer and Loaiciga 1993).

Theoretically, submarine groundwater discharge can occur wherever a coastal aquifer is hydrogeologically connected to the sea (Moore and Shaw 1998; Moore 1999). Artesian or pressurized aquifers can extend for considerable distances from shore, and where the confining units are breached or eroded away, groundwater can flow directly into the sea (Manheim and Paull 1981; Moore 1998). While the magnitude of this submarine groundwater discharge is often less than direct riverine runoff, recent studies have shown that coastal aquifers may contribute significant quantities of fresh water to coastal bottom waters (Zektzer et al. 1973; Moore 1996; Burnett et al. 2001b, 2002). Although it is unlikely that submarine groundwater discharge plays a significant role in the global water budget (Zektzer and Loaiciga 1993), there is strong evidence that suggests that the geochemical signature of many redox sensitive constituents is directly affected by the exchange of subsurface fluids across the sediment-water interface (Corbett et al. 2000a,b,c; Lapointe et al. 1990). This fluid exchange includes direct upward groundwater discharge as well as the reversible exchange at the sediment-water interface (i.e., seawater recirculation) as a result of tidal pumping (Li et al. 1999; Hancock et al. 2000).

Standard seepage meters have traditionally provided physical, time dependent measurements of submarine groundwater discharge, while first order diffusion models have produced elemental flux estimates based on dissolved pore water profiles (Cable et al. 1996b; Corbett et al. 2002b). Both techniques yield evidence for substantial localized submarine groundwater discharge, but these methods fail to provide insight into larger scale, more synoptic estimates for exchange across the sediment/water interface. Due to the patchiness and generally unpredictable nature of submarine groundwater discharges, a tracer capable of integrating the spatial heterogeneities of most coastal bottom sediments is needed. To address this issue, W.S. Moore and W. Burnett and their colleagues have utilized the four naturally occurring isotopes of radium and ^{222}Rn to study both local and regional scaled submarine groundwater discharge.

3.2. Tracing groundwater using ^{222}Rn and the Ra quartet

An ideal submarine groundwater discharge tracer should be highly enriched in groundwater relative to seawater, behave conservatively (i.e., non-reactive) or at least predictably and also be easy to measure. The four isotopes of Ra and ^{222}Rn follow these constraints reasonably well and have recently been utilized to identify and quantify submarine groundwater discharge to various coastal oceans (Krest et al. 2000; Bollinger and Moore 1993; Webster et al. 1995). One strong advantage of these radiotracers over seepage meters is that the coastal water column effectively integrates the submarine groundwater discharge signal over a broad area and time period.

Ra is continually produced in sediments by the decay of insoluble Th parents (see Fig. 7). While in fresh water Ra is bound tightly to particle surfaces, in seawater Ra readily undergoes cation exchange with other dissolved constituents. This provides an additional source of Ra to the water column, which controls the frequently observed non-

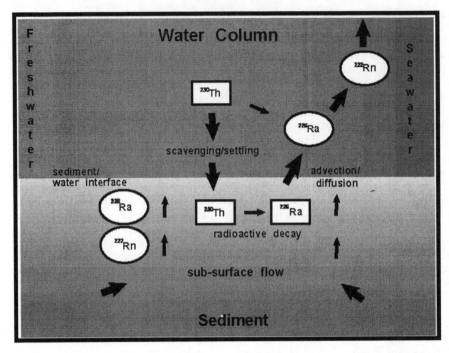

Figure 7. Sediment contains ^{230}Th derived both scavenged from the water column during particle settling and contained in solid material. ^{226}Ra produced in the sediments is highly soluble in pore waters and diffuses into the overlying water or is advected across the sediment-water interface by discharging groundwater. ^{222}Rn is produced within the water column from dissolved ^{226}Ra and within the underlying sediments.

conservative distribution in estuaries (Torgersen et al. 1996). The specific source terms for this additional Ra include: 1) Ra produced from bottom sediments (Bollinger and Moore 1993; Rama and Moore 1996; Hancock et al. 2000), 2) Ra desorbed from suspended particulates (Yang et al. 2002) and 3) Ra advected upward either by the discharge of submarine groundwater (Moore and Shaw 1998) or by reversible exchange across the sediment-water interface; i.e., the flow of recirculated seawater (Hancock and Murray 1996; Moore 1997; Hancock et al. 2000).

The utility of radium as an effective tracer of submarine groundwater discharge is a function of 1) its sedimentary source in fresh waters that typically produces a signal that is highly enriched relative to coastal water, 2) its mobility in brackish waters and in seawater, and 3) the wide range in half-lives of the four Ra isotopes (3.8 days to 1600 years) that corresponds to that of a range in relevant submarine groundwater discharge processes. By developing a simple estuarine mass balance model for Ra, the flux of submarine groundwater and any associated constituents can then easily be derived. Moore (1996) suggested that the discharge and exchange of freshened groundwater across the sediment-water interface had to be the primary source of ^{226}Ra enrichment in a large coastal region of the S. Atlantic Bight. Estimates of submarine groundwater discharge derived only from ^{226}Ra activities need further calibration as the reversible water mass exchange across the sediment-water interface may serve as a "Ra-pump." Corbett et al. (1999) evaluated groundwater discharge into Florida Bay by comparing

^{222}Rn activities with CH$_4$ concentrations and a positive relation between methane and ^{222}Rn was observed. Similar results were found by Cable et al. (1996a) and Kim and Hwang (2002). The location of high groundwater fluxes inferred from Ra and Rn concentrations has been corroborated by direct seepage measurements (Cable et al., 1996a; Corbett et al. 2000b). However, seepage meters provide a measurement of the local water flux across the sediment-water interface, while the chemical tracers provide a more integrated flux estimate to the overlying water. Burnett et al. (2002) brought together a suite of geochemical tracers and physical data in an effort to quantify the links between tracer concentrations and groundwater fluxes. However, further data is required to obtain sufficient coverage across an individual estuary and separate all the possible sources of Ra and Rn isotopes in order to derive a first-order estimate of groundwater fluxes to the entire estuary. While these studies establish that U- and Th-series can be used to identify groundwater discharges in coastal and estuarine systems, further work is required to quantify the associated integrated groundwater fluxes of broad areas.

Recent studies of submarine groundwater discharge into estuaries indicate that select long-lived and short-lived U/Th series isotopes show great promise as new tools to directly examine exchange processes and rates across the sediment-water interface. As new detection techniques and field validation methods develop, e.g., *in situ* ^{222}Rn monitors (Burnett et al. 2001a), coastal scientists will for the first time be able to realistically identify and quantify submarine groundwater discharge. Ideally, a thorough submarine groundwater discharge study should include direct measurements (e.g., seepage meters) and numerical modeling efforts to calibrate the geochemical tracers for quantifying groundwater discharges at the particular estuary. In concert, such an approach provides a powerful diagnostic tool for regional scale submarine groundwater discharge investigations.

4. OPEN ISSUES

Considerable advances continue to be made in the understanding of radionuclide migration in the natural environment. Overall, these models have provided an important framework for understanding the behavior of these nuclides. Clearly, further studies are required to examine different aquifer environments, to further constrain the controlling processes, to refine the analytical tools for calculating retardation rates, and to determine how the data gathered from naturally occurring radionuclides can be extrapolated to the higher concentrations of anthropogenic isotopes. In addition to exploring radionuclide behavior under specific conditions, there are some important broad issues that still require further close attention.

4.1. The effects of well construction and sampling

The question of how well groundwater samples represent the aquifer is generally not discussed in research studies of radionuclides. Contamination studies frequently utilize monitor wells that have been specifically designed and constructed for sample collection. Issues that are considered include the use of non-corroding well materials such as PVC, drilling methods that minimize (but cannot completely avoid) disruption of the surrounding aquifer, extraction of drilling muds by well development, and well screens that target restricted sampling depths. There have been various studies that have examined the impact of sampling procedures on sample integrity. However, other studies that do not have the luxury of installing dedicated wells often utilize other types of wells, such as those for water supply. While there is rarely the opportunity to investigate the potential impact on the sample parameters of interest, it is worth considering some of the factors that may cause deviations in the data. During drilling, a large region around the wells is disrupted, changing stratification, packing, and porosity. A pack of clean sand is generally placed around the well screen; this has very different characteristics from the aquifer, and the time that water spends in this region is dictated by the flow rate in the aquifer, a factor that may be important for the shortest-lived radionuclides. Cement seals used to isolate the screened region from the overlying borehole can contaminate wells if not emplaced properly (Barcelona and Helfrich 1986). Well casing material can also affect water chemistry (e.g., Barcelona and Helfrich 1986). Steel casing, often used for water wells, can release Fe and Fe colloids due to corrosion (Degueldre et al. 1999). Well development generally follows construction, and in removing drilling fluids and muds, fine material in the aquifer may also be redistributed. The rates of purging to remove stagnant water within a well have been found to affect trace element and contaminant concentrations, and more stable concentrations tend to be found when low turbidities are achieved by using low pumping rates in a variety of aquifer conditions (e.g., Puls and Powell 1992; McCarthy and Shevenell 1998; Gibs et al. 2000). Low purge rates have been used to collect U samples using dedicated bladder pumps (Shanklin et al. 1995), although such equipment is generally only practical at continually monitored sites. Waters from different levels or used for drilling may also be drawn into a well, and result in substantial chemical changes (Grenthe et al. 1992).

Overall, it is not possible to evaluate whether any of these factors have compromised the data included in published studies. More research is required to determine how U- and Th-series nuclides are affected by various well construction methods and sampling procedures, how reliable are data from wells not designed fro trace element monitoring, and if there are any factors that have affected the data collected from past studies. Most importantly, radionuclide studies must incorporate greater awareness of the potential problems, summarize whatever details are available regarding the wells used, and consider potential problems where the relevant information is not available.

4.2. Quantification of model parameters

There are various parameters and assumptions defining radionuclide behavior that are frequently part of model descriptions that require constraints. While these must generally be determined for each particular site, laboratory experiments must also be conducted to further define the range of possibilities and the operation of particular mechanisms. These include the reversibility of adsorption, the relative rates of radionuclide leaching, the rates of irreversible incorporation of sorbed nuclides, and the rates of precipitation when concentrations are above Th or U mineral solubility limits. A key issue is whether the recoil rates of radionuclides can be clearly related to the release rates of ^{222}Rn; the models are most useful for providing precise values for parameters such as retardation factors, and many values rely on a reliable value for the recoil fluxes, and this is always obtained from ^{222}Rn groundwater activities. These values are only as well constrained as this assumption, which therefore must be bolstered by clearer evidence.

4.3. Interpreting model-derived information

Evaluations of the behavior of naturally occurring nuclides can provide data on the distributions of radionuclides in groundwater, host minerals, and on sorption sites. However, identifying the causes of variations requires further chemical data and laboratory experiments. Variations due to differences in adsorption that are related to changes in chemistry must be investigated by establishing correlations with water chemistry parameters and confirmation from adsorption experiments that the variations can be explained quantitatively. The effects of colloids on radionuclide transport must also be investigated. Further studies are needed to determine whether radionuclide behavior can be successfully evaluated using approaches integrating field and laboratory investigations.

4.4. Inputs at the water table

Modeling of the transport of the long-lived nuclides, especially U, require knowledge of the input at the water table as a boundary condition for aquifer profiles. There are few studies of the characteristics of radionuclides in vadose zone waters or at the water table. Significant inputs are likely to occur to the aquifer due to elevated rates of weathering in soils, and this is likely to be dependent upon climatic parameters and has varied with time. Soils may also be a source of colloids and so provide an important control on colloidal transport near recharge regions.

4.5. Applications to pollutant radionuclide migration studies

A large number of studies have examined the migration of naturally occurring actinides at potential waste repositories or natural analogue sites. The role of these studies in radionuclide containment performance assessments has been discussed by Ivanovich et al. (1992a) and Smellie et al. (1997). Reservations have been put forward about the applicability of the information obtained from natural radionuclide modeling studies to anthropogenic nuclides that may be released at higher concentrations and in waters that have distinctive characteristics (McKinley and Alexander 1996). It is indeed clear that such studies cannot fully examine all the important characteristics at sites to support concrete predictions about pollutant migration. The magnitudes and patterns of adsorption rates under low concentrations and at present conditions can be defined, and these can provide broad survey information for more targeted exploration and benchmarks for assessing the relevance of laboratory results. Further information is of course required to extrapolate these results to quantify behaviors under different conditions. For example, the extent to which there are changes in sorption or precipitation due to greater radionuclide concentrations (e.g., due to nonlinear adsorption isotherms), or due to accompanying changes in water chemistry or colloid concentrations, must be quantified. Therefore, while predictions cannot be always directly made from the models described here to other conditions, the information provided must undoubtedly constitute an important component in any site study. Overall, the challenge remains to clearly characterize the present behavior of naturally occurring radionuclides and confidently extrapolate this to conditions created by pollutant releases.

4.6. Tracing groundwater discharges

The use of Ra and Rn as quantitative coastal groundwater tracers is still in its infancy. A clearer understanding of comprehensive Ra input and removal processes across the sediment-water interface needs further development in a wide range of hydrogeologic settings and within the context of integrated estuarine studies. Studies are obviously required which combine comprehensive sampling of an estuary along with the associated river and surrounding groundwaters, and consider the gathered data in the context of a well-characterized hydrological framework.

Overall, there are many issues that remain to be studied in greater detail. However, with further understanding of how U, Th, Ra, Rn, and Pb behave in diverse aquifer environments, studies utilizing U- and Th- decay series systematics will become increasingly useful for characterizing groundwater transport of trace elements and tracing groundwater flow.

ACKNOWLEDGMENTS

Reviews by S. Luo, C. Lundstrom, G. Henderson, and two anonymous reviewers are greatly appreciated.

REFERENCES

Adloff JP, Roessler K (1991) Recoil and transmutation effects in the migration behavior of actinides. Radiochim Acta 52/53:269-274

Alexander WR, McKinley IG (1994) Constraints on the use of "in situ distribution coefficients (K_d)" values in contaminant transport modeling. Eclogae Geol Helv 87:321-324

Andrews JN, Wood DF (1972) Mechanism of radon release in rock matrices and entry into groundwaters. Inst Min Metall Trans B81:198-209

Andrews JN, Ford DJ, Hussain N, Trevedi D, Youngman MJ (1989) Natural radioelement solution by circulating groundwaters in the Stripa granite. Geochim Cosmochim Acta 53: 1791-1802

Banner JL, Chen JH, Wasserburg GJ, Moore CH (1990) ^{234}U-^{238}U-^{230}Th-^{232}Th systematics in saline ground waters from central Missouri. Earth Planet Sci Lett 101:296-312

Barcelona MJ, Helfrich JA (1986) Well construction and purging effects on groundwater samples. Environ Sci Tech 20:1179-1184

Benes P (1990) Radium in (continental) surface water. In: The environmental behavior of radium. Vol. 1. Intl Atomic Energy Agency, Vienna, p373-418

Bokuniewicz H (1980) Ground water seepage into Great South Bay, New York. Estuar Coast Mar Sci 10:437-444

Bollinger MS Moore WS (1993) Evaluation of salt marsh hydrology using radium as a tracer. Geochim Cosmochim Acta 57:2203-2212

Bonotto DM, Andrews JN (1993) The mechanism of ^{234}U/^{238}U activity ratio enhancement in karstic limestone groundwater. Chem Geol (Isot Geosci Sect) 103:193-206

Bourdon B, Turner S, Henderson GM, Lundstrom CC (2003) Introduction to U-series geochemistry. Rev Mineral Geochem 52:1-21

Buesseler KO, Bauer JE, Chen RF, Eglinton TI, Gustafsson Ö, Landing W, Mopper K, Moran SB, Santschi PH, Vernon Clark R, Wells ML (1996) An intercomparison of cross-flow filtration techniques used for sampling marine colloids: overview and organic carbon results. Marine Chem 55:1-31

Buffle J, Perret D, Newman M (1992) The use of filtration and ultrafiltration for size fractionation of aquatic particles, colloids, and macromolecules. In: Environmental particles. Buffle J, van Leeuwen HP (eds) Lewis Publishers, Boca Raton FL, p171-230

Bugna GC, Chanton JP, Cable JE, Burnett WC, Cable PH (1996) The importance of ground water discharge to the methane budgets of near shore and continental shelf waters of the northeastern Gulf of Mexico. Geochim Cosmochim Acta 60:4735-4746

Burnett WC, Kim G, Lane-Smith D (2001a) A continuous radon monitor for assessment of radon in coastal ocean waters. J Radioanal Nucl Chem 249:167-172

Burnett WC, Taniguchi M, Oberdorfer J (2001b) Measurement and significance of the direct discharge of groundwater into the coastal zone. J Sea Res 46:109-116

Burnett WC, Chanton J, Christoff J, Kontar E, Krupa S, Lambert M, Moore W, O'Rourke D, Paulsen R, Smith C, Smith L, Taniguchi M (2002) Assessing methodologies for measuring groundwater discharge to the ocean. EOS Trans Am Geophys Un 83:117-123

Cable JE, Bugna GC, Burnett WC, Chanton JP (1996a) Application of ^{222}Rn and CH_4 for assessment of ground water discharge to the coastal ocean. Limnol Oceanogr 41:1347-1353

Cable JE, Burnett WC, Chanton JP, Weatherly GL (1996b) Estimating ground water discharge into the northeastern Gulf of Mexico using radon 222. Earth Planet Sci Lett 144:591-604

Chabaux F, Riotte J, Dequincey O (2003) U-Th-Ra fractionation during weathering and river transport. Rev Mineral Geochem 52:533-576

Chalov PI, Merkulova KI (1968) Effects of oxidation on the separation of uranium isotopes during leaching from minerals. Geochem Int 5 (Suppl):391-397

Chen JH, Edwards RL, Wasserburg GJ (1986) ^{238}U, ^{234}U, and ^{232}Th in seawater. Earth Planet Sci Lett 80:241-251

Cherdyntsev VV (1971) Uranium-234. Israel Program for Scientific Translations, Jerusalem

Cochran JK, Masqué P (2003) Short-lived U/Th-series radionuclides in the ocean: tracers for scavenging rates, export fluxes and particle dynamics. Rev Mineral Geochem 52:461-492

Copenhaver SA, Krishnaswami S, Turekian KK, Shaw H (1992) ^{238}U and ^{232}Th series nuclides in groundwater from the J-13 well at the Nevada test site: implications for ion retardation. Geophys Res Lett 19:1383-1386

Copenhaver SA, Krishnaswami S, Turekian KK, Epler N, Cochran JK (1993) Retardation of ^{238}U and ^{232}Th decay chain radionuclides in Long Island and Connecticut aquifers. Geochim Cosmochim Acta 57:597-603

Corbett, DR, Dillon K, Burnett W (2000a) Tracing groundwater flow on a barrier island in the northeast Gulf of Mexico. Estuar Coast Shelf Sci 51:227-242

Corbett DR, Dillon K, Burnett W, Chanton J (2000b) Estimating the groundwater contribution into Florida Bay via natural tracers ^{222}Rn and CH_4. Limnol Oceanogr 45: 1546-1557

Corbett DR, Kump L, Dillon K, Burnett W, Chanton J (2000c) Fate of wastewater-borne nutrients in the subsurface of the Florida Keys, USA. Mar Chem 69:99-115

Dabous AA, Osmond JK (2001) Uranium isotopic study of artesian and pluvial contributions to the Nubian Aquifer, Western Desert, Egypt. J Hydrol 243:242-253

Davidson MR, Dickson BL (1986) A porous flow model for steady-state transport of radium in ground waters. Water Resour Res 22:34-44

Dearlove JLP, Longworth G, Ivanovich M, Kim JI, Delakowitz B, Zeh P (1991) A study of groundwater colloids and their geochemical interactions with natural radionuclides in Gorleben aquifer systems. Radiochim Acta 52/53:83-89

Degueldre C, Scholtis A, Pearson FJ, Laube A, Gomez P (1999) Effect of sampling conditions on colloids and ground water chemistry. Eclogae Geolog Helv 92:105-114

Dickson BL (1985) Radium isotopes in saline seepages, south-western Yilgarn, Western Australia. Geochim Cosmochim Acta 49:361-368

Dickson BL (1990) Radium in groundwater. *In* The environmental behavior of radium. Vol. 1. IAEA Technical Reports Series No. 310. Intl Atomic Energy Agency, Vienna, p 335-372

Domenico PA, Schwartz FW (1990) Physical and Chemical Hydrogeology. John Wiley and Sons, New York

Dran J-C, Langevin Y, Petit J-C (1988) Uranium isotopic disequilibrium: reappraisal of the alpha-recoil effect. Chem Geol 70:126

Eyal Y, Olander DR (1990) Leaching of uranium and thorium from monazite: I. Initial leaching. Geochim Cosmochim Acta 54:1867-1877

Fjeld RA, DeVol TA, Goff RW, Blevins MD, Brown DD, Ince SM, Elzerman AW (2001) Characterization of the mobilities of selected actinides and fission/activation products in laboratory columns containing subsurface material from the Snake River Plain. Nucl Tech 135:92-108

Fleischer RL (1980) Isotopic disequilibrium of uranium: alpha-recoil damage and preferential solution effects. Science 207:979-981

Fleischer RL (1982) Alpha-recoil damage and solution effects in minerals: uranium isotopic disequilibrium and radon release. Geochim Cosmochim Acta 46:2191-2201

Fleischer RL (1988) Alpha-recoil damage: relation to isotopic disequilibrium and leaching of radionuclides. Geochim Cosmochim Acta 52:1459-1466

Fleischer RL, Raabe RO (1978) Recoiling alpha-emitting nuclei: mechanisms for uranium-series disequilibrium. Geochim Cosmochim Acta 42:973-978

Freeze RA, Cherry JA (1979) Groundwater. Prentice-Hall, Englewood Cliffs, New Jersey

Frissel MJ, Köster HW (1990) Radium in soil. *In:* The environmental behavior of radium. Vol. 1. IAEA Technical Reports Series No. 310. Intl Atomic Energy Agency, Vienna, p323-334

Gascoyne M (1992) Geochemistry of the actinides and their daughters. *In:* Uranium-series disequilibrium. Ivanovich M, Harmon RS (eds) Clarendon Press, Oxford, p 34-61

Giammar DE, Hering JG (2001) Time scales for sorption-desorption and surface precipitation of uranyl on goethite. Environ Sci Technol 35:3332-3337

Giblin AE, Gaines AG (1990) Nitrogen inputs to a marine embayment: the importance of ground water. Biogeochem 10:309-328

Gibs J, Szabo Z, Ivahnenko T, Wilde FD (2000) Change in field turbidity and trace element concentrations during well purging. Ground Water 38:577-588

Gnanapragasam EK, Lewis BA (1991) Elastic strain energy and the distribution coefficient of radium in solid solution with calcium salts. Geochim Cosmochim Acta 59:5103-5111

Grenthe I, Stumm W, Laaksuharju M, Nilsson AC, Wikberg P (1992) Redox potentials and redox reactions in deep groundwater systems. Chem Geol 98:131-150

Hammond DE, Zukin JG, Ku TL (1988) The kinetics of radioisotope exchange between brine and rock in a geothermal system. J Geophys Res 93:13175-13186

Hancock GJ, Murray AS (1996) Source and distribution of dissolved radium in the Bega River Estuary, Southeastern Australia. Earth Planet Sci Lett 138:145-155

Hancock GJ, Webster IT, Ford PW, Moore WS (2000) Using Ra isotopes to examine transport processes controlling benthic fluxes into a shallow estuarine lagoon. Geochim Cosmochim Acta 21:3685-3699

Harvey JW, Odum WE (1990) The influence of tidal marshes on upland groundwater discharge to estuaries. Biogeochem 10: 217-236

Hashimoto T, Aoyagi Y, Kudo H, Sotobayashi T (1985) Range calculation of alpha-recoil atoms in some minerals using LSS-theory. J Radioanal Nucl Chem 90:415-438

Henderson GM, Slowey NC, Naddad GA (1999) Fluid flow through carbonate platforms: constraints from $^{234}U/^{238}U$ and Cl$^-$ in Bahama pore-waters. Earth Planet Sci Lett 169:99-111

Hodge VF, Johannesson KH, Stetzenbach KJ (1996) Rhenium, molybdenum, and uranium in groundwater from the southern Great Basin, USA: evidence for conservative behavior. Geochim Cosmochim Acta 60:3197-3214
Howard AJ, Simsarian JE, Strange WP (1995) Measurements of ^{220}Rn emanation from rocks. Health Phys 69:936-943
Hussain N (1995) Supply rates of natural U-Th series radionuclides from aquifer solids into groundwater. Geophys Res Lett 22:1521-1524
Hussain N, Krishnaswami S (1980) ^{238}U series radioactive disequilibrium in groundwaters: implications to the origin of excess ^{234}U and fate of reactive pollutants. Geochim Cosmochim Acta 44:1287-1291
Hussain N, Lal D (1986) Preferential solution of ^{234}U from recoil tracks and ^{234}U/^{238}U radioactive disequilibrium in natural waters. Proc Indian Acad Sci (Earth Planet Sci) 95:245-263
Hussain N, Church T, Kim G (1999) Use of ^{222}Rn and ^{226}Ra to trace submarine groundwater discharge into the Chesapeake Bay. Marine Chem 65:127-134
Ivanovich M (1991) Aspects of uranium-thorium series disequilibrium applications to radionuclide migration studies. Radiochim Acta 52/53:257-268
Ivanovich M, Fröhlich K, Hendry MJ (1991) Uranium series radionuclides in fluids and solids, Milk River aquifer, Alberta, Canada. Appl Geochem 6:405-418
Ivanovich M, Latham AG, Longworth G, Gascoyne M (1992a) Applications to radioactive water disposal studies. In: Uranium-series disequilibrium. Ivanovich M, Harmon RS (eds) Clarendon Press, Oxford, p583-630
Ivanovich M, Tellam JH, Longworth G, Monaghan JJ (1992b) Rock water interaction timescales involving U and Th isotopes in a Perm-Triassic sandstone. Radiochim Acta 58/59:423-432
Johannes RE (1980) The ecological significance of the submarine discharge of ground water. Mar Ecol Prog Ser 3:365-373
Kigoshi K (1971) Alpha recoil ^{234}Th: dissolution in water and the ^{234}U/^{238}U disequilibrium in nature. Science 173:47-48
Kim JI, Zeh P, Delakowitz B (1992) Chemical interactions of actinide ions with groundwater colloids in Gorleben aquifer systems. Radiochim Acta 58/59:147-154
Kim G, Hwang DW (2002) Tidal pumping of groundwater into the coastal ocean revealed from submarine ^{222}Rn and CH$_4$ monitoring. Geophys Res Lett 29, article no. 1678
King PT, Michel J, Moore WS (1982) Ground water geochemistry of ^{228}Ra, ^{226}Ra, and ^{222}Rn. Geochim Cosmochim Acta 46:1173-1182
Kolodny Y, Kaplan IR (1970) Uranium isotopes in sea-floor phosphorites. Geochim Cosmochim Acta 34:3-24
Krest JM, Moore WS, Gardner LR, Morris JT (2000) Marsh nutrient export supplied by ground water discharge: Evidence from radium measurements. Global Biogeochem Cycles 14:167-176
Krishnaswami S, Seidemann DE (1988) Comparative study of ^{222}Rn, ^{40}Ar, ^{39}Ar, and ^{37}Ar leakage from rocks and minerals- implications for the role of nanopores in gas transport through natural silicates. Geochim Cosmochim Acta 52:655-658
Krishnaswami S, Graustein WC, Turekian KK, Dowd F (1982) Radium, thorium, and radioactive lead isotopes in groundwaters: application to the in-situ determination of adsorption-desorption rate constants and retardation factors. Water Resour Res 6:1663-1675
Krishnaswami S, Bhushan R, Baskaran M (1991) Radium isotopes and ^{222}Rn in shallow brines, Kharaghoda (India). Chem Geol (Isot Geosci) 87:125-136
Kronfeld J, Vogel JC, Talma A (1994) A new explanation for extreme ^{234}U/^{238}U disequilibria in a dolomitic aquifer. Earth Planet Sci Lett 123:81-93
Ku T-L, Luo S, Leslie BW, Hammond DE (1992) Decay-series disequilibria applied to the study of rock-water interaction and geothermal systems. In: Uranium-series disequilibrium. Ivanovich M, Harmon RS (eds) Clarendon Press Oxford, p631-668
Ku TL, Luo S, Leslie BW, Hammond DE (1998) Assessing in-situ radionuclide migration from natural analog studies: response to McKinley and Alexander (1996). Radiochim Acta 80:219-223
Langmuir D (1969) Iron in ground-water of the Magothy and Raritan Formations in Camden and Burlington Counties, NJ. New Jersey Water Res Circ 19
Langmuir D, Herman J (1980) The mobility of Th in natural waters at low temperatures. Geochim Cosmochim Acta 44:1753-1766
Langmuir D, Melchoir D (1985) The geochemistry of Ca, Sr, Ba, and Ra sulfates in some deep brines from the Palo Duro Basin, Texas. Geochim Cosmochim Acta 49:2423-2432
Langmuir D, Reise AC (1985) The thermodynamic properties of radium. Geochim Cosmochim Acta 49:1593-1601
Lapointe BE, O'Connell JD, Garrett GS (1990) Nutrient couplings between on-site sewage disposal systems, ground waters, and nearshore surface waters of the Florida Keys. Biogeochem 10:289-307

Li L, Barry DA, Stagnitti F, Parlange, J-Y (1999) Submarine groundwater discharge and associated chemical input to a coastal sea. Water Resour Res 35:3253-3259

Lieser KH, Hill R (1992) Chemistry of thorium in the hydrosphere and in the geosphere. Radiochim Acta 56:141-151

Luo SD, Ku TL, Roback R, Murrell M, McLing TL (2000) In-situ radionuclide transport and preferential groundwater flows at INELL (Idaho): decay-series disequilibrium studies. Geochim Cosmochim Acta 64:867-881

Mahoney JJ, Langmuir D (1991) Adsorption of Sr on kaolinite, illite, and montmorillonite at high ionic strengths. Radiochim Acta 54:139-144

Manheim FT, Paull, CK (1981) Patterns of ground water salinity changes in a deep continental-oceanic transect off the southeastern Atlantic coast of the U.S.A. J Hydrol 54:95-105

Martin P, Akber RA (1999) Radium isotopes as indicators of adsorption-desorption interactions and barite formation in groundwater. J Environ Radioact 46:271-286

McCarthy J, Shevenell L (1998) Obtaining representative ground water samples in a fractured and karstic formation. Ground Water 36:251-260

McKinley IG, Alexander WR (1996) On the incorrect derivation and use of in-situ retardation factors from natural isotope profiles. Radiochim Acta 74:263-267

Michel J (1984) Redistribution of uranium and thorium series isotopes during isovolumetric weathering of granite. Geochim Cosmochim Acta 48:1249-1255

Miller CW, Benson LV (1983) Simulation of solute transport in a chemically reactive heterogeneous system: model development and application. Water Resourc Res 19:381-391

Moise T, Starinsky A, Katz A, Kolodny Y (2000) Ra isotopes and Rn in brines and ground waters of the Jordan-Dead Sea Rift Valley: enrichment, retardation, and mixing. Geochim Cosmochim Acta 64:2371-2388

Moore WS (1996) Large ground water inputs to coastal waters revealed by ^{226}Ra enrichments. Nature 380: 612-614

Moore WS (1997) High fluxes of radium and barium from the mouth of the Ganges-Brahmaputra River during low river discharge suggest a large ground water source. Earth Planet Sci Lett 150:141-150

Moore WS (1999) The subterranean estuary: a reaction zone of ground water and seawater. Mar Chem 65:111-125

Moore WS (2000) Determining coastal mixing rates using radium isotopes. Cont Shelf Res 20:1995-2007

Moore WS, Shaw T J (1998) Chemical signals from submarine fluid advection onto the continental shelf. J Geophys Res 103:21543-21552

Morawska L, Phillips CR (1992) Dependence of the radon emanation coefficient on radium distribution and internal structure of the mineral. Geochim Cosmochim Acta 57:1783-1797

Neretnieks I (1980) Diffusion in the rock matrix: an important factor in radionuclide retardation? J Geophys Res 88:4379-4397

Nitzsche O, Merkel B (1999) Reactive transport modeling of uranium 238 and radium 226 in groundwater of the Königstein uranium mine, Germany. Hydrogeol J 7:423-430

Ohnuki T, Isobe H, Yanase N, Nagano T, Sakamoto Y, Sekine K (1997) Change in sorption characteristics of uranium during crystallization of amorphous iron minerals. J Nucl Sci Technol 34:1153-1158

Ordonez Regil E, Schleiffer JJ, Adloff JP, Roessler K (1989) Chemical effects of α-decay in uranium minerals. Radiochim Acta 47:177-185

Osmond JK, Cowart JB (1992) Ground water. *In:* Uranium-series disequilibrium. Ivanovich M, Harmon RS (eds) Clarendon Press Oxford, p290-334

Osmond JK, Cowart JB (2000) U-series nuclides as tracers in groundwater hydrology. *In* Environmental tracers in subsurface hydrology, Cook P, Herczeg A (eds). Kluwer Academic Publishers, Boston, p290-333

Osmond JK, Ivanovich M (1992) Uranium series mobilization and surface hydrology. *In* Uranium-series disequilibrium. Ivanovich M, Harmon RS (eds) Clarendon Press, Oxford, p 259-289

Payne TE, Davis JA, Waite TD (1994) Uranium retention by weathered schists—the role of iron minerals. Radiochim Acta 66/67:297-303

Payne TE, Edis R, Fenton BR, Waite TD (2001) Comparison of laboratory uranium sorption data with "in situ distribution coefficients" at the Koongarra uranium deposit, Northern Australia. J Environ Radioact 57:35-55

Petit J-C, Langevin Y, Dran J-C (1985a) Radiation-enhanced release of uranium from accessory minerals in crystalline rocks. Geochim Cosmochim Acta 49:871-876

Petit J-C, Langevin Y, Dran J-C (1985b) ^{234}U/^{238}U disequilibrium in nature: theoretical reassessment of the various proposed models. Bull Minéral 108:745-753

Pliler R, Adams JAS (1962) The distribution of thorium and uranium in a Pennsylvanian weathering profile. Geochim Cosmochim Acta 26:1137-1146

Porcelli D, Andersson PS, Wasserburg GJ, Ingri J, Baskaran M (1997) The importance of colloids and mires for the transport of uranium isotopes through the Kalix River watershed and Baltic Sea. Geochim Cosmochim Acta 61:4095-4113

Porcelli D, Andersson PS, Baskaran M, Wasserburg GJ (2001) Transport of U- and Th-series nuclides in a Baltic Shield watershed and the Baltic Sea. Geochim Cosmochim Acta 65:2439-2459

Puls RW, Powell RM (1992) Acquisition of representative ground-water quality samples for metals. Ground Water Monitor Remediat 12:167-176

Rama, Moore WS (1984) Mechanism of transport of U-Th series radioisotopes from solids into ground water. Geochim Cosmochim Acta 48:395-399

Rama, Moore WS (1990a) Submicronic porosity in common minerals and emanation of radon. Nucl Geophys 4:467-473

Rama, Moore WS (1990b) Micro-crystallinity in radioactive minerals. Nucl Geophys 4:475-478

Rama, Moore WS (1996) Using the radium quartet for evaluating groundwater input and water exchange in salt marshes. Geochim Cosmochim Acta 60:4645-4652

Read D, Andreoli MAG, Knoper M, Williams CT, Jarvis N (2002) The degradation of monazite: implications for the mobility of rare-earth and actinide elements during low-temperature alteration. Europ J Mineral 14:487-498

Reay WG, Gallagher DL, Simmons GM (1992) Ground water discharge and its impact on surface water quality in a Chesapeake Bay inlet. Water Res Bull 28:1121-1134

Roback RC, Johnson TM, McLing TL, Murrell MT, Luo SD, Ku TL (2001) Uranium isotopic evidence for groundwater chemical evolution and flow patterns in the eastern Snake River Plain aquifer, Idaho. Geol Soc Am Bull 113:1133-1141

Roessler K (1983) Uranium recoil reactions. *In:* Uranium, Supplement Vol. A6, Gmelin handbook of inorganic chemistry, Springer-Verlag, Berlin, p 135-164

Roessler K (1989) Thorium recoil reactions. *In:* Thorium, Supplement Vol. 4, Gmelin handbook of inorganic chemistry, Springer-Verlag, Berlin, p 199-246

Rosholt J, Shields WR, Garner EL (1963) Isotopic fractionation of uranium in sandstone. Science 139:224-226

Ryan JN, Elimelech M (1996) Colloid mobilization and transport in groundwater. Colloid Surf A-Physicochem Engin Aspects 107:1-56

Schwartz M, Sharpe J. (2003) Use of ^{222}Rn to trace groundwater discharge into the Delaware River. Estuaries in press

Semkow TM (1990) Recoil-emanation theory applied to radon release from mineral grains. Geochim Cosmochim Acta 54:425-440

Shanklin DE, Sidle WC, Ferguson ME (1995) Micro-purge low-flow sampling of uranium-contaminated ground-water at the Fernald environmental management project. Ground Water Monitor Remediat 15:168-176

Shaw TJ, Moore WS, Kloepfer J, Sochaski MA (1998) The flux of barium to the coastal waters of the Southeastern United States: the importance of submarine ground water discharge. Geochim Cosmochim Acta 62:3047-3052

Sheng ZZ, Kuroda PK (1986) Isotopic fractionation of uranium: extremely high enrichments of ^{234}U in the acid-residues of a Colorado carnotite. Radiochim Acta 39:131-138

Short SA, Lowson RT (1988) ^{234}U/^{238}U and ^{230}Th/^{234}U activity ratios in the colloidal phases of aquifers in lateritic weathered zones. Geochim Cosmochim Acta 52:2555-2563

Simmons GM Jr (1992) Importance of submarine ground water discharge (SGWD) and seawater cycling to material flux across sediment/water interfaces in marine environments. Mar Ecol Prog Ser 84:173-184

Sims R, Lawless TA, Alexander JL, Bennett DG, Read D (1996) Uranium migration through intact sandstone: effect of pollutant concentration and the reversibility of uptake. J Contamin Hydrol 21:215-228

Smellie JAT, Stuckless JS (1985) Element mobility studies of two drill-cores from the Götemar granite (Kråkemåla test site), southeast Sweden. Chem Geol 51:55-78

Smellie JAT, Karlsson F, Alexander WR (1997) Natural analogue studies: present status and performance assessment implications. J Contam Hydrol 26:3-17

Sturchio NC, Bohlke JK, Markun FJ (1993) Radium isotope geochemistry of thermal waters, Yellowstone National Park, Wyoming, USA. Geochim Cosmochim Acta 57:1203-1214

Sturchio NC, Banner JL, Binz CM, Heraty LB, Musgrove M (2001) Radium chemistry of ground waters in Palaeozoic carbonate aquifers, mid-continent, USA. Appl Geochem 16:109-122

Suksi J, Rasilainen K (1996) On the role of α-recoil in uranium migration- some findings from the Palmottu Natural Analogue Site, SW Finland. Radiochim Acta 74:297-302

Suksi J, Ruskeeniemi T, Lindberg A, Jaakkola T (1991) The distribution of natural radionuclides on fracture surfaces in Palmottu Analogue study Site in SW Finland. Radiochim Acta 52/53:367-372

Suksi J, Ruskeeniemi T, Rasilainen K (1992) Matrix diffusion- evidences from natural analogue studies at Palmottu in SW Finland. Radiochim Acta 58/59:385-393

Suksi J, Rasilainen K, Casanova J Ruskeeniemi T, Blomqvist R, Smellie JAT (2001) U-series disequilibria in a groundwater flow route as an indicator of uranium migration processes. J Contam Hydrol 47:187-196

Suzuki Y, Kelly SD, Kemner KM, Banfield JF (2002) Nanometre-size products of uranium bioreduction. Nature 419:134

Swarzenski PW, Reich CD, Spechler RM, Kindinger JL Moore WS (2001) Using multiple geochemical tracers to characterize the hydrogeology of the submarine spring off Crescent Beach, Florida. Chem Geol 179:187-202

Swarzenski PW, Porcelli D, Andersson PS, Smoak JM (2003) The behavior of U- and Th- series nuclides in the estuarine environment. Rev Mineral Geochem 52:577-606

Tadolini T, Spizzico M (1998) Relation between "terra rossa" from the Apulia aquifer of Italy and the radon content of groundwater: experimental results and their applicability to radon occurrence in the aquifer. Hydrogeol J 6:450-454

Tole MP (1985) The kinetics of dissolution of zircon ($ZrSiO_4$). Geochim Cosmochim Acta 49:453-458

Top Z, Brand LE, Corbett RD, Burnett W, Chanton J (2001) Helium and radon as tracers of groundwater input into Florida Bay. J Coast Res 17:859-868

Torgerson T, Turekian KK, Turekian VC, Tanaka N, DeAngelo E, O'Donnell J (1996) ^{224}Ra distribution in surface and deep water of Long Island Sound: Sources and horizontal transport rates. Cont Shelf Res 16: 1545-1559

Tricca A, Porcelli D, Wasserburg GJ (2000) Factors controlling the ground water transport of U, Th, Ra, and Rn. Proc Indian Natl Acad Sci 109:95-108

Tricca A, Wasserburg GJ, Porcelli D, Baskaran M (2001) The transport of U- and Th-series nuclides in a sandy unconfined aquifer. Geochim Cosmochim Acta 65:1187-1210

Valiela I, Costa J, Foreman K, Teal JM, Howes B, Aubrey D (1990) Transport of ground water-borne nutrients from watersheds and their effects on coastal waters. Biogeochem 10:177-197

Webster IA, Hancock GJ Murray, AS (1995) Modeling the effect of salinity on radium desorption from sediments. Geochim Cosmochim Acta 59:2469-2476

Yanase N, Payne TE, Sekine K (1995) Groundwater geochemistry in the Koongarra ore deposit, Australia 2. Activity ratios and migration mechanisms of uranium series nuclides. Geochem J 29:31-54

Yang H-S, Hwang D-W Kim G (2002) Factors controlling excess radium in the Nakdong River estuary, Korea: submarine groundwater discharge versus desorption from riverine particles. Marine Chem 78:1-8

Zektzer IS, Ivanov VA, Meskheteli AV (1973) The problem of direct ground water discharge to the seas. J Hydrol 20:1-36

Zektzer IS Loaiciga H (1993) Groundwater fluxes in the global hydrological cycle: past, present and future. J Hydrol 144:405-427

Zukin JG, Hammond DE, Ku TL, Elders WA (1987) Uranium-thorium series radionuclides in brines and reservoir rocks from two deep geothermal boreholes in the Salton Sea Geothermal Field, southeastern California. Geochim Cosmochim Acta 51:2719-2731

9 Uranium-series Dating of Marine and Lacustrine Carbonates

R.L. Edwards[1], C.D. Gallup[2], and H. Cheng[1]

[1]*Department of Geology and Geophysics*
University of Minnesota, Twin Cities
Minneapolis, Minnesota, 55455, U.S.A.

[2]*Department of Geological Sciences*
University of Minnesota, Duluth
Duluth, Minnesota, 55812, U.S.A.

1. HISTORICAL CONSIDERATIONS

Of the possible uranium-series dating schemes, the most important and most widely applied to marine carbonates is ^{230}Th dating, with ^{231}Pa dating playing an increasingly important role. For this reason, this review will focus on these two methods. ^{230}Th dating, also referred to as U/Th dating or ^{238}U-^{234}U-^{230}Th dating, involves calculating ages from radioactive decay and ingrowth relationships among ^{238}U, ^{234}U, and ^{230}Th. ^{232}Th is also typically measured as a long-lived, essentially stable index isotope (over the time scales relevant to ^{230}Th dating). At present ^{230}Th dating can, in principle, be used to date materials as young as 3 years and in excess of 600,000 years (Edwards et al. 1987a, 1993; Edwards 1988; see Stirling et al. 2001 for an example of dating corals in excess of 600,000 years old). ^{231}Pa dating, also referred to as U/Pa dating, involves calculating ages from the ingrowth of ^{231}Pa from its grandparent ^{235}U. At present ^{231}Pa dating can be used to date materials as young as 10 years and as old 250,000 years (Edwards et al. 1997). ^{230}Th dating covers all of the ^{231}Pa time range and more, with somewhat higher precision, and is therefore the method of choice if a single method is applied. However, the combination of ^{231}Pa and ^{230}Th dating is of great importance in assessing possible diagenetic mobilization of the pertinent nuclides, and thereby, the accuracy of the ages (Allegre 1964; Ku 1968). Even if the primary age exceeds the 250,000 year limit of ^{231}Pa dating, the combined methods can be used to assess the degree to which the samples have remained closed over the past 250,000 years (e.g., Edwards et al. 1997). Thus ^{231}Pa analysis can play an important role in assessing age accuracy. Taken together, ^{230}Th and ^{231}Pa dating cover a critical time period in earth history, which often cannot be accessed with other radiometric dating techniques. Other reviews that may be of interest to the reader include a review of dating of marine sediments by Ku (1976), a review of disequilibrium dating methods (Ivanovich and Harmon 1992), and portions of the lead author's doctoral thesis (Edwards 1988), which includes a review of ^{230}Th dating from the discovery of radioactivity (Becquerel 1896) until 1988, a detailed discussion of mass spectrometric techniques for measurement of ^{230}Th and ^{234}U, and the first in depth discussion of models of marine uranium isotopic composition.

Marine and lacustrine carbonates that have been particularly amenable to ^{230}Th dating (and in some cases ^{231}Pa dating) include coralline aragonite (Barnes et al. 1956), aragonitic ooliths (Veeh 1966), aragonitic carbonate bank sediments (Slowey et al. 1996), and lacustrine tufas (Kaufmann and Broecker 1965) and other lacustrine precipitates (e.g., Ku et al. 1998; Schramm et al. 2000). Because of the large amount of work that has been done with coralline aragonite, this review will focus on dating of corals that live in shallow waters. We will then discuss dating of some of the other materials with regard to the surface coral discussion.

Historically, the most important applications of uranium-series dating techniques of marine and lacustrine carbonates have followed soon after important technical breakthroughs. Two factors have generated this close link. First, such dating methods are important because they can elucidate a broad range of processes including the timing, nature, and causes of climate change, oceanographic processes, and tectonic and seismic processes. ^{230}Th dating has also played a central role in calibration of the radiocarbon time scale and in elucidating atmospheric radiocarbon history. In some cases, ^{230}Th or ^{231}Pa dating has been the only way to characterize some aspect of these processes. Second, for most of the last century the field has been limited by measurement capabilities, which have, in turn limited the degree to which we can characterize these important processes. Thus, technical advances have generally been followed, soon thereafter by uranium-series dating applications.

The reasons for the technical limitations are the vanishingly small concentrations of the key intermediate daughter nuclides. For example, a relatively high concentration material (a 50,000 year old coral) typically has a ^{230}Th concentration of about 100 femtomoles/gram (60 billion atoms/g) and a ^{231}Pa concentration of about 3 femtomoles/gram (2 billion atoms/g). An intermediate concentration material (a 100 year old coral) has a ^{230}Th concentration of about 200 attomoles/gram (100 million atoms/g) and a ^{231}Pa concentration of about 8 attomoles/gram (5 million atoms/gram). At the extreme low end of the concentration range, surface seawater contains about 4 yoctomoles (yocto = 10^{-21}) of ^{230}Th/g (3000 atoms of ^{230}Th/g) and about 80 zeptomoles (zepto = 10^{-24}) of ^{231}Pa/g (50 ^{231}Pa atoms/g). Although we are still limited by analytical capabilities for some applications and are actively pursuing analytical improvements, measurements of ^{230}Th and ^{231}Pa in all of the above materials can now be made with relatively small samples and relatively high precision, despite these low concentrations.

Historically this was not always the case. ^{230}Th was first identified by Boltwood (1907), prior to the discovery of isotopes and was initially called "ionium." However, direct measurements of ^{230}Th concentrations in natural materials were not made until half a century later when Isaac and Picciotto (1953) applied nuclear track techniques to deep-sea sediments. The first demonstration that ^{230}Th could be used to date carbonates (corals) was by Barnes et al. (1956), using alpha-counting techniques to measure the pertinent nuclides. This work spawned a whole field that depended upon alpha-counting uranium-series measurements of natural materials. The field flourished in the 1960's and 1970's. A large portion of this effort focused on ^{230}Th dating of marine carbonates, principally corals (e.g., Veeh 1966; Broecker et al. 1968; Mesolella et al. 1969; Bloom et al. 1974; Chappell 1974; Ku et al. 1974). The field ultimately languished because of the technical limitations of alpha-counting techniques, in terms of sample size and precision (see Edwards 1988, 2000; Wasserburg 2000). The basic problem was a limit on the fraction of atoms that can be detected by decay counting techniques given the large difference between the half-lives of the pertinent nuclides (on the order of 10^5 years or more) and reasonable laboratory counting times (weeks). This problem was solved with the development of mass spectrometric methods for the measurement of ^{234}U (Chen et al. 1986) and ^{230}Th (Edwards et al. 1987a) in natural materials. Mass spectrometric measurements obviate the need to wait for the nuclides of interest to decay as mass spectrometers detect the ions/atoms of interest directly. In this regard, the development of mass spectrometric techniques for ^{230}Th and ^{234}U measurement is analogous to the development of accelerator mass spectrometer techniques for ^{14}C measurement, which improved upon traditional beta-counting techniques. Mass spectrometric methods for measuring ^{230}Th and ^{234}U greatly reduced sample size requirements and improved analytical precision. These technical improvements reinvigorated uranium-series studies and spawned a new era of in the full range of fields discussed in this book. With regard to dating of carbonates, the technique improved the precision of ^{230}Th ages, extended the range of ^{230}Th dating to both older and younger age, improved our ability to detect diagenetic alteration of nuclides used in ^{230}Th dating, and generated substantial activity (see below) in ^{230}Th dating applications to paleoclimatology, paleoceanography, ^{14}C calibration and atmospheric ^{14}C history, and tectonics. Today, the original mass spectrometric technique (Edwards et al. 1987a) is still the preferred method for analyzing ^{238}U, ^{235}U, ^{234}U, ^{230}Th, and ^{232}Th, with a number of modifications applied by different laboratories.

^{231}Pa measurements and their relationship to carbonate dating follow a similar history. Protactinium was discovered by Fajans and Gohring in 1913 in studies of ^{234}Pa. Protactinium was initially called "brevium" reflecting ^{234}Pa's short half-life (6.7 hours). Meitner and Hahn identified ^{231}Pa in 1918 and renamed the element protoactinium, as ^{231}Pa (the longest-lived isotope of protactinium) is the parent of ^{227}Ac, the longest-lived isotope of actinium. The team of Soddy, Cranson, and Fleck and independently, Fajans, also isolated ^{231}Pa at about the same time as Meitner and Hahn. "Protoactinium" was later shortened to "protactinium." Early decay-counting methods for direct measurement of ^{231}Pa in natural materials were presented by Potratz and Bonner (1958) and Sackett (1960). Rosholt (1957), Koczy et al. (1957), and Sackett et al. (1958) present early decay-counting methods for indirect determination of natural levels of ^{231}Pa via measurement of ^{227}Th as a proxy for its grandparent ^{231}Pa. Early applications of ^{231}Pa to the dating of marine carbonates include contributions by Rosholt and Antal (1962), Sakanoue et al. (1967), and Ku (1968). The method was applied in conjunction with ^{230}Th dating to make key tests of open system behavior in marine carbonates (e.g., Ku 1968; Rosholt 1967; Szabo and Rosholt 1969; Kaufmann et al. 1971). Analogous to the development of mass spectrometric techniques for ^{230}Th and ^{234}U, similar techniques were developed for the measurement of ^{231}Pa (Pickett et al. 1994). This method has already begun to fuel activity in the full range of fields discussed in this book. The first application of the Pickett et al. (1994) mass spectrometric techniques to the dating of carbonates was performed by Edwards et al. (1997). As with ^{230}Th, the use of mass spectrometric ^{231}Pa measurement techniques (as opposed to traditional decay counting techniques) improves the precision of ages, reduces sample size requirements, and extends the range of ^{231}Pa dating to both older and younger times. As discussed below, ^{231}Pa dating has and will play a critical role in combination with ^{230}Th dating in testing for diagenesis and age accuracy.

2. THEORY

2.1. Decay chains

^{238}U decay chain. ^{230}Th dating is based on the initial portion of the ^{238}U decay chain (see Bourdon and Turner 2003). ^{238}U decays by alpha emission with a half-life of 4.4683 ± 0.0048 × 10^9 years (Jaffey et al. 1971) to ^{234}Th, which in turn decays (half-life = 24.1 days) by beta emission to ^{234}Pa, which decays (half-life = 6.7 hours) by beta emission to ^{234}U, which decays (half-life = 245,250 ± 490 years, Cheng et al. 2000b) by alpha emission to ^{230}Th (half-life = 75,690 ± 230, Cheng et al. 2000b), which decays through a series of intermediate daughters ultimately to stable ^{206}Pb. Because of their short half-lives compared to the timescales in question, ^{234}Th and ^{234}Pa can be ignored in carbonate dating applications, and from a mathematical standpoint, we can view ^{238}U as decaying directly to ^{234}U. (Despite our slight herein, ^{234}Th does have the important distinction of being one the radionuclides that was originally used to determine the decay law $dN/dt = -\lambda N$ or $N = N°e^{-\lambda t}$, where N is the present number of a parent atom, N° is the initial number of parent atoms, t is time and λ is the decay constant. Rutherford and Soddy 1902). For carbonate dating, the pertinent nuclides are: ^{238}U, ^{234}U, and ^{230}Th.

^{235}U *decay chain.* ^{231}Pa dating is based on the initial portion of the ^{235}U decay chain. ^{235}U decays by alpha emission (half-life = 7.0381 ± 0.0096 × 10^8 years; Jaffey et al. 1971) to ^{231}Th, which decays (half-life = 1.06 days) to ^{231}Pa, which decays (half-life = 32760 ± 220 y, Robert et al. 1969) through a set of intermediate daughters ultimately to ^{207}Pb. Because of its short half-life, ^{231}Th can be ignored for ^{231}Pa dating applications. Thus, the pertinent nuclides for ^{231}Pa dating are ^{235}U and ^{231}Pa.

2.2. Secular equilibrium and uranium-series dating

In decay chains like the ^{238}U and ^{235}U chains, if a system remains closed to chemical exchange for time scales that are long compared to the half-lives of the intermediate daughters, the system reaches a state of secular equilibrium (see Bourdon and Turner 2003) in which the activities of all of the nuclides in a particular decay chain are equal. All U-series dating methods depend on some natural process that fractionates nuclides in the decay chain and thereby disrupts this secular equilibrium state. In the case of ^{230}Th and ^{231}Pa dating of marine and lacustrine carbonates, this event is the extreme fractionation of the parent (uranium, which is generally soluble in natural waters) from the daughter (thorium or protactinium, which generally have low solubilities in natural waters) during weathering and the hydrologic cycle. If one knows isotope ratios immediately after the fractionation event, the subsequent approach to secular equilibrium is a function of time, which can be calculated from the equations of radioactive decay and ingrowth (Bateman 1910).

2.3. ^{230}Th, ^{231}Pa, and ^{230}Th/^{231}Pa age equations

Initial daughter isotope abundances. A fundamental issue in many radiometric dating systems is the ability to quantify the initial amount of daughter isotope, permitting a determination of the relative fractions of radiogenic and initial daughter. The age is then calculated from the amount of radiogenic daughter. Ideally the initial amount of radiogenic daughter is negligible and correction for initial daughter is not necessary. If the initial amount of daughter is small but significant, it may be possible to make the correction by estimating the initial ratio of the daughter isotope to a stable isotope of the same element as the daughter. This ratio is then multiplied by the present abundance of the stable isotope as an estimate of the initial daughter abundance. If the initial amount of daughter is yet higher, then isochron techniques may be required in order to resolve the initial and radiogenic components (see below and Ludwig 2003).

Fractionation of Th and Pa daughter isotopes from U parent isotopes. One of the great advantages of ^{230}Th and ^{231}Pa dating of corals is that initial daughter concentrations are extremely low, in most cases negligible. This results from the extreme fractionation of uranium from both thorium and protactinium during the weathering process. All are actinides, but have different valences under oxidizing conditions. Uranium is typically +6 under oxidizing conditions, protactinium, +5, and thorium, +4. Uranium is soluble as uranyl ion and in various uranyl carbonate forms. Thorium has extremely low solubility in virtually all natural waters (extremely alkaline waters being the exception). Protactinium also has low solubility in natural waters, although its solubility is generally slightly higher than for thorium. A striking illustration of the extreme solubility difference between uranium and the other two elements is the fact that surface sea water has ^{230}Th/^{238}U values 10^5 times lower than secular equilibrium values (Moore 1981), and ^{231}Pa/^{235}U values 10^4 times lower than secular equilibrium (Nozaki and Nakanishi 1985). The initial fractionation of uranium from thorium and protactinium takes place during weathering and soil formation (see Chabaux et al. 2003) where a significant proportion of the uranium tends to dissolve in the aqueous phase and both thorium and protactinium tend to remain associated with solid phases. This process is initially responsible for the relatively high surface water concentrations of uranium and low concentrations of thorium and protactinium (see Porcelli and Swarzenski 2003).

Uranium in solution decays to produce both ^{230}Th and ^{231}Pa. Nevertheless, the aqueous concentrations of ^{230}Th and ^{231}Pa in surface seawater do not build up because of they are continuously removed by adsorption onto solid particles and complexation with organic molecules associated with solid particles (see Cochran and Masque 2003). Most of this ^{230}Th and ^{231}Pa is removed from the water column as the particles with which they are associated settle to the seafloor. However, as the particles settle through the water column, the adsorbed/complexed thorium and protactinium continuously re-equilibrate with seawater (see the reversible exchange model of Bacon and Anderson, 1982). The process is analogous to an ion exchange column (particles equivalent to the solution passing through an exchange column and sea water equivalent to the ion exchange resin). As a result, both ^{230}Th and ^{231}Pa generally increase with depth in the ocean. Thus, with regard to dating of marine carbonates, surface carbonates are less likely to have significant initial ^{230}Th and ^{231}Pa than deep water carbonates. This is the main difference between the dating of surface corals and dating of deep-sea corals (see below).

When coralline aragonite forms from surface seawater, there is potential to fractionate uranium, thorium, and protactinium yet again. As it turns out, however, these elements do not fractionate by large amounts during this process, incorporating the three elements in approximately their proportions in seawater. Molar uranium/calcium ratios of surface corals are within about 30% (Fig. 1, see Shen and Dunbar 1995; Min et al. 1995; see discussion below on primary uranium concentrations in corals and other marine carbonates) of the seawater value of 1.3×10^{-6} (see Broecker and Peng 1982). Molar ^{232}Th/^{238}U values of young surface corals (Fig. 2, Edwards et al. 1987a) are typically somewhat lower than typical seawater values of 3×10^{-5} (Fig. 2, Chen et al. 1986), but within a factor of 3 of this value. ^{231}Pa values for modern corals have not been measured directly because of their extremely low concentrations; however, a measurement on one coral has placed an upper limit on initial ^{231}Pa/^{235}U of 5×10^{-10} (Edwards et al. 1997). This upper limit agrees with the very limited number of surface seawater ^{231}Pa/^{235}U seawater measurements (see Edmonds et al. 1998 and references therein). Thus, existing data suggests that surface corals initially have ^{231}Pa/^{235}U ratios that are similar to or lower than seawater values. Taken together, this data indicates that the main fractionation (several orders of magnitude) leading to low initial ^{231}Pa/^{235}U and ^{230}Th/^{238}U in corals is that between solids and natural waters prior to the precipitation of coralline aragonite. Fractionation during precipitation of aragonite is small (less than a factor of a few).

The ^{230}Th age equation. Because of extremely low initial ^{230}Th/^{238}U ratios in surface corals, we first present the version of the ^{230}Th age equation calculated assuming an initial condition of ^{230}Th/^{238}U = 0. Below, we present tests that indicate that this assumption holds for most surface corals. We then present a variant of this equation, which relaxes the criterion that initial ^{230}Th/^{238}U = 0, but requires some knowledge of initial ^{230}Th/^{232}Th values. It may be necessary to employ this second equation in unusual cases involving surface corals, with deep-sea corals, and in some other marine and lacustrine carbonates. The ^{230}Th age equation, calculated assuming (1) initial ^{230}Th/^{238}U = 0, (2) all changes in isotope ratios are the result of radioactive decay and ingrowth (no chemical/diagenetic shifts in isotope ratios), and (3) ^{238}U (half life of several billion years) has not decayed appreciably over the timescales (several hundred thousand years or less) of interest is:

$$\left[\frac{^{230}\text{Th}}{^{238}\text{U}}\right] - 1 = -e^{-\lambda_{230}t} + \left(\frac{\delta^{234}\text{U}_m}{1000}\right)\left(\frac{\lambda_{230}}{\lambda_{230}-\lambda_{234}}\right)\left(1-e^{-(\lambda_{230}-\lambda_{234})t}\right) \quad (1)$$

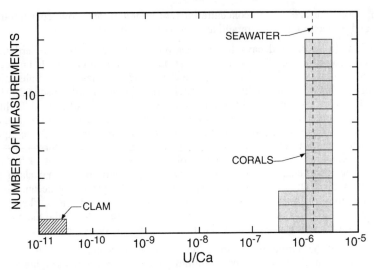

Figure 1. Histogram of measurements of molar U/Ca ratio in a number of samples of reef-building corals and one giant clam sample (after Edwards 1988). Also indicated is the U/Ca ratio of seawater. This illustrates the point that corals do not fractionate U from Ca by large amounts when they make their skeletons. U/Ca ratios of corals are similar to values from inorganically precipitated marine aragonite. Mollusks along with most other biogenic minerals exclude uranium. Note that the horizontal axis is on a log scale and that the U/Ca ratio of the clam is almost 5 orders of magnitude lower than that of the corals. This difference is the fundamental reason why there are difficulties with uranium-series dating of mollusks.

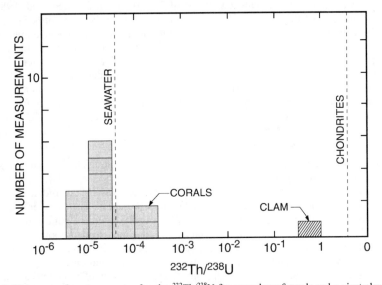

Figure 2. Histogram of measurements of molar $^{232}Th/^{238}U$ for a number of corals and a giant clam sample (after Edwards 1988). Also indicated are the $^{232}Th/^{238}U$ value of sea water and of chondrites. This illustrates the fact that corals do not fractionate thorium from uranium by large amounts during growth. The chondritic value is approximately that of the bulk earth and similar to the crustal value. Thus the large fractionation of uranium from thorium, which makes ^{230}Th dating possible, takes place during the weathering process and hydrologic cycle. Note the log scale of the horizontal axis and the 5 order of magnitude difference between chondrites and sea water. The giant clam has a much higher $^{232}Th/^{238}U$ ratio than corals because clams exclude uranium during growth (see Fig. 1).

The brackets around $^{230}Th/^{238}U$ indicate that this is an activity ratio. λ's are decay constants; t is age; and $δ^{234}U_m$ is the present deviation in parts per thousand (per mil) of the $^{234}U/^{238}U$ ratio from secular equilibrium: $δ^{234}U = ([^{234}U/^{238}U] - 1) \times 1000$. Given measured $^{230}Th/^{238}U$ and $^{234}U/^{238}U$, the only unknown is age, which can be calculated from Equation (1). Because age appears twice, the equation must be solved graphically (Fig. 3) or by iteration (by substituting $δ^{234}U_m$ and then different values of "t" until the measured $^{230}Th/^{238}U$ is calculated). The general form of this equation was first solved by Bateman (1910).

A specific form of this equation was solved by Barnes et al. (1956) when he presented the first decay counting ^{230}Th data for corals. The equation of Barnes et al. (1956) did not include the second term on the right side $\{(δ^{234}U_m/1000)(λ_{230}/(λ_{230} - λ_{234}))(1 - e^{-(λ_{230} - λ_{234})t})\}$ of Equation (1). This term accounts for initial $^{234}U/^{238}U$ values different from equilibrium. At the time of the Barnes et al. (1956) work, it was not known that seawater is out of equilibrium with respect to $^{234}U/^{238}U$, and the discovery that $^{234}U/^{238}U$ can be out of equilibrium in natural waters was only made one year earlier (Cherdyntsev 1955; by analysis of river water). This actually came as a surprise since there is no obvious chemical mechanism for this. ^{234}U is linked to its great-grandparent, ^{238}U by two very short-lived intermediate daughters, and isotopes of the same element, ^{234}U and ^{238}U should not fractionate appreciably from each other. The discovery that natural waters generally have higher $^{234}U/^{238}U$ values than secular equilibrium (and soils generally have lower $^{234}U/^{238}U$ values than secular equilibrium), led to the conclusion that a higher proportion of ^{234}U atoms than ^{238}U atoms

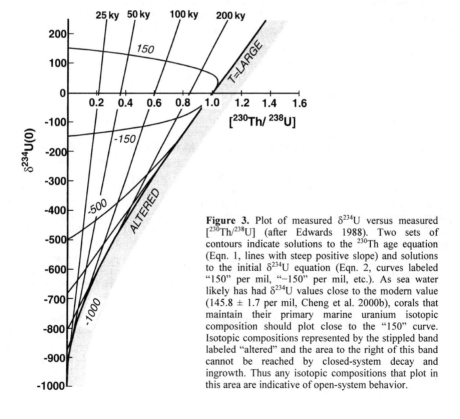

Figure 3. Plot of measured $δ^{234}U$ versus measured $[^{230}Th/^{238}U]$ (after Edwards 1988). Two sets of contours indicate solutions to the ^{230}Th age equation (Eqn. 1, lines with steep positive slope) and solutions to the initial $δ^{234}U$ equation (Eqn. 2, curves labeled "150" per mil, "–150" per mil, etc.). As sea water likely has had $δ^{234}U$ values close to the modern value (145.8 ± 1.7 per mil, Cheng et al. 2000b), corals that maintain their primary marine uranium isotopic composition should plot close to the "150" curve. Isotopic compositions represented by the stippled band labeled "altered" and the area to the right of this band cannot be reached by closed-system decay and ingrowth. Thus any isotopic compositions that plot in this area are indicative of open-system behavior.

are available for leaching from minerals undergoing weathering. The rationale was that ^{234}U atoms are produced from decay of ^{238}U, a process that involves alpha emission. Recoil from alpha-emission can damage chemical bonds that hold the nuclide in a particular site, thereby causing ^{234}U to be more susceptible to leaching into the aqueous phase than ^{238}U. Thurber (1962) first showed that the ^{234}U/^{238}U ratio of seawater was out of secular equilibrium (about 15% higher than secular equilibrium). Immediately thereafter, Broecker (1963) presented the specific equation (Eqn. 1) that includes the term for disequilibrium in initial ^{234}U/^{238}U. Derivations of Equation (1) that include numerous intermediate steps can be found in Edwards (1988) and Ivanovich and Harmon (1992).

Disequilibrium initial ^{234}U/^{238}U ratio is a happy complication. As is clear from Equation (1), this phenomenon requires measurements of both ^{234}U/^{238}U and ^{230}Th/^{238}U to solve for age. However, given both measured values, we can solve for age uniquely given the assumptions presented above. Furthermore, a second equation that relates measured and initial ^{234}U/^{238}U can be calculated from the equations of radioactive production and decay, subject only to the assumption that chemical reactions (diagenesis) involving uranium have not occurred since precipitation of the aragonite:

$$\delta^{234}U_m = \left(\delta^{234}U_i\right) e^{-\lambda_{234} t} \qquad (2)$$

where the subscript "i" refers to the initial value and the decay constant is that of ^{234}U. Thus, Equations (1) and (2) constitute two equations that can be solved for two unknowns (age and initial δ^{234}U). As it turns out, for marine samples, knowledge of initial δ^{234}U is of great importance in assessing dating accuracy. A number of arguments and supporting data suggest that marine δ^{234}U has been constant within fairly tight bounds (see below and Henderson and Anderson 2003). If so, deviations from marine values would indicate diagenetic shifts in uranium and potential inaccuracy in the ^{230}Th age.

Figure 3 shows a graphical solution to both the ^{230}Th age equation (Eqn. 1) and the initial δ^{234}U equation (Eqn. 2). Plotted on the ordinate and abscissa are the two measured quantities $\delta^{234}U_m$ and [^{230}Th/^{238}U]. Contoured with sub-vertical lines is one of the calculated unknowns, age; the ordinate is coincident with the age = zero contour. Emanating from the ordinate are a set of curves (sub-horizontal near the ordinate), which contour the other unknown, $\delta^{234}U_i$. Secular equilibrium is represented by the point (0,1). The age contours get closer together with increasing age and increasing [^{230}Th/^{238}U], reflecting the exponential approach of [^{230}Th/^{238}U] to secular equilibrium. Beyond about 50 ky analytical errors in [^{230}Th/^{238}U] are more or less constant. Thus, as age increases beyond 50 ky, error in age increases because of the compression of age contours. Eventually the age contours become so close that, for a given analytical error, one can no longer distinguish between the isotopic composition of a particular age sample and the isotopic composition of an infinite age sample. This is the upper limit of the ^{230}Th age range, which given current analytical errors is about 700,000 years (Edwards et al. 1987; Edwards et al. 1993; Stirling et al. 2001). Note that there are no age or initial δ^{234}U contours in the right portion of the diagram ("altered region"). Isotopic compositions in this region of the diagram cannot be reached through closed-system decay and ingrowth of materials that initially have zero ^{230}Th/^{238}U. Thus, samples that plot in this region have undergone diagenetic alteration.

Modifications of the ^{230}Th age equation. We show below that surface corals generally satisfy the ^{230}Th/^{238}U = 0 assumption used in calculating Equation (1). However, in unusual cases involving surface corals, most deep-sea corals, and other marine and lacustrine dating applications, correction for non-zero ^{230}Th/^{238}U may be necessary. With some knowledge of the initial ^{230}Th/^{232}Th ratio, age can be calculated with the following equation:

$$\left\{\left[\frac{^{230}Th}{^{238}U}\right]-\left[\frac{^{232}Th}{^{238}U}\right]\left[\frac{^{230}Th}{^{232}Th}\right]_i\left(e^{-\lambda_{230}t}\right)\right\}-1=$$

$$-e^{-\lambda_{230}t}+\left(\frac{\delta^{234}U_m}{1000}\right)\left(\frac{\lambda_{230}}{\lambda_{230}-\lambda_{234}}\right)\left(1-e^{-(\lambda_{230}-\lambda_{234})t}\right) \qquad (3)$$

The second term on the left side ([$^{232}Th/^{238}U$] [$^{230}Th/^{232}Th$]$_i$ ($e^{-\lambda_{230}t}$)) corrects for initial ^{230}Th. If this term is negligible compared to measured [$^{230}Th/^{238}U$], Equation (1) can be used. Included in this term is initial $^{230}Th/^{232}Th$, which must be known independently in order to calculate age. For this reason, Equation (3) is typically used when the magnitude of the second term on the left is small but significant. In this case, even if $^{230}Th/^{232}Th$ is not known precisely, the contribution to error in age is small. Low $^{232}Th/^{238}U$, low initial $^{230}Th/^{232}Th$, or large "t" will contribute to small values for this term. For surface corals there is a limited range of ^{238}U concentrations and initial $^{230}Th/^{232}Th$ values. Thus, this term is only significant if the sample is young (< a few thousand years) and the sample has unusually high ^{232}Th concentrations. For deep sea corals, the same parameters are important; however the term is generally larger and increases with depth as $^{230}Th/^{232}Th$ generally increases with depth in the ocean. For most other materials discussed here, this term should be evaluated on a case by case basis.

To estimate the magnitude of this term, several strategies have been employed. Zero-order estimates have been made by calculating an initial $^{230}Th/^{232}Th$ value (4.4 × 10^{-6} by atom) assuming a bulk earth $^{232}Th/^{238}U$ value (3.8 by atom) and presuming secular equilibrium. Estimates of initial $^{230}Th/^{232}Th$ can also be made by analyzing modern analogues to the samples in question, analyzing samples with ages that are known independently, or by applying development diagrams (see Cheng et al. 2000a) or isochron techniques (see below, Ludwig 2003).

Isochrons. In cases where the initial ^{230}Th term in Equation (3) is large enough that $^{230}Th/^{232}Th$ must be established precisely, isochron methods can be used. Such methods consider the sample to be composed of a mixture of two components. By analyzing sub-samples of the same age with different proportions of the components, one can extrapolate to an endmember that includes only radiogenic daughter and establish an age on the basis of that endmember. The extrapolation also establishes an initial $^{230}Th/^{232}Th$ ratio, assumed to be the same in both components, and the initial $^{234}U/^{238}U$ value for the radiogenic endmember. Isochron methods generally require that the sample only contain two components, although methods that deconvolve more than two components have been developed (see Henderson et al. 2001).

As an example, imagine a carbonate-rich sediment that contains a small but significant fraction of detrital silicate. The sediment might be a lake sediment, a surface coral, a deep-sea coral, a carbonate-rich bank sediment, or a speleothem (see for example, Richards and Dorale 2003). The carbonate has a very high $^{238}U/^{232}Th$ ratio (on the order of 10^4 by atom) and the detrital material has a lower $^{238}U/^{232}Th$ ratio of about 10^0. The carbonate and detrital materials each have specific $\delta^{234}U$ values that differ from each other. Both have the same initial $^{230}Th/^{232}Th$ ratio. Through time, the mixture evolves following the laws of radioactive ingrowth and decay. Sub-samples of the sediment with different proportions of carbonate and detrital material, and therefore different $^{238}U/^{232}Th$, $^{234}U/^{238}U$, and present day $^{230}Th/^{232}Th$ ratios, are measured. If the sub-samples started with the same initial $^{230}Th/^{232}Th$, their isotopic compositions will evolve so that they lie on a line in three-dimensional isotope ratio space. If the points do not lie on a line, there must be an additional thorium or uranium component represented in the mixture or the sediment has been altered. Different choices for the three axes can be made (see Ludwig and Titterington 1994); however, a convenient set for illustrative purposes is ^{230}Th/^{232}Th, ^{238}U/^{232}Th, and ^{234}U/^{232}Th [see Hall and Henderson (2001) for a recent example of use of these types of axes]. Regardless of the choice of axes, the resultant line is termed an isochron. The equation of the line can be used to extrapolate to a hypothetical uranium-free end member, from which initial ^{230}Th/^{232}Th can be calculated, and to a ^{232}Th-free (radiogenic) end member, from which age and initial ^{234}U/^{238}U of the end member can be calculated, essentially from Equations (1) and (2).

The isochron approach is powerful, but ultimately limited by the degree to which the constant initial ^{230}Th/^{232}Th assumption holds in a particular sample. There are indications from different settings (e.g., Lin et al. 1996; Cheng et al. 2000a; Slowey et al. 1996; Henderson et al. 2001) for two distinct sources of initial thorium, a "hydrogenous" source (from solution or colloids) and a "detrital source." The co-linearity (or lack thereof) of data points is an important test of this phenomenon. Even in cases where there are two initial sources of thorium, there may be methods to correct for this additional source (Henderson et al. 2001). Ideally one would physically separate the carbonate-rich fraction from the thorium-rich fraction, but often this cannot be done because of cementation or small grain size. Methods of chemical separation of these components have been attempted (e.g., acid dissolution of the carbonate component leaving the detrital silicate component behind); however, differential adsorption of uranium and thorium onto the solid residue and/or differential leaching of uranium and thorium from the residue have proven to cause problems (Bischoff and Fitzpatrick 1991; Luo and Ku 1991). Thus, in cases where physical separation is not possible, isochron techniques with total sample dissolution are recommended. Further discussion of isochron approaches can be found in Bischoff and Fitzpatrick (1991), Luo and Ku (1991), Ludwig and Titterington (1994) and Ludwig (2003).

The ^{231}Pa age equation. The ^{231}Pa age equation, calculated assuming no chemical shifts in protactinium or uranium and an initial ^{231}Pa/^{235}U = 0, is analogous to the ^{230}Th age equation (Eqn. 1), but simpler. There is no term analogous to the δ^{234}U term because there is no long-lived intermediate daughter isotope between ^{235}U and ^{231}Pa:

$$\left[\frac{^{231}Pa}{^{235}U}\right] - 1 = -e^{-\lambda_{231} t} \tag{4}$$

We will show below that the initial ^{231}Pa/^{235}U = zero assumption holds for a number of corals that have typical low ^{232}Th concentrations. Initial ^{231}Pa/^{235}U values for most other carbonates have not been studied in detail. Furthermore, in contrast to thorium, there is no long-lived isotope of protactinium that can be used as an index isotope although some work has employed corrections for initial ^{231}Pa. Such corrections essentially assume that ^{232}Th is an isotope of protactinium and assume a bulk earth ^{232}Th/^{238}U ratio and secular equilibrium between ^{231}Pa and ^{235}U. The term for applying this correction is analogous to the initial ^{230}Th term in Equation (3).

The ^{231}Pa/^{230}Th age equation. Equation (4) can be divided by Equation (1) to give the following ^{231}Pa/^{230}Th age equation:

$$\left[\frac{^{231}Pa}{^{230}Th}\right] = \frac{\left(1 - e^{-\lambda_{231} t}\right)}{\left[\frac{^{238}U}{^{235}U}\right]\left\{1 - e^{-\lambda_{230} t} + \left(\frac{\delta^{234}U_m}{1000}\right)\left(\frac{\lambda_{230}}{\lambda_{230} - \lambda_{234}}\right)\left(1 - e^{-(\lambda_{230} - \lambda_{234})t}\right)\right\}} \tag{5}$$

[^{238}U/^{235}U] is not a variable because it is constant in nature; our present best estimate of its value is: $(^{238}U/^{235}U)(\lambda_{238}/\lambda_{235}) = 137.88\ (1.5513 \times 10^{-10}\ y^{-1}/9.8485 \times 10^{-10}\ y^{-1}) = 21.718$. As opposed to Equations (1) and (4), Equation (5) does not depend explicitly on ^{238}U or ^{235}U abundance. $^{231}Pa/^{230}Th$ ages are analogous to $^{207}Pb/^{206}Pb$ ages in this respect and many others. $^{231}Pa/^{230}Th$ ages are not sensitive to very recent uranium loss; nor are they sensitive to very recent uranium gain if the added uranium has the same $\delta^{234}U_m$ as the sample. This is clear from Equation (5). On the other hand $^{231}Pa/^{230}Th$ ages are sensitive to uranium gain or loss at earlier times, as these processes affect the subsequent ingrowth and decay of ^{231}Pa and ^{230}Th. Nevertheless, $^{231}Pa/^{230}Th$ can be used in conjunction with ^{230}Th and ^{231}Pa ages to characterize, and potentially "see through" certain types of diagenesis.

2.4. Tests for ^{231}Pa-^{230}Th age concordancy

Perhaps the greatest strength of $^{231}Pa/^{235}U$ data is its use in concert with ^{230}Th-^{234}U-^{238}U data in testing age accuracy (see Cheng et al. 1998). A good method for interpreting this data is through a $^{231}Pa/^{235}U$ vs. $^{230}Th/^{234}U$ concordia diagram (Fig. 4; Allegre 1964; Ku 1968; see Cheng et al. 1998). The ordinate is the key isotopic ratio for ^{231}Pa dating and the abscissa is the key isotopic ratio for ^{230}Th dating. Plotted parametrically is age along the locus of isotopic compositions for which ^{231}Pa and ^{230}Th age are identical: concordia. This plot is analogous the U-Pb concordia plot (Wetherill 1956a; 1956b) except that the $^{231}Pa/^{235}U$ vs. $^{230}Th/^{234}U$ diagram has different concordia for different initial $\delta^{234}U$ values. If a data point plots off of concordia, the sample must have been

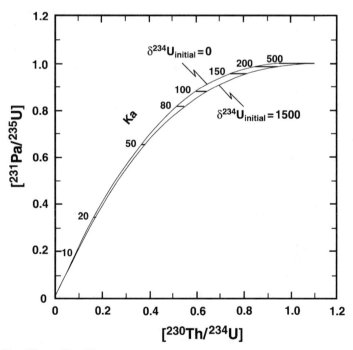

Figure 4. $^{231}Pa/^{235}U$ vs. $^{230}Th/^{234}U$ concordia diagram (after Cheng et al. 1998). Concordia represent the locus of isotopic compositions for which the ^{231}Pa age (see Eqn. 4) and the ^{230}Th age (see Eqn. 1) are equal. Two concordia curves are indicated, one for an initial $\delta^{234}U = 0$ and one for an initial $\delta^{234}U = 1500$. Age is depicted parametrically along the concordia curves (horizontal line segments). The concordia curves are analogous to U/Pb concordia curves used in zircon dating.

altered chemically and one or both of the ages are not accurate. If a data point plots on concordia, then the sample's isotopic composition is consistent with closed-system behavior of the pertinent nuclides. This constitutes a robust test for age accuracy.

Furthermore, should a data point lie off of concordia, it may still be possible to constrain the true age of the sample. Interpretations of points that lie off of concordia are model-dependent. However, there are commonalities among models of a number of likely diagenetic processes (Cheng et al. 1998). Figure 5 shows isotopic compositions generated by one such model. Two separate calculations are illustrated, one for a set of co-genetic materials with a primary age of 80 ky (dashed curve), another for a set of materials with a primary age of 150 ky (thin solid curve). Considering the 80 ky materials first, the intersection of the dashed line with concordia represents a material that has behaved as a closed system and records primary ^{230}Th and ^{231}Pa ages. All other isotopic compositions on the line represent materials that have undergone different degrees of the same type of diagenesis: continuous addition or loss of uranium. The portion of the dashed curve between concordia and the origin represents the isotopic composition of materials that have gained uranium (with a δ^{234}U identical to that of the sample) continuously. The portion immediately below concordia represents material that has gained uranium at a low rate; progressively closer to the origin, the rate of uranium gain increases. The origin represents material that has gained uranium at an infinite rate. Similarly, the portion of the curve above concordia represents the isotopic composition of material that has lost uranium. The portion above and immediately adjacent to the

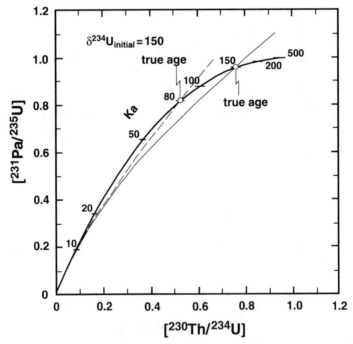

Figure 5. Concordia diagram similar to Figure 4 illustrating the concordia curve for initial δ^{234}U = 150 (appropriate for marine samples), with age in ka depicted parametrically along concordia. Also illustrated are continuous uranium gain/loss model curves for samples with primary ages of 80 ka (dashed) and 150 ka (thin solid curve). See text for discussion of this model and related models (after Cheng et al. 1998).

concordia represents the isotopic composition of material that has lost uranium at a low rate. Points progressively higher on the curve represent progressively higher rates of uranium loss. The thin solid curve is analogous to the dashed curve, but pertains instead to a set of materials with a primary age of 150 ky.

Both of these curves approximate a line in the vicinity of concordia. This is one of the commonalities between this continuous uranium gain/loss model and a number of other open system models (e.g., instantaneous uranium gain/loss models, continuous daughter (thorium and protactinium) gain/loss models, or instantaneous daughter gain/loss models; see Cheng et al. 1998). This has led to the idea that if one can identify a set of equal-age materials affected by the same diagenetic process, but to different extents, the isotopic compositions of this set of materials should approximate a line. By extrapolating or interpolating this line to concordia, one may be able to establish the primary age of the material (see Fig. 6). This possibility exists even if the specific diagenetic process is not known, as long as the diagenetic process includes one of the modeled processes discussed above.

In the model that we illustrate (Fig. 5), points above concordia indicate continuous uranium loss and points below concordia indicate uranium gain. Models of instantaneous uranium loss/gain plot in the same sense relative to concordia. Similarly, models of instantaneous or continuous daughter loss generally (but not always depending on the ratio of lost ^{231}Pa to lost ^{230}Th) yield isotopic compositions below concordia, whereas models of daughter gain generally (but not always) give isotopic compositions above

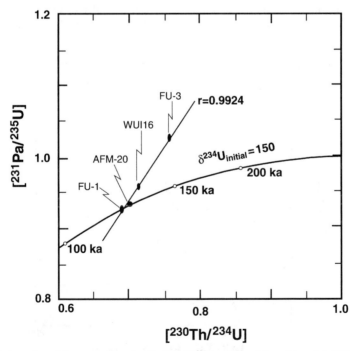

Figure 6. A portion of the concordia curve for initial $\delta^{234}U = 150$ and four data points (Edwards et al. 1997) for coral samples from the last interglacial terrace in Barbados. The points lie along a line, and illustrate the possibility of using best-fit lines through discordant points to extrapolate to concordia and possibly the true age of a set of samples (after Cheng et al. 1998).

concordia. Thus, isotopic compositions above concordia are consistent with uranium loss or daughter gain, whereas those below concordia are consistent with uranium gain or daughter loss. If so, isotopic compositions above concordia yield ^{231}Pa and ^{230}Th ages that are maximum ages, whereas those below concordia yield minimum ages. In sum, isotopic compositions that yield discordant ages place constraints on both diagenetic mechanisms and age. Isotopic compositions that yield concordant ages provide some confidence that the ages are accurate. These two statements, in a nutshell embody the power of combined ^{230}Th-^{231}Pa dating.

3. TESTS OF DATING ASSUMPTIONS

3.1. Are initial ^{230}Th/^{238}U and ^{231}Pa/^{235}U values equal to zero?

One of the keys in ^{230}Th and ^{231}Pa dating is establishing initial ^{230}Th/^{238}U and ^{231}Pa/^{235}U values. Upper limits were originally placed on these ratios in surface corals by analyzing modern corals by alpha-counting techniques (Barnes et al. 1956; Broecker 1963 for ^{230}Th/^{238}U and Ku 1968 for ^{231}Pa/^{235}U). With the advent of mass spectrometric techniques, tighter bounds on initial values were needed. In this regard, the first data that suggested that initial ^{230}Th/^{238}U values of surface corals were extremely low were the ^{232}Th/^{238}U values measured by mass spectrometric techniques (Edwards et al. 1987a). The values were three orders of magnitude lower than the upper limits from the earlier alpha-counting measurements. When multiplied by typical ^{230}Th/^{232}Th ratios of surface sea water, these measurements yielded initial ^{230}Th/^{238}U values equivalent to about 1 year's worth of radiogenic ingrowth (negligible compared to analytical errors of >2 years).

This calculation was further tested by dating portions of a coral with ages known a priori from the counting of annual density bands (Fig. 7). A plot of ^{230}Th age calculated

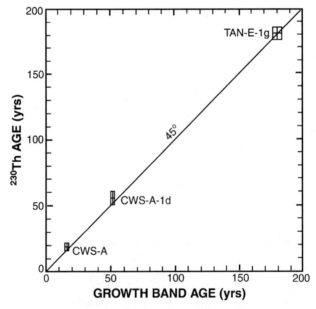

Figure 7. ^{230}Th age vs. growth band age for three coral sub-samples, all younger than 200 years old (after Edwards 1988 and Edwards et al. 1988). All three points lie on a 1:1 line indicating that the ^{230}Th ages are accurate and that initial ^{230}Th/^{238}U is negligible, justifying the use of Equation (1) to determine ^{230}Th age.

with Equation (1) vs. band counting age (Edwards 1988; Edwards et al. 1988) for three such coral sub-samples is shown in Figure 7. Errors are indicated by boxes, with the vertical dimension equivalent to analytical error in ^{230}Th age and the horizontal dimension equivalent to the number of annual bands encompassed by the sub-sample. Equation (1) has no term for correcting for initial ^{230}Th. Thus, if there were significant initial ^{230}Th, the ^{230}Th age would be older than the band counting age. This is not the case as all three points plot within error of a 45° line, indicating that initial ^{230}Th is not significant within errors of about 3 years. All three Figure 7 samples have ^{232}Th concentrations less than 100 pg/g. Thus, this study suggested that initial ^{230}Th levels were negligible in surface corals with ^{232}Th levels below about 100 pg/g. Subsequent work has demonstrated that most surface corals have ^{232}Th levels of about 100 pg/g or less, leading to the conclusion that initial ^{230}Th/^{238}U is negligible for typical surface corals.

However, subsequent work has also shown that a small fraction of surface corals can have elevated ^{232}Th values. ^{232}Th levels of several hundred to 1000 pg/g have been reported from the central Pacific (Cobb et al. in review) and in extreme cases levels of several thousands of pg/g (Zachariasen 1998; Zachariasen et al. 1999) for a small subset of corals from Sumatra. For coralline samples older than several thousand years, even levels of ^{232}Th as high as a few thousand pg/g are not likely to be associated with levels of initial ^{230}Th that are significant compared to analytical error. However, for samples less than several thousand years old, such samples are best avoided, as initial ^{230}Th is likely to be significant. In the very highest ^{232}Th samples from Sumatra, the corals were visibly discolored with a brown stain, and thus could easily be identified. The stain could be partially removed in an ultrasonic bath, a procedure that also lowered ^{232}Th levels. Thus, the thorium is associated with the discoloration, quite likely organic matter. In both the Cobb et al. (in review) and Zachariasen et al. (1999) studies, samples with the highest ^{232}Th levels could be avoided. However, in both studies, it was important to obtain ages on coralline material with somewhat elevated ^{232}Th levels (hundreds of pg/g). In both studies, this was accomplished using Equation (3) with ^{230}Th/^{232}Th determined by analyzing local corals of known age or for which there were constraints on age. In the Sumatra study, initial atomic ^{230}Th/^{232}Th values between 0 and 1.3×10^{-5} were determined and in the central Pacific values covered a similar range from 0 to 2.0×10^{-5}, broadly consistent with values for surface seawater. In each of these studies, corrections for initial ^{230}Th using these isotopic values were on the order of 20 years or less, even for samples with several hundred pg/g of ^{232}Th.

In sum, for surface corals, the assumption that initial ^{230}Th/^{238}U values are negligible holds in most cases, satisfying one of the assumptions used in deriving Equation (1). However, in unusual cases corals younger than several thousand years may have significant initial ^{230}Th/^{238}U compared to analytical errors. In those cases, corrections for initial ^{230}Th/^{238}U can be made accurately with some knowledge of the range of possible ^{230}Th/^{232}Th values.

Initial ^{231}Pa/^{235}U levels are more difficult to assess, primarily because there is no long-lived or stable isotope of protactinium that can be used as an index isotope. Edwards et al. (1997) analyzed a set of surface coral sub-samples younger than 1000 years by both ^{230}Th and ^{231}Pa techniques. For all samples, ^{232}Th concentrations were less than 100 pg/g so that initial ^{230}Th/^{238}U values were negligible. Each sub-sample yielded ^{230}Th and ^{231}Pa ages identical within analytical errors (Fig. 8), indicating that initial ^{231}Pa/^{235}U was negligible. This suggests that surface corals with typical ^{232}Th values do not require corrections for initial ^{231}Pa. Whether or not corals with elevated ^{232}Th require such corrections is an open question.

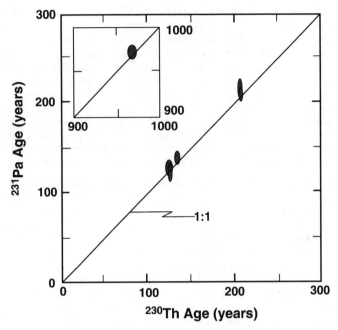

Figure 8. ^{231}Pa age vs. ^{230}Th age for a set of corals all younger than 1000 years old (after Edwards et al. 1997). All points lie along a 1:1 line indicating that the corals record identical ^{230}Th and ^{231}Pa ages, suggesting that the ^{231}Pa ages are accurate and that initial ^{231}Pa/^{235}U is negligible, justifying the use of Equation (4) to determine ^{231}Pa age.

3.2. Tests of the closed-system assumption

The second assumption, used in deriving Equations (1) through (4) is that the system has remained closed to chemical exchange. Testing this assumption is more difficult. A number of different approaches have been used to assess this assumption. Perhaps the most powerful of these is combined ^{231}Pa and ^{230}Th dating, but most of the methods discussed below play a role.

Concerns that diagenetic alteration could affect measured ^{230}Th ages of corals have existed since the early applications using alpha-counting methods. Thurber et al. (1965) offered the following list for reliable coral ^{230}Th ages: a) no evidence of recrystallization, b) uranium concentration of ~3 ppm, c) [^{230}Th/^{232}Th] of > 20, d) [^{234}U/^{238}U] of 1.15 ± 0.02, and e) stratigraphically consistent ages. These phenomena are still applied today, but numerical limits have changed drastically, reflecting higher precision measurement techniques and better understanding of the processes that result in primary and diagenetic values for these parameters.

Initial mineralogy, petrology, and associated elemental concentrations. Corals create an aragonitic skeleton in a seawater environment. Because sea levels are low during glacial and interstadials periods, most fossil corals have been exposed to meteoric waters for a significant portion of their post-depositional history. This exposure creates the potential for these corals to be recrystallized to the more stable calcium carbonate, calcite, by dissolution/reprecipitation reactions associated with percolating waters (Matthews 1968). X-ray diffraction is commonly used to screen for calcite, where the heights of aragonite and calcite peaks of a sample are compared to the peak heights of a series of standards, from 100% calcite and 0% aragonite down to the detection limits of the instrument, usually 1-2% calcite. As Thurber et al. (1965) suggested, a measurable calcite peak indicates that the sample has been recrystallized to some degree, which suggests that the ^{230}Th age is not reliable. However, where samples with and without detectable calcite have been measured that are from the same terrace, and presumably are the same age, there is no clear systematic effect on the ^{230}Th age (e.g., Bloom et al. 1974).

Matthews (1968) documented that the frequency of exposed coral recrystallization generally increases with coral age on Barbados. Chappell and Polach (1972) had similar results for corals from the Huon Peninsula in Papua New Guinea, but with thin section work showed that the microscopic bundles of aragonite that make up the coral skeleton become fused or thickened as recrystallization proceeds (see Figs. 9 and 10). Bar-Matthews et al. (1993) attempted to correlate such petrologic changes (where the thickening is due to secondary aragonite) with shifts in uranium-series isotopes in last interglacial corals from the Bahamas, but could find no clear correlation with the presence of secondary precipitation features and δ^{234}U value or ^{230}Th age. In addition, thin section work can detect precipitation of aragonite cements in pores within the coral skeleton. As aragonite cements initially have similar isotopic characteristics to coralline aragonite, the presence of aragonite cement does not necessarily pose a problem. However, if the cement has a different age than the coral skeleton, the apparent age of the coral-cement mixture will differ from the true age of the coral.

Bloom et al. (1974) measured Sr and Mg concentrations as a possible measure of recrystallization of coral skeletons. Sr concentrations in calcite are less than in aragonite and high-Mg calcite has higher Mg concentrations than aragonite (e.g., Edwards 1988 for Sr and Mg measurements of cements in fossil corals). Consistent with this difference, Bloom et al. (1974) found that samples with detectable calcite had slightly lower Sr concentrations and those with high-Mg calcite had Mg concentrations ~5 times higher than aragonitic corals. Bar-Matthews et al. (1993) did not find a correlation between Sr and Mg concentrations in corals with secondary aragonite and initial δ^{234}U value. However, they did find a distinguishable negative correlation between Na and SO$_3$ concentrations and initial δ^{234}U value, which they interpreted as suggesting diagenetic alteration in a marine environment (many of the samples were within a few meters of sea level). This correlation has not been explored in uplifted corals that have spent most of their history many meters above sea level. Primary coralline aragonite generally has extremely low ^{232}Th concentrations (tens of parts per trillion, Edwards et al. 1987a). Thus, ^{232}Th values significantly higher than this indicate contamination from detrital sources, organic-thorium complexes, or unusual growth settings (e.g., Zachariasen et al. 1999; Cobb et al., in review). In some cases, limiting sample to denser portions of the coral, with little pore space and low macroscopic surface to volume ratio has proven to improve the sample in terms of a number of the measures of diagenesis (Stirling et al. 1995) discussed above and below (e.g., δ^{234}U, ^{232}Th content).

Initial $^{234}U/^{238}U$. Corals incorporate marine uranium into their skeleton without isotopic fractionation. The modern marine δ^{234}U value as measured in marine waters is 140-150 (Chen et al. 1986) and in modern corals is 145.8 ± 1.7‰ (using updated half-lives; Cheng et al. 2000b). The marine δ^{234}U value does not vary more than the analytical error with depth or geographic location (Chen et al. 1986; Cheng et al. 2000a), consistent with the long marine residence time of uranium (200,000 to 400,000 years; Ku et al. 1977) and the long half-life of ^{234}U, compared to the mixing time of the ocean (about 10^3 y). If the uranium isotopic composition of the ocean also remained constant with time, the δ^{234}U value would add a second chronometer in the ^{238}U decay chain for corals

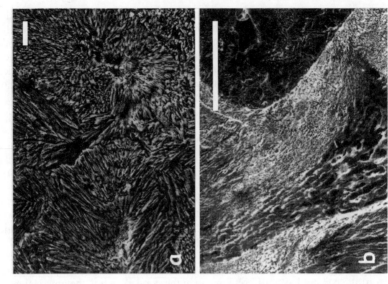

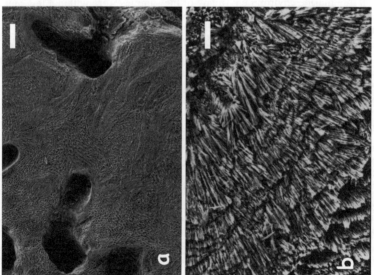

Figure 9. SEM photographs of polished, etched thin sections of modern *Acropora palmata* coral (after Edwards 1988). The scale bar in "a" is 100 microns. Visible in "a" are large macroscopic pores in the skeleton and well as the texture of the very fine aragonite crystals. "b" is the same section as "a" but at higher magnification. The scale bar is 10 microns. Individual aragonite crystal fibers are visible in "b."

Figure 10. SEM photographs of polished, etched thin sections of fossil *Acropora palmata* coral (after Edwards 1988). The scale bar in "a" is 10 microns. "a" depicts sample AFS-12, a last interglacial coral from Barbados. The crystal morphology in this well-preserved sample is indistinguishable from that of a modern sample (see Fig. 9b). The scale bar in "b" is 100 microns. "b" depicts sample PB-5B, a fossil coral collected from North Point Shelf on Barbados. The crystal morphology of this sample shows clear evidence of alteration, including a large calcite crystal filling in a macroscopic pore (dark area in upper right portion of photograph).

(Ku 1965). However, the wide range in riverine $\delta^{234}U$ values (0 to 2000; Cochran 1982) suggests that the elevation of the marine $\delta^{234}U$ value above secular equilibrium results from a complex combination of weathering and alpha-recoil processes (see above). Thus, it is possible that the marine $\delta^{234}U$ value has changed by small amounts with time. In practice, the main impediment to a $\delta^{234}U$ chronometer is the sensitivity of $\delta^{234}U$ to diagenesis (see below).

When dating corals, there are two results when Equations (1) and (2) are solved: the ^{230}Th age and the initial $\delta^{234}U$ value. If all corals remained closed to diagenetic alteration, the initial $\delta^{234}U$ value would reflect changes in the marine $\delta^{234}U$ value with time. Early studies indicated that there is substantial variation in the initial $\delta^{234}U$ value and that, in general, higher initial $\delta^{234}U$ value corresponds with higher ^{230}Th ages (Bender et al. 1979). The question was, did these variations reflect changes in the marine uranium isotopic composition or represent the effects of diagenesis?

Given that the long residence time of uranium should place limits on how much the marine $\delta^{234}U$ value could change over Late Quaternary time scales, several workers have used models to determine what these limits should be (see Henderson and Anderson 2003). Chen et al. (1986) and Edwards (1988) used a simple one-box model and assumed steady state conditions. They showed that:

$$\delta^{234}U_{ss} = \frac{\delta^{234}U_r}{1+(\tau_U/\tau_{234})} \quad (6)$$

where $\delta^{234}U_{ss}$ is the steady state marine $\delta^{234}U$ value, $\delta^{234}U_r$ is the average $\delta^{234}U$ value of the input to the ocean, assumed to be rivers, τ_U is the residence time of uranium in the ocean, and τ_{234} is the mean-life for ^{234}U. To a first approximation, the $\delta^{234}U_{ss}$ is ~1/2 of the $\delta^{234}U_r$, as the residence time of the ocean is roughly equal to the mean-life of ^{234}U, which makes τ_U/τ_{234} ~1. The modern ocean satisfies this equation, as the average riverine $\delta^{234}U$ value is ~300‰ and the modern $\delta^{234}U$ value is ~150‰. To evaluate how much the marine $\delta^{234}U$ value could change, Edwards (1988) assumed that the input would have an instantaneous jump to a new value and showed that the transition to a new steady state marine $\delta^{234}U$ value would be an exponential function (Henderson, 2002, has recently presented a similar equation):

$$\delta^{234}U_{marine} = \delta^{234}U_{ss2} + \left(\delta^{234}U_{ss1} - \delta^{234}U_{ss2}\right)e^{-[(1/\tau_U)+(1/\tau_{234})]t} \quad (7)$$

where the subscripts "1" and "2" refer to steady state marine values before and after the shift in riverine $\delta^{234}U$ (calculated from Eqn. 6). However, for timescales short compared to $1/(1/\tau_U + 1/\tau_{234})$ (~160,000 years) the shift in the marine $\delta^{234}U$ value can be approximated by:

$$\Delta\delta^{234}U_{marine} \approx \Delta\delta^{234}U_r\left(t/\tau_U\right) \quad (8)$$

where the $\Delta\delta^{234}U_{marine}$ and the $\Delta\delta^{234}U_r$ are the shifts from their initial (steady state in the marine case) values. Thus, for a shift of a quarter of the modern-day range of $\delta^{234}U$ for most rivers (a range of about 400‰, $\Delta\delta^{234}U_r$ = 80‰) for 50,000 years, the shift in the $\delta^{234}U_{marine}$ would be ~13‰. If a 50% shift lasted 20,000 years, the shift in the $\delta^{234}U_{marine}$ would be ~10‰. Thus, Edwards (1988) concluded that the $\delta^{234}U_{marine}$ should have remained within 10-20‰ of its modern value during the last several hundred thousand years.

Hamelin et al. (1991) came to similar conclusions regarding the magnitude of shifts in the $\delta^{234}U_{marine}$ value caused by instantaneous shifts in the $\delta^{234}U_r$, using computer simulation as opposed to solving the specific differential equations. They also investigated periodic changes in $\delta^{234}U_r$ and found that with a 30% shift in the value (100‰) with a 100,000 year period the change in the $\delta^{234}U_{marine}$ value was only ~1% (or 10‰ if the modern $\delta^{234}U_{marine}$ is used) and was out of phase with the forcing.

Richter and Turekian (1993) also assumed a simple one-box model for the ocean. They derived a set of equations with the assumption that 1) the volume of the ocean does not change, 2) mass is conserved, i.e., the change in the marine U concentration with time equals the flux in from rivers minus the flux out, and 3) that the marine U concentration does not change with time, which implies that the flux in equals the flux out. This results in the equation:

$$\frac{d\delta A_o}{dt'} + (1+\tau_U \lambda_{234}) \delta A_o = \delta A_r \qquad (9)$$

where δ here represents small changes in A_o, the marine uranium activity, and A_r, the average riverine uranium activity and $t' = t/\tau_U$. At steady state ($d\delta A_o/dt'=0$), this reduces to Edwards' first equation (Eqn. 6). To test the simple system's response to periodic forcing, Richter and Turekian let A_r vary according to:

$$A_{r(t')} = A_{r(0)} + \sum_n a_n \sin\left(\frac{2\pi t'}{P_n}\right) \qquad (10)$$

where $A_{r(0)}$ is the initial riverine uranium activity, a_n is the amplitude of the change in A_r, and P_n is the period. The results of solving Equation (9) for δA_o for various a_n (using the most simple case, where n = 1) are shown in Figure 11 (similar to Richter and Turekian's Fig. 3). As Richter and Turekian concluded, a change of ± 0.1 in A_r (equivalent to a change in $\delta^{234}U_r$ of 100‰, ~30% of the modern value) would produce barely detectable changes (~5‰) in the marine uranium isotopic composition for a period of 100,000 years (similar to the Hamelin et al. results) and undetectable changes at shorter periods. They point out that 1) a change of 0.1 in A_r is equivalent to a ~25% change in the flux of uranium into the ocean, given its current isotopic composition and 2) one could include changes in the flux of uranium into the ocean in the equations as well, but, as the flux and the isotopic composition of uranium in modern rivers is generally inversely correlated, the addition would not significantly change the results. Figure 11 shows that to get clearly detectable changes (~10‰) in the marine uranium isotopic composition would require a change in A_r of 0.2 (or 200‰, ~66% of the modern value) with a period of 100,000 years or longer; to get marine changes from forcing at periods of less than 100,000 years would require even greater changes in A_r. These results are consistent with Edwards' estimate of an outside envelope of a 10-20‰ change in the marine uranium isotopic composition on a 10^5-year time scale.

Within the 10-20‰ envelope, it was less clear if samples at the higher end of this range had been altered or reflected small changes in the marine $\delta^{234}U$ value. Hamelin et al. (1991) made a histogram of initial $\delta^{234}U$ values of apparently well-preserved last interglacial corals and showed that the mean was 160 (157 with new half-lives), ~11‰ above the modern value. However, an extensive data set of corals that grew during the last deglaciation show no deviation from the modern value at a 2σ precision of ± 1‰ (Fig. 12; Edwards et al. 1993), suggesting that glacial/interglacial changes do not affect the marine $\delta^{234}U$ value on a 10^4-year time scale. What was missing was a constraint on how the diagenetic effects that lead to elevated initial $\delta^{234}U$ values in corals affect their ^{230}Th age.

While many models have been proposed for changes in uranium-series isotopic composition with diagenetic alteration (e.g., Hamelin et al. 1991), documentation of a

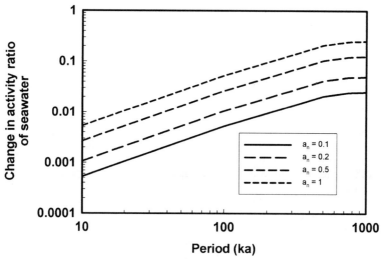

Figure 11. Results from Richter and Turekian (1993) model (Eqns. 9 and 10) for the simplest case, where n = 1. Similar to Figure 3 in Richter and Turekian (1993), but with multiple values for a_n.

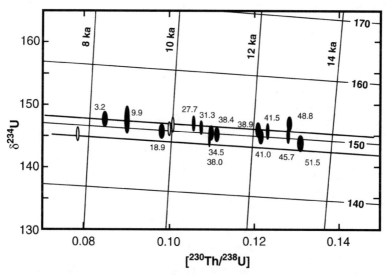

Figure 12. Plot of measured $\delta^{234}U$ vs. measured [$^{230}Th/^{238}U$] for a set of samples collected from the Huon Peninsula, Papua New Guinea (after Edwards et al. 1993). Open ellipses are data for samples collected from outcrop; closed ellipses are data for samples collected from drill core. All points plot along the initial $\delta^{234}U$ =150 contour, indicating that all samples have maintained a primary marine uranium isotopic composition, consistent with closed-system behavior. Relatively young samples such as these are more likely to satisfy the closed-system assumption whereas older corals such as those depicted in Figure 13 are not as likely to satisfy this assumption.

correlation between initial δ^{234}U values and ^{230}Th ages was made possible by the extensive dating that has been done on Barbados corals. Gallup et al. (1994) demonstrated that for corals from the same terrace, and presumably the same age, δ^{234}U values and ^{230}Th ages co-vary. This can best be observed in a plot of measured δ^{234}U vs. [^{230}Th/^{238}U] (similar to Fig. 3). Figure 13 shows TIMS data for Barbados corals available in 1994, coded by terrace. For each terrace, there is a range of initial δ^{234}U values, where the lowest value is within error of the modern marine δ^{234}U value or above it. The data for each terrace are not randomly scattered but form a rough line that goes to higher ^{230}Th age with higher initial δ^{234}U value. Gallup et al. modeled this process assuming continuous addition of ^{230}Th and ^{234}U. In their model, the rate of addition was allowed to vary, but the ratio of ^{230}Th to ^{234}U addition was fixed. The value of this ratio (^{230}Th/^{234}U = 0.71) was chosen so that the model would reproduce the isotopic composition of VA-1 (Fig. 13) an altered coral of known age (from U/He dating, Bender et al. 1979). They integrated the differential form of the age equation, modified to include ^{230}Th and ^{234}U addition terms in the calculated ratio. This yielded an equation that describes the isotopic composition resulting from adding ^{234}U and ^{230}Th in a fixed ratio and to varying degrees to samples of a given true age (dashed lines on Fig. 13). These "addition lines" follow the trends in the data from each terrace, suggesting that addition is broadly continuous and that it involves both ^{234}U and ^{230}Th in a ratio of about 0.71. This ratio of addition corresponds to an increase of ^{230}Th age of 1 ka for every 4‰ rise in initial δ^{234}U value. This provided a semi-quantitative criterion for reliable ^{230}Th ages for corals that have similar trends in isotopic composition. It has been used for corals from many areas, including the Western Atlantic (e.g., Blanchon et al. 2001), the Western Pacific (e.g., Stirling et al. 1995, 1998; Esat et al. 1999), and the Eastern Pacific (e.g., Szabo et al. 1994; Stirling et al. 2001). However, it is often difficult to demonstrate that corals have a similar diagenetic trend where there are a limited number of terraces to sample, or in some cases, such as in Papua New Guinea, corals do not seem to show the same diagenetic trend. Thus, the criterion should be used in conjunction with other measures of diagenesis.

Thompson et al. (in review) have followed up on the Gallup et al. (1994) work with additional analyses and more in depth modeling. The new coral analyses, from the same Barbados terraces, confirm the original Gallup et al. diagenetic trends. Thompson et al. surmised that ^{230}Th and ^{234}U addition could have taken place as ^{230}Th and ^{234}Th adsorption from ground water, prior to decay of ^{234}Th (half-life = 24.1 days to ^{234}U). Although they consider a number of possible sources for aqueous ^{230}Th and ^{234}Th, a likely possibility is recoil ejection from solids into solution during decay of ^{238}U and ^{234}U. This follows a similar idea proposed by Henderson and Slowey (2000) in the context of dating of carbonate bank sediments. Thompson et al. model ^{230}Th and ^{234}U addition on this basis. The model results are a set of calculated addition lines that are numerically similar to the Gallup et al. lines. Their model confirms the Gallup et al. suggestion of broadly continuous addition of both ^{234}U and ^{230}Th. Furthermore, the Thompson et al. model suggests a genetic tie to recoil phenomena. In addition, Thompson et al. suggest that their model can be used to correct ages of samples with non-marine δ^{234}U, and term these "open-system ages." These ages (perhaps better be termed "model ages" as they are model-dependent) essentially use a refined version of the Gallup et al. (1994) addition lines to correct for age as a function of δ^{234}U. This approach may prove useful for certain applications. For example, the empirical trends at Barbados and elsewhere do generally increase in apparent age, with progressive increase in δ^{234}U. Thus, (1) in localities where trends can be demonstrated to be particularly robust and (2) for scientific problems that allow significant error in age, the addition lines may provide the basis for an age correction. However, the trends observed at different localities

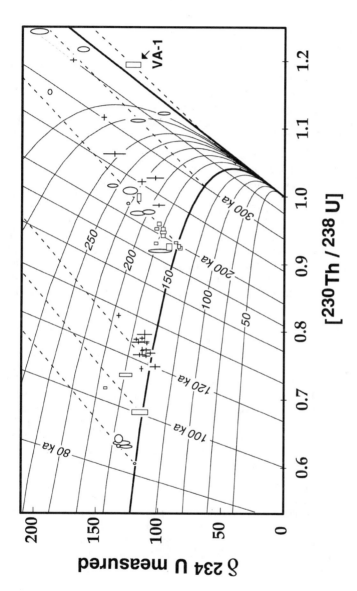

Figure 13. Plot of measured $\delta^{234}U$ versus $^{230}Th/^{238}U$ activity ratio (from Gallup et al. 1994) of all TIMS Barbados data published through 1994 (see Gallup et al. 1994 for references). Data are grouped by terrace, where age and elevation increase from left to right; note symbols are reused for increasingly older terraces. The size of the symbols roughly represents the 2σ error ellipse. The curved lines are contours of initial $\delta^{234}U$ value (the heavy contour is for an initial equal to the modern marine value) and lines with steep positive slope are contours of ^{230}Th age. The area without contours represents isotopic compositions that are inaccessible through closed-system decay. Dashed lines are $^{234}U-^{230}Th$ addition lines, as discussed in the text, calculated with ages of 83, 100, 120, 200, 300, and 520 ka. With additional data from Barbados, Thompson et al. (in review) have confirmed the linear trends originally established by Gallup et al. (1994). Stirling et al. (1998, 2001) have observed similar trends elsewhere.

worldwide are not the same, and in general, the data points from a given terrace are not all co-linear within analytical error (Fig. 13), as required for an exact correction using either the Gallup et al. or Thompson et al. models. Thus, some diagenetic process in addition to the modeled process must affect certain samples. Recent combined ^{231}Pa and ^{230}Th measurements on Barbados corals bear this out (Gallup et al. 2001; Cutler et al. 2003). For example, of 14 corals all with δ^{234}Ui values within 8 per mil of the marine value, 6 had discordant ^{231}Pa and ^{230}Th ages, demonstrating without ambiguity, diagenetic processes that have affected Pa, Th, and/or U, but not δ^{234}U.

In sum, initial δ^{234}U values place important constraints on diagenesis. We are beginning to understand some of the most important processes whereby diagenesis affects δ^{234}U and ^{230}Th age. However, we still do not understand a number of important aspects of this process and δ^{234}U does not appear to respond to all diagenetic processes. This suggests that, in addition to initial δ^{234}U considerations, we need to apply the full range of diagenetic tests.

The above analysis of Barbados corals also places constraints on the history of the marine δ^{234}U value. Empirical evidence suggests that diagenesis generally acts to raise the δ^{234}U value of Barbados fossil corals, which makes the lowest initial δ^{234}U value for each terrace the best estimate of the marine value at the time the terrace was formed. This makes it possible to extend evidence for a stable marine δ^{234}U value beyond the last deglaciation (Fig. 12; Edwards et al. 1993) to Marine Oxygen Isotope Sub-stage 5a (Fig. 13, Gallup et al. 1994), the last interglacial (Henderson et al. 1993) and the penultimate interglacial, 200,000 years ago (Fig. 13; Gallup et al. 1994). Similar arguments can be applied to coral data from East Pacific corals (Stirling et al. 2001) and aragonitic carbonate bank sediments from the Bahamas (Henderson 2002), extending evidence that the marine δ^{234}U value has remained within error of its present value for interglacial periods in the last ~325 ka and ~360 ka, respectively.

Initial U concentration. If uranium concentration has changed as a result of diagenetic reactions, one may, in principle detect this by comparison between uranium concentrations in modern corals and their fossil counterparts. Early work documenting and studying uranium concentrations in corals is extensive (e.g., Barnes et al. 1956; Tatsumoto and Goldberg 1959; Veeh and Turekian 1968; Schroeder et al. 1970; Thompson and Livingston 1970; Gvirtzman et al 1973; Amiel et al. 1973; Swart 1980; Swart and Hubbard 1982; Cross and Cross 1983). This broad body of data shows that primary surface coral uranium concentrations lie between 1.5 and 4 ppm (see Fig. 1). Concentrations appear to be species dependent (Cross and Cross 1983). Furthermore, uranium concentrations vary within individual coral skeletons (Schroeder et al. 1970; Shen and Dunbar 1995; Min et al. 1995).

Shen and Dunbar (1995) and Min et al. (1995) showed that uranium concentrations in individual skeletons and between colonies anti-correlate with sea surface temperature, with a shift in U/Ca ratio of about 4.7% per degree Celsius for *Porites lobata*. At the higher latitudes of the tropics and in the sub-tropics, this may result in variation in uranium concentration in excess of 20% within a single skeleton (Shen and Dunbar (1995); Min et al (1995)). Even in localities with seasonal amplitudes in temperature of one or two degrees, U/Ca varies by several percent seasonally. Thus, primary uranium concentrations in *Porites lobata* vary seasonally by large amounts compared to analytical errors. Because of this temperature-related variation, it is difficult to compare fossil uranium concentrations to primary uranium concentrations in detail, as the true primary concentrations are not known. On the other hand, the Shen and Dunbar (1995) and Min et al. (1995) works suggest that seasonal variation in U/Ca is a general feature of reef-building coral skeletons. Thus, a possible test for diagenetic shifts in uranium is whether a fossil coral maintains its primary annual variation in U/Ca ratio. Although this test has been suggested (Shen and Dunbar 1995; Min et al. 1995), it has yet to be applied, but may well become an important tool in future dating and diagenesis studies.

Concordancy tests. Combined ^{231}Pa and ^{230}Th dating of corals provides perhaps the most rigorous test for closed-system behavior. Because mass spectrometric ^{231}Pa techniques are still fairly new and because the measurements themselves are not easy, there is still limited data of this sort. In coming years, ^{231}Pa measurements will play a major role in assessing the accuracy of ^{230}Th-based chronologies, such as the late Quaternary sea level curve. The few such data sets on reef-building corals include data reported by Edwards et al. (1997), Gallup et al. (2002), and Cutler et al. (2003), and Koetsier et al. (1999).

These data sets have helped in establishing with high probability the accuracy of the ages of key corals that record important aspects of climate change (Gallup et al. 2002; Cutler et al. 2003). They have also demonstrated that coral samples that might otherwise be considered pristine are, in fact, altered. Data from these samples, which plot off of concordia include those that plot above concordia and those that plot below concordia. Shown in Figure 14 (Edwards et al. 1997) are a set of samples from the Huon Peninsula, Papua New Guinea (left graph) and from Barbados (right graph). The younger (around 10 ka old) Huon Peninsula samples are all on or very close to concordia suggesting that these ages are affected little by diagenetic processes. Older Huon Peninsula samples (10's to more than 100 ka old) from the Huon Peninsula generally plot on, below, and above concordia (Cutler et al. 2003). This indicates a variety of diagenetic conditions on the Huon Peninsula can result in daughter loss/parent gain as well as daughter gain/parent loss. In general this contrasts with Barbados samples of about the same age, which have isotopic compositions that generally plot on or above concordia (Fig. 14). Thus, Barbados diagenetic processes tend to be characterized by parent loss/daughter gain.

Combined ^{231}Pa–^{230}Th studies will be important in understanding diagenetic processes, and perhaps using this knowledge to constrain age even in cases where the data are discordant (e.g., Fig. 6). However, given our present state of knowledge regarding these issues, the most immediate benefit of these measurements is the identification, along with other tests, of material that has remained a closed to chemical exchange and therefore records accurate ^{230}Th and ^{231}Pa ages. The relatively small number of analyses that have been made so far indicate that concordant material can be identified and can be used to establish solid chronologies.

4. SOURCES OF ERROR IN AGE

A quick glance at Equations (1) through (5) shows sources of error that contribute to error in age, presuming that the assumptions used in calculating the equations hold (initial condition assumptions and the closed-system assumption). These include errors in the decay constants/half-lives, errors in the measurement of the pertinent isotope ratios, and in the case of Equation (3), the error in our estimate of initial ^{230}Th/^{232}Th. Relationships among error in half-lives, laboratory standardization procedures, and ^{230}Th age are discussed in detail by Cheng et al. (2000b).

4.1. Errors in half-lives and decay constants

Half-lives have typically been determined by measuring the activity (rate of decay) of a sample containing a known number of atoms of the nuclide in question and calculating the decay constant via the equation: $N\lambda = a$, where "a" is the measured activity. The half-lives of all of the nuclides pertinent to ^{230}Th and ^{231}Pa dating have been determined in this fashion. Among those that are known most precisely are those of ^{238}U

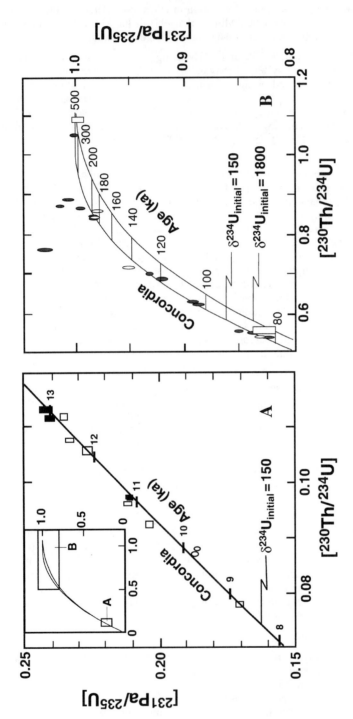

Figure 14. Coral isotopic data from the Huon Peninsula, Papua New Guinea (left diagram) and Barbados (right diagram) plotted in concordia diagrams (after Edwards et al. 1997). The younger Papua New Guinea data all plots close to concordia, suggesting that these samples have behaved as closed-systems. The right diagram includes two concordia curves. The upper curve (initial $\delta^{234}U = 150$) is appropriate for the coral samples (ellipses), whereas the lower curve ($\delta^{234}U = 1800$) is appropriate for Devils Hole samples (Ludwig et al. 1992; depicted by open rectangles). Many of the Barbados corals have concordant isotopic compositions, suggesting closed-system behavior. Those that are discordant plot above concordia, consistent with uranium loss and/or daughter gain. Older samples from Papua New Guinea (not illustrated here) plot on concordia, above concordia, and below concordia (Cutler et al. 2003). Thus diagenetic reactions affecting the Papua New Guinea samples appear to be more complex than those affecting the Barbados samples.

and ^{235}U, with errors close to ± 0.1% at the 2σ level. For this reason an alternate method for determining half-lives of geologically important radio-nuclides has emerged (Renne et al. 1998). This method starts by establishing the age of a material very precisely using U-Pb dating. As the half-lives of ^{238}U and ^{235}U are known very well, they contribute little error to the age. The known-age material can then be "dated" by another method. As the age is known a priori, one can choose the decay constant as an unknown, and in some cases determine its value more precisely than in a pure laboratory study.

Ludwig et al. (1992) and Cheng et al. (2000b) have applied a similar approach to re-determination of the half-lives of some uranium-series nuclides. This method involves identifying materials that have behaved a closed systems for times long compared to the half-lives of uranium-series intermediate daughters. As such materials are in secular equilibrium, $\lambda_{230} = (\lambda_{238})$ ^{238}U/^{230}Th and $\lambda_{234} = (\lambda_{238})$ ^{238}U/^{234}U, where the isotope ratios are atomic ratios. In these two cases the only sources of error in the calculated λ_{230} and λ_{234} values are the error in λ_{238} and the error in the appropriate measured isotope ratio of the secular equilibrium material. As the error in λ_{238} is small, the errors in the calculated decay constants are dominated by errors in the measured isotope ratios, which in turn are dominated by uncertainty in our knowledge of isotope ratios in uranium and thorium isotopic standards. Nevertheless, errors in λ_{234} and λ_{230}, determined in this fashion (Cheng et al. 2000b) are small compared to existing determinations in pure laboratory studies, and are therefore the recommended decay constant (half-life) values given in this volume (Table 1, Bourdon and Turner 2003).

Of note are the values for ^{230}Th and ^{234}U as the revised values postdate the development of mass spectrometric techniques for measurement of ^{234}U and ^{230}Th in natural materials. Data published prior to Cheng et al. (2000b) does use not use the revised values whereas data published subsequently may or may not use the new values. The revised half-lives do have a small, but significant effect on calculated ^{230}Th ages, particularly ages older than about 100 ka. Furthermore, the new value for the ^{234}U half-life changes δ^{234}U values as these are calculated from measured ^{234}U/^{238}U atomic ratios using the secular equilibrium ^{234}U/^{238}U value: $\lambda_{238}/\lambda_{234}$. δ^{234}U values calculated with the new λ_{234} are about 3 per mil lower than those calculated with commonly used λ_{234} values, hence the revised modern sea water δ^{234}U of 145.8 ± 1.7 per mil (Cheng et al. 2000b), compared to earlier values about 3 per mil higher. In general half-lives are now known precisely enough so that their contribution to error in age is comparable to or smaller than typical errors in isotope ratios (determined with mass spectrometric techniques).

Table 1. Preferred half life values.

Isotope	Half Life (years)	Decay Constant (year^{-1})	Reference
^{238}U	4.4683 ± 0.0048 × 10^9	1.5513 ± 0.0017 × 10^{-10}	Jaffey et al. (1971)
^{235}U	7.0381 ± 0.0096 × 10^8	9.8485 ± 0.0134 × 10^{-10}	Jaffey et al. (1971)
^{234}U	245,250 ± 490	2.8263 ± 0.0057 × 10^{-6}	Cheng et al. (2000b)
^{230}Th	75,690 ± 230	9.1577 ± 0.0278 × 10^{-6}	Cheng et al. (2000b)
^{231}Pa	32760 ± 220	2.1158 ± 0.0014 × 10^{-5}	Robert et al. (1969)

4.2. Errors in measurement of isotope ratios

As discussed above, mass spectrometric techniques are the methods of choice for measurement of nuclides pertinent to this study. They supercede earlier decay-counting techniques because of their ability to detect a much larger fraction of the nuclides of interest, thereby reducing sample size and counting statistics error by large amounts. Historical aspects of this transition are discussed by Wasserburg (2000) and Edwards (2000). In terms of precision, the capabilities of mass spectrometric measurements were clear from the very first measurements of the sort (Chen et al. 1986 for ^{234}U; Edwards et al. 1987a for ^{230}Th). Then, as now, the measurements were limited by counting statistics. From a practical standpoint, given typical sample sizes, natural abundances, and ionization efficiencies, measurements of ^{230}Th and ^{234}U are limited to 2σ precisions of about 1 per mil. For this reason, ^{230}Th precisions have not improved by large amounts since the original mass spectrometric measurements. Precisions in ^{234}U have improved modestly from about ± 5 per mil in the original measurements (Chen et al. 1986; Edwards et al. 1987a) to about ± 1 per mil (Edwards et al. 1993). This latter improvement resulted from modest changes to the original procedures, including removal of reflected peaks from the spectrum, changing from the single-filament graphite loading technique to the double-filament technique to stabilize the ion beam, and a modest increase in sample size to increase the number of ions counted.

Mass spectrometric measurements on corals typically result in errors in ^{238}U, ^{234}U, ^{235}U, and ^{230}Th of ± 2 per mil or better (2σ), with the exception that fractional error in ^{230}Th typically increases progressively from this value for samples progressively younger than several ka. This results from the low concentrations of ^{230}Th in very young corals. Errors in ^{231}Pa are typically somewhat larger than those of the other isotopes, with errors of ± several per mil, except for corals younger than a few ka.

Errors age resulting from to analytical errors are given in Figure 15. Considering ^{230}Th ages, samples a few hundred years old have errors in age of about ± 3 years, 10,000 year-old samples have errors in age of about ± 30 years, and 100,000 year-old samples

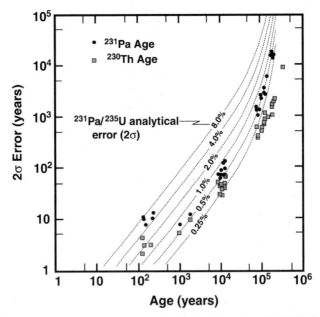

Figure 15. Error in age vs. age, both on log scales (after Edwards et al. 1997). Each data point represents data from a particular sample analyzed by thermal ionization mass spectrometric techniques. Solid circles represent ^{231}Pa ages. Contours of analytical error in ^{231}Pa/^{235}U pertain to the ^{231}Pa data points. Shaded squares represent ^{230}Th ages. See text for discussion.

have errors in age of ± 1 ka. Samples as young as 3 years have ages distinguishable from zero and samples as old as 700 ka have ages distinguishable from infinity. Considering ^{231}Pa ages, samples younger than 1000 years have errors in age of several to 10 years, 10,000 year-old samples have errors in age of several tens to 100 years, and 100,000 year-old samples have errors in age of a few ka. Samples as young as 7 years have ages distinguishable from zero and samples as old as 250 ka have ages distinguishable from infinity. In sum, ^{230}Th ages can be measured precisely for materials younger than about 600,000 years, and ^{231}Pa analyses can be used to test the closed system assumption and age accuracy over the past 250,000 years.

4.3. Error in initial ^{230}Th/^{232}Th

Errors in ^{230}Th/^{232}Th only contribute to error in ^{230}Th if the initial ^{230}Th term in Equation (3) ($[^{232}$Th/^{238}U] $[^{230}$Th/^{232}Th$]_i$ $e^{-\lambda_{230}t}$) is significant compared to measured $[^{230}$Th/^{238}U]. As discussed above, this term is insignificant for most surface corals and can be ignored, along with any error in this term. However, for some of the unusual samples from the central Pacific (Cobb et al. in review) and Sumatra (Zachariasen et al. 1999) this term is significant and errors in age are limited not by analytical error, but instead by the ability to constrain the ^{230}Th/^{232}Th values to a specific range. Although not generally a source of error for surface corals, error in initial ^{230}Th/^{232}Th is more likely to contribute error in ^{230}Th ages of some of the other marine or lacustrine materials discussed below.

5. LATE QUATERNARY SEA LEVELS FROM CORAL DATING

One of the real successes of ^{230}Th dating (and to a lesser degree ^{231}Pa dating) has been its application to the study of sea level change and the causes of the Quaternary glacial cycles. Early work in this area suggested that sea level followed Milankovitch cycles (Broecker et al. 1968; Mesolella et al. 1969). The degree to which this relationship holds and the details of this relationship have been the subject of intense research ever since. The development of mass spectrometric techniques for determining ^{230}Th ages promised to resolve some of these issues. Some of that promise has been realized and more will be forthcoming. Some of the delay has resulted from our incomplete knowledge of diagenetic processes and the fairly recent development of high precision ^{231}Pa techniques. Nevertheless a picture is beginning to emerge from coral dating efforts, in which glacial cycles are partly forced by changes in orbital and rotational geometry, but also respond to other factors, including those responsible for millennial-scale change and perhaps atmospheric CO_2 values. In addition, the timing and rates of sea level change during both glaciation and melting have been established for certain times, and bear on the causes of glaciation and melting. A more extensive review of late Quaternary sea level studies can be found in Edwards et al. (2003).

5.1. Deglacial sea level

One of the major successes of high-precision ^{230}Th dating has been the establishment of a detailed deglacial sea level history (Fig. 16). This has been accomplished directly and indirectly through ^{230}Th dating of corals. One of the main impediments has been the difficulty in recovering sequences of corals that formed during times of low sea level. Fairbanks (1989) first overcame this problem by drilling offshore near Barbados, recovering the full deglacial sequence back to the Last Glacial Maximum (LGM), and dated the corals with ^{14}C techniques. Because the ^{14}C timescale was not yet calibrated in the deglacial time range, the curve was not yet on an absolute timescale. Bard et al. (1990a) dated the same sequence using ^{230}Th techniques. Combined with the ^{14}C data, this yielded an absolute deglacial sea level curve and a ^{14}C calibration. Chappell and Polach (1991) and

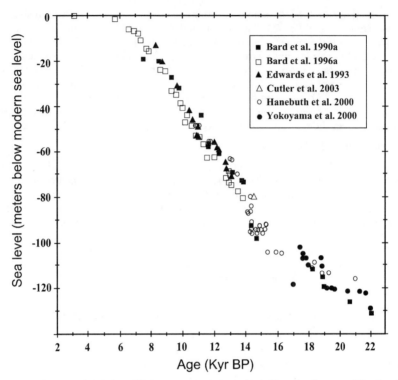

Figure 16. Direct sea level data (23 ky to present) from the indicated references. The data from different sites generally agree, suggesting that different far-field sites have not been affected differentially by isostatic effects. Note also the apparent rapid rises (melt water pulses) in sea level at about 19 ky, 14.6 ky, and 11 ky. The latter two correspond to times of abrupt rise in Greenland temperature (Dansgaard et al. 1993) and East Asian Monsoon intensity (Wang et al. 2001), supporting the idea that sea level can be affected by millennial-scale events. The general rise in sea level observed in this figure corresponds in time to an interval of relatively high insolation at high northern latitudes during summer (see Fig. 17), relatively low insolation at low northern latitudes during winter, and relatively low northern latitudinal gradients during winter. Thus, this curve demonstrates correlations between orbital geometry and sea level, as well as millennial-scale climate change and sea level.

Edwards et al. (1993) obtained a similar sequence of corals by drilling on land into the uplifted deglacial sequence on the Huon Peninsula, Papua New Guinea. Corals were recovered from the portion of deglacial sequence subsequent to 14 ky B.P. ^{230}Th and ^{14}C dating confirmed and increased the resolution of the Barbados results, both in terms of sea level and ^{14}C calibration (Edwards et al. 1993). Similar studies from Tahiti (Bard et al. 1996a) and additional work from the Huon Peninsula (Cutler et al. 2003) have largely confirmed the earlier work. In addition, based in large part upon the ^{14}C calibration work in the ^{230}Th-based coral studies, additional ^{14}C-based sea level studies have contributed further sea level information (Hanebuth et al. 2000; Yokoyama et al. 2000).

The combined data (Fig. 16) show that sea level reached a minimum during the LGM between 20 and 22 ky ago. The rate of subsequent sea level rise was not constant, with relatively rapid rise (melt water pulses) at about 19 ky, between 15 and 14 ky and at about 11 ky, with intervals of little or no rise in sea level immediately before each of these pulses. The latter two of these low sea level rise-melt water pulse pairs correspond to distinct climate events recorded as temperature change in Greenland ice (Dansgaard et al. 1993) and as change in the intensity of the East Asian Monsoon in Chinese Caves (Wang et al. 2001). The most recent pair corresponds to the Younger Dryas and the warming at the end of the Younger Dryas whereas the earlier of the two pairs corresponds to the interval prior to the Bolling-Allerod and the transition into the Bolling-Allerod. Most the most recent deglaciation took place during a time of increasing summer insolation in the high northern latitudes. Thus, the detailed deglacial sea level curve demonstrates sea level response to both insolation and to millennial-scale climate change (e.g., Younger Dryas, Bolling-Allerod).

5.2. Sea level during the last interglacial/glacial cycle and earlier

Figure 17 summarizes coral sea level data since the last interglacial period and illustrates the current state of sea level reconstructions over the timescale of a full glacial-interglacial cycle. The inset in part b shows all available ^{230}Th data coded so that solid points have initial δ^{234}U values within 8 per mil of the marine value and open points have non-marine δ^{234}U. There are clear inconsistencies for the complete dataset, as well as some inconsistencies for the subset of data with marine δ^{234}U values, likely caused by diagenesis. If the dataset is screened further, largely on the basis of ^{231}Pa-^{230}Th concordancy (see Cutler et al. 2003 for details), one obtains a self-consistent dataset, albeit with a limited number of points (Fig. 17b, main portion). The curve shows a clear correlation with high northern latitude summer insolation, supporting the idea that a significant portion of sea level variability results from changes in orbital geometry. The curve also constrains the rate of sea level drop for specific times during the last glacial cycle. The largest and most rapid drop took place between 76 and 71 ky, when sea level dropped 60 m at a rate in excess of 10 m/ky. This and other times of sea level fall correspond to times of low summer insolation at the high northern latitudes, supporting the conventional view of the Milankovitch Theory. However, these intervals also correspond to times of high winter insolation at low northern latitudes and high winter insolation gradients between low and high northern latitudes. These relationships suggest that glacial growth may have been facilitated by high water vapor pressures in the atmospheric source areas for glacial ice, coupled with efficient moisture transport. Thus, insolation clearly has played a role in controlling sea level, from a number of different perspectives.

Further support for insolation control of sea level comes from a large number of studies employing mass spectrometric ^{230}Th dating of corals, carbonate bank sediments, and speleothems, including dating of Marine Oxygen Isotope Stage 5a (e.g., Edwards et al. 1987b, Li et al. 1991; Gallup et al. 1994; Richards et al. 1994; Ludwig et al. 1996; Edwards et al. 1997; Toscano and Lundberg 1999; Cutler et al. 2003), the Last Interglacial high stand (e.g., Edwards et al. 1987a,b, 1997; Li et al. 1989; Bard et al. 1990b, Chen et al. 1991; Henderson et al. 1993; Zhu et al. 1993; Collins et al. 1993a,b; Stein et al. 1993; Muhs et al. 1994, 2002; Szabo et al. 1994; Gallup et al. 1994; Stirling et al. 1995, 1998; Slowey et al. 1996; Eisenhauer et al. 1996; Esat et al. 1999; McCulloch et al. 2000; Cutler et al. 2003), the Penultimate Interglacial high stands (e.g., Li et al. 1989; Bard et al. 1991; Gallup et al. 1994; Lundberg and Ford 1994; Richards 1995; Slowey et al. 1996; Bard et al. 1996b; Edwards et al. 1997; Robinson et al. 2002; Bard et al. 2002), and earlier interglacial periods (Richards 1995; Stirling et al. 2001). However, there is mounting evidence for sea level change that cannot be caused directly by insolation forcing. In addition to the deglacial evidence for millennial-scale changes in sea level, Chappell (2002) has demonstrated millennial-scale variation in sea level, with an amplitude of about 40 m, during Marine Isotope Stages 3 and 4. The frequency of these changes is much higher than orbital frequencies and cannot result directly from insolation changes. Furthermore, there is now mounting evidence that Termination II sea level rise preceded high northern latitude summer insolation rise (Stein et al. 1993; Broecker and Henderson 1998; Henderson and Slowey 2000; Gallup et al. 2002). The cause of this apparent early rise is not clear; however, possibilities include insolation changes other than high northern latitude changes, indirect insolation forcing modulated by glacio-isostatic factors, a response to atmospheric CO_2 rise, or a tie to millennial-scale changes. In sum, uranium-series dating of corals and other marine carbonates have demonstrated that sea level responds directly to changes in insolation resulting from changes in orbital geometry. There is also strong evidence that sea level responds to other processes, such as those responsible for millennial-scale climate change, changes in atmospheric CO_2 levels, and/or insolation forcing modulated in some fashion.

6. DATING OF OTHER MARINE AND LACUSTRINE MATERIALS

6.1. Deep sea corals

Deep sea corals are known to live at depths ranging from 60 to 6000 m. Deep sea corals with aragonitic skeletons have uranium concentrations similar to surface corals. Although the literature on uranium-series dating of deep-sea corals is very limited compared to that on surface corals, both mass spectrometric ^{230}Th ages (Smith et al. 1997; Adkins et al. 1998; Mangini et al. 1998; Lomitschka and Mangini 1999; Cheng et al. 2000a; and Goldstein et al. 2001) and ^{231}Pa ages (Goldstein et al. 2001) have been determined on deep-sea corals. Deep sea corals are of great importance because they are

---

Figure 17 (*on facing page*). Coral sea-level record (after Cutler et al. 2003) compared to insolation (Berger 1978) and benthic $\delta^{18}O$ records (Shackleton et al. 1983). **a.)** 65°N summer half-year insolation curve. Gray bars delineate periods of rapid sea-level fall. (*left inset*) Winter half-year insolation gradients: top curve is 35-50°N, middle is 50-65°N, and bottom is 20-35°N. (*right inset*) 15°N winter half-year insolation curve. **b.)** Coral sea-level record. Circles in inset are data from Edwards 1988; Bard et al. 1990a,b; 1996a; Chen et al. 1991; Edwards et al. 1993; 1997; Stein et al 1993; Collins et al. 1993; Szabo et al. 1994; Gallup et al. 1994; Stirling et al. 1995; 1998; Ludwig et al. 1996; Chappell et al. 1996; Esat et al. 1999; Toscano and Lundberg 1999; Cabioch and Ayliffe 2001; Yokoyama et al. 2001; and Cutler et al 2003. Data presented in main portion of (b) are a subset of the samples from the inset, which show no evidence of diagenesis (see Cutler et al. 2003). Triangles are from Cutler et al. 2003; circles from earlier publications. Papua New Guinea samples are upright and Barbados samples inverted. Light gray symbols are data that satisfy both concordancy and $\delta^{234}U$ criteria (see Cutler et al. 2003). Solid black symbols are a subset of the data that satisfy one of the two criteria (see Cutler et al. 2003). Replicate samples were combined using a weighted average and error bars are 2σ (if not visible, they are smaller than the symbol). Boxes enclose samples older than 15 ky of the genus *Acropora*, known to track sea level closely. Black bars give the duration of MIS 5e sea-level high according to the $\delta^{234}U_i$-screened, coral data sets of Stirling et al. (1995, 1998; upper) and Chen et al. (1991; lower). The dark gray curve provides Cutler et al.'s best estimate of past sea-level change. The portion of the curve immediately before the Last Interglacial (no data shown) is based on data from Gallup et al. (2002). Numbers give average rates of sea-level fall for each of 4 sea level drops. The inset shows all coral sea-level data presently available for times subsequent to the Last Interglacial period, measured with high-precision ^{230}Th dating methods. Solid symbols represent samples having $\delta^{234}U_i$ values that match the modern marine value, open symbols represent samples that do not. **c.)** Benthic $\delta^{18}O$ record for Carnegie Ridge core V19-30 (Shackleton et al. 1983). Numbers are marine oxygen isotope sub-stages. Dotted tie lines match similar features in the coral and $\delta^{18}O$ records. Note that sea-level broadly correlates with 65°N summer insolation and broadly anti-correlates both with winter insolation at low latitudes (15°N) of the northern hemisphere and with latitudinal insolation gradient during northern hemisphere winters. This observation supports the idea that a number of aspects of orbitally-controlled insolation change affect sea level. Not shown are high-resolution, high-frequency sea level changes during Marine Isotope Stage 3 and 4 (Chappell 2002), indicating response of sea level to millennial-scale events. Note also high sea level at about 135 ky, preceding the rise in insolation during Termination II (see Henderson and Slowey 2000 and Gallup et al. 2002). This suggests that the beginning of Termination II was not forced directly by insolation changes.

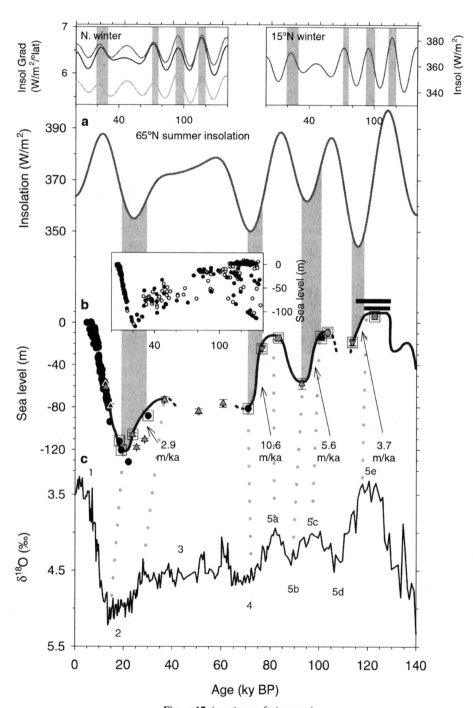

Figure 17. (*caption on facing page*)

one of a very limited number of archives that record deep-sea conditions at very high resolution. For example, individuals of the deep sea coral species, Desmophyllum cristagalli live for on the order of a century or two (Cheng et al. 2000a) and can be sampled at sub-decadal resolution for many proxies. One of the goals of all of the deep-sea coral studies referenced above has been to characterize some aspect of deep ocean circulation and the timing of changes in deep ocean circulation.

The principles of ^{230}Th and ^{231}Pa dating of deep-sea corals are the same as those of surface corals. However, the correction for initial ^{230}Th/^{238}U and ^{231}Pa/^{235}U is generally significant for deep sea corals and the error in the age is generally limited by our ability to quantify initial values for these isotope ratios. This stems from the fact that (1) ^{230}Th/^{232}Th values, ^{230}Th concentrations, and ^{231}Pa concentrations generally increase with depth in the oceans and (2) deep sea corals generally age in the presence of sediments with high ^{230}Th and ^{231}Pa contents, having scavenged these nuclides from the water column. Often the surfaces of deep sea coral skeletons are covered with Mn oxide-rich crusts mixed with detrital sediments. Chemical cleaning techniques (modifications of those used by Shen and Boyle 1988) can remove almost all of this material (Cheng et al. 2000a). Nevertheless, typically fossil deep sea corals have ^{232}Th concentrations (several hundred to thousands of pg/g) significantly higher than their surface counterparts. As the ^{230}Th/^{232}Th ratio of this thorium is also significantly higher than that for surface corals, the initial ^{230}Th term ($\{[^{232}$Th/^{238}U] $[^{230}$Th/^{232}Th]$_i$ $(e^{-\lambda_{230}t})\}$) in Equation (3) is generally significant. Thus, all of the principles and issues discussed for surface corals apply to deep-sea corals, with the added requirement that initial levels of ^{230}Th and ^{231}Pa must be carefully quantified.

The following strategies for minimizing the error in the $\{[^{232}$Th/^{238}U] $[^{230}$Th/^{232}Th]$_i$ $(e^{-\lambda_{230}t})\}$-term have been applied. Chemical cleaning techniques can remove much of the thorium and protactinium on surfaces, lowering ^{232}Th/^{238}U dramatically. Potentially, modifications of cleaning procedures can further reduce ^{232}Th levels. ^{230}Th/^{232}Th values have been determined with isochron techniques and by analyzing young samples. Cheng et al. (2000a) determined a ^{230}Th/^{232}Th value of $(85 \pm 80) \times 10^{-6}$ for a suite of corals, largely from the Atlantic. Using isochron techniques, Goldstein et al. (2001) determined a ^{230}Th/^{232}Th value within the Cheng et al. range for a Southern Ocean sample. Goldstein et al. (2001) also applied a two component mixing model to correct for initial ^{231}Pa/^{235}U, in essence using ^{232}Th as an index isotope of protactinium and isochron techniques. After correcting for initial ^{231}Pa/^{235}U and ^{230}Th/^{238}U, they were able to obtain a concordant age on a 16 ka sample from the Southern Ocean. In sum, by using cleaning techniques and carefully characterizing initial isotope ratios, ages with errors approaching those of surface corals can be obtained for some deep sea coral samples (e.g., ± several to 10 years for samples that are few hundred years old and ± 100 to 400 years for samples that are between 10 and 20 ka old).

6.2. Carbonate bank sediments

Given the high uranium and generally low Th concentrations of aragonite precipitated biogenically from shallow seawater, carbonate bank sediments, which are dominated by fine aragonite needles formed by *Halimeda* algae, are good targets for U-series dating. *Halimeda* lives on shallow banks and breaks into fine aragonite needles after death, which then accumulate on the shallow banks and are also transported to nearby basins (Glaser and Droxler 1991). Early efforts to date such sediments were mixed (Slowey et al. 1995; Gallup et al. 1995). Attempts to date sediment from the Nicaragua Rise in the Walton Basin were unsuccessful (Gallup 1997) because the sediments were deposited in 900 to 1100 m of water (Droxler et al. 1991), resulting in a long settling path that resulted in high Th concentrations (^{232}Th from 0.2 to 0.6 ppm) due to adsoption of Th in the water column. Washing and leaching techniques were unable to remove this adsorbed thorium and it did not appear to be of a uniform composition that would allow a correction to be made (Gallup 1997).

However, the Slowey et al. (1995) work on shallower carbonate bank sediments (off of the Bahamas) was successful and led to the dating of a number of important events in the marine oxygen isotope record (including the timing of the Last Interglacial (Slowey et al. 1996), Termination II (Henderson and Slowey 2000) and portions of the Penultimate Interglacial (Slowey et al. 1996, Robinson et al. 2002). The shallower depths obviate the most serious of the initial thorium problems encountered in the Nicaragua Rise work.

However, the quantification of initial ^{230}Th values was still the most important hurdle in establishing precise and accurate ages. Carbonate deposition rates are much higher during interglacial periods as compared to glacial periods. Thus, the fraction of detrital material (along with the ^{230}Th contained in the detrital component) is higher during glacial periods and lower during interglacial periods. Because of the relatively low initial ^{230}Th concentrations in interglacial sediments, bulk sediment dating is possible (Slowey et al. 1996; Robinson et al. 2002). In this case initial ^{230}Th is quantified using a variant of Equation (3), using two initial ^{230}Th terms, one for ^{230}Th in the detrital fraction, one for ^{230}Th in the marine (also referred to as "hydrogeneous" or "adsorbed") fraction. The fraction of ^{232}Th associated with the detrital component is determined from the aluminum concentration and reasonable estimates of the Al/^{232}Th ratio in the detrital component. The detrital ^{230}Th is then determined using estimates of the ^{230}Th/^{232}Th ratio in the detrital component. The ^{232}Th in the marine component is determined from the difference between measured bulk ^{232}Th and calculated detrital ^{232}Th. The marine or hydrogenous ^{230}Th is then calculated using an estimate of the marine ^{230}Th/^{232}Th ratio. Using this method, fractional errors in initial ^{230}Th are large. However, for interglacial sediments with low initial ^{230}Th, the contribution to error in age is small. Therefore, this bulk sediment approach has been successful in dating Last Interglacial (Slowey et al. 1996) and Penultimate Interglacial (Slowey et al. 1996; Robinson et al. 2002) bank sediments.

Because of higher initial ^{230}Th contents, different techniques were necessary to date Termination II carbonate bank sediments (Henderson and Slowey 2000). In this case, the detrital fraction was quantitatively removed with washing techniques. Aragonite and calcite fractions were then separated with heavy liquid techniques. The carbonate fractions were then used to obtain ages using isochron techniques. However, patterns in the uranium-isotopic composition of the sediments made it clear that an additional correction was necessary. While the bulk sediment gave a δ^{234}U value consistent with the isochron age, subsamples upon which the isochron were based did not. This suggested mobility of uranium-series isotopes within the sediment column, which the authors attributed to recoil processes. A model of this process made it possible to correct for this mobilization by assuming that the subsample should have a δ^{234}U value consistent with the isochron age. The resulting corrected isochron ages have a precision of ~3 ka and agree with ages for the penultimate deglaciation determined by combined ^{231}Pa and ^{230}Th methods on Barbados corals (Gallup et al. 2002).

6.3. Mollusks and foraminifera

Whereas surface corals, deep sea corals, inorganically precipitated marine aragonite, and algal-precipitated aragonite all have U/Ca ratios within a factor of a few of sea water (e.g., Fig. 1 for surface corals), and consequently uranium concentrations in the ppm range, most biologically precipitated minerals have much lower U/Ca ratios. Mollusks and forminifera are good examples of the latter. Low U/Ca ratios in and of themselves do not preclude ^{230}Th or ^{231}Pa dating. Present analytical techniques have sufficient sensitivity to date low U/Ca materials precisely. However, it has been demonstrated that both mollusks and forams pick up and exchange uranium diagenetically, precluding their use for ^{230}Th or ^{231}Pa dating, unless the history of uranium uptake and exchange can be determined.

Broecker (1963) first documented low uranium concentrations in modern mollusks, reporting values in the range of 10 ng/g. Edwards et al. (1987b) published a value about a hundred times lower (75 pg/g) from the aragonitic portion of a giant clam shell (see Fig. 1). The Broecker (1963) values and the Edwards et al. (1987b) value are all well below uranium concentrations of fossil mollusks, which typically lie in the 100's of ng/g to μg/g range. Thus, uranium in fossil mollusks is virtually all diagenetic, violating one of the main assumptions used in solving the ^{230}Th and ^{231}Pa age equations. Two key early studies (Szabo and Rosholt 1969; Kaufman et al. 1971) evaluated ^{230}Th and ^{231}Pa dating of mollusks. Both documented a diagenetic origin for uranium. Szabo and Rosholt (1969) proposed an open system model for determining molluscan ages using combined ^{230}Th-^{231}Pa analyses. Kaufman showed that even with combined ^{230}Th-^{231}Pa analyses, general models for uranium uptake could not be used to accurately determine the true age. Given these studies, there is little possibility that even combined ^{230}Th-^{231}Pa analyses will yield ages similar in precision and accuracy to ages of most of the other materials discussed here. However, there may be a glimmer of hope for using combined ^{230}Th-^{231}Pa analyses to at least constrain ages in some fashion. The sensitivity of mass spectrometric methods may allow focused ^{230}Th-^{231}Pa studies that constrain local diagenetic processes and age. Such studies might require numerous analyses of groups of shells from the same stratigraphic horizon or multiple analyses of the same shell, and would only be worthwhile if broad age constraints on the particular deposit would be of importance.

Dating of forams is perhaps even less viable than dating of mollusks. Forams are similar to mollusks in that primary uranium concentrations are low. Ku (1965) showed that calcitic forams have uranium concentrations of about 20 ng/g, about 100 times lower than corals. Delaney and Boyle (1983), Henderson and O'Nions (1995), and Russell et al. (1996) obtained more extensive data, all broadly within this range of tens of ng/g. In addition to their low uranium concentrations, forams have the disadvantage of being associated with deep sea sediments that have high ^{230}Th concentrations scavenged from the water column above. Delaney and Boyle (1983) showed that even after chemical cleaning techniques were applied, foram samples had [^{230}Th/^{234}U] values in excess of 1 (plot in the "altered" portion of Fig. 3), and therefore had undergone diagenetic addition of ^{230}Th, violating the closed system assumption. Henderson and O'Nions (1995) showed that chemically cleaned foram samples had both elevated uranium concentrations and elevated δ^{234}U values, indicating diagenetic addition and likely exchange of uranium, again violating the closed system assumption. Thus, direct ^{230}Th dating of forams is not presently possible.

6.4. Lacustrine carbonates

Lacustrine carbonates with relatively high uranium concentrations include tufas and inorganically precipitated calcite and aragonite sediments, sometimes rich in organic matter. As with deep sea corals and aragonite bank sediments, the basic dating principles are the same as for surface corals. As with deep sea corals and aragonite bank sediments the initial ^{230}Th term ([^{232}Th/^{238}U] [^{230}Th/^{232}Th]$_i$ ($e^{-\lambda_{230}t}$) term in Eqn. 3) is generally significant. This results from the general association of the carbonates with detrital sediments containing significant levels of thorium. As a consequence isochron techniques are often necessary when dating lacustrine carbonates. Because uranium residence times in lakes are much shorter than in the ocean, lacustrine uranium concentrations and uranium isotopic compositions vary from lake to lake and temporally. Thus, neither initial δ^{234}U values nor uranium concentrations yield information about possible diagenetic reactions, as they would in the marine realm. Because these tools are not available for testing the closed system assumption, future work may well move toward combined ^{230}Th-^{231}Pa dating for tests of the sort. At present only ^{230}Th dating, typically using isochron techniques, has been applied to this sort of sediment. Early work on lacustrine tufas was done by Kaufman and Broecker (1965). This work characterized times of relatively moist climate in the western U.S., and is also one of the first papers to place constraints on Pleistocene atmospheric ^{14}C levels. Lin et al. (1996) followed up on this work using modern techniques and carefully characterizing two different initial thorium components, detrital thorium with low ^{230}Th/^{232}Th values and hydrogenous thorium with higher ^{230}Th/^{232}Th values. Israelson et al. (1997) dated Holocene bulk lake sediment, which was composed of organic-rich carbonate. They were able to obtain ^{230}Th ages without large initial ^{230}Th corrections because of extremely high uranium concentrations in their samples (up to 76 ppm). Schramm et al. (2000) analyzed an extensive sequence of aragonitic lacustrine sediments deposited from Lake Lisan, the last glacial analogue of the Dead Sea. This study is one of a very limited number of studies that constrains atmospheric ^{14}C levels over the first half of the radiocarbon time range. Hall and Henderson (2001) dated early Holocene carbonate lake sediments from Antarctica using detailed isochron techniques. They showed that ^{14}C dating in these lakes was not viable because of large reservoir effects. In addition to dating of lacustrine carbonates, some recent studies have applied ^{230}Th techniques to the dating of lacustrine salts, using the same principles discussed above. As with lacustrine carbonates, isochron approaches are necessary. One such study elucidates the timing of climate change over the last two glacial cycles from Death Valley sediments (Ku et al. 1998), whereas another study characterizes climate change in northern Chile over the last glacial cycle (Bobst et al. 2001). Additional studies such as these are needed in order to characterize and understand climate change on glacial-interglacial time scales at continental sites.

7. CONCLUSIONS

As ^{230}Th and ^{231}Pa dating are among few methods available for dating natural materials that formed in the past 600,000 years, these methods play an important role in the Earth Sciences. In addition to the ongoing efforts to characterize sea level, these methods have been important in, calibrating the ^{14}C time scale, characterizing recent tectonic movement, characterizing ocean circulation changes, and characterizing climate change. The ^{14}C calibration literature can be accessed through references cited in Beck et al. (2001). Although the ^{230}Th and ^{231}Pa dating literature is extensive and this field has made major contributions to the Earth Sciences, major questions remain. For example, there are still major questions regarding the Late Quaternary sea level curve that may require combined ^{230}Th–^{231}Pa techniques to resolve (see Cutler et al. 2003). There are major questions regarding ^{14}C calibration, which again will likely required combined ^{230}Th–^{231}Pa techniques to resolve (see Beck et al. 2001). These and other important problems will likely be tackled using combined ^{230}Th–^{231}Pa dating, and to some degree relying on new inductively-coupled plasma mass spectrometric techniques for increasing throughput and in some instances modestly enhancing precision (see Stirling et al. 2001; Shen et al. 2002; Stirling and Goldstein 2003).

ACKNOWLEDGMENTS

We thank G. Henderson and C. Stirling for careful reviews of an earlier version of this manuscript. This work was supported in part by National Science Foundation Grant ESH 9809459.

REFERENCES

Adkins JF, Cheng H, Boyle EA, Druffel ERM, Edwards RL (1998) Deep-sea coral evidence for rapid change in ventilation of the deep north Atlantic at 15.4 ka. Science 280:725-728

Allegre MC (1964) De l'extension de la méthode de calcul graphique Concordia aux mesures d'âges absolus effectués à l'aide du déséquilibre radioactif. Cas des minéralisations secondaires d'uranium. Note (*) de. C. R. Acad. Sc. Paris, t. 259 p. 4086-4089

Amiel AJ, Miller SD, Friedman GM (1973) Incorporation of uranium in modern corals. Sedimentology 20:523-528

Bacon MP, Anderson RF (1982) Distribution of thorium isotopes between dissolved and particulate forms in the deep sea. J Geophys Res 87:2045-2056

Bar-Matthews M, Wasserburg GJ, and Chen JH (1993) Diagenesis of fossil coral skeletons - correlation between trace-elements, textures, and U-234/U-238. Geochim Cosmochim Acta 57(2):257-276

Bard E, Antonioli F, Silenzi S (2002) Sea-level during the penultimate interglacial period based on a submerged stalagmite from Argentarola Cave (Italy) Earth Planet Sci Lett 196:135-146

Bard E, Fairbanks RG, Hamelin B, Zindler A, Chi Trach H (1991) Uranium-234 anomalies in corals older than 150,000 years. Geochim Cosmochim Acta 55:2385-2390

Bard E, Hamelin B, Fairbanks RG (1990b) U-Th ages obtained by mass spectrometry in corals from Barbados: sea level during the past 130,000 years. Nature 346:456-458

Bard E, Hamelin B, Arnold M, Montaggioni L, Cabioch G, Faure G, Rougerie F (1996a) Deglacial sea-level record from Tahiti corals and the timing of global meltwater discharge. Nature 382:241-244

Bard E, Hamelin B, Fairbanks RG, Zindler A(1990a) Calibration of ^{14}C time scale over the last 30,000 years using mass spectrometric U-Th ages from Barbados corals. Nature 345:461-468

Bard E, Jouannic C, Hamelin B, Pirazzoli P, Arnold M, Faure G, Sumosusastro P, Syaefudin (1996b) Pleistocene sea levels and tectonic uplift based on dating of corals from Sumba Island, Indonesia. Geophys Res Lett 23:1473-1476

Barnes JW, Lang EJ, Potratz HA (1956) The ratio of ionium to uranium in coral limestone. Science 124:175-176

Bateman H (1910) The solution of a system of differential questions occurring in the theory of radioactive transformations. Proc Cambridge Phil Soc 15:423-427

Bender ML, Fairbanks RG, Taylor FW, Matthews RK, Goddard JG, and Broecker WS (1979) Uranium-series dating of the Pleistocene reef tracts of Barbados, West Indies. Geol Soc Amer Bull 90:577-594

Berger AL (1978) Long-term variations of caloric insolation resulting from the earth's orbital elements. Quat Res 9:139-167

Beck JW, Richards DA, Edwards RL, Smart PL, Donahue DJ, Hererra-Osterheld S, Burr GS, Calsoyas L, Jull AJT, Biddulph D (2001) Extremely large variations of atmospheric C-14 during the last glacial period. Science 292:2453-2458

Bischoff JL, Fitzpatrick JA (1991) U-series dating of impure carbonates: An isochron technique using total-sample dissolution. Geochim Cosmochim Acta 55:543-554

Blanchon P, Eisenhauer A (2001) Multi-stage reef development on Barbados during the last interglaciation. Quat Sci Rev 20(10):1093-1112

Bloom AL, Broecker WS, Chappell JMA, Matthews RK, Mesolella KJ (1974) Quaternary sea level fluctuations on a tectonic coast: new ^{230}Th/^{234}U dates on the Huon Peninsula, New Guinea. Quat Res 4:185-205

Boltwood BB (1907) Note on a new radio-active element. Amer J Sci 24:370-372

Broecker WS (1963) A preliminary evaluation of uranium series inequilibrium as a tool for absolute age measurements on marine carbonates. J Geophys Res 68:2817-2834

Broecker WS, Henderson GM (1998) The sequence of events surrounding Termination II and their implications for the cause of glacial-interglacial CO_2 change. Paleoceanography 13:352-364

Broecker WS, Peng T-H (1982) Tracers in the Sea. Eldigio, Palisades, NY, 1982

Broecker WS, Thurber DL, Goddard J, Ku TK, Matthews RK, Mesolella KJ (1968) Milankovich hypothesis supported by precise dating of coral reefs and deep sea sediments. Science 159:297-300

Cabioch G, Ayliffe, LK (2001) Raised coral terraces at Malakula, Vanuatu, southwest Pacific, indicate high sea level during Marine Isotope Stage 3. Quat Res 56:357-365

Chabaux F, Riotte J, Dequincey O (2003) U-Th-Ra fractionation during weathering and river transport. Rev Mineral Geochem 52:533-576

Chappell J (1974) Geology of coral terraces, Huon Peninsula, New Guinea: A study of Quaternary tectonic movements and sea-level changes. Geol Soc Am Bull 85:553-570

Chappell J (1996) Reconciliation of late Quaternary sea levels derived from coral terraces at Huon Peninsula with deep sea oxygen isotope records. Earth Planet Sci Lett 141:227-236

Chappell J (2002) Sea level changes forced ice breakouts in the Last Glacial cycle: new results from coral terraces. Quat Sci Rev 21:1229-1240

Chappell J Polach H (1991) Post-glacial sea-level rise from a coral record at Huon Peninsula, Papua New Guinea. Nature 349:147-149

Chappell JC, Polach H (1972) Some effects of partial recrystallization on ^{14}C dating Late Pleistocene corals and mollusks. Quat Res 2:244-252

Chen JH, Edwards RL, Wasserburg GJ (1986) U-238, U-234, and Th-232 in seawater. Earth Planet Sci Lett 80:241-251

Chen JH, Curran HA, White B, Wasserburg GJ (1991) Precise chronology of the last interglacial period: ^{238}U/^{230}Th data from fossil coral reefs in the Bahamas. Geol Soc Amer Bull 103:82-97

Cheng H, Adkins JF, Edwards RL, Boyle EA (2000a) ^{230}Th dating of deep-sea corals. Geochim Cosmochim Acta 64:2401-2416

Cheng H, Edwards RL, Hoff J, Gallup CD, Richards DA, Asmerom Y (2000b) The half-lives of uranium-234 and thorium-230. Chem Geol 169:17-33

Cheng H, Edwards RL, Murrell MT, Goldstein S (1998) The systematics of uranium-thorium-protactinium dating. Geochim Cosmochim Acta 62(21):3437-3452

Cherdyntsev VV (1955) Transactions of the third session of the commission for determining the absolute age of geological formations. Izv Akad Nauk SSSR, Moscow, p 175-182

Cobb KM, Charles CD, Cheng H, Edwards RL, Kastner M (2002) U/Th-dating living and young fossil corals from the central tropical Pacific. Earth Planet Sci in review

Cochran JK (1982) The oceanic chemistry of U and Th series nuclides. *In:* Uranium series disequilibrium: Application to environmental problems. Ivanovich M, Harmon RS (eds) Clarendon, Oxford, p 384-430

Cochran JK, Masqué P (2003) Short-lived U/Th-series radionuclides in the ocean: tracers for scavenging rates, export fluxes and particle dynamics. Rev Mineral Geochem 52:461-492

Collins LB, Zhu ZR, Wyrwol KH, Hatcher BG, Playford PE, Chen JH, Eisenhauer A, Wasserburg GJ (1993) Late Quaternary evolution of coral reefs on a cool-water carbonate margin; the Abrolhos carbonate platforms, Southwest Australia. Marine Geol 110:203-212

Cross TS, Cross BW (1983) U, Sr and Mg in Holocene and Pleistocene corals A. palmata and M. annularis. J Sed Petr 53:587-594

Cutler KB, Edwards RL, Taylor FW, Cheng H, Adkins J, Gallup CD, Cutler PM, Burr GS, Chappell J, Bloom AL (2003) Rapid sea-level fall and deep-ocean temperature change since the last interglacial. Earth Planet Sci Lett, in press

Dansgaard W, Johnson SJ, Clausen HB, Dahl-Jensen D, Gundestrup NS, Hammer CU, Hvidberg CS, Steffensen JP, Sveinbjornsdottir AE, Jouzel J, Bond G (1993) Evidence for general instability of past climate from a 250-kyr ice-core record. Nature 364:218-220

Delaney ML, Boyle EA (1983) Uranium and thorium isotope concentrations in foraminiferal calcite. Earth Planet Sci Lett 62:258-262

Droxler AW, Morse JW, Glaser KS, Haddad GA, Baker PA (1991) Surface sediment carbonate mineralogy and water column chemistry: Nicaragua Rise versus the Bahamas. Marine Geol 100:277-289

Edmonds HN, Moran BS, Hoff JA, Smith JN, Edwards RL (1998) Pa-231 and Th-230 abundances and high scavenging rates in the western Arctic Ocean. Science 280:405-407

Edwards RL (1988) High-precision Th-230 ages of corals and the timing of sea level fluctuations in the late Quaternary. PhD Dissertation, California Institute of Technology, Pasadena, California

Edwards RL (2000) C.C. Patterson Award Acceptance Speech. Geochim Cosmochim Acta 64:759-761

Edwards RL, Beck JW, Burr GS, Donahue DJ, Druffel ERM, Taylor FW (1993) A large drop in atmospheric ^{14}C/^{12}C and reduced melting during the Younger Dryas, documented with ^{230}Th ages of corals. Science 260:962-968

Edwards RL, Chen JH, Wasserburg GJ (1987a) ^{238}U, ^{234}U, ^{230}Th, ^{232}Th systematics and the precise measurement of time over the past 500,000 years. Earth Planet Sci Lett 81:175-192

Edwards RL, Chen JH, Ku T-L, Wasserburg GJ (1987b) Precise timing of the last interglacial period from mass spectrometric analysis of ^{230}Th in corals. Science 236:1547-1553

Edwards RL, Cheng H, Murrell MT, Goldstein SJ (1997) Protactinium-231 dating of carbonates by thermal ionization mass spectrometry: Implications for the causes of Quaternary climate change. Science 276:782-786

Edwards RL, Gallup CD, Cutler KB, Cheng H (2003) Geochemical evidence for Quaternary Sea Level Changes. *In:* Treatise on Geochemistry. Oceans and Marine Geochemistry. Vol 6. Turekian KK, Holland H (eds.) Elsevier Science, New York, in press

Edwards RL, Taylor FW, Wasserburg GJ (1988) Dating earthquakes with high precision thorium-230 ages of very young corals. Earth Planet Sci Lett 90:371-381

Eisenhauer A, Zhu ZR, Collins LB, Wyrwoll K, Eichstätter (1996) The Last Interglacial sea level change: new evidence from the Abrolhos islands, West Australia. Geologische Rundschau 85:606-614

Esat TM, McCulloch MT, Chappell J, Pillans B, Omura A (1999) Rapid fluctuations in sea level recorded at Huon Peninsula during the Penultimate Deglaciation. Science 283:197-201

Fairbanks RG (1989) A 17,000-year glacio-eustatic sea-level record: influence of glacial melting rates on the Younger Dryas event and deep-ocean circulation. Nature 342:637-642

Gallup CD, Edwards RL, Haddad GA, Droxler AW (1995) Constraints on past changes in the marine $\delta^{234}U$ value from northern Nicaragua Rise, Caribbean Sea cores. EOS 76(46):291-292

Gallup CD (1997) High-precision uranium-series analyses of fossil corals and Nicaragua Rise sediments: The timing of high sea levels and the marine $\delta^{234}U$ value during the past 200,000 years. PhD Dissertation, University of Minnesota, Twin Cities, Minnesota

Gallup CD, Cheng H, Taylor FW, Edwards RL (2002) Direct determination of the timing of sea level change during termination II. Science 295:310-313

Gallup CD, Edwards RL, Johnson RG (1994) The timing of high sea levels over the past 200,000 years. Science 263:796-800

Glaser KS, Droxler AW (1991) High production and highstand shedding from deeply submerged carbonate banks, Northern Nicaragua Rise. J Sed Petrol 61:128-142

Goldstein SJ, Lea DW, Chakraborty S, Kashgarian M, Murrell MT (2001) Uranium-series and radiocarbon geochronology of deep-sea corals: implications for Southern Ocean ventilation rates and the oceanic carbon cycle. Earth Planet Sci Lett 193:167-182

Gvirtzman G, Friedman GM, Miller DS (1973) Control and distribution of uranium in coral reefs during diagenesis. J Sed Petrol 43:985-997

Hall BL, Henderson GM (2001) Use of TIMS uranium-thorium dating to determine past reservoir effects in lakes: Two examples from Antarctica. Earth Planet Sci Lett 193:565-577

Hamelin B, Bard E, Zindler A, Fairbanks RG (1991) $^{234}U/^{238}U$ mass spectrometry of corals: How accurate is the U-Th age of the last interglacial period. Earth Planet Sci Lett 106:169-180

Hanebuth T, Stattegger K Grootes PM (2000) Rapid flooding of the Sunda Shelf: a late-glacial sea-level record. Science 288:1033-1035

Hearty P, Kindle, P, Cheng H, Edwards RL (1999) Evidence for a +20 m sea level in the mid-Pleistocene. Geology 27:375-378

Henderson GM (2002) Seawater ($^{234}U/^{238}U$) during the last 800 thousand years. Earth Planet Sci Lett 199:97-110

Henderson GM, Anderson RF (2003) The U-series toolbox for paleoceanography. Rev Mineral Geochem 52:493-531

Henderson GM, Cohen AS, O'Nions RK (1993) $^{234}U/^{238}U$ ratios and ^{230}Th ages for Hateruma Atoll corals: implication for coral diagenesis and seawater $^{234}U/^{238}U$ ratios. Earth Planet Sci Lett 115:65-73

Henderson GM, O'Nions RK (1995) $^{234}U/^{238}U$ ratios in Quaternary planktonic foraminifera. Geochim Cosmochim Acta 59:4685-4694

Henderson G, Slowey N (2000) Evidence from U-Th dating against Northern Hemisphere forcing of the penultimate deglaciation. Nature 404:61-66

Henderson GM, Slowey NC, Fleisher MQ (2001) U-Th dating of carbonate platform and slope sediments. Geochim Cosmochim Acta 65:2757-2770

Isaac N, Picciotto E (1953) Ionium determination in deep-sea sediments. Nature 171:742-743

Israelson C, Bjorck S, Hawkesworth CJ, Possnert G (1997) Direct U-Th dating of organic- and carbonate-rich lake sediments from southern Scandinavia. Earth Planet Sci Lett 153:251-263

Ivanovich M, Harmon, RS (eds) (1992) Uranium-series Disequilibrium: Application to Earth, Marine, and Environmental Sciences. Oxford Science Publications, Oxford

Jaffey AH, Flynn KF, Glendenin LE, Bentley WC, Essling AM (1971) Precision measurement of Half-Lives and specific activities of ^{235}U and ^{238}U. Phys Rev C 4:1889-1906

Kaufman A, Broecker WS (1965) Comparison of ^{230}Th and ^{14}C ages for carbonate materials from Lakes Lahontan and Bonneville. J Geophys Res 70:4039-4054

Kaufman A, Broecker WS, Ku T-L, Thurber D L (1971) The status of U-series methods of mollusk dating. Geochim. Cosmohcim. Acta 35:1155- 1183

Koczy FF, Picciotto E, Poulaert G, Wilgain S (1957) Mesure des isotopes du thorium dans l'eau de mer. Geochim Cosmochim Acta 11:103-129

Koetsier G, Elliott T, Fruijtier C (1999) Constraints on diagenetic age disturbance: combined uranium-protactinium and uranium-thorium ages of the key Largo Formation, Florida Keys, USA. Ninth Annual V. M. Goldschmidt Conf 157-158

Ku T-L (1965) An evaluation of the U^{234}/U^{238} method as a tool for dating pelagic sediments. J Geophys Res 70:3457-3474

Ku T-L (1968) Protactinium-231 method of dating coral from Barbados Island. J Geophys Res 73:2271-2276

Ku TL (1976) The uranium-series methods of age determination. Ann Rev Earth Planet Sci 4:347-380

Ku T-L, Kimmel MA, Easton WH, O'Neil TJ (1974) Eustatic sea level 120,000 years ago on Oahu Hawaii. Science 183:959-962

Ku TL, Knauss KG, Mathieu GG (1977) Uranium in the open ocean: Concentration and isotopic composition. Deep-sea Res 24:1005-1017

Ku T-L, Luo S, Lowenstein TM, Li J (1998) U-series chronology of lacustrine deposits in Death Valley, California. Quat Res 50:261-275

Li WX, Lundberg J, Dickin AP, Ford DC, Schwarcz HP, McNutt R, Williams D (1989) High precision mass-spectrometric uranium-series dating of cave deposits and implications for paleoclimate studies. Nature 339:534-536

Lin JC, Broecker WS, Anderson RF, Hemming S, Rubenstone JL, Bonani G (1996) New ^{230}Th/U and ^{14}C ages from Lake Lahontan carbonates, Nevada, USA, and a discussion of the origin of initial thorium. Geochim Cosmochim Acta 60:2817-2832

Lomitschka M, Mangini A (1999) Precise Th/U-dating of small and heavily coated samples of deep sea corals. Earth Planet Sci Lett 170:391-401

Ludwig KR (2003) Mathematical–statistical treatment of data and errors for ^{230}Th/U geochronology. Rev Mineral Geochem 52:631-636

Ludwig KR, Muhs D, Simmons K, Halley R, Shinn EA (1996) Sea-level records at ~80 ka from tectonically stable platforms: Florida and Bermuda. Geology 24(3):211-214

Ludwig KR, Simmons KR, Szabo BJ, Winograd IJ, Landwehr JM, Riggs AC, Hoffman RJ (1992) Mass-spectrometric ^{230}Th-^{234}U-^{238}U dating of the Devils Hole calcite vein. Science 258:284-287

Ludwig KR, Titterington DM (1994) Calculation of ^{230}Th/U isochrons, ages, and errors. Geochim Cosmochim Acta 58:5031-5042

Lundberg J, Ford DC (1994) Late Pleistocene sea level change in the Bahamas from mass spectrometric U-series dating of submerged speleothem. Quat Sci Rev 13:1-14

Luo SD, Ku T-L (1991) U-series isochron dating: A generalized method employing total-sample dissolution. Geochim Cosmochim Acta 55:555-564

Mangini A, Lomitschka M, Eichstadter R, Frank N, Vogler S, Bonani G, Hajdas I Patzold J (1998) Coral provides way to age deep water. Nature 392:347-348

Matthew RK (1968) Carbonate diagenesis: Equilibration of sedimentary mineralogy to the subaerial environement; coral cap of Barbados, West Indies. J Sed Petrol 38:1110-1119

McCulloch MT, Esat T (2000) The coral record of last interglacial sea levels and sea surface temperatures. Chem Geol 169:107-129

Mesolella KJ, Matthews RK, Broecker WS, Thurber DL (1969) The astronomical theory of climatic change: Barbados data. J Geol 77:250-274

Min GR, Edwards RL, Taylor FW, Recy J, Gallup CD, Beck JW (1995) Annual cycles of U/Ca in corals and U/Ca thermometry. Geochim Cosmochim Acta. 59:2025-2042

Moore WS (1981) The thorium isotope content of ocean water. Earth Planet Sci Lett 53:419-426

Moran SB, Hoff JA, Edwards RL, Landing WM (1997) Distribution of Th-230 in the Laborador Sea and its relation to ventilation. Earth Planet Sci Lett 150:151-160

Muhs DR, Simmons KR, Steinke B (2002) Timing and warmth of the Last Interglacial period: new U-series evidence from Hawaii and Bermuda and a new fossil compilation for North America. Quat Sci Rev 21:1355-1383

Muhs DR, Kennedy G, Rockwell TK (1994) Uranium-series ages of marine terrace corals from the Pacific coast of North America and implications for last-interglacial sea level history. Quat Res 42:72-87

Nozaki Y, Nakanishi T (1985) ^{231}Pa and ^{230}Th profiles in the open ocean water column. Deep-Sea Res 32:1209-1220

Pickett D, Murrell MT, Williams RW (1994) Determination of femtogram quantities of protactinium in geologic samples by thermal ionization mass spectrometry. Anal Chem 66:1044-1049

Porcelli D, Swarzenski PW (2003)The behavior of U- and Th- series nuclides in groundwater. Rev Mineral Geochem 52:317-361

Potratz HA, Bonner NA (1958) U.S. Energy Comm Rept No. La-1721

Renne PR, Karner DB, Ludwig KR (1998) Radioisotope dating: Enhanced: Absolute ages aren't exactly. Science 282:1840-1841

Richards DA (1995) Pleistocene sea levels and paleoclimate of the Bahamas based on Th-230 ages of speleothems. PhD Dissertation. University of Bristol, Bristol, United Kingdom

Richards DA, Smart PL, Edwards RL (1994) Maximum sea levels for the last glacial period from U-series ages of submerged speleothems. Nature 367:357-360

Richter FM, Turekian KK (1993) Simple models for the geochemical response of the ocean to climatic and tectonic forcing. Earth Planet Sci Lett 119:121-131

Robert J, Miranda CF, Muxart R (1969) Mesure de la période du protactinium-231 par microcalorimétrie. Radiochim Acta 11:104-108

Robinson LF, Henderson GM, Slowey NC (2002) U-Th dating of marine isotope stage 7 in Bahamas slope sediments. Earth Planet Sci Lett 196:175-187

Robst AL, Lowenstein TK, Jordan TE, Godfrey LV, Ku T-L, Luo S (2001) A 106 ka paleoclimate record from drill core of the Salar de Atacama, northern Chile. Paleogeo Paleoclim Paleoecol 173:21-42

Rosholt JN (1957) Quantitative radiochemical methods for determination of the sources of natural radioactivity. Anal Chem 29:1398-1408

Rosholt NJ (1967) Open system model for uranium-series dating of Pleistocene samples. *In:* Radioactive Dating and Methods of low-level Counting. I. A. E. A. Proc Ser Publ, SM-87/50, p 299-311

Rosholt JN, Antal PS (1962) Evaluation of the $Pa^{231}/U-Th^{230}/U$ method for dating Pleistocene carbonate rocks. US Geol Survey Prof Paper 450-E:108-111

Russell AD, Emerson S, Mix AC, Peterson LC (1996) The use of foraminiferal U/Ca as an indicator of changes in seawater uranium content. Paleoceanography 11:649-663

Rutherford E, Soddy F (1902) The cause and nature of radioactivity: Part II. Phil Mag Ser 6 4:569-585

Sackett WM (1960) Protactnium-231 content of ocean water and sediments. Science 132:1761-1762

Sackett WM (1958) Ionium-uranium ratios in marine deposited calcium carbonates and related materials. PhD Dissertation, Washington University, St. Louis, Missouri

Sackett WM, Potratz HA, Goldberg ED (1958) Thorium content of ocean water. Science 128:204-205

Sakanoue M, Konishi K, Komura K (1967) Stepwise determinations of thorium, protactinium, and uranium isotopes and their application in geochronological studies. *In:* Proc. Symp. Radioactive Dating and Methods of Low-Level Counting, Monaco, p 313-329

Schramm A, Stein M, Goldstein SL (2000) Calibration of the ^{14}C time scale to >40 ka by $^{234}U-^{230}Th$ dating of Lake Lisan sediments (last glacial Dead Sea). Earth Planet Sci Lett 175:27-40

Schroeder JH, Miller DS, Friedman GM (1970) Uranium distributions in recent skeletal carbonates. J Sed Petrol 40:672-681

Shackleton NJ, Imbrie J, Hall MA (1983) Oxygen and carbon isotope record of East Pacific core V19-30: implications for the formation of deep water in the late Pleistocene North Atlantic. Earth Planet Sci Lett 65:233-244

Shen C-C, Edwards RL, Cheng H, Dorale JA, Thomas RB, Moran SB, Weinstein S, Edmonds HN (2002) Uranium and thorium isotopic and concentration measurements by magnetic sector inductively coupled plasma mass spectrometry. Chem Geol 185:165-178

Shen GT, Dunbar RB (1995) Environmental controls on uranium in reef corals. Geochim Cosmochim Acta 59:2009-2024

Shen GT, Boyle EA (1988) Determination of lead, cadmium and other trace metals in annually-banded corals. Chem Geol 67:47-62

Slowey NC, Henderson GM, Curry WB (1995) Direct dating of the last two sea-level highstands by measurements of U/Th in marine sediments from the Bahamas. EOS 76(46):296-297

Slowey NC, Henderson GM, Curry WB (1996) Direct U–Th dating of marine sediments from the two most recent interglacial periods. Nature 383:242–244

Smith JE, Risk MJ, Schwarcz HP, McConnaughey TA (1997) Rapid climate change in the North Atlantic during the Younger Dryas recorded by deep-sea corals. Nature 386:818-820

Stein M, Wasserburg G., Aharon P, Chen JH, Zhu ZR, Bloom AL, Chapell J (1993) TIMS U series dating and stable isotopes of the last interglacial event at Papua New Guinea. Geochim Cosmochim Acta 57:2541-2554

Stirling CH, Esat TM, Lambeck K, McCulloch MT (1998) Timing and duration of the last interglacial; evidence for a restricted interval of widespread coral reef growth. Earth Planet Sci Lett 160:745-762

Stirling CH, Esat TM, Lambeck K, McCulloch MT, Blake SG, Lee D-C, Halliday AN (2001) Orbital forcing of the marine isotope stage 9 interglacial. Science 291:290-293

Stirling CH, Esat TM, McCulloch MT, Lambeck K (1995) High-precision U-series dating of corals from Western Australia and implications for the timing and duration of the Last Interglacial. Earth Planet Sci Lett 135:115-130

Swart PK (1980) The environmental geochemistry of scleractinian corals. An experimental study of trace element and stable isotope incorporation. PhD Dissertation, University of London, London, United Kingdom

Swart PK, Hubbard JAEB (1982) Uranium in scleractinian coral skeletons. Coral Reefs 1:13-19

Szabo BJ, Ludwig KR, Muhs DR, Simmons KR (1994) Thorium-230 ages of corals and duration of the last interglacial sea-level high stand on Oahu, Hawaii. Science 266:93-96

Szabo BJ, Rosholt JN (1969) Uranium-series dating of Pleistocene molluscan shell from southern California-An open system model. J Geophys Res 74:3253-3260

Tatsumoto M, Goldberg ED (1959) Some aspect of the marine geochemistry of uranium. Geochim Cosmochim Acta 17:201-208

Thompson G, Livingston HD (1970) Strontium and uranium concentrations in aragonite precipitated by some modern corals. Earth Planet Sci Lett 8:439-442

Thurber DL (1962) Anomalous $^{234}U/^{238}U$ in Nature. J Geophys Res 67:4518-4520

Thurber DL, Broecker WS, Blanchard RL, Potratz HA (1965) Uranium-series ages of Pacific atoll coral. Science 149:55-58

Toscano MA, Lundberg J (1999) Submerged Late Pleistocene reefs on the tectonically-stable S.E. Florida margin: high-precision geochronology, stratigraphy, resolution of Substage 5a sea-level elevation, and orbital forcing. Quat Sci Rev 18:753-767

Veeh HH (1966) $^{230}Th/^{238}U$ and $^{234}U/^{238}U$ ages of Pleistocene high sea level stand. J Geophys Res 71:3379-3386

Veeh HH, Turekian KK (1968) Cobalt, silver and uranium concentrations of reef building corals in the Pacific Ocean. Limol Oceanogr 13:304-308

Wasserburg GJ (2000) Citation for presentation of the 1999 C.C. Patterson Award to R. Lawrence Edwards. Geochim Cosmochim Acta 64:755-757

Wetherill GS (1956a) An interpretation of the Rhodesia and Witwatersrand age patterns. Geochim Cosmochim Acta 9:290-292

Wetherill GS (1956b) Discordant uranium-lead ages. I. Trans. Am Geophys Union 37:320-326

Zachariasen JA (1998) Paleoseismology and Paleogeodesy of the Sumatran Subduction Zone: A Study of Vertical Deformation Using Coral Microatolls. PhD Dissertation, California Institute of Technology, Pasadena, California

Zachariasen J, Sieh K, Taylor FW, Edwards RL, Hantoro WS (1999) Submergence and uplift associated with the giant 1833 Sumatran subduction earthquake: Evidence from coral microatolls. J Geophys Res 104(B1):895-919

Zhu ZR, Wyrwoll KH, Chen L, Chen J, Wasserburg GJ, Eisenhauer A (1993) High-precision U-series dating of Last Interglacial events by mass spectrometry: Houtman Abrolhos Islands, western Australia. Earth Planet Sci Lett 118:281-293

10 Uranium-series Chronology and Environmental Applications of Speleothems

David A. Richards

School of Geographical Sciences
University of Bristol
Bristol, BS8 1SS, United Kingdom

Jeffrey A. Dorale

Department of Geological Sciences
University of Missouri - Columbia
Columbia, Missouri, 65211, U.S.A.

1. INTRODUCTION

An increasing number of scientists recognize the value of speleothems[1] as often extremely well-preserved archives of information about past climate, vegetation, hydrology, sea level, nuclide migration, water-rock interaction, landscape evolution, tectonics and human action. Well-constrained data are required to document past changes, reconstruct past patterns and predict future responses of the Earth system at a wide range of spatial and temporal scales. Speleothems are particularly useful in this regard because they can be found in many locations of the globe, sampled at high-resolution and reliably dated using high-precision uranium-series techniques.

Speleothems are bodies of mineral material formed in caves as the result of chemical precipitation from groundwater flowing or dripping in a cave. Most speleothems are composed of calcite formed by slow degassing of CO_2 from supersaturated groundwater, but aragonite and gypsum forms are also common, particularly near cave entrances where evaporative effects are important. A host of different speleothem types decorate the walls, ceilings and floors of caves, and their mineralogy and morphology is a function of fluid flow and chemistry of waters feeding a particular location as well as the ambient conditions (temperature, chemistry, light) in the air or water-filled void. Subaerial forms include the familiar stalagmites, stalactites, draperies, flowstones. Subaqueous forms include rimstone pools, "rafts," mammillary calcite wall-coatings and "dog-tooth" spar. For an extensive review of the types of speleothem that have been observed, see Hill and Forti (1997).

Speleothems are used in a multitude of ways to explore past environmental conditions, perhaps the most fundamental of which is their very presence or absence. Deposition of speleothems relies on sufficient water supply and soil CO_2 to enable dissolution and transport of reactants in the vadose zone to underlying caves. In arid or glacial times, conditions may not have been favorable for speleothem formation. Thus, speleothem absence/presence or growth frequency can be used as paleoclimatic indicator. The presence of subaerial speleothem forms such as flowstones, stalactites and stalagmites indicates that open passages existed and were not flooded for long intervals during the time of their formation. Age determinations on suitable samples, therefore, provide valuable constraints on the opening of passages and water-table fluctuations, the latter influenced by sea level variation or regional base-level lowering. In addition to

---

[1] In addition to their tremendous value as archives of information about the past state of the Earth's environment, speleothems have an aesthetic value that satisfies the curiosity of all cave visitors. For this reason, we promote conservation guidelines that have been set in place to preserve the quality of cave environments. In many regions of the world, collection of speleothems is prohibited by legislation. Even if permission is obtained, the utmost effort should be made to minimize the impact of sampling.

presence/absence, a suite of potential indicators are contained within the solid and liquid phase (fluid inclusions) of speleothems, such as trace elements, luminescence, δD, $\delta^{18}O$, $\delta^{13}C$, pollen and internal stratigraphy (growth rate and hiatuses). Also, where speleothems display a clear stratigraphic relationship with sedimentary material that has been deposited, disrupted or altered, valuable age constraints can be obtained for events of interest, such as flooding, seismicity, or human occupation. Attempts have also been made to investigate past records of geomagnetic and/or heliomagnetic intensity by analyzing magnetic minerals and concentrations of cosmogenic nuclides within speleothems.

Essential to all the above applications are constraints on the timing of speleothem growth. Initial attempts to obtain chronological information from speleothems by radiocarbon (e.g., Broecker et al. 1960; Geyh 1970) or uranium-series (e.g., Rosholt and Antal 1962; Cherdyntsev et al. 1965) techniques were poorly constrained, but better understanding of chronological techniques in the late 1960s and 1970s, coupled with the use of stable isotopes of hydrogen, oxygen and carbon to obtain paleoclimatic information (Hendy and Wilson 1968; Duplessy et al. 1972; Schwarcz et al. 1976; Harmon et al. 1978; Gascoyne et al. 1978) indicated that speleothems offered considerable potential. Uranium-series methods now provide by far the most reliable and precise method for dating speleothems <0.5 Ma. Electron spin resonance (Ikeya 1975; Bassiakos 2001) and paleomagnetic (Latham and Ford 1993) techniques can be used to extend the dateable range, but suffer from relatively poor precisions.

A surge in interest in speleothems as archives of geological and archaeological information during the past two decades can be attributed in part to improvements in analytical techniques and also to a better understanding of Earth surface processes at the local to global scale that contribute to variation in proxy evidence contained within speleothems. Improvements in technology are readily illustrated by comparison of paleoclimate information obtained from speleothems 20 years ago with that of most recent studies. Through the mid-1980s, the highest resolution geochemical records (e.g., $\delta^{18}O$ and $\delta^{13}C$) were obtained at mm-scale resolution with age control based on sub-samples >10 g with 2σ age uncertainties of the order of 2-10% at 10 ka. In contrast, a recent study of a stalagmite from the west coast of Ireland using an ion microprobe (Baldini et al. 2002) illustrates annual variation in trace elemental (P and Sr) concentration at 10 µm spatial resolution (Fig. 1) for calcite deposited during a period known as the "8.2 ka cold event." Such trace elemental data can be used as proxy evidence for paleo-recharge because they are related to water-soil-rock residence times. A parallel study using laser-ablation gas-chromatography isotope-ratio mass spectrometry to analyze sub-samples of <250 µm diameter from the same stalagmite section (McDermott et al. 2001) highlights a negative $\delta^{18}O$ anomaly for the same period (Fig. 1), which is ascribed to a combination of cold temperatures and freshening of ocean surface waters in the North Atlantic, the dominant source region for precipitation in western Ireland, and is linked with anomalies in marine and ice core records of the high latitude North Atlantic and Greenland. The timing of the cold event is constrained by mass-spectrometric uranium-series ages for sub-samples of calcite <1 g with precisions of ± 80 years. This dramatic event was probably caused by a catastrophic draining of the glacial lakes Agassiz and Ojibway into the North Atlantic. This study highlights a vast improvement on sample sizes required for analysis and, hence, temporal resolution, which has enabled construction of records with resolutions approaching those of tree-ring, coral and ice core records. In many ways, the technical developments of the past decade have had particular influence on speleothem studies in comparison to other carbonate deposits because U concentrations and growth rates are typically limiting in speleothems.

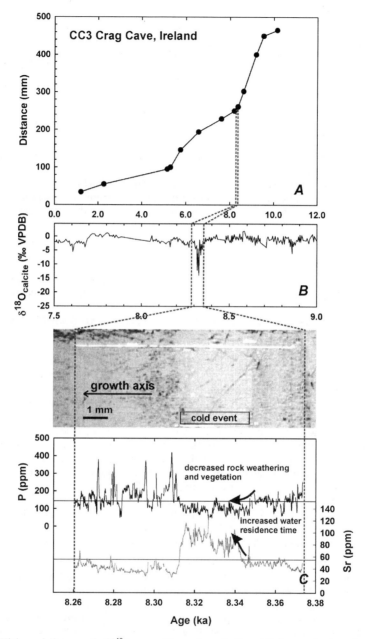

Figure 1. High resolution record of $\delta^{18}O_{calcite}$ and trace elements, Sr and P, for a speleothem from western Ireland (McDermott et al. 2001; Baldini et al. 2002). A) ^{230}Th age vs. distance relationship. Symbols larger than 2σ errors, which range from ± 0.2 to 0.8 ka. B) Negative $\delta^{18}O_{calcite}$ anomaly related to changes in $\delta^{18}O_{water}$ of rainfall during cold event triggered by meltwaters from Laurentide ice sheet. Sample resolution of ~250μm using laser-ablation gas-chromatography isotope-ratio mass spectrometry. C) P and Sr concentrations at ~10 μm resolution for period encompassing cold event. Trace element variation is related to water-rock residence times and vegetation changes in the vadose zone above the sample.

We do not intend to provide an exhaustive review of the development of uranium-series techniques and environmental applications in this chapter because there have been many notable reviews written during the past twenty years (Gascoyne et al. 1978; Gascoyne 1992a; Schwarcz 1986; Latham and Schwarcz 1992; Schwarcz and Blackwell 1992; Atkinson and Rowe 1992; Gascoyne and Harmon 1992; Ford 1997; Ku 2000). Here, we outline considerations that must be made by practicing geochronologists when analyzing speleothems, and illustrate the wide range of recent applications that are informed by ages and geochemistry of these deposits. We focus on the applications of uranium-series dating to speleothems and environmental change of the past million years. There are obvious reasons for this:

(1) Uranium-series dating is by far the most widely-used dating technique applied to speleothems, and is only applicable to material that is currently in a state of disequilibrium. Given the rates of ingrowth of daughter isotopes of ^{231}Pa and ^{230}Th, conditions of uranium-series disequilibrium in systems which have remained isotopically-closed are generally restricted to materials <~500 ka (^{230}Th) and 200 ka (^{231}Pa).

(2) Comparable evidence for environmental change is available from high resolution ocean cores, loess sequences, ice cores and other archives.

We do, however, also include brief discussion of the principles and application of U-Th-Pb dating which can theoretically be applied to material up to the age of the Earth.

2. BASIC GEOCHRONOLICAL PRINCIPLES AND ASSUMPTIONS

2.1. General principles of ^{230}Th-^{234}U-^{238}U and ^{231}Pa-^{235}U dating

A state of secular equilibrium between parent and daughter nuclides in the ^{238}U and ^{235}U radioactive decay-series will have been established in any naturally-occurring material that has remained undisturbed for several million years because the half-life of the parent isotopes is much greater than that of the intermediate daughters in the decay chains. The decay-series in secondary deposits formed from the dissolution and subsequent precipitation of such material, however, will be in a state of disequilibrium at time of formation with either an excess or deficiency of intermediate nuclides because of one or a combination of chemical, physical or nuclear fractionation processes. The extent to which the uranium-series decay chains in the secondary deposit have returned to secular equilibrium in a closed system from an initial state of disequilibrium can be expressed by a straightforward function of time using the decay constants of the radioactive isotopes involved if the following criteria are satisfied: (1) intermediate and daughter decay products at time of formation were absent, or if present, can be corrected for; (2) no gain or loss of the parent nuclide or daughter products occurred since the time of formation.

Uranium-series dating of speleothems is based on the extreme fractionation of the parent U isotopes (^{238}U, ^{235}U and ^{234}U) from their long-lived daughters ^{231}Pa and ^{230}Th in the hydrosphere. The average abundances of U and Th in the Earth's continental crust are 1.7 and 8.5 μg g^{-1} (Wedepohl 1995). Their relative abundances in the hydrosphere, however, are markedly different, principally because of the different solubility of U and Th species in the surface and near-surface environments [see http://earthref.org/GERM for database of elemental abundances in major reservoirs of the Earth].

Uranium is readily mobilized in the meteoric environment, principally as the highly soluble uranyl ion (UO$_2^{2+}$) and its complexes, the most important of which are the stable carbonate complexes that form in typical groundwaters (pH > 5, pCO$_2$ = 10^{-2} bar) (Gascoyne 1992b; Grenthe et al. 1992; see also Langmuir (1997) for review). Uranium is also known to be strongly associated with organic matter, such as fulvic and humic acids, and inorganic colloids. Uranium concentrations in meteoric waters are highly variable (0.01 to 100 µg L^{-1}; Osmond and Cowart 1982) depending on ionization potential, pH, U concentration and solubility of mineral phase in the source rock, water-rock interaction time, and presence of complexing ligands. Groundwaters with the highest U contents are associated with mineralized areas or organic rich deposits such as black shales. Most analyzes have been conducted on rivers and groundwaters, while relatively few have been conducted on vadose drip waters that feed speleothems.

In contrast to U, the long-lived daughter products Th and Pa typically exist in +4 and +5 oxidation states, respectively, and are readily hydrolyzed and either precipitated or adsorbed on detrital particulates (inorganic or organic), clay minerals and iron (oxy)hydroxides. Thorium and Pa can be readily transported, therefore, in association with calcite colloids, humic acids and goethite, for example, and while turbid waters can be expected to have higher Th concentrations, waters feeding secondary calcite and aragonite deposits will generally have negligible Th concentrations. Protactinium and Th associated with airborne particulates can be incorporated into secondary deposits. Also, intermittent flooding events can transport a supply of detrital material.

U and Th concentrations in secondary deposits precipitated from solution generally reflect relative abundances in the hydrosphere. Uranium is co-precipitated with $CaCO_3$ in subaerial environments on exsolution of CO_2 (or evaporation), while the immediate daughter products are essentially absent. This represents extreme chemical fractionation of parent and daughter isotopes within the hydrosphere.

In a closed system, the extent to which the $(^{230}Th/^{238}U)_A$ and $(^{231}Pa/^{235}U)_A$ activity ratios (denoted by subscript A) have returned to unity is a function of time (T) and governed respectively by the following equations, assuming $(^{234}U/^{238}U)_A = 1$ and an initial state of $^{230}Th = ^{231}Pa = 0$.

$$\left(\frac{^{230}Th}{^{238}U}\right)_A = 1 - e^{-\lambda_{230}T} \tag{1}$$

$$\left(\frac{^{231}Pa}{^{235}U}\right)_A = 1 - e^{-\lambda_{231}T} \tag{2}$$

While it is expected that the source rocks for the radionuclides of interest in many environments were deposited more than a million years ago and that the isotopes of uranium would be in a state of radioactive equilibrium, physical fractionation of ^{234}U from ^{238}U during water-rock interaction results in disequilibrium conditions in the fluid phase. This is a result of (1) preferential leaching of ^{234}U from damaged sites of the crystal lattice upon alpha decay of ^{238}U, (2) oxidation of insoluble tetravalent ^{234}U to soluble hexavalent ^{234}U during alpha decay, and (3) alpha recoil of ^{234}Th (and its daughter ^{234}U) into the solute phase. If initial $(^{234}U/^{238}U)_A$ in the waters can be reasonably estimated *a priori*, the following relationship can be used to establish the time T since deposition,

$$\delta^{234}U(T) = \delta^{234}U(0)e^{-\lambda_{234}T} \tag{3}$$

where $\delta^{234}U = 1000*[(^{234}U/^{238}U)_m/(^{234}U/^{238}U)_{eq} - 1]$, $(^{234}U/^{238}U)_m$ is the measured mass ratio and $(^{234}U/^{238}U)_{eq}$ is the mass ratio at secular equilibrium. Reasonable age estimates have been derived for subaqueous speleothem deposits from groundwaters that exhibit minimal long-term variation in $\delta^{234}U(0)$ (Winograd et al. 1988; Ludwig et al. 1992). However, secular variation of $\delta^{234}U(0)$ within individual samples is commonly observed (see Section 2.4), thus limiting the applicability of this technique.

Nearly all natural waters contain ^{234}U and ^{238}U in a state of disequilibrium (i.e., $(^{234}U/^{238}U)_A \neq 1$). Generally $(^{234}U/^{238}U)_A > 1$, and in some cases as high as 30 (Osmond and Cowart 1992), hence, a second term must be added to Equation (1), which gives the standard ^{230}Th/^{238}U age equation (from Kaufman and Broecker 1965):

$$\left(\frac{^{230}Th}{^{238}U}\right)_A = 1 - e^{-\lambda_{230}T} + \left(\frac{\delta^{234}U(T)}{1000}\right)\left(\frac{\lambda_{230}}{\lambda_{230} - \lambda_{234}}\right)\left(1 - e^{(\lambda_{234} - \lambda_{230})T}\right) \quad (4)$$

Figure 2 shows $(^{230}Th/^{238}U)_A$ plotted as a function of T for different values of $\delta^{234}U(0)$. As T becomes large, $(^{230}Th/^{238}U)_A$ approaches unity and the age limit of the technique is reached. The actual limit depends on the precision of the isotopic determinations and $\delta^{234}U(0)$. Equations (2) and (4) can be combined to give an expression of $^{231}Pa/^{230}Th$

$$\left(\frac{^{231}Pa}{^{230}Th}\right)_A = \frac{\left(\frac{^{235}U}{^{238}U}\right)_A \left(1 - e^{-\lambda_{231}T}\right)}{1 - e^{-\lambda_{230}T} + \left(\frac{\delta^{234}U(T)}{1000}\right)\left(\frac{\lambda_{230}}{\lambda_{230} - \lambda_{234}}\right)\left(1 - e^{(\lambda_{234} - \lambda_{230})T}\right)} \quad (5)$$

2.2. Initial conditions

The equations governing the age of secondary carbonate deposits stated above assume that all ^{230}Th or ^{231}Pa present in the mineral is formed *in situ* by radioactive decay of co-precipitated U. Thorium and Pa content at time of formation can often be considered to be negligible in the pure authigenic phase of calcite or aragonite

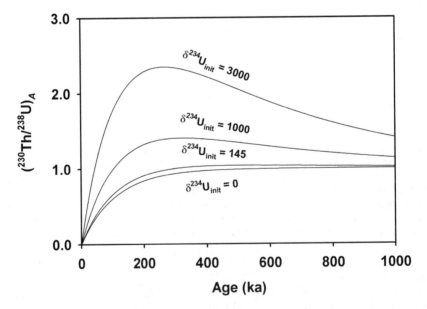

Figure 2. $(^{230}Th/^{238}U)_A$ versus time for different $\delta^{234}U(0)$ values assuming closed system and initial $^{230}Th = 0$ (Eqns. 1 and 2). For those cases where $(^{230}Th/^{238}U)_A > 1$, a unique combination of $(^{230}Th/^{238}U)_A$ and $\delta^{234}U_m$ defines the age T. $\delta^{234}U(0) > 6000$ have been observed in speleothems (Plagnes et al. 2002) but are generally −100 to 1000.

speleothems, however, this component is often accompanied by allochtonous material with a significant amount of Th and Pa. Various methods are available to account for the presence of inherited daughter isotopes but the uncertainty of the final estimate is greater than would be expected for pure material.

Initial ^{230}Th and ^{231}Pa are generally considered to be associated with a detrital component that becomes cemented, or occluded, within the speleothem. This component may be composed of clays, alumino-silicates or Fe-oxyhydroxides (Fig. 3) with strongly adsorbed Th^{4+} and Pa^{5+}. Th and Pa incorporated in speleothems and similar deposits may also have been transported in colloidal phases (Short et al. 1998; Dearlove et al. 1991), attached to organic molecules (Langmuir and Herman 1980; Gaffney et al. 1992) or as carbonate complexes in solution (Dervin and Faucherre 1973a, b; Joao et al. 1987).

A priori estimation of $^{230}Th/^{232}Th_0$ *(or R_0).* The extent of initial Th contamination can be monitored by measurement of ^{232}Th, which is the most abundant, extremely long-lived isotope of Th with a half-life of 1.401×10^{10} a. Where ^{232}Th content is high, initial ^{230}Th concentration is also expected to be significant. To a first approximation, correction can be made for the detrital component of ^{230}Th by using the ^{232}Th concentration as an index of contamination and assuming an appropriate value of ^{230}Th/^{232}Th for the sedimentary context. By assuming that ^{232}Th was incorporated at the time of formation, and that the excess ^{230}Th associated with this ^{232}Th decayed over time T since then, $(^{230}Th*/^{238}U)_A$ (*in situ* ingrowth within authigenic carbonate phase) can be calculated using the following equation:

$$\left(\frac{^{230}Th^*}{^{238}U}\right)_A = \left(\frac{^{230}Th}{^{238}U}\right)_{Am} - \left(\frac{^{232}Th}{^{238}U}\right)_A R_0 e^{-\lambda_{230}T} \tag{6}$$

where R_0 is the ^{230}Th/^{232}Th activity ratio in the detritus at time of formation and is estimated from other sources, subscript *m* refers to the measured value of the dissolved

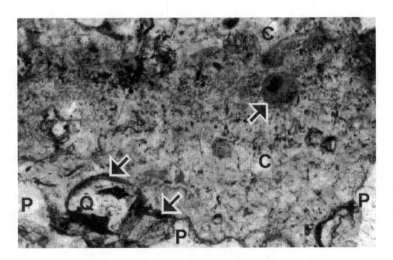

Figure 3. Grain of clay and Fe oxide (upper arrow) and quartz grain (Q) with coating of similar material (lower arrows) amidst calcite (C) in a speleothem from Drotsky's Cave, Botswana. "P" indicates pores. Energy-dispersive X-ray analysis suggests that clay is Mg-bearing smectite and confirms presence of Fe (Railsback et al. 1999). Plane-polarized light; field of view 0.7 mm wide. Stalagmite DS87; thin section BDS1.5-2; sample collected by Drs. G. A. Brook and J. Cooke. [see Atlas of Speleothem Fabrics http://www.gly.uga.edu/railsback/speleoatlas/SAintro1.html]

authigenic and detrital fraction. The concentrations of ^{234}U and ^{238}U in the authigenic carbonate are assumed to equal their concentrations in the leachate that includes the portion of detritus that has entered solution. Kaufman and Broecker (1965) established a mean value for R_0 of 1.7 ± 0.17 (1σ) for Lake Lahontan lacustrine sediments, which was confirmed by subsequent analyzes by Lin et al. (1996). Kaufman (1993) reviews the average values obtained for detritus in carbonate materials (shells, molluscs, tufas) from different locations of the globe and considers the value of 1.7 with a more conservative uncertainty of ± 0.7 (1σ) to be a reasonable *a priori* estimate in the absence of any pertinent data for the region or context of study. Others have generally adopted a value that reflects the Th/U ratio in the terrestrial upper continental crust, which has a range of 3.6 to 3.8 (Taylor and McClennan 1995; Wedepohl 1995). Assuming secular equilibrium between ^{230}Th and ^{238}U, this would give $(^{230}$Th/^{232}Th$)_A$ = 0.83-0.87. If the full range of Th/U values for sediments is used (see Rogers and Adams 1969), a conservative mean value would be 0.8 ± 0.8 (2σ). There is a significant difference between the arbitrary values adopted in published studies. It is accepted that where ^{232}Th contamination is minimal, the effect of using the incorrect ^{230}Th/^{232}Th initial ratio will be negligible, however, caution must be adopted because anomalous values of ^{230}Th/^{232}Th$_0$ may be have been inherited in some environments

Isochron methodology. To thoroughly check for, and accommodate, initial ^{230}Th in bulk carbonate material, we advocate total dissolution and isochron methodologies and application of rigorous statistical treatment (Luo and Ku 1991; Bischoff and Fitzpatrick 1991; Ludwig and Titterington 1994; Ludwig 2003). While various leach-leach, leach-residue methodologies have been promoted to distinguish between the isotopic ratios in the authigenic and detrital phases (Osmond et al. 1970; Ku and Liang 1984; Schwarz and Latham 1989), only total dissolution techniques deal with both lattice-bound and adsorbed Th. Bischoff and Fitzpatrick (1991) demonstrated that these two components could not be adequately separated by selective leaching techniques. Also, ^{230}Th released from the authigenic phase during etching can reabsorb to detrital material present during incomplete dissolution. Henderson et al. (2001) also advocate whole-sample dissolution because of the problems associated with post-depositional mobility of nuclides as a result of α-recoil. For samples with a major component of detrital material, however, there may be insufficient variation in the Th/U ratio of totally dissolved sub-samples of coeval material to arrive at meaningful isochrons, in which case selective dissolution techniques must be employed.

The construction of isochrons is standard methodology in radiometric geochronology where two or more cogenetic mineral phases can be separated from bulk sample and analyzed separately to distinguish between the amount of ingrown radiogenic end-member and its initial abundance (Rb-Sr, U-Pb, Sm-Nd, etc.) Measured U and Th isotopic data for two or more sub-samples of coeval carbonate material with different ^{232}Th/^{238}U should lie on a straight mixing line in ^{230}Th/^{232}Th-^{238}U/^{232}Th and ^{234}U/^{232}Th-^{238}U/^{232}Th space, or ^{230}Th/^{238}U-^{232}Th/^{238}U and ^{234}U/^{238}U-^{232}Th/^{238}U space if the material (1) has behaved as closed system and (2) is a two-component system (i.e., variable quantities of authigenic $CaCO_3$ and homogenous detrital phase). The ^{230}Th/^{238}U and ^{234}U/^{238}U isotopic ratios of the ^{232}Th-free endmember (used to calculate the ^{230}Th-^{234}U-^{238}U age) and ^{230}Th/^{232}Th$_0$ can calculated from intercept and gradient of the straight line that best estimates the relationship in the data.

Weighted regression of ^{238}U-^{234}U-^{230}Th-^{232}Th isotope data on three or more coeval samples provides robust estimates of the isotopic information required for age calculation. Ludwig (2003) details the use of maximum likelihood estimation of the regression parameters in either coupled *XY-XZ* isochrons or a single three dimensional *XYZ* isochron, where *X*, *Y* and *Z* correspond to either (1) ^{238}U/^{232}Th, ^{230}Th/^{232}Th and

^{234}Th/^{232}Th or (2) ^{232}Th/^{238}Th, ^{230}Th/^{238}U and ^{234}U/^{238}U, respectively.

Stalagmites often have observable differences in the quantity of detrital material along growth layers because of the nature of fluid flow on the surface of the sample. Commonly, the central portion at the locus of drip contact is the cleanest material, presumably because the energy of drip impact is greatest at this point. The extent of homogeneity of initial ^{230}Th/^{232}Th in a single layer from a stalagmite (SVC-98-3.B) from Spring Valley Caverns, Minnesota was analyzed using isochron methodology by Dorale et al. (2003). Variation in detrital Th content was confirmed by ^{238}U/^{232}Th measurements, which vary by two orders of magnitude from the core to the periphery of the growth layer. Isotopic data show a linear relationship in ^{230}Th/^{238}U-^{232}Th/^{238}U and ^{234}U/^{238}U-^{232}Th/^{238}U space, however there is excess scatter with a mean square of weighted deviates value of 7.9 (Fig. 4). The resolved ^{230}Th/^{232}Th is 0.801 ± 0.152, which is very similar to the bulk Earth value.

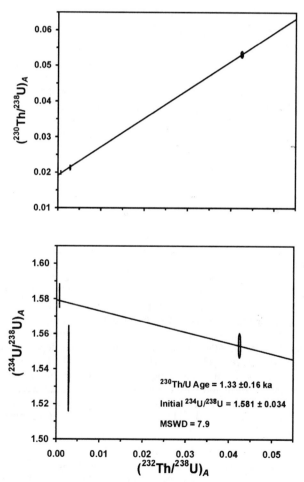

Figure 4. Osmond type II isochron plot for sub-samples from single growth layer in stalagmite SVC-98-3.B from Spring Valley Caverns (Dorale et al. 2003).

The most likely cause of scatter in isochrons is a combination of more than one source of initial Th. Lin et al. (1996) identify three sources of initial ^{230}Th in lacustrine carbonates from Lake Lahontan, USA: (1) detrital Th, (2) hydrogenous Th adsorbed to detritus and (3) hydrogenous Th directly incorporated by carbonates. Detrital Th is assumed to have a Bulk Earth value and be composed of aluminosilicates, however the total initial ^{230}Th/^{232}Th signal is possibly offset to high values by an abundant "hydrogenous" component, which has an additional source of ^{230}Th from *in situ* decay of uranium. Schramm et al. (2000) also identify a hydrogenous component in carbonates deposited in Lake Lisan (last glacial Dead Sea). Here, a sample of modern aragonite material, essentially free of insoluble detritus, had a finite age of 2556 ± 30 years indicating that substantial corrections must be made to ages to account for initial conditions.

A significant hydrogenous component of initial ^{230}Th was also discovered in a speleothem from the Bahamas by Beck et al. (2001). Using a correction for initial ^{230}Th based on measured ^{232}Th concentration and assumed Bulk Earth $(^{230}Th/^{232}Th)_A$ of 0.8 ± 0.8, they observed that samples with high ^{232}Th concentrations had significantly older ^{230}Th ages than proximal samples with lower ^{232}Th concentrations, indicating that a greater value of $(^{230}Th/^{232}Th)_0$ was required. In all cases, no insoluble residue or "detrital component" remained after dissolution in dilute acid and, for this reason, they considered the initial component of Th in meteoric waters to be dissolved or scavenged by colloids and co-precipitated with U in the calcite. Variation in $^{238}U/^{232}Th$ was substantial; two orders of magnitude range and sufficiently different between sub-samples of the same approximate age that ^{230}Th/^{232}Th of the scavenged component could be estimated using isochron techniques. Estimates of the present value of scavenged $(^{230}Th/^{232}Th)_A$ (or $(^{230}Th/^{232}Th)_0 \cdot e^{-\lambda_{230}T}$) were obtained by plotting the isotopic results for sets of 2 or more sub-samples that could be considered as coeval in $(^{238}U/^{232}Th)$-$(^{230}Th/^{232}Th)$-$(^{234}U/^{232}Th)$ space (Fig. 5). A mean value for $(^{230}Th/^{232}Th)_0 \cdot e^{-\lambda_{230}T}$ of 15.8 ± 2.3 (2σ_{mean}, n = 10) was obtained and used to correct measured ^{230}Th/^{238}U prior to calculation of ages (Fig. 6). This value is at least an order of magnitude higher than commonly used values for initial Th corrections and careful consideration should be given to previously published analyzes where the measured ^{230}Th/^{232}Th activity ratio is <~200 and contamination appeared to be insignificant based on lack of insoluble detritus.

One might expect the ^{230}Th/^{232}Th ratio of meteoric waters in the shallow vadose zone of a middle Pleistocene age carbonate platform such as Grand Bahama to be high because the carbonate sediments in the dune ridges that house the cave system are relatively free of silicate material and of sufficient age that ^{230}Th and ^{238}U are expected to be near secular equilibrium. Carbonate material dominates, with a much greater ^{230}Th/^{232}Th than Bulk Earth. Dissolution of this material and subsequent adsorption of hydrogenous Th by Fe-oxyhydroxides and colloids results in transport of Th with elevated ^{230}Th/^{232}Th values to the speleothem. Similar middle Pleistocene oolitic-peloidal deposits from Eleuthera, Bahamas yield ^{230}Th/^{232}Th of 3.43 × 10^{-4} (Hearty et al. 1999). It should be noted that even though a high initial ^{230}Th/^{232}Th value was derived for the Bahamas speleothem GB-89-24-1 (Beck et al. 2001) and corrections of as much as 3 ka were applied to some sub-samples, many of the sub-samples had such low ^{232}Th concentrations that corrections were negligible. The resulting initial ^{230}Th corrected ages present a remarkably coherent chronology (Fig. 6).

Zero(?) age deposits and waters. Perhaps the most sensitive test of initial conditions for speleothems and analogous carbonate deposits is to undertake U-series measurements of the most recent calcite in actively growing samples such as straw stalactites and the waters that feed them. Surprisingly, few such measurements have been published.

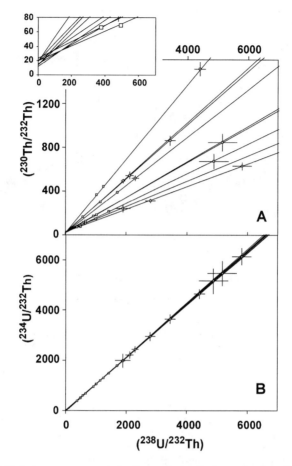

Figure 5. U and Th isotope measurements for 9 coeval sets of sub-samples from GB-89-24-1, Grand Bahama (Beck et al. 2001) are plotted in (A) $(^{230}\text{Th}/^{232}\text{Th})_A$-$(^{238}\text{U}/^{232}\text{Th})_A$ and (B) $(^{234}\text{U}/^{232}\text{Th})_A$-$(^{238}\text{U}/^{232}\text{Th})_A$ space (Rosholt Type-II' diagram). Intercepts of isochrons with abscissa in A are equivalent to $(^{230}\text{Th}/^{232}\text{Th}_{\text{init}})\, e^{-\lambda 230 t}$ of the scavenged Th component, while the gradient is equivalent to $(^{230}\text{Th*}/^{238}\text{U})$ and can be used to calculate the age T. Ages calculated using isochrons are in correct stratigraphic order. Using T for each isochron, a weighted mean $(^{230}\text{Th}/^{232}\text{Th})_0$ activity ratio of 18.67 ± 2.9 ($2\sigma_m$) was obtained, equivalent to an atomic ratio of 1.04×10^{-4} and significantly greater than bulk earth atomic ratio of 4.4×10^{-6}. This value was subsequently used to correct the measured $(^{230}\text{Th}/^{238}\text{U})$ in all sub-samples (Fig. 6) The gradient in B is equivalent to the present value of $(^{234}\text{U}/^{238}\text{U})_A$. Correction to initial values, using T, yields $\delta^{234}\text{U}(0) = 59 \pm 6$ ($2\sigma_m$), which is indistinguishable from the mean value of all other sub-samples; 59 ± 3 ($2\sigma_m$, n = 56). No significant relationship between T and $\delta^{234}\text{U}(0)$ was found.

Whitehead et al. (1999) present finite alpha-spectrometric ^{230}Th-^{238}U and ^{231}Pa-^{235}U ages >2 ka for two straw stalactites considered to be actively forming at the time of sampling and declare that ^{230}Th and ^{231}Pa contamination at zero age is a possible limitation of U-series dating methodology, however, these samples were uncorrected for initial Th, which is known to be relatively high ($^{232}\text{Th}/^{238}\text{U}$ = 0.0283 and 0.0126), at least two orders of magnitude higher than typical surface corals ($^{232}\text{Th}/^{238}\text{U}$ < 0.0003), for example.

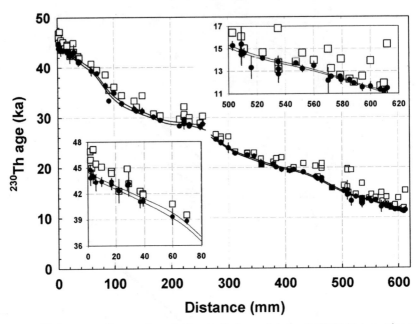

Figure 6. ^{230}Th age vs. distance along the longitudinal growth axis of GB-89-24-1, a submerged speleothem from Grand Bahama (Beck et al. 2001). Estimates of ^{230}Th age are corrected for initial Th using $(^{230}Th/^{232}Th)_0$ activity ratio with Bulk Earth value of 0.8 ± 0.8 ppm (open squares) and significantly higher value of 18.67 ± 2.9 based on isochron results illustrated in Figure 5 (filled squares). The latter correction results in a smooth, monotonic distance-age relationship.

Richards (1995) analyzed U and Th isotopic concentrations in a straw stalactite (TC-92-3) and an unfiltered sample of dripwater from Middle Caicos, British West Indies (Table 1). Further analyzes of waters during the summer of 1999 confirmed the elevated value of $(^{230}Th/^{232}Th)_A$ in drip waters (average = 15.0 ± 7.7, 1σ, n = 19), which is significantly higher than the value of Bulk Earth (~0.8). This value is similar to that predicted using isochrons for a speleothem from Grand Bahama and indicates that $(^{230}Th/^{232}Th)_A$ is high in pure carbonate terrain and initial ^{230}Th must be investigated in detail. In a $^{230}Th/^{232}Th$ vs. 1/[Th] plot for the Turks and Caicos dripwaters (Fig. 7) it can be seen that the points sit on or above a mixing plane between two or more components, Bulk Earth and various radiogenic "hydrogenous" phases.

In a detailed study of numerous suites of co-eval sub-samples from speleothems from Soreq Cave, Israel, Kaufman et al. (1998) demonstrate that Th is positively correlated with Si, Fe and Al, which suggest that ^{232}Th is associated primarily with

Table 1. U, Th data for active straw stalactite and drip water from Conch Bar Cave, Middle Caicos

Sample	^{238}U (ng g^{-1})	^{232}Th (ng g^{-1})	$(^{230}Th/^{232}Th)_A$	Th/U	$\delta^{234}U(0)$ (‰)	$(^{230}Th/^{238}U)_A$	Uncorrected Age (ka)
TC-92-3B	724.3	0.10	18.3±2.2	.000138	24.3±1.5	80(±1)×10^{-4}	0.09±0.01
TC-92-H20	0.3884	2.7×10^{-4}	5.18±0.85	.000695	97.4±2.0	120(±2)×10^{-4}	0.12±0.02

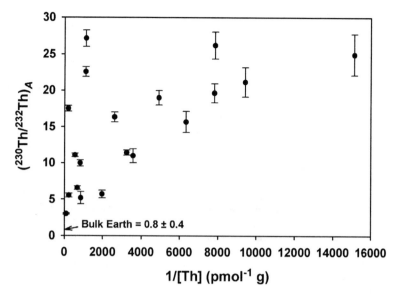

Figure 7. $(^{230}Th/^{232}Th)_A$ vs. 1/[Th] of dripwaters from Conch Bar Caves, Middle Caicos, British West Indies. In all cases dripwaters have $(^{230}Th/^{232}Th)_A > 0.8$ (bulk Earth value).

detrital silicates and ferric-oxyhydroxides. Isochron analyzes reveals a R_0 value of 1.3 to 2.9. Assuming secular equilibrium, this suggests a $^{232}Th/^{238}U$ atomic ratio of 1.08 to 2.4, which is below that of the average crustal value. Analysis of filtered and unfiltered cave drip waters from the same location indicate that ^{232}Th in this environment is associated with detrital particulates, but they also discovered that such material has adsorbed large quantities of U, as has been found for colloids and ferric-oxyhydroxides in rivers and estuaries (Porcelli and Swarzenski 2003; Swarzenski et al. 2003).

2.3. Closed system decay

One of the major assumptions of the ^{230}Th and ^{231}Pa dating schemes presented above is that the U-series radionuclide system should be closed to post-depositional migration or addition constituent nuclides. Authigenic material that shows evidence of weathering should be avoided. Also, any sign of recrystallization, such as conversion from aragonite to calcite, is indicative of potential nuclide migration, as is alteration from "length-fast" palisade low magnesium calcite to blocky, equant low magnesium calcite (Bar Matthews et al. 1997). Dense, macrocrystalline calcite provides the most reliable material.

After material has been analyzed, an extra degree of confidence can be ascribed to the ages if a combination of the following criteria are met:

(1) Age determinations are consistent with their stratigraphic position, either within the sample itself or in the context of the surrounding deposits and sample location.

(2) Combined ^{230}Th-^{238}U and ^{231}Pa-^{235}U dating techniques yield concordant results.

(3) Agreement is found for ages determined on the same material by independent dating methods

In addition to the above, one should also look for internal consistency of the derived values of initial $^{234}U/^{238}U$ and also demonstrate that secular equilibrium conditions have been established and maintained in material >1 Ma, if available, from the same location.

Stratigraphic consistency. The advent of high-resolution U-series techniques has brought about the standard practice of measuring multiple sub-samples along the growth axis of speleothems to provide detailed chronologies and growth rate information. Such data provide an opportunity to look for internal consistency in the derived ages. Beck et al (2001) made 81 U and Th measurements on sub-samples from the growth axis of a speleothem from the Bahamas, which is composed of clear dense calcite with no evidence of post-depositional alteration. After correction of the U and Th isotopic results for the presence of initial ^{230}Th contamination (see above), a smooth relationship is observed between longitudinal position and age, with no statistically significant age reversals (Fig. 6).

The most susceptible material for post-depositional loss or addition of radionuclides is the outer layer of samples that have been exposed to moisture for a long duration. Stratigraphic consistency between ages of the outermost material and that deposited prior to this provides valuable constraints on the technique. Four ages were derived for a band of clear, white calcite deposited on a stalactite from 53.6 m below sea level in a blue hole of Andros, Bahamas (Richards et al. 1994). Isotopic data for the outermost surface, which had been exposed to sea water for at least 8 ka was indistinguishable from the internal material (Fig. 8).

Combined ^{231}Pa-^{235}U ***and*** ^{230}Th-^{234}U-^{238}U ***dating.*** For secondary carbonate deposits less than about 200 ka, finite ^{230}Th and ^{231}Pa ages can be derived using the Equations (4) and (2), respectively. A combined ^{230}Th/^{231}Pa age can also be derived (Eqn. 5). Based on the assumptions that negligible Pa and Th was incorporated in the secondary deposit at

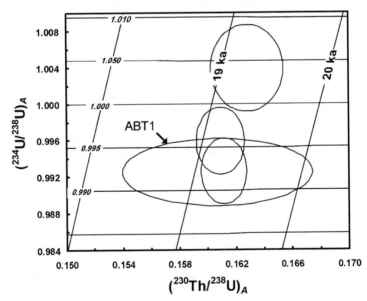

Figure 8. TIMS ^{230}Th ages of latest phase of growth of GB-89-23-2 (Richards et al. 1994). Outermost sample (ABT1; n = 4) indistinguishable from ages of older material up to 10 mm inside outer surface of this stalactite. GB-89-23-2 was collected from a depth of 53.6 m below present sea level in Stargate, Andros, Bahamas, and was submerged since inundation by rising sea levels during the last deglaciation.

the time of formation and decay has proceeded as a closed system, one should expect the ages to agree within analytical precision. If there has been significant episodic or continuous loss or gain of U, Th and Pa since deposition, discordant ages will result. Isotopic data for material with initial ^{230}Th/^{238}U and ^{231}Pa/^{235}U = 0 and closed system decay will plot on an evolution curve, which has a form dictated by initial ^{234}U/^{238}U. Combined measurements of U, Th and Pa measurements can be used to investigate open system behaviour and diagenesis in much the same way as has been used in U-Th-Pb system (Allègre 1964; Ku et al. 1974; Szabo 1979; Kaufman et al. 1995). Prior to the development of high-precision mass-spectrometric techniques, however, the precision of alpha-spectrometric techniques was insufficient to investigate many dating problems using Pa.

Early attempts to demonstrate concordance between speleothem ages derived from each of the ^{238}U and ^{235}U decay chains were presented by Thompson et al. (1976) and Gascoyne (1985), who used ^{227}Th to deduce the activity of ^{231}Pa. This is possible because the half-life of the intermediate daughter, ^{227}Ac, is only 22 yr and material older than a few hundred years has ^{227}Th activity in equilibrium with that of ^{231}Pa. They demonstrated reasonable concordance in some samples, but overall results were equivocal, possibly as a result of coprecipitation of ^{231}Pa during formation, and thus excess ^{227}Th, in some samples (see also discussion in Whitehead et al. 1999). Shen (1996) was more successful in applying the technique to speleothem samples from caves of China and demonstrates reasonable concordance for a suite of sub-samples with ages up to and beyond 200 ka. The analytical uncertainties, however, are generally greater than 10%.

Now that ^{231}Pa measurements can be obtained with precisions approaching that of ^{230}Th measurements on material deposited during the past 50 ka, researchers are in a much better position to verify their U-Th chronologies by testing for age concordancy and diagenetic processes. Using a modified technique of that developed by Pickett et al. (1994), Edwards et al. (1997) tested the accuracy of ages that had been obtained using the ^{230}Th chronometer. They obtained U, Th and Pa isotopic data for numerous corals to demonstrate closed system uranium-series decay from an initial state of ^{231}Pa = ^{230}Th = 0 and δ^{234}U$_{init}$ = 140‰. Also, two sub-samples from Devils Hole calcite vein, which had a ^{230}Th chronology suggesting an earlier timing for the penultimate deglaciation of (Ludwig et al. 1993) than that suggested by orbitally-tuned marine cores, provided concordant ages and similar δ^{234}U$_0$ of 1800‰.

Edwards et al. (1997) demonstrate that for samples with similar δ^{234}U(0), such as corals that grew in open marine water, and remained closed system since time of formation, isotopic data plot on or close to an ideal concordia in ^{231}Pa-^{235}U vs. ^{230}Th/^{234}U space. Concordia represent the loci of points for which ^{231}Pa and ^{230}Th ages are identical for a given δ^{234}U(0), and are more useful for systems with constant δ^{234}U(0) than deposits formed from meteoric water where it is well known that secular variation of δ^{234}U(0) is exhibited because of the changing nature of water-rock interaction and chemistry of percolating waters. Where δ^{234}U(0) varies significantly in individual samples or between samples, concordance is better demonstrated on a ^{231}Pa versus ^{230}Th age plot. The suitability of speleothem calcite for uranium-series dating is well-demonstrated by Figure 9, where isotopic data and ages for material analyzed in the past few years at the University of Minnesota (Beck et al. 2001, Musgrove et al. 2001, Dorale et al. 2003) plot on the ^{230}Th = ^{231}Pa line (Fig 9B). The same data are plotted in ^{231}Pa-^{235}U vs. ^{230}Th/^{234}U space in Fig 9A. Generally, data for individual samples plot on or near concordia indicating limited secular variation of δ^{234}U(0), however, two sub-samples from a speleothem from Crevice Cave, Missouri have different δ^{234}U(0) of 3007 at 89.7 ka and 2589 at 78.9 ka.

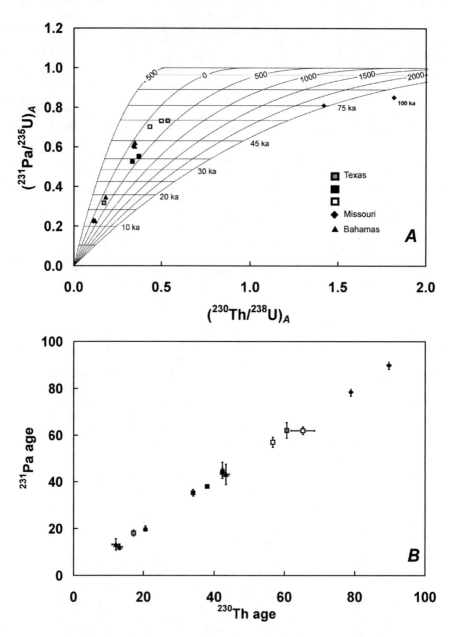

Figure 9. U, Th and Pa isotopic data and ages for speleothems (Beck et al. 2001; Musgrove et al. 2001, Dorale et al. 2003). A.) $(^{230}Th/^{238}U)_A$ vs. $(^{231}Pa/^{235}U)_A$ concordia plot. Evolution curves for different $\delta^{234}U(0)$ are shown. Horizontal lines represent isochrons. B.) ^{230}Th vs. ^{231}Pa ages for the same samples. All data plot within error of the 1:1 line, illustrating age concordance.

A unique strength of concordia diagrams is their ability to reveal modes of diagenetic disturbance. If U-Th-Pa isotopic data for numerous co-eval sub-samples that have experienced different degrees of the same diagenetic process are plotted in $^{231}Pa/^{235}U$ or $^{230}Th/^{238}U$ vs. $^{234}U/^{238}U$ space, they should fall on a curve or straight line that can be used to estimate the true crystallization age and the mode of disturbance. This is similar to methods used in U-Pb geochemistry and is described in detail by Cheng et al. (2000), who illustrate the use of combined U-Th-Pa data to investigate open-system behavior in corals from a last interglacial terrace on Barbados. To the best of our knowledge similar methods have yet to be applied to speleothems.

Comparison with independent "dating" methods. Advances in technology during the past have made uranium-series dating potentially the most precise and/or highest resolution method for dating carbonate material. Alternative radiometric methods are often fraught with additional complications. Only AMS ^{14}C dating can achieve comparable precisions, but as we discuss in Section 4.2, *a priori* assumptions of initial ^{14}C concentrations in speleothem are poorly constrained and accuracy is less reliable. However, confidence can be ascribed to a uranium-series age determination if consistent timing is observed between regional or global events recorded by trace elements or stable isotopes in a speleothem and that observed in other archives, such as ocean cores or ice cores. For example, McDermott et al. (2001) and Baldini et al. (2002) observe a $\delta^{18}O$ and trace element anomaly (~8.3 ka) in a speleothem from western Ireland (Fig. 1) that is synchronous, within 2σ dating uncertainties, with an abrupt and short duration shift to cold climate in the high latitude Northern Hemisphere as recorded in Greenland ice cores (Grootes et al. 1993, Alley et al. 1997).

Dramatic shifts in ocean or atmospheric circulation can result in shifts in $\delta^{18}O$ of precipitation falling above the cave location. Speleothems, therefore, can be used to infer timing of regional or global scale changes in past climate. A high degree of synchroneity between the timing of abrupt shifts of $\delta^{18}O$ records in speleothem calcite and marine and ice cores for the last interglacial-glacial period, therefore, has been also used to indicate the reliability of uranium-series ages speleothems (Dorale et al. 1998, Bar-Matthews et al. 1998; McDermott et al. 2001, Burns et al. 2001 Spötl and Mangini 2002; Wang et al. 2002 *inter alia*; see Section 4.2). We should stress, however, that observation of apparent synchroneity between events in speleothems and other records does not provide unequivocal constraints on the age determination because there exists the possibility of diachronous response to the same forcing events. Researchers should also be aware that in many cases the marine cores and ice cores records are themselves reliant on dating methods that are subject to debate, such as ice flow models, radiocarbon and orbital tuning.

2.4. $^{234}U/^{238}U$ dating methodology

^{230}Th-^{234}U-^{238}U and ^{231}Pa-^{235}U techniques are only applicable to deposits typically less than ~500 ka and ~200 ka respectively. The half-life of ^{234}U is such that the dateable age range can be extended to at least 1 Ma (using Eqn. 3) if robust estimates of $\delta^{234}U(0)$ can be made for a particular cave system. However, speleothems exhibit significant secular and intra-regional variation of $\delta^{234}U(0)$ in most settings because of changes in water-rock interaction times, soil pCO_2, pH and other climate- and hydrology-related factors, thus precluding the use of this technique for dating purposes in most cases (Harmon et al. 1975). Some have advocated the use of a regional best estimate to obtain low precision ages where numerous modern drip waters and material with resolvable U/Th ages from the same cave or region show little variation in $\delta^{234}U(0)$. For example, Gascoyne et al. (1983) showed that the standard deviation of $\delta^{234}U(0)$ for numerous dated speleothems from Victoria Cave, northwest England was sufficiently low that reasonable age estimates could be obtained for material up to 1.5 Ma. Successful attempts have been made to use this technique under marine (Bender et al. 1979; Ludwig et al. 1991), groundwater (Winograd et al. 1988; Ludwig et al. 1992), lacustrine (Kaufman 1971) and soil (Ludwig and Paces 2002) settings where $\delta^{234}U(0)$ is more likely to remain reasonably stable for long periods of time. Ludwig et al. (1992) adopted a mean $\delta^{234}U(0)$ value of 1750 ± 100 to calculate ages >360 ka for sub-samples from a calcite vein from Devils Hole, Nevada. The uncertainty propagated through the age equation encompasses the full range of variation of $\delta^{234}U(0)$ for ages younger <360 ka.

2.5. U-Th-Pb dating of secondary carbonates of Quaternary age

Standard ^{230}Th-^{234}U-^{238}U dating techniques can only be applied usefully to material that satisfy the above criteria and have activity ratios that are distinguishable from unity. For speleothems, this means that the upper age limit is <~500 ka, unless the calcite precipitated from waters with extremely high initial values of ^{234}U/^{238}U (see Fig. 2). Speleothems and other secondary carbonate deposits such as tufa, travertines, lake carbonates and vein calcites have the potential to provide valuable information about past climate, hydrogeochemistry, landscape development and hominid evolution during the early Quaternary and Tertiary periods. Electron-spin resonance (Grün 1989, Rink 1997) and ^{234}U-^{238}U disequilibrium methods (Ludwig et al. 1992) have proved to be useful in extending the dateable range beyond 0.5 Ma but suffer from lack of precision or poorly constrained initial conditions. An alternative geochronological technique is U-Th-Pb dating, which uses the ingrowth of the stable Pb isotopes ^{206}Pb and ^{207}Pb from the decay of their respective long-lived parent isotopes, ^{238}U and ^{235}U. U-Th-Pb dating techniques are more commonly used for much older deposits, indeed up to the age of the Earth. Since the work of Moorbath et al. (1987), U-Pb and Pb-Pb methods have been applied to various ancient carbonates (Paleozoic and Mesozoic), such as limestones (Smith and Farquhar 1989), metamorphosed marbles (Jahn and Cuvellier 1994), calcite concretions (Israelson et al. 1996), paleosol calcite (Rasbury et al. 1997) and scalenohedral calcite (Lundberg et al. 2000). For deposits less than a few million years, the amount of radiogenic Pb present is very small because of the extremely long half-lives of the parent isotopes and, until recently, it has been considered impractical to obtain ages with reasonable precision. Getty and DePaolo (1995) presented a novel method for dating Quaternary volcanic rocks using U-Th-Pb techniques and Richards et al. (1998) demonstrated the feasibility of using U-Th-Pb isochron methodology to date young carbonate materials with high U concentrations, very low Pb concentrations, and sufficient range in U/Pb ratios. More recently, preliminary U-Th-Pb ages have been obtained for Quaternary sedimentary carbonates from Olduvai Gorge that are associated with hominid remains (Cole et al. 2001), speleothems with extremely high U concentrations (> 300 ppm) from Spannagel Cave, Austria (Cliff and Spötl 2001) and corals of middle Pleistocene age from Costa Rica (Getty et al. 2001). The most extensive U-Th-Pb dating study on deposits of Quaternary age is that conducted by Neymark et al. (2000) on sub-millimeter-thick opal (30 to 313 ppm U) that coats fractures and cavities in Tertiary Tuffs at Yucca Mountain Nevada.

Criteria for successful U-Th-Pb dating are similar to those of standard U-series techniques, but in addition (1) sufficient amounts of ^{206}Pb and ^{207}Pb must have accumulated by radiogenic ingrowth to be distinguished from the initial Pb that is always present at the time of formation and (2) the decay scheme must be closed for all intermediate daughters from ^{230}Th and ^{231}Pa to stable ^{206}Pb and ^{207}Pb, respectively.

Isochrons can be used for coeval samples with a range of U-Pb ratios to determine the age and initial Pb ratio in a similar manner to that used in standard U-Th techniques where initial Th is significant. Ideally, sub-samples of pristine calcite from the same growth layer and exhibiting a considerable range of U/Pb ratios can be used to define isochrons in either $^{206}Pb/^{204}Pb$-$^{238}U/^{204}Pb$ or $^{207}Pb/^{204}Pb$-$^{235}U/^{204}Pb$ space. The gradient of the isochrons will define $^{206}Pb*/^{238}U$ and $^{207}Pb*/^{235}U$, where $^{206}Pb*$ and $^{207}Pb*$ are the radiogenic Pb components. Alternatively, three-dimensional isotope plots, which contain complete information about concordance between the two decay schemes and common Pb (Wendt 1984) can be used. We suggest the use of the variant developed by Wendt and Carl (1985) from Tera and Wasserburg (1972), as implemented in ISOPLOT 2.92 (Ludwig 1994, 1999).

For most geochronological applications of U-Pb methodology, departures from initial secular equilibrium of the U-series decay chain at time of precipitation are generally considered to be insignificant relative to subsequent radiogenic ingrowth. For young samples, however, this becomes a significant consideration (Ludwig 1977; Wendt and Carl 1985) (Figs. 10, 11). It should be noted that with increasing precision of U and Pb measurements even mineral ages as great as 380 Ma can be significantly affected by initial ^{230}Th excess (Amelin and Zaitsev 2002). While this can present a problem to the geochronologist, Oberli et al. (1996) recognize that initial disequilibria can be used to provide detailed information about the nature of chemical fractionation processes involved at the time of formation and perhaps the subsequent metamorphic history.

We have considered initial uranium-series conditions in Section 2.2 above in relation to standard U-Th-Pa techniques. Intermediate daughters such as ^{230}Th, ^{231}Pa are insoluble and strongly adsorbed on suspended sediments such that their incorporation in calcite is generally negligible ($^{230}Th_0 = ^{231}Pa_0 = ^{227}Ac_0 = 0$). In nearly all cases, ^{234}U is precipitated in excess of that expected for secular equilibrium with ^{238}U. This must be taken into account by using an independent estimate of $\delta^{234}U(0)$. The effects of $^{234}U/^{238}U$ disequilibrium on calculated age can be seen in Figures 10 and 11, which illustrates the situation for a range of $\delta^{234}U(0)$ typical of calcite deposited from freshwater. For a given $^{238}U/^{206}Pb$ and $^{207}Pb/^{206}Pb$ the corresponding concordant age can vary significantly depending on initial conditions. It is likely that there will be some ^{226}Ra initially coprecipitated with U because excesses of ^{226}Ra over its parent ^{230}Th have been reported for soils (Dickson and Wheller 1992; Olley et al. 1997), however, the effect of any initial abundance is minor because of the short half-life of 1600 years and is likely to be irrelevant for the applicable age range of this technique.

In addition to the assumptions of initial conditions, the validity of U-Pb methodology relies on closed system behavior of U, Pb and intermediate nuclides in the decay chain. Concordance between the two U-series decay chains is most likely to be compromised by Rn loss because Rn is the only gas in the decay chains and has a high diffusivity. Radon-222 in the ^{238}U decay chain has a half life of 3.8 days. This is much longer than the half-life of ^{219}Rn (3.96 s) in the ^{235}U decay chain. Therefore, partial loss of Rn will give rise to an apparent $^{206}Pb*/^{238}U$ age younger than the true age, whereas the $^{207}Pb*/^{235}U$ will remain unaffected. There is evidence that Rn loss is negligible from speleothems composed of dense calcite with columnar crystal fabric (Lyons et al. 1989). Richards et al. (1998) investigated the possibility of Rn loss in a stalactite from the Peak District by measuring the activity of ^{230}Th and ^{210}Po, which are either side of ^{222}Rn in the ^{238}U decay chain and expected to be in secular equilibrium for material much greater than 20 ka. The measured ($^{210}Po/^{230}Th$) activity ratio was 0.942 ± 0.064, which is within error of unity. They obtained a $^{206}Pb*/^{238}U$ age indistinguishable from a $^{230}Th/^{238}U$ age for this stalactite using three-dimensional isochron methodology (Ludwig 2003), indicating that U-Th-Pb dating of Quaternary age samples is possible (Figs. 12, 13). Not all calcite is suitable for U-Th-Pb dating, however, because in many cases too much initial Pb is coprecipitated. Ideal material has very high U and low Pb, so that radiogenic Pb can be distinguished from initial Pb. A survey of selected pristine calcite from various locations of the globe

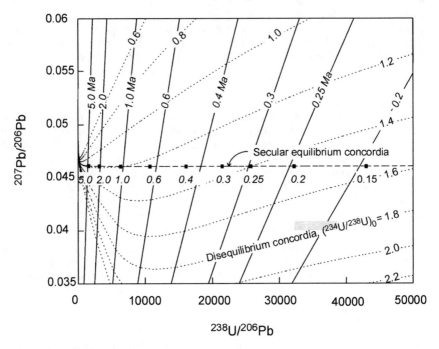

Figure 10. Modified Tera-Wasserburg (^{207}Pb/^{206}Pb-^{238}U/^{206}Pb) plot with disequilibrium concordia (after Wendt and Carl 1985). The curvilinear trajectories (dotted lines), or concordia, show the change in ^{238}U/^{206}Pb and ^{207}Pb/^{206}Pb as age increases for different values of δ^{234}U(0) and ^{231}Pa$_0$ = ^{230}Th$_0$ = ^{226}Ra$_0$ = 0. The near vertical lines (solid) are isochrons. Ages reported in Ma. The secular equilibrium case, where the activity ratios of all parent-daughter pairs in the U-series decay chains are unity is shown by a dashed line. Ages (in Ma) are represented by squares.

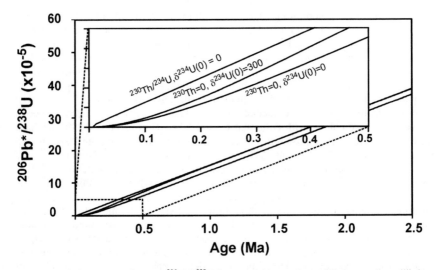

Figure 11. Evolution curves of age vs. ^{206}Pb*/^{238}U for material in secular equilibrium or disequilibrium.

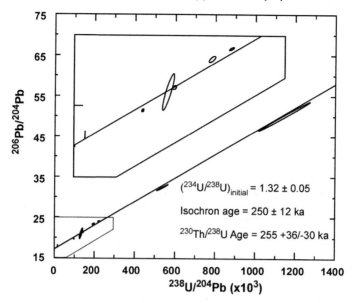

Figure 12. Uranium-lead (^{238}U/^{204}Pb-^{206}Pb/^{204}Pb) isochron plot for the WHC1 stalactite from Peak District, England (Richards et al. 1998).

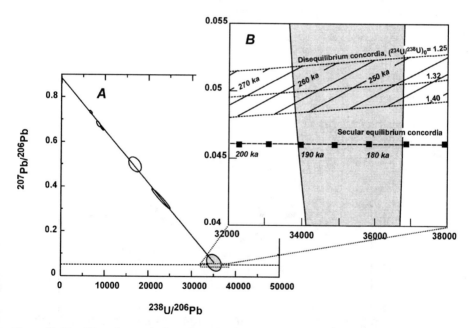

Figure 13. Tera-Wasserburg style three-dimensional concordia plot for the WHC1 stalactite (Richards et al. 1988). Shown is the projection onto the ^{238}U/^{206}Pb- ^{207}Pb/^{206}Pb plane (see Fig. 10 caption). The intercept of the three-dimensional fit with this plane is shown by the shaded error ellipse. A) Three-dimensional regression, disequilibrium concordia (δ^{234}U(0) = 320; dashed line], and plane intercept. B) Detail of intercept with ^{238}U/^{206}Pb-^{207}Pb/^{206}Pb plane, disequilibrium concordia (2σ error), regression line, and the secular equilibrium concordia.

using standard ICPMS coupled with data from the literature reveals an extremely large range in U/Pb ratios of eight orders of magnitude (Fig. 14). Clearly, pre-screening for elemental abundance is recommended prior to isotopic analysis.

3. SPELEOTHEM GEOCHRONOLOGY IN PRACTICE

The morphology, crystallography, mineralogy, purity and growth rates of speleothems vary tremendously in relation to their location and the chemistry of waters feeding them. The potential suitability of speleothems is dependent on the intended application. Paleoclimate records, for example, are best derived from samples from the deep-interior of caves that exhibit continuous growth of dense calcite with a simple internal morphology (flowstones and stalagmites), whereas speleothem intercalated with hominid remains are likely to be found near cave entrances and rock shelters. A review of potential applications can be found below in Section 4. Here, we present field sampling strategies, sub-sampling methodology, and post-analytical treatment of ages. Analytical procedures for chemical separation of uranium-series nuclides and instrumental procedures for obtaining high-precision estimates of isotopic abundance are discussed by Goldstein and Stirling (2003). Statistical treatment of isotopic data is also dealt with in Ludwig (2003).

3.1. Speleothem sampling strategy

At the outset here, we reiterate that collection of material from caves is a highly sensitive issue and at all times researchers must be aware of conservation issues regarding cave deposits. To quote from the National Speleological Society (USA) Code of Conduct, scientific collection should be "professional, selective, and minimal."

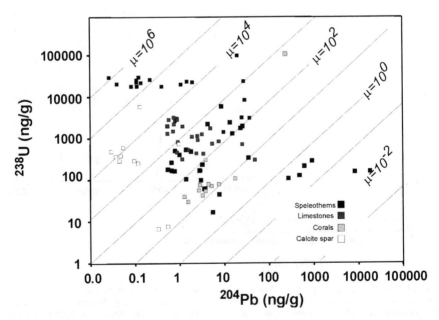

Figure 14. Comparison of U, Pb concentrations in speleothems from the Bristol archive with published data of limestones, corals and calcite spar (Smith et al. 1991; Jones et al. 1995). Eight orders of magnitude variation in $^{238}U/^{204}Pb$ ratio is observed.

***A priori* considerations.** An extensive review of the form and mineralogy of the secondary cave deposits is provided by Hill and Forti (1997), but also see Ford and Williams (1989). By far the most useful speleothems for geological purposes are the most common forms; stalactites, stalagmites and flowstones composed of calcite, and occasionally aragonite. Other deposits that have been used to derive useful information include travertines, gypsum crusts, calcite veins and scalenohedral calcite (dog-tooth spar).

In a discussion of the suitability of speleothems for dating purposes, Gascoyne and Schwarcz (1982) describe the various types of speleothem that form in the different "zones" of a cave from the entrance to the deep interior. The entrance (or "twilight") zone contains sporadic, fast-growing speleothems that are the often the result of evaporative processes. Alteration by microbial action in biofilms (Jones 1995) is often displayed within this zone, and samples have high, detrital and organic content. Here, speleothems often display a degree of porosity and are contaminated with detritus. Samples from the deep-interior, where humidity levels are at their highest and air circulation is non-existent, are the slowest growing, and are generally composed of the dense calcite and free from detritus, unless the site is subject to periodic flooding. To these subaerial zones could be added the subaqueous zones where phreatic calcite is formed under low temperature conditions or associated with hydrothermal fluids (e.g., Devils Hole vein calcite).

The suitability of speleothems for analysis is determined by their mineralogy and fabric. Material displaying unaltered or primary fabric are required if chronological or paleoenvironmental information is sought. Such mineral fabrics include coarse columnar (or palisade) calcite, equant calcite, microcrystalline calcite, botryoidal clusters of elongate (acicular or fibrous) aragonite crystals. Kendall and Broughton (1978) present an overview of the common calcite fabrics (see also González et al. 1992; Frisia et al. 2000). We suggest routine observation of thin sections because they can reveal the extent of porosity, primary fabric, contaminants, and diagenetic history (Railsback 2000).

Sub-sampling of calcite for dating purposes. Sub-sampling for chronological purposes is not a trivial matter. The primary consideration is quality of material, but also important is the requisite mass and dimensions of sub-samples, which is based on a priori consideration of desired precision. Ideally, the spatial resolution of sampling is based on a compromise between the growth period represented by the sub-sample and the achievable precision. For U-Th ages this is a function of U concentration, $\delta^{234}U(0)$, sample age, chemical yield and the ionization/transmission efficiency of the adopted instrumental procedure. The most efficient strategy here is to measure the U concentration of selected sub-samples prior to isotopic analysis. While U concentration can vary dramatically between samples (Fig. 14), there is much less variation along the growth axis of individual samples. It is only after subsequent dating that growth rate can be determined, at which point adjustment of the growth distance represented for further sub-samples can be made. Improvements in ionization and transmission efficiency afforded by plasma source technology (Goldstein and Stirling 2003) have served to further reduce sample sizes, enabling dating of thin cement overgrowths and slow-growing speleothems. Also, Stirling et al. (2000) have demonstrated the feasibility of analyzing speleothems with greater than 100 ppm U by direct *in situ* laser ablation with a spatial resolution of 10-50 μm.

Sub-samples of calcite of between 0.01 and 1 g are usually adequate for most dating applications using mass-spectrometry. Powders can be obtained using micro-drills or wafers can be cut from a slab using a diamond wire saw (0.17-0.3 mm kerf). Milling is generally less destructive to the formation and specific growth layers can be precisely traced. The advantage of sawing is that there is less chance for contamination, and sub-samples can be ultrasonicated to remove cutting debris and detrital material. Growth layers can be followed by adjusting the cutting axis using a micrometer or goniometer stage.

Many samples exhibit growth layers. These may be a result of linear arrays of fluid inclusions, changes in crystal fabric, entrapment of surface contamination such as clays, or changing organic/mineral content of solutions feeding the sample. In many cases, these can provide a useful guide to sub-sampling because they define the past record of growth morphology. We note here that growth layers have been observed at annual intervals or less for some samples using visual (Railsbeck et al. 1994), elemental (e.g., Roberts et al. 1998, 1999; Huang et al. 2000; Baldini et al. 2002) or fluorescence (e.g., Baker et al. 1993, Shopov et al. 1994, Polyak and Asmerom 2001). This could prove to be extremely valuable in terms of correlation with coral, tree-ring and ice core records; however, Betancourt et al. (2002) cautions that the manifestation of a consistent long-term annual signal is likely to be rare because of variability in groundwater residence times and seasonal pattern of soil moisture conditions.

The number of sub-samples that must be obtained for individual speleothems depends on the specific application. When ages are to be used to constrain the periods of continuous growth or hiatuses, sub-samples from the base and outer surface may suffice. For applications based on high-resolution records of environmental change, a suite of ages must be obtained to estimate the distance-age relationship along the axis of growth.

3.2. Treatment of U-series ages

Speleothems offer considerable potential for obtaining high-resolution records of paleoenvironmental change; in fact sub-annual resolution can now be achieved for various proxy indicators. It is therefore important that robust chronologies are also obtained. In many cases speleothem chronologies rely on interpolated estimates of ages for calcite deposited between sub-sample locations. The strategy adopted to obtain a high-resolution chronology for continuous growth periods is often an iterative procedure that considers the variability in growth rate, achievable precision, amount of material sampled and the age derived. Where sample growth is extremely slow, such as opals, models need to be constructed based on integral equations of continuous deposition because standard ages often over-estimate the true age (Neymark and Paces 2000).

A survey of the most recent studies that use multiple uranium-series ages for individual speleothems reveals a range of practices in dealing with interpolation and extrapolation of derived ages. Linear regression, stepwise linear fits, polynomial fits and non-parametric smoothing splines are used to define the most likely age vs. distance relationship for continuous phases of growth. There are advantages with each of these approaches but weighted non-parametric smoothing splines are, perhaps, the most "natural" and "automatic" of curve fitting routines. Non-parametric smoothing methodology consists of techniques that make few assumptions about the form of an underlying function that needs to be estimated. Essentially, these techniques allow the dataset to "speak for itself" in determining the final estimate which is constructed by weighted local averaging, the extent of which is controlled by a smoothing parameter. Unfortunately, interpolation is not trivial because of the non-parametric nature of the function, and we recognize that low-order polynomial fits produce simple-to-use, or readily transferable, formulae for interpolating between dated sub-sample locations. Also, non-parametric smoothing techniques are computationally intensive, but standard statistics software such as S-PLUS® (Venables and Ripley 1999) can be easily used to derive of the age-distance relationship.

Demand for high-resolution records of environmental change from speleothems has led to a dramatic increase in the number of sub-samples required from individual samples to define the distance-age relationships along the axis of growth. Twenty years ago, the age determination of top and bottom sub-samples was generally considered adequate to constrain the growth period. In more recent times, however, many more sub-samples are utilized: Dorale et al. (1998) reported 38 ages (74.0 to 21.4 ka) for sub-samples from 4 stalagmites from Crevice Cave, Missouri, one of which is supplemented by an additional 13 ages in Dorale et al. (2003); Wang et al. (2001) determined ages for 59 sub-samples from 5 samples from Hulu Cave, China (74.9 to 10.9 ka); Zhao et al. (2001) determined 17 sub-sample ages along the axis of a speleothem from Newdegate Cave, Tasmania (154.5 to 100.3 ka); Musgrove et al. (2001) produced ages for a total of 45 sub-samples from 4 samples from central Texas (7.6 to 71 ka); Beck et al. (2001) reported 81 ages (44 to 11 ka) from the growth axis of a single submerged speleothem sample from Grand Bahama. In most of the above examples, the age-distance relationship is represented by a piecewise linear fit. While this is convenient for interpolation purposes, growth rates cannot be adequately represented because of the stepwise nature of the resultant function. For this reason, we advocate the use of curvilinear fits, in particular smoothing splines. Researchers should be aware of the implicit assumptions that are adopted in the various approaches used for interpolation, or more strictly approximation, and extrapolation. Particular caution must be adopted if interpolated results are to be examined in the frequency domain, as is often the case in palaeoclimate research.

4. SPELEOTHEM CHRONOLOGY AND ENVIRONMENTAL CHANGE

There is a critical need to assess natural environmental change at increasingly higher temporal and spatial resolution to elucidate the pattern of response to internal and external forcings and feedbacks in the Earth's system. Speleothems and other secondary carbonate deposits are one of a host of different archives that can be used to provide such information for the past 0.5 Ma, a time period of dramatic and often abrupt change. Speleothem applications can be broadly classified into those based on (1) presence/absence or rate of speleothem growth, or (2) proxy information for environmental change derived from the isotope and chemical composition of the mineral deposits themselves. Essential to both of these classes is the determination of accurate chronologies using the methods described above. By obtaining reliable and high-precision chronologies of environmental and geophysical change for many sites across the globe using speleothems, comparison can be made with independent records from the terrestrial environment, oceans and cryosphere, so that the amplitude and spatial patterning of response to different forcing can be investigated. Below, we illustrate the wide range of possible applications focussing on the most recent literature. We deliberately avoid reviewing the fundamental controls on proxy evidence in each of the categories and concentrate on chronological aspects. For further information on the controls of paleoclimate evidence, we suggest that readers look at the upcoming review by Lauritzen (in press) (see also Gascoyne 1992).

4.1. Applications based on the presence/absence or growth rate of speleothems

Speleothem growth is obviously dependent on the continued supply of seepage water supersaturated with respect to calcite to a point location in a cave. Initiation and cessation of growth can be caused by a range of different factors such as changes in vadose water chemistry, aridity, flooding events, permafrost conditions and random shifts in flow routing. Reliable ages of the period of growth, therefore, can provide valuable spatial and temporal constraints on environmental factors. The spatial extent of the principal factor controlling presence or absence may be global (e.g., eustatic sea level change), regional (influence of advancing ice sheet during glacial periods) or local (e.g., back flooding of cave passage after sediment blockage downstream).

Palaeo-sea (and -water-table) fluctuations. Edwards et al. (2003) describe methods used to obtain valuable information about past sea level elevations and tectonics using U-series ages of well-preserved shallow marine carbonates from uplifted coral reef terraces (e.g., Barbados, Papua New Guinea and Vanuatu) and submerged coral sequences (e.g., Barbados and Tahiti). Most of the information derived from corals is related to either periods of sea level stasis or rise. Well-preserved speleothems found below present sea level can be used as complimentary evidence about the timing and elevation of *low* sea levels in the past because they could only have formed when cave passages were air-filled.

Alpha-spectrometric U-series ages determined for submerged speleothems from the stable platforms of Bermuda (Harmon et al. 1978, 1981, 1983) and Bahamas (Spalding and Mathews 1972; Gascoyne et al. 1979, Gascoyne 1984) provided valuable constraints for the timing of sea level fluctuations above −20 m during glacial and interglacial periods of the past 200 ka. Age uncertainties, however, were typically >5 %. The advent of mass-spectrometric techniques enabled much higher precision chronologies using slower growing samples, such as flowstones. Li et al. (1989) and Lundberg and Ford (1994), for example, presented high-precision ages for a flowstone sequence from −15 m in a cave on Grand Bahama that exhibited interruptions in deposition from 235-230 ka, 220-212 ka, 133-110 ka, 100-97 ka and <30 ka. These hiatuses in growth were caused by submergence during interglacial high sea stands during oxygen isotope stages 7, 5 and 1. Richards et al. (1994) and Smart et al. (1998) present results from a comprehensive suite of samples to depths of −60 m in numerous "blue holes" of the Bahamas that constrain the maximum elevation of sea-level fluctuations during the last and penultimate glacial periods. Continuous growth is observed from 70 to 16 ka at an elevation of −20 m, indicating that maximum sea level during all of isotope stage 3 parts of stages 2 and 4 were no higher than −20 m. Periods of submergence are represented by breaks in deposition and crystal terminations, followed by re-nucleation of calcite growth after sea level regression leaves cave passages air filled. In most cases, material is extremely well-preserved, showing no evidence of dissolution despite submergence for up to 50 ka (Fig. 15A,B). In some cases, however, outer surface may be truncated by dissolution in the aggressive waters of the mixing zone (Fig. 15C,D). Here, top ages for phases of growth may be represent maximum age constraints. Researchers must also be aware that initiation of growth after a hiatus may not be immediate because of paleoclimate control (e.g., regional aridity) or random shifts in flow routes.

Generally, submerged speleothems from inland caves provide the most suitable material for dating purposes because they are less susceptible to mechanical and chemical bioerosion by epilithic and endolithic organisms. Gascoyne et al. (1979) successfully obtained ages of poorly preserved samples by separating primary and altered components of carbonate. Borings and encrustation by marine organisms can, however, provide unambiguous evidence for submergence by rising sea levels. Antonioli and Oliverio (1996) obtained AMS ^{14}C ages for well-preserved fossil shells of the boring mussel *Lithophaga lithophaga* from within a submerged stalagmite from −48 m in a cave off the west coast of Italy. The oldest age obtained for one of the marine colonists was 9.6 ^{14}C kyr BP (or 10.3 ka) and is used to constrain the rate of sea level rise during the last deglaciation by comparison with the sea level curve derived from submerged corals in Barbados (Fairbanks 1989; Bard et al. 1990). Earliest colonization by serpulid worm communities on submerged speleothems from the Tyrrhenian coast of Italy has also been dated by radiocarbon techniques to constrain a regional sea level curve for the Mediterranean region (Antonioli et al. 2001). Recently, Bard et al. (2002a) reported high-precision TIMS ^{230}Th ages for calcite from the core of an encrusted stalagmite from the same coastline. Two continuous growth phases of dense calcite are exhibited with biogenic crust overgrowths formed during high sea-stands. This submerged sample from −18.5 m constrains the timing of the high sea-stand associated with marine isotope stage (MIS) 7.1 (Martinson et al. 1987) to between 201.6 ± 1.8 and 189.7 ± 1.5 ka, which

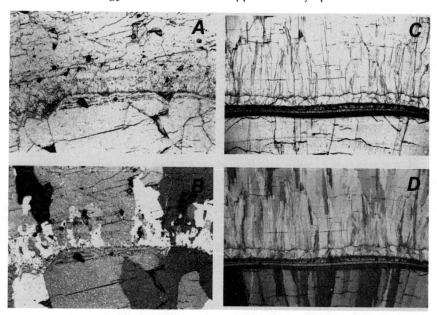

Figure 15. Thin section photomicrographs of hiatuses in Bahamas flowstone and stalagmite sequence from Sagittarius blue hole, Grand Bahama. A,B) GB-89-25-5A (plane and cross-polarized light, respectively; field of view = 2.6 mm). Hiatus represents stage 5 high sea level during which sample was submerged for at least 50 ka. Note well-preserved steep-sided crystal terminations. C,D) GB-89-27-1 (plane and cross-polarized light respectively; field of view = 2.6 mm). Hiatus in growth represents stage 9 high sea levels. Iron-rich red/orange flocculate is observed at hiatus (lepidocrosite and hematite) and crystal termination of lower phase of calcite growth is truncated.

agrees with an estimate of 190 ± 5 ka for sea level fall after stage 7 derived from a flowstone sequence from the Bahamas (Richards et al. 1992) and the youngest ages (200.1 ± 1.2 ka; 200.8 ± 1.0 ka) for the penultimate interglacial (MIS 7) coral reef terrace in Barbados (Gallup et al. 1994). Robinson et al. (2002) report ages as young as 178.4 ka for well-preserved aragonitic material from Bahamas slope sediments considered to be of MIS stage 7 age based on $\delta^{18}O$ data and cite confirmatory evidence of speleothem ages from Oman for an interglacial period from 210 to 180 ka (Burns et al. 2001).

Clearly, U-series ages of speleothems can provide valuable constraints for the timing of past sea levels and can be used in conjunction with marine evidence to test the hypothesis that glacial-interglacial fluctuations are related to solar forcing, but there remain some discrepancies and much attention is currently focused on the timing of the penultimate deglaciation, or Termination II. It was the long-held belief that increases in summer insolation in the high latitude regions of the Northern hemisphere were the major cause of the onset of deglaciation and oxygen-isotope records were tuned to changes in northern hemisphere insolation in the frequency domain. The predicted age of the midpoint of penultimate deglaciation (~128 ka) on the basis of such theory is significantly younger than suggested by uranium-series ages on speleothems and marine aragonite deposits. Bard et al. (2002a) report latest growth in a continuous phase of speleothem growth from the submerged Argentarola Cave prior to the penultimate deglaciation of 145.2 ± 1.1 ka, which constrains the maximum age of Termination II, while compilation of minimum ages for latest speleothem growth during the penultimate glaciation in the

Bahamas suggest a younger maximum constraint of >130 ka (Gascoyne et al. 1979; Richards et al. 1994; Lundberg and Ford 1994; Lundberg 1997). Ages for Termination II as recorded in the $\delta^{18}O$ record of Devils Hole calcite vein from continental USA predated estimates from Milankovitch theory by some 6 ka at 140 ± 3 ka (Winograd et al. 1992; Ludwig et al. 1992). An earlier termination has also been indicated by marine evidence: Henderson et al. (2000) dated aragonite deposits from the slopes of the Bahama Banks and obtained a well-constrained ^{230}Th age of 135 ± 2.5 ka for Termination II; Gallup et al. (2002) reported ^{230}Th and ^{231}Pa ages of 135.8 ± 0.8 ka for coral samples that were 18 ± 3 m below sea level at time of formation; Esat et al. (1999) obtained ages of corals the Huon Peninsula that suggest sea levels was at least as high as 14 m below present at 135 ka.

Most studies using speleothems to constrain past sea-levels rely on U-series ages of calcite that was deposited at times of low sea level. Vesica et al. (2000) and Fornós et al. (2002), however, present mass-spectrometric ^{230}Th ages for phreatic calcite overgrowths that form on previously vadose speleothems in Mallorca that were submerged by brackish water during high sea stands. Phreatic overgrowths have been documented for each of the high stands associated with MIS 9 or older, 7, 5e, 5c and 5a and conform broadly to the expected time ranges suggested by the Milankovitch theory. With increased precision afforded by the latest techniques, it is envisaged that such material will be re-analyzed to improve upon precisions and investigate phase relationships between Milankovtch forcing and sea level at these much earlier times. One of the key issues is whether or not phase relationships between insolation and sea level and duration of high sea stands are consistent throughout the Quaternary period (Winograd et al. 1997)

Not only does flooding by rising sea levels cause cessation of calcite growth, but also flooding by regional water table fluctuations in areas distant from the sea. Auler and Smart (2001) obtained ^{230}Th ages for water table and subaqueous speleothems formed during last glacial maximum and MIS 6 in Toca de Boa Vista, northeastern Brazil, 13 m above the present water table, indicating that climate was wetter than present during these glacial periods, in marked contrast to other records from lowland Brazil sites and general circulation model results. Regional fluctuations in paleoclimate are also inferred from water table fluctuations documented in Browns Room, an air-filled chamber that is part of the Devils Hole fissure system, Nevada (Szabo et al. 1994). Uranium-series ages of calcite with morphological and petrographical features that can be related to former water levels provide reliable evidence for the timing of former high water tables 5 m above present from 116 to 53 ka, and fluctuations between 5 and 9 m between 44 and 20 ka. This data is in broad agreement with effective moisture records from lacustrine, soil and travertine deposits of the Great Basin, USA.

Paleoclimate. One of the disadvantages of speleothems as paleosea-level indicators is the fact that growth may cease for reasons other than submergence by rising sea levels. For example, Richards et al. (1994) suggest that regional aridity in the Bahamas may have been associated with the onset of deglaciation and in North America, thus halting supply of dripwater to the caves up to 5 ka prior to submergence in some cases. In many studies, however, the recognition that changing environmental conditions control speleothem growth frequency, periods and rates has provided valuable regional paleoclimate information that can be compared with independent evidence such as lake level, paleovegetation, marine and ice core evidence.

Growth frequency. In high latitude regions of the globe, periodic advance of glaciers dramatically affected karst regions; dripwater flow routes to the cave became permanently frozen, rate of production of soil CO_2 was reduced and, in some cases, soil would have been stripped from the surface by ice masses. Harmon et al. (1977) recognized four distinct periods of deposition in the Rocky Mountains of North America based on alpha-spectrometric ages of many U-series ages of speleothems that could be related to periods of low continental ice volume. Since this early work, many regional compilations of alpha-spectrometric ^{230}Th ages have been produced with a view to discerning changing environmental conditions, among them, Gascoyne et al. (1983)—northern England; Hennig et al. (1983)—Europe; Lively (1983)—Minnesota, USA; Gordon et al. (1989)—United Kingdom; Baker et al. (1993)—northwest Europe; Lauritzen (1991, 1993)—Norway; Onac and Lauritzen (1996)—Romania; Hercman (2000)—central Europe; Lauritzen and Mylroie (2000)—New York, USA; Ayliffe and Veeh (1988), Ayliffe et al. (1997)—Naracoorte, Australia. The most suitable method for graphical representation of speleothem growth frequency is by probability density functions of age frequency, where each age determination represented by a normal distribution and normalized to equal area (Gordon and Smart 1984) (see Fig. 16). Smooth density functions constructed for compilations of alpha-spectrometric ^{230}Th ages can be compared with other proxy records of climate change. Kashiwaya et al. (1991) analyzed the speleothem frequency curve of Gordon et al. (1989), which is based on several hundred British speleothems, to demonstrate that temporal change in frequency was related to Milankovitch parameters. While speleothem growth generally ceased in high latitudes during glacial conditions, growth in temperate mid-latitude regions continued but at a reduced frequency under conditions of changing vegetation, soil CO_2 levels and soil moisture availability and temperature.

Speleothem frequency distributions have provided a useful tool for broad comparisons, but they suffer from the problem of biased sampling strategies and low resolution at times of known abrupt change. The increased precision afforded by mass-spectrometric techniques will result in fewer studies using this approach to assess of growth frequency and, more often, records of continuous deposition and growth rate studies will be graphically illustrated.

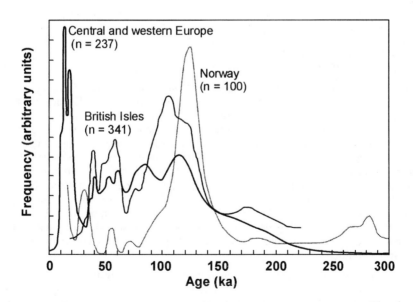

Figure 16. Probability density function of compilations of alpha-spectrometric U-series ages for central and western Europe (Hercmann 2000), British Isles (Gordon et al. 1989) and Norway (Lauritzen 1991, 1993) (redrawn from Hercmann 2000).

Growth periods. Growth phases in individual speleothems, or more ideally suites of speleothems from individual caves, can be used to constrain periods of time when regional environmental conditions were conducive for calcite growth. Mass-spectrometric ages of sub-sample wafers or powders representing < 1 mm growth from close to the base and outer surface of continuous phases of growth in stalagmites, flowstones and stalactites precisely define the growth periods.

Baker et al. (1996) compare the pattern of growth in two flowstone sequences from Yorkshire, England. Six growth periods of both short duration (1 to 3 ka) and fast growth rate are separated by non-depositional hiatuses in a sequence from Lancaster Hole. The ages of these phases were determined to be 128.8 ± 2.7, 103.1 ± 1.8, 84.7 ± 1.2, 57.9 ± 1.5, 49.6 ± 1.3, and 36.9 ± 0.8 ka by mass-spectrometric techniques. There is a remarkably good correlation between the periods of active speleothem growth and the timing of Northern hemisphere solar insolation maxima, which they consider to be related to recharge rates or volume. This sample, they suggest, only grew at times when precipitation was sufficiently high to overcome a threshold and supply dripwater to this location. A flowstone sequence from a nearby cave, Stump Cross Caverns, also exhibits numerous phases of growth. In common with the Lancaster Hole sequence, no growth was found at times of glaciation but there appear to be an additional control on calcite growth because no growth is exhibited during MIS 5. The authors suggest that although conditions were warm and moist enough for flowstone growth as evidenced by hippopotamus faunal remains in hiatus material (Gascoyne et al. 1981) and calcite deposition elsewhere in the cave (Sutcliffe et al. 1985), the location of this flowstone sequence must have been flooded for a long duration. Clearly, this is a site-specific effect and researchers must be careful not to infer regional paleoclimate information on the basis of individual samples.

High-precision TIMS ^{230}Th ages of a flowstone samples from Spannagel Cave, Zillertal Alps (Austria) at 2500 m above sea level demonstrates that this site is highly sensitive to climate changes because growth is restricted to interglacial periods between 207 to 180 ka and 135 to 116 ka (Spötl et al. 2002). The estimated age of initiation of calcite growth after the termination of the penultimate glacial is 135 ± 1.8 ka, which is in remarkable agreement with estimates from the marine realm (Esat et al. 1999; Henderson and Slowey 2000; Gallup et al. 2002) discussed above in Section 4.1.

In contrast to regions in high latitudes, carbonate cave deposits in desert locations might be expected to exhibit growth during the cooler periods of the late Pleistocene. Vaks et al. (2001), for example, determined 41 ^{230}Th ages for material from 13 samples from a cave located at the current desert boundary in Israel. Hiatuses in growth between 160 and 80 ka and the Holocene are ascribed to increased evaporation/precipitation ratio from either higher temperatures or reduced rainfall during interglacial periods. Abundant samples are also observed in tropical regions at times of intensified monsoon activity. Burns et al. (1988, 2001) report fifty-one ^{230}Th ages of speleothems from Hoti Cave, Oman, that are restricted to the intervals 325-300 ka, 210-180 ka, 130-117 ka and 82-78 ka (one sample records three short separate growth periods during MIS 9, 7a and 5e, separated by long intervening periods of non-deposition marked by detrital material). Each growth period corresponds to a continental pluvial period coincident with peak interglacial conditions.

Growth rates. Where many age determinations have been made along the growth axis of individual samples, accumulation rates of calcite can be calculated and related to changing conditions in the vadose zone above the location of drip source. Theoretical models of speleothem growth have been developed based on chemical kinetics of calcite precipitation (Dreybrodt 1980; Buhmann and Dreybrodt 1985a 1985b, Dreybrodt 1997), but the inter-relationships between the various factors are complex. Calcite precipitation rates are dependent on calcium concentration, temperature and drip rates, all of which are likely to change with climate and vegetation changes. Calcium concentration is related to porosity and permeability of limestone, soil pCO_2 levels, moisture and temperature, while drip rates are related excess soil moisture and routing of water in the hydrological system. The nature of flow at the solid-solution interface is also important, with turbulent flow conditions giving rise to greater precipitation rates than stagnant or laminar flow conditions (Dreybrodt 1997). Relatively few studies have compared empirical observations of growth rates within stalagmites collected in artificial and natural "cave" settings with theoretical estimates: Baker and Smart (1995) used flowstone samples that had grown in excavated caves or mines, where it was possible to estimate the maximum time since opening of the void; A. Baker et al. (1998) used recent samples from natural and artificial settings that exhibit annual laminae. In both cases, reasonable agreement was found between empirical and theoretical estimates of mass accumulation, but complicating factors such as seasonality of water supply and variations in calcite porosity were implicated (A. Baker et al. 1998). Positive correlations have also been observed in comparisons between thickness of annual laminae and historical records of effective precipitation (Railsback et al. 1994; Genty and Quinif 1996; Brook et al. 1999) and environmental change based on archaeological records (Polyak and Asmerom 2001). However, it is recognized that the factors involved are complex and not fully determined for sites that are well-monitored and have detailed historical records of climate and vegetation change and only broad conclusions can be deduced from observed variations in growth rate using U-series ages where sub-samples are typically analyzed at intervals of 10^2 to 10^4 years.

Musgrove et al. (2001) reported calcite growth rates that varied by three orders of magnitude during the period 70 to 7 ka based on ^{230}Th ages for 4 speleothems from caves up to 130 km apart in central Texas. The correspondence of fast and slow growth rates in the four samples suggest that a common controlling mechanism is involved at the regional scale. Fastest growth rates are observed at times of rapid Laurentide ice sheet growth in North America (MIS 2 and 4).

In a detailed study of a single stalagmite from Tasmania, Australia, that grew from 155 to 100 ka, Zhao et al. (2001) show that the fastest growth rates (~61.5 mm ka^{-1}) occurred between 129.2 ± 1.6 and 122.1 ± 2.0 ka. They suggest that highest rates of growth are associated with greater moisture supply during the height of the last interglacial because of latitudinal shifts in the band of subtropical highs. In contrast to this pattern, further to the north, in the Naracoorte region of South Australia, speleothems exhibit enhanced growth during stadials rather than interglacials (Ayliffe et al. 1998). Here, effective moisture in the past must have been related to evaporation rates or air temperature.

Landscape evolution. Atkinson and Rowe (1992) review the application of uranium-series dating to investigations of regional uplift, denudation and valley downcutting. The datable range of U-Th techniques is such that rates of change can be established for the past 500,000 years and extrapolated with reasonable confidence for the past few million years. While it is envisaged that future applications of U-Th-Pb dating techniques will extend the range of possibilities (Richards et al. 1998), most studies have relied on alternative dating strategies such as palaeomagnetism, ESR and ^{40}Ar/^{39}Ar to extend the time frame of study.

Speleothems have been most usefully applied to investigations of the rate of valley downcutting, where base level changes are associated with paleo-watertable lowering and draining of cavernous voids. Uranium-series ages of the oldest material found at a particular elevation in a cave system defines the minimum age of change from phreatic to vadose conditions. By plotting the age vs. elevation of numerous speleothems samples, estimates of maximum valley incision rates can be derived. This approach has been used to deduce valley downcutting rates in Yorkshire, England (Atkinson 1978, Gascoyne et al. 1983) of 0.12 m ka^{-1}. Rowe et al. (1988) used a combination of paleomagnetic data and U-series ages of speleothem from the Peak District, England to demonstrate that mean downcutting rates were 0.055 m ka^{-1}. Farrant et al. (1995) deduce a long-term (~2 Ma.) base-level lowering rate of 0.19 ± 0.04 m ka^{-1} based on uranium series, electron spin resonance, and paleomagnetic dating of an extensive sequence of speleothems and limestone caves in Sarawak, Malaysia. Polyak et al. (1998) extend the timeframe for analysis of water table lowering in the Guadalupe Mountains of Mexico, which was originally based on U-series ages (Ford and Hill 1989), by $^{40}Ar/^{39}Ar$ dating fine-grained alunite that forms during cave genesis by sulfuric acid dissolution. Alunite found up to 1100 m above the present valley floor had $^{40}Ar/^{39}Ar$ ages of 12 Ma, implying a downcutting rate of ~0.092 m ka^{-1}.

Palaeoseismicity. Uranium-series ages constraining deformation, breakage and differential growth of stalagmites and stalactites in caves can used to provide data on the occurrence of historic and prehistoric earthquakes, and hence return periods of seismic action. Postpichl et al. (1991), for example, dated different generations of tilted and collapsed speleothems attributed to paleoearthquakes of central Italy, and Lemeille et al. (1999) dated either side of offsets in speleothems to confirm that disruption in Bätterloch and Dieboldslöchli caves, Switzerland, was associated with the 1356 AD Basel seismic event. Kagan et al. (2002) documents uranium-series ages and carbon and oxygen isotope records of phases of speleothem growth precipitated directly on severed stalagmites, collapsed pillars, ceiling blocks, and various stalactites from caves near the Dead Sea Transform Fault System. Eighteen separate seismic events are recognized, which can in some cases be correlated with archaeologically- and geologically- recorded earthquakes. In a similar way, travertines in neotectonic regions can be dated by U-series isochron methods to calculate time-averaged dilation and lateral propagation rates for individual fissures (Hancock et al. 1999).

Archaeology. Some of the most important archaeological and faunal remains have been found in caves and rock shelters and uranium-series ages of cave deposits therein have been instrumental in providing chronological control. An extensive review of the wide range of geochronological approaches that have been applied specifically to cave deposits is provided by Schwarcz and Rink (2001), while Schwarcz and Blackwell (1992) focus on U-series applications. In most cases, ^{230}Th ages have been used to constrain the age of detrital layers that have been washed, blown or carried into caves. Calcite material, such as straw stalactites that have spalled from the cave roof, associated with archaeologically valuable finds in detrital layers can be dated to obtain a maximum, or *terminus post quem* age constraint. Flowstones overlying detrital layers can provide minimum ages. Thin calcite coatings can now be dated using high-precision techniques: Frank et al. (2002), for example, determined mass-spectrometric U-series ages on sub-samples representing < 3 mm of individual calcite layers that grew on walls of artificial water-supply tunnels at Troy/Ilios. Variable quantities of detritus in the samples demanded correction for initial ^{230}Th, and first-order estimates of the actual age of deposition based on an *a priori* estimate for R_0 (Eqn. 6) of 0.75 ± 0.1 were demonstrated to be reasonable after careful evaluation of other sub-samples using isochron methodology. The ^{232}Th-free estimates of $^{230}Th/^{238}U$ activity ratio and $\delta^{234}U$ using the error-weighted regression slope for data in $^{230}Th/^{232}Th$ vs. $^{238}U/^{232}Th$ and $^{234}U/^{232}Th$ vs. $^{238}U/^{232}Th$ space (Section 2.2), respectively, yielded the same age (4200 ± 750 yr) as the first order estimates (~4350 ± 570 yr) for the oldest overgrowth material. In this way,

Frank et (2002) demonstrate that the tunnels must have been built during the Bronze Age (Troy I-II) by Anatolian and south-eastern European cultures.

One of the prime motivators for improved precision of mass-spectrometric U-series ages (and decay constants) is extension of the dateable age range to beyond 500 ka. Numerous important archaeological finds are poorly constrained by only minimum alpha-spectrometric ages of "> 350 ka" and reappraisal of earlier estimates is ongoing for many sites. The Zhoukoudian hominid specimens, from near Beijing, China, for example, have widely been recognized as representative of *Homo erectus*, but attempts to resolve the issue of whether this species was a direct ancestor of later eastern Asian populations or a side branch of human evolution have been confounded by conflicting geochronological results. Prior to the recent study by Shen et al. (2001), ages for the youngest *H. erectus* in the sequence at Zhoukoudian ranged from a time span between 290 and 230 ka, based on ^{230}Th dating of fossils (Yuan et al. 1991), to 414 ± 14 ka based on multiple TIMS ^{230}Th ages of flowstone samples (Shen et al. 1996). Additional ages by Shen et al. (2001) confirm the older age constraint, but cannot be used to limit the age of the youngest skull to a finite range > 400 ka. Fossil remains of *H. erectus* have also been discovered at Tangshan Cave, Nanjing and are morphologically correlated with specimens from Zhoukoudian. High-precision mass-spectrometric U-series ages of flowstone samples overlying the Nanjing Man fossils range from 563 +50/−36 ka to 600 +60/−50 ka (Cheng et al. 1996, Wang et al. 1999, Zhao et al. 2001). It should be borne in mind that these ages are heavily dependent on the precision and accuracy of the decay constants used and careful calibration strategies need to be adopted. Zhao et al. (2001) rely on standardization to a secular equilibrium material (HU-1), which minimizes the fractional errors in age but relies on the fact that the standard is indeed at equilibrium.

4.2. Applications based on proxy evidence for environmental change contained within speleothems

Speleothems and similar deposits contain trace substances and isotopic information that can be used to derive information about the past state of the vadose system overlying the site of deposition. Variations in oxygen and carbon isotopes in both the solid and fluid phase, trace elements such as Mg, Sr, P, Ba, U and Fe, organic acids, pollen and fossil mites, combined with accurate and precise chronology, can be used to construct temporal records of past regional environmental, hydrological or geophysical conditions for the past 500 ka, at resolutions as high as sub-annual in some cases. Secondary calcite deposits are particularly useful because they are often well-preserved, continuous and provide direct radiometric ages. Ocean cores, on the other hand, are subject to bioturbation and can only be directly dated by uranium-series methods in rare circumstances, such as well-preserved aragonitic deposits shed from carbonate platforms (Henderson.and Slowey 2000). Also, ice cores provide multi-proxy records but cannot be directly-dated by radiometric methods.

$\delta^{18}O$ in calcite. The distribution of ^{18}O between calcite and water during the precipitation of speleothems and similar deposits is dependent on temperature alone if the system remains in isotopic equilibrium, which is often the case in deeper sections of the cave, where ventilation is minimal and humidity is high. Fractionation between ^{18}O and ^{16}O on precipitation of calcite is about −0.22‰/°C at 20°C, increasing to −0.24‰/°C at 10°C based on experimental studies of inorganic calcites (O'Neil et al. 1969; Friedman and O'Neil 1977). Hendy and Wilson (1968) optimistically declared that changes in mean annual temperatures as small as 0.2°C could be achieved based on their preliminary investigations of speleothems from New Zealand. However, since this early work, extensive research has demonstrated that the derivation of paleotemperatures from $\delta^{18}O_{\text{calcite}}$ is complex and few reliable quantitative estimates of past temperature have

been derived, principally because of the much greater expected temporal amplitude of $\delta^{18}O$ variation in the waters ($\delta^{18}O_{water}$) from which the calcite has precipitated. Variation in $\delta^{18}O_{calcite}$ along the axis of growth, then, is a function of temperature and the isotopic composition of rainfall and percolating vadose waters, the latter attributed to a combination of factors related to the temperature of formation of water in the atmosphere [~0.7‰/°C for oceanic sites (Dansgaard 1964); generally lower gradient for continental sites (Rozanski et al. (1993)], changes in $\delta^{18}O$ of the sea water, which fluctuates with changing ice volume, and changes in the path of water from source to site of precipitation. The present day geographical distribution of mean annual $\delta^{18}O$ is reasonably well-known, but past spatial patterns must be inferred from models of past atmospheric circulation, ocean $\delta^{18}O$ changes and temperature distributions (Cole et al. 1999; Jouzel et al. 2000; Werner et al. 2000).

In comparing the temporal relationship of $\delta^{18}O$ in speleothem calcite with other proxy records of climate change, both positive and negative relationships with expected mean annual temperature have been observed, dependent on the relative influence of additional factors. For oceanic sites and some continental sites, the temperature dependence of $\delta^{18}O_{water}$ can outweigh the effects of $\delta^{18}O$ of sea water and fractionation at the calcite-water interface, such that calcites deposited during interglacials are less negative (Gascoyne et al. 1981; Goede et al. 1996, Lauritzen 1995, Dorale et al. 1992 1998). In other locations, negative shifts are related to a combination of factors such as glacial-interglacial changes in ocean water $\delta^{18}O$, intensification of rainfall, the "amount effect," seasonal patterns of monsoon rainfall and anomalous shifts in surface water $\delta^{18}O$ in source regions (Bar-Matthews et al. 1997, 1999; Hellstrom et al. 1998, Frumkin et al. 1999 Burns et al. 2001, Wang et al. 2001, Neff et al. 2001, Bard et al. 2002b). It is clear that in some cases, amplitudinal range of $\delta^{18}O_{calcite}$ is too great to be ascribed to temperature variation alone (e.g., −11.65 to −0.82‰ for Crag Cave, south west Ireland; McDermott et al. 2001) and dramatic shifts in $\delta^{18}O$ of water vapour accentuate the apparent relationships. In most recent studies, paleodata are supported by detailed survey of the modern scenario and confirmation that calcite is deposited under isotopic equilibrium conditions. It is recognized that many speleothem samples are likely to be unsuitable for the investigation of the $\delta^{18}O_{calcite}$-$\delta^{18}O_{water}$ relationship because of kinetic effects related to large differences in pCO_2 of drip waters and ambient air in the cave, slow drip rates and evaporation. It is now routine to measure $\delta^{18}O$ along distinctive growth layers to determine whether significant variation occurs from the central axis outwards, the so called "Hendy test" (Hendy 1971). While the appearance of kinetic fractionation of the isotopic signal is generally a problem it must be borne in mind that in certain circumstances it may be possible to recognize periodic shifts between equilibrium and non-equilibrium conditions and relate this to regional, not locally-specific, environmental changes (Niggemann et al. in press).

Although the apparent causal mechanisms for $\delta^{18}O_{calcite}$ variation are complicated, consistent regional patterns have been observed with a high degree of correlation with high-resolution $\delta^{18}O$ records from ice and marine cores for compilations of data based on many mass-spectrometric [230]Th ages (Winograd et al. 1992; Dorale et al. 1998, Bar-Matthews et al. 1998; McDermott et al. 2001, Burns et al. 2001 Spötl and Mangini 2002; Wang et al. 2002). Where observed patterns are replicated in many samples from the same cave, it is unlikely that non-equilibrium fractionation has occurred because each drip site is likely to be influenced to a different degree by flow path, pCO_2, residence times and degassing history (Dorale et al. 1998, 2002).

Perhaps one of the most widely cited $\delta^{18}O_{calcite}$ records is that of the Devils Hole vein calcite (Winograd et al. 1992) which records glacial-interglacial shifts in 1 to 2‰ from

566 to 60 ka. The record is reliably dated by ^{230}Th-^{234}U-^{238}U and ^{234}U-^{238}U methods (Ludwig et al. 1992). The variation in $\delta^{18}O_{calcite}$ exhibited most likely reflects changing $\delta^{18}O_{water}$ in atmospheric precipitation that recharges the groundwaters of the surrounding area. The local T-$\delta^{18}O_{calcite}$ relationship is considered to be positive at all times, such that highest $\delta^{18}O$ occurs at times of interglacial periods and is similar to the empirical relationship derived for mid- to high latitude precipitation (e.g., Johnsen et al. 1989) or cave waters (Yonge et al. 1985).

Recently published speleothem $\delta^{18}O_{calcite}$ records (Dorale et al. 1998; Bar-Matthews et al. 1998; Wang et al. 2001; Spotl and Mangini 2002) improve upon the resolution of Devils Hole and extend the investigation of regional variation in $\delta^{18}O_{water}$ and paleorecharge across the globe. These records are of sufficient resolution that leads and lags in the climate system can be investigated by comparison with high resolution polar ice core records. This can only be done, however, if both records are accurate in their timing. Ice core records beyond the last glacial maximum rely on ice flow models because annual layers at such depths no longer exist. Speleothems, on the other hand are directly dated by radiometric methods. Spötl and Mangini (2002) postulate minor differences in the timing of large and abrupt events in the $\delta^{18}O_{water}$ and $\delta^{18}O_{ice}$ from 60 ka to 44 ka. These may possibly be leads and lags, but are more likely to be due to either inaccuracies in the chronologies involved or the correlations between records. A comparison with a similarly dated speleothem record from Hulu Cave, China illustrates the complexities (Fig. 17). Unambiguous markers that are displayed in different paleoenvironmental archives, including speleothems, are required before absolute chronologies can be determined and terrestrial-ocean-cryosphere correlations established with confidence.

Temporal resolution approaching that of the ice deposited during the last glacial period in Greenland can commonly be obtained for stalagmites and stalactites using a drill bit <0.5 mm diameter. To investigate decadal to centennial sensitivity of the ocean-atmosphere-hydrosphere-vegetation system, comparable resolution to ice cores for the Holocene, where annual layers are preserved, is required. McDermott et al. (2001) turned to laser ablation stable isotope measurements using a beam diameter of ~150 μm and recognized features such as the 8200 year cold event that were not apparent in coarser resolution studies (Fig. 1). By matching the resolution of measurements for ice cores, ocean cores and speleothems, researchers in the paleoclimate field are presented with some interesting challenges. Most records at the centennial to millennial scale resolution have a component of global and regional variation. To match these records, one must identify common causal mechanisms that manifest themselves synchronously. If this can be done, leads and lags in the climate system can be investigated. One of the principal difficulties with comparison, however, is that generally only the speleothem records are directly datable.

$\delta^{13}C$ in calcite. Dissolved carbon in seepage waters is derived from three sources: atmospheric CO_2, decaying organic matter and root respiration in the soil zone and dissolved limestone along the flow route to the cave void. The net isotopic composition of dissolved HCO_3^- and the precipitated $CaCO_3$ depends, amongst other factors, on the $\delta^{13}C$ of reactants in the system, kinetic fractionation factors in the H_2O-$CaCO_3$-CO_2 system, saturation state with respect to $CaCO_3$ (and $CaMg(CO_3)_2$), exchange with gaseous phase and rate of precipitation. The system is complex (see reviews by Wigley et al. 1978; Salomans and Mook 1986; Dulinski and Rozanski 1990) and, while rainfall and temperature variation can cause temporal and spatial variation in $\delta^{13}C$ of precipitated calcite by influencing the chemical kinetics of dissolution and precipitation, the most useful information derived from $\delta^{13}C$ of speleothems relates to vegetation changes.

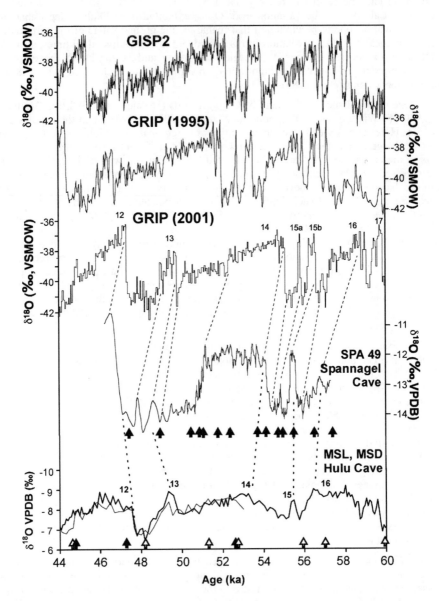

Figure 17. Last glacial period $\delta^{18}O_{calcite}$ for speleothem records from China (Wang et al. 2001) and Austria (Spötl and Mangini 2002), and Greenland $\delta^{18}O_{ice}$ records (Grootes et al. 1993; Johnsen et al. 1995; Johnsen et al. 2001). Ice core records are based on annual band counting and ice flow models. Speleothem records are based on TIMS ^{230}Th ages (shown by arrows). Numbers refer to Dansgaard-Oeschger (D-O) events in ice cores. Climate teleconnections may cause synchronous shifts in atmospheric circulation in different areas of the globe. The abrupt shifts in climate recorded in $\delta^{18}O_{ice}$ in the GRIP and GISP2 ice cores may be synchronous with changes in monsoonal rainfall patterns in China and temperature shifts above the Central Alps recorded in the speleothem records. While broad agreement is demonstrated, apparent discrepancies indicate exceptions to synchronous responses, inaccurate chronologies, or erroneous correlations.

Numerous studies of secondary pedogenic carbonates have successfully interpreted changes in $\delta^{13}C$ as a response to changing vegetation. Geochemical models (Dreybrodt 1980) indicate that the expected range of $\delta^{13}C$ in secondary carbonates precipitated in soils in regions dominated by C_3 (Calvin cycle) plants is −14 to −6‰, whereas those in C_4-dominated regions will have a range of −6 to +2‰ (C_3 plants have $\delta^{13}C$ value of −30 to −24; C_4 plants, −16 to −10‰; Vogel 1993). The differences between these two end-members is sufficiently large than that any temporal variation in their relative abundance should be discernible in secondary carbonates, although vegetation density (Amundson et al. 1988) and soil respiration rates (Quade et al. 1989) are also likely to play a role. Speleothems are likely to display a similar range of $\delta^{13}C$ to pedogenic carbonates under the same conditions.

Studies based on speleothems from the midwestern USA exhibit significant variation in $\delta^{13}C$ that has been related to shifts in the location of prairie and forest ecotones during the last glacial and Holocene periods (Dorale et al. 1992 1998; Denniston et al. 1999a,b, 2000, 2001). If samples can be obtained from locations that experienced the passage of ecotonal boundaries, rapid shifts in $\delta^{13}C$ are observed. A clear example of the relationship between speleothem $\delta^{13}C$ values and the C_3/C_4 biomass ratio was provided by R. G. Baker et al. (1998), who compared well-dated Holocene profiles from Northeast Iowa. Pollen, plant macrofossils, and sedimentary organic matter $\delta^{13}C$ values from alluvial deposits along Roberts Creek were compared to stalagmite $\delta^{13}C$ values from nearby Cold Water Cave. The pollen and plant macrofossil evidence (which in many cases allowed species-level plant identification) showed that a C_4-inclusive prairie replaced a C_3-rich deciduous forest around 6,300 years ago, which in turn was replaced by oak savanna (and a more intermediate C_3-C_4 mixture) approximately 3,500 years ago. The sedimentary organic matter $\delta^{13}C$ values showed a corresponding trend toward heavier isotopic values. The similarity of both the timing and isotopic trend of two speleothem records from Cold Water Cave with the vegetation record from Roberts Creek argues strongly that the speleothems are recording long-terms changes in the isotopic composition of the soil organic matter resulting from regional changes in the vegetation (Fig. 18).

Although changes in $\delta^{13}C$ can be attributed to changing vegetation, there are samples that demonstrate secular variation in regions that have been continuously dominated by either C_3 and C_4 plant types. In these areas, factors other than the distribution of C_3 and C_4 plants must be involved. In New Zealand, no plants are found to use the C_4 photosynthetic pathway except two species (both salt-tolerant lowland species), yet a speleothem from the mountainous region of north-western South Island exhibits substantial variation in $\delta^{13}C$ from 31 ka to present that is reasonably correlated with both $\delta^{18}O$ in the same sample and ice core records of temperature change (Hellstrom et al. 1998). The authors suggest that variability, here, is related to soil moisture conditions with vegetation density. During the last glacial maximum, drier conditions and sparser vegetation produced lower soil atmospheric CO_2 levels and increase in $\delta^{13}C$ (Dulinski and Rozanski 1992).

In a study that ends with a cautionary note, Baker et al. (1997) studied the spatial variation of $\delta^{13}C$ in actively-growing and late Quaternary speleothems from the British Isles, where native C4 plants are non-existent. Despite sampling material predicted to have been deposited under equilibrium conditions (i.e where humidity was high and ventilation minimal), $\delta^{13}C$ values greater than expected for the C_3 pathway (−14 to −6‰) were observed in many cases. They suggest that in some cases, where residence times in the soil are short, waters may not have equilibrated with soil CO_2. Also, kinetic fractionation may occur during degassing of fluids at the location of the sample or *en route* to the cave void. Researchers must study contemporaneous samples from the same cave (or region) to establish whether factors influencing secular variation of $\delta^{13}C$ are regional or local (Dorale et al. 1998, 2002).

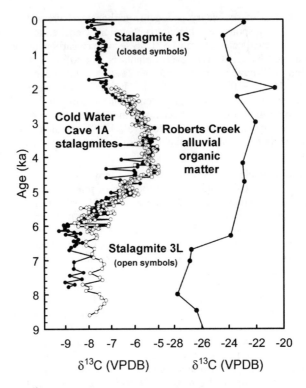

Figure 18. Holocene $\delta^{13}C$ profiles for stalagmites 1S and 3L from Cold Water Cave, Iowa and alluvial sedimentary organic matter from Roberts Creek (~60 km from Cold Water Cave—figure modified from Baker et al. 1998). Note the general similarity in timing and trend of the different $\delta^{13}C$ records. The $\delta^{13}C$ values of the stalagmites reflect the sources of carbon, namely soil CO_2 ultimately derived from the vegetation, for which the Roberts Creek record provides representative data, and the overlying limestone, which typically has $\delta^{13}C$ values close to 0‰ VPDB. In principle, these two sources contribute 50:50 to the dissolved carbonate species, although complex isotopic exchange scenarios are possible that may modify this ratio by the time seepage waters reach the caves. Nonetheless, the carbon which originates as soil CO_2 undergoes isotopic fractionation as it converts from dissolved CO_2 to aqueous CO_2 to HCO_3^- to CO_3^{-2} to solid $CaCO_3$; under equilibrium conditions, this sequence represents an approximate 10‰ enrichment (Hendy 1971). Thus, when soil CO_2 with a $\delta^{13}C$ value of −28‰ (representing a typical C_3 plant value) fractionates approximately 10‰ and mixes 50:50 with bedrock carbonate with a $\delta^{13}C$ value of ~ 0‰, the resultant speleothem carbonate would have a $\delta^{13}C$ value of ~ −9‰, consistent with the observations at Cold Water Cave and Roberts Creek. Because of the 50:50 contributions and the invariant nature of the bedrock component, shifts in the vegetation $\delta^{13}C$ values should cause speleothem shifts that are roughly half the magnitude of the vegetation shifts, again consistent with the observations. Note the factor of two differences in the $\delta^{13}C$ scales.

$\Delta^{14}C$ in calcite. Radiocarbon dating is the most widely-used technique for derivation of timescales for terrestrial and oceanic records of climate change. However, its use in speleothem chronology is considered problematical, principally because of potential variability in the contribution of "dead carbon" from the host limestone (Hendy 1970). While HCO_3^- in the soil zone in many locations may be found to be in equilibrium with soil gas, precipitating waters contain a proportion of carbon inherited from the dissolution of ^{14}C-free limestone in the vadose zone that varies according to the extent of open system conditions and incongruent dissolution (Wigley et al. 1978). Where organic matter turnover rates are relatively short (often <10 years), soil $^{14}CO_2$ can be considered to be effectively in equilibrium with the atmosphere, given the sampling resolution of calcite. Fractionation of carbon isotopes also occurs on precipitation of calcite depending on environmental conditions (see above). Because of these problems, relatively few speleothem chronologies have been based on ^{14}C ages, but where independent estimates of the age were obtained, the dead carbon proportion (dcp) has generally been found to have a range in dcp of 10-20 % (Bastin and Gewelt 1986; Genty et al. 1998, Holmgren et al. 1994), although much greater values are also observed (Vogel and Kronfeld 1997). In a detailed study of ^{230}Th dated speleothems from western Europe, Genty et al. (1999) demonstrate that temporal and spatial changes of dcp can be significant and dependent not only on the extent of open-system conditions, but also the age of the soil organic matter. Uncorrected ^{14}C ages, then, will over estimate the true age of deposition in most cases.

Despite the inherent problems associated with ^{14}C in speleothems, recent work by Beck et al. (2001) has focused attention on this topic because they demonstrated that in certain circumstances the dcp correction may be sufficiently well constrained that calibration of the radiocarbon timescale might be possible. Speleothems have an advantage over marine corals (e.g., Bard et al. 1990, 1998; Edwards et al. 1993), which have been used to extend the calibration curve, because they provide continuous records and are less likely to be affected by post-depositional alteration. Beck et al. (2001) obtained an extensive suite ^{230}Th and ^{231}Pa ages for a submerged speleothem (GB-89-24-1) from the Bahamas that grew continuously from 44 to 11 ka (see Fig. 19). Over 270 AMS ^{14}C ages were obtained for sub-samples along the axis of growth and where the timing of speleothem deposition overlapped with the earliest section (15.6 to 11.1 ka) of the existing radiocarbon calibration curve (Stuiver et al. 1998), a constant dcp of 1450 ± 450 (or 16 ± 5 %) was observed.

AMS ^{14}C ages for the rest of the sample were dcp-corrected and a surprising amount of structure was found in the data that suggests dramatic shifts in atmospheric ^{14}C occurred in the past (Fig. 19). Estimated atmospheric ^{14}C concentrations ($\Delta^{14}C$) were also much greater during the last glacial period, possibly reflecting increased ^{14}C production at times of reduced geomagnetic intensity. Variations in atmospheric ^{14}C concentrations are also related to the distribution of carbon in the Earth's major reservoirs. Detailed study of ^{14}C variation in speleothems could provide valuable information about the timing of changes and nature of carbon exchange between these reservoirs, which is likely to be caused by changes in the pattern of ocean circulation (Stocker and Wright 1996).

$\delta^{18}O$ and δD in fluid inclusions in calcite. Speleothems commonly contain between 0.001 and 0.3 wt. % water trapped within the calcite (Schwarcz et al. 1976). To obtain unambiguous paleoclimate information from the oxygen isotope signal preserved in the speleothem calcite, it would be extremely advantageous to directly measure the isotopic composition of relict drip water that is preserved within fluid inclusions. In this way, the problem of estimating past patterns of $\delta^{18}O_{water}$ in precipitation might be circumvented. Early attempts (Schwarz et al. 1976, Harmon et al. 1979) focused on the hydrogen isotope composition of fluids because it was considered that the $\delta^{18}O_{water}$ in the inclusions could have been altered after exchange with calcite since the time of entrapment. Such an approach is feasible because of the following general relationship observed in meteoric waters (Craig et al. 1961)

$$\delta D_{water} = 8\ \delta^{18}O_{water} + 10 \qquad (6)$$

However, few studies have been conducted because of analytical difficulties during extraction and measurement such as re-adsorption or incomplete removal, and the low temporal resolution that can be achieved Also, it is considered likely that the global meteoric water relationship in Equation (6) does not apply consistently across the globe

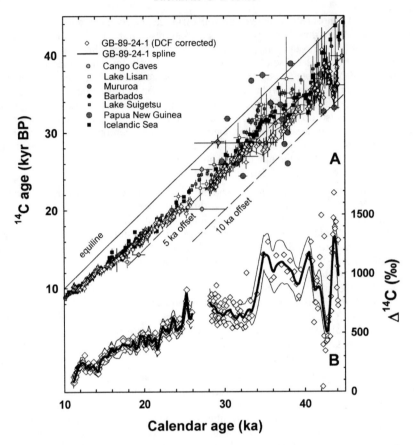

Figure 19. A) Radiocarbon ages vs. calendar ages (45 to 11 ka) for speleothems, coral, lake sediments and a marine core. GB-89-24-1—speleothem from Bahamas (Beck et al. 2001); Cango Caves—speleothem from South Africa (Vogel and Kronfeld 1997); Lake Lisan—lacustrine sediments from paleo-Dead Sea (Schramm 2000) Mururoa, Barbados—corals (Bard et al. 1998); Lake Suigetsu—lacustrine sediments (Kitagawa and van der Plicht 2000), Papua New Guinea (Yokoyama et al. 2000); Icelandic Sea -marine sediments (Voelker at al. 2000). Note increasing scatter in data with increasing and also apparent structure in some of the archives. B) Variations in atmospheric $\Delta^{14}C$ based on dcp-corrected ^{14}C age vs. ^{230}Th ages for GB-89-24-1. Both raw data and smoothed spline with 95% confidence bands are plotted. A sharp peak of approximately 1 ka in duration is observed in GB-89-24-1 at circa 44 ka, reaching levels of circa 1300‰—nearly twice as high as the 20th century bomb pulse. This may be related to changes in the atmosphere-ocean exchange of carbon associated with changing mode of ocean circulation during one of the meltwater events that influenced the North Atlantic Ocean during the last glacial period.

or between time periods of with temperature and precipitation regimes (Fritz et al. 1987, Gat and Carmi 1970; Harmon and Schwarz 1981).

Despite the potential problems, two recent studies provide positive results. Matthews et al. (2000) analyzed the δD of fluid inclusions in present-day cave carbonates from Soreq cave (Israel) using the method of thermal vacuum extraction and standard techniques on cave pool water from the same cave to estimate isotopic fractionation during measurement ($\Delta_{ex} = \delta D_{extracted\ water} - \delta D_{cave\ water}$). After subtracting Δ_{ex} from δD

measurements on extracted water from speleothem calcite that is known to be of last glacial age and has previously been analyzed for $\delta^{18}O_{calcite}$ (Bar-Matthews et al. 1997), they demonstrate, by comparing their results with the $\delta^{18}O$ and δD relationship expressed by the Mediterranean Meteoric Water Line (Gat and Carmi 1970), that sea-surface and atmospheric temperatures were indeed cooler during glacial times in this region and closely approximate estimates derived from the global Meteoric Water Line. Dennis et al. (2001) detail robust techniques using a crushing cell, cryogenic vacuum extraction and thermal desorption to analyze artificial and natural samples. Precisions of ± 3‰ for δD and ± 0.4‰ for $\delta^{18}O$ measurements were achieved on samples of < 1μL (which is equivalent to 1 g calcite for samples studied). Gross change of mid-latitude precipitation between glacials and interglacials is only expected to be 5-6 times greater than achievable precisions at present but overall the results are encouraging.

Trace elements. The partitioning of Mg in calcite-water system is known to temperature dependent (Mucci and Morse 1990; Morse and Bender 1990) and early attempts were made by Gascoyne (1983) and Goede and Vogel (1991) to use the secular variation of Mg content in speleothems to derive paleotemperature information. Mg concentrations in speleothems are sufficiently high that high-resolution techniques can be used to investigate its distribution in the calcite. Roberts et al. (1998), for example, used SIMS analysis to investigate the spatial pattern of Mg at 1 μm resolution in SU-80-11, a speleothem from NW Scotland previously dated by TIMS ^{230}Th ages (Baker et al. 1993). This resolution permits observation of annual bands and Roberts et al. suggest that regular fluctuations in Mg/Ca along the axis of growth (mean amplitude 0.00135 ± 0.00045) might be related to a combination of seasonal temperature changes (the sample was collected 50 m from the cave entrance where seasonal variation would have been experienced, which is in contrast to deep sections of a caves where temperatures are constant). Longer term trends of Mg/Ca in SU-80-11 cannot be explained by temperature effects because the amplitude of variation is to large

Information about the environmental conditions in the vadose system above the cave can be derived from the supply of trace elements to the cave void by percolating waters. Concentration of Sr, Mg, Ba, U, P and other constituents is likely to be affected by supply effects and groundwater residence times. Temporal fluctuations of Mg and Sr have been related to evaporative concentration in drip waters (Railsbeck et al. 1994), variation in residence times affecting extent of dolomite dissolution relative to calcite dissolution (Fairchild et al. 1996) and selective leaching of Sr and Mg (Fairchild et al. 2000) and enhanced calcite precipitation along flow route prior to location of drip site in cave void (Fairchild et al. 2000). Each of these factors can be related to paleohydrological condition in the past, and hence well-dated speleothems could provide valuable proxy high-resolution information for past regional climates. The achievable resolution is well-demonstrated by a recent study by Baldini et al. (2002), who combine laser ablation $\delta^{18}O$ and SIMS elemental analysis to obtain a multi-proxy record at sub-annual resolution from a speleothem from Crag Cave, Ireland (Fig. 1).

Organic acid fluorescence. In a similar manner to trace constituents, such as Mg, Sr and P, concentrations of organic acids present in speleothem calcite are sufficient to observe variation at temporal scales of less than annual in some cases (e.g., Baker et al. 1993, Shopov et al. 1994). Organic acids (humic and fulvic) are formed in the soil by humification, and transported to the cave void by percolating waters where they are entrapped in precipitating carbonates. Under certain circumstances, where precipitation patterns are strongly seasonal and the nature of vadose percolation is such that seasonal mixing is incomplete, bands with different luminescent intensities can be differentiated after excitation with UV radiation. In other cases, bands are not observable but secular variation in intensity is found. Such features can be used to derive a variety of paleoenvironmental information (McGarry and Baker 2000).

If annual bands can be distinguished, annual growth rates can be estimated and related to changing environmental conditions to a combination of factors that include drip water rates, soil pCO_2, Ca concentration in percolating waters (see above). Baker et al. (1993) demonstrated that the frequency of banding in a sample from Scotland was annual based on TIMS ^{230}Th ages along the axis of growth. Interpretation of growth rates in terms of palaeoclimatic change is difficult, however, and independent confirmatory records are generally required for sections of the speleothem record. Perhaps more useful are luminescence intensity records, which have been related to past moisture levels in overlying soils and peats: Baker et al. (1999) suggest that 90-100 year oscillations in intensity of luminescence two speleothems from NW Scotland is caused by fluctuations in the extent of humification associated with bog wetness during the past 2500 years. Charman et al. (2001) strengthen this argument for luminescence fluctuation in NW Scotland by analyzing humification, using degree of decay estimated by colorimetry, testate amoebae counts, which indicate water levels, and pollen records in overlying peat cores dated by AMS ^{14}C methods and find a good degree of correlation with the ^{230}Th dated speleothem record.

Secular variation of $\delta^{234}U$ in calcite. Although it has been demonstrated in many studies that there is a wide range in measured $\delta^{234}U$ for speleothems from a single cave site, past environmental conditions can be inferred from the variation of $\delta^{234}U$ in secondary calcite in certain vadose systems using individual samples or sequences. Elevated $\delta^{234}U(0)$, for example, is expected to be related to accumulation of ^{234}U in damaged sites during a period of reduced leaching, either as a result of lower soil $pCO2$ or reduced soil moisture excess. Richards et al. (1997) relate the saw-tooth pattern of $\delta^{234}U(0)$ in speleothem calcite in a submerged sequence on the Bahamas (Fig. 20) to a combination of factors: Uranium in the secondary calcite is primarily inherited from the marine carbonates overlying the cave which were deposited with $\delta^{234}U(0)$ similar to present sea water (~145‰). The first-order trend (A) towards lower values along the axis of growth can be explained by radioactive decay of U in the overlying carbonates towards secular equilibrium. Superimposed on this is a saw-tooth, or step-like, second order trend. This is likely to be caused by (B) deposition of additional marine carbonates with marine $\delta^{234}U$ during high sea-stands and accumulation of ^{234}U in damaged sites in the vadose zone under dry conditions, which is then preferential leached, and (C) dissolution of increasingly lower and older material during glacials because of stabilization, fissure development and input of organic material.

The pattern of decreasing $\delta^{234}U(0)$ at a greater rate than that expected simply from radioactive decay is common in continuous phases of speleothem growth and has been interpreted in terms of paleohydrology in a recent study by Plagnes et al. (2002). Kaufman et al. (1998) recognize a rapid fall in $\delta^{234}U$ at 19 ka in speleothems from Soreq cave, Israel, synchronous with shifts in $\delta^{18}O_{calcite}$. They interpret this as reflecting the transition from glacial to interglacial conditions in the eastern Mediterranean region, when dramatic increases in precipitation rates elevated the preferential leaching rate of ^{234}U. Also, in a multi-proxy study by Hellstrom and McCulloch (2000), consistent temporal relationships are found between $\delta^{234}U(0)$ variation and that of trace elemental concentrations (Sr, Ba, Mg and U), stable isotope data ($\delta^{13}C$ and $\delta^{18}O$) and luminescence intensity along the axis of growth of a speleothem from South Island, New Zealand. They relate this to changing effective moisture and vegetation density in this region from the last glacial maximum and the present day.

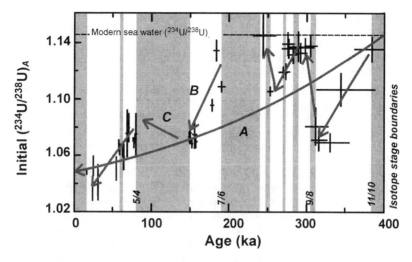

Figure 20. Secular variation in $\delta^{234}U(0)$ for Bahamas flowstone sequence. Changes in $\delta^{234}U(0)$ are related to uranium-series disequilibrium conditions in host limestone, periodic addition of new material with elevated (marine) $\delta^{234}U(0)$, alpha recoil effects and variation in recharge, and hence water-rock interaction times (see text for details).

5. CONCLUDING REMARKS

There is little doubt on the basis of the above review that speleothems are useful in an extremely wide range of applications and the frequency of publications devoted to speleothem-based archives is increasing rapidly. Fundamental to all applications is a robust and high-precision chronology and uranium-series techniques currently set the standard in this regard for speleothems. Reliable ages can be obtained for most samples where material appears to be well-preserved. However, it will become increasingly important to demonstrate reliability as the spatial resolution of analysis improves further.

Most early studies were conducted by researchers from research groups that specialized in karst research. The field is now populated by researchers from a whole host of geochemical and geophysical backgrounds and it can be viewed as truly inter-disciplinary. This is one of the strengths of speleothem research and we envisage even greater collaborative efforts between research teams in the future. Essential to the success of speleothem research, however, is the adherence to a few simple guidelines (see also Smart et al. 1996) that are already routinely observed by many current researchers:

(1) Geochemical evidence from precipitated calcite should be supplemented with study of the present-day processes in the vadose system above the location of the sample.

(2) Analyzes should be conducted on more than one sample, when available, to determine whether inferred environmental factors are indeed regional or global, and not related to specific local effects.

(3) Reliable chronologies should be based on multiple ^{230}Th age determinations at a sampling interval that is appropriate for the spatial resolution of geochemical analysis along the axis of growth and accounts for the possibility of non-linear or discontinuous growth.

(4) Multi-proxy evidence should be obtained along the axis of growth for individual samples. If trace elemental anomalies, related to water-rock interaction, correspond with $\delta^{18}O$ anomalies, related to regional atmospheric circulation or "amount effect," for example, then regional to global scale environmental change is implicated and can be studied at very high resolution.

Speleothems have many advantages over other archives of past geochemical and geophysical information, and we expect their use to become even more widespread. Their value is readily exemplified by recent high-resolution $\delta^{18}O_{calcite}$ records for the last glacial cycle, which have comparable resolution to the polar ice cores. The latter are currently the most widely cited archives for evidence of dramatic and abrupt environmental change for the late Pleistocene and Holocene, yet, in contrast to speleothems, cannot be directly dated for much of their length. Speleothems will play a complimentary role by providing chronological information by way of uranium-series dating and correlation, and more global coverage of regional $\delta^{18}O_{water}$ information. Such information can then be used to verify coupled isotope-atmosphere-ocean models results. We also expect speleothems to be more widely used to provide environmental information prior to the last interglacial-glacial cycle because remarkably well-preserved samples, exhibiting no geochemical or physical alteration, can be obtained. Currently, one of the limiting factors in using speleothems close to limit of uranium-series dating techniques is the precision and accuracy of the half-lives of the shorter-lived nuclides in the uranium-series decay chain. Continued attention must be focused on reducing the precision of these constants if we are to investigate the phasing of environmental change and internal and external forcing during the past 0.5 Ma. Also, more feasibility work should be undertaken in the field of U-Th-Pb dating, because this technique offers the possibility of extending the dateable range to beyond 0.5 Ma. While some researchers will be looking to older deposits, others will be devoting effort to the construction of environmental archives with, perhaps, sub-annual resolution. This is an exciting prospect because such records can be used to test models of the sensitivity of environmental change at much shorter timescales by comparison with historical records and geological evidence from the Holocene. The future then is bright, at least for those stages of research projects that will be undertaken above ground!

ACKNOWLEDGMENTS

Many of the views presented here have been informed by valuable discussions with colleagues at the Universities of Bristol, Leeds, Iowa and Minnesota. The list of names is too long to place here, but special mention should be reserved for the contributions of Peter Smart and Larry Edwards. We have drawn upon numerous examples from our own work in this review and acknowledge the support of the Natural Environment Research Council (UK) and the National Science Foundation (USA). We are grateful for the constructive comments of G. Henderson, A. Mangini and two anonymous reviewers.

REFERENCES

Allègre CJ (1964) De l'extension de la méthode de calcul graphique Concordia aux measures d'âges absolus effctués à l'aide du déséquilibre radioactif. Cas ds minéralisations secondaires d'uanium. C R Acad Sci Paris 259:4086-4089

Alley RB, Mayewski PA, Sowers T, Stuiver M, Taylor KC, Clark PU (1997) Holocene climatic instability: A prominent, widespread event 8200 yr ago. Geology 25:483-486

Amelin Y, Zaitsev AN (2002) Precise geochronology of phoscorites and carbonatites: The critical role of U-series disequilibrium in age interpretations. Geochim Cosmochim Acta 66:2399-2419

Amundson RG, Chadwick OA, Sowers JM, Doner HE (1988) Relationship between climate and vegetation and the stable carbon isotope chemistry of soils in the eastern Mojave Desert, Nevada. Quat Res 29:245-254

Antonioli F, Oliverio M (1996) Holocene sea-level rise recorded by a radiocarbon-dated mussel in a submerged speleothem beneath the Mediterranean Sea. Quat Res 45:241-244

Antonioli F, Silenzi S, Frisia S, (2001) Tyrrhenian Holocene palaeoclimate trends from spelean serpulids Fabrizio. Quat Sci Rev 20:1661-1670

Atkinson TC, Lawson TJ, Smart PL, Harmon, RS, Hess JW (1978) Paleoclimate and geomorphic implications of $^{230}Th/^{234}U$ dates on speleothem from Britain. Nature 272:24-8

Atkinson TC, Rowe PJ (1992) Applications of dating to denudation chronology and landscape evolution. *In:* Uranium-series disequilibrium: Applications to Earth, Marine and Environmental Sciences, Ivanovich M, Harmon RS (eds) Oxford University Press, Oxford, p 669-703

Auler AS, Smart PL (2001) Late Quaternary paleoclimate in semiarid Northeastern Brazil from U-series dating of travertines and water table speleothems. Quat Res 55:159-167

Ayliffe LK, Marianelli PC, Moriarty KC, Wells RT, McCulloch MT, Mortimer GE, Hellstrom JC (1998) 500 ka precipitation record from southeastern Australia: Evidence for interglacial relative aridity. Geology 26:147-150

Ayliffe LK, Veeh HH (1988) Uranium-series dating of speleothems and bones from Victoria Cave, Naracoorte, South Australia. Chem Geol 72:11-234

Baker A, Barnes WL, Smart PL (1996) Speleothem luminescence intensity and spectral characteristics: Signal calibration and a record of paleovegetation change. Chem Geol 130:65-76

Baker A, Genty D, Dreybrodt W, Barnes WL, Mockler HJ, Grapes J (1998) Testing theoretically predicted stalagmite growth rate with recent annually laminated samples: Implications for past stalagmite deposition. Geochim Cosmochim Acta 62:393-404

Baker A, Genty D, Smart PL (1998) High-resolution records of soil humification and paleoclimatic change from variations in speleothem luminescence excitation and emission wavelengths. Geology 26:903-906

Baker A, Ito E, Smart PL, McEwan R (1997) Elevated ^{13}C in speleothem and implications for palaeo-vegetation studies. Chem Geol 136:263-270

Baker A, Smart PL (1995) Recent flowstone growth rates: Field measurements in comparison to theoretical predictions. Chem Geol 122:121-128

Baker A, Smart PL and Ford DC (1993a) Northwest European palaeoclimate as indicated by growth frequency variations of secondary calcite deposits. Palaeogeog Palaeoclim Palaeoecol 100:291-301

Baker A, Smart PL Edwards RL (1995) Paleoclimate implications of mass spectrometric dating of a British flowstone. Geology 23:309-312

Baker A, Smart PL, Edwards RL (1996) Mass spectrometric dating of flowstones from Stump Cross Caverns and Lancaster Hole, Yorkshire: palaeoclimate implications. J Quat Sci 11:107-115

Baker A, Smart PL, Edwards RL Richards DA (1993b) Annual growth banding in a cave stalagmite. Nature 364:518-520

Baker RG, Gonzalez LA, Raymo M, Bettis EA, Reagan MK, and Dorale JA (1998) Comparison of multiple proxy records of Holocene environments in midwestern USA. Geology 26:1131-1134

Baldini JUL, McDermott F, Fairchild IJ (2002) Structure of the 8200-year Cold Event revealed by a speleothem trace element record. Science 296:2203-2206

Bard E, Antonioli F, Silenzi S (2002a) Sea-level during the penultimate interglacial period based on a submerged stalagmite from Argentarola Cave (Italy). Earth Planet Sci Lett 196:135-146

Bard E, Arnold M, Hamelin B, Tisnerat-Laborde N, Cabioch G (1998) Radiocarbon calibration by means of mass spectrometric Th-230/U-234 and C-14 ages of corals: An updated database including samples from Barbados, Mururoa and Tahiti. Radiocarbon 40:1085-1092

Bard E, Arnold M, Hamelin B, Tisnerat-Laborde N, Cabioch G (1998) Radiocarbon calibration by means of mass spectrometric $^{230}Th/^{234}U$ and ^{14}C ages of corals. An updated base including samples from Barbados, Mururoa and Tahiti. Radiocarbon 40:1085-1092

Bard E, Delaygue G, Rostek F, Antonioli F, Silenzi S, Schrag DP (2002b) Hydrological conditions over the western Mediterranean basin during the deposition of the cold sapropel 6 event. Earth Planet Sci Lett 202:481-494

Bard E, Hamelin B, Fairbanks RG, Zindler A (1990) Calibration of the ^{14}C timescale over the past 30,000 years using mass spectrometric U-Th ages from Barbados corals. Nature 345:405-410

Bar-Matthews M, Avalon A, Matthews A, Sass E, Halicz L (1996) Carbon and oxygen isotope study of the active water-carbonate system in a karstic Mediterranean cave: Implications for paleoclimate research in semiarid regions. Geochim Cosmochim Acta 60:337-347

Bar-Matthews M, Ayalon A, Kaufman A (1997) Late Quaternary paleoclimate in the eastern Mediterranean region from stable isotope analysis of speleothems at Soreq Cave, Israel. Quat Res 47:155-168

Bar-Matthews M, Matthews A, Avalon A (1991) Environmental controls of speleothem mineralogy in a karstic dolomitic terrain (Soreq Cave, Israel). J Geol 99:189-207

Bassiakos Y (2001) Assessment of the lower ESR dating range in Greek speleothems. J Radioan Nucl Chem 247:629-633

Bastin B, Gewelt M (1986) Analyze pollinique et datation ^{14}C des concretions stalagmitiques Holocenes: Apports complementaires des deaux methods. Geographie Physique et Quaternaire 40:185-196

Beck JW, Richards DA, Edwards RL, Silverman BW, Smart PL, Donahue DL, Hererra-Osterheld S, Burr GS, Calsoyas L, Jull AJT, Biddulph D (2001) Extremely large variations of atmospheric ^{14}C concentration during the last glacial period. Science 29:2453-2458

Bender ML, Fairbanks RG, Taylor FW, Matthews RK, Goddard JG, Broecker WS (1979) Uranium-series dating of the Pleistocene reef tracts of Barbados, West Indies. Geol Soc Am Bull 90:577-594

Betancourt JL, Grissino-Mayer HD, Salzer MW, Swetnam TW (2002) A test of "annual resolution" in stalagmites using tree rings. Quat Res 58:197-199

Bischoff JL, Fitzpatrick JA (1991) U-series dating of impure carbonates: An isochron technique using total-sample dissolution. Geochim Cosmochim Acta 55:543-554

Broecker WS, Olson EA, Orr PC (1960) Radiocarbon measurements and annual rings in cave formations. Nature 185:93-94

Brook GA Sheen S-W, Rafter MA, Railsback LB, Lundberg J (1999) A high-resolution proxy record of rainfall and ENSO since AD 1550 from layering in stalagmites from Anjohibe Cave, Madagascar. Holocene 9:695-705

Buhmann D, Dreybodt W (1985a) The kinetics of calcite dissolution and precipitation in geologically relevant situations of karst areas, 1. Open system. Chem Geol 48:189-211

Buhmann D, Dreybodt W (1985b) The kinetics of calcite dissolution and precipitation in geologically relevant situations of karst areas, 1. Closed system. Chem Geol 53:109-124

Burns SJ, Fleitman D, Matter A, Neff U, Mangini A (2001) Speleothem evidence from Oman for continental pluvial events during interglacial periods. Geology 29:623-626

Burns SJ, Matter A, Frank N, Mangini A (1998) Speleothem-based paleoclimate record from northern Oman. Geology 26:499-502

Charman DJ, Caseldine CJ, Baker A, Gearey B, Hatton J (2001) Palaeohydrological records from peat profiles and speleothems in Sutherland, NW Scotland. Quat Res 55:223-234

Cheng H, Adkins JF, Edwards RL, Boyle EA (2000) ^{230}Th dating of deep-sea corals. Geochim Cosmochim Acta 64:2401-2416

Cheng H, Edwards RL Wang Y (1996) U/Th and U/Pa dating of Nanjing Man. EOS (Trans, Am Geophys Union) 78:F787

Cheng H, Edwards RL, Murrell MT, Benjamin TM (1998) Uranium-thorium-protactinium dating systematics. Geochim Cosmochim Acta 62:3437-3452

Cherdyntsev VV, Kazachevskii IV, Kuz'mina YA (1965) Age of carbonate determined from the isotopes of thorium and uranium. Geochem Int 2:749-756

Chung GS, Swart PK (1990) The concentration of uranium in freshwater vadose and phreatic cements in a Holocene ooid clay: a method of identifying ancient water tables. J Sediment Petrol 60:735-746

Cliff RA, Spötl C (2001) U-Pb dating of speleothems from the Spannagel Cave, Austria. XI EUG Conference, Strasbourg, France. J Conf Abstr 6:601

Cole JE, Rind D, Webb RS, Jouzel J, Healy R (1999) Climatic controls on interannual variability of precipitation δ18O: Simulated influence of temperature, precipitation amount, and vapor source region. J Geophys Res 104:14223-14235

Craig H (1961) Isotopic variations in meteoric waters. Science 133:1702-1703

Crowley TJ, Kim K-Y (1994) Milankovitch forcing of the last interglacial sea level. Science 265:1566-1568

Dansgaard W (1964) Stable isotopes in precipitation. Tellus 16:436-468

Dennis PF, Rowe PJ, Atkinson TC (2001) The recovery and isotopic measurement of water from fluid inclusions in speleothems. Geochim Cosmochim Acta 65:871-884

Denniston RF, González LA, Asmerom Y, Baker RG, Reagan MK, Bettis EA (1999a) Evidence for increased cool season moisture during the middle Holocene. Geology 27:815-818

Denniston RF, González LA, Asmerom Y, Polyak V, Reagan MK, Saltzman MR (2001) A high-resolution speleothem record of climatic variability at the Allerod-Younger Dryas transition in Missouri, central United States. Palaeogeog Palaeoclim Palaeoecol 176:147-155

Denniston RF, González LA, Asmerom Y, Reagan MK (2000) Speleothem records of Holocene paleoenvironmental change in the Ozark Highlands, USA. Quat Int 67:21-28

Denniston RF, González LA, Semken HA, Jr, Baker RG, Recelli-Snyder H, Reagan MK, Bettis, EA (1999b) Integrating stalagmite, vertebrate, and pollen sequences to investigate Holocene vegetation and climate change in the southern Midwest. Quat Res 52:381-387

Dervin J, Faucherre J (1973a) Étude des carbonates complexes de thorium et de cérium. II- Constitution des complexes en colution. Bull Soc Chim Fr 11:2926-2929

Dervin J, Faucherre J (1973b) Étude des carbonates complexes de thorium et de cérium. I- Solubilité et nature des ions complexes en solution. Bull Soc Chim Fr 11:2930-2933

Dickson BL, Wheller GE (1992) Uranium-series disequilibrium in exploration geology *In*: Uranium-Series Disequilibrium: Applications to Earth, Marine, and Environmental Sciences, 2nd ed. Ivanovich M, Harmon RS (eds) Clarendon Press, Oxford. p 704-729

Dorale JA, Edwards RL, Alexander CA, Shen CC, Richards DA, Cheng H (2003) Uranium-series dating of speleothems: Current techniques, limits and applications. *In*: Studies of Cave Sediments, Sasowsky ID, Mylroie JE (eds) Kluwer Academic/Plenum Publishers, New York. (in press)

Dorale JA, Edwards RL, Ito E, González LA (1998) Climate and vegetation history of the Midcontinent from 75 to 25 ka: a speleothem record from Crevice Cave, Missouri, USA. Science 282:1871-1874

Dorale JA, Edwards RL, Onac BP (2002) Stable isotopes as environmental indicators in speleothems. *In*: Karst Processes and the Carbon Cycle. Yuan, D-X (ed) Geological Publishing House, Beijing, China. p 107-120

Dorale JA, González LA, Reagan MK, Pickett DA, Murrell MT, Baker RG (1992) A high-resolution record of Holocene climate change in speleothem calcite from Cold Water Cave, northeast Iowa. Science 258:1626-1630

Dreybrodt W (1980) Deposition of calcite from thin films of natural calcareous solutions and the growth of speleothems. Chem Geol 29:89-105

Dreybrodt W (1997) Chemical kinetics, speleothem growth, and climate. *In:* Climate Change: The Karst Record. S-E Lauritzen (ed) Karst Waters Institute Special Publications 2: Charles Town, West Virginia p 23-25

Dulinski M, Rozanski K (1990) Formation of $^{13}C/^{12}C$ isotope ratios in speleothems: A semi-dynamic model. Radiocarbon 32:7-16

Duplessy JC, Labeyrie L, Lalou C, Nguyen HV (1970) Continental climate variations between 130,000 and 90,000 years BP Nature 226:631-632

Edwards RL, Beck JW, Burr GS, Donahue DJ, Chappell JMA, Bloom AL, Druffel ERM, Taylor FW (1993) A large drop in atmospheric C-14/C-12 and reduced melting in the Younger Dryas documented with Th-230 ages of corals. Science 260:962-968

Edwards RL, Cheng H, Murrell MT, Goldstein SJ (1997) Protactinium-231 dating of carbonates by thermal ionization mass spectrometry: Implications for Quaternary climate change. Science 276:782-786

Edwards RL, Gallup CD, Cheng H (2003) Uranium-series dating of marine and lacustrine carbonates. Rev Mineral Geochem 52:363-405

Esat TM, McCulloch MT, Chappell J, Pillans B, Omura A (1999) Rapid fluctuations in sea level recorded at Huon Peninsula during the Penultimate Deglaciation. Science 283:197-201

Fairbanks RG (1989) A 17,000 year glacio-eustatic sea level record: Influence of glacial melting rates on the Younger Dryas event and deep-ocean circulation. Nature 342:637-642

Fairchild IJ, Borsato A, Tooth AF, Frisia S, Hawkesworth CJ, Huang Y, McDermott F, Spiro B (2000) Controls on trace element (Sr-Mg) compositions of carbonate cave waters: implications for speleothem climatic records. Chemical Geology 166:255-269

Fairchild IJ, Tooth AF, Huang Y, Borsato A, Frisia S, McDermott F (1996) Spatial and temporal variations in water and stalactite chemistry in currently active caves: a precursor to interpretations of past climate. *In:* Proc. 4th International Symposium on Geochemistry of the Earth's Surface. Bottrell SH (ed) Ilkley, Yorkshire, University of Leeds, UK p 229-233

Farrant AR, Smart PL, Whitaker FF (1995) Long-term Quaternary uplift rates inferred from limestone caves in Sarawak, Malaysia. Geology 23:357-360

Ford DC (1997) Dating and paleo-environmental studies of speleothems. *In:* Cave Minerals of the World (2nd edn). Hill CA, Forti P (eds) National Speleological Society of America Press, Huntsville, Alabama. p 271-284

Ford DC, Hill CA (1989) Dating results from Carlsbad Cavern and other caves in the Guadalupe Mountains, New Mexico. Isochron/West 54:3-7

Ford DC, Williams PW (1989) Karst geomorphology and hydrology. Unwin Hyman, London

Fornós JJ, Gelabert B, Ginés A, Ginés J, Tuccimei P, Vesica P (2002) Phreatic overgrowths on speleothems: a useful tool in structural geology in littoral karstic landscapes. The example of eastern Mallorca (Balearic Islands). Geodinimica Acta, 15:113-125

Frank N, Mangini A, Korfmann M (2002) $^{230}Th/U$ dating of the Trojan "Water Quarries." Archaeometry 44:305-314

Friedman I, O'Neil JR (1977) Compilation of stable isotope fractionation factors of geochemical interest. U S Geol Surv Prof Paper 440-KK

Frisia S, Borsato A, Fairchild IJ, McDermott F (2000) Calcite fabrics, growth mechanisms, and environments of formation in speleothems from the Italian Alps and southwestern Ireland. J Sediment Res 70:1183-1196

Fritz P, Drimmie RJ, Frape SK, O'Shea K (1987) The isotopic composition of precipitation and groundwater. *In:* Canada International Symposium on the Use of Isotope Techniques in Water Resources Development. IAEA Symposium, Vienna 299:539-550

Frumkin A, Ford DC, Schwarcz HP (1999) Continental oxygen isotopic record of the last 170,000 years in Jerusalem. Quat Res 51:317-327

Gallup CD, Cheng H, Taylor FW, Edwards RL (2002) Direct determination of the timing of sea level change during Termination II. Science 295:310-313

Gallup CD, Edwards RL, Johnson RG (1994) The timing of high sea levels over the past 200,000 years. Science 263:796-800

Gascoyne M (1983) Trace element partition coefficients in the calcite-water system and their paleoclimatic significance. J Hydrol 61:213-222

Gascoyne M (1984) Uranium Series ages of Speleothems from Bahaman Blue Holes and their significance. Trans British Cave Res Assoc 11:45-49

Gascoyne M (1985) Application of the $^{227}Th/^{230}Th$ method to dating Pleistocene carbonates and comparisons with other dating methods. Geochim Cosmochim Acta 49:1165-1171

Gascoyne M (1992a) Paleoclimate determination from cave deposits. Quat Sci Rev 11:609-632

Gascoyne M (1992b) Geochemistry of the actinides and their daughters. *In:* Uranium-Series Disequilibrium: Applications to Earth, Marine, and Environmental Sciences, 2nd ed. Ivanovich M, Harmon RS (eds) Clarendon Press, Oxford. p 34-61

Gascoyne M, Benjamin GJ, Schwarcz HP, Ford DC (1979) Sea-level lowering during the Illinoian glaciation: evidence from a Bahama "Blue Hole." Science 205:806-808

Gascoyne M, Currant AP, Lord TC (1981) Ipswichian fauna of Victoria Cave and the marine palaeoclimatic record. Nature 294:652-4

Gascoyne M, Ford DC, Schwarcz HP (1981) Late Pleistocene chronology and paleoclimate of Vancouver Island determined from cave deposits. Ca J Earth Sci 18:1643-1652

Gascoyne M, Harmon RS (1992) Palaeoclimatology and palaeoselevels. *In:* Uranium-series disequilibrium: Applications to Earth, Marine and Environmental Sciences. Ivanovich M, Harmon RS (eds) Oxford University Press, Oxford. p 553-582

Gascoyne M, Schwarcz HP (1982) Carbonate and sulphate precipitates. *In:* Uranium series disequilibrium: Applications to environmental problems (1st edn). Ivanovich M, Harmon RS (eds) Calrendon Press, Oxford. p 268-301

Gascoyne M, Schwarcz HP, Ford DC (1978) Uranium-series dating and stable-isotope studies of speleothems. Part 1. Theory and techniques. Trans Brit Cave Res Assoc 5:91-112

Gascoyne M, Schwarcz HP, Ford DC (1983) Uranium-series ages of speleothem from northwest England: correlation with Quaternary climate. Phil Trans Royal Soc London B301:143-164

Gat JR, Carmi I (1970) evolution of the isotopic composition of stamospheric waters in the Mediterranan Sea area. J Geophys Res 75:3039-3048

Genty D, Baker A, Massault M, Proctor C, Gilmour M, Pons-Branchu E, Hamelin B (2001) Dead carbon in stalagmites: Carbonate bedrock paleodissolution vs. ageing of soil organic matter. Implications for ^{13}C variations in speleothems. Geochim Cosmochim Acta 65:3443-3457

Genty D, Massault M (1999) Carbon transfer dynamics from bomb-^{14}C and $\delta^{13}C$ time series of a laminated stalagmite from SW France - Modeling and comparison with other stalagmite records. Geochim Cosmochim Acta 63:1537-1548

Genty D, Quinif Y (1996) Annually laminated sequences in the internal structure of some Belgian stalagmites - importance for paleoclimatology. J Sediment Res 66:275-288

Genty D, Vokal B, Obelic B, Massault M (1998) Bomb ^{14}C time history recorded in two modern stalagmites - importance for soil organic matter dynamics and bomb ^{14}C distribution over continents. Earth Planet Sci Lett 160:795-809

Getty SR, Asmerom Y, Quinn TM (2001) Accelerated Pleistocene coral extinctions in the Caribbean Basin from uranium-lead (U-Pb) dating. Geology 29:639-642

Getty SR, DePaolo DJ (1995) Quaternary geochronology using the U-Th-Pb method. Geochim Cosmochim Acta 59:3267-3272

Geyh M (1970) Isotopenphysikalische Untersuchungen an Kalksinter, ihre Bedeutung fur die 14C-Alterbestimmung von Gründwasser und die erforshung des paläoklimas. Geol Jahrb 88:149-159

Goede A, Green DC, Harmon RS (1986) Late Pleistocene palaeotemperature record from a Tasmanian speleothem. Aust J Earth Sci 33:333-342

Goede A, Vogel JC (1991) Trace element variations and dating of a late Pleistocene Tasmanian speleothem. Palaeogeog Palaeoclim Palaeoecol 88:121-131

Goldstein SJ, Stirling CH (2003) Techniques for measuring uranium-series nuclides: 1992-2002. Rev Mineral Geochem 52:23-57

González LA, Carpenter SJ, Lohmann KC (1992) Inorganic calcite morphology: roles of fluid chemistry and fluid flow. J Sediment Petrol 62:382-399

Gordon D, Smart PL (1984) Comments on "Speleothems Travertines and Paleoclimates" by G.J. Hennig, R. Grün and K. Brunnacker. Quat Res 22:144-147

Gordon D, Smart PL, Ford DC, Andrews JN, Atkinson TC, Rowe PJ, Christopher NSJ (1989) Dating of late Pleistocene interglacial and interstadial periods in the United Kingdom from speleothem growth frequency. Quat Res 31:14-26

Grenthe I, Fuger J, Konings RJM, Lemire RJ, Muller AB, Nguyen-Rung C, Wanner H (1992) Chemical Thermodynamics of Uranium. Nuclear Energy Agency. North-Holland, Amsterdam

Grootes PM, Stuiver M, White JWC, Johnsen S, Jouzel JJ (1993) Comparison of oxygen isotope records from the GISP2 and GRIP Greenland ice cores. Nature 366:552-554

Grün R (1989) Electron spin resonance (ESR) dating. Quat Int 1:65-109

Hancock PL, Chalmers RML, Altunel E, Cakir Z (1999) Travitonics: using travertines in active fault studies. J Struct Geol 21:903-916

Harmon RS, Ford DC, Schwarcz HP (1977) Interglacial chronology of Rocky and Mackenzie Mountains based upon ^{230}Th-^{234}U dating of speleothems. Can J Earth Sci 14:2543-2552

Harmon RS, Land LS, Mitterer RM, Garrett P, Schwarcz HP, Larson GJ (1981) Bermuda sea levels during the last interglacial. Nature 289:481-3

Harmon RS, Mitterer RM, Kriausakul N, Land LS, Schwarcz HP, Garrett P, Larson GJ, Vacher HL, Rowe M (1983) U-Series and amino-acid racemization geochronology of Bermuda: Implications for eustatic sea-level fluctuation over the past 250,000 years. Palaeogeog Palaeoclim Palaeoecol 44:41-70

Harmon RS, Schwarcz HP (1981) Changes of ^{2}H and ^{18}O enrichment of meteoric water and Pleistocene glaciation. Nature 290:125-128

Harmon RS, Schwarcz HP, O'Neil JR (1979) D/H ratios in speleothem fluid inclusions: A guide to variations in the isotopic compositions of meteoric precipitation? Earth Planet Sci Lett 42:254-266

Harmon RS, Thompson P, Schwarcz HP, Ford DC (1975a) Late Pleistocene paleoclimates of North America as inferred from stable isotope studies of speleothems. Quat Res 9:54-70

Harmon RS, Thompson P, Schwarcz HP, Ford DC (1975b) Uranium-series dating of speleothems. Nat Speleological Soc Bull 37:21-33

Hearty PJ, Kindler P, Cheng H, Edwards RL (1999). Evidence for a +20 m middle Pleistocene sea-level highstand (Bermuda and Bahamas) and partial collapse of Antarctic ice. Geology 27:375-378

Hellstrom J, McCulloch M, Stone J (1998) A detailed 31,000-year record of climate and vegetation change from the isotope geochemistry of two New Zealnd speleothems. Quat Res 50:167-178

Hellstrom JC, McCulloch MT (2000) Multi-proxy constraints on the climatic significance of trace element records from a New Zealand speleothem. Earth Planet Sci Lett 179:287-297

Henderson GM, Slowey NC (2000). Evidence against northern-hemisphere forcing of the penultimate deglaciation from U-Th dating. Nature 402:61-66

Henderson GM, Slowey NC, Fleisher MQ (2001) U-Th dating of carbonate platform and slope sediments. Geochim Cosmochim Acta 65:2757-2770

Hendy CH (1970) The use of ^{14}C in the study of cave processes. In: Twelfth Nobel Symposium, Radiocarbon Variations and Absolute Chronology. Olsson IU (ed) Almqvist and Wiksell, Stockholm, and Wiley and Sons, New York, p 419-443

Hendy CH (1971) The isotopic geochemistry of speleothems. I. The calculation of the effects of different modes of formation on the composition of speleothems and their applicability as paleoclimatic indicators. Geochim Cosmochim Acta 35:801-824

Hendy CH, Wilson AT (1968) Paleoclimatic data from speleothems. Nature 219:48-51

Hennig GJ, Grün H, Brunnacker K (1983) Speleothems, travertines, and paleoclimates. Quat Res 20:1-29

Hercman H (2000) Reconstruction of palaeoclimatic changes in Central Europe between 10 and 200 thousand years BP, based on analysis of growth frequency of speleothems. Studia Quat 17:35-70

Hill CA, Forti P (1997) Cave Minerals of the World (2nd edn). National Speleological Society, Huntsville

Holmgren K, Lauritzen S-E, Possnert G (1994) ^{230}Th/^{234}U and ^{14}C dating of a Late Pleistocene stalagmite in Lobatse II Cave, Botswana. Quat Sci Rev 13:111-119

Huang Y, Fairchild IJ (2001) Partitioning of Sr^{2+} and Mg^{2+} into calcite under karst-analog experimental conditions. Geochim Cosmochim Acta 65:47-62

Huang Y, Fairchild IJ, Borsato A, Frisia S, Cassidy NJ, McDermott F, Hawkesworth CJ (2001) Seasonal variations in Sr, Mg and P in modern speleothems (Grotta di Ernesto, Italy). Chem Geol 175:429-448

Ikeya M (1975) Dating a stalactite by electron paramagnetic resonance. Nature 255:48-50

Israelson C, Halliday AN Buchardt B (1996) U-Pb dating of calcite concretions from Cambrian black shales and the Phanerozoic time scale. Earth Planet Sci Lett 141:153-159

Jahn B-m, Cuvellier H (1994) Pb-Pb and U-Pb geochronology of carbonate rocks: An assessment. Chem Geol 115:125-151

Joao A, Bigot S, Fromage F (1987) Study of the carbonate complexes of IVB elements: I. Determination of the stability constant of Th(IV) pentacarbonate. Bull Soc Chim Fr, Issue 1, p. 42-44

Johnsen SJ, Dahl-Jensen D, Dansgaard W, Gundestrup N (1995) Greenland palaeotemperatures derived from GRIP bore hole temperature and ice core isotope profile. Tellus 47B:624-629

Johnsen SJ, Dahl-Jensen D, Gundestrup N, Steffensen JP, Clausen HB, Miller H, Masson-Delmotte V, Sveinbjörnsdottir AE, White J (2001) Oxygen isotope and palaeotemperature records from six Greenland ice-core stations: Camp Century, Dye-3, GRIP, GISP2, Renland and NorthGRIP J Quat Sci 16:299-307

Johnsen SJ, Dansgaard W, White JWC (1989) The origin of arctic precipitation under present and glacial conditions. Tellus 41B:452-468

Jones B (1995) Processes associated with microbial biofilms in the twilight zone of caves: examples from the Cayman Islands. J Sediment Res A65:552-560

Jones CE, Halliday AN, Lohmann KC (1995) The impact of diagenesis on high-precision U-Pb dating of ancient carbonates: An example from the Late Permian of New Mexico. Earth Planet Sci Lett 134:409-423

Jouzel J, G. Hoffmann G, Koster RD, Masson V (2000) Water isotopes in precipitation: data/model comparison for present-day and past climates. Quat Sci Rev 19:363-379

Kagan E, Agnon A, Bar-Matthews M, Ayalon A (2002) Cave deposits as recorders of paleoseismicity: A record from two caves located 60 km west of the Dead Sea Transform (Jerusalem, Israel). *In:* Environmental Catastrophes and Recoveries in the Holocene (online abstract http://atlas-conferences.com/cgi-bin/abstract/caiq-38)

Kashiwaya K, Atkinson TC, Smart PL. (1991). Periodic variations in late Pleistocene speleothem abundance in Britain. Quat Res 35:190-196

Kaufman A (1993) An evaluation of several methods for determining ^{230}Th/U ages in impure carbonates. Geochim Cosmochim Acta 57:2303-2317

Kaufman A, Broecker WS (1965) Comparison of ^{230}Th and ^{14}C ages for carbonate materials from Lakes Lahontan and Bonneville. J Geophys Res 70:4039-4054

Kaufman A, Ku T-L, Luo S (1995) Uranium-series dating of carnotites: Concordance between Th-230 and Pa-231 ages. Chem Geol 120:175-181

Kaufman A, Wasserburg GJ, Porcelli D, Bar-Matthews M, Ayalon A, Halicz L (1998) U-Th isotope systematics and U-series ages of speleothems from Soreq Cave, Israel and climatic correlations. Earth Planet Sci Lett 156:141-155

Kendall AC, Broughton PL (1978) Origin of fabrics in speleothems composed of columnar calcite crystals. J Sediment Petrol 48:519-538

Kitagawa H; van der Plicht J (2000) Atmospheric radiocarbon calibration beyond 11,900 cal BP from Lake Suigetsu laminated sediments. Radiocarbon 42:369-380

Ku T-L (2000) Uranium-Series Methods. *In:* Quaternary Geochronology: Methods and Applications. Stratton Noller J, Sowers JM, Lettis WR (eds) American Geophysical Union, Washington DC, p 101-114

Ku T-L, Kimmel MA, Easton WH, O'Neil TJ (1974) Eustatic sea level 120,000 years ago on Oahu Hawaii. Science 183:959-962

Ku T-L, Liang ZC (1984) The dating of impure carbonates with decay series isotopes. Nucl Instr Meth 223:563-571

Langmuir D (1997) Aqueous Environmental Geochemistry. Prentice Hall, New Jersey

Langmuir D, Herman JS (1980) The mobility of thorium in natural waters at low temperatures. Geochim Cosmochim Acta 44:1753-1766

Latham AG, Ford DC (1993) The paleomagnetism and rock magnetism of cave and karst deposits. *In:* Applications of Paleomagnetism to Sedimentary Geology. Aissaoui DM, McNeill DF, Hurley NF (eds). SEPM Special Publications. 49:149-155

Latham AG, Schwarz HP (1992) Carbonate and sulphate precipitates. *In:* Uranium-series disequilibrium: Applications to Earth, Marine and Environmental Sciences, Ivanovich M, Harmon RS (eds) Oxford University Press, Oxford, p 423-459

Lauritzen S-E (1991) Karst resources and their conservation in Norway. Norsk Geografisk Tidskrift 45:119-142

Lauritzen S-E (1993) Natural environmental change in karst: The quaternary record. Catena Supp 25:21-40

Lauritzen S-E (1995) High-resolution paleotemperature proxy record for the last interglaciation based on Norwegian speleothems. Quat Res 43:133-146

Lauritzen S-E (2003) Reconstructing Holocene climate records from speleoethems. *In:* Global Change in the Holocene. Mackay AW, Battarbee RW, Birks HJB (eds) Oldfield F. Arnold, London (in press)

Lauritzen S-E, Mylroie JE Results of a Speleothem U/Th Dating Reconnaissance from the Helderberg Plateau, New York. J Cave Karst Studies 62(1):20-26

Lauritzen S-E, Onac BP (1995) Uranium-series dating of speleothems from Romanian caves. Theor Appl Karst 8:25-36

Lemeille F, Cushing M, Carbon D, Grellet B, Bitterli T, Flehoc C (1999) Co-seismic ruptures and deformations recorded by speleothems in the epicentral zone of the Basel earthquake. Geodinam Acta 12:179-191

Li W-X, Lundberg J, Dickin AP, Ford DC, Schwarcz HP, McNutt R, Williams D (1989) High-precision mass-spectrometric uranium-series dating of cave deposits and implications for paleoclimate studies. Nature 339:534-536

Lin JC, Broecker WS, Anderson RF, Hemming S, Rubenstone JL, Bonani G (1996) New Th-^{230}/U and C-14 ages from Lake Lahontan carbonates, Nevada, USA, and a discussion of the origin of initial thorium. Geochim Cosmochim Acta 60:2817-2832

Lively RS (1983) Late Quaternary U-series speleothem growth record from southeastern Minnesota. Geology 11:259-262

Lomitschka M, Mangini A (1999) An ascorbic acid/Na$_2$EDTA cleaning procedure for precise Th/U-dating of small samples of deep sea corals. Earth Planet Sci Lett 170:391-401

Ludwig KR (1977) Effect of initial radioactive-daughter disequilibrium on U-Pb isotope apparent ages of young minerals. J Res United States Geol Surv 5:663-667

Ludwig KR (1999) Using Isoplot/Ex, Version 2.01: A geochronological toolkit for Microsoft Excel. Berkeley Geochronology Center Special Publication 1a:47

Ludwig KR (2003) Mathematical-statistical treatment of data and errors for ^{230}Th/U geochronology. Rev Mineral Geochem 52:631-636

Ludwig KR, Paces JB (2002) Uranium-series dating of pedogenic silica and carbonate, Crater Flat, Nevada. Geochim Cosmochim Acta 66:487-506

Ludwig KR, Simmons KR, Szabo BJ, Winograd IJ, Landwehr JM, Riggs AC, Hoffman RJ (1992) Mass-spectrometric ^{230}Th-^{234}U-^{238}U dating of the Devils Hole calcite vein. Science 258:284-287

Ludwig KR, Simmons KR, Winograd IJ, Szabo BJ, Riggs AC (1993) Dating of the Devils Hole calcite vein. Science 259:1626-1627

Ludwig KR, Szabo BJ, Moore JG, Simmons KR (1991) Crustal subsidence rate off Hawaii determined from ^{234}U/^{238}U ages of drowned coral reefs. Geology 19:171-174

Ludwig KR, Titterington DM (1994) Calculation of ^{230}Th/U isochrons, ages, and errors. Geochim Cosmochim Acta 58:5031-5042

Lundberg J (1997) Paleoclimatic reconstruction and timing of sea level rise at the end of the Penultimate Glaciation, from detailed stable isotopic study and TIMS dating of submerged Bahamian speleothem. Proceedings of the 12th Int. Confr. Speleology, Switzerland 1:101

Lundberg J, Ford DC (1994) Late Pleistocene sea level change in the Bahamas from U-series dating of speleothem by mass spectrometry. Quat Sci Rev 13:1-14

Lundberg J, Ford DC, Hill CA (2000) A preliminary U-Pb date on cave spar, Big Canyon, Guadalupe Mountains, New Mexico, USA. J Cave Karst Studies 62:144-148

Luo S, Ku T-L (1991) U-series isochron dating: A generalized method employing total sample dissolution. Geochim Cosmochim Acta 55:555-564

Lyons RG, Crossley PC, Ditchburn RG, McCabe WJ, Whitehead N (1989) Radon escape from New Zealand speleothems. Appl Radiat Isot 40:1153-1158

Martinson DG, Pisias NG, Hays JD, Imbrie J, Moore TC, Shackleton NJ (1987) Age dating and the orbital theory of the Ice Ages: Development of a high-resolution 0 to 300,000 yr chronostratigraphy. Quat Res 27:1-29

Matthews A, Ayalon A, Bar-Matthews M (2000) D/H ratios of fluid inclusions of Soreq cave (Israel) speleothems as a guide to the Eastern Mediterranean Meteoric Line relationships in the last 120 ky. Chem Geol 166:183-191

McDermott F, Mattey DP, Hawkesworth CJ (2001) Centennial-scale Holocene climate variability revealed by a high-resolution speleotherm δ^{18}O record from SW Ireland. Science 294:1328-1331

McGarry SF, Baker A (2000) Organic acid fluorescence: applications to speleothem paleoenvironmental reconstruction. Quat Sci Rev 19:1087-1101

Moorbath S, Taylor PN, Orpen JL, Treloar P, Wilson JF (1987) First direct radiometric dating of Archaen stromatolitic limestone. Nature 326:865-867

Morse JW, Bender ML (1990) Partition coefficients in calcite: Examination of factors influencing the validity of experimental results and their application to natural systems. Chem Geol 82:265-277

Mucci A, Morse JW (1990) The chemistry of low temperature abiotic calcites: Experimental studies on coprecipitation, stability and fractionation. Rev Aquatic Sci 3:217-254

Musgrove ML, Banner JL, Mack LE, Combs DM, James EW, Cheng H, Edwards RL (2001) Geochronology of late Pleistocene to Holocene speleothems from central Texas: Implications for regional paleoclimate. Geol Soc Am Bull 113:1532-1543

Neff U, Burns SJ, Mangini A, Mudelsee M, Fleitmann D, Matter A (2001) Strong coherence between solar variability and the Monsoon in Oman between 9 and 6 kyrs ago. Nature 411:290-293

Neymark LA, Amelin YV, Paces JB (2000) ^{206}Pb-^{230}Th-^{234}U-^{238}U and ^{207}Pb-^{235}U geochronology of Quaternary opal, Yucca Mountain, Nevada. Geochim Cosmochim Acta 64:2913-2928

Neymark LA, Paces JB (2000) Consequences of slow growth for 230Th/U dating of Quaternary opals, Yucca Mountain, Nevada, USA. Chem Geol 164:143-160

Niggemann S, Mangini A, Richter DK, Wurth G (in press) A paleoclimate record of the last 17,600 years in stalagmites from the B7-cave, Sauerland, Germany. Quat Sci Rev

Oberli F, Meier M, Berger A, Rosenberg C, Gieré R (1996) ^{230}Th/^{238}U disequilibrium systematics in U-Th-Pb dating: Nuisance or powerful tool in geochronology. 6th V.M. Goldschmidt Conference, Heidelberg, Germany. J Conf Abstr 1:439

Olley JM, Roberts RG, Murray AS (1997) A novel method for determining residence times of river and lake sediments based on disequilibrium in the thorium decay series. Water Resour Res 33:1319-1326

Onac BP, Lauritzen S-E (1996) The climate of the last 150,000 years recorded in speleothems: preliminary results from north-western Romania. Theor Appl Karstology 9:9-21

O'Neil JR., Clayton RN, Mayeda TK (1969) Oxygen isotope fractionation in divalent metal carbonates. J Chem Phys 51:5547-5558

Osmond JK, Cowart JB (1982) Groundwater *In:* Uranium series disequilibrium: Applications to environmental problems (1st edn). Ivanovich M, Harmon RS (eds) Calrendon Press, Oxford, p 202-245

Osmond JM, May JP, Tanner WF (1970) Age of the Cape Kennedy barrier-and-lagoon complex. J Geophys Res 75:469-479

Pickett DA, Murrell MT, Williams RW (1994) Determination of femtogram quantities of protactinium in geologic samples by thermal ionization mass spectrometry. Anal Chem 66:1044-1049

Plagnes V, Causse C, Genty D, Paterne M, Blamart D (2002) A discontinuous climatic record from 187 to 74 ka from a speleothem of the Clamouse Cave (south of France). Earth Planet Sci Lett 201:87-103

Polyak VJ, Asmerom Y (2001) Late Holocene climate and cultural changes in the southwestern United States. Science 294:148-151

Polyak VJ, McIntosh WC, Provencio P, Güven N (1998) Age and Origin of Carlsbad Caverns and related caves from ^{40}Ar/^{39}Ar of alunite. Science 279:1919-1922

Porcelli D, Swarzenski PW (2003)The behavior of U- and Th- series nuclides in groundwater. Rev Mineral Geochem 52:317-361

Postpichl D, Agostini S, Forti P, Quinif Y (1991) Paleoseismicity from karst sediments: "Grotta del Cervo" cave, Central Italy. Tectonophysics 193:33-44

Quade J, Cerling TE, Bowman JR (1989) Systematic variations in the carbon and oxygen isotopic composition of pedogenic carbonate along elevation transects in the southern Great Basin. United States Geol Soc Am Bull 101:464-475

Railsback LB (2000) An Atlas of Speleothem Microfabrics: *http://www.gly.uga.edu/speleoatlas/SAIndex1.html*

Railsback LB, Brook GA, Webster JW (1999) Petrology and paleoenvironmental significance of detrital sand and silt in a stalagmite from Drotsky's Cave, Botswana. Phys Geography 20:331-347

Railsback LB, Brook, GA, Chen J, Kalin R, Fleisher CJ (1994) Environmental controls on the petrology of a Late Holocene speleothem from Botswana with annual layers of aragonite and calcite: J Sediment Res A64:147-155

Rasbury ET, Hanson GN, Meyers WJ, Saller AH (1997) Dating of the time of sedimentation using U-Pb ages for paleosol calcite. Geochim Cosmochim Acta 61:1525-1529

Richards DA (1995) Pleistocene sea levels and paleoclimate of the Bahamas based on ^{230}Th ages of speleothems. PhD dissertation. University of Bristol, Bristol

Richards DA, Bottrell SH, Cliff RA, Ströhle K, Rowe PJ (1998) U-Pb dating of a Quaternary-age speleothem. Geochim Cosmochim Acta 62:3683-3688

Richards DA, Smart PL Edwards RL (1994) Maximum sea levels for the last glacial period from U-series ages of submerged speleothems. Nature 367:357-360

Richards DA, Smart PL, Borton CJ, Edwards RL, Roberts MS (1997) Uranium-series disequilibria in speleothems form the Bahamas: Sea levels, carbonate deposition and diagenesis. Seventh Annual V. M. Goldschmidt Conference, LPI Contribution 921, Lunar and Planetary Institute, Houston. p 173

Richards DA, Smart PL, Edwards RL (1992) Late Pleistocene sea-level change based on high-precision mass-spectrometric ^{230}Th ages of submerged speleothems from the Bahamas. EOS Trans, Am Geophys Union 73:172

Rink WJ (1997) Electron spin resonance (ESR) dating and ESR applications in Quaternary science and archaeometry. Rad Meas 5-6:975-1025

Roberts MS, Smart PL, Baker A (1998) Annual trace element variations in a Holocene speleothem. Earth Planet Sci Lett 154:237-246

Roberts MS, Smart PL, Hawkesworth CJ, Perkins WT, Pearce NJP (1999). Trace element variations in coeval Holocene speleothems from GB Cave, southwest England. The Holocene 9:138-139

Robinson LF, Henderson GM, Slowey NC (2002) U-Th dating of marine isotope stage 7 in Bahamas slope sediments. Earth Planet Sci Lett 196:175-187

Rogers JJW, Adams JAS (1969) Uranium. In: Handbook of Geochemistry. Wedepohl, KH (ed) Spinger-Verlag, Berlin

Rosholt JN, Antal PS (1962) Evaluation fo the $Pa^{231}/U-Th^{230}/U$ method for dating Pleistocene carbonate rocks. US Geol Surv Pro Paper 450-E:108-111

Rowe PJ, Austin T, Atkinson TC (1988) The Quaternary evolution of cave deposits at Cresswell Crags gorge, England. Cave Science. Trans Brit Cave Res Assoc 16:3-17

Rozanski KL, Araguas-Araguas L, Gonfiantini R (1993) Isotopic patterns in modern global precipitation. In: Climatic Change in Continental Isotope Records. Swart PK, Lohmann KC, MacKenzie J, Savin S (eds) American Geophysical Union, Washington D.C., p 1-36.

Salomans W, Mook WG (1986) Isotope geochemistry of carbonates in the weathering zone In: Handbook of Environmental Isotope Geochemistry: Vol 2 The Terrestrial Environment. Fritz P, Fontes JC (eds), Elsevier, Amsterdam, p 239-269

Schramm A, Stein M, Goldstein SL (2000) Calibration of the ^{14}C time scale to 50 kyr by $^{234}U-^{230}Th$ dating of sediments from Lake Lisan (the paleo-Dead Sea). Earth Planet Sci Lett 175:27-40

Schwarcz HP (1986) Geochronology and isotope geochemistry of speleothems. In: Handbook of Environmental Isotope Geochemistry: Vol 2 The Terrestrial Environment. Fritz P, Fontes JC (eds) Elsevier, Amsterdam, p 271-303

Schwarcz HP, Blackwell BA (1992) Archaeological applications. In: Uranium-series disequilibrium: Applications to Earth, Marine and Environmental Sciences. Ivanovich M, Harmon RS (eds) Oxford University Press, Oxford, p 513-552

Schwarcz HP, Harmon RS, Thompson P, Ford DC (1976) Stable isotope studies of fluid inclusions in speleothems and their paleoclimatic significance. Geochim Cosmochim Acta 40:657-665

Schwarcz HP, Rink WJ (2001) Dating methods for sediments of caves and rock shelters. Geoarchaeology 16:355-372

Shen GJ (1996) $^{227}Th/^{230}Th$ dating method: Methodology and application to Chinese speleothem sample. Quat Sci Rev 15:699-707

Shen GJ, Ku T-l, Cheng H, Edwards RL, Yuan ZX, Wang Q (2001) High-precision U-series dating of Locality 1 at Zhoukoudian, China. J Human Evol 41:679-688

Shen GJ, Ku T-L, Gahleb B, Yuan ZX (1996) Preliminary results on U-series dating of Peking Man site with high precision TIMS. Acta Anthropol Sinica 15:210-217 (in Chinese)

Shopov YY, Ford DC, Schwarcz HP (1994) Luminescent micro-banding in speleothems: high-resolution chronology and paleoclimate. Geology 22:407-410

Short SA, Lowson RT, Ellis J (1988) $^{234}U/^{238}U$ and $^{230}Th/^{234}U$ activity ratios in the colloidal phases of aquifers in lateritic weathered zones. Geochim Cosmochim Acta 52:2555-2563

Smart PL, Richards DA, Edwards RL (1998) Uranium-series ages of speleothems from South Andros, Bahamas; implications for Quaternary sea-level history and palaeoclimate. Cave Karst Sci 25:67-74

Smart PL, Roberts MS, Baker A, Richards DA (1996) Palaeoclimate determination from speleothems. A critical appraisal of the state of the art. In: Climate Change: The Karst Record. Lauritzen S-E (ed) Karst Waters Institute Special Publications 2. Charles Town, West Virginia, p 157-159

Smith PE, Brand U, Farquhar RM (1994) U-Pb systematics and alteration trends of Pennsylvanian-aged aragonite and calcite. Geochim Cosmochim Acta 58:313-322

Smith PE, Farquhar RM (1989) Direct dating of Phanerozoic sediments by the $^{238}U/^{206}Pb$ method. Nature 341:7-20

Smith PE, Farquhar RM Hancock RG (1991) Direct radiometric age determination of carbonate diagenesis using U-Pb in secondary calcite. Earth Planet Sci Lett 105:474-491

Spalding RF, Mathews TD (1972) Submerged stalagmites from caves in the Bahamas: Indicators of low sea level stand. Quat Res 2:470-472

Spötl C, Mangini A, Frank N, Eichstädter R, Burns SJ (2002) Start of the last interglacial period at 135 ka: Evidence from a high alpine speleothem. Geology 30:815-818

Spötl C, Mangini A. (2002) Stalagmite from the Austrian Alps reveals Dansgaard-Oeschger events during isotope stage 3: Implications for the absolute chronology of Greenland ice cores. Earth Planet Sci Lett 203:507-518

Stirling CH, Lee DC, Christensen JM, Halliday AN (2000) High-precision in situ U-238-U-234-Th-230 isotopic analysis using laser ablation multiple-collector ICPMS. Geochim Cosmochim Acta 64:3737-3750

Stocker TF, Wright DG (1996) Rapid changes in ocean circulation and atmospheric radiocarbon. Paleoceanogaphy 11:773-796

Stuiver M, Reimer PJ, Bard E, Beck JW, Burr GS, Hughen KA, Kromer B, McCormac G, van der Plicht J, Spurk M (1998) INTCAL98 radiocarbon age calibration, 24,000-0 cal BP Radiocarbon 40:1041-1083

Sutcliffe AJ, Lord TC, Harmon RS, Ivanovich M, Rae A, Hess JW (1985) Wolverine in northern England about 83,000 years BP: Faunal evidence for climatic change during Isotope stage 5. Quat Res 24:73-86

Swarzenski PW, Porcelli D, Andersson PS, Smoak JM (2003) The behavior of U- and Th- series nuclides in the estuarine environment. Rev Mineral Geochem 52:577-606

Szabo BJ (1979) ^{230}Th, ^{231}Pa and open system dating of fossil corals and shells. J Geophys Res 84:4927-4930

Szabo BJ, Kolesar PT, Riggs AC, Winograd IJ, Ludwig KR (1994) Paleoclimatic inferences from a 120,000-year calcite record of water-table fluctuations in Browns Room of Devils Hole, Nevada. Quat Res 41:59-69

Taylor SR, McLennan SM (1995) The geochemical evolution of the continental crust. Rev Geophys 33:241-265

Tera F, Wasserburg GJ (1972) U-Th-Pb systematics in three Apollo 14 basalts and the problem of initial Pb in lunar rocks. Earth Planet Sci Lett 14:281-304

Thompson P, Schwarcz HP, Ford DC (1976) Stabke isotope geochemistry, geothermometry, and geochronology of speleothems from West Virginia. Geol Soc Am Bull 87:1730-1738

Vaks A, Ayalon A. Gilmour M, Frumkin A, Kaufman A, Matthews A, Bar-Matthews M. (2001) Pleistocene paleoclimate evidences from speleothem record of a karstic cave located at the desert boundary - Maale-Efraim, Eastern Shomron, Israel. PAGES - PEPIII: Past Climate Variability Through Europe and Africa Conference Abstr (*http://atlas-conferences.com/c/a/g/c/54.htm*)

Venables WN, Ripley BD (1999) Modern Applied Statistics with S-PLUS. Springer-Verlag, New York

Vesica PL, Tuccimei P, Turi B, Fornós JJ, Ginés A, Ginés J (2000) Late Pleistocene paleoclimates and sea-level change in the Mediterranean as inferred from stable isotope and U-series studies of overgrowths on speleothems, Mallorca, Spain. Quat Sci Rev 19:865-879

Voelker AHL, Grootes PM, Nadeau M-J, Sarnthein M (2000) Radiocarbon levels in the Iceland Sea from 25-53 kyr and their link to the Earth's magnetic field intensity. Radiocarbon 42:437-452

Vogel JC (1993) Variability of carbon isotope fractionation during photosynthesis. *In:* Stable isotopes and plant carbon-water relations. Ehleringer JR, Hall AE, Farquhar GD (eds) Academic Press, San Diego, p 29-38

Vogel JC, Kronfeld J (1997) Calibration of radiocarbon dates for the late Pleistocene using U/Th dates on stalagmites. Radiocarbon 39:27-32

Wang YG, Cheng H, Edwards RL, An ZS, Wu JY, Shen C-C, Dorale JA (2001) A high-resolution absolute-dated late Pleistocene monsoon record from Hulu Cave, China. Science 294:2345-2348

Wang YJ, Hai, C, Luo CL, Xia, YF, Wu JY, Chen J (1999) TIMS U-series ages of speleoethem from Tangshan caves, Nanjing. Chinese Sci. Bull. 44:1987-1991

Wedepohl KH (1995) The composition of the continental crust. Geochim Cosmochim Acta 59:1217-1239

Wendt I (1984) A three-dimensional U-Pb discordia plane to evaluate samples with common lead of unknown isotopic composition. Chem Geol 2:1-12

Wendt I, Carl C (1985) U/Pb dating of discordant 0.1 Ma old secondary U minerals. Earth Planet Sci Lett 73:278-284

Werner M, Mikolajewicz U, Hoffmann G, Heimann M (2000) Possible changes of δ^{18}O in precipitation caused by a meltwater event in the North Atlantic. J Geophys Res 10:10161-10167

Whitehead NE, Ditchburn RG, Williams PW, McCabe WJ (1999) ^{231}Pa and ^{230}Th contamination at zero age: a possible limitation on U/Th series dating of speleothem material. Chem Geol 156:359-366

Wigley TML, Plummer LN, Pearson FJ (1978) Mass transfer and carbon isotope evolution in natural water systems. Geochim Cosmochim Acta 42:1117-1140

Winograd IJ, Coplen, TB, Landwehr JM, Riggs AC, Ludwig KR, Szabo BJ, Kolesar PT, Revesz KM (1992) Continuous 500,000-year climate record from vein calcite in Devils Hole, Nevada. Science 258:255-260

Winograd IJ, Landwehr JM, Ludwig KR, Coplen TB, Riggs AC (1997) Duration and structure of the past four interglaciations. Quat Res 48:141-154

Winograd IJ, Szabo BJ, Coplen TB, Riggs AC (1988) A 250,000-year climatic record from Great Basin vein calcite: implications for Milankovitch theory. Science 242:1275-1280

Yokoyama Y, Esat TM, Lambeck K, Fifield LK (2000) Last ice age millennial scale climate changes recorded in Huon Peninsula corals. Radiocarbon 42:383-401

Yonge CJ, Ford DC, Gray JP, Schwarcz HP (1985) Stable isotope studies of cave seepage water. Chem Geol 58:97-105

Yuan SX, Chen TM, Gao SJ, Hu YQ (1991) Study on uranium series dating of fossil bones from Zhoukoudian sites Acta Anthropol. Sinica 10:189-193 (in Chinese)

Zhao J-X, Hu K, Collerson KD, Xu H-K (2001) Thermal ionization mass spectrometry U-series dating of a hominid site near Nanjing, China. Geology 29:27-30

11 Short-lived U/Th Series Radionuclides in the Ocean: Tracers for Scavenging Rates, Export Fluxes and Particle Dynamics

J. K. Cochran
Marine Sciences Research Center
State University of New York
Stony Brook, New York 11794-5000, U.S.A.

P. Masqué
Departament de Física
Institut de Ciències i Tecnologies Ambientals
Universitat Autònoma de Barcelona, 08193 Bellaterra, Spain

1. INTRODUCTION

The uranium and thorium decay series have numerous radionuclides with half-lives ranging from fractions of a second to a few years. The former include many of the isotopes of radon, polonium, bismuth, lead and thallium and these are generally too short to show appreciable disequilibrium with respect to their precursors in the decay series. Others, however, have half-lives that make them useful in tracing a variety of oceanic processes that occur on short time scales. This group includes isotopes of thorium: ^{234}Th (half-life = 24.1 d), ^{227}Th (18.6 d), ^{228}Th (1.9 y), radium: ^{224}Ra (3.64 d), ^{223}Ra (11.1 d), radon: ^{222}Rn (3.8 d) and polonium: ^{210}Po (138 d). Of these, the Th isotopes and ^{210}Po are particle-reactive, while the radium isotopes and radon tend to remain in solution. As a consequence, Th and Po are useful in quantifying the rate of scavenging from solution onto particles and the processes governing the dynamics of particles in the ocean, while Ra and Rn are appropriate tracers for fluid processes such as advection and diffusion (Cochran 1992). Other chapters in this volume highlight recent developments in the use of short-lived Ra isotopes and Rn as tracers of estuarine mixing and submarine groundwater discharge (Swarzenski et al. 2003; Porcelli and Swarzenski 2003). Likewise, the oceanic chemistries of the longer-lived particle reactive radionuclides in the U/Th series, such as ^{230}Th, ^{231}Pa and ^{210}Pb are treated elsewhere in this volume (Henderson and Anderson 2003). Accordingly our emphasis here is with recent developments in the application of the short-lived Th isotopes (^{234}Th, ^{227}Th, ^{228}Th) and ^{210}Po as tracers for particle-associated processes. These radionuclides have in common the fact that each is produced by a longer lived parent which is stably dissolved or tends to remain in solution longer than its progeny. In the case of Th, the nascent atom can quickly hydrolyze and sorb onto available particle surfaces. Indeed, Th has one of the strongest affinities for particle surfaces of all the elements, with a K_d of ~10^6-10^7 (IAEA 1985), and as a consequence it is useful in determining the rates of a variety of particle-associated oceanic processes. Of the short-lived Th isotopes we consider in this chapter, ^{234}Th (half-life 24.1 d) has been the focus of greatest attention and application in the past ten years. The production of ^{234}Th from ^{238}U, coupled with the conservative behavior of ^{238}U in seawater, makes the source of ^{234}Th easy to characterize. Moreover, the half-life of ^{234}Th is sufficiently short to make it sensitive to short-term (e.g., seasonal) changes that occur in the upper water column of the open ocean or in the sediments or water column of coastal areas. In contrast, the sources of ^{227}Th, ^{228}Th and ^{210}Po, (decay of ^{227}Ac, ^{228}Ra and ^{210}Pb, respectively) show substantial depth variations in the water column.

2. MEASUREMENT TECHNIQUES

The longer lived radionuclides of the U/Th series (e.g., ^{238}U, ^{235}U, ^{232}Th, ^{230}Th, ^{231}Pa) are amenable to measurement by classical decay counting or by mass spectrometric techniques (Goldstein and Stirling 2003). The short-lived radionuclides that are the subject of this chapter are not characterized by atom abundances sufficient to permit their measurement by mass spectrometry, and counting of beta, alpha or gamma emissions remains the method of choice.

^{227}Th, ^{228}Th and ^{210}Po, all decay by alpha emission and are thus measurable by isotope dilution and alpha spectrometry (Ivanovich and Murray 1992). However, ^{234}Th is produced by the alpha decay of ^{238}U and in turn decays by beta emission to ^{234}U via the short-lived intermediate ^{234}Pa (half-life 1.18 m):

$$^{238}U \rightarrow \, ^{234}Th \rightarrow \, ^{234}Pa \rightarrow \, ^{234}U \rightarrow \, ...$$

and must be determined by beta counting or gamma spectrometry (or liquid scintillation, Pates et al. 1996). Moreover, the short half-life of ^{234}Th requires prompt processing and counting (i.e., within a few weeks) to obtain the highest quality data.

Early measurements of ^{234}Th were on seawater samples and Th was co-precipitated from 20-30 L of seawater with iron hydroxide (Bhat et al. 1969). This procedure may not recover all of the ^{234}Th in the sample, and an alpha emitting Th isotope (e.g., ^{230}Th or ^{229}Th) is added as a yield monitor. Following chemical purification of the Th fraction by ion exchange chromatography, the Th is electrodeposited onto platinum or stainless steel planchets. The planchets are then counted in a low background gas-flow beta detector to measure the beta activity and subsequently with a silicon surface barrier detector to determine the alpha activity of the yield monitor. The ^{234}Th activity is thus determined as:

$$^{234}Th = \frac{count\,rate_{beta}}{\varepsilon \cdot Y} \qquad (1)$$

where ε is the beta counting efficiency determined from standards and Y is the Th yield determined by the comparison of the alpha activity of the recovered tracer relative to the tracer added to the sample. In preparing samples for beta counting, the planchet is usually covered with a Mylar or aluminum film that effectively blocks the weak ^{234}Th betas. As a consequence, it is the ^{234}Pa betas that are measured and used to determine the ^{234}Th activity in Equation (1).

Measurements of ^{234}Th in sediment samples (Aller and Cochran 1976; Cochran and Aller 1979) used much the same approach as outlined above. In this case, the dried sediment sample (~10 g) was leached with strong mineral acid (HCl) in the presence of a yield monitor (generally ^{229}Th, an artificial Th isotope resulting from the decay of ^{233}Th that is produced by neutron capture on ^{232}Th). Thorium was separated from U and purified by ion exchange chromatography, and electrodeposited onto stainless steel planchets. Counting and determination of ^{234}Th activity followed the procedure outlined above.

Modifications to the procedure for both seawater and sediments were made in the 1990's in order to measure ^{234}Th by detection of its gamma emission (Buesseler et al. 1992a; Feng et al. 1999a). The branching ratio for the ^{234}Th gamma emission is small (approximately 4.8% at 63.3 keV), and coupled with the low counting efficiency for gamma counting using a planar intrinsic germanium detector, made it necessary to increase sample size. For seawater samples, this necessitated the application of pumping systems that filter large volumes of seawater (at least several hundred liters) through a series of prefilters and manganese oxide-impregnated filter cartridges to remove the particulate and dissolved Th respectively. The filter cartridges were either melted (Buesseler et al. 1992a), compressed (Buesseler et al. 1995; Cochran et al. 2000) or ashed (Cochran et al. 1995; Bacon et al. 1996) to put them into a geometry suitable for counting. In some cases, filter cartridges were also used as prefilters to separate particulate Th, and these were processed in the same fashion as the Mn-treated cartridges (e.g., Buesseler et al. 1992a; Cochran et al. 1995).

The extraction efficiency of the Mn-treated cartridges for Th from seawater is determined on each sample by measuring the activity of two cartridges in series, assuming the two cartridges have identical extraction efficiencies (Livingston and Cochran 1987). The extraction efficiency (E) is then:

$$E = 1 - \frac{MnB}{MnA} \qquad (2)$$

where MnA is the ^{234}Th activity on the first cartridge in the series and MnB is the activity on the second cartridge. Extraction efficiencies may depend on factors such as the rate of flow through the system (Cochran et al. 1995) but are typically in excess of 70% (Livingston and Cochran 1987; Buesseler et al. 1992b).

The dissolved activity (^{234}Th) is given by:

$$^{234}Th = \frac{cpm_A}{E \cdot \varepsilon \cdot V} \qquad (3)$$

where cpm_A is the count rate on the first Mn cartridge, E is the Mn cartridge extraction efficiency, ε is the gamma counting efficiency for the sample and V is the sample volume. If a prefilter cartridge (particulate ^{234}Th) is being counted, E drops out of Equation (3). The counting efficiency includes both sample geometry and the branching ratio for ^{234}Th gamma decay (~4.8% for 63.3 keV). There is some uncertainty as to the value of the ^{234}Th branching ratio, however this is not critical because the value of ε is determined for a given detector by adding a known quantity of a standard solution of ^{238}U (with ^{234}Th in equilibrium) to a filter cartridge and preparing it for counting in the same manner as a sample cartridge. Generally well-type gamma detectors have higher counting efficiencies based on geometry than do planar detectors, and thus with their use the sample volumes can be decreased (Charette and Moran 1999).

For determination of activities of ^{234}Th in sediments by gamma spectrometry, the sediment (~100 g) is dried, ground and sealed in sample jars for counting. Activities are determined using a modification of Equation (3) in which the E term is not included, and V is replaced by the sample mass. Counting efficiencies are determined by counting sediment standards of known ^{238}U (and equilibrium ^{234}Th) activity.

Measuring ^{234}Th activity by gamma spectrometry requires consideration of the self-absorption of gamma rays by the sample itself. In the case of the filter cartridges used for seawater ^{234}Th determinations, the standards are identical to the samples in terms of matrix. In sediment samples however, self-absorption can easily change within a single core, among cores, or between samples and standard sediment as a consequence of compositional changes. The effect of self-absorption on counting efficiency can be evaluated using either a theoretical relationship between density and self-absorption (Cutshall et al. 1983) or an empirical relationship determined by preparing standards with different densities (Cochran et al. 1993). Density is evaluated by counting a ^{234}Th source through the sample and comparing it with the same source counted through an empty sample container. The relationship between sample density as measured by source transmission and counting efficiency, once established, can be used to select the appropriate counting efficiency for any given sample (once the transmission of a ^{234}Th source through it has been measured).

Modifications to these procedures include measurement of particulate ^{234}Th on membrane-type prefilters (e.g., Millipore or glass fiber) by direct beta counting. In this case, plugs are taken from the filters and stacked in sample holders before being beta counted (Buesseler et al. 1995; Bacon et al. 1996). Larger pore size filters (53 μm Nitex of 70 μm Teflon) have been added to *in situ* pumping systems to separate large and small particles (retained on the 0.7 μm GFF or Microquartz filter). In order to determine particulate organic carbon (POC) on the same samples as ^{234}Th, Particles were rinsed off of these large pore size filters and re-filtered onto Microquartz or silver filters for beta counting (Bacon et al. 1996; Buesseler et al. 1995, 1998). Recently, total ^{234}Th activities have been assayed in small volume samples (~2 to 4 L) by precipitation of ^{234}Th along with manganese oxide, filtration of the precipitate and non-destructive beta counting (Buesseler et al. 2001). Because one is measuring the ^{234}Th (^{234}Pa) beta emissions directly, and the beta counting efficiency is generally greater than that for gamma counting, this approach has the advantage of small sample size and thus permits increased depth resolution in water column profiles. An alternate method for measuring ^{234}Th on discrete water samples (20 L) has been developed by Pates et al. (1996) and involves co-precipitating ^{234}Th with Fe(OH)$_3$ in the presence of ^{230}Th as a yield tracer. The ^{234}Th (and ^{234}Pa) are measured by liquid scintillation spectrometry. With both small volume approaches, generally only total ^{234}Th (particulate + dissolved) is measured in a water sample and information on particulate ^{234}Th must be gained by other means such as *in situ* pumps. Most recent studies of ^{234}Th/^{238}U disequilibrium in open ocean surface waters have counted samples at sea using a planar intrinsic germanium gamma detector or a low-level beta counting system. These detectors permit the analysis of ~2 (gamma) to >10 (beta) samples per day and data are developed in near-real time. As a consequence sampling can be adjusted to take advantage of trends in the data.

The short-lived alpha emitting Th isotopes require chemical processing of the samples prior to measurement. Following measurement of ^{234}Th, samples are dissolved in mineral acids (HCl, HNO$_3$) in the presence of ^{229}Th as a yield monitor, chemically purified by ion exchange chromatography and plated onto stainless steel to prepare a source for alpha counting (Buesseler et al. 1992a). Neither ^{210}Po nor ^{210}Pb are retained on the manganese cartridges used for Th in seawater, and discrete water samples (4 – 20 L) are taken for measurement of these radionuclides. They are commonly measured together because ^{210}Pb is the source of ^{210}Po. After addition of yield monitors (^{209}Po or ^{208}Po for ^{210}Po and stable Pb for ^{210}Pb), Po and Pb are co-precipitated from a seawater sample (4-20 L) with iron hydroxide (Nozaki 1986) or Co-APDC (Chung et al. 1983). The precipitates are recovered from the solution, dissolved in 1.0-1.5 N HCl, Po is plated onto a silver disk and counted by alpha spectrometry (Flynn 1968). The Pb fraction is purified by ion exchange chromatography and stored for ingrowth of additional ^{210}Po. After sufficient time (~6 months) the ingrown ^{210}Po is plated and counted as an index of the ^{210}Pb in the sample. Pb yields for this procedure are determined by measurement (usually by atomic absorption spectrophotometry) of the stable Pb carrier that was added initially. Simultaneous determination of ^{210}Po and ^{210}Pb by liquid scintillation recently has been proposed by Biggin et al. (2002). This procedure, although requiring more elaborate radiochemical treatment of the sample, avoids the delay of several months in obtaining the results.

Applications of Short-lived Radionuclides in the Ocean

3. SCAVENGING FROM SEAWATER

3.1. Early observations of Th scavenging

Impetus for measurement of Th in seawater came from its use in describing the process of scavenging of reactive chemical species. As well the high reactivity of Th made it an ideal tracer for demonstrating how scavenging rates of reactive species varied from place to place in the ocean and in predicting the fate of particle-reactive pollutants (Broecker et al. 1973). ^{228}Th and ^{234}Th were viewed as especially useful in this regard. ^{234}Th was first measured in seawater by Bhat et al. (1969), who observed deficiencies in ^{234}Th relative to ^{238}U in the upper 100 m of the water column in the Indian Ocean. Bhat et al. (1969) also observed occasional excesses of ^{234}Th at depths between 100 and 200 m and speculated that scavenging of ^{234}Th onto sinking biogenic material, coupled with remineralization of this material and release of ^{234}Th to solution at depth, was responsible for the profiles. In addition, Bhat et al. (1969) documented a strong trend of decreasing ^{234}Th/^{238}U activity ratios towards the coast, suggesting that scavenging rates were appreciably greater in coastal waters.

The results of Matsumoto (1975) reinforced the pattern of ^{234}Th/^{238}U disequilibrium observed by Bhat et al. (1969). His results for the Pacific Ocean showed ^{234}Th deficiencies in the upper 150-200 m of stations throughout the Pacific Ocean: the median ^{234}Th/^{238}U activity ratio for all samples in the upper 100 m was 0.8. Matsumoto (1975) evaluated the mean residence time of Th with respect to scavenging using a simple scavenging model that balances ^{234}Th production with decay and scavenging. Neglecting vertical advective transport (Fig. 1),

$$P = (\lambda + k)N \tag{4}$$

where P is the atomic production rate (per liter of seawater) of ^{234}Th from ^{238}U, N is the total ^{234}Th atoms per liter, λ is the ^{234}Th decay constant and k is the scavenging rate constant (assumed to be first-order) for ^{234}Th removal from the water column. Equation (4) can be rewritten in activity units to yield an expression for k:

$$k = \frac{\lambda(P - A_{Th})}{A_{Th}} \tag{5}$$

where A_{Th} is the total ^{234}Th activity ($A = \lambda N$) in the water sample. The inverse of k is the mean residence time (τ_{Th}) of ^{234}Th with respect to scavenging.

Matsumoto's (1975) data yielded a τ_{Th} of ~0.4 years. This value was consistent with previous results by Broecker et al. (1973), who calculated τ_{Th} for surface waters from ^{228}Th/^{228}Ra disequilibrium. Kaufman et al. (1981) used both ^{234}Th/^{238}U and ^{228}Th/^{228}Ra disequilibrium to calculate Th residence times in the New York Bight. They established a "concordia" or trend of isotope ratios that yielded consistent results from both radionuclide pairs and showed that occasionally residence times disagreed. Such disagreement was attributed to water from a different scavenging regime (e.g., farther offshore) that had recently moved into the Bight. Because the ^{228}Th/^{228}Ra ratio takes longer to "adjust" to the new scavenging environment than does the ^{234}Th/^{238}U ratio, the observation of different residence times using the two isotopes of Th suggests that such non-steady state processes are operating.

Documentation of disequilibrium between ^{227}Th and its parent ^{227}Ac has been quite limited (Nozaki and Horibe 1983; Geibert et al. 2002) and the two isotopes have not generally been measured on the same samples. As with ^{234}Th, ^{227}Th must be measured promptly after sample collection. Indeed most studies of the ^{227}Th/^{227}Ac pair have

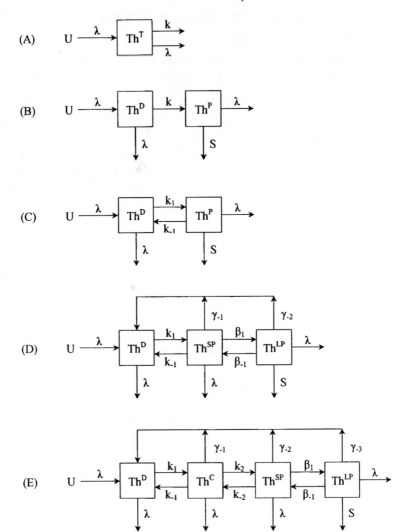

Figure 1. Schematic diagrams of the cycling of thorium in the ocean. The figure shows how scavenging models have evolved over time, with increasing complexity. In Model A, total Th activity (Th^T) is produced from the decay of dissolved U (or Ra), λ is the Th decay constant and k is the scavenging rate constant for thorium removal from the water column. In Model B, thorium activity is divided between the dissolved (Th^D) and the particulate (Th^P) phases, k is the first-order rate constant for uptake of dissolved thorium onto particle surfaces and S accounts for the removal of particles from the water column. Reversible scavenging equilibrium between the dissolved and particulate phases is included in Model C, where k_1 and k_{-1} are the adsorption and desorption rate constants. Size fractionation is considered in Model D, with suspended small particles (Th^{SP}) and sinking large particles (Th^{LP}) reservoirs. The first-order aggregation and disaggregation rate constants between them are β_1 and β_{-1}, respectively. Remineralization rates (γ_1 and γ_2) are also taken into account. Finally, the scavenging of thorium by the colloidal pool (Th^C) is included in Model E. Of all the rate constants, only λ is known. Each Th box generates a mass balance equation, yet as model complexity increases, the number of unknowns exceeds the number of equations. Using multiple Th isotopes (e.g., ^{234}Th and ^{228}Th) can help solve for more of the unknown parameters, but often simplifying assumptions must be made. (Adapted from Matsumoto 1975, Bacon and Anderson 1982, Clegg and Whitfield 1991, 1992, Murnane et al 1994 and Quigley et al 1996).

focused on distributions of ^{227}Ac, which is measured by purifying and storing it to allow ^{227}Th to grow into equilibrium (Nozaki 1984; 1993; Nozaki and Yang 1987; Nozaki et al. 1990; Geibert et al. 2002). ^{227}Ac has a geochemistry similar to Ra and is added to seawater from marine sediments, where it is produced from ^{231}Pa. Migration of ^{227}Ac into the overlying water results in gradients away from the seafloor that can be used to quantify near-bottom mixing (Nozaki 1993; Nozaki and Yang 1987; Geibert et al. 2002). In principle, ^{227}Th/^{227}Ac disequilibrium in near-bottom waters can be used to study near-bottom scavenging, in a fashion similar to the use of ^{234}Th/^{238}U disequilibrium in this zone (Bacon and Rutgers van der Loeff 1989), however clear evidence of ^{227}Th/^{227}Ac disequilibrium in the benthic boundary layer remains to be established.

3.2. Development of scavenging models based on Th

The scavenging model described in Equation (4) is robust and has been applied in a number of instances (e.g., Kaufman et al. 1981). However, a more detailed description of Th scavenging results from a model that treats the dissolved and particulate phases separately (e.g., Krishnaswami et al. 1976; Fig. 1). For dissolved ^{234}Th:

$$\lambda P = \lambda A_{Th}^d + kA_{Th}^d \qquad (6)$$

and for particulate ^{234}Th:

$$kA_{Th}^d = \lambda A_{Th}^p + \psi A_{Th}^p \qquad (7)$$

where k is the first-order rate constant for uptake of dissolved ^{234}Th onto particle surfaces and ψ is the rate constant pertaining to the removal of particulate ^{234}Th from the water column. Equations (6) and (7) provide the foundation for explicit consideration of particle-solution interactions and particle sinking, although they do require an operational definition of what constitutes "dissolved" and "particulate" Th. For example, the "particulate" fraction is often defined as that retained on a 0.4 or 0.45 μm membrane filter.

Efforts to apply Equations (6) and (7) to distributions of Th isotopes in the oceans showed that the situation was more complex. For example, Bacon and Anderson (1982) measured vertical distributions of ^{230}Th in the deep sea and found that both the particulate and "dissolved" fractions increased linearly with depth. While the former observation is predictable from Equation (7) if sinking particles continue to scavenge Th during their descent, the latter is inconsistent with Equation (6). Bacon and Anderson (1982) suggested that the data could best be explained by a reversible scavenging equilibrium maintained between dissolved and particulate Th. Thus Equation (6) must be modified to:

$$\lambda P + k_{-1} A_{Th}^p = \left(\lambda + k_1 \right) A_{Th}^d \qquad (8)$$

where k_1 and k_{-1} are the first order rate constants pertaining to uptake of dissolved Th from solution onto particle surfaces and release from particle surfaces to solution. Reversibility of Th scavenging was further supported by Nozaki et al. (1981), Moore and Hunter (1985) and Moore and Millward (1988). Applications of ^{230}Th (and ^{231}Pa) scavenging to estimating oceanic sedimentation rates and paleoproductivity are reviewed in this volume by Henderson and Anderson (2003).

Further developments of scavenging models included the notion of reversible scavenging from the dissolved phase, but incorporated more than one particulate phase and linked the particle phases through terms representing aggregation and disaggregation (Fig. 1; Clegg et al. 1991; Cochran et al. 1993; Murnane et al. 1994; 1996; Burd et al. 2000; Cochran et al. 2000). The measurement of multiple isotopes of Th can help constrain the rates of aggregation and disaggregation and their variation with depth.

The values of k_1 calculated by Bacon and Anderson (1982), and used in most models of Th scavenging, varied with particle concentration and ranged from 0.2 to 1.2 y^{-1}. Such values are appreciably longer than expected from sorption rates onto particle surfaces. The discrepancy can be explained if dissolved Th is initially sorbed to surfaces of very small particles (colloids) that pass through the typical filters (0.1-0.4 μm) used to separate "dissolved" from particulate fractions (Santschi et al. 1986).

3.3. The role of colloids in Th scavenging

The potential importance of colloids in scavenging had been pointed out in 1978 by Tsunogai and Minagawa (1978). Some ten years later, Honeyman and Santschi (1989) elaborated a theory of scavenging involving colloids that implicated a process termed "Brownian" or colloidal pumping. In this view, dissolved metal species first rapidly associate with surface sites on both colloidal and filterable particles and the colloidal fraction slowly aggregates or coagulates to form filterable particles. Among the mechanisms discussed by Honeyman and Santschi (1989), Brownian coagulation relies on the random collisions of particles. Brownian pumping is then the mechanism by which these particles move up the particle size spectrum and enter the "filterable" fraction.

Initial tests of Brownian pumping required the measurement of Th in colloids separated from seawater samples. ^{234}Th proved to be an especially useful tracer of colloidal uptake of metal species because of its constant source and relative abundance. Baskaran et al. (1992) and Moran and Buesseler (1992) used cross-flow filtration to separate the colloidal fraction and both studies reported significant (up to 78% of total) ^{234}Th in this fraction. Subsequent work largely supported these observations (Moran and Buesseler 1993; Huh and Prahl 1995) and suggested the importance of colloidal organic matter in scavenging Th (Niven et al. 1995).

However, as more studies of colloidal Th were made, methodological concerns associated with the techniques for separating colloids from seawater samples were raised. The development of high-temperature catalytic techniques for measuring dissolved organic carbon in seawater focused research on carbon cycling through the colloidal organic carbon (COC) pool and raised issues concerning the role of COC in scavenging reactive chemical species. The possibility of experimental artifacts in quantifying the colloidal fraction of organic carbon and other constituents using cross-flow filtration was addressed in an intercalibration exercise termed the "Colloid Cookout" (Buesseler et al. 1996). The results showed large differences in the COC separated by different types of ultrafilters, as well as differences within a single filter type. Buesseler et al. (1996) concluded that it was critical to carefully evaluate blanks, characterize cross-flow filtration systems using standard compounds and compare the results with other methods to validate them.

Recent studies of the interaction of Th with colloids suggest that the association of Th with colloidal organic matter is stronger than for known mineral adsorbents and essentially irreversible (Quigley et al. 2001; 2002). This is somewhat at variance with the notion of reversible scavenging presented by Bacon and Anderson (1982), reinforced by Moore and Hunter (1985) and Moore and Millward (1988) and incorporated into most current models of Th scavenging (e.g., Clegg et al. 1991; Murnane et al. 1994; 1996; Burd et al. 2000). However Quigley et al. (2001) argue that colloidal disaggregation could produce the appearance of sorption reversibility in comparisons of the filterable (particulate) and "dissolved" fractions. Whether scavenging is treated as a reversible process or not, colloidal pumping is a mechanism that permits transfer of Th into the filterable fraction. Estimates of the residence time of Th in the colloidal fraction have been made and are short, on the order of 10 days (Moran and Buesseler 1992), suggesting that colloidal pumping is effective.

Applications of Short-lived Radionuclides in the Ocean 469

One of the issues surrounding reversibility of Th scavenging onto colloidal organic matter is the exact nature of the surface. Excretion of polysaccharides by phytoplankton is an important pathway in the production of marine colloids and this process has been proposed to be important for Th scavenging onto colloidal organic carbon (Niven et al. 1995). Quigley et al. (2002) documented the strong binding of Th onto polysaccharide and suggested that variation in polysaccharide content of particles plays an important role in governing the ratio of particulate organic carbon (POC) to ^{234}Th of the particles. As we shall see below, this parameter is important in estimating the sinking flux of POC using ^{234}Th deficits in the water column.

3.4. Scavenging of Po

In much the same fashion as the use of ^{234}Th/^{238}U or ^{228}Th/^{228}Ra disequilibria, the extent and rate of ^{210}Po scavenging from seawater can be estimated by comparison of its distribution with that of its grandparent, ^{210}Pb. Early results documenting distributions of ^{210}Po and ^{210}Pb in the oceans (Bacon et al. 1976; 1988; Thomson and Turekian 1976; Nozaki et al. 1976; Nozaki and Tsunogai 1976; Cochran et al. 1983; Chung and Finkel 1988; see also Cochran 1992) showed that ^{210}Po was more rapidly removed from surface waters than ^{210}Pb (^{210}Po/^{210}Pb in solution < 1). At depth, however, ^{210}Po can be released to solution from sinking particles such that ^{210}Po/^{210}Pb in solution is >1 (Cochran et al. 1983; Bacon et al. 1988; Cochran 1992).

Clues to this behavior lie in the fact that while Pb and Th, as well as Po, can be scavenged onto particle surfaces, Po is enriched in the organic fraction (Kharkar et al. 1976; Heyraud and Cherry 1979) and is assimilated into cells, possibly as an analog of sulfur (Fisher et al. 1983; 1987). Indeed Stewart and Fisher (2002) showed that Po uptake in marine phytoplankton cultured in the laboratory is explained by both a surface bound fraction and a cellular fraction associated with protein. Such behavior helps account for observed links between chlorophyll a (denoting living organic matter), productivity, POC concentration and the rate of Po scavenging (Nozaki et al. 1997; 1998; Hong et al. 1999; Sarin et al. 1999). As well, if some of the Po taken up by phytoplankton is assimilated, it may be released as the organic matter decomposes during sinking. Thus Po may behave more like carbon than Th or Pb with respect to its scavenging on particulate organic matter and involvement in marine food webs. Such a possibility is supported by the recent work of Masqué et al. (2002), who observed residence times of ^{210}Po that were greater than those of ^{210}Pb in the surface waters of the northwestern Mediterranean Sea. They explained these results by the more efficient recycling of organic mater and ^{210}Po in the surface waters or by the preferential uptake of ^{210}Po onto low density, buoyant particles that do not sink. Similar observations on the residence time of ^{210}Po in the Mediterranean were made by Radakovitch et al. (1998; 1999). Moreover the ^{210}Po/^{210}Pb activity ratio on particles caught in sediment traps can help determine the provenance of the particles: values significantly greater than 1.0 indicate biogenic particles, while values ~1 are consistent with detrital particles in which the radionuclide pair are in equilibrium. Heussner et al (1990) used this difference to determine the origin of particles in sediment traps in the Gulf of Lions.

4. THORIUM AND POLONIUM AS TRACERS FOR ORGANIC CARBON CYCLING IN THE OCEANS

4.1. Basis, approach and early results

Although estimates vary, about ⅓ of the CO_2 that is released to the atmosphere by human activities (e.g., deforestation and burning of fossil fuels) is taken up by the oceans (Feely et al. 2001). The extent to which this carbon is incorporated into biological production in the ocean has become an area of intense interest. It has long been known that a certain fraction of the production of marine organic matter escapes the euphotic zone (approximately the upper 100 m) and sinks to depth, either as fecal pellets or as sinking aggregates termed "marine snow." This sinking or export flux of POC removes C from the surface water and adds it to the deep oceans, where some of it is decomposed and a small fraction reaches the bottom. The magnitude of the POC export has been estimated from nutrient balances in the euphotic zone, such that the production supported by the supply of "new" nutrients via upwelling for example—termed new production—is balanced by the export of POC (Dugdale and Goering 1967; Eppley and Peterson 1979). Other approaches for estimating POC export include the use of sediment traps that collect sinking particles, although these are often subject to significant problems such as biases associated with the flow of water over the trap face.

The scavenging of ^{210}Po and ^{234}Th onto particles in the ocean suggests the possibility of using these radionuclides as tracers of POC export and their short half-lives make them sensitive to seasonal changes in the processes of POC production and export. Indeed, this is one of the most significant applications of ^{234}Th/^{238}U disequilibrium within the past 20 years. The approach builds on the observations of ^{234}Th deficiencies in the upper ocean described above (e.g., Matsumoto 1975) and argues that such deficiencies are created by the scavenging of ^{234}Th onto biogenic particles and the sinking of such particles out of the euphotic zone. Coale and Bruland (1985; 1987) explicitly showed that the pattern of ^{234}Th/^{238}U disequilibrium in the upper 100 m of the North Pacific gyre and the California Coastal Current was linked to profiles of nutrients and dissolved O_2, such that the production and cycling of organic matter were causing the ^{234}Th disequilibrium (Fig. 2).

Eppley (1989) proposed that if the residence time of Th with respect to sinking (as calculated from Eqn. 7, for example) could be applied to the standing crop of POC in the euphotic zone, as seems reasonable if sinking POC is responsible for ^{234}Th removal, then the sinking flux of POC could be calculated. He equated this flux to the new production and argued that ^{234}Th disequilibrium would be a useful approach to determining new production in the oceans. Murray et al. (1989) determined residence times of both POC and particulate ^{234}Th in the eastern Equatorial Pacific and found different values, seemingly calling into question Eppley's (1989) suggestion. In part the disparity may be due to the way the particulate ^{234}Th residence time is typically calculated (via Eqn. 7) but, because the chemistries of C and Th are substantially different with respect to their incorporation and cycling in organic matter, it is likely that their residence times would be different in the euphotic zone.

Indeed Kim et al. (1999) have argued that better agreement is obtained from a one-box model approach, in which the particulate Th residence time is calculated as the difference between the residence times of total Th and dissolved Th (effectively the difference in the $1/k$ values calculated from Eqns. 5 and 6). However, Buesseler and Charette (2000) argued in a response to Kim et al. (1999) that there is abundant evidence to support the notion that residence times of POC and Th are different in the euphotic zone.

Buesseler et al. (1992b) proposed a method to circumvent these difficulties in comparing residence times. They argued that the deficiency in total ^{234}Th with respect to ^{238}U indicates a flux of ^{234}Th in association with particles sinking out of the euphotic zone. If the POC (or particulate organic nitrogen, PON)/^{234}Th ratio of these sinking particles is known, a POC (or PON) flux can be calculated as:

$$POC\ flux = \left[\lambda \int (A_U - A_{Th}) dz \right] \times \frac{POC}{^{234}Th} \qquad (9)$$

where $\int (A_U - A_{Th}) dz$ is the integrated ^{234}Th deficit (dpm/m^2) in the euphotic zone,

Applications of Short-lived Radionuclides in the Ocean

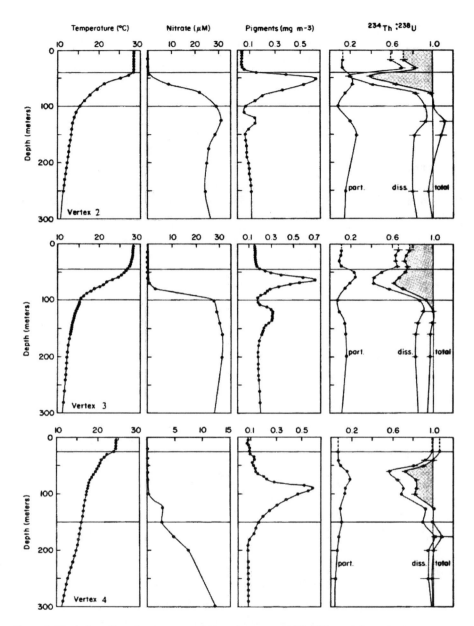

Figure 2. Vertical profiles of temperature, nitrate, pigments, and ^{234}Th/^{238}U activity ratios in the eastern Pacific Ocean. Upper horizontal lines delineate the base of the mixed layer and lower horizontal lines delineate the base of the euphotic zone. The vertical line at ^{234}Th/^{238}U = 1 represents radioactive equilibrium. The hatched area represents total ^{234}Th deficiency with respect to its parent ^{238}U, in correspondence to the profiles of nitrate and pigments and thus pointing to the role of organic matter in scavenging Th from the upper water column However, the removal of ^{234}Th is not uniform throughout the euphotic zone but is enhanced near depths corresponding to the active production of biogenic particles, as seen in the pigment and particle maxima. [Used by permission of the American Society of Limnology and Oceanography, Inc., © 1987, *Limnology and Oceanography*, Vol. 32, Coale and Bruland, pp. 189-200.]

$POC/^{234}Th$ (mol C/dpm ^{234}Th) is the ratio on sinking particles and λ is the decay constant of ^{234}Th (0.029 d^{-1}). This approach makes no assumptions about residence times, although it implicitly assumes that sinking biogenic particles are the principal carriers of ^{234}Th atoms, that the POC/^{234}Th ratio on sinking particles can be measured, that steady state applies and that horizontal and vertical transport of ^{234}Th via advection of water are negligible.

4.2. Results of the past decade: JGOFS and other studies

A major opportunity to test the use of ^{234}Th as a proxy for POC flux arose with the Joint Global Ocean Flux Study (JGOFS). JGOFS had as a central goal a better understanding of the ocean carbon cycle, including the flux of POC leaving the euphotic zone. Process studies were carried out in the Atlantic Ocean, Pacific Ocean, Arabian Sea and Southern Ocean. ^{234}Th profiles were obtained as a part of each process study.

The first results came from a study of the 1989 spring phytoplankton bloom in the North Atlantic (Buesseler et al. 1992b). The approach of Buesseler et al. (1992b) to applying Equation (9) involved measurement of ^{234}Th on filterable particles (>1 um) and in solution. Because the particles were retained on a polypropylene filter cartridge, the POC/^{234}Th ratio could not be measured directly on these particles and instead the ratio measured in sediment trap samples was used to estimate the sinking POC flux using Equation (9). These results showed that the development of $^{234}Th/^{238}U$ disequilibrium during the bloom clearly tracked the production and sinking of biogenic particles (Fig. 3).

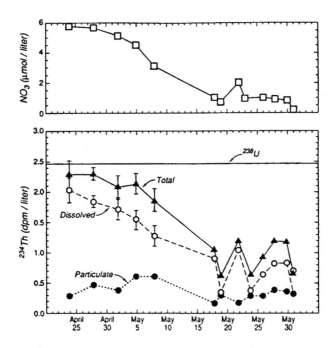

Figure 3. Time series of nitrate (Slagle and Heimerdinger 1991) and dissolved, particulate, and total ^{234}Th in surface water at 47°N, 20°W (Atlantic Ocean) in April-May 1989. ^{238}U activity calculated as 0.0686·salinity (Chen et al. 1986). The production of biogenic particles during the bloom enhances the scavenging of Th, resulting in growing $^{234}Th/^{238}U$ disequilibrium with time due to sinking of particles. [Reprinted from Buesseler et al., *Deep-Sea Research I*, Vol. 39, pp. 1115-1137, © 1992, with permission from Elsevier Science.]

However, the rapid change of the ^{234}Th profiles over the course of the bloom, a time scale comparable to half-life of ^{234}Th, necessitated the use of a non-steady state model. Anticipating this, Buesseler et al. (1992b) collected profiles to permit these calculations (Fig. 4). The results showed a fundamental disagreement between sinking POC fluxes derived from water column ^{234}Th deficits and directly collected in floating sediment traps at the same time, with lower fluxes observed in the traps. Indeed, Buesseler (1991) compiled measurements of ^{234}Th fluxes from water column and sediment data and found frequent offsets, both positive and negative, that he attributed to biases in sediment trapping (Fig. 5). (Not surprisingly, advocates of sediment trapping argued that it was the estimates of ^{234}Th flux from the water column data that were flawed). Similar offsets between traps and water column Th fluxes were observed by Buesseler et al. (1994) on the basis of a compilation of available data from the Equatorial Pacific, the Bermuda time series site, the North Atlantic, the Arabian Sea and the Arctic and the Southern Oceans.

In the Equatorial Pacific (EqPac) JGOFS program, refinements to the sampling scheme included separating particles via a series of prefilters that permitted POC and ^{234}Th to be determined directly on the particulate fraction (Buesseler et al. 1995; Bacon et al. 1996). This scheme involved collecting large particles (>53 um) on a Nitex screen. After collection, the particles were rinsed off the screen and re-filtered onto a Microquartz filter for measurement of both ^{234}Th and POC. Bacon et al. (1996) argued that the large size fraction (>53 um) most closely approximated the sinking fraction and used the POC/^{234}Th ratio in this fraction in Equation (9) to calculate POC fluxes. The results at a station on the equator showed that a relatively small fraction (~2%) of the primary production was exported as sinking POC in this region. Buesseler et al. (1995) obtained similar data along a transect across the equator (12°N to 10°S) and found that export peaked near the equator, but in all cases was <10% of the primary production. In both cases the process of equatorial upwelling had to be taken into account, and was different under El Niño and non-El Niño conditions.

Murray et al. (1996) also collected ^{234}Th profiles from the Equatorial Pacific and compared water column ^{234}Th fluxes and POC fluxes derived from them with ^{234}Th and POC fluxes in floating sediment traps. These results showed that trap ^{234}Th fluxes ranged from 0.8 to 3.5 times those derived from the water column deficiency, with greater offsets in the shallowest trap (100 m). Possible reasons for this discrepancy include hydrodynamic effects on trapping efficiency, the different time scales over which the fluxes are measured (floating traps are deployed for ~2 days while water column ^{234}Th integrates over the mean life of ^{234}Th), and the assumptions made to calculate ^{234}Th fluxes from water column data (see below). Nevertheless Murray et al. (1996) argued that agreement was good except for the shallow traps. Murray et al. (1996) documented a latitudinal variation in POC flux derived from ^{234}Th data similar to that observed by Buesseler et al. (1995), as well as a decrease in export during El Niño conditions.

The EqPac studies also provided data on the POC/^{234}Th ratios of different sized particles as well as on material collected in sediment traps (Buesseler et al. 1995; Bacon et al. 1996; Murray et al. 1996). In general the EqPac results show that the POC/^{234}Th ratio decreases as: small particles > large particles ~ sediment trap material. This trend is consistent with preferential loss of carbon due to decomposition in large aggregates or sinking material, however a coupled adsorption/aggregation model (without decomposition) also simulates a decrease in POC/^{234}Th ratios with increasing particle size (Burd et al. 2000). In the EqPac data, the decrease in POC/^{234}Th ratio of large filterable particles with depth in the water column (Bacon et al. 1996) reinforces the notion of progressive loss of POC as particles sink. In contrast, other studies have shown that the POC/^{234}Th ratio is greater in larger particles (Buesseler et al. 1996; 1998;

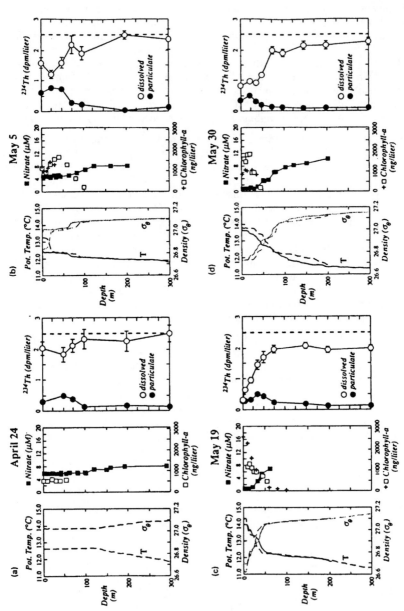

Figure 4. Time series profiles of ^{234}Th and temperature, potential density, Chl a, and nitrate (Slagle and Heimerdinger 1991) at 47°N, 20°W (Atlantic Ocean) in April-May 1989. Dashed vertical line represents estimated ^{238}U activity (Chen et al. 1986). The evolution of ^{234}Th/^{238}U disequilibrium with time follows that of Chl a and nitrate, confirming the observations illustrated in Figure 3. The series of profiles taken approximately one week apart permits application of a non-steady state model to the data. [Reprinted from Buesseler et al., *Deep-Sea Research I*, Vol. 39, pp. 1115-1137, © 1992, with permission from Elsevier Science.]

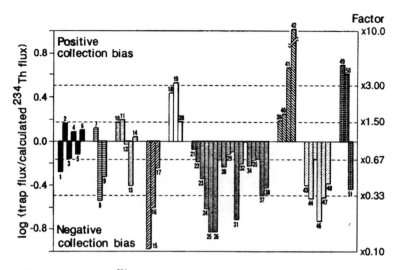

Figure 5. Comparison between ^{234}Th fluxes measured in shallow sediment traps and those calculated for the same depth and time from the ^{234}Th deficit in the overlying water column. The data were compiled from many studies and the right-hand scale shows the factor of positive or negative offset between the two data sets. Significant differences are common and are linked to the various assumptions and constraints of the two approaches for measuring POC flux. [Reprinted from *Nature*, Vol. 353, Buesseler, pp. 420-423, © 1991, Macmillan Publishers Ltd.]

Charette and Moran 1999; Charette et al. 1999; Amiel et al. 2002). Such observations often coincide with the collection of significant quantities of large phytoplankton during the growth phase of a bloom such that the volume-to-surface area ratio of the cells is increasing and the POC/^{234}Th ratio, representing the ratio of an assimilated element to an adsorbed one, follows.

Further insight on the relationship between Th and C and the POC/^{234}Th ratio of different size fractions is likely to be gained by measurements of ^{210}Po in the same samples. Indeed, given the ^{210}Po and ^{210}Pb half-lives, the use of this pair leads to estimates of export of particles (and carbon) over longer timescales than with ^{234}Th/^{238}U alone. Friedrich and Rutgers van der Loeff (2002) were recently the first to apply this approach to determine the nature of the material settling out of the mixed layer in terms of organic carbon and biogenic silica content in the Southern Ocean, and thus the fluxes of POC and biogenic silica. These authors noted that particulate ^{210}Po was correlated with POC while particulate ^{210}Pb and ^{234}Th were correlated with both POC and biogenic silica. Friedrich and Rutgers van der Loeff (2002) used ^{210}Po-^{210}Pb and ^{210}Po-^{234}Th as coupled tracers to estimate the sinking velocities of biogenic silica and POC. This allowed them, for instance, to corroborate that biogenic silica preferentially settles when heavily silicified diatoms dominate the phytoplankton.

Recent efforts to use ^{234}Th/^{238}U disequilibrium to determine POC export from the euphotic zone largely have used the sampling scheme applied by Buesseler et al. (1995) and Bacon et al. (1996) in the JGOFS EqPac program, namely *in situ* filtration through a series of membrane filters followed by manganese impregnated cartridge filters to scavenge dissolved Th. The modification of this approach for small volume samples (~2 L) (Buesseler et al. 2001; Benitez-Nelson et al. 2001a) permits faster processing of samples and greater resolution in depth profiles. As a result, the calculation of the ^{234}Th deficit is more reliable, yet application of Equation (9) still requires pumping or sediment traps at discrete depths to measure the POC/^{234}Th ratio.

The fluxes of POC determined by the ^{234}Th method applied to the world's oceans are summarized in Table 1. Where possible we have tabulated the ratio of Th-derived POC export to independent estimates of primary production. As noted above, this ratio, termed the "*ThE*" ratio (Buesseler 1998), is important in the euphotic zone carbon balance as it represents the "leakage" of POC out of the euphotic zone (The *ThE* ratio is so named to evoke the "*e*" ratio, which is defined as the ratio of POC flux measured with sediment traps to primary production).

Several aspects of the data assembled in Table 1 stand out. First is the generally low fraction (<10%) of primary production that is exported from the euphotic zone (Fig. 6), according to the ^{234}Th method. These low *ThE* ratios imply efficient remineralization of organic matter and a low leakage of POC from the euphotic zone. A second observation is the high export seen under bloom conditions and in high latitude systems. These areas of high export often have high POC/^{234}Th ratios, and Buesseler (1998) has suggested that primary production in such areas is often characterized by large diatom cells that are likely to sink rapidly following a bloom. Areas that have abundant dissolved N and P yet low production (termed High Nutrient-Low Chlorophyll or HNLC areas) are characterized by low *ThE* ratios, and in some cases show marked increases in Th-estimated POC export upon fertilization with Fe (Bidigare et al. 1999). In other cases, the export is not significantly greater after fertilization (Charette and Buesseler 2000). Clearly more must be known about the relationships between food web structure and POC export to completely explain the data presented in Table 1. A more complete account of some of the regional studies that produced these data is given in Buesseler (1998).

4.3. Unresolved issues

Given the extent to which the ^{234}Th method has been adopted to determine POC export fluxes, it is imperative that uncertainties in the approach be fully evaluated or at least considered. The advantages of the method are clear: the source of ^{234}Th is uniform and well characterized, Th strongly associates with particles and is thus a good tracer for particle transport processes, ^{234}Th deficits can be reliably and accurately measured and the half-life of ^{234}Th is long enough to permit the profiles to integrate over time yet short enough to permit resolution of seasonal variations in POC flux associated with blooms. However, there remain issues with the application of the method that must be addressed and form the basis for future research. We consider these below.

Steady state vs. non-steady state ^{234}Th profiles. Determination of ^{234}Th deficits is relatively simple. Production is balanced against decay and export, with the expectation that in the absence of the latter, ^{234}Th should be in radioactive equilibrium with ^{238}U and the flux is given by the integral term in Equation (9):

$$\lambda \int (A_U - A_{Th}) dz \tag{10}$$

It is important to note however that Equation (10) assumes steady state in the Th distribution so that production truly is balanced by decay and export. It is easy to imagine a scenario after a phytoplankton bloom, when the export of POC (and ^{234}Th) has decreased or even ceased, such that the water column ^{234}Th profile would still show a deficit with respect to ^{238}U caused by prior high flux events. This "relict" deficit will disappear as ^{234}Th grows into equilibrium with ^{238}U on a time scale set by the ^{234}Th half-life. The magnitude by which the Th flux is over- or under-estimated depends on whether deficits are increasing or decreasing and at what rate.

It is straightforward to correct for non-steady state profiles if a station is occupied repeatedly over a period of weeks to months. However, this has been done in relatively few instances (e.g., Buesseler et al. 1992b; Bacon et al. 1996; Cochran et al. 2000; Benitez-Nelson et al. 2001b; Friedrich and Rutgers van der Loeff 2002). More often, oceanographic cruises attempt to cover large areas or occupy large numbers of stations, few of which are re-occupied in a systematic manner.

Advective effects. The vertical and horizontal transport of water through a sampling area may contribute to temporal changes in water column ^{234}Th profiles and alter estimates of the flux required to support the observed ^{234}Th deficit. The simple box model that produces Equation (10) assumes that there are no influences on ^{234}Th profiles related to advection of different water masses. For example, Bacon et al. (1996) collected four ^{234}Th profiles under non-El Niño conditions at a Pacific equatorial station and found the profiles relatively constant with time. Yet upwelling was transporting water from depth (with higher ^{234}Th activity) into the euphotic zone and depressing the apparent deficits. Bacon et al. (1996) took this advection into account to more accurately estimate the sinking flux of ^{234}Th from the area. Bacon et al. (1996) also sampled during El Niño conditions and found slightly greater deficits than during non-El Niño conditions. Had they not taken upwelling into account, they would have concluded that ^{234}Th export was greater during the lower production of an El Niño period than during non-El Niño conditions.

Horizontal advection also may bias the interpretation of ^{234}Th profiles. Cochran et al. (1997) sampled stations along the axis of circulation in the Northeast Water Polynya on the East Greenland Shelf. ^{234}Th deficits increased along the circulation path, suggesting progressively increasing export of ^{234}Th. Yet the data also could be interpreted in a Lagrangian fashion, with a regionally constant rate of Th removal and increasing deficits as the water mass "ages." This interpretation is supported by progressive decreases in dissolved nutrients along the path of circulation.

The role of advection also was shown clearly by Friedrich and Rutgers van der Loeff (2002) in the Southern Ocean, especially when attempting to use ^{210}Po as a proxy for the estimation of POC and biogenic silica export. These authors pointed out that the ^{210}Po-^{210}Pb pair could be used to constrain POC and biogenic silica export fluxes if advection could be properly characterized.

Effects on ^{234}Th profiles also can be caused by the passage of an eddy through a sampling area. Eddies can have distinctly different scavenging histories for reactive elements than the water through which they pass, and if not recognized can bias the interpretation of Th deficits. Horizontal effects can be at least partly taken into account through spatial sampling that is designed to reveal them if they occur, through monitoring hydrographic parameters at stations where ^{234}Th profiles are taken and through prior knowledge of the circulation of an area. Such a multifaceted approach was used by Buesseler et al. (1994) in an attempt to compare ^{234}Th fluxes derived from water column deficits and sediment traps.

Sampling artifacts. The use of *in situ* pumps to collect water samples for ^{234}Th analysis permits simultaneous collection (and separation) of different particle fractions as well as dissolved Th. As pumping systems have been modified to permit determination of POC on the pump filters, it became possible to compare POC determined from the pump samples with conventional POC determinations made on small volume samples (0.5 - 2 L) taken from hydrocasts. The JGOFS data from multiple studies show large discrepancies between these two sample collection methods, with pump POC values 3 to 100 times lower than bottle POC values. Possible artifacts with each approach have been identified. For example Moran et al. (1999) have suggested that DOC is adsorbed onto

Table 1. ^{234}Th-derived POC export fluxes from the euphotic zone.

Region	Method* POC	^{234}Th	^{234}Th flux (dpm m^{-2} d^{-1})	POC/^{234}Th (μmol C dpm^{-1})	POC flux (mmol C m^{-2} d^{-1})	Primary production (mmol C m^{-2} d^{-1})	ThE ratio (%)	Ref.
Atlantic								
NABE, 47°N, 20°W, Apr-May 1989, 0-35 m	B	IC	1050 - 2610	14 - 22	5 - 41	87 - 98	5 - 42	(1)
NABE, 47°N, 20°W, Apr-May 1989, 0-75 m	B	IC	1720 - 3600	4 - 23	7 - 77	87 - 98	8 - 79	(1)
Bermuda Atlantic Time Series (BATS), Mar-Oct 1993-1995, 0-150 m	ST	IM	0 - 2500	2 - 9	<1 - 6	6 - 67	<1 - 56	(2)
Atlantic Equator 35°S-10°N, 25°W-5°E, May-June 1996, 0-100 m	B	B	605	5 - 90	3.2 - 43	140§	13§	(3)
Gulf of Maine, Mar-Apr, July-Aug 1997, 0-10 m	B	IM	<280	5 - 110	0 - 19	24 - 30	9 - 21	(4)
Northern Iberian Margin, May-June 1997, Dec 1997-Jan 1998, 0-50 m	B	B	500 - 1750	10 - 60	1 - 56	43 - 162	19	(5)
Gulf of Maine, Mar, June, Sept 1995, Aug-Sept 1997, 0-50 m	B	IM, IC	600 - 2400	10 - 300	15 - 34	25 - 400	11 - 25	(6)
Pacific								
Equatorial Pacific, 12°N-12°S, 95°W-170°W, Spring 1992, 0-100 m	IM	IM	1000 - 2800	1.0 - 3.5	1 - 4.5	20 - 100	1 - 8	(7)
Equatorial Pacific, 12°N-12°S, 95°W-170°W, Fall 1992, 0-100 m	IM	IM	1000 - 3500	0.8 - 4.0	1 - 8	40 - 120	3 - 10	(7)
Equatorial Pacific, 140°W, Spring 1992, 0-120 m	IM	IM	400 - 2600	0.3 - 3	0.5 - 3	90	2	(8)
Equatorial Pacific, 140°W, Fall 1992, 0-120 m	IM	IM	200 - 5000	0.4 - 6	1 - 4	130	2	(8)
Equatorial Pacific 12°N-12°S, 140°W, Winter 1992, 0-120 m	B, ST, IM	B, ST, IM	500 - 2400		1 - 6	20 - 80	3 - 11	(9)
Equatorial Pacific 12°N-12°S, 140°W, Summer 1992, 0-120 m	B, ST, IM	B, ST, IM	1300 - 3600	1 - 12.5	2 - 30	20 - 150	4 - 23	(9)
Subarctic NE Pacific, 125°W-145°W, 50°N, Feb, May, Aug 1996, Feb 1997, 0-200 m	IM	IM	895 - 1500	0.90 - 3.1	2.8 - 7.6	42.3 - 115	6 - 13	(10)
Western & Central Equatorial Pacific, Equator, 165°E-150°W, Oct 1994, 0-120 m	B, ST	B, ST	700 - 2800	2 - 5	4 - 14	52 - 94	8 - 14	(11)
Western & Central Equatorial Pacific, Equator, 165°E-150°W, April 1996	B, ST	B, ST	1200 - 4000	2 - 4	3 - 13	80	4 - 16	(11)
North Pacific Subtropical Gyre, 22°45'N, 158°W, Apr 1999-Mar 2000, 0-150 m	B, IM, ST	B, IM, ST	300 - 2300	2 - 8	0.4 - 4	15 - 60	4 - 22	(12)
Mediterranean								
North-western Mediterranean, 43°25'N, 7°51'E, Spring 1995, 0-40 m	ST	B, ST	200 - 2000	5 - 8	1 - 9	17 - 50	3 - 55	(13)
Arabian Sea								
Arabian Sea, Jan-Aug 1995, 0-100 m	IM	IM	<200 - 5500	1 - 4	1 - 17	50 - 130	1 - 10	(14)
Arabian Sea, Aug-Sept (Late SW Monsoon) 1995, 0-100 m	IM	IM	3000 - 5500	3 - 8	11 - 26	50 - 150	17 - 27	(14)

Table 1 continued.

Region	Method* POC	Method* ^{234}Th	^{234}Th flux (dpm m^{-2} d^{-1})	POC/^{234}Th (μmol C dpm^{-1})	POC flux (mmol C m^{-2} d^{-1})	Primary production (mmol C m^{-2} d^{-1})	ThE ratio (%)	Ref.
Antarctic								
Bellinghausen Sea, Dec 1992, 0-100 m	IM	IM	1600	14	21	60	35	(15)
Polar Front, Oct-Nov 1992, 0-100 m	B	B	0 - 3200	6 - 12	13 - 26	110	12 - 24	(16)
Southern Antarctic Circumpolar Current, Oct-Nov 1992, 0-100 m	B	B	0 - 1600		3 - 5	25	12 - 24	(16)
Ross Sea, Spring 1996, 0-100 m	IM	IM	0 - 500	2 - 40	0 - 4	5 - 35	<10	(17)
Ross Sea, Summer 1997, 0-100 m	IM	IM	800 - 2600	10 - 250	7 - 91	45 - 90	25 - 70	(17)
Ross Sea, Fall 1997, 0-100 m	IM	IM	800 - 2500	5 - 300	3 - 22	3		(17)
Southern Ocean, W. Pacific sector, Transect 170°W, Oct 1997-Mar 1998, 0-100 m	IM	IM	1800 - 3800	3 - 14	5 - 50	20 - 150	up to 50	(18)
Polar Front, Oct-Nov bloom 1992, 0-200 m	B	B	1800 - 3200		14 - 84	110	46	(19)
Southern Antarctic Circumpolar Current, Oct-Nov bloom 1992, 0-200 m	B	B	160 - 1700		17 - 39	25	100	(19)
Arctic								
Northeast Water Polynya, Greenland, 0-50 m	B	IC	350 - 1600	13 - 120	12 - 56	20 - 80	30 - 60	(20)
Chuckchi Shelf, August-Sept 1994, 0-30 m	IM	IM	550	67	38	214	18	(21)
Chuckchi Slope and Interior Arctic, Aug-Sept 1994, 0-30 m	IM	IM	55 - 420	5 - 21	0.3 - 7	1 - 28	4 - 100	(21)
Beaufort Sea, shelf-slope, Aug-Sept 1995, 0-50 m	IM	IM	250 - 1400	10 - 12	1 - 7		nd	(22)
Beaufort Sea, offshore, Aug-Sept 1995, 0-50 m	IM	IM	650	8.5	2		nd	(22)
North Water Polynya, Greenland, May, July 1998, Aug-Sept 1999, 0-100 m	IM	IM	300 - 2400	4 - 45	4 - 89	10 - 150	10 - 90	(23)

* B: Bottle; IM: *In situ* pump/membrane filter (53, 70 μm); IC: *In situ* pump/cartridge filter (0.5, 1 μm); ST: sediment trap

§ 5°S - 5°N and 25°W - 5°E

References: (1) Buesseler et al. 1992, (2) Buesseler 1998, (3) Charette and Moran 1999a, (4) Benitez-Nelson et al. 2000, (5) Hall et al. 2000, (6) Charette et al. 2001, (7) Buesseler et al. 1995, (8) Bacon et al. 1996, (9) Murray et al. 1996, (10) Charette et al. 1999, (11) Dunne et al. 2000, (12) Benitez-Nelson et al. 2001b, (13) Schmidt et al. 2002, (14) Buesseler et al. 1998, (15) Shimmield et al. 1995, (16) Rutgers van der Loeff et al. 1997, (17) Cochran et al. 2000, (18) Buesseler et al. 2001, (19) Friedrich and Rutgers van der Loeff 2002, (20) Cochran et al. 1995; 1997, (21) Moran et al. 1997, (22) Moran and Smith 2000, (23) Amiel et al. 2002

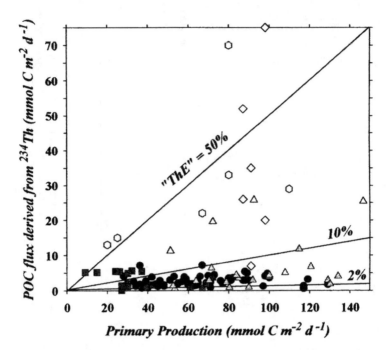

Figure 6. Plot of POC flux derived from ^{234}Th against primary production (*ThE* ratio) using data compiled from several case studies (EqPac: Equatorial Pacific Process Study, BATS: Bermuda Atlantic Time Series, NABE: U.S. JGOFS North Atlantic Bloom Experiment). This use of ^{234}Th/^{238}U disequilibrium in the surface ocean provides insight into the workings of the "biological pump," or the leakage of POC from the surface ocean via sinking particles. Typically, less than 10% of the primary production is exported from the euphotic zone, and only in high latitude areas and under bloom conditions is the export significantly larger. (From Buesseler et al 1994 and references therein)

the glass fiber filters used to collect POC samples and that this artifact is proportionately more important for small volume (i.e., bottle) samples. In contrast, Gardner et al. (2002) have suggested that the pressure difference across the filter is greater in a pump sample than under the laboratory conditions in which bottle samples are filtered. This would break up aggregates and force particles through the filters used in the pumps. This bias thus would lead to underestimates of the POC and particulate ^{234}Th fractions, but not the POC/^{234}Th ratios or total ^{234}Th values on which Equation (9) depends. However, if phytoplankton cells are being lysed by the larger pressure difference in the pumps, intercellular carbon may be forced through the filter unaccompanied by ^{234}Th, which remains associated with the residual material. This might lead to underestimates of the POC/^{234}Th ratio in the particles and possibly to underestimates of POC flux.

Determination of the POC/^{234}Th ratio on sinking particles. Application of Equation (9) requires knowledge of the POC/^{234}Th ratio on sinking particles. As noted above, this has often been taken as the POC/^{234}Th ratio in the large-size particulate fraction, with the assumption that this fraction is more likely to be sinking.

An alternative is to use the POC/^{234}Th ratio in sediment trap material. Sediment traps do unquestionably capture sinking material, but hydrodynamic biases associate with the traps may lead them to sample only a fraction of the sinking material. As well, the POC/^{234}Th ratio in the large particle fraction as collected on Nitex or Teflon screens used in *in situ* pumps does not always equal that in traps, in instances where such comparisons can be made (Armiel et al. 2002). The fact that the POC/^{234}Th ratios used to calculate the sinking POC fluxes in Table 1 were determined in several different ways (e.g., measured in filtered large particles >53 or >70 μm or sediment traps, combining bottle POC data with pump ^{234}Th) makes rigorous comparison among regions difficult. What is needed to resolve this issue is the measurement of POC/^{234}Th ratios on material known to be sinking at different speeds, and a comparison of these values with the sampling approaches listed above. This can be approached by comparing POC/^{234}Th ratios on particles separated according to sinking velocity by centrifugation or an elutriator with the large particle fractions sampled with pumps.

Comparison of water column-derived ^{234}Th and POC fluxes with sediment trap fluxes. Sediment traps have contributed greatly to our understanding of the magnitudes and modifications of particle fluxes through the oceanic water column (see Gardner 2000 for a review). It has long been recognized that trapping efficiency may not be 100%. Indeed, the flux of ^{230}Th measured in deep traps (>1000 m), relative to its known production in the overlying water column, has been a prime means of calibrating and verifying trap efficiency. (Owing to its long half-life and high reactivity, ^{230}Th essentially is completely scavenged from the overlying water column by sinking particles. See Henderson and Anderson (2003) for additional details). In the case of ^{234}Th, fluxes derived from deficits in the euphotic zone are compared with surficial trap arrays. These are often deployed as depth arrays on floating moorings, with the shallowest trap at generally at ~50-100 m. Comparison of the ^{234}Th (and POC) fluxes in the traps with those determined from the water column ^{234}Th deficit often shows large discrepancies as pointed out by Buesseler (1991). Most commonly, the water column fluxes are greater than the sediment traps.

It is thought that one of the primary sources of error in sediment traps is hydrodynamic effects at the trap mouth (e.g., Gust and Kozerski 2000). One way of evaluating hydrodynamic biases on fixed-depth floating traps is to compare them with neutrally buoyant traps that move with the water and sample a single isopycnal surface. Such a comparison by Buesseler et al. (2000) in the waters off Bermuda showed that material caught in the two types of traps differed significantly in components such as ^{234}Th and fecal pellet fluxes. These results suggest that trapping efficiencies likely vary for different types of particles and that the POC/^{234}Th ratio measured in trap material may not be representative of that in sinking particles. Moreover, the ^{234}Th flux measured with the neutrally buoyant trap was lower than that in the floating trap and more consistent with prior water column results for the comparable time of year. Considerable additional research is needed to reconcile the different estimates of flux.

Time scales of integration. The various parameters of the carbon system integrate over different time scales depending on measurement technique. Water column profiles of ^{234}Th, for example are based on samples collected within a few hours, but the profiles integrate flux events that occurred some time in the past. The "memory" of the system is limited by the half-life of ^{234}Th. Moored sediment traps are often deployed for months and record detailed temporal trends of fluxes, yet floating traps usually are deployed for intervals of 1 to 3 days before recovery. Thus, the recent ^{234}Th flux recorded in a floating trap may be compared with the temporally integrated flux derived from the water column ^{234}Th deficit. The temporal factor may not be significant under steady state conditions in which the flux is changing little with time, but is likely to be far more significant under bloom conditions.

Differences in integration time scales may also affect our perception of key derived parameters such as the *ThE* ratio (Cochran et al. 2000). This ratio (see above) compares the POC flux derived from water column ^{234}Th profiles (and thus integrating into the past) with present primary production. As classically measured using ^{14}C incubation techniques, primary production is an instantaneous measurement representing the phytoplankton community as sampled at a single time. Under bloom conditions, the export of POC may lag the production of fresh organic matter and *ThE* ratios calculated late in a bloom may be overestimates.

5. ^{234}Th AS A TRACER FOR PARTICLE TRANSPORT AND SEDIMENT PROCESSES IN THE COASTAL OCEAN

In addition to the use of ^{234}Th to estimate POC fluxes from the euphotic zone of the open ocean, high particle concentrations and fluxes near the bottom of the open ocean and in estuarine and coastal areas produce enhanced scavenging and effective transfer of ^{234}Th to bottom sediments in these areas. Although early measurements of ^{234}Th/^{238}U disequilibrium in near-bottom waters of the Pacific (Amin et al. 1974) failed to document ^{234}Th scavenging, areas with a well-developed bottom nepheloid (or high turbidity) layer indeed show ^{234}Th removal (Bacon and Rutgers van der Loeff 1989; Turnewitsch and Springer 2001). An extreme example of such an environment is the High Energy Benthic Boundary Layer Experiment (HEBBLE) site in the Atlantic, where the strong bottom currents ensure substantial resuspension of sediment leading to scavenging of ^{234}Th (DeMaster et al. 1991).

Measurements of ^{234}Th/^{238}U disequilibrium in coastal and estuarine waters (e.g., Aller and Cochran 1976; McKee et al. 1986; Cochran et al. 1995) show rapid ^{234}Th removal. Mean residence times with respect to scavenging are often less than a few days, and ^{234}Th produced from ^{238}U decay in the overlying water column is present in the sediments, where it is in "excess" of the sediment ^{238}U activity. Excess ^{234}Th profiles in sediments can be used to estimate time scales of remobilization of trace metal as well as the effects of episodic events such as storms on near-interface sediments, but perhaps their greatest application is in evaluating the rates of sediment mixing by the benthic fauna. A knowledge of the reworking rates of sediments is important in deciphering the chronology of sediment cores because the distributions of longer lived radionuclides such as ^{210}Pb are affected both by sediment accumulation and mixing. The latter can produce overestimates of sediment accumulation if ignored, and ^{234}Th helps to sort out these effects.

5.1. Sediment mixing rates

The processes of sediment accumulation and perturbation of the surface sediments by organisms or physical processes are responsible for distributing ^{234}Th to depth in the sediments. The short half-life of ^{234}Th ensures that it typically does not penetrate deep into the sediment column by sediment accumulation alone, and bioturbation or physical mixing are the dominant processes distributing ^{234}Th to several centimeters depth. The importance of mixing on ^{234}Th profiles was initially documented by Aller and Cochran (1976) and Cochran and Aller (1979) through a comparison of excess ^{234}Th profiles with X-radiographs and faunal samples that showed faunal activity (Fig. 7). Sediment profiles of excess ^{234}Th typically show quasi-exponential decreases in the upper 3 to5 cm (Fig. 7). Particle mixing by benthic fauna is often parameterized as an eddy diffusion-like process, such that mixing of ^{234}Th into the sediments is balanced by radioactive decay:

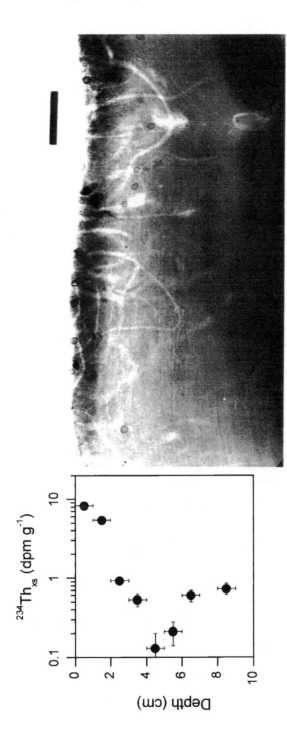

Figure 7. Excess ^{234}Th activity versus depth (left) and X-radiograph (right) in a sediment core collected from the New York Bight, showing the importance of mixing by benthic fauna in the upper part of the seabed. Abundant individuals of the small bivalve *Nucula proxima* may be seen in the X-radiograph near the sediment-water interface, and the light-colored areas represent burrows of *Nephtys* sp. and *Ceriantheopsis* sp. Reprinted from *Estuarine Coastal and Shelf Science* (formerly Estuarine and Coastal Marine Science) Vol. 9, Cochran and Aller, pp. 739-747, © 1979, with permission from Elsevier Science.

$$\frac{\partial A}{\partial t} = D_b \frac{\partial^2 A}{\partial x^2} - \lambda A \tag{11}$$

where A is the excess ^{234}Th activity (dpm/g), D_b is the particle mixing coefficient (cm^2/y) and x is depth (cm) in the sediment. Equation (11) properly includes a sediment accumulation term, but this is commonly ignored for ^{234}Th as it is smaller than the mixing and decay terms. However porosity changes in the upper few cm's can affect the value of A, and the activities are often converted to dpm/cm^3 wet sediment (by multiplying A by the dry bulk density (g dry/cm^3 wet sediment). Equation (11) can be solved for steady state conditions ($\partial A/\partial t = 0$) with the boundary conditions $A = A_0$ at $x = 0$ and $A \rightarrow 0$ as $x \rightarrow \infty$:

$$A = A_0 \exp(-\sqrt{\frac{\lambda}{D_b}} x) \tag{12}$$

Values of D_b determined by ^{234}Th profiles commonly range from ~1 to 50 cm^2/y in estuarine and slope sediments (Aller and Cochran 1976; Aller et al. 1980; Sun et al. 1994; Gerino et al. 1998; Green et al. 2002) to ≤ 10 cm^2/y in the deep sea (Aller and DeMaster 1984; Pope et al. 1996). Although there is some overlap, mixing rates in the deep sea tend to be lower than those in nearshore sediments. This trend has been documented with ^{210}Pb profiles as well (e.g., Henderson et al. 1999).

Mixing coefficients derived from ^{234}Th generally agree with those determined by other short-lived radionuclides (e.g., ^{210}Pb; see Henderson and Anderson 2003) although the ^{234}Th-derived values may be greater due the fact that the short half-life of ^{234}Th limits it to tracing recently deposited material, which may contain fresher organic matter and therefore may be preferentially consumed by deposit feeding organisms (Smith et al. 1993). Particle selectivity and particle processing rates of bathyal deposit feeders have been evaluated by comparing ^{234}Th activities (and chl-a and phaeopigments) in the gut contents of deep-sea deposit feeders to surface sediments and material collected with sediment traps (Miller et al. 2000). A decrease in mixing intensity with depth in the sediment column also may contribute to differences in mixing rates calculated with different tracers. ^{234}Th-derived mixing coefficients also are generally consistent with those determined from the distributions of both natural and added tracers of mixing. For example, a collection of time series sediment profiles of ^{234}Th and chlorophyll-a at a station in Long Island Sound showed similar distributions due to mixing (Gerino et al. 1998).

5.2. ^{234}Th as a tracer of particle transport in shelf and estuarine environments

Because the sources of particle-reactive natural radionuclides are well characterized, they serve as good indicators of the relative importance of lateral vs. vertical processes in delivering them to the sea floor. This is especially true for ^{234}Th, the source of which is the integrated U activity in the overlying water column. The supply of ^{234}Th arising from vertical transport through the water column to bottom sediments thus increases linearly with total water depth. Comparison of inventories of ^{234}Th in bottom sediments with production in the overlying water column thus provides useful information about particle redistribution processes that may affect other particle-associated chemical species as well as ^{234}Th. Aller et al. (1980) used this approach in considering the spatial distribution of inventories of ^{234}Th in the sediments of Long Island Sound, USA. They observed a pattern that ranged between direct vertical delivery from the overlying water column and complete homogenization of ^{234}Th inventories in sediments via lateral transport.

The water column distribution of particulate ^{234}Th in partially mixed estuaries aids in assessing the transport of particles throughout the system, as a consequence of tidal mixing or the estuarine circulation. Feng et al. (1999a) took advantage of the fact that the source of ^{234}Th (dissolved ^{238}U) increases seaward such that particles from the lower Hudson River estuary have proportionately more ^{234}Th than particles in the fresher, up-estuary portion of the system. They normalized the ^{234}Th activities to ^{7}Be, a cosmogenic radionuclide whose input to the estuary is via direct atmospheric deposition and is constant over this spatial scale, to resolve the effects of particle source and varying grain size (Fig. 8). The results confirmed that particles are transported a distance comparable to the tidal excursion. The ^{234}Th/^{7}Be ratio in the Hudson River estuary is also useful in resolving the importance to the turbidity maximum of local, resuspended sediment vs.

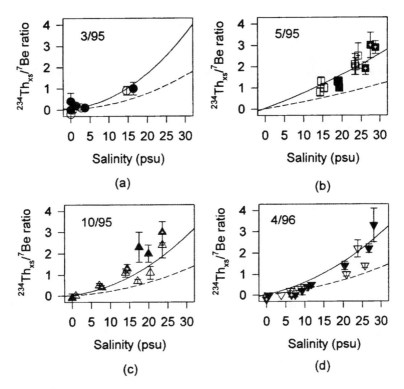

Figure 8. ^{234}Th/^{7}Be activity ratio in suspended particles vs. salinity in the partially mixed Hudson River estuary. Normalization of ^{234}Th activities to ^{7}Be eliminates the influence of particle grain size on the radionuclide activities and permits identification of the provenance of the particles in the estuary. Both Th and Be are rapidly scavenged onto particles in thus turbid estuary and the ^{234}Th/^{7}Be ratio on particles thus increases toward the estuary mouth because the increase in ^{238}U with salinity leads to greater production of ^{234}Th down-estuary. In comparison the addition of ^{7}Be to the estuary from the atmosphere is spatially constant. Open and filled symbols represent surface and bottom water, respectively. Samples collected on flood tide are designated with a +. The solid curves denote upper limits on ^{234}Th/^{7}Be activity ratios, estimated from a model assuming in situ production of ^{234}Th, atmospheric supply of ^{7}Be, rapid scavenging, and short residence time of suspended particles in water column. The dashed curves are estimated lower limits set by resuspension of bottom sediments. During periods of low flow (e.g., October 1995) the greater-than-expected ^{234}Th/^{7}Be ratios on particles in salinities of 17-25 psu indicate that particles are transported significant distances up-estuary in association with the estuarine circulation. Conversely, during times of high river flow (e.g., April 1996), the ^{234}Th/^{7}Be activity ratios trend below the solid curve, indicating enhanced resuspension of bottom sediments or transport of particles through the estuary toward the ocean. Such patterns can be used to trace the rate and scale of the transport of particle-reactive contaminants through estuaries. [Reprinted from *Geochimica et Cosmochimica Acta*, Vol. 63, Feng et al., pp. 2487-2505, © 1999, with permission from Elsevier Science.]

material brought in with the estuarine circulation (Feng et al. 1999b). In urbanized estuaries such as the Hudson, which have significant sources of particle-reactive contaminants near the mouth of the estuary, the ^{234}Th/^{7}Be ratio also serves to trace the dispersion of these contaminants up-estuary (Feng et al. 1999c).

On a larger scale, ^{234}Th gradients may be used to trace lateral transport on continental shelves. In the inner Yangtze continental shelf, McKee et al. (1984) observed sediment inventories that were 30 times the supply from local overlying water column. Their data provide evidence of the transport of particulate ^{234}Th and deposition in the inner shelf of the Yangtze. Water column and sediment ^{234}Th measurements in the Gulf of Maine also indicate that horizontal transport delivers ^{234}Th into Inner Casco Bay (Gustafsson et al. 1998). Gustafsson et al.'s (1998) observation of the gradient in total ^{234}Th with distance away from the Bay mouth was similar to the lateral gradients observed by Bhat et al. (1969) in the Arabian Sea and Kaufman et al. (1981) in the New York Bight. In the case of Casco Bay, the gradient in ^{234}Th supported inventories in inner Bay sediments that were greater than supply from the local overlying water column (Gustafsson et al. 1998). Such ^{234}Th data provide the basis for calibrating transport models that clarify the dispersion of particle-reactive contaminants such as hydrophobic organics away from sources (Gustafsson et al. 1998).

6. CONCLUDING REMARKS

Applications of natural radionuclides to understanding ocean processes have increased tremendously in the last decade, and this is especially so for applications of ^{234}Th/^{238}U disequilibrium. Interestingly this is in spite of the fact that, unlike the longer lived thorium isotopes, it has not been possible to measure ^{234}Th by techniques other than decay counting (e. g. thermal ionization mass spectrometry, inductively coupled plasma-mass spectrometry). Measurement of total ^{234}Th in seawater samples essentially has come full circle from relatively small samples (~20 L) to large volumes sampled by *in situ* pumps, to a combination of pumping and small volume samples. A great deal of the impetus driving the measurement of ^{234}Th has been its use as a means of calibrating the flux of POC sinking out of the euphotic zone. This application has tremendous potential because it is relatively easy to occupy a station, collect a ^{234}Th depth profile and obtain a POC flux integrated over the past ~1 to 3 months. Yet, considerable additional research is necessary to fully understand the constraints of this approach and what it is really telling us about the oceanic carbon cycle. Some of the potential research directions have been discussed earlier in this chapter, but among the most important are:

- controls on the POC/^{234}Th ratio of different sized particles
- the relationship between the POC/^{234}Th ratio of filterable particles and that of sinking particles
- better use of ^{210}Po and ^{234}Th as coupled tracers for export fluxes of POC as well as biogenic silica or carbonate
- use of ^{234}Th and ^{228}Th as coupled tracers to extract information on particle dynamics such as aggregation/disaggregation
- more detailed comparisons of ^{234}Th fluxes derived from water column profiles and sediment traps of different designs

Resolving some of the issues surrounding the use of ^{234}Th as a tracer for POC flux, as well as collecting simultaneous ^{210}Po and ^{210}Pb data, will enable a better characterization of other components of the particulate flux, such as biogenic silica (Buesseler et al. 2001; Friedrich and Rutgers van der Loeff 2002) and particle-associated contaminants (Gustafsson et al. 1997a,b; 1998).

The short-lived particle reactive radionuclides of the U/Th series also have enormous potential for tracking particle source and transport in ocean margins. Mass balances comparing inventories in sediments with supply can be used to determine import or export of particles to an area. Such approaches are increasingly important in understanding the fates of particle-reactive contaminants whose sources are often enhanced in the coastal ocean. Studies of ^{234}Th, especially when supplemented by other radionuclides such as ^{7}Be, ^{210}Po and ^{210}Pb, will enable us to refine our view of how vertical and lateral transport of particles and particle-associated chemical species occurs in the oceans, as well as how solutes interact with particle surfaces.

ACKNOWLEDGMENTS

The authors' involvement in some of the research described in this chapter has been supported by the US National Science Foundation, the National Oceanic and Atmospheric Administration and its Sea Grant Program, the US Department of Energy, the Hudson River Foundation and the European Union. Financial support in the form of postdoctoral fellowship to PM from the Government of Spain and the Fulbright Commission for is gratefully acknowledged. We also thank Gideon Henderson and two anonymous reviewers for their comments on the manuscript.

REFERENCES

Aller RC, Cochran JK (1976) Th-234/U-238 disequilibrium in nearshore sediments: particle reworking and diagenetic time scales. Earth Planet Sci 29:37-50
Aller RC, Benninger LJ, Cochran JK (1980) Tracking particle associated processes in nearshore environments by use of ^{234}Th/^{238}U disequilibrium. Earth Planet Sci Lett 47:161-175
Aller RC, DeMaster DJ (1984) Estimates of particle-flux and reworking at the deep-sea floor using Th-234/U-238 disequilibrium. Earth Planet Sci Lett 67:308-318
Amiel D, Cochran JK, Hirschberg DJ (2002) ^{234}Th/^{238}U disequilibrium as an indicator of the seasonal export flux of POC in the North Water (NOW) Polynya. Deep-Sea Res II 49:5191-5209
Amin BS, Krishnaswami S, Somayajulu BLK (1974) ^{234}Th/^{238}U activity ratios in Pacific Ocean bottom waters. Earth Planet Sci Lett 21:342-344
Bacon MP, Spencer DW, Brewer PG (1976) Pb-210/Ra-226 and Po-210/Pb-210 disequilibria in seawater and suspended particulate matter. Earth Planet Sci Lett 32:277-296
Bacon MP, Anderson RF (1982) Distribution of thorium isotopes between dissolved and particulate forms in the deep sea. J Geophys Res 87:2045-2056
Bacon MP, Belastock RA, Tecotzky M, Turekian KK, Spencer DW (1988) Lead-210 and polonium-210 in ocean water profiles of the continental shelf and slope south of New England. Cont Shelf Res 8:841-853
Bacon MP, Rutgers van der Loeff MM (1989) Removal of Thorium-234 by scavenging in the bottom nepheloid layer of the ocean. Earth Planet Sci Lett 92:157-164
Bacon MP, Cochran JK, Hirschberg DJ, Hammar TR, Fleer AP (1996) Export flux of carbon at the equator during the EqPac time-series cruises estimated from ^{234}Th measurements. Deep-Sea Res II 43:1133-1153
Baskaran M, Santschi PH, Benoit G, Honeyman BD (1992) Scavenging of Th isotopes by colloids in seawater of the Gulf of México. Geochim Cosmochim Acta 56:3375-3388
Benitez-Nelson CR, Buesseler KO, Crossin G (2000) Upper ocean carbon export, eddy diffusivity and horizontal transport in the southwestern Gulf of Maine. Cont. Shelf Res 20:707-736
Benitez-Nelson C, Buesseler KO, Rutgers van der Loeff M, Andrews J, Ball L, Crossin G, Charette MA (2001a) Testing a new small-volume technique for determining thorium-234 in seawater. J Radioanal Nucl Chem 248:795-799
Benitez-Nelson C, Buesseler KO, Karl D, Andrews J (2001b) A time-series study of particulate matter export in the North Pacific Subtropical Gyre based upon ^{234}Th:^{238}U disequilibrium. Deep-Sea Res I 48:2595-2611
Bhat SG, Krishnaswami S, Lal D, Rama, More WS (1969) ^{234}Th/^{238}U ratios in the ocean. Earth Planet Sci Lett 5:483-491

Bidigare RR, Hanson KL, Buesseler KO, Wakeham SG, Freeman KH, Pancost RD, Millero FJ, Steinberg P, Popp BN, Latasa M, Landry MR, Laws EA (1999) Iron-stimulated changes in C-13 fractionation and export by equatorial Pacific phytoplankton: Toward a paleogrowth rate proxy. Paleoceanography 15 (5):589-595

Biggin CD, Cook GT, MacKenzie AB, Pates JM (2002) Time-efficient method for the determination of Pb-210, Bi-210, and Po-210 activities in seawater using liquid scintillation spectrometry. Anal Chem 74:671-677

Broecker WS, Kaufman A, Trier RM (1973) The residence time of Thorium in surface sea-water and its implications regarding the fate of reactive pollutants. Earth Planet Sci Lett 20:35-44

Buesseler KO (1991) Do upper-ocean sediment traps provide an accurate record of particle flux? Nature 353:420-423

Buesseler KO, Cochran JK, Bacon MP, Livingston HD, Casso SA, Hirschberg DJ, Hartman, MC, Fleer, AP (1992a) Determination of thorium isotopes in seawater by non-destructive and radiochemical procedures. Deep-Sea Res I 39:1103-1114

Buesseler KO, Bacon MP, Cochran JK, Livingston HD (1992b) Carbon and nitrogen export during the JGOFS North Atlantic Bloom Experiment estimated from ^{234}Th:^{238}U disequilibria. Deep-Sea Res I 39:1115-1137

Buesseler KO, Michaels AF, Siegel DA, Knap AH (1994) A three dimensional time-dependent approach to calibrating sediment trap fluxes. Global Biogeochem Cycl 8:179-193

Buesseler KO, Andrews JA, Hartman MC, Belastock R, Chai F (1995) Regional estimates of the export flux of particulate organic carbon derived from thorium-234 during the JGOFS EqPac program. Deep-Sea Res II 42:777-804

Buesseler KO, Bauer J, Chen R, Eglinton T, Gustafsson Ö, Landing W, Mopper K, Moran SB, Santschi PH, Vernon Clark R, Wells M (1996) An intercomparison of cross-flow filtration techniques used for sampling marine colloids: overview and organic carbon results. Mar Chem 55:1-31

Buesseler KO (1998) The de-coupling of production and particulate export in the surface ocean. Glob Biogeochem Cycl 12:297-310

Buesseler KO, Ball L, Andrews J, Benitez-Nelson C, Belastock R, Chai F, Chao Y (1998) Upper ocean export of particulate organic carbon in the Arabian Sea derived from thorium-234. Deep-Sea Res II 45:2461-2487

Buesseler KO, Charette MA (2000) Commentary on "How accurate are the ^{234}Th based particulate residence times in the ocean?" by G. Kim, N. Hussain and T. Church. Geophys Res Lett 27:1939-1940

Buesseler KO, Steinberg DK, Michaels AF, Jonson RJ, Andrews JE, Valdes JR, Price JF (2000) A comparison of the quantity and quality of material caught in a neutrally buoyant versus surface-tethered sediment trap. Deep-Sea Res I 47:277-294

Buesseler KO, Ball L, Andrews J, Cochran JK, Hirschberg DJ, Bacon MP, Fleer A, Brzezinski M (2001) Upper ocean export of particulate organic carbon and biogenic silica in the Southern Ocean along 170°W. Deep-Sea Res II 48:4275-4297

Burd AB, Moran SB, Jackson, GA (2000) A coupled adsorption-aggregation model of the POC/^{234}Th ratio of marine particles. Deep-Sea Res I 47:103-120

Charette MA, Moran SB (1999) Rates of particle scavenging and particulate organic carbon export estimated using ^{234}Th as a tracer in the subtropical and equatorial Atlantic Ocean. Deep-Sea Res II 46:885-906

Charette MA, Moran SB, Bishop JKB (1999) ^{234}Th as a tracer of particulate organic carbon export in the subarctic Northeast Pacific Ocean Deep-Sea Res II 46(11-12):2833-2861

Charette MA, Buesseler KO (2000) Does iron-fertilization lead to rapid carbon export in the Southern Ocean? Geochem Geophys Geosyst 1:Paper number 2000GC000069

Charette MA, Moran BS, Pike SM, Smith JN (2001) Investigating the carbon cycle in the Gulf of Maine using the natural tracer thorium 234. J Geophys Res 106:11553-11579

Chen JH, Edwards GJ, Wasserburg GJ (1986) ^{238}U, ^{234}U and ^{232}Th in sea water. Earth Planet Sci Lett 80:241-251

Chung Y, Finkel R, Bacon MP, Cochran JK, Krishnaswami S (1983) Intercomparison of ^{210}Pb measurements at GEOSECS station 500 in the northeast Pacific. Earth Planet Sci Lett 65:393-405

Chung Y, Finkel R (1988) Po-210 in the Western Indian-ocean – distributions, disequilibria and partitioning between the dissolved and particulate phases. Earth Planet Sci Lett 88:232-240

Clegg SL, Whitfield M (1991) A generalized model for the scavenging of trace metals in the open ocean – I. Particle cycling. Deep-Sea Res 37:809-837

Clegg SL, Whitfield M (1992) A generalized model for the scavenging of trace metals in the open ocean – II. Thorium scavenging, Deep-Sea Res 38:91-120

Clegg SL, Bacon MP, Whitfield M (1991) Application of a generalized scavenging model to thorium isotope and particle data at equatorial and high-latitude sites in the Pacific Ocean. J Geophys Res 96:20655-20670

Coale KH, Bruland KW (1985) Th-234:U-238 disequilibria within the California Current. Limnol Oceanogr 30:22-33

Coale KH, Bruland KW (1987) Oceanic stratified euphotic zone as elucidated by Th-234:U-238 disequilibria. Limnol Oceanogr 32:189-200

Cochran JK, Aller RC (1979) Particle reworking in sediments from the New York Bight apex: evidence from ^{234}Th/^{238}U disequilibrium. Estuar Coast Mar S 9:739-747

Cochran JK, Bacon MP, Krishnaswami S, Turekian KK (1983) ^{210}Po and ^{210}Pb distributions in the central and eastern Indian Ocean. Earth Planet Sci Lett 65:433-445

Cochran JK, Livingston HD, Hirschberg DJ, Surprenant LD (1987) Natural and anthropogenic radionuclide distributions in the northwest Atlantic-ocean. Earth Planet Sci Lett 84:135-152

Cochran JK (1992) The oceanic chemistry of the uranium and thorium-series nuclides *In*: Uranium-series disequilibrium: applications to earth, marine, and environmental sciences. Ivanovich M, Harmon RS (eds) Oxford University Press, New York, p 334-395

Cochran JK, Buesseler KO, Bacon MP, Livingston HD (1993). Thorium isotopes as indicators of particle dynamics in the upper ocean: Results from the JGOFS North Atlantic Bloom Experiment. Deep-Sea Res I 40(8):1569-1595

Cochran JK, Barnes C, Achman D, Hirschberg DJ (1995) Thorium-234/Uranium-238 disequilibrium as an indicator of scavenging rates and particulate organic carbon fluxes in the Northeast Water Polynya, Greenland. J Geophys Res 100(C3):4399-4410

Cochran, JK, Frignani M, Hirschberg DJ, Barnes C (1995) Thorium isotopes as indicators of scavenging rates in the Venice Lagoon. Mar. Freshwater Res. 46:215-221

Cochran JK, Roberts KA, Barnes C, Achman D (1997) Radionuclides as indicators of particle and carbon dynamics of the East Greenland Shelf. Radioprot – Coll 32:129-136

Cochran JK, Buesseler KO, Bacon MP, Wang HW, Hirschberg DJ, Ball L, Andrews J, Crossin G, Fleer A (2000) Short-lived thorium isotopes (^{234}Th, ^{228}Th) as indicators of POC export and particle cycling in the Ross Sea, Southern Ocean. Deep-Sea Res II 47(15-16):3451-3490

Cutshall NH, Larsen IL, Olsen CR (1983) Direct analysis of Pb-210 in sediment samples – self-absorption corrections. Nucl Intr Meth Phys Res 206:309-312

DeMaster DJ, Brewster DC, McKee BA, Nittrouer CA (1991) Rates of particle scavenging, sediment reworking, and longitudinal ripple formation at the HEBBLE site based on measurements of ^{234}Th and ^{210}Pb. Mar Geol 99:423-444

Dugdale RC, Goering JJ (1967) Uptake of new and regenerated forms of nitrogen in primary productivity. Limnol Oceanogr 12:196-206

Dunne JP, Murray JW, Rodier M, Hansell DA (2000) Export flux in the western and central equatorial Pacific: zone and temporal variability. Deep-Sea Res I 47:901-936

Eppley RW, Peterson BJ (1979) Particulate organic matter flux and planktonic new production in the deep ocean. Nature 282:670-680

Eppley RW (1989) New production: history, methods, problems. *In*: Productivity of the ocean: Present and past. Berger WH, Smetacek VS, Wefer G (eds) John Wiley & Sons, Chichester, p 85-97

Feely RA, Sabine CL, Takahashi T, Wanninfhof R (2001) Uptake and storage of carbon dioxide in the ocean: The Global CO_2 Survey. Oceanogr 14:18-32

Feng H, Cochran JK, Hirschberg DJ (1999a) ^{234}Th and ^{7}Be as tracers for the transport and dynamics of suspended particles in a partially mixed estuary. Geochim Cosmochim Acta 63(17):2487-2505

Feng H, Cochran JK, Hirschberg DJ (1999b) ^{234}Th and ^{7}Be as tracers for the sources of particles to the turbidity maximum of the Hudson River estuary. Estuar Coast Shelf Sci 49:629-645

Feng H, Cochran JK, Hirschberg DJ (1999c) ^{234}Th and ^{7}Be as tracers for transport and sources of particle-associated contaminants in the Hudson River Estuary. Sci Total Environ 237/238:401-418

Fisher NS, Burns KA, Cherry RD, Heyraud M (1983) Accumulation and cellular distribution of ^{241}Am, ^{210}Pb and ^{210}Po in two marine algae. Mar Ecol Prog Ser 11:233-237

Fisher NS, Teyssie JL, Krishnaswami S, Baskaran M (1987) Accumulation of Th, Pb, U, and Ra in marine-phytoplankton and its geochemical significance. Limnol Oceanogr 32 (1):131-142

Flynn WW (1968) Determination of low levels of polonium-210 in environmental materials. Anal Chim Acta 43 (2):221-227

Friedrich J, Rutgers van der Loeff MM (2002) A two-tracer (^{210}Po – ^{234}Th) approach to distinguish organic carbon and biogenic silica export flux in the Antarctic Circumpolar Current. Deep-Sea Res I 49:101-120

Gardner WD (2000) Sediment trap technology and sampling in surface waters. In: The changing ocean carbon cycle: A midterm synthesis of the Joint Global Ocean Flux Study. Hanson RB, Ducklow HW, Field JG (eds) Cambridge University Press, Cambridge, p 240-281

Gardner WD, Richardson MJ, Carlson CA, Hansell DA (2002) Determining POC: bottles, pumps and transmissometers. Limnol. Oceanogr. (submitted)

Geibert W, Rutgers van der Loeff MM, Hanfland C, Dauelsbergf HJ (2002) Actinium-227 as a deep-sea tracer: sources, distribution and applications. Earth Planet. Sci Lett. 198:147-165

Gerino M, Aller RC, Lee C, Cochran JK, Aller JY, Green MA, Hirschber D (1998) Comparison of different tracers and methods used to quantify bioturbation during a spring bloom: 234-Thorium, luminospheres and chlorophyll a. Estuar Coast Shelf S 46:531-547

Goldstein SJ, Stirling CH (2003) Techniques for measuring uranium-series nuclides: 1992-2002. Rev Mineral Geochem 52:23-57

Green MA. Aller RC, Cochran JK, Lee C, Aller JY (2002). Bioturbation in shelf/slope sediments off Cape Hatteras, North Carolina: The use of ^{234}Th, Chl-a, and Br$^-$ to evaluate rates of particle and solute transport. Deep-Sea Res II 49(20):4627-4644

Gust G, Kozerski HP (2000) In situ sinking-particle flux from collection rates of cylindrical traps. Mar Ecol Prog Ser 208:93-106

Gustafsson Ö, Gschwend PM, Buesseler KO (1997a) Using ^{234}Th disequilibria to estimate the vertical removal rates of polycyclic aromatic hydrocarbons from the surface ocean. Mar Chem 57(1-2):11-23

Gustafsson Ö, Gschwend PM, Buesseler KO (1997b) Settling removal rates of PCBs into the northwestern Atlantic derived from ^{238}U-^{234}Th disequilibria. Environ Sci Tech 31:3544-3550

Gustafsson Ö, Buesseler KO, Geyer WR, Moran SB, Gschwend PM (1998) On the relative significance of horizontal and vertical transport of chemicals in the coastal ocean: application of a two-dimensional Th-234 cycling model. Cont Shelf Res 18:805-829

Hall IR, Schmidt S, McCave IN, Reyss J-L (2000) Suspended sediment distribution and Th-234/U-238 disequilibrium along the Northern Iberian Margin: implication for particulate organic carbon export. Deep-Sea Res I 47:557-582

Henderson GM, Lindsay FN, Slowey NC (1999) Variation in bioturbation with water depth on marine slopes: a study on the Little Bahamas Bank. Mar Geol 160:105-118

Henderson GM, Anderson RF (2003) The U-series toolbox for paleoceanography. Rev Mineral Geochem 52:493-531

Heussner S, Cherry RD, Heyraud M (1990) Po-210 and Pb-210 in sediment trap particles on a Mediterranean continental margin. Cont. Shelf Res 10:989-100

Heyraud M, Cherry RD (1983) Correlation of Po-210 and Pb-210 enrichments in the sea-surface microlayer with neuston biomass. Cont Shelf Res 1:283-293

Honeyman BD, Santschi PH (1989)The role of particles and colloids in the transport of radionuclides and trace metals in the oceans. In: Environmental particles. Buffle J, van Leewen HP (eds) Lewis Publishers, Boca Raton, p 379-423

Hong GH, Park SK, Baskaran M, Kim SH, Chung CS, Lee SH (1999) Lead-210 and polonium-210 in the winter well-mixed turbid waters in the mouth of the Yellow Sea Cont Shelf Res 19 (8):1049-1064

Huh C-A, Prahl FG (1995) Role of colloids in upper ocean biogeochemistry in the northeast Pacific Ocean elucidated from ^{238}U-^{234}Th disequilibria. Limnol Oceanogr 40:528-532

International Atomic Energy Agency (IAEA) (1985) Sediment K_d's and concentration factors for radionuclides in the marine enviroment. Technical Reports Series 247. IAEA, Vienna, Austria

Ivanovich M, Murray A (1992) Spectroscopic methods. In: Uranium-series disequilibrium: applications to earth, marine, and environmental sciences. Ivanovich M, Harmon RS (eds) Oxford University Press, New York, p 127-173

Kaufman A, Li Y-H, Turekian KK (1981) The removal rates of ^{234}Th and ^{228}Th from waters of the New York Bight. Earth Planet Sci Let 54:385-392

Kim G, Hussain N, Church TM (1999) How accurate are the ^{234}Th based particulate residence times in the ocean?. Geophys Res Lett 26:619-622

Kharkar DP, Thomson J, Turekian KK, Forster WO (1976) Uranium and thorium series nuclides in plankton from the Caribbean. Limnol Oceanogr 21:294-299

Krishnaswami S, Lal D, Somayajulu BLK, Weiss R, Craig H (1976) Large-volume in situ filtration of deep Pacific waters: mineralogical and radioisotope studies. Earth Planet Sci Lett 32:420-429

Livingston HD, Cochran JK (1987) Determination of transuranic and thorium isotopes in ocean water: in solution and in filterable particles. J Radioanal Nucl Chem 115:299-308

Masqué P, Sanchez-Cabeza JA, Bruach JM, Palacios E, Canals M (2002) Balance and residence times of ^{210}Pb and ^{210}Po in surface waters of the northwestern Mediterranean Sea. Cont Shelf Res 22:2127-2146

Matsumoto E (1975) Th-234-U-238 radioactive disequilibrium in the surface layer of the oceans. Geochim Cosmochim Acta 39:205-212

McKee BA, DeMaster DJ, Nittrouer CA (1984) The use of Th-234/U-238 disequilibrium to examine the fate of particle-reactive species on the Yangtze continental shelf. Earth Planet Sci Lett 68:431-42

McKee BA, DeMaster DJ, Nittrouer CA (1986) Temporal variability in the portioning of thorium between dissolved and particulate phases on the Amazon shelf: Implications for the scavenging of particle-reactive species. Cont Shelf Res 6:87-106

Miller RJ, Smith CR, DeMaster DJ, Fornes WL (2000) Feeding selectivity and rapid particle processing by deep-sea megafaunal deposit feeders: A ^{234}Th tracer approach. J Mar Res 58:653-573

Moore RM, Hunter KA (1985) Thorium adsorption in the ocean – reversibility and distribution amongst particle sizes. Geochim Cosmochim Acta 49:2253-2257

Moore RM, Millward GE (1988) The kinetics of reversible Th reactions with marine particles. Geochim Cosmochim Acta 52:113-118

Moran SB, Buesseler KO (1992) Short residence time of colloids in the upper ocean estimated from ^{238}U:^{234}Th disequilibria. Nature 359:221-223

Moran SB, Buesseler KO (1993) Size-fractionated ^{234}Th in continental shelf waters off New England: implications for the role of colloids in oceanic trace metal scavenging. J Mar Res 51:893-922

Moran, SB, Ellis KM, Smith JN (1997) ^{234}Th/^{238}U disequilibrium in the central Arctic Ocean: implications for particulate organic carbon export. Deep-Sea Res II 44(8):1593-1606

Moran SB, Charette MA, Pike SM, Wicklund CA (1999) Differences in seawater particulate organic carbon concentration in samples collected using small- and large-volume methods: the importance of DOC adsorption to the filter blank. Mar Chem 67:33-42

Moran SB, Smith JN (2000) ^{234}Th as a tracer of scavenging and particle export in the Beaufort Sea. Cont Shelf Res 20(2):153-167

Murnane RJ, Cochran JK, Sarmiento JL (1994) Estimates of particle- and thorium-cycling rates in the northwest Atlantic Ocean. J Geophys Res 99:3373-3392

Murnane RJ, Cochran JK, Buesseler KO, Bacon MP (1996) Least-squares estimates of thorium, particle and nutrient cycling rate constants from the JGOFS North Atlantic Bloom Experiment. Deep-Sea Res I 43(2):239-258

Murray JW, Downs JN, Stroms S, Wei C-L, Jannasch HW (1989) Nutrient assimilation, export production and ^{234}Th scavenging in the eastern equatorial Pacific. Deep-Sea Res 36:1471-1489

Murray JW, Young J, Newton J, Dunne J, Chapin T, Paul B, McCarthy JJ (1996) Export flux of particulate organic carbon from the Central Equatorial Pacific determined using a combined drifting trap- ^{234}Th approach. Deep-Sea Res II 43 (4-6):1095-1132

Niven SEH, Kepkay PE, Boraie A (1995) Colloidal organic carbon and colloidal ^{234}Th dynamics during a coastal phytoplankton bloom. Deep-Sea Res II 42:257-273

Nozaki Y, Thomson J, Turekian KK (1976) The distribution of Pb-210 and Po-210 in the surface waters of the Pacific Ocean. Earth Planet Sci Lett 32:304-312

Nozaki Y, Tsunogai S (1976) Ra-226, Pb-210 and Po-210 disequilibria in western North-Pacific. Earth Planet Sci Lett 32 (2):313-321

Nozaki Y, Horibe Y, Tsubota H (1981) The water column distributions of thorium isotopes in the western North Pacific. Earth Planet Sci Lett 54 (2):203-216

Nozaki Y, Horibe Y (1983) Alpha-emitting thorium isotopes in northwest Pacific deep waters. Earth Planet Sci Lett 65:39-50

Nozaki Y (1984) Excess ^{227}Ac in deep ocean water. Nature 310:486-488

Nozaki Y (1986) ^{226}Ra-^{222}Rn-^{210}Pb systematics in seawater near the bottom of the ocean. Earth Planet Sci Lett 80:3-4

Nozaki Y, Yang HS, Yamada M (1987) Scavenging of thorium in the ocean. J Geophys Res 92 (C1):772-778

Nozaki Y, Yang HS (1987) Th and Pa isotopes in the waters of the western margin of the Pacific near Japan: evidence for release of ^{228}Ra and ^{227}Ac from slope sediments. J Oceanogr Soc Japan 43:217-227

Nozaki Y, Yamada M, Nikaido H (1990) The marine geochemistry of actinium-227: evidence for its migration through sediment pore water. Geophys Res Lett 17:1933-1936

Nozaki Y (1993) Actinium-227: a steady state tracer for the deep-se basin wide circulation and mixing studes. In: Deep Ocean Circulation, Physical and Chemical Aspects. Teramoto T (ed) Elsevier p 139-155

Nozaki Y, Zhang J, Takeda A (1997) ^{210}Pb and ^{210}Po in the equatorial Pacific and Bering Sea: the effects of biological productivity and boundary scavenging. Deep-Sea Res II 44:2203-2220

Nozaki Y, Dobashi F, Kato Y, Yamamoto Y (1998) Distribution of Ra isotopes and the ^{210}Pb and ^{210}Po balance in surface waters of the mid Northern Hemisphere. Deep-Sea Res. I 45:1263-1284

Pates JM, Cook GT, MacKenzie AB, Anderson R, Bury SJ (1996) Determination of Th-234 in marine samples by liquid scintillation spectrometry. Anal Chem 68:3783-3788

Pope RH, DeMaster DJ, Smith CR, Seltman Jr H (1996) Rapid bioturbation in equatorial Pacific sediments: evidence from excess ^{234}Th measurements. Deep-Sea Res II 43:1339-1364.

Porcelli D, Swarzenski PW (2003)The behavior of U- and Th- series nuclides in groundwater. Rev Mineral Geochem 52:317-361

Quigley MS, Honeyman BD, Santschi PH (1996) Thorium sorption in the marine environment: equilibrium partitioning at the Hematite/water interface, sorption/desorption kinetics and particle tracing. Aquat Geochem 1:277-301

Quigley MS, Santschi PH, Guo L, Honeyman BD (2001) Sorption irreversibility and coagulation behaviour of ^{234}Th with marine organic matter. Mar Chem 76:27-45

Quigley MS, Santschi PH, Hung C-C, Guo L, Honeyman BD (2002) Importance of acid polysaccharides for ^{234}Th complexation to marine organic matter. Limnol Oceanogr 47:367-377

Radakovitch O, Cherry RD, Heyraud M, Heussner S (1998) Unusual ^{210}Po/^{210}Pb ratios in the surface water of the Gulf of Lions. Oceanol Acta 21:459-468

Radakovitch O, Cherry RD, Heussner S (1999) Particulate fluxes on the Rhone continental margin (NW Mediterranean). Part III: ^{210}Po and ^{210}Pb data and the particle transfer scenario. Deep-Sea Res I 46:1539-1563

Rutgers van der Loeff MM, Friedrich J, Bathmann UV (1997) Carbon export during the spring bloom at the Antarctic Polar Front, determined with the natural tracer ^{234}Th. Deep-Sea Res II 44:457-478

Santschi PH, Nyffeler UP, Li Y-H, O'Hara P (1986) Radionuclide cycling in natural waters: relevance of scavenging kinetics. *In*: Sediments and water interactions. Sly P (ed) Springer-Verlag, New York, p 183-191

Sarin MM, Kim G, Church TM (1999) ^{210}Po and ^{210}Pb in the South-equatorial Atlantic: distribution and disequilibrium in the upper 500 m. Deep-Sea Res II 46:907-917

Schmidt S, Andersen V, Belviso S, Marty JC (2002) Strong seasonality in particle dynamics of north-western Mediterranean surface waters as revealed by ^{234}Th/^{238}U. Deep-Sea Res I 49:1507-1518

Shimmield GB, Ritchie GD, Fileman TW (1995) The impact of marginal ice zone processes on the distribution of ^{210}Pb, ^{210}Po and ^{234}Th and implications for new production in the Bellinghausen Sea, Antarctica. Deep-Sea Res II 42:1313-1335

Slagle R, Heimerdinger G (1991) North Atlantic Bloom Experiment, April-July 1989. Process Study Data Report P-1, NODC/U.S. JGOFS Data Management Office, Woods Hole, MA, USA

Smith CR, Pope RH, DeMaster DJ, Magaard L (1993) Age-dependent mixing of the deep-sea sediments. Geochim Cosmochim Acta 57:1473-1488

Stewart G, Fisher NS (2002) Experimental studies on the accumulation of polonium-210 by marine phytoplankton. Limnol Oceanogr, submitted

Sun MY, Aller RC, Lee C (1994) Spatial and temporal distributions of sedimentary chloropigments as indicators of benthic processes in Long-Island Sound. J Mar Res 52:149-176

Swarzenski PW, Porcelli D, Andersson PS, Smoak JM (2003) The behavior of U- and Th- series nuclides in the estuarine environment. Rev Mineral Geochem 52:577-606

Thomson J, Turekian KK (1976) Polonium-210 and Lead-210 distributions in ocean water profiles from the eastern South Pacific. Earth Planet Sci Lett 32:297-303

Tsunogai S, Minagawa M (1978) Settling model for the removal of insoluble chemical elements in seawater. Geochem J 12:47-56

Turnewitsch R, Springer BM (2001) Do bottom mixed layers influence ^{234}Th dynamics in the abyssal near-bottom water column?. Deep-Sea Res I 48:1279-1307

12 The U-series Toolbox for Paleoceanography

Gideon M. Henderson
Department of Earth Sciences
Oxford University, South Parks Road
Oxford, OX1 3PR, ENGLAND
gideonh@earth.ox.ac.uk

Robert F. Anderson
Lamont-Doherty Earth Observatory of Columbia University
Route 9W
Palisades, New York, 10964, USA
boba@ldeo.columbia.edu

1. INTRODUCTION

The geochemistry of marine sediments is a major source of information about the past environment. Of the many measurements that provide such information, those of the U-series nuclides are unusual in that they inform us about the rate and timescales of processes. Oceanic processes such as sedimentation, productivity, and circulation, typically occur on timescales too short to be assessed using parent-daughter isotope systems such as Rb-Sr or Sm-Nd. So the only radioactive clocks that we can turn to are those provided by cosmogenic nuclides (principally ^{14}C) or the U-series nuclides. This makes the U-series nuclides powerful allies in the quest to understand the past ocean-climate system and has led to their widespread application over the last decade.

As in other applications of the U-series, those in paleoceanography rely on fractionation of the nuclides away from secular equilibrium. In the oceanic setting, this fractionation is generally due to differences in the solubility of the various nuclides. The behavior of the U-series nuclides in the ocean environment was widely researched in the middle decades of the twentieth century. This work established knowledge of the concentrations of the nuclides in the various compartments of the ocean system, and of their fluxes between these compartments. Such understanding was comprehensively summarized in the Ivanovich and Harmon U-series volume (1992), particularly by Cochran (1992). Understanding of the behavior of the U-series nuclides has not advanced very dramatically in the decade since that summary but a major theme of research has been the use of this geochemical understanding to develop U-series tools to assess the past environment (Table 1).

This chapter summarizes the use of U-series nuclides in paleoceanography. It starts with a brief summary of the oceanic U budget and an introduction to important features of the behavior of U-series nuclides in the marine realm. It then discusses the various U-series tools which have proved useful for paleoceanography, starting at ^{238}U (and ^{235}U) and progressing down the decay chain towards Pb. One tool that will not be discussed is U/Th dating of marine carbonates which has seen sufficient application to merit a chapter on its own (Edwards et al. 2003). The use of U-series nuclides to assess rates of processes in the modern ocean will also not be discussed in depth here but are dealt with elsewhere in this volume (Cochran and Masque 2003).

2. U-SERIES ISOTOPES IN THE OCEAN ENVIRONMENT

2.1. The ocean uranium budget

Uranium has a reasonably constant seawater concentration in both space and time,

Table 1. Summary of seawater data for U-series nuclides with paleoceanographic applications. S - soluble, I - insoluble. Full descriptions of the paleoceanographic uses and references are provided in the text. Further details of the half lives are in Bourdon et al. (2003).

Nuclide	Average Concentration		Residence Time Including Decay (years)	Residence Time Excluding Decay (years)	Halflife (years)	Sol.	Paleoceanographic Uses
	(g/g)	(dpm/1000l)					
^{238}U decay series							
^{238}U	3.3×10^{-9}	$2.5 \times 10^{+3}$	$4.0 \times 10^{+5}$	$4.0 \times 10^{+5}$	$4.46 \times 10^{+9}$	S	authigenic U - paleoproductivity and deepwater oxygen
^{234}Th	4.8×10^{-20}	$2.5 \times 10^{+3}$	9.6×10^{-2}	$2.0 \times 10^{+1}$	6.60×10^{-2}	I	sediment mixing, trapping efficiency
^{234}U	2.0×10^{-13}	$2.9 \times 10^{+3}$	$1.9 \times 10^{+5}$	$4.0 \times 10^{+5}$	$2.45 \times 10^{+5}$	S	continental weathering
^{230}Th	1.1×10^{-17}	5.0×10^{-1}	$2.0 \times 10^{+1}$	$2.0 \times 10^{+1}$	$7.57 \times 10^{+4}$	I	U/Th chronology, constant flux proxy, sediment focusing
^{226}Ra	8.4×10^{-17}	$1.9 \times 10^{+2}$	$2.2 \times 10^{+3}$	$7.2 \times 10^{+4}$	$1.60 \times 10^{+3}$	S	Holocene dating
^{210}Pb	7.6×10^{-19}	$1.3 \times 10^{+2}$	$2.0 \times 10^{+1}$	$4.8 \times 10^{+1}$	$2.26 \times 10^{+1}$	I	sediment mixing
^{235}U decay series							
^{235}U	2.4×10^{-11}	$3.5 \times 10^{+3}$	$4.0 \times 10^{+5}$	$4.0 \times 10^{+5}$	$2.34 \times 10^{+7}$	S	
^{231}Pa	2.8×10^{-18}	3.0×10^{-1}	$1.3 \times 10^{+2}$	$1.3 \times 10^{+2}$	$3.25 \times 10^{+4}$	I	rate of ocean circulation, opal paleoproductivity

varying only in line with changes in salinity (Ku et al. 1977) (Table 1). This feature simplifies the use of many proxies making, for instance, the supply rate of Th and Pa isotopes constant. The ocean U budget was summarized by Cochran (1992) and, despite work in several areas since then, does not need dramatic revision. There remains, however, significant uncertainty about the sizes of some of the fluxes in the budget.

The dominant supply of U to the oceans is from the continents by river runoff. Palmer and Edmond (1993) measured dissolved U concentrations in a number of rivers and summarized existing literature to arrive at a total flux close to that of Cochran (1992) of 11×10^9 g/year. This flux is uncertain by about 35% due to inadequate sampling of rivers with large seasonal cycles (Palmer and Edmond 1993).

Additional uncertainty about the size of the riverine flux arises from the behavior of U in the estuarine zone. The Amazon is one of the few major world rivers where estuarine U behavior has been studied. Early work suggested that significant U was released from Amazon shelf-sediments as salinity approached its open-ocean value during mixing. From the Amazon alone, this release would represent an additional source of U of up to 15% of the global riverine flux and the flux would obviously be larger if U behaves in a similar fashion in other estuaries. Further work, however, has demonstrated that U is removed from Amazon waters by colloidal coagulation at salinities of <10 psu (Swarzenski et al. 1995) and by reducing conditions in the underlying sediments (Barnes and Cochran 1993). This removal may balance the release of U at greater salinity so that the Amazon provides neither a net sink nor input of U to the oceans. Other rivers, notably the Ganges-Brahmatuptra, appear to act as a net sink of U (Carroll and Moore 1993). The behavior of U in estuaries clearly varies depending on river chemistry and the style of mixing and is fully summarized elsewhere in this volume (Swarzenski and Porcelli 2003). At this stage, the poor knowledge of U behavior in estuaries represents a large uncertainty in the ocean U budget.

Additional sources of U to the oceans are dissolution of wind-blown dust and groundwater discharge. The former is unlikely to be significant and a 10% release of U from dust would amount to only about 3% of the riverine U flux. Groundwater may introduce a more significant flux of U but this is extremely difficult to estimate as both the total flux of groundwater to the ocean and the average U concentration of such waters are poorly known (Dunk et al. 2002).

U is removed from the oceans principally into marine sediments and by alteration of ocean basalts (Table 2). The removal into marine sediments is dominated by reducing sediments where pore-water U concentrations are lowered as U becomes insoluble. This lowering leads to a diffusion gradient which draws more seawater U into the pore-waters to be removed (Barnes and Cochran 1990; Klinkhammer and Palmer 1991). Recent estimates for each of the removal terms are given in Table 2, but significant uncertainty remains over most of these fluxes (Dunk et al. 2002). It is probably fair to say that the total removal flux may be incorrect by a factor of 2 and, at such a level of uncertainty, the U budget is in balance. This budget, together with the seawater U concentration of 3.3 ppb (Chen et al. 1986a), indicates an oceanic residence time for U of $\approx$400 kyr.

This long residence time suggests that changes in seawater U concentration during the 100 kyr glacial-interglacial cycle are unlikely. Calculations using the expected area of reducing sediments during glacials also suggest that seawater U concentration does not change during the cycle (Rosenthal et al. 1995a). These calculations make the assumption that U incorporated in anoxic sediments during glacial periods is not released to the ocean during interglacials when the deep sea is more oxygenated. This release was suggested (Emerson and Huested 1991) but is now thought to be unlikely as U mobilized by re-

Table 2. Known inputs and outputs of U to the oceans. Units are 10^9 g yr^{-1}, or thousand tons per year. References are the most recent primary studies of that flux. Some other fluxes (e.g., groundwater input and input or removal at estuaries) are so poorly known that they cannot be realistically included. Significant uncertainty remains in most of the fluxes listed above so that, within these uncertainties, the U budget can be considered to be in balance.

		Flux (10^9 g yr^{-1})	Reference
Sources	Rivers	11.0	(Palmer and Edmond 1993)
	Dust dissolution	0.3	
	TOTAL	**11.3**	
Sinks	Oxic sediments	0.8	(Sacket et al. 1973)
	Metalliferous sediments	0.2	(Mills et al. 1994)
	Suboxic sediment	1.3	(Barnes and Cochran 1990)
	Anoxic sediments	2.8	(Morford and Emerson 1999)
	Shallow water carbonates	0.8	(Sacket et al. 1973)
	Salt marshes	1.0	(Church et al. 1996)
	MOR - low T	0.9	(Staudigel et al. 1996)
	MOR - high T	0.4	(Chen et al. 1986b)
	TOTAL	**8.2**	

oxygenation tends to move downward to be immobilized in reducing sediment at greater depth (Rosenthal et al. 1995b; Thomson et al. 1998).

2.2. Chemical behavior of U-series nuclides in the oceans

The chemical behavior of U and its daughter nuclides in the ocean environment was extensively studied in the 1960s and 1970s and has been well summarized (Cochran 1992). The most important mechanism by which nuclides are separated from one another to create disequilibrium is their differing solubility. For U, this solubility is in turn influenced by the redox state. The process of alpha-recoil can also play an important role in producing disequilibrium.

In oxidizing aqueous conditions, such as those found in most seawater, U forms the soluble uranyl-carbonate species. In anoxic or suboxic conditions, however, U is reduced from its hexavalent to its tetravalent state and becomes insoluble. This process is not instantaneous and does not occur in anoxic deep-waters of, for instance, the Black Sea (Anderson et al. 1989). Reduction to the tetravalent state does, however, frequently occur in marine sediments and causes them to act as one of the principal sinks for U from the oceans (Barnes and Cochran 1990; Klinkhammer and Palmer 1991). Indeed, the removal of U into reducing sediments has been used as a proxy for both bottom-water oxygen concentrations (Francois et al. 1997, Frank et al. 2000) and past productivity (Kumar et al. 1995). These proxies rely on the fact that the redox state of sediment is controlled by the competition between the supply of oxygen from deep water, and the supply of organic material which consumes oxygen during its decay. Authigenic U concentrations are therefore high when deep-water oxygen concentrations are low, or organic material supply is high. The delivery of organic matter to the sea bed at any particular site depends both on the productivity-driven flux of particulate organic matter sinking from surface waters above, and on the lateral redistribution of particles by deep-sea currents. Discriminating among the factors controlling authigenic U formation can be difficult but the spatial pattern of authigenic U deposition over a wide region can sometimes be used to infer the relative importance of surface productivity and bottom-water oxygen levels (Chase et al. 2001).

Thorium generally exists as a neutral hydroxide species in the oceans and is highly insoluble. Its behavior is dominated by a tendency to become incorporated in colloids and/or adhere to the surfaces of existing particles (Cochran 1992). Because ocean particles settle from the water column on the timescale of years, Th isotopes are removed rapidly and have an average residence time of ≈20 years (Fig. 1). This insoluble behavior has led to the common assertion that Th is always immobile in aqueous conditions. While this is generally true in seawater, there are examples of Th being complexed as a carbonate (e.g., Mono Lake waters, Anderson et al. 1982; Simpson et al. 1982) in which form it is soluble.

As with ^{230}Th, ^{231}Pa is produced by decay of U in the water column. While decay of ^{234}U produces 2.52×10^{-2} dpm m^{-3} yr^{-1} of ^{230}Th, decay of ^{235}U produces 2.33×10^{-3} dpm m^{-3} yr^{-1} of ^{231}Pa leading to a production activity ratio of $(^{231}$Pa/^{230}Th$) = 0.093$ (note that here, and throughout this chapter, round brackets are used to denote an activity or activity ratio rather than an atom ratio). Like ^{230}Th, ^{231}Pa is removed from the water column by sorbtion to the surfaces of settling particles. However, Pa is slightly less insoluble than Th leading to a longer residence time than ^{230}Th of ≈130 years (Fig. 1; Table 1). This can lead to the spatial separation of ^{231}Pa and ^{230}Th after their production in the water column and before their burial in sediments. This is reflected by $(^{231}$Pa/^{230}Th$)$ ratios in recent modern sediments ranging from 0.03 to 0.3.

Radium, like most other group II metals, is soluble in seawater. Formation of ^{226}Ra and ^{228}Ra by decay of Th in marine sediments leads to release of these nuclides from the sediment into the deep ocean. Lead, in contrast, is insoluble. It is found as a carbonate or dichloride species in seawater (Byrne 1981) and adheres to settling particles to be removed to the seafloor.

The energy involved in the α-decay of some nuclides within the decay chain leads to the largely physical fractionation process of α-recoil (Osmond and Ivanovich 1992). As the α particle is ejected, the daughter nuclide recoils in the opposite direction and moves a distance of ≈550 angstroms in a typical mineral (Kigoshi 1971). This recoil causes a fraction of the daughter nuclide produced during α decay to be ejected from the host mineral into the surrounding medium. An additional fraction of the daughter is left residing in damaged crystallographic sites within the mineral, from where it can be more readily mobilized. α-recoil therefore gives daughter nuclides of α-decay a tendency to leave their host mineral by a process which is independent of their chemistry. Alpha-recoil is most important in preferentially releasing ^{234}U from minerals over ^{238}U, but also plays a role in mobilizing other nuclides.

3. HISTORY OF WEATHERING – $(^{234}$U/^{238}U$)$

Seawater $(^{234}$U/^{238}U$)$ is higher than secular equilibrium due to α-recoil on the continents and in marine sediments. The history of seawater $(^{234}$U/^{238}U$)$ may provide information about the history of weathering during the Pleistocene.

Decay of ^{238}U on the continents causes α-recoil of the immediate daughter, ^{234}Th, which quickly decays to ^{234}U. This process leads to $(^{234}$U/^{238}U$)$ above the secular

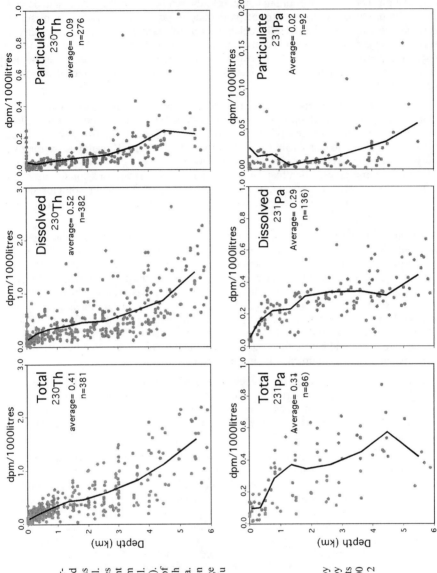

Figure 1. A summary of water-column measurements of ^{230}Th and ^{231}Pa. Most of this data was compiled by Henderson et al. (1999a) and a full list of references is contained therein. More recent data have been added from Edmonds et al. (1998), Moran et al. (2001) and Vogler et al. (1998). Averages and the total number of measurements are shown on each panel for the various types of data. The average ^{230}Th concentration can be used to calculate the average ocean residence time for ^{230}Th (Yu et al. 1996) using the equation:

$\tau_{Th} = (^{230}Th)/(^{234}U)*(1/\lambda_{230})$
$= 0.5/2750*1.09 \times 10^5$
$= 20$ years.

Similarly,
$\tau_{Pa} = (^{231}Pa)/(^{235}U)*(1/\lambda_{231})$
$= 0.3/108*4.61 \times 10^4$
$= 130$ years.

Average profiles are shown by black lines and were calculated by averaging nuclide measurements and depth data in 9 layers (0-100 m, 100-500 m, 500 m layers to 2 km, and 1 km layers to 6 km).

equilibrium value of 1.0 in most continental surface waters. Rivers, for example, have ($^{234}U/^{238}U$) values which range from 0.9 to 3.0 and the average riverine U flux to the oceans has a ($^{234}U/^{238}U$) lying somewhere close to 1.25 (Chabaux et al. 2003). Given the long residence time of U in the oceans (≈400 kyr) there is time for a portion of the excess ^{234}U which enters the oceans to decay before U is again removed from the oceans. A simple calculation suggests that, if rivers were the only source of excess ^{234}U, the oceans would have a ($^{234}U/^{238}U$) value of ≈1.08. Observed values are higher than this at 1.146 ± 0.002 (Chen et al. 1986a, Cheng et al. 2000, Robinson et al. Press). This difference requires there to be an additional source of ^{234}U to the oceans, probably related to α-recoil in marine sediments (Ku 1965). Alpha-recoil leads to high ($^{234}U/^{238}U$) in marine pore waters (Cochran and Krishnaswami 1980; Henderson et al. 1999c; Russell et al. 1994) and thus to diffusion of ^{234}U into bottom waters. This process operates in addition to the net flux of seawater U into reducing sediments so that, although both ^{238}U and ^{234}U are removed from seawater, some ^{234}U is also returned by diffusion from pore-waters. Although the size of the bottom-water ^{234}U flux has not yet been independently assessed, the mismatch between riverine inputs of excess ^{234}U and observed seawater ($^{234}U/^{238}U$) suggest that it must be of approximately equal importance to riverine inputs in supporting the high ($^{234}U/^{238}U$) values observed in seawater (Henderson 2002a).

Changes in seawater ($^{234}U/^{238}U$) with time are most likely to be induced by changes in the riverine input and may therefore provide information about past continental weathering. In particular, riverine ($^{234}U/^{238}U$) is thought to be increased by physical weathering (Kronfeld and Vogel 1991) because the grinding of rocks increases the surface area from which ^{234}U can be directly recoiled, and releases ^{234}U held in damaged sites within mineral grains.

The history of seawater ($^{234}U/^{238}U$) was initially investigated in order to assess the reliability of U/Th ages of corals (Gallup et al. 1995; Henderson et al. 1993) and found to be within error of its present value during the last two interglacials. Further work has indicated constant interglacial values for the last 400 kyr (Henderson 2002a) (Fig. 2). This suggests that the rate of physical weathering on the continents, averaged over a glacial-interglacial cycle, has not varied during the last 400 kyr.

The presence of changes in seawater ($^{234}U/^{238}U$) within a single glacial to interglacial cycle has been suggested based on compilations of coral data (Esat and Yokoyama 1999; Yokoyama et al. 2001). Last glacial corals have ($^{234}U/^{238}U$) up to 0.020 lower than Holocene corals and the history of change mimics the history of sea level. It is possible that these changes reflect differential diagenesis of corals, dependant on when they grew within the sea level cycle and therefore their duration of subaerial versus submarine exposure. If such diagenesis could be discounted, however, the glacial-interglacial changes in seawater ($^{234}U/^{238}U$) would require dramatic changes in the U budget related to climate or sea level. The increase in ($^{234}U/^{238}U$) from the glacial to the Holocene would require a very large pulse of high ($^{234}U/^{238}U$) The possibility of such glacial-interglacial changes in seawater ($^{234}U/^{238}U$) remains controversial and is an active area of research.

4. SEDIMENTATION RATE – $^{230}Th_{xs}$

4.1. The downward flux of ^{230}Th

^{230}Th is extremely insoluble and adheres to the surface of particles in the ocean soon after it forms from the decay of ^{234}U. Because these particles continuously settle from the water column, ^{230}Th is rapidly removed from the oceans to the seafloor. The combined process of surface adsorption, followed by particle settling, is termed scavenging. Measurement of the very low ^{230}Th concentrations in seawater that result from this

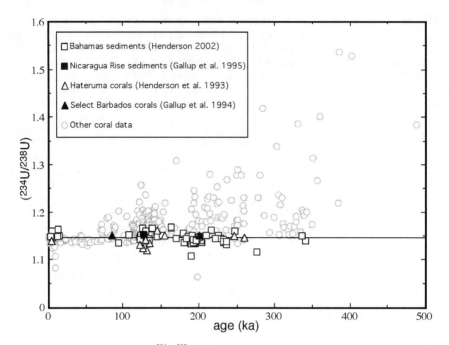

Figure 2. The history of seawater $(^{234}U/^{238}U)$ for the last 500 kyr. Data shown in black are from studies specifically investigating the U isotope history of seawater and demonstrate that $(^{234}U/^{238}U)$ has remained close to the modern value of 1.146 (shown by black line) for at least the last 400 kyr. Data is mostly from the Bahamas slope sediment (Henderson 2002) and is supplemented by similar sediment from the Nicaragua slope (Gallup et al. 1995) and by two coral studies which focused on unaltered corals (Gallup et al. 1994; Henderson et al. 1993). Other corals have frequently suffered alteration, as shown by their high initial $(^{234}U/^{238}U)$ values (gray circles).

scavenging are challenging but were successfully performed in the particulate phase (Krishnaswami et al. 1976; Krishnaswami et al. 1981) and in total seawater (Moore 1981; Moore and Sackett 1964; Nozaki et al. 1981). Those studies showed a broadly linear increase in ^{230}Th concentration with water depth, and total concentrations ≈10 times the particulate concentration. These features of the ^{230}Th profile are best explained by a reversible scavenging model in which ^{230}Th on the surface of settling particles continues to exchange with dissolved ^{230}Th as the particles settle through the water column (Bacon and Anderson 1982; Nozaki et al. 1987). The tendency for ^{230}Th to adhere to the surface, rather than remain dissolved, can be described by the use of a distribution coefficient, K_d, defined as the concentration of ^{230}Th per mass of particles divided by the concentration of ^{230}Th per mass of water. The K_d for ^{230}Th in the open ocean is typically 10^7, indicating the extreme insolubility of thorium (Fig 3).

This low solubility leads to the theoretical expectation that ^{230}Th should be removed to the seafloor as soon as it forms because there is insufficient time for advection or diffusion to occur (Bacon and Anderson 1982; Bacon and Rosholt 1982). This would make the ^{230}Th flux to the seafloor dependant only on the depth of overlying water and on the concentration of ^{234}U in that water. In numerical terms, the production rate of ^{230}Th within a column of water, in dpm m^{-2} yr^{-1} is given by:

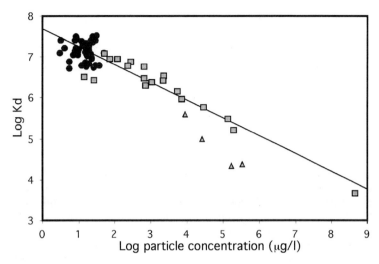

Figure 3. Distribution coefficient (K_d) versus particle concentration for Th. Note that, for typical open-ocean particle concentrations, Th is about 10^7 times more likely to adhere to a mass of particles than to remain in the same mass of water. This tendency to be found in the particulate phase decreases with particle concentration, probably due to the presence of a larger number of colloids which, because they pass through filters, appear to be in the dissolved phase (Honeyman et al. 1988). Grey squares are ^{234}Th data from Honeyman et al. (1988); gray triangles are ^{234}Th data from the continental shelf from McKee et al. (1986) and black circles are a compilation of open ocean ^{230}Th data from Henderson et al. (1999a).

$$P_{Th} = Z \left(^{234}U \right) \lambda^{230} \tag{1}$$

Where Z is the water depth in meters, (^{234}U) is the activity of ^{234}U in seawater (= 2750 dpm m^{-3}) and λ^{230} is the decay constant of ^{230}Th. If the assumption of immediate removal is correct then the downward flux of ^{230}Th, V_{Th}, is equal to P_{Th}, and $V_{Th}/P_{Th} = 1$.

It is worth mentioning at this stage that the actual downward flux of ^{230}Th in the water column consists of not only the ^{230}Th derived from ^{234}U decay in the water column, but also a small component of ^{230}Th contained in detrital material and supported by decay of ^{234}U. Measured ^{230}Th values must therefore be corrected for this small portion of detrital ^{230}Th to assess ^{230}Th$_{xs}$, i.e., the unsupported ^{230}Th derived directly from water-column ^{234}U decay (see Appendix).

If $V_{Th}/P_{Th} = 1$ then the accumulation of ^{230}Th$_{xs}$ in marine sediments would provide an assessment of their sedimentation rate. For instance, if P_{Th} is N dpm m^{-2} yr^{-1}, and N dpm are found in the upper 1 cm of 1 m^2 of seafloor, then the sedimentation rate must be 1 cm yr^{-1}. Sedimentation rate is an important variable in paleoceanographic reconstruction as it provides the timescale for the continuous record of environmental change recorded in marine sediments. Sedimentation rate is also a key geochemical variable as sediments are the major sink for most chemical species in the ocean. A tool allowing assessment of past sedimentation rates is therefore an appealing prospect.

The use of ^{230}Th$_{xs}$ to assess sedimentation rate relies on the assumption that the downward flux of ^{230}Th$_{xs}$, V_{Th}, equals the production of ^{230}Th$_{xs}$, P_{Th}. The simple fact that ^{230}Th exists in the water column at all (Fig. 1) suggests that some advection must occur and that V_{Th} may not always exactly equal P_{Th}. Similarly, the spatial variability in ^{230}Th concentration in the water column (Henderson et al. 1999a) indicates that its rate of removal is not uniform and therefore that advection of ^{230}Th is not a constant. A good example of ^{230}Th advection has been the recognition that unusually low ^{230}Th concentrations found in deep waters of the North Atlantic reflect movement of surface waters into the deep ocean by North Atlantic Deep Water (NADW) formation (Moran et al. 1997; Moran et al. 1995; Vogler et al. 1998). Similarly, the high ^{230}Th observed in intermediate-depth waters of the Weddell Sea partially reflects upward advection of high ^{230}Th deep waters (Rutgers van der Loeff and Berger 1993). It is important to assess the degree to which such advection invalidates the assumption that V_{Th} equals P_{Th}. Such an assessment has been made using two approaches—sediment-trap measurements and modeling.

Using sediment traps to estimate the downward flux of ^{230}Th$_{xs}$ through the water column is made difficult by the problems of incomplete or over-efficient trapping of falling material. Sediment traps in the open ocean frequently collect less ^{230}Th$_{xs}$ than is produced in the water overlying them (Anderson et al. 1983b; Bacon et al. 1985; Colley et al. 1995; Taguchi et al. 1989). These low fluxes may be explained by either incomplete trapping of ^{230}Th$_{xs}$-bearing material as it falls through the water column, or by advection of ^{230}Th$_{xs}$ away from the production site before it adheres to particles. This question can be investigated by measurement and modeling of shorter lived U-series nuclides, particularly ^{234}Th (Cochran and Masque 2003). Such studies indicate that traps frequently do not capture the entire downward particle flux, particularly when they are situated in the upper ocean (Buesseler 1991; Clegg and Whitfield 1993). Unless the degree of undertrapping can be accurately assessed, the question of whether $V_{Th} = P_{Th}$ cannot be answered with trap studies. In fact, some studies have turned the question around and, by assuming that $V_{Th} = P_{Th}$, have used ^{230}Th$_{xs}$ fluxes to assess sediment trap efficiency (Brewer et al. 1980). While such an approach can be useful to correct sediment-trap particle fluxes, it obviously precludes the use of sediment traps to assess the underlying assumption that $V_{Th} = P_{Th}$.

An accurate assessment of sediment trap efficiencies can be provided by coupling ^{230}Th$_{xs}$ measurements with those of ^{231}Pa$_{xs}$—a nuclide with a similar formation and chemistry in the oceans. This approach was suggested by Anderson et al. (1983a) who used it to compare downward ^{230}Th fluxes in the open ocean with those at the ocean margin. It was also used by Bacon et al. (1985) to assess trapping efficiency of a single trap near Bermuda and has more recently been used to assess trap efficiency in a wide range of oceanographic settings (Scholten et al. 2001; Yu et al. 2001a; Yu et al. 2001b). In this approach, only an assumption about the ratio of ^{231}Pa$_{xs}$ to ^{230}Th$_{xs}$ in seawater is required in order to assess sediment-trap efficiency. The downward flux of ^{230}Th$_{xs}$ is equal to the production of ^{230}Th$_{xs}$ plus the net horizontal flux of ^{230}Th$_{xs}$ (Scholten et al. 2001):

$$V_{Th} = P_{Th} + H_{Th} \tag{2}$$

Where H_{Th} is the net horizontal flux of ^{230}Th$_{xs}$. Similarly:

$$V_{Pa} = P_{Pa} + H_{Pa} \tag{3}$$

The vertical flux ratio of the radionuclides, R_v, is measured in the trap:

$$R_v = \frac{V_{Th}}{V_{Pa}} \tag{4}$$

And the ratio of the nuclides in the net horizontally transported portion is assumed to equal that observed in the water column:

$$R_h = \frac{H_{Th}}{H_{Pa}} \tag{5}$$

The four Equations (2) to (5) therefore include only four unknowns (V_{Th}, V_{Pa}, H_{Th}, H_{Pa}) and can be solved for V_{Th} (Yu et al. 2001b):

$$V_{Th} = \frac{((P_{Th} - R_h P_{Pa})R_v)}{(R_v - R_h)} \quad (6)$$

V_{Th} can then be compared to P_{Th} to assess the magnitude of lateral $^{230}Th_{xs}$ advection, or with the measured sediment-trap $^{230}Th_{xs}$ flux to assess the trapping efficiency.

Making an assumption about the $(^{231}Pa_{xs}/^{230}Th_{xs})$ ratio in the horizontal component is not entirely straightforward as the value varies significantly in seawater both with depth and from region to region. Unfortunately, seawater measurements are also presently rather sparse so this pattern of variability is not particularly well constrained, limiting the ability to assess sediment trap efficiencies. There is also a question as to whether the seawater ratio is really an appropriate ratio to use. If advection transports the nuclides, then they will be transported at their seawater ratio. But if they are transported by diffusive processes then the concentration gradient of the two nuclides will control their rate of transport independently of one another and the seawater ratio may not be appropriate. A sensitivity analysis which addresses some of these issues indicated that sediment-trap efficiency can be assessed reasonably precisely in much of the open ocean, although the ocean margins and Southern Ocean are more problematic (Yu et al. 2001b). Efficiencies assessed with $(^{231}Pa_{xs}/^{230}Th_{xs})$ in this way were found to be close to 100% in the deep ocean (>1200 m) but are less good in the shallow ocean (Yu et al. 2001b). This is not simply related to current speeds, but probably related to changes in the size, density, and cohesiveness of particles with depth in the water column (Yu et al. 2001b)

Such sediment-trap studies indicate, V_{Th}/P_{Th} slightly below 1 for much of the open oceans (0.90 ± 0.06), suggesting slight lateral advection of ^{230}Th from these regions. In regions of high particle flux, V_{Th}/P_{Th} is slightly greater than 1 (e.g., 1.2 in the Panama basin) demonstrating horizontal addition of ^{230}Th to these regions (Yu et al. 2001a). Deviations from the assumption that $V_{Th}/P_{Th} = 1$ are therefore reasonably small, supporting the use of sedimentary $^{230}Th_{xs}$ concentrations to calculate sedimentation rates.

An alternative method of assessing the spatial and temporal variability of V_{Th}/P_{Th} for $^{230}Th_{xs}$ is to introduce the nuclide into an ocean general circulation model (OGCM). By incorporating a single class of particles settling at 3 m day^{-1} into an existing OGCM, Henderson et al., (1999a) demonstrated that ocean particle concentrations could be well reproduced. Important features such as the typical particle concentration, the near-surface particle maximum, and the greater particle concentrations in equatorial and high-latitude regions were all successfully mimicked by the model. Decay of ^{234}U was then incorporated into the model and the resulting $^{230}Th_{xs}$ allowed to scavenge reversibly to the particles. The distribution coefficient for ^{230}Th was tuned to give a good fit between the model $^{230}Th_{xs}$ and the ≈900 literature measurements of water column $^{230}Th_{xs}$ concentration. A distribution coefficient within error of that constrained by observations provided a good fit to the data, except in the Southern Ocean, and enabled the rate and pattern of $^{230}Th_{xs}$ removal from most of the world's oceans to be assessed. This model demonstrated that ≈70% of the ocean floor is expected to have V_{Th}/P_{Th} within 30% of 1.0 (Fig. 4). The model results compare well with sediment-trap measurements of V_{Th}/P_{Th}, being within error in all but 1 of the 14 traps investigated by Yu et al. (2001a).

Such OGCM modeling also suggests the importance of ice-cover in controlling the amount of $^{230}Th_{xs}$ advection (Henderson et al. 1999a). Low particle fluxes beneath sea-ice may lead to low scavenging rates in these areas, particularly where ice cover is permanent. In these areas, $^{230}Th_{xs}$ may be advected to the edge of the ice sheet where it is

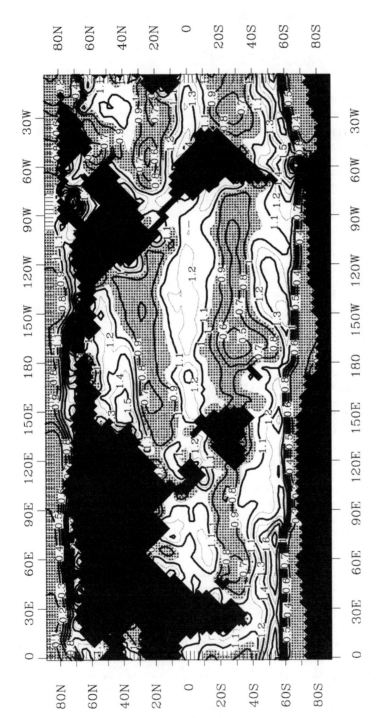

Figure 4. A model of the ^{230}Th flux to the sediment normalized to its production in the water column (Henderson et al. 1999a). Values less than 1 signify that ^{230}Th is advected away from that region prior to removal to the sediment, and those greater than 1 that ^{230}Th is advected into that region to be removed. This model result is in agreement with 13 of 14 sediment trap measurements of the downward ^{230}Th flux (Yu et al. 2001a). These observations and modeling suggest that the ^{230}Th flux to the seafloor is within 30% of its production in the overlying water column over about 70% of the area of the open ocean.

scavenged by the higher particle fluxes. OGCM modeling may allow the correction of V_{Th} in these areas, and elsewhere where V_{Th} is not equal to P_{Th}, so that the use of ^{230}Th to assess sedimentation rates can be extended.

In summary, sediment trap and modeling studies have now constrained reasonably well the precision and limitations of the assumption that ^{230}Th$_{xs}$ is removed to the seafloor immediately where it forms. In much of the open ocean, the assumption is good to better than 30% and the flux of ^{230}Th$_{xs}$ to the sediment is close to its production in the overlying water column. This result supports the use of ^{230}Th$_{xs}$ to assess sedimentation rates in ocean sediments for these regions. More care is needed in using ^{230}Th$_{xs}$ to assess sedimentation rates at ocean margins, and in regions close to permanent sea-ice either now or in the past. More care may also be needed in regions underlying hydrothermal plumes as Mn-Fe oxyhydroxide particles have been shown to be particularly good scavengers of ^{230}Th (Shimmield and Price 1988).

4.2. Seafloor sediments

The predictable flux of ^{230}Th$_{xs}$ to the seafloor means that the flux of other components into marine sediments can be assessed by simply measuring their concentration relative to that of ^{230}Th (Fig. 5). This approach, termed ^{230}Th$_{xs}$ profiling, has seen widespread use in the last decade and has become a standard technique for measuring accumulation rates of many chemical species and sedimentary components. ^{230}Th$_{xs}$ provides possibly the best constraint on such accumulation rates for late Pleistocene sediments and is therefore an important tool. It is the best constrained of the "constant flux proxies" which include other chemical species such as Ti (Murray et al. 2000) and ^{3}He (Marcantonio et al. 1995). As with these other proxies, ^{230}Th$_{xs}$ is not mobilized during sediment dissolution because of its extreme insolubility so that ^{230}Th$_{xs}$ profiling assesses the final sedimentary burial flux, rather than the flux that initially arrives at the seafloor.

In downcore records, measured ^{230}Th requires correction not just for the presence of supported ^{230}Th but also for ingrowth of ^{230}Th from authigenic U incorporated in the sediment, and for decay of excess ^{230}Th with time (see Appendix for details). These corrections require the measurement of ^{232}Th and ^{234}U in the sediment, and some knowledge of the age-to-depth relationship for the core being studied. Because the half life of ^{230}Th is 76 kyr, ingrowth and decay of ^{230}Th are relatively slow compared to the sedimentological or oceanographic processes being investigated which occur on timescales of thousands of years. The age-to-depth model therefore does not need to be particularly precise and sufficient precision is provided by stratigraphic markers such as changing species assemblages or oxygen-isotope stratigraphy. ^{230}Th measurements corrected for detrital ^{230}Th, ingrown ^{230}Th, and the decay of excess ^{230}Th are denoted ^{230}Th$_{xs}^0$.

Having corrected measured ^{230}Th to derive ^{230}Th$_{xs}^0$, the flux of a component, i, into the sediment is given by (Bacon 1984; Suman and Bacon 1989):

$$F_i = Z\left(^{234}U\right)\lambda^{230} \frac{f_i}{^{230}Th_{xs}^0} \qquad (7)$$

Where F_i is the normalized flux of i to the sediment in g yr^{-1} m^{-2}; Z is the water depth at the core location in meters, f_i is the weight fraction of i in the sediment, (^{234}U) is the seawater activity of ^{234}U in dpm.m^{-3}, and ^{230}Th$_{xs}$ is the ^{230}Th activity of the sediment in dpm.g^{-1}.

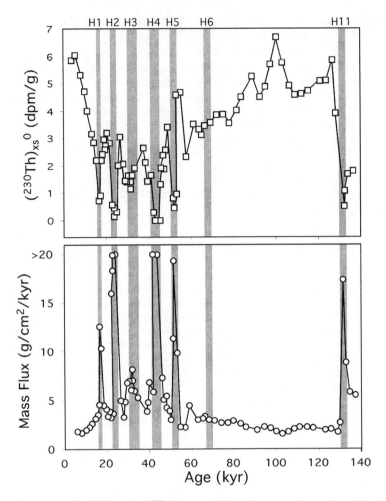

Figure 5. An example of the use of ^{230}Th concentrations to assess changes in sediment mass accumulation taken from McManus et al. (1998). The upper panel shows the measured ^{230}Th$_{xs}^{0}$, calculated from measured ^{230}Th concentrations by correction for detrital ^{230}Th, and for the effects of age using δ^{18}O stratigraphy (see Appendix). Because the supply of ^{230}Th to the sediment is a constant, low ^{230}Th represent times of rapid sediment mass accumulation. The calculated mass flux is shown in the lower panel. Dramatic increases in mass flux are observed during all but one of the Heinrich events, shown by the gray bands.

Since accumulation rates of ^{230}Th were first assessed (Bacon and Rosholt 1982; Ku 1965) ^{230}Th profiling has seen many applications. One of the first uses was to ascertain the cause of regular cycles in carbonate concentration in North Atlantic sediments during the Pleistocene (Bacon 1984). These changes might have been caused by changes in the detrital flux while the carbonate flux remained constant, or by changes in the carbonate flux while the detrital flux remained constant. Only by assessing the flux of one or both of these components to the sediment can this question be answered. Through the use of ^{230}Th profiling, Bacon concluded that the major cause of the cycles was changes in the supply rate of detrital material which varied by a factor of 2-5 while changes in calcite accumulation changed by less than a factor of 2.

This work was pursued by Francois et al. (1990) who assessed both spatial and temporal changes in calcite deposition rates in the tropical Atlantic and demonstrated a rapid increase in carbonate preservation at the onset of the last deglaciation, followed by an increase in carbonate dissolution as deglaciation continued. Only through the use of a precise accumulation tool such as ^{230}Th could this level of resolution be achieved and the rapid change in carbonate burial be recognized. Previous attempts to assess sedimentation rate had relied on recognition of marine oxygen isotope stages, or on multiple ^{14}C ages, neither of which can provide the high time resolution of ^{230}Th. For instance, the use of δ^{18}O curves can assess average sedimentation rates over a marine isotope stage (10-20 kyr), but over shorter periods runs into problems due to the uncertainty in the timescale of the SPECMAP δ^{18}O curve and in the precise identification of marine isotopes stages in the core of interest.

Another application of ^{230}Th$_{xs}$ profiling has been to assess sedimentation rates during Heinrich events (Francois and Bacon 1994; Thomson et al. 1995) (Fig. 5). Abrupt increases in the coarse detrital content of North Atlantic sediments during Heinrich events might be caused by a lowering of biogenic particle flux to the seafloor, or by an increase in the flux of ice-rafted detrital material. By measuring ^{230}Th$_{xs}$ concentrations, the sedimentation rate was found to be up to 7 times higher during Heinrich layers, indicating an increase in the supply of ice-rafted detritus (Thomson et al. 1995). ^{230}Th$_{xs}$ profiling has been used more recently to assess the duration of each Heinrich layer in Labrador Sea sediments (Veiga-Pires and Hillaire-Marcel 1999) and has indicated typical durations of ≈1.2 kyr.

Several studies have also used ^{230}Th$_{xs}$ to identify and quantify the redistribution of sediments on the seafloor by bottom currents. This has become an important field as paleoclimatologists seek to increase the resolution of marine climate records. To achieve high resolution, many climate records have been measured on "drift" deposits in which fine-grained sediment has been transported and deposited by deep-ocean circulation. Suman and Bacon (1989) first recognized that ^{230}Th$_{xs}$ profiling enabled the past vertical flux of sediment from the water column to be assessed even in areas where horizontal redeposition dominates sediment supply. They applied the term "sediment focusing" for such areas where redeposition causes sediment, together with its complement of ^{230}Th, to accumulate rapidly. Because the ^{230}Th removed by the downward sediment flux remains with the sediment as it is redistributed, the ^{230}Th concentration of the sediment still provides information about the downward fluxes. And, if an independent age model is available from, say, δ^{18}O stratigraphy, then the accumulation rate of ^{230}Th$_{xs}$ provides information about the extent of sediment focusing. ^{230}Th$_{xs}$ has, for instance, been used to quantify sedimentary accumulation of opal and biogenic Ba up to 10 times their vertical rain rate in the Subantarctic zone of the Indian Ocean (Francois et al. 1993, 1997); in the Weddell Sea (Frank et al. 1995); and in the Atlantic sector of the Southern Ocean (Asmus et al. 1999; Frank et al. 1996, 1999). Similarly, sedimentation rates up to 4 times the vertical rain-rate were assessed on the Iberian margin (Hall and McCave 2000; Thomson et al. 1999). And systematic changes in sediment focusing with climate change have been recognized for the Equatorial Pacific (Marcantonio et al. 2001b).

Other applications of ^{230}Th$_{xs}$ profiling to assess accumulation rates of sedimentary components include carbonate accumulation in the Western Equatorial Atlantic (Rühlemann et al. 1996); biogenic and terriginous particle accumulation on the Australian continental margin (Veeh et al. 2000); sedimentation rates in the North East Atlantic (McManus et al. 1998; Thomson et al. 1993) (Fig. 5); sedimentation rates during key changes in thermohaline circulation (Adkins et al. 1997); the constancy of the micrometeroritic ^{3}He flux to sediments during the Pleistocene (Marcantonio et al. 1999); and the rate of ^{10}Be deposition as a guide to past variations in cosmogenic radionuclide production (Frank et al. 1997). The ^{230}Th approach has also frequently been used to normalize ^{231}Pa fluxes in order to assess ocean productivity or circulation as discussed in sections 5 and 6.

4.3. Mn crusts

^{230}Th incorporation has also been used to assess the growth rates and ages of Mn crusts and nodules growing on the seafloor (Chabaux et al. 1995; Krishnaswami et al. 1982; Ku and Broecker 1969). These deposits have proved useful for the long-term records of marine geochemistry and paleoceanography that they contain (Frank 2002) but have proved difficult to date. Unlike marine sediments, Mn deposits cannot be assumed to collect the entire flux of ^{230}Th from the water column above them and the incorporation rate of ^{230}Th is therefore generally poorly known. This limits the accuracy of the technique, although assuming a constant incorporation rate of ^{230}Th$_{xs}$ with time seems appropriate in at least some crusts because they contain exponential decreases in ^{230}Th$_{xs}$ with depth (Chabaux et al. 1995; Henderson and Burton 1999; Ku et al. 1979) (Fig. 6). Some workers have preferred to normalize ^{230}Th against ^{232}Th on the assumption that seawater ^{230}Th/^{232}Th is less likely to vary with time than is the rate of ^{230}Th$_{xs}$ incorporation (Chabaux et al. 1995; Neff et al. 1999). Variation in past ocean ^{232}Th/^{230}Th is to be expected, however, due to climate-related changes in dust input of ^{232}Th with time (Henderson et al. 2001). The isotope ratio approach is therefore not necessarily any more reliable than the assumption of a constant ^{230}Th$_{xs}$ incorporation rate, although using the two approaches together can at least serve to improve confidence in resulting growth rates (Krishnaswami et al. 1982, Claude-Ivanaj et al. 2001) (Fig. 6).

Additional concern has been raised about the possibility of diffusion of Th isotopes within the Mn deposits (Ku et al. 1979; Mangini et al. 1986). This seemed possible given the very slow growth rate of such material (typically millimeters per million years) and therefore the small length scales required for diffusion to perturb ^{230}Th chronology. Agreement between ^{230}Th$_{xs}$ and ^{10}Be chronologies argue against the importance of diffusion, however (Krishnaswami et al. 1982, Claude-Ivanaj et al. 2001). Diffusion of U within Mn crusts has been observed in several studies (Henderson and Burton 1999, Neff et al. 1999). This diffusion prevents the use of ^{234}U$_{xs}$ to date such crusts but allows an assessment of the diffusion rates of other elements in crusts (Henderson and Burton 1999). This approach confirms that Th diffusion is too slow to perturb Mn crust chronology, due largely to its extremely high concentration in crusts relative to its concentration in seawater.

Despite the uncertainty about the rate of incorporation of ^{230}Th$_{xs}$ into Mn deposits, Th chronology can still provide some useful information about their growth rates. For instance, ^{230}Th$_{xs}$ has been used to suggest short-term changes in growth rates of Mn crusts (Eisenhauer et al. 1992) and to check the ^{10}Be chronologies of crusts used to reconstruct the radiogenic isotope history of seawater (Abouchami et al. 1997; Frank 2002).

5. PAST EXPORT PRODUCTIVITY – (^{231}Pa$_{xs}$/^{230}Th$_{xs}$)

Paleoceanographers have long sought to reconstruct past changes in the biological productivity of the ocean. Much of the motivation for this work has come from two related objectives: (1) understanding the sensitivity of ocean ecosystems to perturbation by climate change and, (2) establishing the role of the ocean's biological pump in regulating the partitioning of carbon dioxide, a potent greenhouse gas, between the

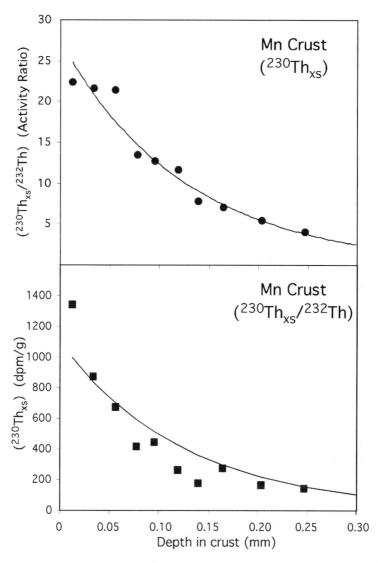

Figure 6. An example of the use of ^{230}Th to assess the growth rate of Mn nodules taken from Krishnaswami et al. (1982). Both panels show the same ^{230}Th$_{xs}$ data from nodule R/V Vitiaz from the Southern Indian Ocean. Errors on the activities are within symbol size. The lower panel shows the ^{230}Th$_{xs}$ activity, while the upper panel shows the same data normalized to the ^{232}Th activity. Note that both profiles show a general exponential decrease which can be used to assess the growth rate using the relationship that ^{230}Th$_{xs}^{now}$ = ^{230}Th$_{xs}^{initial} \cdot e^{-\lambda 230 t}$. Curves shown on both panels are for a steady growth rate of 1.15 mm/Myr. Note, however, that this calculation involves an assumption that either the ^{230}Th$_{xs}$ and/or the ^{230}Th$_{xs}/^{232}$Th remains constant at the outer surface of the nodule throughout its growth. That this is not exactly true is shown by the fact that the growth rate history for the crust is not identical using the two chronometers. For instance, the outermost three ^{230}Th$_{xs}$ data imply a slow growth rate, while the same three points, when normalized to ^{232}Th, imply a rapid growth rate. The problem of variable initial conditions limits the usefulness of ^{230}Th$_{xs}$ dating of Mn nodules and crusts.

atmosphere and the deep sea. Reconstructing past changes in biological productivity of the oceans is difficult because biogenic phases that would provide the most direct measure of ocean productivity (e.g., organic carbon) are poorly preserved in marine sediments. Preservation of these phases is also highly variable and cannot be estimated from first principles. Consequently, a suite of indirect approaches have been developed to evaluate past changes in ocean productivity including barite or Ba accumulation, authigenic U concentrations, and ($^{231}Pa_{xs}/^{230}Th_{xs}$) (Henderson 2002b; Lochte et al., in press).

A correlation exists between the flux of particles collected by sediment traps and the ($^{231}Pa_{xs}/^{230}Th_{xs}$) of these particles (Kumar et al. 1995). In pelagic regions, particulate material settling through the deep sea is almost entirely of biogenic origin. A geochemical proxy that records particle flux, such as ($^{231}Pa_{xs}/^{230}Th_{xs}$), might therefore serve as a valuable measure of past changes in biological productivity. In this section we review the rationale for using ($^{231}Pa_{xs}/^{230}Th_{xs}$) as such a productivity proxy, together with more recent information that leads to a somewhat different view of the factors regulating particulate ($^{231}Pa_{xs}/^{230}Th_{xs}$) in the ocean.

5.1. Chemical fractionation and boundary scavenging

Early studies demonstrated that ($^{231}Pa_{xs}/^{230}Th_{xs}$) in particulate material caught in sediment traps (Anderson et al. 1983a,b) and in marine sediments (Shimmield et al. 1986; Yang et al. 1986) deviated from the production ratio (0.093), with lower values found in open-ocean sediments and higher values found in near-shore sediments. These observations were interpreted to reflect the combined effects of chemical fractionation during scavenging of ^{230}Th and ^{231}Pa by open-ocean particles and the intensified scavenging of particle-reactive substances that occurs near ocean margins. The tendency for insoluble nuclides such as ^{231}Pa (together with ^{10}Be, ^{26}Al, ^{210}Pb, etc.) to be removed at the margins of the ocean basins has been termed boundary scavenging (Spencer et al. 1981). Boundary scavenging must also influence the spatial pattern of ^{230}Th burial in marine sediments to some extent but the effect is believed to be small because ^{230}Th is removed from the open ocean too quickly to permit significant lateral redistribution (see Section 4.1). There are several causes for the increased scavenging in the near-shore environment, including the supply of particles eroded from nearby continents, the enhanced production of biogenic particles due to upwelling, and the greater redox cycling of Fe and Mn (Cochran 1992).

The differential removal of ^{230}Th and ^{231}Pa from the open ocean can be expressed in terms of a fractionation factor (Anderson et al. 1983b), F(Th/Pa), where:

$$F\left(\frac{Th}{Pa}\right) = \frac{\left(^{230}Th_{xs}/^{231}Pa_{xs}\right)^{particles}}{\left(^{230}Th_{xs}/^{231}Pa_{xs}\right)^{dissolved}} = \frac{K_d(Th)}{K_d(Pa)} \tag{8}$$

Particles in open-ocean regions have an affinity for Th that is roughly an order of magnitude greater than their affinity for Pa; i.e., F(Th/Pa) ≈10. Preferential removal of ^{230}Th therefore leads to dissolved ($^{231}Pa/^{230}Th$) in open-ocean seawaters of ≈0.3 to 0.4 (i.e., higher than the production ratio of 0.093). Such ^{230}Th removal also leads to a correspondingly low ($^{231}Pa_{xs}/^{230}Th_{xs}$) of ≈0.03 to 0.04 in the sediments underlying open-ocean seawaters (Anderson et al. 1983b; Nozaki and Yang 1987).

The intensified scavenging that occurs at ocean boundaries lowers the concentrations of dissolved ^{230}Th and ^{231}Pa in deep waters several-fold relative to those found in open-ocean regions (Anderson et al. 1983b). Consequently, eddy diffusion and advection of deep waters from the open ocean toward ocean margins causes a net supply of dissolved ^{231}Pa and ^{230}Th at a ($^{231}Pa_{xs}/^{230}Th_{xs}$) much greater than 0.093, and potentially as high as the dissolved $(^{231}Pa_{xs}/^{230}Th_{xs})$ ratio in open-ocean waters (0.3 to 0.4). Early studies concluded that the intense scavenging at ocean margins of ^{231}Pa and ^{230}Th supplied by lateral transport, with a dissolved $(^{231}Pa_{xs}/^{230}Th_{xs})$ typical of open-ocean deep water, was responsible for the high $(^{231}Pa_{xs}/^{230}Th_{xs})$ ratios found in ocean-margin sediments (Anderson et al. 1983a; Nozaki and Yang 1987).

5.2. $(^{231}Pa_{xs}/^{230}Th_{xs})$ ratios as a paleoproductivity proxy

The principles of boundary scavenging were later invoked to explain the empirical correlation between $(^{231}Pa_{xs}/^{230}Th_{xs})$ and the annual mass flux of particles collected by sediment traps (Kumar et al. 1995; Yu 1994). Whereas ^{230}Th is scavenged so intensely that its flux to the sea bed is everywhere nearly in balance with its production in the overlying water column (Section 4), the residence time of dissolved ^{231}Pa is sufficiently long to permit substantial lateral redistribution, by eddy diffusion or advection, from regions of low particle flux (low productivity) to regions of high particle flux (Fig. 7). Early assessments of $(^{231}Pa_{xs}/^{230}Th_{xs})$ as a paleo flux (productivity) proxy seemed to support this view, in that sedimentary $(^{231}Pa_{xs}/^{230}Th_{xs})$ was found to be well correlated with ^{230}Th-normalized accumulation rates of biogenic opal at sites in the Southern Ocean (Francois et al. 1993; Kumar et al. 1993, 1995). As diatoms are a primary component of the phytoplankton in the Southern Ocean, the good correlation between $(^{231}Pa_{xs}/^{230}Th_{xs})$ ratios and opal accumulation rate seemed to support the basis for using $(^{231}Pa_{xs}/^{230}Th_{xs})$ as a paleoproductivity proxy. The use of $(^{231}Pa_{xs}/^{230}Th_{xs})$ as a paleoproductivity proxy has

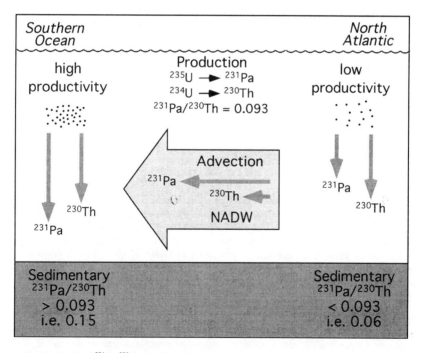

Figure 7. Schematic of ^{231}Pa-^{230}Th fractionation in the oceans. Both nuclides are formed from decay of U throughout the water column. The length of gray arrows represents the size of the fluxes illustrating that ^{230}Th is rapidly scavenged everywhere, while ^{231}Pa can be advected or transported by diffusion (not shown) from areas of low to high productivity. Sedimentary $^{231}Pa/^{230}Th$ is therefore a function of both the productivity and the net lateral transport of ^{231}Pa by ocean circulation.

the disadvantage that the relationship between $(^{231}Pa_{xs}/^{230}Th_{xs})$ and particle flux observed in the modern ocean is influenced by factors other than the flux of particles at the site studied. Rates of lateral transport (advection and mixing) as well as fluxes of particles in surrounding regions also influence the $(^{231}Pa_{xs}/^{230}Th_{xs})$. However, $(^{231}Pa_{xs}/^{230}Th_{xs})$ has an advantage over biogenic phases as a paleoproductivity proxy in that $(^{231}Pa_{xs}/^{230}Th_{xs})$ is insensitive to dissolution of biogenic phases during early diagenesis (i.e., the ratio is insensitive to the degree of preservation of biogenic material).

The Southern Ocean has been a particular focus for the application of $(^{231}Pa_{xs}/^{230}Th_{xs})$ as a paleoproductivity proxy. This region is believed to play a critical role in climate change because it serves as the "window" through which deep waters of the world's oceans exchange gases, including the greenhouse gas CO_2, with the atmosphere. Furthermore, the nutrients upwelled in the Southern Ocean today are used inefficiently by phytoplankton. Greater nutrient utilization efficiency in the past may have contributed to the lower atmospheric CO_2 levels that existed during glacial times. For these reasons, a substantial body of paleoceanographic research has been devoted to understanding past changes in biological productivity, circulation, and nutrient utilization efficiency of the Southern Ocean (see reviews by Elderfield and Rickaby 2000; Sigman and Boyle 2000; Anderson et al. 2002).

$(^{231}Pa_{xs}/^{230}Th_{xs})$ has contributed to the view that the zone of maximum biological productivity in the Southern Ocean was located equatorward of its present position during glacial periods. Compared to Holocene conditions, Kumar et al. found $(^{231}Pa_{xs}/^{230}Th_{xs})^0$ and ^{230}Th-normalized opal accumulation rates to have been greater during the Last Glacial maximum in the Subantarctic zone (located just north of the Antarctic Polar Front; APF), but lower in the Antarctic zone (located south of the APF) (Kumar et al. 1993, 1995). These results were interpreted to reflect a shift of productivity northwards, probably in response to a greater extent of sea-ice cover in the south. The amplitude of the changes was furthermore interpreted to indicate an increase in the overall productivity of the Southern Ocean, possibly due to an increased supply of the limiting micronutrient, Fe (Kumar et al. 1995). A later study substantiated the observation of higher $(^{231}Pa_{xs}/^{230}Th_{xs})^0$ in the glacial Subantarctic zone (Asmus et al. 1999), and a comparison of several different paleoproductivity proxies supported the northward shift in productivity during glacial periods, while suggesting that total glacial productivity may not have been significantly higher than at present (Frank et al. 2000).

Until recently, most of the work on paleoproductivity of the Southern Ocean has been conducted in the Atlantic sector. In a recent study of the Pacific sector (Chase et al., press-a), $(^{231}Pa_{xs}/^{230}Th_{xs})^0$ was used, together with other proxies, to show that the glacial-interglacial change in productivity differed fundamentally from the pattern observed in the Atlantic sector. Whereas the Pacific and Atlantic sectors both showed evidence for much lower productivity south of the APF during the last glacial period, the glacial Pacific saw little Holocene increase in productivity north of the APF, in contrast to the increase observed in the Atlantic. Overall, it appears that productivity of the Pacific sector of the Southern Ocean was lower than today during the last glacial period.

$(^{231}Pa_{xs}/^{230}Th_{xs})^0$ ratios have also been used to assess changes in productivity in the Arabian Sea in response to changing monsoon strength (Marcantonio et al. 2001a). Abrupt changes in $(^{231}Pa_{xs}/^{230}Th_{xs})^0$ were found to coincide with climate events such as the Younger Dryas which are normally associated with changes in the North Atlantic region. Sediments from the glacial and Younger Dryas exhibited low $(^{231}Pa_{xs}/^{230}Th_{xs})^0$ while those from the Bolling Allerod and Holocene had high $(^{231}Pa_{xs}/^{230}Th_{xs})^0$. The increased productivity indicated by high $(^{231}Pa_{xs}/^{230}Th_{xs})^0$ were assumed to reflect greater monsoon intensity and hence increased upwelling during warm periods.

5.3. The role of particle composition

Early views of boundary scavenging relied on differential scavenging rates coupled with lateral mixing to drive the net flux of ^{231}Pa from the open ocean to ocean margins (Bacon 1988). It was recognized that manganese oxides scavenge ^{231}Pa and ^{230}Th without detectable fractionation (i.e., at F(Th/Pa)≈1) (Anderson et al. 1983a,b), and that the redox cycling of Fe and Mn, which generates oxide-rich particles in ocean-margin waters, might contribute to the boundary scavenging of dissolved nuclides (Anderson et al. 1983a; Cochran 1992). These phases were, however, ignored in the interpretation of open-ocean results because it was assumed that the Fe and Mn oxides would settle out of the water column near their source. Early studies did not consider that variability in F(Th/Pa) might contribute significantly to the spatial pattern of $(^{231}Pa_{xs}/^{230}Th_{xs})$ in marine sediments.

Evidence for regional variability in F(Th/Pa), even in the open ocean, began to mount as investigators examined the scavenging of ^{230}Th and ^{231}Pa in different regions. Taguchi et al. (1989) noted that particulate $(^{231}Pa_{xs}/^{230}Th_{xs})$ in North Pacific sediment-trap samples increased with increasing opal content of the samples. Subsequent studies showed that F(Th/Pa) in the Atlantic sector of the Southern Ocean decreases southwards and is close to 1.0 south of 60°S (Rutgers van der Loeff and Berger 1993; Walter et al. 1997). These results led to the hypothesis that biogenic opal scavenges $^{230}Th_{xs}$ and $^{231}Pa_{xs}$ without significant fractionation. That hypothesis was confirmed by the recent comparison of $^{231}Pa_{xs}$ and $^{230}Th_{xs}$ concentrations in sediment trap samples with the corresponding concentrations of ^{230}Th and ^{231}Pa in the surrounding water column (Chase et al. 2002). Effective K_d values computed from these results (Fig. 8) demonstrated that the affinity of marine particles for ^{231}Pa increases with increasing opal content of particles, and that this change controls the previously observed decrease in F(Th/Pa) for opal-rich particles. Consequently, particle composition, as well as particle flux, influences the $(^{231}Pa_{xs}/^{230}Th_{xs})$ of marine sediments.

A surprising feature of the particle reactivity of ^{230}Th is that its K_d decreases with increasing opal content of particles (Fig. 8). The underlying chemical cause for this decrease in K_d(Th) is not known, but it is consistent with laboratory studies. Osthols (1995) found that Th sorbs weakly to amorphous silica and lab experiments by Guéguen and Guo (2002) are qualitatively consistent with the results of Chase et al., showing that Th preferentially sorbs on $CaCO_3$ surfaces, whereas Pa preferentially sorbs on opal. The decrease of K_d(Th) contributes as much to the absence of fractionation during scavenging by opal-rich particles as does the increase of K_d(Pa). Another surprise is that the K_d(Th) of the calcite-rich end member reaches values as large as observed anywhere in the ocean, including ocean-margin environments ($\sim 1 \times 10^7$). This result precludes the possibility that lithogenic phases have K_d(Th) orders of magnitude greater than K_d(Th) of biogenic phases, as had previously been suggested by Luo and Ku (1999). Because open-ocean particles with less than 1% lithogenic content have as high a K_d(Th) as ocean-margin particles containing more than 50% lithogenic material, preferential scavenging of ^{230}Th onto lithogenic material cannot occur (Chase et al. 2002).

Much of the geographic variability in sedimentary $(^{231}Pa_{xs}/^{230}Th_{xs})$ observed in modern sediments may be explained by variability in the composition of biogenic particles arising from variability in the structure of the planktonic ecosystem. This can be inferred from the composition-dependence of F(Th/Pa) (Fig. 8), and is shown explicitly by the relationship between sediment trap $(^{231}Pa_{xs}/^{230}Th_{xs})$ and the opal/calcite ratio of the trapped particles (Fig. 9). Sediment trap $(^{231}Pa_{xs}/^{230}Th_{xs})$ also exhibits a positive relationship with the mass flux of particles, but the correlation is poorer than that with particle composition (Fig. 9). Indeed, the relationship between particulate $(^{231}Pa_{xs}/^{230}Th_{xs})$

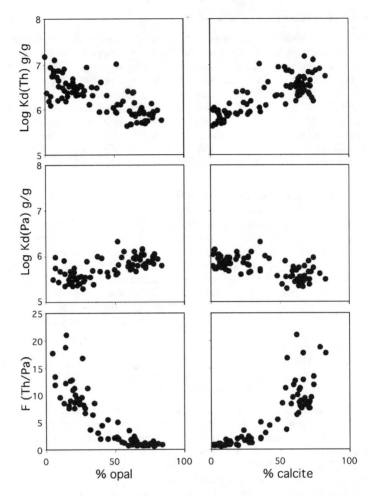

Figure 8. Partition coefficients (K_d) for Th and Pa and the fractionation factor (F) between Th and Pa plotted as a function of the opal and calcium carbonate percentage in settling particulate material. Note the tendency for the K_d for Th to increase with increasing carbonate fraction and decrease with increasing opal fraction. Pa shows the opposite behavior so that F increases with low opal fraction or high carbonate fraction. This plot is modified from Chase et al. (in press-b) but excludes the continental margin data also shown in that study and instead focuses exclusively on open-ocean sites.

and particle flux is partly due to the fact that pelagic systems with high particle flux also tend to have abundant diatoms and therefore a high opal content in the resulting particles (Buesseler 1998).

5.4. Prospects for future use

The recognition that particle composition is of prime importance in controlling fractionation of ^{231}Pa from ^{230}Th during scavenging casts doubt over the use of $(^{231}Pa_{xs}/^{230}Th_{xs})^0$ to assess past productivity. Certainly, previous results must be reconsidered in the light of this knowledge (Chase et al., in press-b). The good news,

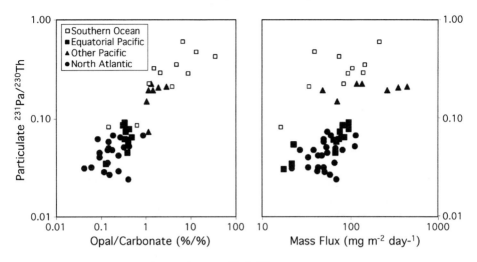

Figure 9. A log-log plot of the annual average $(^{231}Pa_{xs}/^{230}Th_{xs})$ as a function of sediment trap particle composition, and as a function of total mass flux. Note the importance of particle composition on the $(^{231}Pa_{xs}/^{230}Th_{xs})$ of trapped material, with a high opal fraction leading to higher ratios. Note also the poor relationship between $(^{231}Pa_{xs}/^{230}Th_{xs})$ and mass flux. This data was compiled by Chase et al. (in press-b) and includes data from that study, as well as from Lao et al. (1993), Scholten et al. (2001), and Yu et al. (2001a).

though, is that $(^{231}Pa_{xs}/^{230}Th_{xs})^0$ may provide information about the composition of settling particles in the past and therefore about the nature of the past phytoplankton communities (Chase et al. 2002).

Sedimentary $(^{231}Pa_{xs}/^{230}Th_{xs})^0$ can also be used to interpret downcore records of ^{230}Th-normalized opal accumulation rates. Opal accumulation depends both on production and on preservation, which is in turn variable and difficult to evaluate (Sayles et al. 2001). By combining opal accumulation rates with $(^{231}Pa_{xs}/^{230}Th_{xs})^0$ and with other paleoproductivity proxies such as excess barium, however, it is possible to discriminate between a production control and a preservation control when interpreting ^{230}Th-normalized opal accumulation rates. Such a use is illustrated by the strong similarities between the downcore pattern of opal burial and the corresponding records of excess barium accumulation and $(^{231}Pa_{xs}/^{230}Th_{xs})^0$ in sediments from a site in the SW Pacific sector of the Southern Ocean (Fig. 10). Excess barium is widely used as a proxy for export production because fluxes of excess barium collected by sediment traps have been shown to correlate with fluxes of organic carbon (Dymond et al. 1992; Francois et al. 1995). Whereas all three proxies are similarly responsive to changes in diatom production, they have distinct differences in their sensitivity to regeneration at the sea bed, with $(^{231}Pa_{xs}/^{230}Th_{xs})^0$ ratios being completely insensitive to regeneration of biogenic phases (Chase et al., in press-b). Consequently, the strong similarity among the three proxy records provides robust evidence that the pattern of opal burial at this site reflects climate-related changes in diatom production rather than regeneration. A similar use for $(^{231}Pa_{xs}/^{230}Th_{xs})^0$ was suggested in the Atlantic sector of the Southern Ocean where low opal accumulation rates are coincident with low $(^{231}Pa_{xs}/^{230}Th_{xs})^0$ confirming that the low opal values are a reflection of low opal productivity rather than low opal preservation (Frank et al. 2000). When combined with complementary proxies, and used with an awareness of the particle composition-dependent fractionation between ^{230}Th and ^{231}Pa,

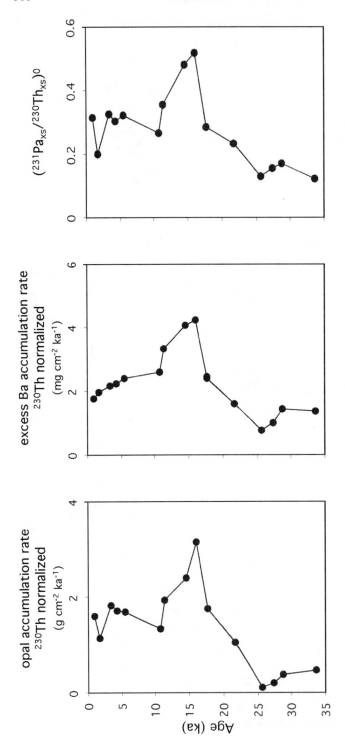

Figure 10. An example of the use of (^{231}Pa$_{xs}$/^{230}Th$_{xs}$) to reconstruct past productivity taken from Chase et al. (in press-a). Data is from a core in the Southern Pacific (AESOPS Station 6 at 61° 52.5'S, 169° 58.3'W). The downcore record of (^{231}Pa/^{230}Th)$_{xs,0}$ indicates an increase in opal productivity centered at 15 ka. This productivity increase is also seen in the records of two other productivity proxies—the opal accumulation and Ba accumulation rates. Downcore records of these sediment constituents are also shown and have been converted from concentrations to accumulation fluxes by normalizing to the known ^{230}Th$_{xs}$ flux.

sediment $(^{231}Pa_{xs}/^{230}Th_{xs})^0$ will continue to provide valuable information about past changes in ocean productivity.

6. RATES OF PAST OCEAN CIRCULATION – $(^{231}Pa_{xs}/^{230}Th_{xs})$

The oceanic fractionation of ^{231}Pa from ^{230}Th has also been used to assess the rate of past deep-ocean circulation. Such ocean circulation is fundamental to climate, transporting heat and greenhouse gases around the globe. There are several tools with which to assess past circulation, but the most commonly used of these ($\delta^{13}C$ and Cd/Ca) assess the distribution of water masses in the past, rather than the rate of flow. While the distribution of water masses is significant, it is their rate of flow which ultimately controls heat transport and is therefore of critical importance. Four proxies have been used to reconstruct past flow rates: surface versus deep water ^{14}C analyses (Adkins and Boyle 1997); stable-isotope geostrophy (Lynch-Stieglitz et al. 1999); the grain-size of deep-sea sediments (McCave et al. 1995); and sedimentary $(^{231}Pa_{xs}/^{230}Th_{xs})^0$ (Yu et al. 1996).

The use of $(^{231}Pa_{xs}/^{230}Th_{xs})^0$ to assess circulation rates again relies on the difference in solubility of ^{230}Th and ^{231}Pa (Fig. 7). When these nuclides form from U decay, the short residence time of ^{230}Th (≈20 years) does not allow it to be advected far before it is removed to the sediment. The longer residence time of ^{231}Pa (≈130 years), on the other hand, allows it to be advected so that, if the water mass is moving, ^{231}Pa may be transported downstream to be removed in regions of high particle flux. This residence time is appropriate for the assessment of circulation in areas of reasonably rapidly moving water, such as the north Atlantic, where deepwaters move several thousand kilometers in 100 years and have a total residence time in the basin of ≈250 years. Sediments underlying waters with such active transport of dissolved ^{231}Pa by ocean circulation, and with a low scavenging efficiency for ^{231}Pa, are expected to have $(^{231}Pa_{xs}/^{230}Th_{xs})^0$ below the seawater production ratio. The degree of lowering below the production ratio will be proportional to the speed of the current (Marchal et al. 2000). Sediments down-stream of such areas, particularly in regions of high particle flux, are expected to have $(^{231}Pa_{xs}/^{230}Th_{xs})^0$ correspondingly higher than the production ratio.

This approach has been applied to the Atlantic to assess changes in the flow rate of NADW between the last glacial and the Holocene (Yu et al. 1996). Holocene sediments from 68 cores situated north of the Southern-Ocean opal belt had $(^{231}Pa_{xs}/^{230}Th_{xs})^0$ below the production ratio at 0.060 ± 0.004, indicating net advection of ^{231}Pa out of the Atlantic. Holocene sediments in the Southern-Ocean opal belt had corresponding high $(^{231}Pa_{xs}/^{230}Th_{xs})^0$ of 0.17 ± 0.04 and clearly act as the sink for the advected ^{231}Pa. Glacial-age sediments from the same cores showed a similar pattern with $(^{231}Pa_{xs}/^{230}Th_{xs})^0$ of 0.059 ± 0.007 in most of the Atlantic, and of 0.15 ± 0.02 in the opal belt. The similar distribution of $(^{231}Pa_{xs}/^{230}Th_{xs})^0$ between the glacial and the Holocene strongly suggests that there was no change in the rate of NADW flow between these periods. This result came as a surprise because previous water-mass studies had demonstrated that less of the Atlantic region was influenced by NADW during the glacial (e.g., $\delta^{13}C$ (Duplessy et al. 1988) and Cd/Ca (Boyle and Keigwin 1982)).

The suggestion that glacial and Holocene NADW flow rates were similar has been sufficiently controversial that the use of $(^{231}Pa_{xs}/^{230}Th_{xs})^0$ to assess the flow rate has been carefully assessed. The recognition that opal was an effective scavenger of ^{231}Pa (Walter et al. 1997; Section 5.3) suggested that Southern Ocean $(^{231}Pa_{xs}/^{230}Th_{xs})^0$ would be high due to the high opal productivity in the region, regardless of the rate of NADW flow,. This suggestion has been confirmed by the discovery that sedimentary $(^{231}Pa_{xs}/^{230}Th_{xs})^0$ in the Pacific sector of the Southern Ocean, far removed from the inflow of NADW, are as high as those observed within the Atlantic sector of the Southern Ocean which does experience NADW flow (Chase et al., in press-b). It has also been demonstrated that, in the specific oceanographic regime of the Southern Ocean, scavenging from surface waters plays an unusually important role in the behavior of ^{231}Pa and ^{230}Th thereby reducing the ability to use $(^{231}Pa_{xs}/^{230}Th_{xs})^0$ in this area to assess deep-water flow out of the Atlantic. (Chase et al., in press-b; Walter et al. 2001). These Southern-Ocean problems of particle composition and regional oceanography do not exist for the rest of the Atlantic, however. A modeling study has indicated that existing $(^{231}Pa_{xs}/^{230}Th_{xs})^0$ data in the North Atlantic constrain the glacial NADW flow rate to be no more than 30% lower than that during the Holocene (Marchal et al. 2000).

Another suggested flaw with the use of $(^{231}Pa_{xs}/^{230}Th_{xs})^0$ to assess glacial rates of NADW flow is that the spatial coverage of cores in the Yu et al. study is not complete and may miss some important sink regions for $^{231}Pa_{xs}$. Additional cores have been analyzed, particularly in the mid-latitude North Atlantic, but this region was not found to be a significant or changeable sink of ^{231}Pa (Anderson et al. 2001) so that the conclusion of unchanged NADW flow appears robust.

One issue which has not yet been addressed is the depth at which the southward flow of deep water occurred in the past. It is possible, for instance, that true NADW did weaken at the LGM, as is suggested by the nutrient-like tracers of water masses, but that this decrease was largely compensated by increases in flow higher in the water column. This issue could be investigated using $(^{231}Pa_{xs}/^{230}Th_{xs})^0$ if cores from a variety of depths were analyzed.

The successful use of $(^{231}Pa_{xs}/^{230}Th_{xs})^0$ to assess glacial-interglacial changes in ocean circulation rates has led to its use to assess circulation changes across more rapid climate events. Early work has assessed the changes in circulation during Heinrich event H1 and the Younger Dryas (McManus et al. 2002). By working on cores at different water depths in the region close to deep-water formation, flow rates of both deep and intermediate waters were reconstructed and found to work in opposition. When deep circulation rates decreased, intermediate circulation rates increased, but not by the same magnitude. Net flow rates therefore varied and were found to be lower during cool periods—by ≈50% in the Younger Dryas, and by more than this during H1 (McManus et al. 2002).

The use of $(^{231}Pa_{xs}/^{230}Th_{xs})^0$ to assess past circulation rates only works well where the residence time of deep waters is low and advection therefore dominates over the removal of ^{231}Pa by particle scavenging (Yu et al. 2001a). The Atlantic Ocean fits these criteria well but, in many other areas $(^{231}Pa_{xs}/^{230}Th_{xs})^0$ is more strongly influenced by particle flux and composition and the ratio is then better suited for assessing changes in past productivity or ecosystem structure (Yu et al. 2001a). Nevertheless, in regions which do have rapidly flowing deepwaters $(^{231}Pa_{xs}/^{230}Th_{xs})^0$ is a powerful tool which will doubtless see further application.

7. HOLOCENE SEDIMENT CHRONOLOGY – ^{226}Ra

^{226}Ra is soluble and therefore tends to be released to deep waters when it is formed by ^{230}Th decay in marine sediments. Substrates which capture the resulting excess of ^{226}Ra found in seawater can potentially be dated using the decay of this ^{226}Ra excess ($^{226}Ra_{xs}$). Unfortunately there is no stable isotope of Ra with which to normalize measured ^{226}Ra values but the marine chemistry of Ba is sufficiently close to that of Ra that it can be used as a surrogate for a stable Ra isotope and seawater ^{226}Ra/Ba ratios are constant throughout the oceans, except in the deep North Pacific (Chan et al. 1976). The half life of ^{226}Ra is only 1600 years so $^{226}Ra_{xs}$/Ba chronology is limited to the Holocene but it nevertheless has potential for use in several regions.

Given the chemical similarity between Ra and Ba it is not surprising that $^{226}Ra_{xs}$ is high in marine barites that form in the upper water column, particularly in regions of high productivity. Such barite forms a minor constituent of marine sediment but can be separated from other sediment components by dissolution because barite is extremely resistant to chemical attack. Barite separated in this way from two Pacific cores had ^{226}Ra activities at least ten times higher than those of the parent ^{230}Th and ranging up to 1400 dpm ^{226}Ra g^{-1} (Paytan et al. 1996). ^{226}Ra activities decreased exponentially with depth in the sediment at a rate consistent with ^{14}C-derived sedimentation rates and were in secular equilibrium with ^{230}Th at depths equivalent to >8 kyr (Fig. 11). These observations suggest that such marine barite behaves as a closed system for ^{226}Ra and initially incorporates ^{226}Ra at a reasonably constant rate. Marine barite might therefore be used to assess the age of Holocene marine sediments. A more detailed study further demonstrated the suitability of the system for chronology by more accurately assessing the nature of supported and ingrown ^{226}Ra within marine sediment (Van Beek and Reyss 2001). The first application of this technique has been to marine barite from the Southern Ocean in order to assess the ^{14}C reservoir age of surface waters in this region during the Holocene (van Beek et al. 2002). This work has indicated a reasonably constant reservoir age of 1100 years in the late Holocene, but a higher value of ≈1900 years in the early Holocene.

^{226}Ra has also proved useful to date some Holocene Mn crusts (Liebetrau et al. 2002). Most Mn crusts grow extremely slowly and the distribution of ^{226}Ra within them is dominated by diffusion—both out of the crust and inward from underlying sediments (Moore et al. 1981). Unusual crusts from the Baltic Sea, however, accumulate sufficiently quickly that the $^{226}Ra_{xs}$ distribution is controlled by initial incorporation of $^{226}Ra_{xs}$ and subsequent decay. Liebetrau et al. (2002) used the decrease of $^{226}Ra_{xs}/Ba$ from the edge to the center of such fast-growing Baltic crusts to calculate growth rates ranging from 0.02

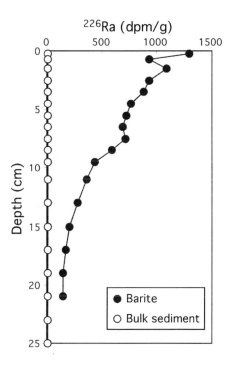

Figure 11. ^{226}Ra in barite and bulk sediment from Core MC82 in the equatorial Pacific (Paytan et al. 1996). Note the considerably higher ^{226}Ra concentration in the barite, and the general exponential decrease in barite ^{226}Ra with depth. This decrease allows barite ^{226}Ra to be to used to provide Holocene chronology in sediments devoid of carbonates for ^{14}C dating.

to 1 mm yr^{-1}. They found ^{226}Ra$_{xs}$/Ba values over an order of magnitude higher than those in Baltic seawater suggesting a dominant source of ^{226}Ra from the underlying sediments. At the rapid growth rates observed, the effects of diffusion were deduced to be negligible and the initiation of crust growth found to coincide with the stabilization of sea level at modern values. These nodules may provide a good archive to investigate changes in weathering inputs to the Baltic Sea during the Holocene, and of anthropogenic contamination of the area.

Marine carbonates may also prove datable using ^{226}Ra$_{xs}$/Ba. Modern Antarctic mollusks have been found to have ^{226}Ra/Ba equal to that of seawater (Berkman et al. 1992) and a significant excess of ^{226}Ra over ^{230}Th. Turning this observation into a useable chronometer requires assessment of whether mollusks are closed systems for Ra and Ba; correction for ingrown and detrital ^{226}Ra; and assumptions about the constancy of seawater ^{226}Ra/Ba through time (Staubwasser et al., in press). Such a chronometer, however, would prove useful for assessing the Holocene history of sea surface ^{14}C, particularly in upwelling regions where no other suitable dating tool presently exists.

8. SEDIMENT MIXING – ^{210}Pb

Mixing of the upper surface of marine sediments by benthic animals limits the resolution at which past climate and oceanography is recorded by marine sediments. Typical marine sediments experience such mixing to a depth of 10 ± 5 cm (Boudreau 1994) and have sedimentation rates of only a few cm kyr^{-1}. Any climate event shorter than ≈1 kyr is, therefore, all but lost from the sediment record. Constraining the rate and depth of mixing in diverse oceanographic settings is therefore important as it allows the limits of paleoclimate resolution to be assessed, and allows inverse modeling to be performed to attempt reconstruction of the full amplitude of past climate events (Bard et al. 1987). The rate of mixing is also important in models of sediment diagenesis and geochemistry.

The ^{210}Pb distribution in marine sediments provides a method of assessing this mixing. ^{210}Pb is produced in the water column by decay of ^{226}Ra, and is also added to the surface of the ocean by ^{222}Rn decay in the atmosphere and subsequent rainout (Henderson and Maier-Reimer 2002). Pb is insoluble in seawater so, once formed, ^{210}Pb adheres to particles and is removed to the seafloor. ^{234}Th is also removed in this way when it forms as the immediate daughter of ^{238}U decay in seawater (Cochran and Masque 2003). In the absence of mixing, these nuclides would only be found on the uppermost surface of the sediment because their half lives are short compared to typical sedimentation rates (^{210}Pb = 22 years; ^{234}Th = 24 days). The presence of these nuclides at depth in the sediment therefore reflects their downward mixing by bioturbation. The depth and pattern of penetration of these isotopes into sediments has been used to assess sediment mixing rates in a large number of studies since the 1970s (e.g., Nozaki et al. 1977).

Because mixing consists of many small events which progressively move the sediment grains it is akin to diffusion and can be modeled as such following the mathematical approach of Guinasso and Schink (1975). ^{210}Pb and ^{234}Th decay as they are mixed downwards which leads to an activity profile in the sediment which decreases exponentially with depth (Fig. 12). The activity of the nuclide, A, is given by (Anderson et al. 1988):

$$A = A_0 e^{\left(\frac{w - \sqrt{(w^2 + 4D\lambda)}}{2D}\right) z} \tag{9}$$

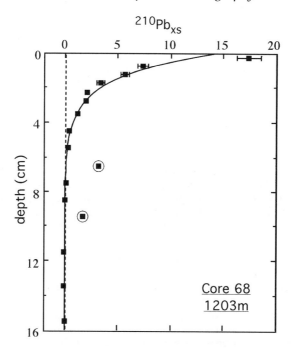

Figure 12. An example of use of ^{210}Pb to assess the rate and depth of sediment mixing from a core on the slopes of the Bahamas (Henderson et al. 1999b). The exponential decrease in ^{210}Pb$_{xs}$ seen in the upper 6 cm of the sediment reflects decay of ^{210}Pb as it is mixed downward. The diffusional model of mixing described in the text indicates a mixing rate, D, of 51 cm^2 kyr^{-1} for this core. The two circled points at greater depth reflect rapid injection of surface material to depth in a process known as "conveyor-belt" feeding (Robbins 1988; Smith et al. 1997).

Where A$_0$ is the activity at the sediment surface, w is the sedimentation rate (cm yr^{-1}), D is the mixing rate (cm^2 yr^{-1}), λ is the decay constant for the nuclide of interest (yr^{-1}); and z is the depth in the sediment (cm). In some near-shore environments both sedimentation and bioturbation must be considered. But in most open marine environments the sedimentation rate is sufficiently slow that it can be ignored and the equation simplifies to:

$$A = A_0 \exp\left(-\left(\lambda/D\right)^{0.5} z\right) \quad (10)$$

D, the mixing rate, varies widely depending on environment. Boudreau (1994) comprehensively summarized measurements of D from some 36 papers including 22 using ^{210}Pb and 7 using ^{234}Th. These studies indicated mixing rates varying from 0.02 to 370 cm^2 yr^{-1}. Several environmental variables appear to be responsible for this large range in mixing rate and a recent thrust of research has been to establish general rules for mixing rate which enable it to be assessed for a new environment without direct measurement. Examples include that of Boudreau who, based on ^{210}Pb data from over 130 sites, suggested a power-law relationship between mixing rate and sedimentation rate (Boudreau 1994):

$$D = 15.7 w^{0.69} \quad (11)$$

Where D is in cm^2 yr^{-1} and w in cm yr^{-1}. Other studies have used ^{210}Pb to demonstrate a relationship between D and water depth (Henderson et al. 1999b; Soetaert et al. 1996):

$$D = 6.33 \, \text{Depth}^{-1.65} \tag{11}$$

Where D is in $cm^2 \, yr^{-1}$ and Depth in m. Although these two relationships explain some of the variability in D, it is clear that other environmental factors are also important, including sediment grain-size (Wheatcroft 1992) and the organic carbon flux (Trauth et al. 1997).

The different half lives of the nuclides that can be used to assess mixing indicate that mixing is more frequent close to the sediment-water interface. In the ≈100 days that ^{234}Th survives before decay, it is only mixed a short distance into the sediment. Its profile suggests values of D higher than those calculated with ^{210}Pb, which can survive for ≈100 years and is mixed to greater depth. This decrease in biological mixing rate probably reflects the greater supply of labile and nutritious organic matter close to the sediment surface (Fornes et al. 2001; Smith et al. 1993). The use of ^{234}Th and other short-lived nuclides to assess mixing is also discussed in this volume by Cochran and Masque (2003).

In addition to the rate of mixing, these nuclides provide an indication of the depth of mixing (L) on various timescales. The ≈100 yrs required for ^{210}Pb decay is sufficiently long to give a reasonable idea of the total depth of mixing that occurs before sediment is buried to sufficient depth to escape further disturbance. This makes ^{210}Pb a particularly useful nuclide to constrain mixing depths for paleoceanographic studies, where the total depth of mixing is the key variable in the smoothing of climate records. ^{210}Pb measurements of L in the deep sea range from 4 to 20 cm (Boudreau 1994) but indicate deeper mixing in near shore environments (Henderson et al. 1999b; Kim and Burnett 1988). Below this well-mixed zone, subsurface peaks in ^{210}Pb concentration are sometimes observed and provide evidence for rapid injection of surface material to depth (Fig. 12). This process, referred to as "conveyor-belt" feeding (or mixing) has been modeled (Robbins 1988; Smith et al. 1997) and incorporated into models of diagenesis (Sayles et al. 2001).

Short-lived U-series nuclides such as ^{210}Pb and ^{234}Th thus provide key information about the rate and depth of sediment mixing in marine records. This information is critical if the resolution of down-core paleoceanographic records are to be assessed.

9. CONCLUDING REMARKS

This chapter has outlined the uses of U-series nuclides to investigate the past ocean environment. The most widely used tools are presently ^{230}Th$_{xs}^0$ (to assess sedimentary fluxes and sediment focusing) and short lived-nuclides such as ^{210}Pb (to assess sediment mixing rates). The fractionation of ^{231}Pa from ^{230}Th in seawater has also been widely investigated to assess past productivity and ocean circulation rates. The controls on sedimentary $(^{231}Pa_{xs}/^{230}Th_{xs})^0$ are only now being fully understood, however, calling into question the conclusions reached in some early studies using this proxy. Armed with this better understanding of $(^{231}Pa_{xs}/^{230}Th_{xs})^0$, this tool has significant promise for assessing circulation rates in areas of rapid flow such as the Atlantic, and for assessing particle fluxes and composition elsewhere. Seawater $(^{234}U/^{238}U)$ has also been used to assess the past U budget of the oceans and the extent of physical weathering on the continents. And ^{226}Ra$_{xs}$/Ba dating has potential to improve chronologies for the Holocene in a variety of settings.

The U-series nuclides used in these tools range in half life from 4.5×10^9 yrs (^{238}U) to 22 yrs (^{210}Pb). The next longest-lived nuclides in the U and Th series are ^{227}Ac (22yrs),

^{228}Ra (5.8 yrs) and ^{228}Th (1.9 yrs). While these nuclides provide information about modern ocean processes, their half lives are probably too short for use in paleoceanography. The range of U-series isotopes with which to investigate the past environment has therefore probably been exhausted. But there are doubtless new applications for these longer-lived nuclides which will go beyond the uses described above. Such new applications will continue to make use of the fact that, together with the cosmogenic nuclides, U-series isotopes provide quantitative estimates of the rates of past processes. This attribute will ensure that U-series geochemistry continues to be widely applied to questions of paleoceanography.

ACKNOWLEDGMENTS

We thank Martin Frank, Michiel Rutgers van der Loeff, and Bernard Bourdon for helpful and thorough reviews. GMH thanks the Leverhulme foundation for financial support.

REFERENCES

Abouchami W, Goldstein SL, Galer SJG, Eisenhauer A, Mangini A (1997) Secular changes of lead and neodymium in central Pacific seawater recorded by a Fe-Mn crust. Geochim Cosmochim Acta 61(18):3957-3974

Adkins JF, Boyle EA (1997) Changing atmospheric $\Delta^{14}C$ and the record of deep water paleoventilation ages. Paleoceanography 12(3):337-344

Adkins JF, Boyle EA, Keigwin LD, Cortijo E (1997) Variability of the North Atlantic thermohaline circulation during the last interglacial period. Nature 390:154-156

Anderson RF (1982) Concentration, vertical flux and remineralization of particulate uranium in seawater. Geochim Cosmochim Acta 46:1293-1299

Anderson RF, Bacon MP, Brewer PG (1982) Elevated concentrations of actinides in Mono Lake. Science 216:514-516

Anderson RF, Bacon MP, Brewer PG (1983a) Removal of ^{230}Th and ^{231}Pa at ocean margins. Earth Planet Sci Lett 66:73-90

Anderson RF, Bacon MP, Brewer PG (1983b) Removal of ^{230}Th and ^{231}Pa from the open ocean. Earth Planet Sci Lett 62:7-23

Anderson RF, Bopp RF, Buesseler KO, Biscaye PE (1988) Mixing of particles and organic constituents in sediments from the continental shelf and slope off Cape Cod: SEEP-1 results. Cont Shelf Res 8(5-7):925-946

Anderson RF, Chase Z, Fleisher MQ, Sachs JP (2002) The Southern Ocean's biological pump during the last glacial maximum. Deep-Sea Research II 49:1909-1938

Anderson RF, Fleisher MQ, Kubik P (2001) Boundary scavenging in the North Atlantic ocean. EOS Trans AGU (Fall Meeting Supplement): OS21D-11

Anderson RF, Fleisher MQ, Lehuray AP (1989) Concentration, oxidation state, and particulate flux of uranium in the Black Sea. Geochim Cosmochim Acta 53:2215-2224

Asmus T, Frank M, Koschmieder C, Frank N, Gersonde R, Kuhn G, Mangini A (1999) Variations of biogenic particle flux in the southern Atlantic section of the subantarctic zone during the late Quaternary: Evidence from sedimentary ^{231}Pa$_{ex}$ and ^{230}Th$_{ex}$. Marine Geo 159:63-78

Bacon MP (1984) Glacial to interglacial changes in carbonate and clay sedimentation in the Atlantic ocean estimated from ^{230}Th measurements. Isot Geosci 2:97-111

Bacon MP (1988) Tracers of chemical scavenging in the ocean: Boundary effects and large-scale chemical fractionation. Phil Trans R Soc London A 325:147-160

Bacon MP and Anderson RF (1982) Distribution of thorium isotopes between dissolved and particulate forms in the ocean. J Geophys Res 87(C3):2045-2056

Bacon MP, Huh C-A, Fleer AP, Deuser WG (1985) Seasonality in the flux of natural radionuclides and plutonium in the deep Sargasso Sea. Deep-Sea Res 32(3):273-286

Bacon MP, Rosholt JN (1982) Accumulation rates of Th-230, Pa-231, and some transition metals on the Bermuda Rise. Geochim Cosmochim Acta 46:651-666

Bard E, Arnold M, Duprat J, Moyes J, Duplessey J-C (1987) Reconstruction of the last deglaciation: Deconvolved records of $\delta^{18}O$ profiles, micropaleontological variations and accelerator mass spectrometric ^{14}C dating. Clim Dyn 1:102-112

Barnes CE, Cochran JK (1990) Uranium removal in oceanic sediments and the oceanic U balance. Earth Planet Sci Lett 97:94-101

Barnes CE, Cochran JK (1993) Uranium geochemistry in estuarine sediments: Controls on removal and release processes. Geochim Cosmochim Acta 57:555-569

Berkman PA, Foreman DW, Mitchell JC, Liptak RJ (1992) Scallop shell mineralogy and crystalline characteristics: Proxy records for interpreting Antarctic nearshore marine hydrochemical variability. Contrib Antarctic Research 57:27-38

Bourdon B, Turner S, Henderson GM, Lundstrom C (2003) An introduction to U-series geochemistry, Rev Miner Geochem XX xxx-xxx

Boudreau BP (1994) Is burial velocity a master parameter for bioturbation? Geochim Cosmochim Acta 58(4):1243-1249

Boyle EA, Keigwin LD (1982) Deep Circulation of the North Atlantic over the last 200,000 years: geochemical evidence. Science 218:784-787

Brewer PG, Nozaki Y, Spencer DW, Fleer AP (1980) Sediment trap experiments in the deep North Atlantic: isotopic and elemental fluxes. J Mar Res 38(4):703-728

Buesseler KO (1991) Do upper-ocean sediment traps provide an accurate record of particle flux? Nature 353:420-423

Buesseler KO (1998) The decoupling of production and particulate export in the surface ocean. Global Biogeochem Cycles 12(2):297-310

Byrne RH (1981) Inorganic lead complexation in natural seawater determined by UV spectroscopy. Nature 290:487-489

Carroll J, Moore WS (1993) Uranium removal during low discharge in the Ganges-Brahmaputra mixing zone. Geochim Cosmochim Acta 58:4987-4995

Chabaux F, Cohen AS, O'Nions RK, Hein JR (1995) ^{238}U-^{234}U-^{230}Th chronometry of Fe-Mn crusts: Growth processes and recovery of thorium isotopic ratios of seawater. Geochim Cosmochim Acta 59:633-638

Chabaux F, Riotte J, Dequincey O (2003) U-Th-Ra fractionation during weathering and river transport. Rev Mineral Geochem 52:533-576

Chan LH, Edmond JM, Stallard RF, Broecker WS, Chung YC, Weiss RF, Ku TL (1976) Radium and barium at GEOSECS stations in the Atlantic and Pacific. Earth Planet Sci Lett 32:258-267

Chase Z, Anderson RF, Fleisher MQ (2001) Evidence from authigenic uranium for increased productivity of the glacial subantarctic ocean. Paleoceanography 16(5):468-478

Chase Z, Anderson RF, Fleisher MQ, Kubik PW (2002) The influence of particle composition and particle flux on scavenging of Th, Pa and Be in the ocean.: Earth Planet Sci Lett 204:215-229

Chase Z, Anderson RF, Fleisher MQ, Kubik PW (In press-a) Accumulation of biogenic and lithogenic material in the Pacific sector of the Southern Ocean during the past 40,000 years. Deep-Sea Res II

Chase Z, Anderson RF, Fleisher MQ, Kubik PW (In press-b) Scavenging of ^{230}Th, ^{231}Pa and ^{10}Be in the Southern Ocean (SW Pacific sector): The importance of particle flux and advection. Deep-Sea Res II

Chen JH, Edwards RL, Wasserburg GJ (1986a) ^{238}U-^{234}U-^{232}Th in seawater. Earth Planet Sci Lett 80: 241-251

Chen JH, Wasserburg GJ, Von Damm KL, Edmond JM (1986b) The U-Th-Pb systematics in hot springs on the East Pacific Rise at 21°N and Guaymas Basin. Geochim Cosmochim Acta 50:2467-2479

Cheng H, Edwards RL, Hoff J, Gallup CD, Richards DA, Asmerom Y (2000) The half lives of uranium-234 and thorium-230. Chemical Geology 169:17-33.

Church TM, Sarin MM, Fleisher MQ, Ferdelman TG (1996) Salt marshes: An important coastal sink for dissolved uranium. Geochim Cosmochim Acta 60(20):3879-3889

Claude-Ivanaj C, Hofmann AW, Vlastelic I, Koschinsky A (2001) Recording changes in ENADW composition over the last 340 ka using high-precision lead isotopes in a Fe-Mn crust. Earth Planet Sci Lett 188:73-89

Clegg SL, Whitfield M (1993) Application of a generalized scavenging model to time series ^{234}Th and particle data during the JGOFS North Atlantic bloom experiment. Deep-Sea Res 40(8):1529-1545

Cochran JK (1992) The oceanic chemistry of the uranium and thorium series nuclides. In: Uranium-series disequilibrium: Applications to earth, marine, and environmental sciences. Ivanovich M, Harmon RS (eds) Oxford University Press, Oxford p 334-395

Cochran JK, Krishnaswami S (1980) Radium, thorium, uranium and ^{210}Pb in deep-sea sediments and sediment pore waters from the north equatorial Pacific. Am J Sci 280:849-889

Cochran JK, Masqué P (2003) Short-lived U/Th-series radionuclides in the ocean: tracers for scavenging rates, export fluxes and particle dynamics. Rev Mineral Geochem 52:461-492

Colley S, Thomson J, Newton PP (1995) Detailed ^{230}Th, ^{232}Th and ^{210}Pb fluxes recorded by the 1989/90 BOFS sediment trap time-series at 48°N, 20°W. Deep-Sea Res 42(6):833-848

Dunk RM, Jenkins WJ, Mills RA (2002) A re-evaluation of the oceanic uranium budget. Chem Geol 190:45-67

Duplessy JC, Shackleton NJ, Fairbanks RG, Labeyrie L, Oppo D, Kallel N (1988) Deepwater source variations during the last climatic cycle and their impact on the global deepwater circulation. Paleoceanography 3(3):343-360

Dymond J, Suess E, Lyle M (1992) Barium in deep-sea sediment: A geochemical proxy for paleoproductivity. Paleoceanography 7(2):163-181

Edmonds HN, Moran SB, Hoff JA, Smith JN, Edwards RL (1998) Protactinium-231 and Thorium-230 abundances and high scavenging rates in the Western Arctic ocean. Science 280:405-407

Edwards RL, Gallup CD, Cheng H (2003) Uranium-series dating of marine and lacustrine carbonates. Rev Mineral Geochem 52:363-405

Eisenhauer A, Gögen K, Pernicka E, Mangini A (1992) Climatic influences on the growth rates of Mn crusts during the Late Quaternary. Earth Planet Sci Lett 109:25-36

Elderfield H, Rickaby REM (2000) Oceanic Cd/P ratio and nutrient utilization in the glacial Southern Ocean. Nature 405:305-310

Emerson SR, Huested SS (1991) Ocean anoxia and the concentrations of molybdenum and vanadium in seawater. Marine Chem 34:177-196

Esat TM, Yokoyama Y (1999) Rapid fluctuations in the uranium isotope composition of the oceans (abstract). Eos Transactions AGU 80(46(Fall meeting supl)):581

Fornes WL, DeMaster DJ, Smith CR (2001) A particle introduction experiment in Santa Catalina Basin sediments: Testing the age-dependent mixing hypothesis. J Marine Res 59(1):97-112

Francois R, Bacon MP (1994) Heinrich events in the North Atlantic: radiochemical evidence. Deep-Sea Res I 41:315-334

Francois R, Bacon MP, Altabet MA (1993) Glacial/interglacial changes in sediment rain rate in the SW Indian sector of subantarctic waters as recorded by ^{230}Th, ^{231}Pa, U and δ^{15}N. Paleoceanography 8(5):611-629

Francois R, Bacon MP, Suman DO (1990) Thorium 230 profiling in deep-sea sediments: high resolution records of flux and dissolution of carbonate in the equatorial Atlantic during the last 24,000 years. Paleoceanography 5(5):761-787

Francois R, Honjo S, Manganini SJ, Ravizza GE (1995) Biogenic barium fluxes to the deep-sea—Implications for paleoproductivity reconstruction. Global Biogeochem Cycles 9(2):289-303

Francois R, Altabet MA, Yu E-F, Sigman DM, Bacond MP, Frank M, Bohrmann G, Bareille G, Labeyrie LD (1997) Contribution of Southern Ocean surface-water stratification to low atmosphere CO_2 concentrations during the last glacial period. Nature 389:929-935

Frank M (2002) Radiogenic isotopes: tracers of past ocean circulation and erosional input. Rev Geophys 40(1):10.1029/2000RG000094

Frank M, Eisenhauer A, Bonn WJ, Walter P, Grobe H, Kubik PW, Dittrich-Hannen B, Mangini A (1995) Sediment redistribution versus paleoproductivity change: Weddell Sea margin sediment stratigraphy and biogenic particle flux of the last 250,000 years deduced from ^{230}Th$_{ex}$, ^{10}Be and biogenic barium profiles. Earth Planet Sci Lett 136:559-573

Frank M, Gersonde R, Rutgers van der Loeff MM, Kuhn G, Mangini A (1996) Late quaternary sediment dating and quantification of lateral sediment redistribution applying ^{230}Th$_{ex}$: A study from the eastern Atlantic sector of the Southern Ocean. Geol Rundsch 85:554-566

Frank M, Schwarz B, Baumann S, Kubik PW, Suter M, Mangini A (1997) A 200 kyr record of cosmogenic radionuclide production rate and geomagnetic field intensity from ^{10}Be in globally stacked deep-sea sediments. Earth Planet Sci Lett 149:121-129

Frank M, Gersonde R, Mangini A (1999) Sediment redistribution, ^{230}Th$_{ex}$-normalization and implications for reconstruction of particle flux and export productivity. In: Use of Proxies in Paleoceanography: Examples from the South Atlantic. Fischer G, Wefer G (eds) Springer-Verlag, p 409-426.

Frank M, Gersonde R, Rutgers van der Loeff MM, Bohrmann G, Nürnberg CC, Kubik PW, Suter M, Mangini A (2000) Similar glacial and interglacial export bioproductivity in the Atlantic sector of the Southern Ocean: Multiproxy evidence and implications for glacial atmospheric CO_2. Paleoceanography 15(6):642-658

Gallup CD, Edwards RL, Haddad GA, Droxler AW (1995) Constraints on past changes in the marine δ^{234}U value from Northern Nicaragua Rise, Caribbean sea cores. EOS 76(46):291-292

Gallup CD, Edwards RL, Johnson RG (1994) The timing of high sea levels over the past 200,000 years. Science 263:796-800

Gueguen C, Guo LD (2002) Thorium and protactinium sorption on silica. Geochim Cosmochim Acta 66(S1):A295

Guinasso NL, Schink DR (1975) Quantatative estimates of the biological mixing rates in abyssal sediments. J Geophys Res 80:3032-3043

Hall IR, McCave IN (2000) Palaeocurrent reconstruction, sediment and thorium focussing on the Iberian margin over the last 140 ka. Earth Planet Sci Lett 178:151-164

Henderson GM (2002a) Seawater ($^{234}U/^{238}U$) during the last 800 thousand years. Earth Planet Sci Lett 199:97-110

Henderson GM (2002b) New oceanic proxies for paleoclimate. Earth Planet Sci Lett 203:1-13

Henderson GM, Burton KW (1999) Using ($^{234}U/^{238}U$) to assess diffusion rates of isotopic tracers in Mn crusts. Earth Planet Sci Lett 170(3):169-179

Henderson GM, Cohen AS, O'Nions RK (1993) $^{234}U/^{238}U$ ratios and ^{230}Th ages for Hateruma Atoll corals: implications for coral diagenesis and seawater $^{234}U/^{238}U$ ratios. Earth Planet Sci Lett 115:65-73

Henderson GM, Heinze C, Anderson RF, Winguth AME (1999a) Global distribution of the ^{230}Th flux to ocean sediments constrained by GCM modelling. Deep Sea Res 46(11):1861-1893

Henderson GM, Lindsay F, Slowey NC (1999b) Variation in bioturbation with water depth on marine slopes: A study on the slopes of the Little Bahamas Bank. Marine Geo 160(1-2):105-118

Henderson GM, Maier-Reimer E (2002) Advection and removal of ^{210}Pb and stable Pb isotopes in the oceans: A GCM study. Geochim Cosmochim Acta 66(2):257-272

Henderson GM, Slowey NC, Fleisher MQ (2001) U-Th dating of carbonate platform and slope sediments. Geochim Cosmochim Acta 65(16):2757-2770

Henderson GM, Slowey NC, Haddad GA (1999c) Fluid flow through carbonate platforms: Constraints from $^{234}U/^{238}U$ and Cl- in Bahamas pore-waters. Earth Planet Sci Lett 169:99-111

Honeyman BD, Balistrieri LS, Murray JW (1988) Oceanic trace metal scavenging: the importance of particle concentration. Deep-Sea Res 35(2):227-246

Imbrie J, Hays JD, Martinson DG, McIntyre A, Mix AC, Morley JJ, Pisias NG, Prell W L, Shackleton NJ (1984) The orbital theory of Pleistocene climate : Support from a revised chronology of the marine $\partial^{18}O$ record. In: Milankovitch and Climate, Vol. 1. Berger A, Imbrie J, Hays J, Kukla G, Saltzman B (eds) D. Reidel, Boston p 269-305

Ivanovich M, Harmon RS (1992) Uranium-series disequilibrium: Applications to earth, marine, and environmental sciences. Oxford University Press, Oxford.

Kigoshi K (1971) Alpha-recoil ^{234}Th: dissolution into water and the $^{234}U/^{238}U$ disequilibrium in nature. Science 173:47-48

Kim DH, Burnett WC (1988) Accumulation and biological mixing of Peru margin sediments. Mar Geol 80:181-194

Klinkhammer GP, Palmer MR (1991) Uranium in the oceans: where it goes and why. Geochim Cosmochim Acta 55:1799-1806

Krishnaswami S, Lal D, Somayajulu BLK (1976) Large-volume in-situ filtration of deep Pacific waters: Mineralogical and radioisotope studies. Earth Planet Sci Lett 32:420-429

Krishnaswami S, Mangini A, Thomas JH, Sharma P, Cochran JK, Turekian KK, Parker PD (1982) ^{10}Be and Th isotopes in manganese nodules and adjacent sediments: nodule growth histories and nuclide behavior. Earth Planet Sci Lett 59:217-234

Krishnaswami S, Sarin MM, Somayajulu BLK (1981) Chemical and radiochemical investigations of surface and deep particles of the Indian Ocean. Earth Planet Sci Lett 54:81-96

Kronfeld J, Vogel JC (1991) Uranium Isotopes in surface waters from southern Africa. Earth Planet Sci Lett 105:191-195

Ku T-L (1965) An evaluation of the $^{234}U/^{238}U$ method as as tool for dating pelagic sediments. J Geophys Res 70(14):3457-3474

Ku T-L, Broecker WS (1969) Radiochemical studies on manganese nodules of deep-sea origin. Deep-Sea Res 16:625-637

Ku TL, Kuass KG, Mathieu GG (1977) Uranium in the open ocean : Concentrations and isotopic composition. Deep Sea Res 24:1005

Ku TL, Omura A, Chen PS (1979) Be^{10} and U-series isotopes in manganese nodules from the central north Pacific. In Marine geology and oceanography of the Pacific manganese nodule province (ed. J. L. Bishoff and Z. Piper), pp. 791-804. Plenum

Kumar N, Anderson RF, Mortlock RA, Froelich PN, Kubik P, Dittrich-Hannen B, Suter M (1995) Increased biological productivity and export production in the glacial Southern Ocean. Nature 378:675-680

Kumar N, Gwiazda R, Anderson RF, Froelich PN (1993) $^{231}Pa/^{230}Th$ ratios in sediments as a proxy for past changes in Southern Ocean productivity. Nature 362:45-48

Lao Y, Anderson RF, Broecker WS, Hofmann HJ, Wolfi W (1993) Particulate fluxes of ^{230}Th, ^{231}Pa, and ^{10}Be in the northeastern Pacific Ocean. Geochim Cosmochim Acta 57:205-217

Liebetrau V, Eisenhauer A, Gussone N, Wörner G, Hansen BT, Leipe T (2002) $^{226}Ra_{excess}/Ba$ growth rates and U-Th-Ra-Ba systematic of Baltic Mn/Fe crusts. Geochim Cosmochim Acta 66(1):73-84

Lochte K, Anderson RF, Francois R, Jahnke RA, Shimmield G, Vetrov A (in press) Benthic processes and the burial of carbon. *In*: Ocean Biogeochemistry: A JGOFS Synthesis. Fasham M, Zeitzschel B, Platt T (eds)

Luo S, Ku T-L (1999) Oceanic $^{231}Pa/^{230}Th$ ratio influenced by particle composition and remineralization. Earth Planet Sci Lett 167:183-195

Lynch-Stieglitz J, Curry WB, Slowey NC (1999) A geostrophic transport estimate for the Florida Current from the oxygen isotope composition of benthic foraminifera. Paleoceanography 14(3):360-373

Mangini A, Segl M, Kudrass H, Wiedicke M, Bonani G, Hofmann HJ, Morenzoni E, Nessi M, Suter M, Wolfli W (1986) Diffusion and supply rates of ^{10}Be and ^{230}Th radioisotopes in two manganese encrustations from the South China Sea. Geochim Cosmochim Acta 50:149-156

Marcantonio F, Anderson RF, Higgins S, Fleisher MQ, Stute M, Schlosser P (2001a) Abrupt intensification of the SW Indian Ocean monsoon during the last deglacition: Constraints from Th, Pa, and He isotopes. Earth Planet Sci Lett 184:505-514

Marcantonio F, Anderson RF, Higgins S, Stute M, Schlosser P, Kubik PW (2001b) Sediment focusing in the central equatorial Pacific Ocean. Paleoceanography 16(3):260-267

Marcantonio F, Kumar N, Stute M, Anderson RF, Seidl MA, Schlosser P, Mix A (1995) A comparative study of accumulation rates derived by He and Th isotope analysis of marine sediments. Earth Planet. Sci. Letters 133:549-555

Marcantonio F, Turekian KK, Higgins S, Anderson RF, Stute M, Schlosser P (1999) The accretion rate of extraterrestrial ^3He based on oceanic ^{230}Th flux and the relation to Os isotope variation over the past 200,000 years in an Indian Ocean core. Earth Planet Sci Lett 170:157-168

Marchal O, Francois R, Stocker TF, Joos F (2000) Ocean thermohaline circulation and sedimentary $^{231}Pa/^{230}Th$ ratio. Paleoceanography 15(6):625-641

McCave IN, Manighetti B, Robinson SG (1995) Sortable silt and fine sediment size/composition slicing: Parameters for palaeocurrent speed and palaeoceanography. Paleoceanography 10(3):593-610

McKee BA, DeMaster DJ, Nittrouer CA (1986) Temporal variability in the partitioning of thorium between dissolved and particulate phases on the Amazon shelf: Implications for the scavenging of particle-reactive species. Cont Shelf Res 6:87-106

McManus JF, Anderson RF, Broecker WS, Fleisher MQ, Higgins SM (1998) Radiometrically determined sedimentary fluxes in the sub-polar North Atlantic during the last 140,000 years. Earth Planet Sci Lett 129:29-43

McManus JF, Francois R, Gherardi J, Nuwer JM (2002) Variable rates of ocean circulation from sedimentary $^{231}Pa/^{230}Th$ during the last ice-age termination. Geochim Cosmochim Acta 66(S1):A503

Mills RA, Thomson J, Elderfield H, Hinton RW. and Hyslop E (1994) Uranium enrichment in metalliferous sediments from the mid-Atlantic ridge. Earth Planet Sci Lett 124:35-47

Moore WS (1981) The thorium isotope content of ocean water. Earth Planet Sci Lett 53:419-426

Moore WS, Ku T-L, Macdougall JD, Burns VM, Burns R, Dymond J, Lyle MW, Piper DZ (1981) Fluxes of metals to a manganese nodule: Radiochemical, chemical, structural, and mineralogical studies. Earth Planet Sci Lett 52:151-171

Moore WS, Sackett WM (1964) Uranium and thorium series inequilibrium in seawater. Journal of Geophysical Research 69:5401-5405

Moran SB, Charett MA, Hoff JA, Edwards RL, Landing WM (1997) Distribution of ^{230}Th in the Labrador Sea and its relation to ventilation. Earth Planet Sci Lett 150:151-160

Moran SB, Hoff JA, Buesseler KO, Edwards RL (1995) High precision ^{230}Th and ^{232}Th in the Norwegian Sea and Denmark by thermal ionization mass spectrometry. Geophys Res Letters 22(19):2589-2592

Moran SB, Shen C-C, Weinstein SE, Hettinger LH, Hoff JH, Edmonds HN, Edwards RL (2001) Constraints on deep water age and particle flux in the Equatorial and Southern Atlantic ocean based on seawater ^{231}Pa and ^{230}Th data. Geophysical Research Letters 28(18):3437-3440

Morford JL, Emerson S (1999) The geochemistry of redox sensitive trace metals in sediments. Geochim Cosmochim Acta 63(11/12):1735-1750

Murray RW, Knowlton C, Leinen M, Mix AC, Polsky CH (2000) Export production and carbonate dissolution in the central equatorial Pacific Ocean over the past 1 Myr. Paleoceanography 15(6):570-592

Neff U, Bollhöfer A, Frank N, Mangini A (1999) Explaining discrepant depth profiles of $^{234}U/^{238}U$ and $^{230}Th_{exc}$ in Mn-crusts. Geochim Cosmochim Acta 63(15):2211-2218

Nozaki Y, Cochran JK, Turekian KK, Keller G (1977) Radiocarbon and ^{210}Pb distribution in submersible-taken deep-sea cores from Project Famous. Earth Planet Sci Lett 34:167-173

Nozaki Y, Horibe Y, Tsubota H (1981) The water column distributions of thorium isotopes in the western North Pacific. Earth Planet Sci Lett 54:203-216

Nozaki Y, Yang H-S (1987) Th and Pa isotopes in the waters of the western margin of the Pacific near Japan: Evidence for release of ^{228}Ra and ^{227}Ac from slope sediments. J Oceanograph Soc Jap 43:217-227

Nozaki Y, Yang H-S, Yamada M (1987) Scavenging of thorium in the ocean. J Geophys Res 92(C1):772-778
Osmond JK, Ivanovich M (1992) Uranium-series mobilization and surface hydrology. In Uranium-series disequilibrium - applications to earth, marine, and environmental sciences. (ed. M. Ivanovich and R. S. Harmon), pp. 259-289. Oxford University Press, Oxford.
Osthols E (1995) Thorium sorption on amorphous silica. Geochim Cosmochim Acta 59(7):1235-1249
Palmer MR, Edmond JM (1993) Uranium in river water. Geochim Cosmochim Acta 57:4947-4955
Paytan A, Moore WS, Kastner M (1996) Sedimentation rate as determined by ^{226}Ra activity in marine barite. Geochim Cosmochim Acta 60(22):4313-4320
Robbins JA (1988) A model for particle-selective transport of tracers in sediments with conveyor-belt deposit feeders. Journal of Geophysical Research 91:8542-8558
Robinson LF, Belshaw NS, Henderson GM (in press) U and Th isotopes in seawater and modern carbonates from the Bahamas. Geochim Cosmochim Acta
Rosenthal Y, Boyle EA, Labeyrie L, Oppo D (1995a) Glacial enrichments of authigenic Cd and U in subantarctic sediments: A climatic control on the elements' oceanic budget? Paleoceanography 10(3):395-413
Rosenthal Y, Lam P, Boyle EA, Thomson J (1995b) Authigenic cadmium enrichments in suboxic sediments: Precipitation and postdepositional mobility. Earth Planet Sci Lett 132:99-111
Rühlemann C, Frank M, Hale W, Mangini A, Mulitza S, Müller PJ, Wefer G (1996) Late quaternary productivity changes in the western equatorial Atlantic: Evidence from ^{230}Th-normalized carbonate and organic carbon accumulation rates. Mar Geo 135:127-152
Russell AD, Edwards RL, Hoff JA, McCorkle D, Sayles F (1994) Sediment source of ^{234}U suggested by δ^{234}U in North Pacific pore waters. EOS 75(44) Fall Meeting Supplement:332
Rutgers van der Loeff MM, Berger GW (1993) Scavenging of ^{230}Th and ^{231}Pa near the Antarctic Polar front in the South Atlantic. Deep-Sea Res 40(2):339-357
Sacket WM, Mo T, Spalding RF, Exner ME (1973) A revaluation of the marine geochemistry of uranium. Symposium on the interaction of radioactive containments with the constituents of the marine environment, 757-769
Sayles FL, Martin WR, Chase Z, Anderson RF (2001) Benthic remineralization and burial of biogenic SiO_2, $CaCO_3$, organic carbon and detrital material in the Southern Ocean along a transect at 170° west. Deep-Sea Research II 48:4323-4383
Scholten JC, Fietzke J, Vogler A, Rutgers van der Loeff MM, Mangini A, Koeve W, Waniek J, Stoffers P, Antia A, Kuss J (2001) Trapping efficiencies of sediment traps from the deep Eastern North Atlantic: The ^{230}Th calibration. Deep-Sea Res II 48:2383-2408
Shimmield GB, Murray JW, Thomson J, Bacon MP, Anderson RF, Price NB (1986) The distribution and behaviour of ^{230}Th and ^{231}Pa at an ocean margin, Baja California, Mexico. Geochim Cosmochim Acta 50:2499-2507
Shimmield GB, Price NB (1988) The scavenging of U, ^{230}Th, and ^{231}Pa during pulsed hydrothermal activity at 20°S, East Pacific Rise. Geochim Cosmochim Acta 52:669-677
Sigman DM, Boyle EA (2000) Glacial/Interglacial variations in atmospheric carbon dioxide. Nature 407:859-869
Simpson HJ, Trier RM, Toggweiler JR, Mathieu G, Deck BL, Olsen CR, Hammond DE, Fuller C, Ku TL (1982) Radionuclides in Mono Lake, California. Science 216:512-514
Smith CR, Berelson W, Demaster DJ, Dobbs FC, Hammond D, Hoover DJ, Pope RH, Stephens M (1997) Latitudinal variations in benthic processes in the abyssal equatorial Pacific: control by biogenic particle flux. Deep-Sea Res Part II-Topical Studies in Oceanography 44(9-10):2295
Smith CR, Pope RH, Demaster DJ, Magaard L (1993) Age-dependent mixing of deep-sea sediments. Geochim Cosmochim Acta 57(7):1473-1488
Soetaert K, Herman PMJ, Middelburg JJ, Heip C, deStigter HS, van Weering TCE, Epping E, Helder W (1996) Modelling ^{210}Pb-derived mixing activity in ocean margin sediments: Diffusive versus nonlocal mixing. J Mar Res 54:1207-1227
Spencer DW, Bacon MP, Brewer PG (1981) Models of the distribution of ^{210}Pb in a section across the North Equatorial Atlantic Ocean. J Mar Res 39(1):119-137
Staubwasser M, Henderson GM, Berkman PA, Hall BL (in press) Ba, Ra, Th and U in marine mollusc shells and the potential of ^{226}Ra/Ba dating of Holocene marine carbonate shells. Geochim Cosmochim Acta
Staudigel H, Plank T, White B, Schmincke H-U (1996) Geochemical fluxes during seafloor alteration of the basaltic upper oceanic crust: DSDP Sites 417 and 418. In: Subduction: Top to Bottom. Bebout GE, Scholl DW, Kirby SH, Platt JP(eds), AGU, Washington DC, p 19-37
Suman DO, Bacon MP (1989) Variations in Holocene sedimentation in the North American Basin determined by ^{230}Th measurements. Deep Sea Res 36:869-787

Swarzenski PW, McKee BA, Booth JG (1995) Uranium geochemistry on the Amazon shelf: Chemical phase partioning and cycling across a salinity gradient. Geochim Cosmochim Acta 59:7-18

Swarzenski PW, Porcelli D, Andersson PS, Smoak JM (2003) The behavior of U- and Th- series nuclides in the estuarine environment. Rev Mineral Geochem 52:577-606

Taguchi K, Harada K, Tsunogai S (1989) Particulate removal of ^{230}Th and ^{231}Pa in the biologically productive northern North Pacific. Earth Planet Sci Lett 93:223-232

Thomson J, Colley S, Anderson R, Cook GT, MacKenzie AB, Harkness DD (1993) Holocene sediment fluxes in the northeast Atlantic from ^{230}Th$_{excess}$ and radiocarbon measurements. Paleoceanography 8:631-650

Thomson J, Higgs NC, Clayton T (1995) A geochemical criterion for the recognition of Heinrich events and estimation of their depositional fluxes by the (^{230}Th$_{excess}$)^0 profiling method. Earth Planet Sci Lett 135:41-56

Thomson J, Jarvis I, Green DRH, Green DA, Clayton T (1998) Mobility and immobility of redox-sensitive elements in deep-sea turbidites during shallow burial. Geochim Cosmochim Acta 62(4):643-656

Thomson J, Nixon S, Summerhayes CP, Schonfeld J, Zahn R, Grootes P (1999) Implications for sedimentation changes on the Iberian margin over the last two glacial/interglacial transitions from (^{230}Th$_{excess}$)^0 systematics. Earth Planet Sci Lett 165:255-270

Trauth MH, Sarnthein M, Arnold M (1997) Bioturbidational mixing depth and carbon flux at the seafloor. Paleoceanography 12(3):517-526

van Beek P, Reyss JL (2001) ^{226}Ra in marine barite: New constraints on supported ^{226}Ra. Earth Planet Sci Lett 187:147-161

van Beek P, Reyss JL, Paterne M, Gersonde R, Rutgers van der Loeff MM, Kuhn G (2002) ^{226}Ra in barite: Absolute dating of Holocene Southern Ocean sediments and reconstruction of sea-surface reservoir ages. Geology 30(8):731-734.

Veeh HH, McCorkle DC, Heggie DT (2000) Glacial/interglacial variations of sedimentation on the West Australian continental margin: Constraints from excess ^{230}Th. Mar Geol 166:11-30

Veiga-Pires CC and Hillaire-Marcel C (1999) U and Th isotope constraints on the duration of Heinrich events H0-H4 in the southeastern Labrador Sea. Paleoceanography 14(2):187-199

Vogler S, Scholten J, Rutgers van der Loeff M, Mangini A (1998) Th-230 in the eastern North Atlantic: The importance of water mass ventilation in the balance of Th-230. Earth Planet Sci Lett 156(1-2): 61-74

Walter HJ, Geibert W, Rutgers van der Loeff MM, Fischer G, Bathman U (2001) Shallow vs. deep-water scavenging of ^{231}Pa and ^{230}Th in radionuclide enriched waters of the Atlantic sector of the Southern Ocean. Deep-Sea Res I 48:471-493

Walter HJ, Rutgers van der Loeff MM, Hoeltzen H (1997) Enhanced scavenging of ^{231}Pa relative to ^{230}Th in the South Atlantic south of the Polar front: Implications for the use of the ^{231}Pa/^{230}Th ratio as a paleoproductivity proxy. Earth Planet Sci Lett 149:85-100

Wheatcroft RA (1992) Experimental tests for particle size-dependant bioturbation in the deep oceans. Limnol Oceanogr 37(1):90-104

Yang H-S, Nozaki Y, Sakai H (1986) The distribution of ^{230}Th and ^{231}Pa in the deep-sea surface sediments of the Pacific Ocean. Geochim Cosmochim Acta 50:81-89

Yokoyama T, Esat TM, Lambeck K (2001) Coupled climate and sea-level changes deduced from Huon Peninsula coral terraces of the last ice age. Earth Planet Sci Lett 193:579-587

Yu E-F, Francois R, Bacon MP, Fleer AP (2001a) Fluxes of ^{230}Th and ^{231}Pa to the deep sea: Implications for the interpretation of excess ^{230}Th and ^{231}Th profiles in sediments. Earth Planet Sci Lett 191(3-4):219-230

Yu E-F, Francois R, Bacon MP, Honjo S, Fleer AP, Manganini SJ, Rutgers van der Loeff M. M., Ittekot V. (2001b) Trapping efficiency of bottom tethered sediment traps estimated from the intercepted fluxes of ^{230}Th and ^{231}Pa. Deep-Sea Res Part 1 48:865-889

Yu E-F (1994) Variations in the particulate flux of ^{230}Th and ^{231}Pa and paleoceanographic applications of the ^{231}Pa/^{230}Th ratio. PhD Dissertation. Woods Hole Oceanographic Institute & Massachusetts Institute of Technology, Boston, MA

Yu E-F, Francois R, Bacon M (1996) Similar rates of modern and last-glacial ocean thermohaline circulation inferred from radiochemical data. Nature 379:689-694

Zheng Y, Anderson RF, van Geen A, Fleisher MQ (2002) Preservation of particulate non-lithogenic uranium in marine sediments. Geochim Cosmochim Acta 66(17):3085-3092.

APPENDIX

Measured concentrations of ^{230}Th and ^{231}Pa in marine sediments consist of three components: that scavenged from seawater; that supported by U contained within lithogenic minerals; and that produced by radioactive decay of authigenic U. Most of the proxies described in this paper make use of only the scavenged component. Measured ^{230}Th and ^{231}Pa must therefore first be corrected for the presence of the other two components. To make this correction to the measured ^{230}Th or ^{231}Pa, it is generally assumed that the U decay series are in secular equilibrium in lithogenic phases, and that the formation of authigenic U is contemporary with the deposition of the sediments. Accepting those assumptions, unsupported (scavenged from seawater) ^{230}Th is calculated from its measured value as:

$$^{230}Th_{xs} = {}^{230}Th_{meas} - \left\{(0.6 \pm 0.1)^{232}Th_{meas}\right\} - \\ \left\{\begin{bmatrix} ^{238}U_{meas} - (0.6 \pm 0.1)^{232}Th_{meas} \end{bmatrix} \times \\ \begin{bmatrix} (1 - e^{-\lambda_{230}t}) + \frac{\lambda_{230}}{\lambda_{230} - \lambda_{234}}(e^{-\lambda_{234}t} - e^{-\lambda_{230}t})\left(\left(\frac{^{234}U}{^{238}U}\right)_{init} - 1\right) \end{bmatrix}\right\} \quad (A1)$$

and unsupported ^{231}Pa as:

$$^{231}Pa_{xs} = {}^{231}Pa_{meas} - \left\{0.046(0.6 \pm 0.1)^{232}Th_{meas}\right\} - \\ \left\{\begin{bmatrix} 0.046\,^{238}U_{meas} - (0.6 \pm 0.1)^{232}Th_{meas} \end{bmatrix}\begin{bmatrix} 1 - e^{-\lambda_{231}t} \end{bmatrix}\right\} \quad (A2)$$

Where all nuclide measurements are expressed in activities. In both expressions, the first term in curly brackets corrects for ^{230}Th (or ^{231}Pa) supported by U in lithogenic material, and the second term in curly brackets corrects for ^{230}Th (or ^{231}Pa) ingrown from authigenic U. For samples that are known to be young, such as sediment trap or core-top samples, t = 0 and this second term equals zero reflecting the lack of time for decay of authigenic U in such samples.

The concentration of lithogenic ^{238}U (or ^{235}U) is estimated from the measured concentration of ^{232}Th, which is entirely of lithogenic origin (Brewer et al. 1980), together with an appropriate lithogenic ^{238}U/^{232}Th (or ^{235}U/^{232}Th) ratio. Although there has never been a formal compilation of lithogenic U/Th ratios in marine sediments, based on our own experience as well as results reported in the literature (e.g., Walter et al. 1997), appropriate lithogenic ^{238}U/^{232}Th ratios (expressed as activity ratios) are 0.6 ± 0.1 for the Atlantic Ocean, 0.7 ± 0.1 for the Pacific Ocean, and 0.4 ± 0.1 for regions south of the Antarctic Polar Front in the Southern Ocean. In the equations above, 0.6 ± 0.1 represents the applied lithogenic ^{238}U/^{232}Th activity ratio, and 0.046 is the natural ^{235}U/^{238}U activity ratio.

Authigenic U is precipitated within chemically-reducing marine sediments (Klinkhammer and Palmer 1991). Particulate non-lithogenic U is also formed in surface ocean waters (Anderson 1982). Although much of this particulate non-lithogenic U is regenerated prior to burial in pelagic sediments (Anderson 1982), a substantial fraction (tens of percent) survives to be buried in ocean-margin sediments, particularly in regions where an intense oxygen minimum zone impinges on the sediments (Zheng et al. 2002).

Authigenic U is assumed to have an initial $^{234}U/^{238}U$ ratio equivalent to that of U dissolved in seawater (= 1.146). To correct for ingrowth from U using the above equations, an estimate of the age of the sediment, t, is required. This is generally derived from stratigraphy (e.g., $\delta^{18}O$) and by assuming the SPECMAP timescale (Imbrie et al. 1984).

Age correction is also generally required in order to correct for the decay of excess ^{230}Th (or ^{231}Pa) since the formation of the sediment. Only by such a correction can the conditions at the time of sediment formation be determined. For $^{230}Th_{xs}$, this correction is given by:

$$^{230}Th_{xs}^{\,0} = {}^{230}Th_{xs} e^{\lambda_{230} t} \tag{A3}$$

And for $^{231}Pa_{xs}$, by:

$$^{231}Pa_{xs}^{\,0} = {}^{231}Pa_{xs} e^{\lambda_{231} t} \tag{A4}$$

Again, a stratigraphic estimate of the age of the sediment is used to give t in both equations. The ratio of age-corrected excess ^{230}Th to ^{231}Pa, denoted $(^{231}Pa_{xs}/^{230}Th_{xs})^0$ is that most widely used to investigate the past environment.

13 U-Th-Ra Fractionation During Weathering and River Transport

F. Chabaux, J. Riotte and O. Dequincey

Centre de Géochimie de la Surface (CNRS/ULP)
Ecole et Observatoire des Sciences de la Terre
1 rue Blessig 67084 Strasbourg Cedex, France

1. INTRODUCTION

The potential of radioactive disequilibria as tracers and chronometers of weathering processes has been recognised since the 1960's (e.g., Rosholt et al. 1966; Hansen and Stout 1968). This interest results from the dual property of the nuclides of the U and Th radioactive series (1) to be fractionated during water-rock interactions and (2) to have radioactive periods of the same order of magnitude as the time constants of many weathering processes and chemical transfers to ground and river waters. Therefore, the study of radioactive disequilibria in surface environments should help to bring information about the nature, the intensity but also the time-scale of the water-rock interactions produced during weathering and related chemical transfers. These different properties have justified many of the studies on U-Th series in weathering profiles and river waters.

Rosholt (1982), Scott (1982) and Osmond and Ivanovich (1992) gave a synthesis of the studies of U-series in weathering and surface hydrology up to the 1980's and the 1990's respectively. Here, we present the main directions taken in these domains over the last decade. They were partly stimulated by the analytical developments made in the measurement of the medium half-life nuclides of ^{238}U series (i.e., ^{234}U-^{230}Th-^{226}Ra) in the mid-1980's, namely, the use of thermal ionisation mass spectrometry (TIMS) (e.g., Chen et al. 1986; Edwards et al. 1987; Cohen et al. 1991; Chabaux 1993;Chabaux et al. 1994) and more recently the use of MC ICP-MS (Turner et al. 2001; Robinson et al. 2002) for U and Th isotope analysis. Details of these developments are given in this volume (Goldstein and Stirling 2003). Compared to the radioactive counting methods previously used, the new techniques permit (1) a reduction by one order of magnitude or more of the size of the sample required for the measurement and (2) an improvement of the analytical precision by a factor of 5 to 10, two improvements that significantly increase the usefulness of these geochemical tools. Compared to other U-series applications such as magma genesis or palaeoclimatological studies, the progress in the domain of the weathering and surface hydrology could appear less spectacular. However, real and significant progress has been made in fields such as (a) the dating of pedogenic formations, (b) the geochemical tracing and the time-constant determination of the chemical mobility within weathering profiles and surface waters, or (c) the calculation of weathering budgets from the analysis of U-Th series in river waters. These studies have opened up new ways which now have to be explored, and which could prompt new research in the field over the coming years.

In the following pages, we will summarise the main processes controlling the fractionation of radionuclides during weathering and transfers into surface waters. Subsequently, we will present the main results obtained on surface weathering and transport in the river waters. Throughout this chapter, we will use parentheses to denote activity ratios.

2. ORIGIN OF RADIONUCLIDE FRACTIONATION DURING WEATHERING AND TRANSFERS INTO SURFACE WATERS

Nuclides of the radioactive U-Th series are fractionated from each other during rock weathering, as attested by the general occurrence of radioactive disequilibria among the nuclides of these series in soils and surface waters. As now well acknowledged and already detailed (Gascoyne 1992), two main processes create radioactive disequilibria. The first one is chemical fractionation among nuclides belonging to different chemical elements. The second is alpha recoil.

2.1. Chemical fractionation and mobilization factors

Thermodynamics of radionuclides in solution. During weathering processes and transfers into river waters, fractionation among nuclides of different chemical elements are largely controlled by the differences in chemical properties of these elements in solution. The properties are controlled by the classical thermodynamic parameters of aqueous solutions: T, P and solution composition (pH, redox potential, ionic strength and occurrence of complex forming ligands). The thermodynamic properties of actinides in aqueous solutions have been largely studied. Many reviews about U are available (e.g., Grenthe et al. 1992; Langmuir 1997; Shock et al. 1997; Murphy and Shock 1999). Langmuir and Herman (1980) and Langmuir and Riese (1985) give an overview of the thermodynamics of Th and Ra in solution.

Uranium thermodynamic data indicate that in environmental conditions the predominant species stable with H_2O are in either U(IV) or U(VI) oxidation state (Fig. 1). Under reducing condition, U(IV) is very insoluble and tends to precipitate as insoluble uraninite. This solubility however increases at low pH in presence of fluoride by formation of uranous fluoride complexes, and above pH 7-8 by complexation of U^{4+} with hydroxyl ions (formation of the $U(OH)_5^-$ complex) (Fig. 2). In oxic conditions, U is under its most oxidized state U^{6+}, which forms in aqueous environments the linear uranyl ion UO_2^{2+}, easily complexed with carbonate and hydroxide, and also with phosphate and fluoride (Fig. 3). Uranium(VI) is considerably more soluble than U(IV) and the formation of uranyl complexes, such as carbonate and phosphate complexes, significantly increases the solubility of U minerals and the mobility of U in surface and ground waters (Fig. 4). Organic ligands can also strongly bind U in environmental systems (e.g., Gayscone 1992; Lenhart et al. 2000; Montavon et al. 2000), and significantly influence its transport in natural medium. Compared to U, Th is found in nature only as a tetravalent cation and is considered to be very insoluble and immobile, at least for pH > 3-4. Mobility and solubility of Th may however be significantly increased by complexation with both inorganic and organic ligands as illustrated in Langmuir and Herman (1980) with the computed solubility of thorianite (Fig. 5). The main inorganic species that complex with Th in surface and ground waters are sulfate, fluoride phosphate and hydroxide (Fig. 6) (see also Anderson et al. 1982, Lin et al. 1996). Alike U, Th forms strong organic complexes with humic and fulvic acids. In freshwaters, Ra solubility is generally low except in specific contexts such as reduced and saline waters, by formation of complexes with chlorides ($RaCl^+$) (e.g., Dickson 1985; Langmuir and Melchior 1985; Sturchio et al. 1993).

The thermodynamic properties of U-Th series nuclides in solution are important parameters to take into account when explaining the U-Th-Ra mobility in surface environments. They are, however, not the only ones controlling radionuclide fractionations in surface waters and weathering profiles. These fractionations and the resulting radioactive disequilibria are also influenced by the adsorption of radionuclides onto mineral surfaces and their reactions with organic matter, micro-organisms and colloids.

Figure 1. Oxidation potential-pH diagram for the system U-C-H-O at 25°C and 1 atm, with $\Sigma U = 1$ μmol / L and $P_{CO_2} = 10^{-2}$ atm. [Used with permission of Elsevier Science, from Langmuir (1978) *Geochim Cosmochim Acta*, Vol. 42, Fig. 2, p. 554].

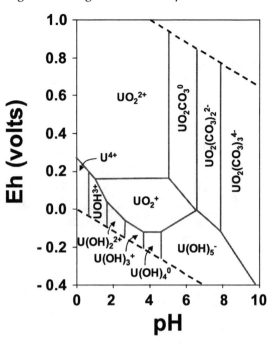

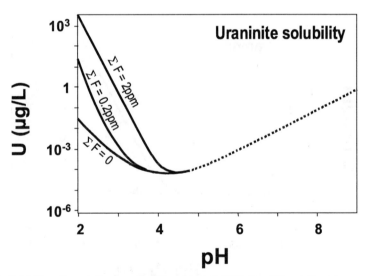

Figure 2. Solubility of uraninite as a function of pH at 25°C for different fluoride concentrations. Formation of uranous fluoride complexes greatly enhances uraninite solubility below pH 3-4. The increase of uraninite solubility at higher pH results from the formation of uranous hydroxyl complexes. [Used with permission of Elsevier Science, from Langmuir (1978) *Geochim Cosmochim Acta*, Vol. 42, Fig. 4, p. 555].

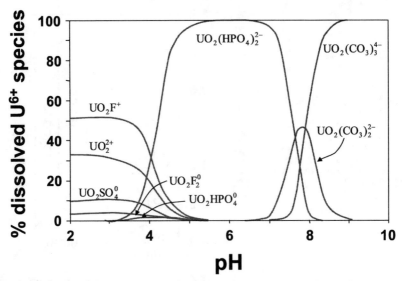

Figure 3. Distribution of uranyl complexes as a function of pH at 25°C in presence of typical ligands in surface and ground waters ($P_{CO_2} = 10^{-2.5}$ atm, $\Sigma F = 0.3$ ppm, $\Sigma Cl = 10$ ppm, $\Sigma SO_4 = 100$ ppm, $\Sigma PO_4 = 0.1$ ppm, $\Sigma SiO_2 = 30$ ppm). Below pH 4-5 uranyl (UO_2^{2+}) ion and uranyl fluoride complexes predominate, at intermediary pHs (4.5 < pH < 7.5) $UO_2(HPO_4)_2^{2-}$ is the predominant species, whereas at higher pH uranyl is complexed with carbonates. [Used with permission of Elsevier Science, from Langmuir (1978) *Geochim Cosmochim Acta*, Vol. 42, Fig. 11, p. 558].

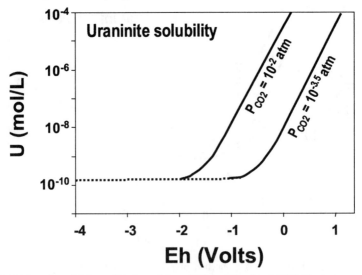

Figure 4. Solubility of uraninite as a function of Eh and P_{CO_2} at pH = 8 and 25°C. The increase of uraninite solubility at high P_{CO_2} results from the formation of uranyl carbonate complexes. [Used with permission of Elsevier Science, from Langmuir (1978) *Geochim Cosmochim Acta*, Vol. 42, Fig. 15, p. 561].

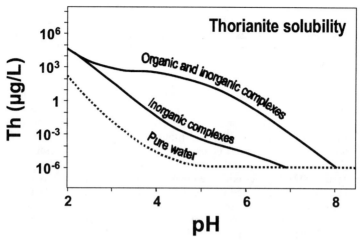

Figure 5. Solubility of thorianite as a function of pH at 25°C in pure water, and in presence of inorganic and organic ligands (Σ F = 0.3 ppm, Σ Cl = 10 ppm, Σ PO$_4$ = 0.1 ppm, Σ SO$_4$ = 100 ppm, Σ NO$_3$ = 2,5 ppm, Σ oxalate = 1 ppm, Σ citrate = 0.1 ppm, Σ EDTA = 0.1 ppm). Thorianite is a very insoluble mineral in pure water, except at very low pH. Inorganic ligands significantly affect ThO$_2$ solubility below pH 6, whereas organic ligands up to pH 8. [Used with permission of Elsevier Science, from Langmuir and Herman (1980) *Geochim Cosmochim Acta*, Vol. 44, Fig. 11, p. 1763].

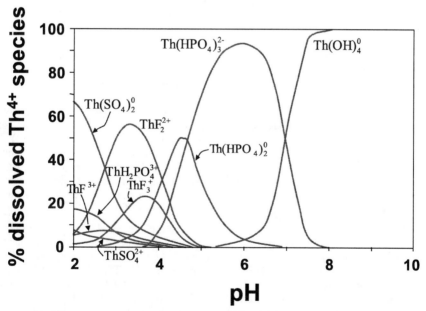

Figure 6. Distribution of thorium complexes as a function of pH at 25°C in the presence of typical inorganic ligands in surface and ground waters (Th = 0.01µg/L, Σ F = 0.3 ppm, Σ Cl = 10 ppm, Σ PO$_4$ = 0.1 ppm, Σ SO$_4$ = 100 ppm). The predominant species are: Th(SO$_4$)$_2^0$, ThF$_2^{2+}$ and Th(HPO$_4$)$_2^0$ for pH < 4-5; Th(HPO$_4$)$_3^{2-}$ for pH between 4.5 to 7.5 and Th(OH)$_4^0$ for pH > 7.5. [Used with permission of Elsevier Science, from Langmuir and Herman (1980) *Geochim Cosmochim Acta*, Vol. 44, Fig. 7, p. 1761].

Radionuclide interactions with minerals. The adsorption of radionuclides onto mineral surfaces is so important that it can become a limiting factor for the mobility of radionuclides in surface and ground-water. Among the parameters that control elemental adsorption onto mineral surfaces are physical parameters such as temperature, cationic exchange capacity and specific area of minerals (Borovec 1981; Prikryl et al. 2001), but also the chemical characteristics of the solution: pH, ionic strength, organic and inorganic ligand concentrations.

Generally speaking, the capacity of U sorption onto mineral surfaces decreases from Fe-oxides and silica gels, to clays and micas and to opals (e.g., Ames et al. 1983a,c,d; Allard et al. 1999). For clay minerals, U retention is low for kaolinites and up to two orders of magnitude higher for illite and montmorillonite (Borovec 1981; Shirvington 1983). Bentonite (a smectitic clay) is also a good U-fixing clay mineral which seems to act as a catalyst for the reduction of adsorbed U(VI) to insoluble U(IV) by organic matter (Giaquinta et al. 1997). For Fe-oxides, a decreasing U retention capacity is reported from the amorphous Fe-oxyhydroxides to goethite and to hematite (e.g., Hsi and Langmuir 1985; Murphy et al. 1999). From this observation, Murphy et al. (1999) proposed that the crystallisation of amorphous Fe-oxides into better crystallised minerals could release part of the adsorbed U. This scenario would be compatible with mechanisms involved during U sorption onto amorphous Fe-oxides (e.g., Bruno et al. 1995; Murakami et al. 1997). The pH of the solution highly influences the retention capacity of U by minerals. This is illustrated by Figure 7 for muscovite and hematite, which fix nearly no U at pH < 3-4 and pH > 9, while they have a maximum U retention for pH 6-7 (see also Pabalan and Turner 1997; Pabalan et al. 1998). The presence of organic and inorganic ligands into the solution also modifies the fixation of radionuclides onto mineral surfaces. The occurrence of carbonate in solution inhibits, for example, U retention onto minerals such as biotite and Fe-oxyhydroxides (Ames et al. 1983a,c,d; Hsi and Langmuir 1985; Duff and Amrhein 1996). By contrast, humic acids significantly enhance U sorption onto hematite and clay minerals at acidic pH (e.g., Geckeis et al. 1999; Lenhart and Honeyman 1999; Murphy et al. 1999; Schmeide et al. 2000) (Fig. 7). These few examples highlight two properties of the behavior of U in the environment. This chemical element, acknowledged to be a very soluble element in oxidising conditions may look totally immobile in oxidizing waters because of its removal from water through concentration onto mineral surfaces. It is by such a mechanism that Kronfeld et al. (1994) explain the low U concentrations in waters of carbonate aquifers, namely a UO_2^{2+} adsorption onto carbonates through the formation of hydroxyl-carbonate species (Morse et al. 1984). The U immobilization has nevertheless to be viewed as a reversible state, in that U fixation onto mineral is very dependent of the physical and chemical conditions of the surrounding medium and can be totally modified if these conditions change. These two characteristics are not restricted to U alone. They are also found, but at different levels, for the other radionuclides, including Ra and, in some cases, Th.

Compared to U, Th and Ra are also very reactive chemical elements, since they are easily adsorbed onto mineral surfaces. Th adsorption onto Fe-hydroxides is higher than that of U, and can be strong, even at pH as low as 2. As for U, it is enhanced at acidic pH by organic matter (Cromières et al. 1998; Geckeis et al. 1999; Murphy et al. 1999). Ra is generally slightly less retained than U by Fe-oxides but better by micas and secondary clay minerals (Ames et al. 1983b,c,d). It can be also easily adsorbed onto Mn oxides (Herczeg et al. 1988) and zeolites (Sturchio et al. 1989). The relative affinity of U-Th-Ra nuclides for the Fe oxyhydroxides were indirectly confirmed by the results on the Kalix river samples (Andersson et al. 1998; Porcelli et al. 2001). In these water samples, the calculated partition coefficient between Fe oxyhydroxides and water of the three nuclides decreases in the following order: Th>U>Ra.

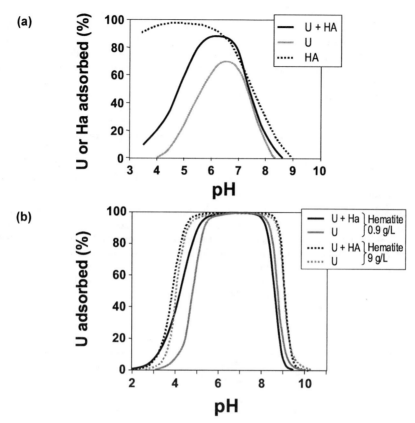

Figure 7. U(VI) sorption onto muscovite (7a, Schmeide et al. 2000) and hematite (7b, Lenhart and Honeyman 1999) in the absence (U) and in the presence of humic acid (U+HA). 7a: $[UO_2^+] = 1$ μmol/L, [HA] = 5 mg/L, muscovite content of about 1.2g/L. Complexation of U with HA in solution and onto mineral surface may influence U sorption. For instance, U sorption onto muscovite is enhanced in presence of HA at low pH. 7b: [U] = 1 μmol/L, [HA] = 10 mg/L. Hematite content in solution = 0.9 and 9g/L. Uptake of U increases with increasing hematite content. In presence of hematite, an increase of U sorption onto hematite is observed at low pH, especially at low hematite content.

Impacts of organic matter, micro-organisms and colloids. Many studies confirm the strong affinity of organic matter as a ligand for U and Th (e.g., Halbach et al. 1980; Nagy et al. 1991; Czerwinski et al. 1994; Zuyi and Huanxin 1994; Zuyi et al. 1996; Lenhart et al. 2000; Schmeide et al. 2000). Among the organic acids of soils, citric acid is a weaker ligand than fulvic and humic acids. The sorption of humic acid onto minerals at low pH can explain, at least partly, the increase in U retention onto minerals in the presence of organic matter.

Biological activity, especially microbiological activity in soils and waters, could also play an important role on the mobility of U. The geo-microbiology of U is today a fast-growing discipline. A recent review is given by Suzuki and Banfield (1999). Many of these studies have tried to characterize the interactions between microbiology and U. It was shown, that U(VI) can be reduced to U(IV) by iron-reducing and sulfate-reducing bacteria (e.g., Fredrikson et al. 2000; Lovley et al. 1991; Lovley and Phillips 1992; Spear et al. 2000) and that numerous mechanisms of U accumulation by microbes exist. Following Suzuki and Banfield (1999), they are sub-divided into metabolism-dependent and metabolism-independent mechanisms. In the first group, one can cite the enzymatic precipitation of uranyl phosphate, the enzymatic reduction of U(VI), or the complexation of U by chelating molecules such as siderophores, produced in response to metal stress. Metabolism-independent mechanisms reflect physico-chemical interactions between the cationic forms of U species and negatively charged microbial sites, and occur, whether the cells are or are not living (references in Suzuki and Banfield 1999). Some of these mechanisms, such as the fixation and the reduction of U, may decrease U mobility, whereas dispersion of micro-organisms in the environment or the complexation of U by chelating agents may increase its mobility in water. All these studies highlight the non-negligible, indeed central, role that the micro-organisms could have in cycling of U(-Th-Ra) in the surface environment. Its real impact on the mobility of radio nuclides in natural systems remains however to be quantified.

The fixation of radio-nuclides by colloids is another important mechanism influencing the fractionation of the U-Th series nuclides. Colloids have sizes ranging from 1 to 1000 nm. They are not retained by a classical frontal filtration membrane (Ø between 0.1 to 0.45 µm), and encompass a large variety of organic and inorganic components (Fig. 8). The inorganic colloids are generally composed of Fe and Al oxyhydroxides, clays and silica, while the organic ones are essentially humic and fulvic acids, as well as viruses and fragments of micro-organism such as bacteria and algae. The retention properties of colloids highly depend on their mineralogical and chemical composition, explaining that colloidal

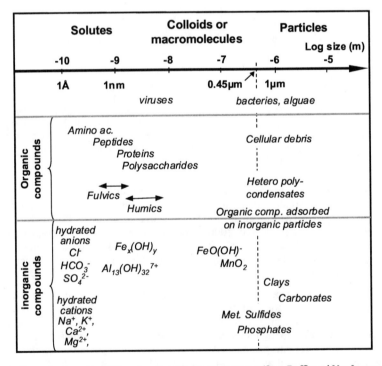

Figure 8. Size and nature of colloids and particles in aqueous systems (from Buffle and Van Leewen 1992).

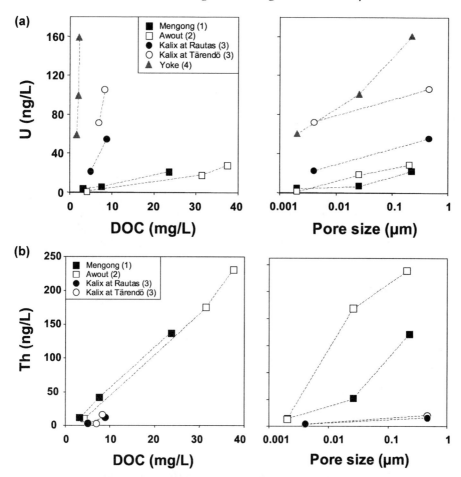

Figure 9. Variations of uranium (a) and thorium (b) contents in the filtrates of a sample as a function of the filtration size, and relation with the variations of the dissolved organic carbon (DOC). Data sources: (1) Viers et al. (1997), (2) Dupré et al. (1999), (3) Porcelli et al. (1997, 2001), (4) Riotte et al. (2003). Filtrates are recovered by tangential ultra-filtration. Low filtration sizes are usually given in Dalton—a molecular weight unit of 1 g/mol—and are ranging between 3 and 300 kD. These filtration sizes have been converted here into an approximate µm pore size.

The results highlight the important role played by organic colloids on Th transport in organic-rich waters (Mengong and Awout). They also point out various U concentration trends as a function of DOC depending of the filtrated sample. These variations were not discussed at this time. They could indicate the diversity of organic colloids involved in U transport, various contribution of inorganic colloids in these different water samples, or various U complexation properties according to the chemical composition of the waters. Future works are required to clarify this point.

retention can be very variable from one water sample to another. Impact of colloids on the mobility of radionuclides in natural waters has been detailed in numerous studies (e.g., Dearlove et al. 1991; Moulin and Ouzounian 1992). Their effect on U and Th transport has been illustrated by ultrafiltration experiments (Fig. 9) and will be more detailed in Section 4. It is important to stress here that the colloidal transport can totally modify the apparent mobility of a chemical element compared to what is expected from thermodynamic data. This is well illustrated with Th, a well known highly insoluble element, which may be yet significantly mobilised under colloidal form.

2.2. Alpha recoil

Alpha recoil is another mechanism that creates radioactive disequilibria. In this case, the radioactive daughter is mobilised from its initial position by the energy of alpha-decay. Besides the displacement of the recoil atom (20 nm or more, Kigoshi 1971), alpha-recoil can induce changes in the physico-chemical properties of the recoil atom and damage the host mineral. Alpha recoil can be an important way for releasing Th, Ra and Rn into groundwater (e.g., Sun and Semkow 1998). It also accounts for the ^{234}U- ^{238}U radioactive disequilibria, a general feature of the surficial processes, initially pointed out by Cherdyntsev et al. (1955). Damages to the crystal lattice and/or the oxidation of the ^{238}U(IV) to ^{234}U (VI) during the alpha decay, cause the ^{234}U atom to preferentially go into solution. Compared to the parent nuclide ^{238}U, the recoil atom ^{234}U is loosely held in its mineralogical site. The direct ejection of the recoiling nucleus from the solid phase to the fluid one is another process that accounts for ^{234}U enrichment in waters. This has been shown by Kigoshi (1971), who demonstrated the occurrence of direct ejection of ^{234}Th from zircon into an aqueous phase and its subsequent decay to ^{234}U. This mechanism is certainly important for explaining high activity ratios in groundwater (Osmond and Cowart 1982; Kronfeld et al. 1994; Tricca et al. 2001). Ejection of the recoiling nucleus from the solid phase is also called upon to account for the high (^{234}U/^{238}U) activity ratios observed in some mineral phases. Such high ratios have first been observed by Sheng and Kuroda (1984) in a sequence of leaching experiments conducted on a sample of U ore from the Oklo uranium deposit. Subsequently, very high U activity ratios have been reported for acid residues of a Colorado carnotite (Sheng and Kuroda 1986a,b). They would result from the recoil of ^{234}U from a U-rich phase into adjacent U-poor phases. Lowson et al. (1986) came to a similar conclusion on the basis of a study of U and Th activity ratios in mineral phases of a lateritic profile. This internal reorganisation of ^{234}U due to alpha recoil in rocks is certainly a process which will have to be considered in the near future to describe radioactive disequilibria in sedimentary rocks (Henderson et al. 2001) or in weathering profiles (von Gunten et al. 1999; Ludwig and Paces 2002). It could, for instance, significantly influence the operating procedure for dating these formations (e.g., Henderson et al. 2001; Ludwig and Paces 2002). The influence of this internal redistribution by alpha recoil on the other nuclides of U-series has now to be assessed.

In summary, alpha recoil, thermodynamic properties of the radionuclides in solution, sorption of radionuclides onto the mineral surfaces, and their complexation by organic matter in soils and by colloids in surface waters are important processes controlling the mobility of, and hence the fractionations between, the U-Th series nuclides within weathering profiles and in river waters. This explains why radioactive disequilibria are excellent tracers of water-rock interactions occurring during weathering and transfers in rivers and soils, and should help to study these interactions. However, quantifying weathering and transfer processes by radioactive disequilibria implies recognition of the different mechanisms that fractionate U-Th series nuclides in these surface environments. This stage, as shown here, is far from being complete, and certainly will require many studies in the coming years. Sound progress however has been made over the past 10 years.

3. RADIOACTIVE DISEQUILIBRIA IN WEATHERING PROFILES: DATING AND TRACING OF CHEMICAL MOBILITY

Uranium- and thorium-series disequilibria have been analysed in soils and weathering profiles in order to date pedogenic formations and to characterise the chemical mobilization processes within these profiles as well as their time constants.

3.1. Dating of pedogenic concretions

Uranium-series dating of soils mainly relies upon the application of ^{230}Th/U chronometry to pedogenic carbonates and silica (e.g., Ku and al. 1979; Rowe and Maher 2000; Ludwig and Paces 2002, and references therein) with only a few studies on ferruginous concretions (Short et al. 1989). The U-Th series dating methods are presented in this volume (Edwards et al. 2003; Ludwig 2003). Here we will focus on specific problems related to the dating of pedogenic concretions.

An initial Th-free material evolving in a closed-system after its deposition can be dated by measuring (^{230}Th/^{234}U) and (^{234}U/^{238}U). This is the classical method used to date carbonate rocks. However, its direct application to pedogenic carbonates is difficult because of the presence of initial detrital U and Th in the sample (see Edwards et al. 2003). Several methods have been proposed for correcting the detrital contaminant. All these methods rely on the hypothesis that the pedogenetic precipitates (carbonates, silica) are a simple mixture of two isotopically homogeneous end-members, with one containing only U isotopes at the time of deposition. Correction techniques try to determine the ^{230}Th/^{234}U and ^{234}U/^{238}U ratios of the Th-free end-member, and to use these ratios to calculate the age of the sample. Standard approaches based on sample leaching procedures developed for dating impure carbonates have been applied to pedogenic precipitates. The mathematical justification for these procedures as well as their range of applications have been presented by Ku and Liang (1984) and by Schwarcz and Latham (1989), and discussed by Kaufman (1993). Several studies have demonstrated the possibility of reliably dating pedogenic concretions by applying such correction methods (e.g., Ku and al. 1979; Rowe and Maher 2000). Nevertheless, it has also been shown that these correction methods could be affected by artefacts due to U and Th isotope fractionations during the leaching procedures, making it difficult to use such data for age determination. The in-sample ^{234}U redistribution from U-rich to U-poor phases via alpha recoil is probably another reason for obtaining U-Th data by leaching procedures without any age significance (Henderson et al. 2001; Ludwig and Paces 2002). Isochron technique using total-sample dissolution, initially proposed by Luo and Ku (1991) and Bischoff and Fitzpatrick (1991), is an alternative approach to date impure carbonates; its application to dating soil concretions is certainly possible (Peterson et al. 1995) but would not be systematic. Such techniques require the use of a series of cogenetic samples that vary in their relative proportion of detrital fraction (Fig. 10). As outlined by Rowe and Maher (2000) the latter two constrains are not systematically encountered in pedogenic concretions: many pedogenic carbonates have quite homogeneous distribution of detritus, and, on the scale of the sample, they are often formed by several generations of carbonate and silicate. The introduction of new analytical techniques including TIMS and ICP-MS could help solving such problems. The sample-size reduction that it implies should enable the analysis of very small samples with high enough U/Th ratios to calculate an age from the raw U-Th data without correction, on the basis of a single analysis per sample. This method is illustrated by Ludwig and Paces (2002) for pedogenic silica and carbonates. It could initiate new perspectives for soil age dating, which have to be evaluated in detail.

3.2. Characterization and time scale of chemical mobility in weathering profiles

A different approach for recovering weathering time information from U-Th series nuclides is to interpret depth variations of radioactive disequilibria in weathering profiles by simple but realistic modelling of U-Th series nuclides during formation and evolution

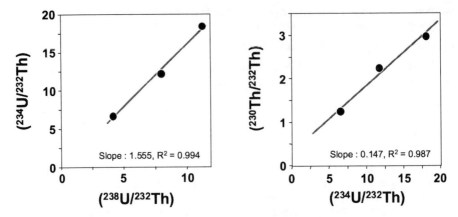

Figure 10. "Isochron plots" for caliche sample YM-U3 (Luo and Ku 1991). This sample is part of a laminar layer of carbonate and opal from an illuvial piemont unit in the Yucca Mountain. The data points represent results of the analysis of three subsamples of the same caliche. If these subsamples are assumed (a) to be coeval, (b) to be a mixture in variable proportion of an authigenic phase without Th and a detrital phase bearing Th, (c) to have the same initial $(^{230}Th/^{232}Th)_0$ ratio and (d) to evolve in closed-system, then they have to plot along a straight line in a plot of $(^{230}Th/^{232}Th)$ vs. $(^{234}U/^{232}Th)$ and $(^{234}U/^{232}Th)$ vs. $(^{238}U/^{232}Th)$. The slope of this arrays gives respectively the $(^{230}Th/^{234}U)$ ratios and $(^{234}U/^{238}U)$ of the pure authigenic end-member, which are the two parameters required to calculate age of the authigenic U (see equations and demonstration in Luo and Ku 1990, or discussion in Edwards et al. 2003).

of the profiles. Such an approach requires, as a preliminary, to recognise the main processes controlling the distribution of radionuclides in weathering profiles.

Distribution of U-Th-Ra in weathering profiles. The first U-Th studies (Pliler and Adams 1962; Rosholt et al. 1966; Hansen and Stout 1968) generally showed a U loss relative to Th at the base of the profiles, and an enrichment in the uppermost horizons and/or in some accumulation layers. The development of weathering studies, however, point out that this situation is not to be generalized and that reverse trends can be observed even at the scale of a single toposequence (Fig. 11).

During weathering processes Th is often supposed to be immobile, as illustrated for instance in the lower part of a syenite lateritic profile studied by Braun et al. (1993). Two main reasons have been proposed to explain the Th immobility: (1) Th is mainly contained in primary and secondary resistates, including relict accessory minerals, and possibly in clay minerals (e.g., Middleburgh et al. 1988; Braun et al. 1993); (2) Th is strongly complexed by or adsorbed onto neoformed minerals, such as clay particles and/or Fe-oxyhydroxides (e.g., Middleburgh et al. 1988; Gueniot et al. 1988a,b). However, such an immobility is not systematic. Migration and redistribution of Th within weathering profiles has been reported. From this point of view, the study by Kurtz et al. (2000) on Hawaiian soils is quite demonstrative (Fig. 12). The reported Th migration would be a consequence of the strong affinity of Th for organic matter, and might be controlled by the migration of the latter within the profile, an assumption in agreement with a scenario initially given by Hansen and Stout (1968).

Compared to Th, U is generally much more mobile. In weathering profiles, however, its behavior is different from that of the purely soluble/weatherable elements such as Na, K (e.g., Dequincey et al. 2002). Processes involved in the partial immobilization of U in

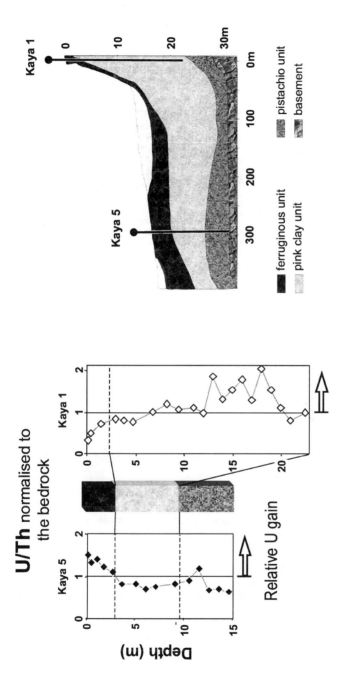

Figure 11. Distribution with depth of U/Th normalized to the bedrock in two lateritic profiles of the Kaya toposequence, about 300m apart (Burkina Faso) (Dequincey et al. 2002; submitted). One profile is located downhill (Kaya 5) and the other one at the top of a residual hill (Kaya 1). The laterite consists of an uppermost ferruginous hardtop, an intermediate pink clay unit and a lowest "pistachio" unit. For Kaya 5 profile, U/Th distribution shows a relative enrichment of U in the uppermost horizon and depletion in the lower part of the profile. This kind of distribution is quite common in weathering profiles but is not systematic as illustrated by the Kaya 1 profile. In the latter, a relative depletion of U is observed in the upper part and a U-enriched level in the intermediate horizon. This lateral difference in U distribution is explained by vertical redistribution of U from the ferruginous top to the underlying horizons, whose intensity is controlled by the evolution of the iron oxides from the uppermost horizons (Dequincey et al. submitted).

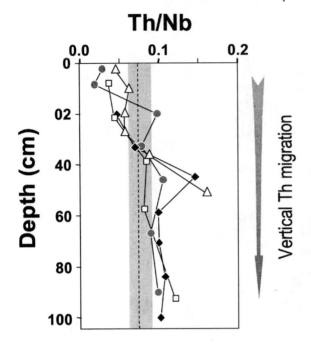

Figure 12. Profiles of Th to Nb ratios for four Hawaiian soils developed on lava flows with ages ranging from 20 ka to 4100 ka (Kurtz et al. 2000). In these soils, Nb is assumed to be an immobile element. Variations of Th/Nb ratios, with lower values than those of basalt (greyed area) in the upper part of the profile and higher values in the lower part indicate an internal downward migration of Th in these four weathering profiles. Circles = 20 ka Laupahoehoe site, triangles = 150 ka Kohala site, squares = 1400 ka Molokai site, diamonds = 4100 ka Kauai site.

soils and weathering horizons include: (a) concentration in relict accessory minerals and (b) adsorption or coprecipitation by amorphous Fe-oxyhydroxides, clays minerals and/or organic matter (e.g., Moreira-Nordemann and Sieffermann 1979; Tieh et al. 1980; Gueniot et al. 1988 a,b; Middleburgh et al. 1988). A case of such U immobilization can be found, for example, in Gueniot et al. (1988a,b). These authors studied aerated and hydromorphic profiles, and explained U accumulation by retention with Fe-oxyhydroxides in oxidised environments or by adsorption onto organic matter in reduced zones. The role of Fe-oxides is also outlined by Dequincey et al. (2002, in prep) in the case of the accumulation of U and Th in the indurated ferruginous cap of African laterites, and of their secondary redistribution into the profile (see Fig. 11).

The distribution of Ra in weathering profiles is less well documented than those of U and Th. However, high Th enrichments relative to U and Ra are observed at the scale of the profile (Greeman et al. 1990, 1999). In the A horizon of soils, the cycling of Ra by vegetation and its retention by humified organic matter explains the observed high Ra accumulation relative to U and Th (Fig. 13). In the other horizons, Ra behaves mainly like U, as observed by Bonotto (1998) for a Brazilian weathering profile.

The distribution of U, Th and Ra in weathering profiles depends on many parameters such as their initial localisation in the parent material, the retention capacity of the primary relict and secondary minerals. It also depends on the presence of organic and inorganic ligands (e.g., humic acids, carbonates) in the environment. Weathering of the bedrock is undoubtedly a major process of U-Th-Ra fractionation in soils. However, it is not the only one. As now well recognised, the occurrence of secondary remobilization and mineral recrystallization also induces redistribution of these elements in the profiles, so that the observed distribution may reflect the superimposition of all these fractionation processes. If these processes are "recent" (i.e., <1.2Ma) they will induce radioactive disequilibria among U-Th series nuclides, which, in return, can potentially be used as a

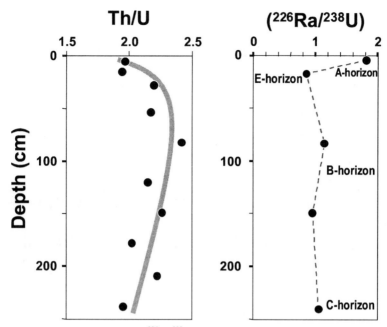

Figure 13. Variation of Th/U ratio and (^{226}Ra/^{238}U) activity ratio in a Central Pennsylvania soil (site 14-80) derived from dolomite (Greeman et al. 1990). The relative enrichment of Ra in the top horizon of the soil is attributed to the retention of Ra by humified soil organic matter.

tracer and chronometer of the processes. Such an approach has mainly been developed for ^{238}U-^{234}U-^{230}Th disequilibria.

Modelling of recent mobility by ^{234}U-^{238}U-^{230}Th disequilibria. Variations of ^{238}U-^{234}U-^{230}Th radioactive disequilibria with depth in weathering profiles are, generally, not interpretable in terms of a simple U-Th fractionation front at the base of the profile and a return to secular equilibrium above (closed system behavior). Instead, these variations indicate that geochemical fractionation takes place in all or many of the profile levels and affects the U-Th radioactive series (Fig. 14). The chronometric interpretation of U-Th disequilibrium variations with depth then requires the use of U mobility models.

Assuming that (^{234}U/^{238}U) fractionation is negligible relative to (^{230}Th/^{238}U) variations, Boulad et al. (1977) proposed a simple mathematical modelling of (^{230}Th/^{238}U) variations with depth in a laterite from Cameroon. These authors assumed the occurrence of two main U-Th fractionation fronts: one at the base of the weathering profile with U loss, and one in the upper part of the profile where U is released and redeposited deeper in the profile (Fig. 15). By fitting theoretical curves to the data, they estimated weathering rates ranging from 50 to 70 mm/ka. An estimate of 50 mm/ka was also obtained with a similar approach by Mathieu et al. (1995) for a Brazilian laterite.

When both (^{234}U/^{238}U) and (^{230}Th/^{238}U) activity ratios are studied, discussion and interpretation of the data are often presented as a plot of (^{234}U/^{238}U) against (^{230}Th/^{238}U) activity ratios (Thiel et al. 1983; Osmond and Ivanovich 1992) (Fig. 16). In such a diagram, instantaneous U gains and losses are represented by straight-line vectors and the radioactive decays by curved lines. Due to the relative decay constants of the nuclides, a

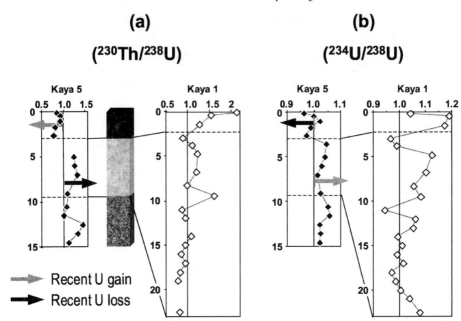

Figure 14. Variation of $(^{230}Th/^{238}U)$ and $(^{234}U/^{238}U)$ activity ratios with depth in the two lateritic profiles of the Kaya toposequence (Burkina Fasso) (see description in Fig. 11). These variations cannot be interpreted by a simple model with a closed-system evolution of ^{238}U-^{234}U-^{230}Th nuclides above a fractionation zone. $(^{230}Th/^{238}U)$ and $(^{234}U/^{238}U)$ activity ratios indicate the occurrence of recent U mobilizations in all or many of the profile levels. It can be noted that levels where $(^{234}U/^{238}U)$ a.r. would indicate quite systematically U gains (resp. U loss) are levels where $(^{230}Th/^{238}U)$ indicate U losses (resp. U gain), suggesting that each level of the profile is affected by both recent U gains and losses. A model with two concomitant and antagonist U fluxes was therefore proposed to account for the ^{238}U-^{234}U-^{230}Th variations in the Kaya toposequence (Dequincey et al. 2002; submitted) (see also Fig. 18).

sample in this diagram cannot cross the $(^{234}U/^{238}U)$ and $(^{234}U/^{230}Th)$ equilines by radioactive decay. The diagram can be therefore subdivided into four different quadrants. A system affected only by U gains or losses relative to Th can only plot in two of these four quadrants, the accumulation quadrant with $(^{234}U/^{238}U)>1$ and $(^{230}Th/^{234}U) <1$, and the leaching quadrant with reverse ^{238}U-^{234}U-^{230}Th characteristics, respectively. No sample can plot or evolve in the two other quadrants, which are therefore called the forbidden zones or complex zones (Fig. 16). It clearly appears from the U-Th studies on weathering profiles that data points can plot in all of the quadrants of the $(^{234}U/^{238}U)$ vs. $(^{230}Th/^{238}U)$ diagram (Fig. 17a,b,c). This implies that ^{238}U-^{234}U-^{230}Th disequilibria in soils and weathering profiles cannot generally be interpreted in terms of single U accumulation or leaching processes. More complex scenarios have to be involved.

A first approach for interpreting the pattern of evolution of radioactive disequilibria in sedimentary and weathering formations has been to use qualitative and empirical laws for the various processes that fractionate ^{238}U-^{234}U-^{230}Th nuclides in sub-surface environments. This approach is known as the U-trend dating method (see Szabo and Rosholt 1982; Rosholt 1985). A summary of this approach and related results can be found in Rosholt (1982) and in Osmond and Ivanovich (1992). This method was mainly applied for dating sedimentary formations, rather than soils or weathering profiles.

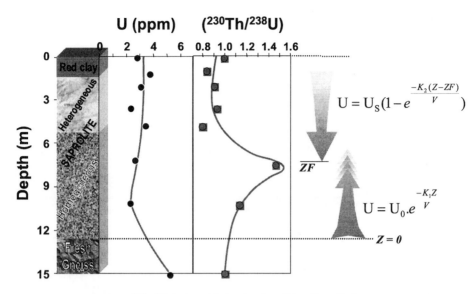

Figure 15. Interpretation of ^{238}U-^{230}Th disequilibrium in a ferrallitic soil profile from Cameroon. The model assumes the occurrence of two U fluxes (Boulad et al. 1977): a progressive U loss from the bedrock to the top and a progressive U accumulation from the top to a maximum depth ZF. These gain and loss functions are described as first order kinetic laws with kinetic coefficient K_1 and K_2 respectively. U content is corrected assuming that Th is conservative ($U_{corr} = (U/Th)_{gneiss}.Th$). Z = distance from weathering front to horizon, ZF = maximum depth of U accumulation, U_s = maximum concentration of adsorbed U.

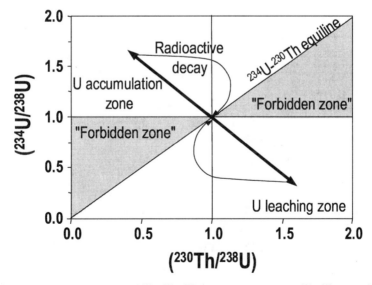

Figure 16. Theoretical evolution of the ^{238}U-^{234}U-^{230}Th disequilibria in the (^{234}U/^{238}U) vs. (^{230}Th/^{238}U) activity ratio diagram, for a simple U mobilization event (see text).

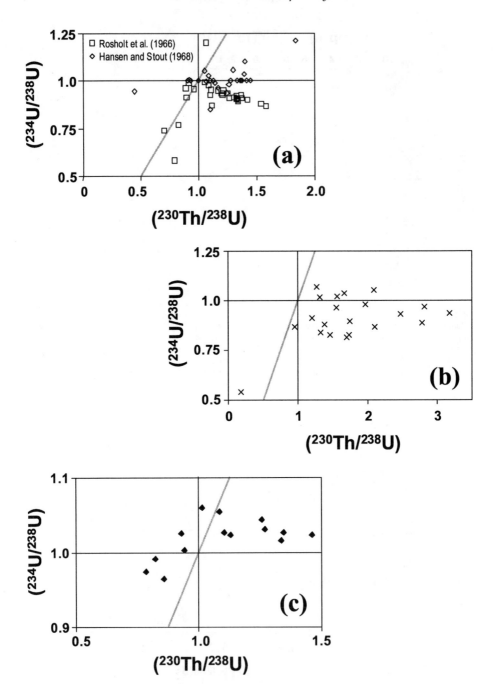

Figure 17. ^{234}U/^{238}U and ^{230}Th/^{238}U activity ratios in different weathering profiles: (a) Rosholt et al. (1966) and Hansen and Stout (1968), soils from USA; (b) Mathieu et al. (1995), laterite from Brazil; (c) Dequincey et al. (2002), laterite from Burkina Faso.

Thiel et al. (1983) and Scott et al. (1992) have proposed more quantitative approaches, based on mathematical modelling of instantaneous and/or continuous U gain and loss. These models are applicable to weathering and account for the position of the data in the "forbidden" zones of the (^{234}U/^{238}U) and (^{230}Th/^{238}U) diagram. Several studies used this approach to interpret ^{238}U-^{234}U-^{230}Th data in soils and weathering profiles (e.g., Ghaleb et al. 1990; Goetz 1990; Hillaire-Marcel et al. 1990; Dequincey et al. 2002). The models retained in the various studies differ from each other in the relative chronology assumed for U gains and losses. All the models have assumed, however, first order kinetics for U losses, which is supported by a few U leaching experiments on both whole rocks and separated minerals (Latham et al. 1987a; b).

Ghaleb et al. (1990) and Hillaire-Marcel et al. (1990) pointed out the potential of the modelling of U recycling in weathering profiles for estimating physical erosion rates. The authors studied the behavior of U in the soils of an endoreic basin of Syria, where U is brought into the soil, remobilised and redeposited deeper in the profile. The modelling of this U cycle and the fit with the observed (^{230}Th/^{238}U) variations with depth constrain the residence time of U in the profile and gives a current erosion rate of 10-20 mm/ka. Dequincey et al. (2002) have proposed to explain the position of the Kaya laterite samples in the complex regions of the (^{234}U/^{238}U) vs. (^{230}Th/^{238}U) diagram (Fig. 17c) with a scenario assuming the occurrence at each level of the laterite sections of continuous U gains and losses. The scenario and the related equations are presented in Figure 18. Its application to two lateritic profiles of the Kaya toposequence about 300 m apart (Burkina Faso) (data presented in Fig. 14 and 17) shows that the weathering profile is not presently at steady state with respect to the ^{238}U-^{234}U-^{230}Th disequilibria. It also indicates that the model coefficients cannot be considered as constant in any part of the profile (Dequincey et al. 2002, in prep). The U mobilization parameters would in fact vary with time and depth. Estimation of time constants for erosion and weathering processes (e.g., propagation rate of the basal weathering front) would then require selecting realistic variation laws for U gain and loss parameters.

The study of separate mineral phases or of granulometric fractions is another approach which can be used to recover temporal information from ^{238}U-^{234}U-^{230}Th radioactive disequilibria in weathering profiles. Such approaches rely on the assumption that the fractions only contain or concentrate minerals phases specific of a single or of few stages of formation and evolution of weathering profiles, and hence can help to characterise the time constants of the corresponding stages.

Using a procedure of chemical sequential separation, Lowson et al. (1986) and von Gunten et al. (1999) separated three main phases in laterites from Australia: amorphous Fe-oxides, crystalline Fe-oxides and resistates. These authors show that Fe-oxides are the main U-bearing phases and that amorphous Fe-oxides are in adsorption/desorption equilibrium with the local waters. Moreover, they observed a progressive enrichment in the nuclides produced by their parent nuclide by alpha-decay, from amorphous to crystalline Fe-oxides, and from Fe-oxides to resistates. They proposed that this could result from a daughter nuclide incorporation into the "enriched" phases during alpha-recoil. These studies confirm therefore that the different phases of weathering profiles could behave independently from each other, with respect to U-Th disequilibria, and could record different weathering steps or processes. Unfortunately, the secondary redistribution of U series nuclides by alpha recoil certainly highly restrains the chronometrical potential of chemically separated phases.

Physical separation of granulometric fractions by sedimentation and ultracentrifugation could also help in constraining weathering rates at local scale. The fine fractions are useful when they exclusively contain secondary minerals, that is, when they

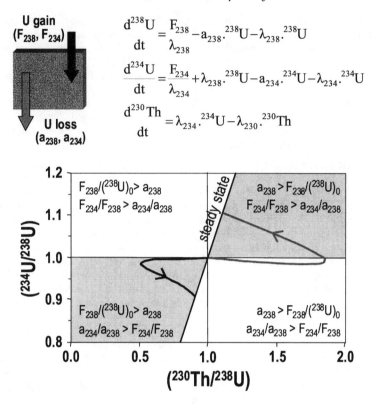

Figure 18. Modelling of the ^{238}U-^{234}U-^{230}Th nuclide evolution in a system affected by two continuous and antagonist U mobilization processes, with different U fractionation intensities (Dequincey et al., submitted). F_{238} and F_{234} are the activity inputs of ^{238}U and ^{234}U per time unit; a_{238} and a_{234} are the time constants of ^{238}U and ^{234}U output. Such a scenario can account for the data points in all parts of the U-Th diagram. The evolution of the system relies on two main parameters: the intensity of U fractionations (F_{234}/F_{238} and a_{234}/a_{238}) and the intensity of the U input flux ($F_{238}/(^{238}U)_0$) relative to the output flux (a_{238}). When time increases, the system eventually reaches a steady state, i.e., a point along the equiline (^{234}U = ^{230}Th).

are "not polluted" by inherited U. In this case the fractions will only bring information related to the weathering processes. The study of the Kaya toposequence in Burkina Faso, underscores the interest of the fine fractions (<0.2 μm) for the characterisation of the more recent events which affected the profiles (Dequincey et al. 2002, in prep). The study of another African profile, the Goyoum latosol, Cameroun, suggests, at least in this example, that the <0.2 μm granulometric fractions could evolve as a closed system above fronts where U fractionates and which migrate downwards. A migration rate of 8 mm/ka, assumed to correspond to the weathering front of the bedrock, is inferred from the observed progressive (^{234}U/^{238}U) return to equilibrium (Dequincey et al. 1999).

The distribution and mobility of the ^{238}U-series nuclides in weathering profiles not only stems from the distribution and properties of its constitutive mineral phases, but also from the U fluxes controlled by climatic conditions, groundwater level variations or local drainage conditions. Many models can then be proposed to account for the variations of ^{238}U-^{234}U-^{230}Th disequilibria in weathering profiles. Fitting a model with the observed disequilibria constrains some parameters of the model, such as weathering rates or elemental residence time in soils. It is obvious that the same data set can generate different estimates depending on the model used. The choice of model is therefore crucial for calculating realistic weathering parameters from U-Th series nuclides. Correct recognition and characterisation of the various mobilization processes affecting weathering profile becomes then a fundamental prerequisite for correctly interpreting U-Th radioactive disequilibria in soils and weathering profiles. More often than not such information will be acquired by combining different data sets including mineralogical and chemical data.

4. TRANSPORT OF U-Th-Ra ISOTOPES IN RIVER WATERS

As discussed elsewhere in this volume, another field where the use of U-series disequilibria has proven to be very useful is the study of chemical transport in waters, either marine (see Cochran and Masque 2003; Henderson and Anderson 2003), estuarine (Swarzenski et al. 2003) or continental waters. In the continental domain, in addition to characterization of transfer processes related to groundwater flows (Porcelli and Swarzenski 2003), radioactive disequilibria have also helped in constraining chemical exchanges between particulate, dissolved and colloidal loads of waters, as well as the origin of chemical fluxes carried by waters.

4.1. Transport of uranium isotopes in river waters

In surface waters, which are nearly always oxygenated, U occurs in its soluble form, i.e., U^{6+}. Uranium often forms complexes with carbonates such as $UO_2(CO_3)_3^{4-}$ or $UO_2(CO_3)_3^{2-}$ and also with hydroxyl and phosphate (see Section 2.1 and reviews in Gascoyne 1992; Langmuir 1997; Murphy et Shock 1999). A significant proportion of U in rivers is, however, transported as suspended particles, recovered by filtration at <0.1-0.45 µm. For instance, in rivers of the Congo basin, 40 to 60% of total river U budget is associated with the suspended sediments (Dupré et al. 1996). Also, in the Kalix river, a particle poor, Fe/Mn-rich river in northern Sweden, particles >0.45 µm contain 20-50% of the riverine U budget (Andersson et al. 1998). In the latter case, the main part of U contained in particles (i.e., 60 to 90%) is in nondetrital phases. It is incorporated in, or complexed with, secondary Fe-oxyhydroxides, which either form secondary particles or form coatings on detrital particles. Isotopically, this uranium is shown to exchange rapidly with the ambient "dissolved" uranium (Andersson et al. 1998). The importance of secondary minerals as carrier of U in riverine particles and the importance of sorption processes in controlling the U series isotopic characteristics of the suspended sediments were also argued by Plater et al. (1992) for English rivers. A large part of U carried by particles could be, therefore, in close relation with "dissolved" U, i.e., with U in the 0.45-µm-filtered waters. Even in the so-called "dissolved" fraction, U is not in a purely dissolved form. Fixation of U by colloids is indeed important and can represent a large part of U in filtered waters. Organic colloids are recognized to have a higher complexation capacity than inorganic ones (e.g., Short et al. 1988; Viers et al. 1997; Porcelli et al. 1997; Riotte et al. 2003). The complexation of U by organic colloids is greater for humic acids than for fulvic ones (Lenhart et al. 2000) and is pH-dependent (Read et al. 1993, Lienert et al. 1994). In organic-rich waters, up to 75-90% of "dissolved" U can be in a colloidal form (e.g., Viers et al. 1997; Dupré et al. 1999) (see also Fig. 9a). Porcelli et al (1997) reported for the Kalix river samples containing around 8-9 mg/L of dissolved organic carbon (DOC), that between 30 to 90% of U is bound to organic colloids but also that at least 10% of the colloidal U fraction is scavenged by Fe-oxyhydroxides. Although low, the inorganic colloidal U fraction might have a major role in the U exchanges between dissolved and solid phases of rivers, as possibly suggested from results on particles (Andersson et al. 1998). Influence of

U colloidal transport in organic-poor surface waters has been far less studied. Riotte et al. (2003) reported U losses from 0 to 70% during ultrafiltration experiments for surface waters of Mount Cameroon without nearly any DOC. Even in the low concentration waters, U can be significantly fractionated from other soluble elements by the occurrence of a colloidal phase, probably inorganic in origin. However, such fractionations are not systematic because of the occurrence of various colloidal phases, characterised by different physical and chemical properties, and hence different sorption and/or complexation capacities (Section 2.1).

The role of the colloidal transport on U mobility in rivers is therefore not negligible, and makes it difficult to precisely determine the truly-dissolved U carried by rivers. Until now, this parameter has not been taken into consideration when estimating the global dissolved U flux carried by rivers to the oceans. Estimates have been made by using U data in water fractions filtrated below 0.45-0.1 μm. From these studies, it appears that U content in rivers ranges from 0.01 to 100 μg/L (e.g., Osmond and Ivanovich 1992). The global average is about 0.3 μg/L (Palmer and Edmond 1993). Mangini et al. (1979) and Scott (1982) observed a correlation between alkalinity and U for several major world rivers. In contrast, Palmer and Edmond (1993), who compiled results from more than 250 rivers, observed that rivers like the Ganges, the Brahmaputra and the Huang Ho have high U contents without any link with alkalinity. Furthermore, in the Amazon, Orinoco and Ganges, U contents are correlated with the sum of cations. For the Amazon and Orinoco, the high U end-member corresponds to chemical fluxes associated with carbonate dissolution. For the Ganges, the high U-concentration end-member has initially been interpreted as a U flux from both oxidation of U-rich phases in black shales and carbonate dissolution (Sarin et al. 1990). The latter assumption was only partly confirmed by studies of Himalayan rivers (Chabaux et al. 2001), since, in these streams, the main lithologies contributing to U fluxes are black shales and high-elevation leucogranites. In contrast, carbonates would only play a minor role in the dissolved U flux of these Himalayan rivers.

^{238}U-^{234}U disequilibria in river waters. The first studies of U-isotope ratios in river waters were mainly undertaken to provide information about the U mass balance in the oceans (e.g., Ku et al. 1977; Martin et al. 1978b; Mangini et al.,1979; Borole et al. 1982; Mac Kee et al. 1987; Caroll and Moore 1993). Subsequently, several studies focused on the source of riverine uranium (e.g., Sarin et al. 1990; Pande et al. 1994; Riotte et al. 1999; Chabaux et al. 2001) and on speciation and transport of radionuclides in rivers (e.g., Andersson et al. 1995; 1998; Porcelli et al 1997, 2001). About 30 main rivers have been analyzed for their ($^{234}U/^{238}U$) ratios (Table 1). The data reveal a wide range of U activity ratios in the world rivers, from 1.02 to 2.59 (see also Dunk et al. 2002).

Comparison of U activity ratios in rain water with variations of U ratios in river and stream waters, collected in the same geographical area and at different rain periods over a yearly cycle, confirm the negligible effect of rain water on the U budget of freshwaters. Such a comparison was made for Himalayan and Mount Cameroon streams (Chabaux et al. 2001; Riotte et al. 2003), and also for a small Vosges Mountain watershed, the Strengbach, France (Fig. 19). In earlier studies, fertiliser-derived uranium has been suggested as a potential source of U in river waters. However, its impact on U budget of surface waters was shown to be negligible (Mangini et al. 1979; Riotte et al. 1999), except in specific areas characterised by a high agricultural activity along with a very low natural concentration of U, as reported by Zielinski et al. (2000) in the Florida Everglades, USA. Moreover, this influence is negligible at the global scale (Mangini et al. 1979, Palmer and Edmond 1993). The immediate consequence of the negligible effect of atmospheric and anthropogenic contributions on U budget of freshwater, is that U activity ratios of dissolved load of rivers can be seen as a specific tracer of chemical fluxes

Table 1. Available data on dissolved U and $(^{234}U/^{238}U)$ ratios in world rivers (from Chabaux et al. 2001). The data are presented in decreasing order of U flux. Analytical precision are ± 2 σ for samples measured by TIMS (references 1 and 13), and ± 1 σ for those obtained by alpha counting.

River	Location	Sample	Discharge 10^9 m³/y	U µg/L	U flux 10^6g/y	$(^{234}U/^{238}U)$	precision	Ref.
Indus	Thatta Bridge	25	238	4.94	1176	1.026	0.0047	(1)
Ganges	Rajshahi	BGP 4-65	450	2.0	900	1.050	0.005	(1)
Yangtse Kiang	Nankui	-	694	1.1	763	1.351	0.0057	(1)
Brahmaputra	Chilmari	BGP 15-79	612	1.0	612	1.034	0.005	(1)
Mississippi	La Place	-	530	0.94	500	1.31	0.07	(2)
Huang Ho	Jimau	-	49	7.5	368	1.318	0.0045	(1)
Amazon	Santarem	-	5400	0.04	217	1.10	0.06	(2)
Rhine	Germany	-	80	2.48	198	1.29	0.07	(3)
Mackenzie	Canada	-	306	0.5	153	1.40	0.10	(4)
Saint Lawrence	Quebec	mean QU -	450	0.29	130	1.18	0.01	(5)
Congo	Zaïre	22	1240	0.08	100	1.09	0.07	(6)
Godavari	Rajahmundry	-	84	0.78	65	1.38	0.03	(7)
Zambezi	Tete	3	224	0.26	43	1.19	0.10	(8)
Krishna	Vijayawada	-	35	1.08	38	1.58	0.04	(7)
Rhône	Arles	R1	52	0.56	29	1.086	0.006	(1)
Wistula	Poland	-	32	0.72	23.2	1.259	0.003	(9)
Garonne	La Réole	7 values	30	0.75	23	1.16	0.005	(1)
Narmada	Broach	mean NB	22	0.84	19	1.39	0.04	(10)
Rio Grande	Del Rio	-	3.9	4.5	18	1.74	0.03	(11)
Mahanadi	Asia	-	67	0.25	17	1.30	0.03	(10)
Limpopo	Messina	11	5	2.8	14	1.50	0.14	(8)
Brazos	America	-	6.75	1.6	11	1.22	0.09	(12)
Loire	Nantes	L1	27	0.37	10	1.145	0.0064	(1)
Orange	Upington	15	12	0.79	9.5	2.59	0.15	(8)
Charente	France	CH 04-33	3	2.0	6	1.22	0.05	(13)
Cauveri	Trichi	-	7.4	0.58	4.3	1.28	0.03	(7)
Seine	Rouen	S2	8	0.53	4.2	1.107	0.0045	(1)
Tapti	Surat	mean TP -	11.4	0.36	4.1	1.26	0.04	(10)
Kalix	Sweden	Kalix 1-4	9	0.17	2	1.659	0.007	(9)-(14)
Suwannee	Live Oak	-	11	0.15	1.7	1.90	0.11	(11)
Susquehannah	Gettysburg	-	32	0.04	1.3	1.31	0.09	(11)
Fenland	UK	-	1.1	0.92	1	1.276	-	(15)
Clyde	UK	-	0.39	0.16	0.06	1.63	0.06	(16)
Forth	UK	-	0.4	0.09	0.04	1.50	0.13	(16)
Tamar	UK	-	0.57	0.04	0.02	1.44	0.36	(16)

	Discharge 10^9 m³/an	U µg/L	U flux 10^6g/an	$(^{234}U/^{238}U)$
Mean	10 734	0.51	5459	1.171

References for Table 1: (1) Chabaux et al. (2001), (2) Moore (1967), (3) Mangini et al. (1979), (4) Vigier et al. (2001), (5) Durand and Hillaire-Marcel (pers. com.), (6) Martin et al. (1978a), (7) Bhat and Krishnaswami (1969), (8) Kronfeld et Vogel (1991), (9) Andersson et al. (1995), (10) Borole et al. (1982), (11) Scott (1982) or ref. in Scott (1982), (12) Sackett et Cook (1969), (13) Martin et al. (1978b), (14) Ingri et al. (2000), (15) Plater et al. (1992), (16) Toole et al. (1987). Discharge data from Meybeck (1984) and Palmer and Edmond (1989).

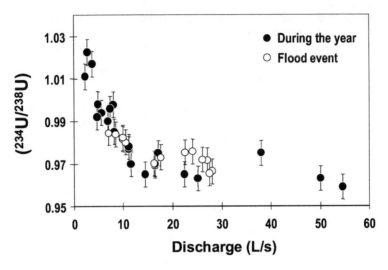

Figure 19. $(^{234}U/^{238}U)$ ratios as a function of discharge in the Strengbach stream (Riotte and Chabaux 1999 and unpub. data). The decrease of $(^{234}U/^{238}U)$ ratios when the discharge increases cannot be explained by the $(^{234}U/^{238}U)$ ratios in local rainwater, which range from 1.16 to 1.19 (unpub. data). This decrease represents contribution of chemical fluxes coming from various levels of the weathering profile.

coming from rocks and soils (e.g., Riotte and Chabaux 1999). This interest is certainly reinforced by the observation that $(^{234}U/^{238}U)$ ratios of dissolved and colloidal fractions of streams, in organic-rich waters (Porcelli et al. 1997) as well as in very dilute and organic-poor waters (Riotte et al. 2003) are similar. $(^{234}U/^{238}U)$ ratios in waters is then independent from filtration size of samples. Sarin et al. (1990), based on their studies on Himalayan rivers, suggested that the intensity of ^{238}U-^{234}U disequilibrium in dissolved load of stream waters depends on the lithologic nature of the formations found in the watersheds. This was well illustrated by Riotte and Chabaux (1999), at the scale of the Strengbach watershed in the Vosges Mountain. The authors observed a systematic variation of the U isotopic ratio in stream waters for each lithology, and a covariation of U and Sr isotope ratios along the stream. For the Himalayan rivers, Chabaux et al. (2001) confirmed the initial suggestion by Sarin et al (1990), and showed how coupling of U with Sr isotopic data permits estimating the contribution of a floodplain component to the U and Sr budgets of the Ganges and the Brahmaputra (Fig. 20). These studies highlight the potential of U isotopes as hydrological tracer of the origin of the chemical fluxes carried by rivers, and the great potential of combined U-Sr investigations for understanding and quantifying these fluxes.

The relationship between U activity ratios and basin geology was first explained in terms of variations of weathering processes with the geology of the drainage basin. As carbonate weathering would be congruent whereas silicate weathering rather incongruent, and hence susceptible of producing a preferential leaching of ^{234}U into the weathering fluids, small U disequilibria would be expected in rivers with dominant carbonate weathering and high U disequilibria in river with dominant silicate weathering. (Sarin et al. 1990; Plater et al. 1992). However, current U data on river waters demonstrate that this model is oversimplified. It was observed, for instance, that waters draining granitic basements may have a small ^{234}U-^{238}U disequilibrium of only a few percent (e.g., Riotte

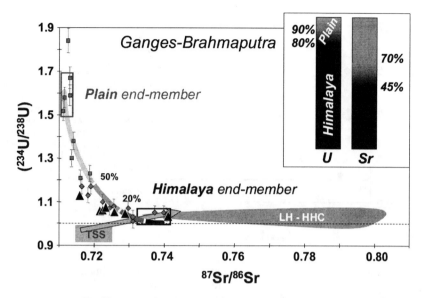

Figure 20. Plot of $(^{234}U/^{238}U)$ ratios against Sr isotope ratios for water samples of Himalayan streams draining the Tethyan Sedimentary Series (TSS) and the Lesser Himalaya-High Himalaya Crystalline (LH-HHC), and for river samples of the Ganges plain. The arrow represents the general trend that the Himalayan streams define between the TSS and the LH-HHC end-members. The elemental U and Sr fluxes carried by the Himalayan rivers at the outflow of the Highlands are homogeneous at the scale of the Himalayan range which permits to define an Himalayan end-member. Rivers flowing on the Indian plain define a different trend from that of the Himalayan rivers, indicating the contribution of a specific floodplain component to the U and Sr budgets of the Ganges and the Brahmaputra. In insert: results of a simple mass balance calculation using these two end-members; the U budget of the Ganges in Bangladesh is strongly dominated by Himalayan contributions whereas the Sr flux is much more affected by the "Indian plain end-member" (details in Chabaux et al. 2001).

and Chabaux 1999; Chabaux et al. 2001). In addition, explaining U disequilibria in river waters in terms of carbonate dissolution and silicate weathering alone should induce negative correlation between U activity ratios and Ca/Na ratios for the river dissolved loads. This is not observed at a global scale: no simple relationship exists between U isotope and Ca/Na ratios (Fig. 21). A reverse correlation is even observed in some streams, such as the Kali Gandaki river in the Himalayas (Chabaux et al. 2001). All these results indicate that the lithological control on $(^{234}U/^{238}U)$ ratios of dissolved loads of rivers cannot be directly related to the petrological nature of the weathered material. Detailed studies of several watersheds have underscored the significant influence of exchanges between river and deep groundwaters on U budget of rivers (Briel 1976, in Osmond and Cowart 1982; Lienert et al. 1994; Porcelli et al. 1997; Riotte and Chabaux 1999; Riotte et al. 2003). Based on these observations, it was suggested that the high ^{238}U-^{234}U disequilibria in river waters were associated with inputs from groundwater which generally have high activity ratios (Riotte and Chabaux 1999; Chabaux et al. 2001). Accordingly, relationship between lithology and $(^{234}U/^{238}U)$ ratios are indirect. A key parameter for high activity ratios in river waters is the capacity of the substratum to store water in aquifers connected with surface waters, which is clearly dependent upon the lithology of the drainage basin. Consequently, the ^{234}U-^{238}U disequilibrium in surface waters might become a relevant and useful tool to distinguish in the global chemical

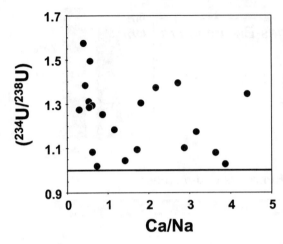

Figure 21. (^{234}U/^{238}U) ratios against Ca/Na molar ratio in the dissolved phase of the major world rivers. Uranium data, see Table 3; Ca and Na data from Gaillardet et al. (1999) and Meybeck (1984). The lack of negative correlation between these two ratios suggests that the lithological control on (^{234}U/^{238}U) ratios of dissolved loads of rivers cannot be directly related to the petrological nature of the weathered material (carbonate dissolution/silicate weathering).

fluxes carried by rivers, those derived from surface water-rock interactions from those induced by groundwater inputs. In the future, it will certainly help to characterise the different levels of water-rock interactions controlling the chemical fluxes of rivers.

The studies of U in river waters has enabled to assess the mean activity ratio of the dissolved U flux carried by rivers to ocean. The value of 1.17, based on 50% of the global exported U flux (Table 1), is in the lower range of older estimates based on smaller data sets (e.g., Mangini et al. 1979; Borole et al. 1982). Such an estimate is too low compared to the (^{234}U/^{238}U) ratios of seawater. The point has been already raised by several authors (e.g., Ku et al. 1977; Borole et al. 1982). It relies on a simple (simplistic?) U oceanic cycle assuming both (1) that the only dissolved U flux to oceans is the river input, and that the hydrothermal systems and the reducing margin sediments are the major sinks, and (2) that the system is at steady state. Under these conditions, assuming a U activity ratio of the riverine flux that is constant with time and using the present-day sea water activity ratio of 1.144 (Henderson et al. 1993; Henderson 2002), the theoretical mean riverine activity ratio should range between 1.25-1.35 instead of the estimated value of 1.17. Several assumptions were presented to explain this apparent contradiction. It was proposed for instance, that there is an additional source of excess-^{234}U such a U supply from groundwater, or a diffusion of ^{234}U from marine sediment pore-waters (e.g., Ku et al. 1977; Borole et al. 1982), or that the mean (^{234}U/^{238}U) activity ratio of rivers is still underestimated (e.g., Kronfeld et Vogel 1991). Alternatively, it could be considered that the U seawater cycle is not at steady state, or that U isotope ratios of world rivers has changed with time in response to climatic variations. The latter scenario has been detailed in Riotte (2001). It is presented in Figure 22 and elsewhere in this volume (Henderson and Anderson 2003).

4.2. Transport of thorium and radium isotopes

Thorium is a highly insoluble element, mainly carried in the particulate form in river waters. This is well shown by Th data for the MacKenzie river (Vigier et al. 2001) and for the Kalix river (Andersson et al. 1995; Porcelli et al. 2001): in both cases, more than 95% of Th is carried by >0.45 μm particles. An important part of this Th is included within detrital material. This is illustrated by sequential extractions performed on sediments from the Witham river (Plater et al. 1992), which show the very low amount of Th in ion-exchangeable and organic-bound fractions compared to Th in Fe-Mn oxides

(a)

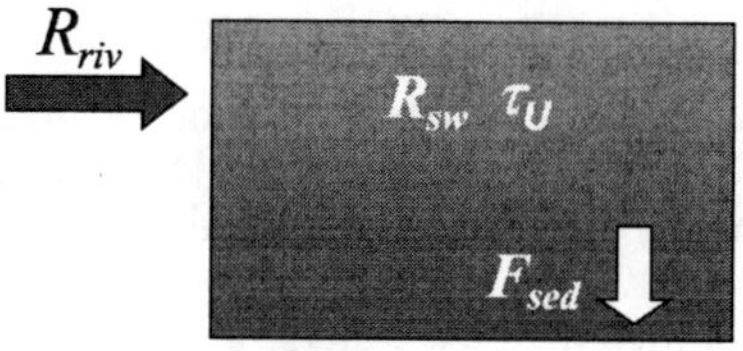

$$\frac{dR_{sw}}{dt} = \frac{R_{riv}}{\tau_U} + \lambda_{234} - R_{sw}\left\{\frac{1}{\tau_U} + \lambda_{234}\right\}$$

(b)

$$\text{If:} \quad R_{riv} = R_0 + R_1 \cdot \sin\left(\frac{2\pi \cdot t}{T}\right)$$

$$\text{Then} \quad (R)_{edm} = \frac{\frac{(R)_0}{\tau_U} + \lambda_{234}}{\lambda_{234} + \frac{1}{\tau_U}} + \frac{\frac{(R)_1}{\tau_U}}{w} \cdot \sin\left(wt + \arctan\left[\frac{-w}{\lambda_{234} + \frac{1}{\tau_U}}\right]\right)$$

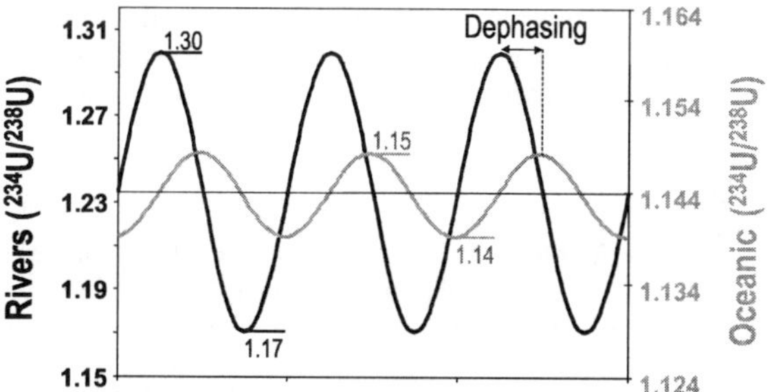

Figure 22. Response of the seawater U activity ratio to a sinusoidal variation of the U activity ratio of world rivers (adapted from Richter and Turekian 1993). Such a scenario could explain the apparent discrepancy between the theoretical mean riverine activity ratio of 1.25-1.35 and the estimated value of 1.17 (see text). The scenario could be supported by the preliminary conclusions from the study of U in Himalayan rivers (Chabaux et al. 2001), which assumed a climatic dependence of the Himalayan U flux, sufficient to induce a periodic variation of the mean U activity ratio of the world rivers on a glacial-interglacial time-scale (T = 10^5 y). The amplitude of variation proposed for the mean (^{234}U/^{238}U) ratios of the world rivers (1.17-1.3) is compatible with a present day sea water value of 1.144 which has remained constant (<0.5‰ variation) over the last 400ka (Henderson 2002), if the U residence time in sea water is higher than 250 ka. In the frame of the above model, the dephasing between the sinusoidal variation of U activity ratio in world river and the variation of the U activity ratio in sea water is of about 25ky. τ_U = U residence time in sea water; $w = 2\pi/T$; T = time period of the sinusoidal variation (here T= 10^5y); R_0 = mean (^{234}U/^{238}U) ratio of the world rivers; R1 amplitude of the sinusoidal variation.

and in resistates. A significant part of non-detrital Th, however, can be carried by the iron-oxyhydroxide fraction of the particles (Porcelli et al. 2001).

The strong influence of colloids on Th mobility is documented by ultra-filtration experiments on a few river samples (e.g., Viers et al. 1997; Dupré et al. 1999; Porcelli et al. 2001) (Fig. 9b) or by geochemical observations such as the similarity of Th/Nd ratios in dissolved and particulate load of river waters (Allègre et al. 1996). Organic colloids play an important role on Th mobility in river waters, as highlighted by concomitant decrease of Th and DOC concentrations in the successive filtrates recovered during ultrafiltration experiments of organic-rich waters (e.g., Viers et al. 1997). The number of ultrafiltration experiments is however too limited to assess the influence of the inorganic colloids on Th mobility. However, from experimental data on the fixation of Th onto mineral phases (see Section 2), a complexation by, or an adsorption onto colloidal Fe-oxyhydroxides is highly probable. From these results, along with the thermodynamic data for Th solubility (Langmuir and Herman 1980), one can conclude that Th measured in 0.45/0.1 μm filtered fractions (Table 2), does not represent truly "dissolved" Th, but Th carried by colloids or fine particles which pass through the filters. This readily accounts for the common observation of higher Th concentrations in organic-rich rivers than in non-organic ones (e.g., Dupré et al. 1996). It also supports the Andersson et al. (1995) interpretation for the correlation observed in the Kalix river waters between Th concentration in the 0.45 μm- filtered fraction and river discharge, i.e., an increase of fine particles carried in the river when the discharge increases.

Presently, the precise determination of the "true dissolved" Th fraction in water samples remains a challenge. Results from ultrafiltration experiments on organic-rich water samples from the Mengong river tend to demonstrate that Th concentration is less than 15 ng/L in absence of DOC (Table 2 and Viers et al. 1997), and that Th is still controlled by organic carbon in the final filtrate of the ultrafiltration experiments. The latter conclusion is also supported by the results obtained for the Kalix river (Porcelli et al. 2001). These results therefore not only raised the question of the determination of the amount of dissolved Th in water but also of the nature of Th chemical speciation.

Table 2. Recent data on dissolved Th isotopes in world rivers.

River	Th (ng/L)	Reference
Amazon catchm.	5-121	Allègre et al., 1996
Congo catchm.	31-128	Allègre et al., 1996
Niger	3-20	Picouet et al., 2002
Wista	2.5	Andersson et al., 1995
Kalix	9-19	Andersson et al., 1995
MacKenzie	0.935	Vigier et al., 2001
Narmada	0.365	Vigier, 2000
Tapti	0.139	Vigier, 2000
Awout	10-239	Dupré et al., 1999
Sanaga	11.6	Viers et al., 1997
Nyong	111	Viers et al., 1997
Jucar	24	Sanchez and Rodriguez-Alvarez, 1999
Taiwan rivers	5-1340	Chu and Wang, 1997

Such questions are not specific to ^{232}Th alone. They are also important for the short-lived Th isotopes (i.e., ^{230}Th, ^{228}Th and ^{234}Th) which are derived from the ^{238}U-^{232}Th series. Isotope distribution among the different phases of water are generally assumed to mostly depend on the behavior of their respective parent nuclides, in particular on their sorption/solubility properties. For instance, according to Sarin et al. (1990), covariation between (^{230}Th/^{238}U) ratios and Th/U ratios in Himalayan rivers reflect the preferential leaching of U during weathering. Andersson et al. (1995) inferred from (^{230}Th/^{232}Th) and U/Th measurements in the Kalix river that the ^{230}Th excess is generated in U-rich, ^{232}Th-poor peatlands and trapped in authigenic particles. Similarly, the observation of (^{228}Th/^{232}Th) ratios far above unity in some waters (e.g., Chu and Wang 1997; Sanchez and Rodriguez-Alvarez 1999) is the result of the higher solubility of ^{228}Ra relative to ^{232}Th, that is the higher solubility of the direct daughter of ^{228}Th compared with its parent: soluble ^{228}Ra indeed produces ^{228}Th which remains in solution. One must note, nevertheless, that all the analyses are performed on the 0.45/0.1 μm filtered water fractions, and therefore give information on Th isotopic characteristics of colloidal fractions rather than of the dissolved ones. Comparison between Th isotope ratios measured in sediments, suspended matters and filtrates (e.g., Sanchez and Rodriguez-Alvarez 1999) provides some information on Th exchanges between sediments particles and colloids and with the dissolved phase. To our knowledge the only study reporting ^{230}Th/^{232}Th isotopic ratios of ultrafiltered fractions is that of Porcelli et al. (2001) on the Kalix river. Their data suggest approximate Th isotopic equilibration between particles, colloids and dissolved species. Such (^{230}Th/^{232}Th) isotopic equilibration, at least between sediments, particles and 0.45 μm filtered-fraction is also reported in Sanchez and Rodriguez-Alvarez (1999). This work nevertheless outlines a significant difference between (^{228}Th/^{232}Th) ratios of these three fractions, which are used to assess the origin of Th in suspended matter (resuspension of sediments rather than exchanges with water). These differences could certainly also give information on exchange rate between colloids and particles. The number of studies on Th isotopes in water still remains scarce. Their development should bring new insights about the speciation of Th, the transport of insoluble element, as well as about the exchange rates among dissolved load, colloids and particles.

Most of Ra studies in river waters focused on the origin and the chemical speciation of ^{226}Ra in rivers. Benes (1983) proposed that Ra in waters could exist as Ra^{2+} or as $Ra(SO_4)$ complexes. Ra adsorption onto particles is an important, and probably a dominant process controlling Ra mobility in rivers (e.g., Michel et al. 1981, Krishnaswami et al. 1982; Herczeg et al. 1988; Sturchio et al. 1989). The Ra adsorption onto particles may be limited by competition with other ions during cation-exchange with iron hydroxides, clays and organic substrates (Benes 1981; Webster et al. 1995). The adsorption processes could account for the apparent heterogeneity of the Ra distribution in river sediments, such as those reported by Burnett et al. (1990) for the Suwannee river, where about 70% of the bedload Ra is located in the <63 μm size fraction. As Ra adsorption significantly decreases with increasing salinity (Kraemer and Reid 1984), part of the adsorbed Ra will be released in estuaries and create an important Ra flux to ocean (see Swarzenski et al. 2003). The importance of the Ra complexation by colloids is still unclear. The chemical similarity between Ra and Ba properties should allow some speculations concerning the Ra affinity for colloids. Thus, on the basis of ultrafiltration experiments (e.g., Viers et al. 1997; Porcelli et al. 2001) a significant fixation of Ra by organic colloids could be anticipated, but with a lower affinity than for U and Th. Unfortunately, such conclusions remain very speculative, as it has been argued that alkali elements retained on ultrafilters from organic-rich waters are largely an artefact of filtration and reflect no bonding by organic colloids (Viers et al. 1997).

On a global scale, the ^{226}Ra concentration in river waters varies widely from 0.004 to 2 dpm/L, i.e., from 1.8 to 900 fg/L (Table 3). The highest Ra concentrations seem to occur in limestone regions and in HCO_3^- rich waters (in Plater et al. 1995). According to Burnett et al. (1990), rivers draining arid areas (Rio Grande or Rio Pecos), or percolating through U-rich lithologies (Texas rivers) and phosphate deposits (the Suwannee river) are enriched in radium 2-3 times over the world average. The lithologic differences among the drained watersheds are therefore certainly an important parameter to explain the Ra concentration in world rivers. Such lithologic control is common. Input of groundwater into river waters can also create a significant Ra source in surface waters. Burnett et al. (1990) explain the downstream ^{226}Ra increase in the Suwannee river (from 0.189 to 0.27 dpm/L) by an increasing contribution of Ra-rich springs related to deep aquifers. An aquifer contribution is also observed in the Dead Sea Rift Valley by Moise et al. (2000). Other parameters proposed to explain the Ra concentrations of stream waters include U-Th contents of the weathered rocks of drainage basin, adsorption-desorption reactions on particle surfaces, the supply of radium by dissolution and recoil, the precipitation of insoluble phases and the mixing of different water masses (e.g., Krishnaswami et al. 1982, Scott 1982, Rama and Moore 1984, Sarin et al. 1990, Plater et al. 1995). More detailed studies will be necessary to better characterise and understand the role of the different parameters on the Ra content in river waters. The use of TIMS should help the development of such studies.

Studies of (^{228}Ra/^{226}Ra) isotope ratios in rivers are scarce. The most comprehensive Ra isotopic data set published for a major river is included in a general study of major and trace element concentration in the Ganges and its tributaries (Sarin et al. 1990). The authors assume, on the basis of major ion chemistry, that the regional variations result from differences in weathering regimes. The origin of differences observed between Himalayan and Indian streams is, however, not clear: according to the authors, dissolved radium contents does not seem to be related to other ions. It is nevertheless possible to make out a correlation by combining Ra data for the Ganges (Sarin et al. 1990) with corresponding Sr isotopic ratios published by Palmer et Edmond (1992). The negative correlation between these two isotopic ratios in waters (Fig. 23) suggests a control of the Ra isotopic ratio of waters by the basin geology. A possible explanation is a control of the (^{226}Ra/^{228}Ra) ratios of rivers by the U/Th ratios of the rocks forming the watershed, which remains to be confirmed by further studies. The lower (^{228}Ra/^{226}Ra) activity ratios found in waters from catchments with a greater proportion of carbonate-rich materials in the Fenland basin, U.K. (Plater et al. 1995) is another example of the control of the catchment geology on the riverine Ra isotope ratios. In this case, however, hydrological parameters are also suspected to play a role in the Ra isotope ratios. The influence of hydrology on Ra isotopic ratios of rivers has been shown by Eikenberg et al. (2001) for the Rhine valley streams. Radium isotope budget is significantly affected by groundwater inputs into surface waters. The authors propose to combine (^{228}Ra/^{226}Ra) ratios and ^{87}Sr/^{86}Sr isotopic ratios to follow groundwater contribution to the streams when water tables vary. The use of the (^{228}Ra/^{226}Ra) ratios as tracer of water masses was also suggested by Porcelli et al. (2001), who have ruled out the contribution of Ra from mires to the Kalix river, as Ra activity ratios in river and source rock are equal (1.2) and very

---

References for Table 3 (*on facing page*): (1) Moore (1967), (2) Moore et al. (1995), (3) Key et al. (1985), (4) Moore and Edmond (1984), (5) Moore and Todd (1993), (6) Sarin et al. (1990), (7) Scott (1982), (8) Krest et al. (1999), (9) Rona and Urry (1952), (10) Bhat and Krishnaswami (1969), (11) Vigier et al. (2001), (12) Eikenberg et al. (2001), (13) Burnett et al. (1990), (14) Nozaki et al. (2001), (15) Li et al. (1977), (16) Li and Chan (1979), (17) Eslinger and Moore (1983), (18) Eslinger and Moore (1980), (19) Miyake et al. (1964), (20) Sanchez and Rodriguez-Alvarez (1999), (21) Plater et al. (1995), (22) Haridasan et al. (2001), (23) Hancock and Murray (1996), (24) Yang et al. (2002), (25) Eslinger et Moore (1984). Discharge data from Meybeck 1984, Palmer and Edmond (1989).

Table 3. Available data on dissolved Ra isotopes in world rivers.

River	Discharge km³/y	[^{226}Ra] dpm/L	[^{228}Ra] dpm/L	[^{224}Ra] dpm/L	(^{228}Ra/^{226}Ra)	Ref.
Amazon	5400	0.02-0.07	0.012-0.097	0.099-0.314	12.25-2.15	(1)-(4)
Orinoco	1135	0.0285-0.038	0.063-0.0675	0.08	1.65-2.37	(5)
Yangtse Kiang	694	0.11	0.165		1.45	(25)
Brahmaputra	612	0.07	0.07		1	(6)
Mississippi	530	0.0118-0.111	0.115		1.25	(1) (7) (8)
St Lawrence	450	0.005				(9)
Ganges	450	0.1-0.2	0.098-0.156		0.78-1.5	(10) (6)
MacKenzie	306	0.106				(11)
Atchafalaya	240	0.1128	0.1318		1.17	(8)
Godaveri	84	0.05				(10)
Rhine	80	0.1152	0.0852	0.0312	0.74	(12)
Narmada	41	0.0092				(11)
Krishna	35	0.05				(10)
Apalachicola	23	0.055				(13)
Tapti	18	0.004				(11)
Chao Phraya	13	0.15	0.23		1.53	(14)
Hudson	12	0.009-0.07				(9) (15) (16)
Suwannee	11	0.289-0.4				(7) (13)
Delaware	10.3	0.048-0.053	0.073	0.108	1.45	(17)
Pee Dee	7	0.03-0.06				(18)
Brazos	6.75	0.187				(7)
Rio Grande	3.9	0.256				(7)
Black	1.3	0.05				(18)
Waccamaw	3	0.06				(18)
Ochlockonee	1.5	0.102				(13)
Withlacoochee	1.8	0.27				(13)
Hillsborough	0.6	0.432				(13)
Alafia	0.38	0.593				(13)
Peace	1.9	1.087				(13)
Little Manatee	0.24	2.073				(13)
Caloosahatchee	1.5	2.097				(13)
Sabarati		0.09	0.306		3.4	(10)
10 Japanese rivers		0.082				(19)
Pecos		0.256				(7)
Nueces		0.296				(7)
Gadalupe		0.208				(7)
Colorado		0.201				(7)
Ohio		0.0727				(7)
Miami		0.0278				(7)
Tombigbee		0.0615				(7)
Alabama		0.0523				(7)
Flint		0.0331				(7)
Jucar		0.0792		0.114		(20)
Witham		0.14			1.25	(21)
Welland		0.06-0.2			0.58-1.57	(21)
Nene		0.12-0.18			1.37-1.68	(21)
Great Ouse		0.13			0.95-1.17	(21)
Citrapuzha		0.18-0.36				(22)
Bega		0.0378	0.078	0.066	1.18	(23)
Nakdong		0.0365	0.092		2.56	(24)

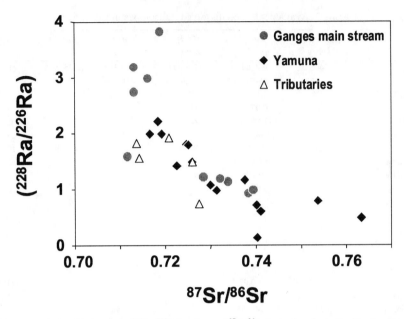

Figure 23. Relationship between (^{228}Ra/^{226}Ra) ratios and ^{87}Sr/^{86}Sr isotopic ratio in the dissolved load of the Ganges main stream and its tributaries. Radium data from Sarin et al. (1990), Sr data from Palmer and Edmond (1992).

different from those of the mires (18). All these studies suggest that Ra isotope ratios of river waters could be used as a hydrological tracers of water masses. The Ra isotope potential as geochemical tracer for the study of chemical transfers in rivers, however, remains largely under-exploited and probably underestimated.

Studies of U,Th and Ra behavior in surface waters have demonstrated the variable affinity of these elements for the different river components. Therefore, the transport of these elements in stream and river waters may induce fractionation among them, and then induce radioactive disequilibria in U-Th series. Analysis of radioactive disequilibria should bring important information on both the dynamics and the rates of transport in rivers. The studies of the Kalix River watershed (Andersson et al. 1995, 1998; Porcelli et al. 1997, 2001) undoubtedly constitute important illustrations of such a potential. In addition to the determination of the different form of nuclide transport in river waters, these studies have shown how U-series nuclides are helpful in constraining exchanges between the different reservoirs of the watershed. This is well illustrated with the study of the impact of mire peats on the U-Th budget of river waters (Porcelli et al. 1997, 2001). By using a simple box model for accumulation of U-radioactive series nuclides in peat and for their transport through the mire, these authors show that the mire waters cannot be the primary source of uranium in the river. In this case bedrock groundwaters must be the dominant source of riverine uranium (Porcelli et al. 1997). By contrast, Th in the filtrated waters of the river would mainly come from mire peat through entrainment of colloidal organic material (Porcelli et al. 2001). This conclusion agrees with the previous interpretation accounting for the large difference between the (^{230}Th/^{232}Th) ratios in fresh and brackish waters from the Baltic Sea (Andersson et al. 1995).

Fractionation During Weathering & River Transport 565

The role of radionuclides as tracer of the chemical transport in river is also reinforced by the fact that each of the U-Th-Ra elements has several isotopes of very different half-lives belonging to the U-Th radioactive series. Thus, these series permit comparison of the behavior of isotopes of the same element which are supposed to have the same chemical properties, but very different lifetimes. These comparisons should be very helpful in constraining time scales of transport in rivers. This was illustrated by Porcelli et al. (2001) who compared (^{234}Th/^{238}U) and (^{230}Th/^{238}U) ratios in Kalix river waters and estimated a transit time for Th of 15 ± 10 days in this watershed. The development of such studies in the future should lead to an important progress in understanding and quantifying of transport parameters in surface waters. This information could be crucial for a correct use of U-series radioactive disequilibria measured in river waters to establish weathering budgets at the scale of a watershed.

5. ESTIMATES OF WEATHERING MASS BALANCE FROM U-SERIES DISEQUILIBRIA IN RIVER WATERS.

The use of radioactive disequilibria for studying erosion rates motivated pioneering studies of U-series nuclides in soils and weathering profiles (Rosholt et al. 1966; Hansen and Stout 1968). As detailed in Section 3, one method is to determine the "local" weathering and erosion rates by using the data obtained from soil profiles. Another approach is based on the comparison between radioactive disequilibria in dissolved load of river waters and their value in associated river particles and/or in soils of the watershed (Osmond et Cowart 1976; Moreira Nordeman 1977, 1980; Plater et al. 1988, 1994; Vigier et al. 2001). The data are used for establishing weathering budgets and determining erosion parameters at a watershed scale, such as erosion rates or sediment yields. The study of Vigier et al. (2001) is undoubtedly the first to give a very convincing demonstration of the potential of U-series disequilibria in river products for exploring chemical and physical weathering processes. It combines the analysis of the three radioactive disequilibria ^{238}U-^{234}U, ^{234}U-^{230}Th and ^{230}Th-^{226}Ra on both the dissolved and suspended loads of river samples and clearly provides the basis for the future studies in this field.

All these studies assume a geochemical complementarity among the products of continental denudation, i.e., between the flux of chemical erosion, usually identified with the dissolved load of river, and the residue of the weathering. Depending on the studies, the latter is assimilated to soils/weathering profiles (Moreira Nordeman 1980), river sediments (Plater et al. 1988; 1994) or suspended particles (Vigier et al. 2001). The above complementarity is a classical concept of surface geochemistry, used, for instance, in recent studies establishing weathering mass balances at a watershed scale based on major and trace elements (e.g., Gaillardet et al. 1995). Compared with the other geochemical tracers, the radioactive disequilibria have the unique advantage of adding a strong initial constraint independent from the source rock composition for the unweathered bedrock, namely initial radioactive equilibrium among ^{238}U-^{234}U-^{230}Th-^{226}Ra, at least for pristine rocks older than 1.2 Ma.

Considering a steady state system, consisting of two phases, i.e., water and weathering product, and neglecting the duration of the chemical erosion relative to the half-life of a nuclide N_i, mass balance involving N_i can be written as:

$$N_{io}M_o = N_{iw}M_w + N_{is}M_s \tag{1}$$

where N_{io}, N_{iw}, N_{is} are the concentrations in activity of the nuclide N_i in the fresh rock, the river water and the residual weathering products, respectively; M_o, the mass of fresh material weathered per time unit, M_w the water discharge and M_s the mass of weathering residue formed per time unit. Using a notation similar to that defined in Gaillardet et al. (1995) this equation can be rewritten as:

$$\tau N_{io} = N_{iw} + N_{is} P \qquad (2)$$

with $\tau = M_o/M_w$, the total flux of fresh rock undergoing the alteration and P the flux of residual material produced during alteration.

τ can be seen as the global weathering rate, and $(\tau - P)$ as the chemical weathering rate. The use of two nuclides, i and j, leads to simple relationships between the parameters. Assuming secular equilibrium for the unweathered material, the fraction of nuclide j in each phase of the system is inferred from the N_i/N_j activity ratios in water and weathering product.

$$\frac{N_{js}}{N_{jw}} = \frac{\left(\frac{N_i}{N_j}\right)_w - 1}{1 - \left(\frac{N_i}{N_j}\right)_s}$$

Also, the knowledge or the estimate of N_i or N_j in the bedrock allows calculation of the two parameters τ and P:

$$\tau = \frac{N_{jw}}{N_{jo}} \times \left\{ 1 - \frac{\left(\frac{N_i}{N_j}\right)_w - 1}{1 - \left(\frac{N_i}{N_j}\right)_s} \right\} \qquad P = \frac{N_{jw}}{N_{js}} \times \frac{\left(\frac{N_i}{N_j}\right)_w - 1}{1 - \left(\frac{N_i}{N_j}\right)_s}$$

If the weathering process is now thought to be in steady-state, i.e., the production of residual material supplied by chemical erosion balances the amount exported by mechanical erosion to the river, and if no storage of weathered material occurs within the watershed, or more precisely, if the duration of such storage remains negligible compared to the half-lives of the involved nuclides, then the riverine sediments represent the mean weathering product formed during bedrock alteration, and the parameter P gives the mean mechanical erosion rate or the mean sediment yield of the watershed. Such an approach neglects any secondary modification of weathering fluxes during transport by rivers.

The above considerations allow to understand the theoretically high potential of radioactive disequilibria in river waters for quantifying weathering processes at a watershed scale.

The first studies, dealing with calculation of weathering rates from radioactive disequilibria, used ^{238}U and ^{234}U. Mainly for illustrating the potential of the method, Osmond and Cowart (1976) used early estimates of (^{234}U/^{238}U) ratios in the average continental run-off and in the mean riverine sediments to calculate a global value of 5 for the ratio of U mobilised by chemical weathering to U associated with mechanical erosion, and to infer a world average erosion rate of 6 g/cm^2/ky (60 tons/km^2/y). Applying the same principle, Moreira-Nordemann (1980) estimated the regional chemical weathering rate in the Petro River Basin (Bahia state, Brazil) by using U activity ratios of soils instead of those of river sediments.

Plater et al. (1988) developed a similar approach to estimate mean sediment yield on a watershed scale. These studies are based on the analysis of both dissolved load and river sediments. They show the theoretical potential of combining (^{238}U-^{234}U) and (^{234}U-^{230}Th) ratios to discuss some basic hypotheses of the mass budget calculations. Indeed, in a plot of (^{234}U/^{238}U) versus (^{230}Th/^{238}U) ratios, the dissolved and particulate loads should fall on a straight line passing through the equipoint (1,1) (details also in Osmond and Ivanovich 1992). Plater et al. (1988) did not find such a theoretical line. This would result from chemical exchanges in rivers between sediments and dissolved phase. The (^{234}U/^{238}U) ratios of sediments would be more significantly affected by the exchanges than the (^{230}Th/^{238}U) ratios. For this reason, only the latter ratio is used by Plater et al. (1994) for computing sedimentation yields of two English watersheds.

Compared with the previous studies, Vigier et al. (2001) is the first study to analyse also the ^{238}U-^{226}Ra disequilibrium on the Mackenzie Basin. Indisputably, it gives the general framework for future studies in the field. It relies on the classical assumption of a geochemical complementarity between the dissolved and suspended loads of rivers. ^{238}U-^{234}U-^{230}Th-^{226}Ra disequilibria confirm this hypothesis, as (^{234}U/^{238}U), (^{230}Th/^{238}U) and (^{226}Ra/^{238}U) ratios in dissolved load of the Mackenzie Basin rivers are about systematically complementary of particle values, with regard to secular equilibrium point (see Fig. 2 in Vigier et al. 2001). With the use of the short period nuclide ^{226}Ra, the duration of weathering processes can no longer be neglected, and data interpretation requires a time dependant model for the behavior of the radionuclides during alteration. Figure 24 presents the model used by Vigier et al. (2001), and the related equations. The model assumes a bedrock initially in secular equilibrium, and a continuous leaching of U-Th-Ra in the weathering profile. It also supposes that the dissolved load represents the present day leaching occurring in the whole profile, while suspended particles integrates the continuous leaching over their residence time in the weathering profile. The model is entirely defined by five constants: the leaching coefficients of ^{238}U, ^{234}U, ^{230}Th, ^{226}Ra, and τ the duration of chemical erosion. These parameters are determined by measurement of (^{234}U/^{238}U) and (^{230}Th/^{238}U) ratios in both dissolved and particulate loads, and of (^{226}Ra/^{238}U) ratios in particles. Application of this model to the Mackenzie Basin indicates a short duration for the chemical erosion in the watershed (9-28 ± 10 ka), and may suggest a recent change in chemical erosion in this northern latitude watershed. This in turn, as noticed by the authors, clearly questions the steady state assumption, for the alteration process in the basin. It is interesting to note that from a different approach, but still using U-Th radioactive disequilibria, a quite similar conclusion is obtained for the watershed of the Kalix River (Porcelli et al. 2001)

All these estimates and conclusions, however, rely on several strong hypotheses. In addition to the steady-state hypothesis and the choice of the erosion model, there is also the strong assumption of a conservative behavior of U, Th, and Ra during their transfer into rivers. Examples of exchange between dissolved load and particles are now well documented (e.g., Plater et al. 1992; Andersson et al. 1998; Porcelli et al. 2001), and are rather the rule than the exception. Moreover, as discussed elsewhere in this volume, the U budget could be significantly affected by deep groundwater inputs without real relation to surface weathering processes (Porcelli and Swarzenski 2003). Taking into account the latter parameters will be certainly necessary in the future to refine estimates of weathering parameters from the radioactive disequilibria in river waters, and also to better understand the real interest of the U and Th nuclides in determining weathering budget.

Continuous leaching model

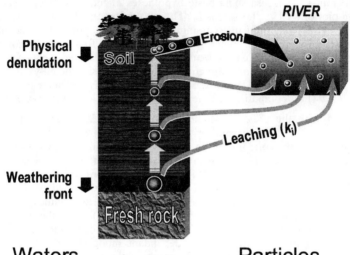

Waters

$$^{238}U_w = \int_0^t k_{238} \cdot {}^{238}U_p \cdot dt$$

$$^{234}U_w = \int_0^t k_{234} \cdot {}^{234}U_p \cdot dt$$

$$^{230}Th_w = \int_0^t k_{230} \cdot {}^{230}Th_p \cdot dt$$

Particles

$$\frac{d\,^{238}U}{dt} = -(k_{238} + \lambda_{238}) \cdot {}^{238}U$$

$$\frac{d\,^{234}U}{dt} = \lambda_{238} \cdot {}^{238}U - (k_{234} + \lambda_{234}) \cdot {}^{234}U$$

$$\frac{d\,^{230}Th}{dt} = \lambda_{234} \cdot {}^{234}U - (k_{230} + \lambda_{230}) \cdot {}^{230}Th$$

$$\frac{d\,^{226}Ra}{dt} = \lambda_{230} \cdot {}^{230}Th - (k_{226} + \lambda_{226}) \cdot {}^{226}Ra$$

Figure 24. Drawing of the continuous leaching model used by Vigier et al. (2001) to estimate the residence time of particles in the soils of the Mackenzie Basin, and the related equations. This model assumes that particles are continuously leached in the soil before leaving to the river. Dissolved load of the river integrates the present leaching of the whole soil profile. k_{238}, k_{234}, k_{230} and k_{226} are the leaching coefficients of ^{238}U, ^{234}U, ^{230}Th and ^{226}Ra nuclides, respectively, and τ is the duration of the chemical erosion.

6. CONCLUDING REMARK

This review highlights the important potential of U-series disequilibria in understanding the continental alteration and related mass transfers. The recent analytical developments, including TIMS and MC ICP MS techniques, for measuring small amounts of U, Th and Ra in geologic samples, offer today new possibilities and new perspectives for analysing U-series disequilibria in weathering profiles and river waters, and could lead to new and, as yet, unanticipated advances in the field of continental alteration.

ACKNOWLEDGMENTS

The manuscript benefited from very careful and constructive reviews by B. Peucker-Ehrenbrink, B. Bourdon and J. Gaillardet, and from a thorough brushing of the English presentation by J. Honnorez, B. Peucker-Ehrenbrink and B. Bourdon. F.C. wishes to acknowledge N. Clauer for his constant support during the development of U-Th series radioactive disequilibria at the CGS, as well as M.C. Pierret and S. Rihs who contribute today to the development of this research in Strasbourg. This is an EOST-CGS contribution.

REFERENCES

Allard T, Ildefonse Ph., Beaucaire C, Calas G (1999) Structural chemistry of uranium associated with Si, Al, Fe gels in a granitic uranium mine. Chem Geol 158:81-103
Allègre CJ, Dupré B, Négrel P, Gaillardet J (1996) Sr-Nd-Pb isotope systematics in Amazon and Congo River systems: Constraints about erosion processes. Chem Geol 131:93-112
Ames LL, McGarrah JE, Walker BA (1983a) Sorption of trace constituents from aqueous solutions onto secondary minerals. I Uranium. Clays Clay Miner 31:321-334
Ames LL, McGarrah JE, Walker BA (1983b) Sorption of trace constituents from aqueous solutions onto secondary minerals. II Radium. Clays Clay Miner 31:335-342
Ames LL, McGarrah JE, Walker BA (1983c) Sorption of uranium and radium by biotite, muscovite and phlogopite. Clays Clay Miner 31:343-351
Ames LL, McGarrah JE, Walker BA, Salter PF (1983d) Uranium and radium sorption on amorphous ferric oxyhydroxide. Chem Geol 40:135-148
Anderson RF, Bacon MP, Brewer PG (1982) Elevated concentration of actinides in Mono Lake. Science 216:514-516
Andersson PS, Wasserburg GJ, Chen JH, Papanastassiou DA, Ingri J (1995) ^{238}U-^{234}U and ^{232}Th-^{230}Th in the Baltic Sea and in river water. Earth Planet Sci Lett 130:217-234
Andersson PS, Porcelli D, Wasserburg GJ, Ingri J (1998) Particle transport of ^{234}U-^{238}U in the Kalix River and in the Baltic Sea. Geochim Cosmochim Acta 62:385-392
Benes P (1983) Physico-chemical forms and migration in continental waters of radium from uranium mining and milling. *In:* Environmental Migration of Long-Lived Radionuclides. IAEA VIENNA, p 3-23
Bhat SG, Krishnaswami S (1969) Isotopes of uranium and radium in Indian rivers. Proc Ind Acad Sci Earth Planet Sci 70:1-17
Bischoff JL, Fitzpatrick JA (1991) U-series dating of impure carbonates: An isochron technique using total-sample dissolution. Geochim Cosmochim Acta 55:543-554
Bonotto DM (1998) Implication of groundwater weathered profile interactions to the mobilization of radionuclides. J South Am Earth Sci 11:389-405
Borole DV, Krishnaswami S, Somayulu BLK (1982) Uranium isotopes in rivers, estuaries and adjacent coastal sediments of western India: Their weathering, transport and oceanic budget. Geochim Cosmochim Acta 46:125-137
Borovec Z (1981) The adsorption of uranyl species by fine clay. Chem Geol 32:45-58
Boulad AP, Muller J-P, Bocquier G (1977) Essai de détermination de l'âge et de la vitesse d'altération d'un sol ferrallitique camerounais à l'aide de la méthode du déséquilibre radioactif uranium-thorium. Sci Géol Bul. 30:175-188
Braun J-J, Pagel M, Herbillon A, Rosin C (1993) Mobilization and redistribution of REEs and thorium in a syenitic lateritic profile: a mass balance study. Geochim Cosmochim Acta 57:4419-4434
Bruno J, de Pablo J, Duro L, Figuerola E (1995) Experimental study and modeling of the U(VI)-Fe(OH)$_3$ surface precipitation/coprecipitation equilibria. Geochim Cosmochim Acta 59:4113-4123
Buffle J, Van Leeuwen HP (1992) Environmental Particles. Lewis Publishers, London
Burnett WC, Cowart JB, Deetae S (1990) Radium in the Suwannee River and Estuary. Biogeochemistry 10:237-255
Caroll JL, Moore WS (1993) Uranium removal during low discharge in the Ganga-Brahmaputra mixing zone. Geochim Cosmochim Acta 57:4987-4995
Chabaux F (1993) Systématique ^{238}U-^{230}Th-^{226}Ra dans les roches volcaniques. PhD Thesis, Univesité Paris 7
Chabaux F, Ben Othman D, Birck JL (1994) A new Ra-Ba chromatographic separation and its application to Ra mass-spectrometric measurement in volcanics rocks. Chem Geol 114:191-197

Chabaux F, Riotte J, Clauer N, France-Lanord Ch (2001) Isotopic tracing of the dissolved U fluxes of Himalayan rivers: implications for present and past U budgets of Ganges-Brahmaputra system. Geochim Cosmochim Acta 65:3201-3217

Chen JH, Edwards LR, Wasserburg GJ (1986) ^{234}U-^{238}U and ^{232}Th in sea water. Earth Planet Sci Lett 80:241-251

Cherdyntsev VV, Chalov PI, Khaidarov GZ (1955) Uranium series disequilibrium dating. *In:* Trans. 3rd Sess. Commission for determining the absolute age of geological formations, Izd. Akad. Nauk, SSSR, p 175-182

Chu TC, Wang JJ (1997) Radioactive disequilibrium of Uranium and Thorium nuclide series in river waters from the Ta-Tun Volcanic group area in Taiwan. Appl Radiat Isot 48:1149-1155

Cochran JK, Masqué P (2003) Short-lived U/Th-series radionuclides in the ocean: tracers for scavenging rates, export fluxes and particle dynamics. Rev Mineral Geochem 52:461-492

Cohen AS, O'Nions RK (1991) Precise determination of femtogram quantities of radium by thermal-ionisation mass spectrometry. Anal Chem 61:2705-2706

Cromières L, Moulin V, Fourest B, Guillaumont R, Giffaut E (1998) Sorption of thorium onto hematite colloids. Radiochim Acta 82:249-255

Czerwinski KR, Buckau G, Scherbaum F, Kim JI (1994) Complexation of the uranyl ion with aquatic humic acid. Radiochim Acta 65:111-119

Dearlove JPL, Longworth G, Ivanovich M, Kim JL, Delakowitz B, Zeh P (1991) A study of ground waters colloids and their geochemical interactions with natural radionuclides in Gorleben aquifer systems. Radiochim Acta 52/53:83-89

Dequincey O, Chabaux F, Clauer N, Liewig N, Muller J-P (1999) Dating of weathering profiles by radioactive disequilibria: contribution of the study of authigenic mineral fractions. CR Acad Sci Paris IIa 328:679-685

Dequincey O, Chabaux F, Clauer N, Sigmarsson O, Liewig N, Leprun J-C (2002) Chemical mobillizations in laterites: Evidence from trace elements and ^{238}U-^{234}U-^{230}Th disequilibria. Geochim Cosmochim Acta 66:1197-1210

Dequincey O, Chabaux F, Leprun J-C, Clauer N (in prep) Origin and modeling of ^{238}U-^{234}U-^{230}Th and lanthanide fractionations in laterites: clues from Kaya toposequence (Burkina Faso). Chem Geol

Dickson BL (1985) Radium isotope in saline seepages, southwestern Yilgarn, western Australia. Geochim Cosmochim Acta 49:361-368

Duff MC, Amrhein C (1996) Uranium (VI) adsorption on goethite and soil in carbonates solutions. Soil Sci Soc Am J 60:1393-1400

Dunk RM, Mills RA, Jenkins WJ (2002) A reevaluation of the oceanic uranium budget for the Holocene. Chem Geol 190:45-67

Dupré B, Gaillardet J, Rousseau D, Allègre CJ (1996) Major and trace elements of river-borne material: The Congo Basin. Geochim Cosmochim Acta 60:1301-1321

Dupré B, Viers J, Dandurand LJ, Polvé M, Bénézeth P, Vervier P, Braun JJ (1999) Major and trace elements associated with colloids in organic-rich river waters: ultrafiltration of natural and spiked solutions. Chem Geol 160:63-80

Edwards RL, Gallup CD, Cheng H (2003) Uranium-series dating of marine and lacustrine carbonates. Rev Mineral Geochem 52:363-405

Edwards LR, Chen JH, Wasserburg GJ (1987) ^{238}U-^{234}U-^{230}Th-^{232}Th systematics and the precise measurement of time over the past 500,000 years. Earth Planet Sci Lett 81:175-192

Eikenberg J, Tricca A, Vezzu G, Stille P, Bajo S, Ruethi M (2001) ^{228}Ra/^{226}Ra/^{224}Ra and ^{87}Sr/^{86}Sr isotope relationships for determining interactions between ground and river water in upper Rhine valley. J Environ Rad 54:133-162

Eslinger RJ, Moore WS (1980) ^{226}Ra behavior in the Pee Dee river-Winyah Bay esturay. Earth Planet Sci Lett 48:239-249

Eslinger RJ, Moore WS (1983) ^{224}Ra, ^{228}Ra, and ^{226}Ra in Winyah Bay and Delaware Bay. Earth Planet Sci Lett 64:430-436

Eslinger RJ, Moore WS (1984) ^{226}Ra and ^{228}Ra in the mixing zones of the Pee Dee River-Winyah Bay, Yangtze River and Delaware Bay Estuaries. Estuar Coast Shelf Sci 18:601-613

Fredrikson JK, Zachara JM, Kennedy DW, Duff MC, Gorby YA, Li SMW, Krupka KM (2000) Reduction of U(VI) in goethite (α-FeOOH) suspensions by a dissimilatory metal-reducing bacterium. Geochim Cosmochim Acta 64:3085-3098

Gaillardet J, Dupré B, CJ Allègre (1995) A global geochemical mass budget applied to the Congo Basin rivers: erosion rates and continental crust composition. Geochim Cosmochim Acta 59:3469-3485

Gaillardet J, Dupré B, Louvat P, CJ Allègre (1999) Global silicate weathering and CO_2 consumption rates deduced from the chemistry of large rivers. Chem Geol 159:3-30

Gascoyne M (1992) Geochemistry of the actinides and their daughters. *In:* Uranium-series disequilibrium: Application to Earth, Marine, and Environmental Sciences. Ivanovich M, Harmon RS (eds), Oxford Sciences Publications, Oxford, p 34-61

Geckeis H, Klenze R, Kim JI (1999) Solid-water interface reactions of actinides and homologues: sorption onto mineral surfaces. Radiochim Acta 87:13-21

Ghaleb B, Hillaire-Marcel C, Causse C, Gariepy C, Vallières S (1990) Fractionation and recycling of U and Th isotopes in a semi-arid endoreeic depression of central Syria. Geochim Cosmochim Acta 54:1025-1035

Giaquinta DM, Soderholm L, Yuchs SE, Wassermann SR (1997) The speciation of uranium in a smectite clay: evidence for catalysed uranyl reduction. Radiochim Acta 76:113-121

Goetz C (1990) Traçage isotopique et chronologie des processus d'altération et de sédimentation par l'étude des déséquilibres U et Th. Application aux systèmes lacustres de Magadi (Kenya) et Manyara (Tanzanie). Doctorat Thesis, Université d'Aix-Marseille, France

Goldstein SJ, Stirling CH (2003) Techniques for measuring uranium-series nuclides: 1992-2002. Rev Mineral Geochem 52:23-57

Greeman DJ, Rose AW, Jester WA (1990) Form and behavior of radium, uranium, and thorium in Central Pennsylvania soils derived from dolomite. Geophys Res Lett 17:833-836

Greeman DJ, Rose AW, Washington JW, Dobos RR, Ciolkosz EJ (1999) Geochemistry of radium in soils of the Eastern United States. Appl Geochem 14:365-385

Grenthe I, Fuger J, Konings RJM, LemireRJ, Muller AB, Nguyen-Trung C, Wanner H (1992) Chemical Thermodynamics of Uranium. Nuclear, North-Holland

Gueniot B, Munier-Lamy C, Berthelin J (1988a) Geochemical behavior of Uranium in soils, part I. Influence of pedogenetic processes on the distribution of uranium in aerated soils. J Geochem Explor 31:21-37

Gueniot B, Munier-Lamy C, Berthelin J (1988b) Geochemical behavior of Uranium in soils, part II Distribution of uranium in hydromorphic soils and soil sequences. Application for surficial prospecting. J Geochem Explor 31:39-55

Halbach P, von Borstel D, Gundermann KD (1980) The uptake of uranium by organic substances in a peat bog environment on a granitic bedrock. Chem Geol 29:117-138

Hancock GJ, Murray AS (1996) Source and distribution of dissolved radium in the Bega River estuary, southeastern Australia. Earth Planet Sci Lett 138:145-155

Hansen RO, Stout PR (1968) Isotopic distribution of uranium and thorium in soils. Soil Sci. 105:44-50

Haridasan PP, Paul AC, Desai MVM (2001) Natural radionuclides in the aquatic environment of a phosphogypsum disposal area. J Environ Rad 53:155-165

Henderson GM (2002) Seawater ($^{234}U/^{238}U$) during the last 800 thousand years. Earth Planet Sci Lett 199:97-110

Henderson GM, Anderson RF (2003) The U-series toolbox for paleoceanography. Rev Mineral Geochem 52:493-531

Henderson GM, Cohen AS, O'Nions RK (1993) $^{234}U/^{238}U$ ratios and ^{230}Th ages for Hateruma Atoll corals: implications for coral diagenesis and seawater $^{234}U/^{238}U$ ratios. Earth Planet Sci Lett 115:65-73

Henderson GM, Slowey NC, Fleisher MQ (2001) U-Th dating of carbonate platform and slope sediments. Geochim Cosmochim Acta 65:2757-2770

Herczeg A, Simpson J, Anderson R, Trier R, Mathieu G, Deck B (1988) Uranium and Radium mobility in groundwaters and brines within the Delaware basin, Southeastern New Mexico, USA Chem Geol 72:181-196

Hillaire-Marcel C, Vallières S, Ghaleb B, Mareschal J-C (1990) Déséquilibres Th/U dans les sols carbonatés en climat subaride: estimation des flux d'uranium et vitesse d'érosion. Le cas du bassin de Palmyre (Syrie). CR Acad Sci Paris IIa 311:233-238

Hsi C, Langmuir D (1985) Adsorption of uranyl onto ferric oxyhydroxides: applications of the surface complexation site-binding model. Geochim Cosmochim Acta 49:1931-1941

Ingri J, Widerlund A, Land M, Gustafsson Ö, Anderson P, Öhlander B (2000) Temporal variation in the fractionation of the rare earth elements in a boreal river: the role of colloidal particles. Chem Geol 166:23-45

Kaufman A (1993) An evaluation of several methods for determining $^{230}Th/U$ ages in impures carbonates. Geochim Cosmochim Acta 57:2303-2317

Key RM, Stallard RF, Moore W S, Sarmiento JL (1985) Distribution and flux of ^{226}Ra and ^{228}Ra in the Amazon River estuary. J Geophys Res 90:6995-7004

Kigoshi K (1971) Alpha-recoil thorium-234: dissolution into water and the uranium-234/uranium-238 disequilibrium in nature. Science 173:47-48

Kraemer T, Reid D (1984) The occurrence and behavior of radium in saline formation water of the US gulf coast region. Isot Geosci 2:153-174

Krest JM, Moore WS, Rama (1999) ^{226}Ra and ^{228}Ra in the mixing zones of the Mississippi and Atchafayala rivers: indicators of groundwater inputs. Mar Chem 64:129-152

Krishnaswami S, Graustein W, Turekian K, Dowd J (1982) Radium, thorium and radioactive lead isotopes in groundwaters: application to the *in-situ* determination of adsorption-desorption rate constants and retardation factors. Water Resour Res 18:1633-1675

Kronfeld J, Vogel JC (1991) Uranium isotopes in surface waters from southern Africa. Earth Planet Sci Lett 105:191-195

Kronfeld J, Vogel JC, Talma AS (1994) A new explanation for extreme ^{234}U/^{238}U disequilibria in a dolomitic aquifer. Earth Planet Sci Lett 123:81-93

Ku T-L, Bull WE, Frieman ST, Knauss KG (1979) ^{230}Th/^{234}U dating of pedogenic carbonates in gravelly desert soils of Vidal Valley, Southeastern California. Geol Soc Am Bull 90:1063-1073

Ku TH, Krauss KG, Mathieu GG (1977) Uranium in open ocean: concentration and isotopic composition. Deep Sea Res 24:1005-1017

Ku T-L, Liang Z-C (1984) The dating of impure carbonates with decay-series isotopes. Nucl Instr Meth 223:563-571

Kurz A, Derry LA, Chadwick OA, Alfano MJ (2000) Refractory element mobility in volcanic soils. Geology 28:683-686

Langmuir D (1978) Uranium solution-mineral equilibria at low temperatures with applications to sedimentary ore deposits. Geochim Cosmochim Acta 42:547-569

Langmuir D, Herman JS (1980) The mobility of thorium in natural waters at low temperatures. Geochim Cosmochim Acta 44:1753-1766

Langmuir D, Melchior D (1985) The geochemistry of Ca, Sr, Ba and Ra sulfates in some brines from the Palo Duro Basin Texas. Geochim Cosmochim Acta 49:2423-2432

Langmuir D, Riese A (1985) The thermodynamics properties of Ra. Geochim Cosmochim Acta 49:1593-1601

Langmuir D (1997) Aqueous Environmental Geochemistry. Prentice Hall, New Jersey, USA

Latham AG, Schwarcz HP (1987a) On the possibility of determining rates of removal of uranium from crystalline igneous rocks using U-series disequilibria – 1: a U-leach model, and its applicability to whole-rock data. Appl Geochem 2:55-65

Latham AG, Schwarcz HP (1987b) On the possibility of determining rates of removal of uranium from crystalline igneous rocks using U-series disequilibria – 2: applicability of a U-leach model to mineral separates. Appl Geochem 2:67-71

Lenhart JJ, Honeyman BD (1999) Uranium (VI) sorption to hematite in the presence of humic acid. Geochim Cosmochim Acta 63:2891-2901

Lenhart JJ, Cabanis SE, McCarthy P, Honeymann BD (2000) Uranium (VI) complexation with citric, humic and fulvic acids. Radiochim Acta 58:5455-5463

Lienert C, Short SA, Von Gunten HR (1994) Uranium infiltration from a river to a shallow groundwater. Geochim Cosmochim Acta 58:5455-5463

Li YH, Mathieu G, Biscaye P, Simpson HJ (1977) The flux of ^{226}Ra from estuarine and continental shelf sediments. Earth Planet Sci Lett 37:237-241

Li YH, Chan LH (1979) Desorption of Ba and ^{226}Ra from river-borne sediments in the Hudson estuary. Earth Planet Sci Lett 43:343-350

Lin JC, Broecker WS, Anderson RF, Hemming S, Rubenstone JL, Bonani G (1996) New ^{230}Th/U and ^{14}C ages from lake Lahontan carbonates, Nevada, USA and a discussion of the origin of initial thorium. Geochim Cosmochim Acta 60:2817-2832

Lovley D, Phillips EJP (1992) Reduction of uranium by Desulfovibrio desulfuricans. Appl Environ Microbio 58:850-856

Lovley D, Phillips EJP, Gorby YA, Landa ER (1991) Microbial reduction of uranium. Nature 350:413-416

Lowson RT, Short SA, Davey BG, Gray DJ (1986) ^{234}U/^{238}U and ^{230}Th/^{234}U activity ratios in mineral phases of a lateritic weathered zone. Geochim Cosmochim Acta 50:1697-1702

Ludwig KR (2003) Mathematical–statistical treatment of data and errors for ^{230}Th/U geochronology. Rev Mineral Geochem 52:631-636

Ludwig KR, Paces JB (2002) Uranium-series dating of pedogenic silica and carbonate, Crater Flat, Nevada. Geochim Cosmochim Acta 66:487-506

Luo S, Ku T-L (1991) U-series isochron dating: A generalized method employing total-sample dissolution. Geochim Cosmochim Acta 55:555-564

Mac Kee BA, Demastre DJ, Nittrouer CA (1987) Uranium geochemistry on the Amazon Shelf: Evidence for uranium release from bottom sediments. Geochim Cosmochim Acta 51:2779-2786

Mangini A, Sonntag C, Bertsch G, Müller E (1979) Evidence for a higher natural uranium content in world rivers. Nature 278:337-339

Martin JM, Meybeck M, Pusset M (1978a) Uranium behavior in Zaire Estuary. Neth J Sea Res 12:338-344

Martin JM, Nijampurkar V, Salvadori F (1978b) Uranium and Thorium isotope behavior in estuarine systems. *In:* Biogeochemistry of estuarine sediments. UNESCO, p 111-127

Mathieu D, Bernat M, Nahon D (1995) Short-lived U and Th isotope distribution in a tropical laterite derived from Granite (Pitinga river basin, Amazonia, Brazil): application to assessment of weathering rate. Earth Planet Sci Lett 136:703-714

Meybeck M (1984) Les fleuves et le cycle géochimique des éléments. Thèse d'état, Université Paris 6, France

Michel J, Moore W, King P (1981) γ-ray spectrometry for determination of radium-228 and radium-226 in natural waters. Anal Chem 53:1885-1889

Middelburg JJ, van der Weijden CH, Woittez JRW (1988) Chemical processes affecting the mobility of major, minor and trace elements during weathering of granite rocks. Chem Geol 68:253-273

Miyake Y, Sugimura Y, Tsubota H (1964) Content of uranium, radium, and thorium in river waters in Japan. *In:* The Natural Radiation Environment. Adams JAS, Lowder WM (eds) Univ. Chicago Press, Chicago, USA, p 219-225

Moise T, Starinsky A, Katz A, Kolodny Y (2000) Ra isotopes and Rn in brines and groundwaters of the Jordan-Dead Sea Rift Valley: Enrichment, retardation, and mixing. Geochim Cosmochim Acta 64:2371-2388

Montavon G, Mansel A, Seibert A, Keller H, Kratz JV, Trautmann N (2000) Complexation studies of UO_2^{2+} with humic acid at low metal ion concentrations by indirect speciation methods. Radiochim Acta 88:17-24

Moore DG (1967) Amazon and Mississippi River concentrations of uranium, thorium, and radium isotopes. Earth Planet Sci Lett 2:231-234

Moore DG, Edmond (1984) Radium and barium in the Amazon River system. J Geophys Res 89:2061-2065

Moore DG, Todd JF (1993) Radium isotopes in Orinoco estuary and eastern Caribbean Sea. J Geophys Res 98:2233-2244

Moore WS, Astwood H, Lindstrom C (1995) Radium isotopes in coastal waters on the Amazon shelf. Geochim Cosmochim Acta 59:4285-4298

Moreira-Nordemann LM (1977) Etude de la vitesse d'altération des roches au moyen de l'uranium utilisé comme traceur naturel. Application à deux bassins du Nord Est du Brésil. Thèse, Université Paris VI, France

Moreira-Nordemann LM (1980) Use of $^{234}U/^{238}U$ disequilibrium in measuring chemical weathering rate of rocks. Geochim Cosmochim Acta 44:103-108

Moreira-Nordemann LM, Sieffermann G (1979) Distribution of uranium in soil profiles of Bahia state, Brazil. Soil Science 127:275-280

Morse JW, Shanbhag PM, Saito A, Choppin GR (1984) Interaction of uranyl ions in carbonate media. Chem Geol 42:85-99

Moulin V, Ouzounian G (1992) Role of colloids and humic substances in the transport of radio-elements through the geosphere. Appl Geochem (Suppl. Issue I:179-186

Murakami T, Ohnuki T, Isobe H, Sato T (1997) Mobility of uranium during weathering. Am Mineral 82:888-899

Murphy RJ, Lenhart JJ, Honeyman BD (1999) The sorption of thorium (IV) and uranium (VI) to hematite in the presence of natural organic matter. Colloid Surf A 157:47-62

Murphy WM, Shock EL (1999) Environmental aqueous geochemistry of actinides. Rev Min 38:221-253

Nagy B, Gauthier-Lafaye F, Holliger P, Davis DW, Mossman DJ, Leventhal JS, Rigali MJ, Parnell J (1991) Organic matter and containment of uranium and fissiogenic isotopes at the Oklo natural reactors. Nature 354:472-475

Nozaki Y, Yamamoto Y, Manaka T, Amakawa H, Snidvongs A (2001) Dissolved barium and radium isotopes in the Chao Phraya river estuarine mixing zone in Thailand. Cont Shelf Res 21:1435-1448

Osmond JK, Cowart JB (1976) The theory and uses of natural uranium isotopic variations in hydrology. Atomic Energy Rev 14:621-679

Osmond JK, Ivanovitch M (1992) Uranium-series mobilization and surface hydrology. *In:* Uranium series disequilibrium: Application to Earth, Marine, and Environmental Sciences. Ivanovich M, Harmon RS (eds), Oxford Sciences Publications, Oxford, p 259-289

Osmond JK, Cowart JB (1982) Groundwater. *In:* Uranium series disequilibrium – Applications to environmental problems. Ivanovich M, Harmon RS (eds) Oxford Science Publications, Oxford, p 202-245

Pabalan RT, Turner DR (1997) Uranium(6+) sorption on montmorillonite: experimental and surface complexation modeling study. Aquat Geochem 2:203-226

Pabalan RT, Turner DR, Bertetti FP, Prikryl JD (1998) Uranium(VI) sorption onto selected mineral surfaces. *In:* Adsorption of Metals by Geomedia. Jenne EA (ed) Academic Press, San Francisco, p 99-130

Palmer MR, Edmond JM (1989) The strontium isotope budget in the modern ocean. Earth Planet Sci Lett 92:11-26
Palmer MR, Edmond M (1992) Controls over the strontium isotope composition of river water. Geochim Cosmochim Acta 56:2099-2111
Palmer MR, Edmond JM (1993) Uranium in river water. Geochim Cosmochim Acta 57:4947-4955
Pande K, Sarin MM, Trivedi JR, Krishnaswami S, Sharma KK (1994) The Indus River system (India-Pakistan): Major-ion chemistry, uranium and strontium isotopes. Chem Geol 116:245-259
Peterson FF, Bell JW, Dorn RI, Ramelli AR, Ku T-L (1995) Late Quaternary geomorphology and soils in Crater Flat, Yucca Mountain area, southern Nevada. Geol Soc Am Bull 107:379-395
Plater AJ, Dugdale RE, Ivanovich M (1988) The application of uranium series disequilibrium concepts to sediment yield determination. Earth Surf. Process. Landforms 13:171-182
Plater AJ, Ivanovich M, Dugdale RE (1992) Uranium series disequilibrium in river sediments and waters: the significance of anomalous activity ratios. Appl Geochem 7:101-110
Plater AJ, Dugdale RE, Ivanovich M (1994) Sediment yield determination using uranium-series radionuclides: the case of the Wash and Fenland drainage basin, eastern England. Geomorphology 11:41-56
Plater AJ, Ivanovich M, Dugdale RE (1995) ^{226}Ra contents and ^{228}Ra/^{226}Ra activity ratios of the Fenland rivers and the Wash, eastern England: spatial and seasonal trends. Chem Geol 119:275-292
Pliler R, Adams JAS (1962) The distribution of thorium and uranium in a Pennsylvanian weathering profile. Geochim Cosmochim Acta 26:1137-1146
Porcelli D, Swarzenski PW (2003)The behavior of U- and Th- series nuclides in groundwater. Rev Mineral Geochem 52:317-361
Porcelli D, Andersson PS, Wasserburg GJ, Ingri J, Baskaran M (1997) The importance of colloids and mires for the transport of uranium isotopes through the Kalix River watershed and Baltic Sea. Geochim Cosmochim Acta 61:4095-4113
Porcelli D, Andersson PS, Baskaran M, Wasserburg GJ (2001) Transport of U- and Th-series in a Baltic Shield watershed and the Baltic Sea. Geochim Cosmochim Acta 65:2439-2459
Prikryl JD, Jain A, Turner DR, Pabalan RT (2001) Uranium (VI) sorption behavior on silicate mixtures. J Cont Hydrol 47:241-253
Rama, Moore WS (1984) Mechanism of transport of U-Th series radioisotopes from solids into groundwater. Geochim Cosmochim Acta 48:395-399
Read D, Bennett DG, Hooker PJ, Ivanovich M, Longworth G, Milodowski AE, Noy DJ (1993) The migration of uranium into peat-rich soils at Broubster, Caithness, Scotland, UK. J Contaminant Hydrol 13:291-308
Richter FM, Turekian KK (1993) Simple models for the geochemical response of the ocean to climatic and tectonic forcing. Earth Planet Sci Lett 119:121-131
Riotte J (2001) Etude du déséquilibre ^{234}U-^{238}U dans les eaux de rivières. Cas du Strengbach, du Mont Cameroun et de l'Himalaya. PhD dissertation, Université Strasbourg 1
Riotte J, Chabaux F (1999) (^{234}U/^{238}U) activity ratios in freshwaters as tracers of hydrological processes: The Strengbach watershed (Vosges, France). Geochim Cosmochim Acta 63:1263-1275
Riotte J, Chabaux F, Benedetti M, Dia A, Gérard M, Boulègue J, Etamé J (2003) U colloidal transport and origin of the ^{234}U-^{238}U fractionation in surface waters: New insights from Mount Cameroon. Chem Geol (in press)
Robinson LF, Henderson GM, Slowey NC (2002) U-Th dating of marine isotope stage 7 in Bahamas slope sediments. Earth Planet Sci Lett 196:175-187.
Rona E, Urry WD (1952) Radium and uranium content of ocean and river waters. Am J Sci 250:241-262
Rosholt J, Doe B, Tatsumoto M (1966) Evolution of the isotopic composition of uranium and thorium in soil profiles. Geol Soc Am Bull 77:987-1004
Rosholt J (1982) Mobilization and weathering. *In:* Uranium series disequilibrium: Application to environmental problems. Ivanovich M, Harmon RS (eds) Oxford Sciences Publications, Oxford, p 167-180
Rosholt BJ (1985) Uranium-trend systematics for dating Quaternary sediments. US Geol Surv Open-file Rep 85-298
Rowe PJ, Maher BA (2000) "Cold" stage formation of calcrete nodules in the Chinese Loess Plateau: evidence from U-series dating and stable isotope analysis. Palaeogeogr Palaeoclimatol Palaeoecol 157:109-125
Sackett WM, Cook G (1969) Uranium geochemistry of the Gulf of Mexico. Trans. Gulf-Coast Ass Geol Soc 19:233-238
Sanchez F, Rodriguez-Alavarez MJ (1999) Effect of pH, conductivity and sediment size on thorium and radium activities along Jucar River (Spain). J Radioanal Nucl Chem 242:671-681

Sarin MM, Krishnaswami S, Somayajulu BLK, Moore WS (1990) Chemistry of U, Th, and Ra isotopes in the Ganga-Brahmaputra river system: Weathering processes and fluxes to the bay of Bengal. Geochim Cosmochim Acta 54:1387-1396

Schmeide K, Pompe S, Bubner M, Heise KH, Bernhard G, Nitsche H (2000) Uranium (VI) sorption onto phyllite and selected minerals in the presence of humic acid. Radiochim Acta 88:723-728

Schwarcz HP, Latham AG (1989) Dirty calcites 1. Uranium series dating of contaminated calcite using leachate alone. Chem Geol 80:35-43

Shock EL, Sassani DC, Betz H (1997) Uranium in geologic fluids: estimates of standard partial molal properties, oxidation potentials and hydrolysis constants at high temperatures and pressures. Geochim Cosmochim Acta 61:4245-4266

Scott MR (1982) The chemistry of U- and Th-series nuclides in rivers. *In:* Uranium series disequilibrium: Application to environmental problems. Ivanovich M, Harmon RS (eds) Oxford Sciences Publications, Oxford, p 181-201

Scott RD, MacKenzie RD, Alexander WR (1992) The interpretation of ^{238}U-^{234}U-^{230}Th-^{226}Ra disequilibria produced by rock-water interactions. J Geochem Explor 45:323-343

Sheng ZZ, Kuroda PK (1984) The α-recoil effects of uranium in the Oklo reactor. Nature 312:535-536

Sheng ZZ, Kuroda PK (1986a) Isotopic fractionation of uranium: extremely high enrichments of ^{234}U in the acid-residues of a Colorado Carnotite. Radiochim Acta 39:131-138

Sheng ZZ, Kuroda PK (1986b) Further studies on the separation of acid residues with extremely high ^{234}U/^{238}U ratios from a Colorado Carnotites. Radiochim Acta 40:95-102

Shirvington PJ (1983) Fixation of radionuclides in the ^{238}U decay series in the vicinity of mineralized zones: 1. The Austatom Uranium Prospect, Northern Territory, Australia. Geochim Cosmochim Acta 47:403-412

Short SA, Lowson RT, Ellis J (1988) ^{234}U/^{238}U and ^{230}Th/^{234}U activity ratios in the colloidal phases of aquifers in lateritic weathered zones. Geochim Cosmochim Acta 52:2555-2563

Short SA, Lowson RT, Ellis J, Price DM (1989) Thorium-uranium disequilibrium dating of late Quaternary ferruginous concretions and rinds. Geochim Cosmochim Acta 53:1379-1389

Spear JR, Figueroa LA, Honeyman BD (2000) Modeling the removal of uranium (VI) under variable sulfate concentrations by sulfate-reducing bacteria. Appl Envir Microbiology 66:3711-3721

Sturchio NC, Bohlke JK, Binz C (1989) Radium-Thorium disequilibrium and zeolite-water exchange in a Yellowstone hydrothermal environment. Geochim Cosmochim Acta 53:1025-1034

Sturchio NC, Bohlke JK, Markun FJ (1993) Radium isotope geochemistry of thermal waters, Yellowstone National Park, Wyoming, USA. Geochim Cosmochim Acta 57:1203-1214

Sun H, Semkow TM (1998) Mobilization of thorium, radium and radon nuclides in ground water by successive alpha-recoils. J Hydrol 205:126-136

Suzuki Y, Banfield JF (1999) Geomicrobiology of uranium. Rev Min Geochem 38:393-432

Swarzenski PW, Porcelli D, Andersson PS, Smoak JM (2003) The behavior of U- and Th- series nuclides in the estuarine environment. Rev Mineral Geochem 52:577-606

Szabo BJ, Rosholt JN (1982) Surficial continental sediments. *In:* Uranium series disequilibrium: Application to environmental problems. Ivanovich M, Harmon RS (eds) Oxford Sciences Publications, Oxford, p 181-201

Thiel K, Vorwerk R, Saager R, Stupp HD (1983) ^{235}U fission tracks and ^{238}U-series disequilibria as a means to study recent mobilization of uranium in archean pyritic conglomerates. Earth Planet Sci Lett 65:249-262

Tieh T, Ledger E, Rowe M (1980) Release of uranium from granitic rocks during in situ weathering and initial erosion (Central Texas). Chem Geol 29:227-248

Tricca A, Wasserburg GJ, Porcelli D, Baskaran M (2001) The transport of U- and Th-series nuclides in a sandy unconfined aquifer. Geochim Cosmochim Acta 65:1187-1210

Turner S, Van Calsteren P, Vigier N, Thomas L (2001) Determination of thorium and uranium isotope ratios in low-concentration geological materials using a fixed multicollector ICP-MS. J Anal Atom Spectrom 16:612-615

Toole J, Baxter MS, Thomson J (1987) The behavior of uranium isotopes with salinity change in three UK estuaries. Est Coast Shelf Sci 25:283-297

Viers J, Dupré B, Polvé M, Schott J, Dandurand L, Braun JJ (1997) Chemical weathering in the drainage basin of a tropical watershed (Nsimi-Zoetele site, Cameroon): comparison between organic-poor and organic-rich waters. Chem Geol 140:181-206

Vigier N (2000) Apport des séries de l'uranium sur les temps caractéristiques des processus géologiques: Différenciation de la chambre magmatique du volcan Ardoukoba (Rift d'Asal), contraintes sur l'érosion continentale (bassins de la Mackenzie et de la Tapti). PhD Thesis, UNiversité Paris VII, France

Vigier N, Bourdon B, Turner S, Allègre CJ (2001) Erosion timescales derived from U-decay series measurements in rivers. Earth Planet Sci Lett 193:549-563

von Gunten HR, Roessler E, Lowson RT, Reid PD, Short SA (1999) Distribution of uranium- and thorium series radionuclides in mineral phases of a weathered lateritic transect of a uranium ore body. Chem Geol 160:225-240

Webster I, Hancock G, Murray A (1995) Modeling the effects of salinity on radium desorption to ferrihydrite: application of a surface complexation model. Geochim Cosmochim Acta 59:2469-2476

Yang HS, Hwang DW, Kim G (2002) Factors controlling excess radium in the Nakdong river estuary, Korea: submarine groundwater discharge versus desorption from riverine particles. Mar Chem 78:1-8

Zielinski RA, Simmons KR, Orem WH (2000) Use of ^{234}U and ^{238}U isotopes to identify fertilizer-derived uranium in the Florida Everglades. Appl Geochem 15:369-383

Zuyi T, G Huanxin (1994) Use of the ion exchange method for determination of stability constants of thorium ions with humic and fulvic acids. Radiochim Acta 65:121-123

Zuyi T, Jinzhou D, Jun L (1996) Use of the ion exchange method for determination of stability constants of uranyl ions with three soil fulvic acids. Radiochim Acta 72:51-54

14   The Behavior of U- and Th-series Nuclides in the Estuarine Environment

Peter W. Swarzenski
Center for Coastal and Regional Marine Studies
US Geological Survey
Saint Petersburg, Florida 33701 U.S.A.
pswarzen@usgs.gov

Donald Porcelli
Department of Earth Sciences
University of Oxford
Parks Rd.
Oxford, OX1 3PR, United Kingdom
don.porcelli@earth.ox.ac.uk

Per S. Andersson
Laboratory for Isotope Geology
Swedish Museum of Natural History
Box 50007
104 05 Stockholm, Sweden
per.andersson@nrm.se

Joseph M. Smoak
Dept of Environmental Science and Policy
Univ. of South Florida
St. Petersburg, Florida 33701, U.S.A.
smoak@bayflash.stpt.usf.edu

1. INTRODUCTION

Rivers carry the products of continental weathering, and continuously supply the oceans with a broad range of chemical constituents. This erosional signature is, however, uniquely moderated by biogeochemical processing within estuaries. Estuaries are commonly described as complex filters at land-sea margins, where significant transformations can occur due to strong physico-chemical gradients. These changes differ for different classes of elements, and can vary widely depending on the geographic location. U- and Th-series nuclides include a range of elements with vastly different characteristics and behaviors within such environments, and the isotopic systematics provide methods for investigating the transport of these nuclides and other analog species across estuaries and into the coastal ocean.

There are numerous types and definitions of estuaries (see Dyer 1973). In this paper, it is simply considered to be a region where freshwater and seawater mix, to emphasize the chemical focus of the issues involved here. This region can either be contained within a river channel or extend onto the shelf, with effects often extending well into an adjoining ocean basin (Fairbridge 1980). These are complex ecological, biogeochemical, and hydrodynamic systems, with strong gradients in the concentrations and composition of micro- and macro particulate matter as well as dissolved organic and inorganic species. Particle-reactive radionuclides and trace elements being transported across estuaries are subject to partitioning between the dissolved, colloidal and particulate phases. The variable chemical nature of the U- and Th-series radionuclides is evident in their distributions in an estuarine environment. Such radionuclides can effectively serve as tracers to identify the sources, fate and transport of particles and colloids as well as other pollutants that behave similarly to these nuclides. These nuclides also have a wide range of half-lives that can be used to examine processes over various time scales. For example, Th and Ra daughter isotopes have half lives of 3.6 days to 75,200 years. Although most of the U- and Th-series radionuclides with half-lives less than ~100 yrs are commonly measured with standard counting techniques, recent advances in mass spectrometry have greatly improved the measurement of the long-lived nuclides, leading to smaller sample size as well as much higher precision. Studies involving high precision determinations of uranium isotopic compositions are now possible using mass spectrometric techniques (Goldstein and Stirling 2003), such as thermal ionization mass spectrometry, TIMS, (Chen et al. 1992) and more recently also multiple collector inductively coupled plasma mass spectrometry, MC-ICP-MS (Halliday et al. 1998; Henderson 2002) and Sector ICP-MS (Shiller and Mao 2000).

Understanding the behavior of radionuclides in estuaries, as the dynamic interface between the continental hydrochemical systems and the ocean basins, requires consideration of broader chemical cycling in the hydrosphere. In this volume, the behavior of U- and Th-series isotopes in rivers is discussed by Chabaux et al. (2003), that in groundwaters by Porcelli and Swarzenski (2003), and that in oceans by Cochran and Masque (2003). General background information is provided by Bourdon et al. (2003).

1.1. Estuarine mixing

When riverine dissolved species (metals, radionuclides, and other organic and inorganic contaminants) are discharged into an estuary, they can behave *conservatively* or *non-conservatively* during mixing. During conservative mixing, there is no net loss or gain of a particular constituent as a function of salinity, so that variations in concentrations are due only to dilution of river water (with ~0 salinity) and ocean water (~35 salinity). However, some elements show non-conservative behavior; that is, they are removed or added in an estuarine system during mixing. When the concentration of a chemical species is plotted against salinity, mixing between two components falls on a straight line if the species is conservative as a function of salinity, with the endmembers defined by river and seawater concentrations (Fig. 1). Where data fall below a line connecting these endmember components, removal occurs, and where data fall above the ideal dilution line, there must be additional inputs, presumably either from underlying sediments, from lateral mixing, or from submarine groundwater discharge. Boyle et al. (1977) pointed out that it is only where curvature in this line occurs that losses or gains take place, while elsewhere mixing is likely to be conservative. Unfortunately, in practice there are sometimes regions where such a distinction cannot be clearly made, due either to limited data coverage or multi-component mixing. Clearly, identification of the regions where inputs or removal occur requires good coverage across the salinity gradients to unambiguously define salinity-concentration relationships. Note that salinity is not exactly defined as it becomes more dilute, as the mixture of major elements is not likely to be the same as in seawater (see discussion in Millero and Sohn 1992). Strictly, conductivity or total dissolved solids (TDS) should be used. However, on the scale of effects seen in the species of interest here, such accurate specification of salinities is unnecessary.

An important issue that influences estuarine behavior is the determination of the riverine component. Variations in the riverine endmember may occur over timescales that are short compared to the residence times of water in the estuary, which result in nonlinear relationships between salinity and trace elements across the estuary as different riverine compositions progress through the estuary (Loder and Reichard 1981; Officer

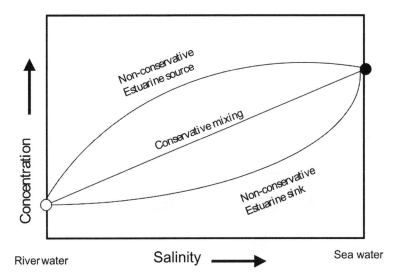

Figure 1. Schematic cartoon for idealized estuarine mixing of a dissolved component versus salinity, which serves as a conservative measure of the degree of mixing between freshwater and seawater. Redrawn after Berner and Berner (1987).

and Lynch 1981). Also, comparison of estuarine profiles with a river sample that was collected at another time may incorrectly define the conservative mixing benchmark. Note that this is a one-dimensional representation and assumes that concentrations normalized to salinity always remain constant, and so the plotted data follow the pattern across the estuary. This therefore does not take into account lateral differences across the estuary. Vertical redistributions of components can occur as well, as discussed in detail by Shiller (1996). This may be due to such mechanisms as the gravitational settling of biogenic particulates that have adsorbed solutes and are then oxidized and at least partially re-dissolved. Such a process can effectively redistribute nutrients and particle-reactive species, and may be overlooked in studies relying on suites of surface samples. Unfortunately, there is limited data available regarding the impact of these reservoirs on some of the U- and Th-series nuclides.

An important element for the cycling of trace elements, and so for considering the background to radionuclide behavior, is Fe. Fe is highly insoluble in oxidized waters, and is generally present in river water either on particles or in colloids (e.g., Kennedy et al. 1974). When discharge into an estuary occurs, Fe flocculates at very low salinities to form rapidly removed particulates (Sholkovitz 1977; Boyle et al. 1977). The result is that Fe concentrations drop off precipitously within the lowest salinity regime of an estuary. Since ferrous oxyhydroxides have highly reactive surfaces, this removal of Fe provides a vehicle for the associated removal of surface-reactive trace elements.

1.2. Estuary fluxes

Estuaries are clearly open systems, and several radionuclide fluxes must be considered to balance an estuarine mass budget (Fig. 2):

- Seawater inflow. The seawater endmember is relatively uniform, although coastal processes and upwelling can generate local variations. This is

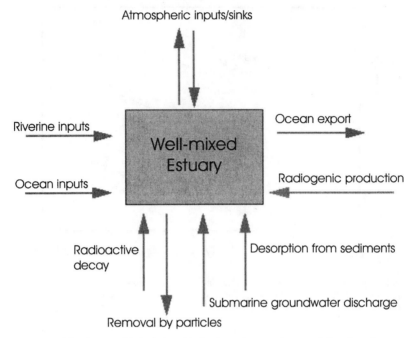

Figure 2. Simplified box model depicting the dominant input and removal functions for a particle reactive radionuclide such as ^{210}Pb in a well-mixed estuarine system.

particularly true of trace elements with short residence times in the ocean, such as Th, that have wide variations in concentration, as opposed to less reactive uranium.

- Atmospheric inputs. The atmosphere can be either a source or sink for select coastal radionuclides; for example, for ^{210}Pb the atmosphere can be a sizeable source, yet for ^{222}Rn the atmosphere is a sink.

- River inputs. The riverine endmember is most often highly variable. Fluctuations of the chemical signature of river water discharging into an estuary are clearly critical to determine the effects of estuarine mixing. The characteristics of U- and Th-series nuclides in rivers are reviewed most recently by Chabaux et al. (2003). Important factors include the major element composition, the characteristics and concentrations of particular constituents that can complex or adsorb U- and Th-series nuclides, such as organic ligands, particles or colloids. River flow rates clearly will also have an effect on the rates and patterns of mixing in the estuary (Ponter et al. 1990; Shiller and Boyle 1991).

- Removal to sediments. Removal of surface-reactive trace elements from the oceans readily occurs by adsorption onto settling particles, and this process is most pronounced in the typically high-energy, particle-rich estuarine environment. Particles are supplied by rivers, augmented by additions of organic material generated within the estuary. Also, flocs are created in estuaries from such components as humic acids and Fe. The interaction between dissolved and colloidal species is enhanced by the continuous resuspension of sediments in turbulent zones. Outputs to sediments by sediment-water interactions are also important. Marshes through which large proportions of river and estuarine water can filter through, and which are rich in organics and particles, provide local environments for trace element complexing and retention.

- Input from sediments. Sediments can release constituents that are produced within sediments, those that were bound onto surfaces during sedimentation, or those within phases that are mobilized by breakdown or transformation of the sediments. Therefore, the evolution of underlying sediments due to changes in such factors as redox conditions, which can redistribute Fe and Mn from phases that may have held radionuclides, must also be considered. Radionuclides can be transported into the overlying water through diffusion from interstitial waters, flushing of sediments, and resuspension of sediments. This includes turbidity maxima that can occur at intermediate salinities, and fluid muds with very high suspended particle concentrations.

- Submarine groundwater inputs to the coastal zone. In some areas, submarine groundwater discharge (SDG) is an important source of nutrients and other dissolved constituents. However, precise estimates of the magnitude and geographic location of submarine groundwater flow are still scarce, as SGD sites are difficult to identify and quantify. The contributions of SGD-derived species are important to an estuarine mass budget if either the discharge rates or the ratio of constituents in groundwater relative to coastal bottom waters is sufficiently large (Moore 1996, 1999).

Due to the complexities involved with identifying and quantifying these fluxes, defining the behavior of elements in an estuary has required more extensive and continuing studies at many locations. Note that the 20 largest world rivers account for only 30% of the global water discharge into the oceans. This makes a comprehensive accounting difficult, and the transfer or extrapolation of knowledge from one river-estuarine system to another, unstudied system uncertain and tenuous.

1.3. Colloids

Many of the earlier studies that were summarized by Moore (1992) did not include a discussion of the role of colloids in the removal and cycling of U- and Th-series radionuclides. Colloids are defined as particles or macromolecules with an equivalent spherical diameter in the size range of 1 nm to 1 μm (Buffle and van Leeuwen 1992). Colloids play a major role in the behavior and cycling of aquatic trace elements and radionuclides. Past studies generally defined all constituents passing through a 0.2 or 0.45 μm filter membrane as "dissolved" while the filter-retained phase was defined as "particulate" and comprised a size class that can effectively undergo gravitational settling. However, the "dissolved" constituents clearly contain not only dissolved ions and small inorganic as well as organic complexes that are traditionally considered in speciation calculations, but also larger constituents such as macromolecules and mineral phases (see discussion by Buffle and van Leeuwen 1992). The mass concentration of colloidal material in the coastal ocean and the open ocean has been found to be significantly higher than in the filter-retained particulate matter. Colloids have been shown to mediate the distribution of Th isotopes between dissolved species and settling particles in the oceans (Honeyman and Santschi 1989; Baskaran et al. 1992), and much of the colloidal materials undergo coagulation and are removed from the water column by settling particles. Once a particle-reactive radionuclide (or trace metal) is introduced to the water column, such as ^{234}Th produced from its dissolved parent, ^{238}U, it is taken up by colloids and removed by coagulation to form sinking particles and sedimentation

(Honeyman and Santschi 1989). Therefore, flocculation of colloids to form settling particles in estuaries is an important mechanism for trace element removal (Sholkovitz 1977). This is particularly true of Fe, which is a ubiquitous colloidal species and is removed at low salinities. Additional removal may occur by adsorption onto flocs, as demonstrated by mixing of organic-rich waters with seawater in the laboratory (Sholkovitz 1977).

Improvements in filtration techniques now allow separation of particles and colloids found in aqueous environments for studies of phase partitioning of U- and Th isotopes (Quigley et al. 2001). This has been crucial for studying the importance of particles and colloids for estuarine geochemistry. Previous studies isolated colloidal material using methods that were dependent upon the nature of the colloidal material, such as by adsorption on XAD resins. Isolation of colloidal material based on the molecular sizes from large volume water samples was developed using cross-flow ultrafiltration cartridges to investigate the size distribution of dissolved organic carbon and of trace metal carriers in coastal water (e.g., Moran and Moore 1989; Whitehouse et al. 1990; Baskaran et al. 1992; Swarzenski et al. 1995; Guo and Santschi 1996; Guo et al. 2000, 2001; Wen et al. 1997). Application of different ultrafiltration techniques that separate particles and colloids from dissolved species have been applied in a variety of hydrological and geochemical settings such as the Amazon and the sub-arctic boreal Kalix River system (Swarzenski et al. 1995; Porcelli et al. 1997; Andersson et al. 2001).

A comprehensive discussion of ultrafiltration methods is given by Buffle and van Leeuwen (1992). Generally, materials >0.45 µm are first separated using conventional filtration techniques and are considered "particles." A commonly used and effective method to further separate the filtrate has been cross-flow ultrafiltration, which uses high flow rates perpendicular to the filter surfaces to minimize buildup of excluded material on membrane surfaces. A portion of the sample passes through the membrane and this ultrafiltrate containing <1 kD or <10 kD material (here nominally referred to as "solutes") is collected separately. The remainder of the sample, along with colloids excluded from passage through the membrane, is recirculated until the ultrafiltered water that has passed through the walls of the fibers is ~90% of the initial sample volume. The remaining circulating sample, the colloid concentrate, contains colloids often enriched in concentration by ~10. After each sample is filtered, acid (and sometimes base) washes are circulated through the system to remove material trapped on the filter. These are retained separately for analysis and are sometimes assumed to contain only colloidal material. An experimental set-up for seawater Th is shown by Baskaran et al. (1992).

Several caveats should be considered in evaluating ultrafiltration data.

- The separation of colloids is set by operational criteria only. Filters are generally rated by the size exclusion of some globular macromolecule that may have a size of 1 kD to 10 kD. However, other molecules with different shapes may not be separated as predicted by their size alone, and this effect may vary between filtering systems and filter materials. A comparison between ultrafiltration systems for seawater analyses (Buesseler et al. 1996) found that there might be significant differences in the fraction of trace constituents that pass through the filters, depending upon the composition of the filter material. Therefore, while qualitative characteristics can be more readily considered, quantitative values for the association with colloids must be considered to be somewhat dependent upon operational conditions.

- The operational separation of colloids does not directly identify the nature of the colloid fraction, which may consist of clays, Fe and Mn oxyhydroxides, or organic compounds such as humic and fulvic acids. Colloids may directly bind to trace elements, or, due to the very large specific surface area, adsorb constituents. The likelihood that the colloid fraction is often a complex mixture of components hampers relating ultrafiltration data to experimental data or thermodynamic calculations (see Gustafsson and Gschwend 1996). For example, Fe not only forms colloids but also combines with humic acids, so that it is often difficult to separate the association of these components with trace elements from field studies.

- The operational distinction between colloids and particles may not accurately describe inherent physico-chemical differences in surface sites and scavenging capabilities. Continuous aggregation/dissaggregation processes produce a particle size continuum (Baskaran et al. 1992; Gustafsson and Gschwend 1996; Gustafsson et al. 2000a,b). For example, settling characteristics will vary considerably; dense mineral phases, such as Fe-Mn oxyhydroxides, will have higher settling velocities while more loosely associated organic aggregates that might be considered part of the colloid pool may remain as non-settling material.

- Comparison between the inventories in the collected fractions (the colloidal and ultrafiltered fractions) and in the starting sample often indicate that there are losses of nuclides on to the ultrafiltration cartridge. These are largely recovered by subsequent acid rinses of the ultrafilters and filtration system. It is not clear whether the recovered abundances should be considered part of the colloids retained by the filter, or solutes that have adsorbed in the system (Gustafsson et al. 1996; Andersson et al. 2001), even though test experiments with colloidally-bound ^{234}Th showed significant losses in the ultrafiltration system (Baskaran et al. 1992.)

Overall, colloids appear to play a fundamental role in the behavior of radionuclides and trace elements, and while ultrafiltration data must be treated with some caution, it provides valuable information. Other methods may soon be developed to directly address some of these difficulties, although for species such as Th, processing sufficient volumes of material for analysis will continue to remain a major challenge.

2. URANIUM

The marine geochemistry of uranium has been studied for several decades, although more recently progress has been made in understanding geochemical reactions in the estuarine environment that greatly impact the terrestrial flux of uranium to the ocean. Some of the first determinations of uranium in the marine environment dates back to the 1930's (Hernegger and Karlik 1935; Föyn et al. 1939). Although the uranium data for the open ocean displayed a large range in concentration it was suggested that "coastal seawater," with lower salinity, showed lower concentration of uranium compared to seawater from the open ocean (Föyn et al. 1939). Holland and Kulp (1954) reviewed the status of the knowledge of uranium in the marine environment and suggested that there was a need for improvement in the determination of uranium concentration in river and ocean water to be able to obtain a quantitative understanding of the geochemical cycling of uranium. However, it was not until the 1960's that an "oceanic average" for the uranium concentration became generally accepted (Burton 1965). In contrast to uranium concentrations, reported (^{234}U/^{238}U) activity ratios for open ocean waters showed less variation and with a value of about 1.15 (Koide and Goldberg 1964), a value earlier suggested from the study of marine carbonates (Thurber 1962). Determination of uranium in river and brackish water from the Baltic Sea showed a general correlation between salinity and U and it was also suggested that U was being removed when uranium (VI) is reduced to uranium (IV) in anoxic waters (Koczy et al. 1957). It was not until the late 1970's that studies of uranium estuarine chemistry continued (Martin et al. 1978a,b).

2.1. U in seawater

Uranium appears to behave conservatively in oxygenated seawater because of the formation of a variety of stable and soluble uranyl (U^{6+}) carbonate complexes (Langmuir 1978; Djogic and Branica 1991), and so there are no strong associations with particles or colloids. The uranium concentration in seawater, normalized to a salinity of 35, is 13.60 nmol/kg (Chen et al. 1986) and the residence time is estimated to be in the range of 0.2 to 0.4×10^6 years (Ku et al. 1977). The activity ratio ($^{234}U/^{238}U$) is higher than secular equilibrium and appears to be constant, with a value of 1.144 (Chen et al. 1986; Cheng et al. 2000). This reflects the process of α-recoil during weathering and preferential release of ^{234}U into continental waters, which keep seawater above secular equilibrium. When examining uranium behavior during estuarine mixing, the seawater component can be considered well defined in uranium concentration as well as ($^{234}U/^{238}U$) ratio.

2.2. River water U inputs

The characteristics of riverine U are reviewed in Chabaux et al. (2003). Continental weathering releases the naturally occurring U isotopes ^{238}U, ^{235}U and ^{234}U to solution in oxidizing environments. Uranium is transported to the ocean by rivers and the riverine uranium concentrations vary considerably between different river systems, with an average value of about 1.3 nmol/L (Bertine et al. 1970). This value was later confirmed by a study of more than 250 rivers (Palmer and Edmond 1993). However, it was concluded that the world average river concentration was biased by very high levels observed in the Ganges-Brahmaputra and Yellow Rivers. In a recent study another 29 rivers were added, but these new data do not significantly change the world average river value of about 1.3 nmol/L (Windom et al. 2000). The ($^{234}U/^{238}U$) activity ratios in river waters also show large variations, mainly related to weathering intensity, with values close to equilibrium when intense weathering prevails (Osmond and Ivanovich 1992). Overall, due to large differences both in U concentration and ($^{234}U/^{238}U$) activity ratio between different rivers, the freshwater component can vary substantially between different estuaries.

A study of the Kalix River in northern Sweden monitored seasonal chemical changes. During times of increased discharge, many inorganic constituents are diluted by factors of up to 4 (Ingri 1996). However, U concentrations fall in a much more restricted range, probably due to buffering of U-bearing organic deposits in the watershed (Porcelli et al. 1997). In contrast, the isotopic composition to U varies significantly (Andersson et al. 1995). It is clear that the river and estuary must be sampled simultaneously in order to define the relevant estuarine river inputs (see Swarzenski and McKee 1998).

2.3. U behavior in estuaries

Thorough reviews of U data from the estuarine and coastal zone by Cochran (1992) and Moore (1992) established that different estuarine systems behave differently and that chemical reactions in the estuary may affect the flux of uranium through the system. Early studies of U behavior from the Zaire (Martin et al. 1978a), Gironde, France (Martin et al. 1978b), the Narbada, Tapti and Godavari estuaries, India (Borole et al. 1977; Borole et al. 1982), and estuaries in the UK (Toole et al. 1987) suggest that U behaves conservatively at these locations. In contrast, examples of substantial removal were found in the Ogeechee and Savannah Rivers in Georgia (USA) (Maeda and Windom 1982), and were attributed to precipitation of iron and manganese and/or flocculation of organic matter during low river discharge. In the low-salinity zone of the Narbada estuary in

India (Borole et al. 1982) and the Forth estuary in the UK (Toole et al. 1987), non-conservative behavior of uranium was also demonstrated. In the Amazon estuary, uranium showed elevated concentrations compared to simple mixing (McKee et al. 1987). Release of uranium from bottom sediments on the shelf was suggested to be a source of dissolved (<0.4 μm) uranium. However, subsequent studies in the Amazon also demonstrated that U removal (Fig. 3) occurred at salinities <12 (Swarzenski et al. 1995, Swarzenski et al. 2003). Overall, it was established that the behavior of U is highly variable; examples have been found of conservative behavior as well as both additions and removal of U by interaction with sediments.

The extensive review by Moore (1992) concluded that a better understanding of U estuarine behavior was central to establishing a quantitative budget for the ocean. It was also stressed that the transport of U on colloids and particles and on authigenic coatings on larger particles might be of importance for the estuarine behavior of U. Improved analytical techniques to measure low uranium concentrations and high precision

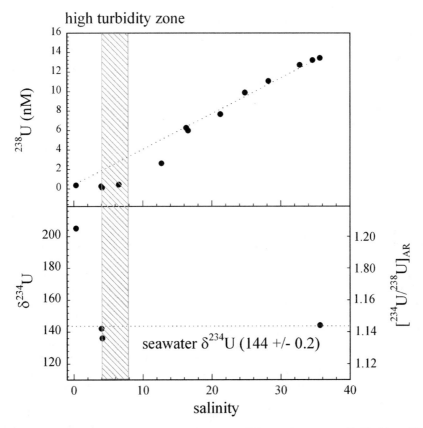

Figure 3. The concentration of uranium (nM) versus salinity on the Amazon Shelf with an ideal dilution line drawn through the riverine and seawater end members. Removal of dissolved U is evident at salinities that range from 0 to 16. The bottom illustration shows δ^{234}U as a function of salinity for the same waters on the Amazon Shelf. A seawater value (144 ± 0.2) is rapidly reached at a salinity of 4 during estuarine mixing. The high turbidity zone of the water column is defined by the greatest suspended particulate concentrations. Data from Swarzenski et al. (2003).

determination of (^{234}U/^{238}U) activity ratios were suggested to be essential for a better understanding of the U estuarine behavior. During the 1990's, studies of uranium during estuarine mixing continued. In the Mahanadi River-estuarine system in India, (Ray et al. 1996) and from the sediment-rich Fly River estuary of Papua New Guinea, uranium appears to exhibit reasonably conservative behavior (Swarzenski et al. 2003).

The Ganges-Brahmaputra River system supplies about 10% of the estimated global river flux of dissolved U. Non-conservative behavior, with removal of U during low discharge, was reported in the low salinity region, with <12 salinity, of the Ganges-Brahmaputra mixing zone (Carroll and Moore 1994) as well as in the nearby Hooghly estuary (Somayajulu 1994). It was suggested that U removal occurs in the organic-rich sediments of the mangrove forest due to redox processes but no detailed removal mechanism was presented (Carroll and Moore 1994).

In the Delaware and Chesapeake estuaries (USA), uranium shows distinctly non-conservative behavior at salinities <5 (Sarin and Church 1994; Church et al. 1996). This was suggested to be due to sedimentary redox processes in the extensive salt marshes in the Delaware and Chesapeake bays. From mass balance calculations it was concluded that almost two-thirds of the uranium in the tidal waters were retained in the sediments. It was also suggested that, extrapolated globally, uranium removal in salt marshes and marine wetlands, including mangroves, are important sinks for U that may responsible for up to 50% of the total marine removal (Church et al. 1996). Removal of U is also observed within the Baltic Sea, related to the association of U with colloids (see Section 2.5).

In contrast, U within surface waters of the Mississippi shelf-break region predominantly displays conservative behavior. U behavior therefore is not foremost controlled by traditional sorption and/or desorption reactions involving metal oxides or colloids. Mixing of the thin freshwater lens into ambient seawater is largely defined by wind-driven rather than physical processes. As a consequence, in the Mississippi outflow region uranium removal is evident only during anomalous river discharge regimes (Swarzenski and McKee 1998).

2.4. Uranium removal to anoxic sediments

There are a few estuarine environments where seasonal anoxia exists and it is important to review the behavior of U in those environments (Cochran et al. 1986; Barnes and Cochran 1990, 1993). Reduction of U(VI) to insoluble U(IV) during diffusion across the sediment-water interface in organic-rich reducing sediments has been suggested to be a major sink of uranium in the ocean (Klinkhammer and Palmer 1991). Accumulation of uranium in organic rich anoxic sediments has been documented from anoxic basins around the world (Cochran 1992). U oxidation/reduction processes are also possible in a water column. Examination of the behavior of uranium isotopes in estuarine waters of the Baltic Sea showed that in the periodically anoxic deep water from a depression in the central Baltic Sea has dissolved uranium concentrations that are depleted by up to 48% compared to those calculated assuming a linear relationship between salinity and U concentration (Löfvendahl 1987; Andersson et al. 1995). This removal is attributed to reduction of U(VI) to insoluble U(IV). Work done in Framvaren Fjord in southern Norway found that U in the water column was depleted in the anoxic waters below the oxic surface layer, but this depletion is initiated not at the redox boundary but at a maximum in microbial activity (McKee and Todd 1993) due to redox cycling of Fe and Mn carriers (Swarzenski et al. 1999a,b). However, direct measurements of U oxidation states (Anderson 1987; Anderson et al. 1989; Swarzenski et al. 1999b) indicated that U remains mostly in the U(VI) form, implying that U reduction likely requires catalysis on sedimentary surfaces.

2.5. Importance of particles and colloids for controlling estuarine uranium

Several studies have examined the partitioning of U on particles and colloids. Results from detailed sampling and particle separation in the Amazon estuary shows that most of the uranium at the Amazon River mouth is associated with particles (>0.4 μm) and that >90% of the uranium in filtered water (<0.4 μm) is transported in a colloidal phases (from a nominal molecular weight of 10 000 MW up to 0.4 μm) (Swarzenski et al. 1995; Moore et al. 1996). Mixing diagrams for uranium in different size fractions in the Amazon estuary reveal that uranium in all size fractions clearly display both removal and substantial input during mixing.

Concentrations of colloidal U comprise up to 92% of the dissolved U fraction at the river mouth and attain highest values in the productive, biogenic region of the Amazon shelf (salinities above approximately 20). Both colloidal and dissolved (i.e., passing through ultrafilters) phases are highly nonconservative relative to ideal dilution of river water and seawater, indicating extensive removal at salinities below approximately 10. Saltwater-induced precipitation and aggregation of riverine colloidal material is most likely the dominant mechanism of U removal in the low salinity region of the Amazon shelf. There is evidence of a substantial colloidal U input (approximately 245% of the riverine colloidal U flux) into surface waters above a salinity of 5. Such enrichment of colloidal U most likely is the result of colloidal U-rich porewater advection, or U diffusion within the porewaters, from the seabed and fluid muds or shelf-wide particle disaggregation. The colloidal fraction rapidly becomes less significant and at salinities approaching 35 only about 15% of the uranium is in the colloidal phase.

The Kalix River drains into the Baltic Sea, which is a broad, shallow estuarine environment, with relatively gradual and stable salinity gradients. The Kalix River estuary and the Baltic Sea have been subject to a comprehensive research program on a variety of uranium series radionuclides (Andersson et al. 1995, 1998, 2001; Porcelli et al. 1997, 2001). At the river mouth the uranium transport is also found to be dominated by particles (>0.45 μm) and >90% of the uranium in the filtered water is also associated with a colloidal phase (3 kD – 0.2 μm) (Porcelli et al. 1997; Andersson et al. 1998; Andersson et al. 2001). Upon entering the estuary, the colloidal fraction still dominates at low salinities, with >80% at about a salinity of 1, but rapidly becomes less significant and at salinities of about 3 the colloid-bound uranium is insignificant (Fig. 4). A closer examination of the uranium concentration in unfiltered water, along with 0.2 μm- and 3 kD- filtered waters from the Kalix River estuary low salinity zone (0 to 3) all show linear trends that seem to reflect simple conservative mixing (Fig. 4), and this is confirmed using U isotope variations (see Section 2.6 below). The uranium concentrations in the filtered water are lower than those in the corresponding unfiltered waters and the uranium concentration increases with salinity but the difference between the different size fractions is less than 5% at a salinity of 3. The rapid decrease of colloid-bound U indicates that U desorbs rapidly from the colloidal carriers due to decreasing colloid concentration and the stabilization of dissolved uranyl carbonate complexes in the estuary (Andersson et al. 2001). The observation that the U concentration in the Kalix River water is above the extrapolated freshwater input to the estuary, implied by the conservative mixing relationships, suggest that there must be removal at very low salinities so that the freshwater input falls upon the mixing lines for low salinity waters. It is likely that during initial estuarine mixing, riverine uranium associated with Fe-organic rich colloids, is removed. This removal is due to colloidal aggregation into larger particles that can sink on a short time-scale. Note that U removal is not likely to occur due to adsorption onto inorganic colloids, as adsorption is diminished in higher ionic strength waters. Rather, U is likely removed as U-humate complexes discharged from the

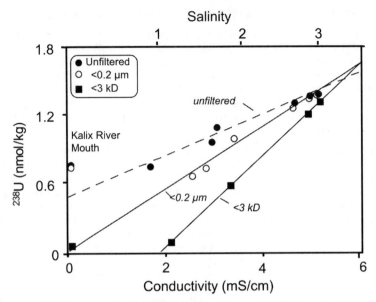

Figure 4. The uranium concentration in unfiltered water, 0.2 μm and 3 kD filtered water in river water from the Kalix River mouth and samples from the low salinity estuarine zone (0-3). Data plotted against conductivity (although the salinity scale is not defined below 2, a tentative scale is indicated). The lines represent the best fit for each fraction in the estuary. The data from the Kalix river mouth represent the river water component, which show <10% annual variation in concentration. The analytical errors are smaller than the symbols. Data from Andersson et al. (2001). Copyright 2001 Elsevier Science.

Kalix. This is consistent with the suggestion of Mann and Wong (1993) that the speciation of U may be important for estuarine behavior, based upon evidence that a significant fraction of U in coastal waters with relatively high concentrations of DOC were associated with organic matter. Concurrently, there is also desorption and re-association to other colloidal carrier phases, probably other organic phases such as humic substances (Andersson et al. 2001), but which do not cause further U removal. The results from the Kalix River estuary show that uranium adsorbs and desorbs from a variety of colloidal carrier substances that alternately dominate at different stages during river – estuarine mixing.

Evidence for the association of U with humic acids has been documented elsewhere. Dearlove et al. (1991) showed that U concentrated by ultrafiltration techniques from organic-rich groundwater samples were associated with humic colloids. Humic and fulvic acids have been shown to strongly complex U. Lienert et al. (1994) modeled the distribution of U species in the Glatt River and concluded that U-humate complexes become important at pH < 6.8. These results reinforce the conclusions in the estuarine studies that U humate and fulvate complexes may account for the association of U with colloids.

2.6. The (^{234}U/^{238}U) activity ratios in estuaries

Measuring the isotopic composition of U in estuaries has the potential for further constraining the interpretations of uranium behavior. However, this has been hampered by large uncertainties in conventional methods using counting techniques. While rivers often display (^{234}U/^{238}U) activity ratios above equilibrium, the ratios generally do not represent a strong contrast to the value for the ocean, and due to the high oceanic U concentration, the isotopic composition of U in the estuaries rapidly approaches the value for the ocean. The development of mass spectrometric methods to obtain highly precise isotopic compositions has made it possible to document more subtle variations, and on smaller samples, than previously possible (Goldstein and Stirling 2003), and so examine U mixing, at least at low salinities. Mass spectromentric data are generally reported as per mil deviations from equilibrium, such that $\delta^{234}U=[((^{234}U/^{238}U)_s/(^{234}U/^{238}U)_{eq}))-1]\times 10^3$, where $(^{234}U/^{238}U_{eq})$ is the secular equilibrium ratio of 1 and $(^{234}U/^{238}U)_s$ represents the measured ratio in the sample.

The first high-precision determination of $(^{234}U/^{238}U)$ ratios in estuarine waters were reported from the Baltic Sea and showed significant deviations from that of seawater detected at salinities as high as 12 (Andersson et al. 1995). Determination of $\delta^{234}U$ in filtered water, particles and colloids has revealed that particles, colloids and dissolved fraction show no isotopic differences between the different size fractions (Porcelli et al. 1997; Andersson et al. 1998; Andersson et al. 2001). The lack of significant variation in isotopic composition between different phases within samples demonstrates that substantial and rapid isotope exchange must occur between U in particles, colloids, and dissolved species in both river and estuarine waters. Note that significant isotopic differences were found by Dearlove et al. (1991) between colloid-bound and "solute" U in groundwaters, indicating that isotopic exchange between these U species was limited. A detailed examination of $\delta^{234}U$ in the Kalix River and estuary revealed that there were indeed small differences in isotopic composition between the different carrier phases, as well as in sinking material collected in sediment traps (Fig. 5). The bulk $\delta^{234}U$ decrease from high values in the river water to those approaching the seawater value at increasing

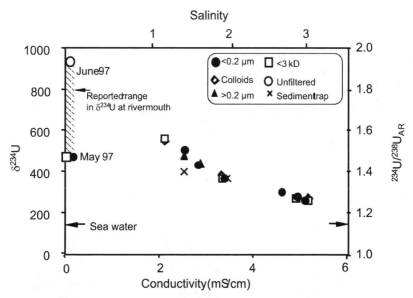

Figure 5. The $\delta^{234}U$ in 0.2μm and 3 kD filtered water and colloids phase (3kD – 0.2μm) and particles (>0.2 μm) as well as material from sediment traps plotted versus conductivity in the low salinity zone (0-3) of the Kalix River estuary. The stippled area marks the reported annual range in $\delta^{234}U$ at the Kalix river mouth, which show a substantial variation compared to the uranium concentration. Data from Andersson et al. (2001). Copyright 2001 Elsevier Science.

salinities (Fig. 5). The $\delta^{234}U$ also indicate removal of uranium at very low salinities and was used to estimate the amount of uranium removal in the inner part of the Kalix River estuary and these data indicated a significant uranium loss of up to 50% (Andersson et al. 2001). The most important finding from the uranium isotope study is that continuous substantial and rapid isotopic exchange must occur between particles, colloids and dissolved phases in river water as well as during estuarine mixing (Andersson et al. 2001; Swarzenski et al. 2003). Such behavior cannot be readily inferred from the U concentration data alone (see Fig. 4).

3. THORIUM

Thorium is considered to be one of the most highly particle reactive radionuclides in natural waters. Therefore, it is efficiently removed from the dissolved (i.e., <0.4 µm) phase onto colloids and particulates during its residence time in aqueous systems. Thorium exhibits this affinity for particle surfaces in both fresh and marine waters. Therefore most of the thorium entering an estuary is already associated with particulates and colloids. Short-lived Th isotopes that do exist in the dissolved phase are largely produced from dissolved parents in an estuarine environment. Dissolved thorium will rapidly adsorb onto particles in the estuary and much of the thorium will be removed from the water column during estuarine mixing. In estuarine environments, thorium is of interest mainly as a tracer of sediment mixing and as a proxy for other particle-reactive species. There are six thorium isotopes in the naturally occurring uranium and thorium decay series. They are: ^{232}Th ($t_{1/2}$ = 1.4 × 10^{10} y), ^{230}Th (7.5 × 10^{4} y), ^{228}Th (1.9 y), ^{234}Th (24.1 d), ^{227}Th (18.7 d) and ^{231}Th (1.1 d). All have been utilized in the estuarine environment with the exception ^{227}Th and ^{231}Th, which not only have relatively short half-lives but also low abundances, making measurement of these isotopes difficult. The most commonly applied estuarine tracer is ^{234}Th.

3.1. ^{234}Th

^{234}Th is produced by ^{238}U, which is present in low concentrations in river waters; therefore most of the ^{234}Th found in estuaries is supplied either in situ or from ocean water. The first ^{234}Th measurements in coastal waters revealed increased removal from dissolved to particulate phase with increasing proximity to shore (Bhat et al. 1969).

The residence time of ^{234}Th, with respect to removal by scavenging, can be calculated by assuming that the dominant source of ^{234}Th is by decay of in situ ^{238}U (Kaufman et al. 1971; Porcelli et al. 2001)

$$^{234Th}\tau_{SED} = {}^{234Th}\tau \Big/ \left[\left(\frac{^{238}U}{^{234}Th}-1\right)\right] \qquad (1)$$

where $^{234Th}\tau = 1/\lambda_{234Th}$ is the mean life of ^{234}Th. Some of the first ^{234}Th estuarine measurements where made in Long Island Sound by Aller and Cochran (1982) to determine sediment mixing rates. These authors further examined the (^{234}Th/^{238}U) activity ratios in the water column of Long Island Sound to derive residence times on the order of 1-10 days.

In more recent studies, Feng et al. (1999) calculated a Th water column residence time of 2 to 12 days in the Hudson River estuary. McKee et al (1986b) determined that ^{234}Th was removed on a time scale of a day or less in the very particle-rich environment of the Yangtze River estuary. In the Amazon River estuary, another particle-rich environment, McKee et al. (1986a) determined that the residence time of dissolved ^{234}Th ranged from 2 to 4 days. McKee et al. (1986a) also calculated apparent distribution coefficients between particulate and water in the Amazon that were much lower than expected. These low distribution coefficients implied that soluble and particulate phases of ^{234}Th are not in equilibrium because particle residence times are short relative to sorption times of ^{234}Th. McKee et al. (1986b) found locations with the shortest dissolved ^{234}Th residence times due to high turbidity had the longest particle residence times due to resuspension. Honeyman and Santschi (1989) suggested that the relatively slow sorption rates are due to a two-step process, sorption onto colloids coupled with colloid aggregation, in which colloid aggregation was the rate-limiting step. Moore et al. (1996) found 40-90% of the soluble phase ^{234}Th in the Amazon shelf waters is in the colloidal fraction. In contrast, in the Baltic, high (^{234}Th/^{238}U) ratios indicate that the residence time of ^{234}Th with respect to particle removal is long (~50 days), largely due to low particle and colloid concentrations (Porcelli et al. 1997).

Baskaran and Santschi (1993) examined ^{234}Th from six shallow Texas estuaries. They found dissolved residence times ranged from 0.08 to 4.9 days and the total residence time ranged from 0.9 and 7.8 days. They found the Th dissolved and total water column residence times were much shorter in the summer. This was attributed to the more energetic particle resuspension rates during the summer sampling. They also observed an inverse relation between distribution coefficients and particle concentrations, implying that kinetic factors control Th distribution. Baskaran et al. (1993) and Baskaran and Santschi (2002) showed that the residence time of colloidal and particulate ^{234}Th residence time in the coastal waters are considerably lower (1.4 days) than those in the surface waters in the shelf and open ocean (9.1 days) of the Western Arctic Ocean (Baskaran et al. 2003). Based on the mass concentrations of colloidal and particulate matter, it was concluded that only a small portion of the colloidal ^{234}Th actively participates in Arctic Th cycling (Baskaran et al. 2003).

^{234}Th has been widely used to determine sediment mixing rates on a time scale of a few months. In an early study, (Aller and Cochran 1976) determined that the top 4-5 cm of sediment were mixed based on the ^{234}Th profile in Long Island Sound. More recently, Fuller et al. (1999) found ^{234}Th in sediments to a depth of up to 10 cm. Other sediment mixing rate calculations were made in the New York Bight, the Yangtze estuary (McKee et al. (1983), the Amazon estuary (McKee et al. 1986b) and Narragansett Bay (Santschi et al. 1979).

McKee et al (1983) used ^{234}Th to determine short-term deposition rates from the Yangtze River. The sediment ^{234}Th activity profiles were determined to largely reflect deposition because radiographs revealed distinct layers that would not exist if the sediment was mixed, and so could be used to calculate that deposition was on the order of 4.4 cm per month. However, this did not agree with long-time scale tracers. Therefore, it was suggested that winter storms periodically erode the sediment and transport it elsewhere reducing the long-term accumulation rates.

In the Amazon River estuary, DeMaster et al. (1986) determined ^{234}Th seabed inventories were in excess of those produced in the overlying water column. They suggest the excess ^{234}Th was probably supplied by lateral transport of offshore water. However, Moore et al. (1996) found that ^{234}Th seabed inventories were only in excess with respect to production on the outer portion of the Amazon shelf in the absence of nearbed high-concentration suspensions (i.e., fluid mud). Smoak et al. (1996) further found a large percentage of the excess inventory in the Amazon estuary is in the thick fluid mud layer and not immediately incorporated into the seabed. The distribution of ^{234}Th in the Amazon estuary therefore appears to be related to the formation and redistribution of the fluid-mud layer (Smoak et al. 1996), which is a function of the spring-neap tidal cycle dynamics (Kineke et al. 1996).

3.2. ^{228}Th

^{228}Th has a half-life of 1.91 years and is produced from ^{228}Ra, with ^{228}Ac as a very short-lived intermediate daughter. The first detailed investigations on the distribution of ^{228}Th in the coastal and open ocean was published by Broecker et al. (1973). They showed that the median activity ratio of (^{228}Th/^{228}Ra) was 0.5, corresponding to a mean removal time with respect to transport into sediments of 2.7 years. ^{228}Th can be used like ^{234}Th for sediment mixing studies, although on a somewhat longer time scale. However, the generation and release of ^{228}Ra ($t_{1/2}$ = 5.7 yr) in an area of sediment mixing complicates interpretations (Cochran 1992). ^{232}Th produces ^{228}Ra within the underlying sediment which in turn is readily dissolved (when recoiled into the pore water) and migrates towards the sediment-water interface where daughter ^{228}Th then accumulates. Therefore, surface sediment has a distinct (^{228}Ra/^{228}Th) ratio. Hancock (2000) used the release of Ra from the seabed to examine resuspended sediment in the Bega River estuary, Australia, where the (^{228}Th/^{232}Th) ratio was used to distinguish resuspended sediment from recent river sediment.

3.3. Long-lived Th isotopes-^{232}Th and ^{230}Th

While various studies have examined the behavior of long-lived Th isotopes in the oceans (see Cochran 1992), there is still comparatively little data for rivers and estuaries. Moore (1967) published the first values for ^{232}Th concentrations in two major rivers, the Amazon and Mississippi. The ^{232}Th concentrations were found to be an order of magnitude higher than that of Atlantic Ocean water, although it was noted that Th on particulates may not have been completely excluded in the analyses. Other data for filtered ^{232}Th in rivers are very limited (Andersson et al. 1995), and it is generally assumed that all the Th is on fine particles. Colloids and ferro-oxyhydroxides can be important for riverine U transport (Porcelli et al. 1997; Andersson et al. 1998), and are also possible carriers of Th (Viers et al. 1997). In general, non-particulate Th has been found in concentrations above the solubility of thorianite (Langmuir and Herman 1980), and so is generally dominantly present on colloids (Porcelli et al. 2001).

There is little data on the estuarine behavior of long-lived, river-supplied Th isotopes, although very little Th is likely to enter the ocean from rivers, as much is removed in the estuary. More recently, mass spectrometric methods have allowed the high precision measurements of ^{232}Th and ^{230}Th in low concentration aqueous samples (Chen et al. 1986; Guo et al. 1995). This method was employed by Andersson et al. (1995) in the Baltic Sea estuarine environment. They found ^{232}Th decreased very rapidly between salinities 0 and 5 and at a salinity of about 7 approaches the level found in seawater. A large portion of ^{232}Th transported by rivers is removed as soon as the river water enters the estuarine environment. Andersson et al. (1995) also found unusually high (^{230}Th/^{232}Th) ratios that suggested different behavior between ^{230}Th and ^{232}Th, although lower ratios were found in samples from the same locations in a later study (Porcelli et al. 1997). The reasons for the differences between these two studies is unknown. At higher salinities further into the Baltic, ^{232}Th concentrations are much lower, and the ^{232}Th budget is dominated by eolian inputs, with river-borne ^{232}Th largely lost to sediments at discharge points. Sedimentation rates calculated from the water column data are compatible with measured sedimentation rates (Porcelli et al. 1997). It was also found that a substantial fraction of the ^{232}Th was on particles. Values of (^{234}Th/^{232}Th) activity ratios in unfiltered water and on particles are similar, indicating that there is isotopic equilibrium of Th in particles and "dissolved" phases.

The association of Th with colloids may be an important precursor to incorporation in sedimenting particles (Honeyman and Santschi 1989). There is limited data available regarding the distribution of ^{232}Th on colloids. In the Baltic, only a small fraction of the ^{232}Th appears associated with colloids at low salinities, since 85% of the ^{232}Th not on particles passed through the ultrafilter (Porcelli et al. 2001). In contrast, Baskaran et al. (1992), found that a dominant fraction of the Th in coastal waters of the Gulf of Mexico are associated with colloids. Mass balance calculations indicated that the terrigenous silicate-derived colloids (which are not expected to readily exchange Th with surrounding species) accounted for only <6% of the total colloidal mass concentration, implying that >94% is of biogenic origin that may have Th on more exchangeable surface sites. Kersten et al. (1998) used model calculations to estimate that 98% of the ^{234}Th in waters of Mecklenburg Bay in the southern Baltic is colloid-bound. Overall, it appears that the extent to which Th is associated with colloids can vary substantially.

A survey of available ^{232}Th data for the ocean basins demonstrated that the highest concentrations are found nearer to the coasts, and it was concluded that while eolian inputs likely dominated the budget in the open ocean and could account for increases near the coast, fluvial inputs may be more important in coastal regions. This implies that some a mechanism causes recycling of Th that has been removed to estuarine sediments (Huh et al. 1989). A study of an ice-covered region of the western Arctic Ocean found that significant amounts of ^{230}Th and ^{232}Th were advected into the basin (Edmonds et al. 1998). Therefore, it appears that while long-lived Th isotopes are rapidly removed into estuarine sediments, transport into the ocean basins may continue.

4. RADIUM

In the U- and Th-decay series there are four naturally occurring radium isotopes: ^{223}Ra, ^{224}Ra, ^{228}Ra and ^{226}Ra with half-lives that coincide well with the time scales of many coastal and oceanic processes (Swarzenski et al. 2001). The historic discovery of radium in 1898 by Marie Curie initiated a remarkable use of this element as an early oceanic tracer. Less than ten years after its discovery, Joly (1908) observed elevated ^{226}Ra ($t_{1/2}$ = 1600 years) activities in deep-sea sediments that he attributed to be the result of water column scavenging and accumulation of Ra in sediments. This hypothesis was later challenged with the first seawater ^{230}Th measurements, which indicated that ^{226}Ra in the sediments was due to production by scavenged parent ^{230}Th (Broecker and Peng 1982), and these new results confirmed that radium was instead actively migrating across the sediment-water interface into the overlying water. This seabed source stimulated much activity to use radium as a tracer for ocean circulation. Unfortunately, the utility of Ra as a deep ocean circulation tracer never came to full fruition, since biological cycling has been repeatedly shown to have a strong and unpredictable effect on the vertical distribution of this isotope. In addition to diffusion across the sediment – water interface into the ocean, Ra also has a riverine source, although in most marine environments the riverine flux is insignificant in the overall mass balance (Cochran, 1982). The short-lived Ra isotopes of the ^{232}Th series, which are highly depleted in the ocean basins due to their rapid decay and the strong depletion in parent Th isotopes, have been used to track advection from the coasts. Because of the 5.7 year half-live of ^{228}Ra, this isotope has been used effectively to estimate oceanic horizontal eddy diffusivities and coastal water residence times over timescales of less than 30 years. By using both ^{228}Ra and ^{224}Ra ($t_{1/2}$ = 3.7 days), timescales of less than 10 days can be investigated.

4.1. Ra behavior in estuaries

Two important geochemical characteristics make Ra isotopes potentially useful as an estuarine tracer: 1) having highly particle reactive Th isotopes residing largely in sediments as its direct radiogenic parents, which ties Ra directly to bottom sediments, and 2) exhibiting vastly different environmental behavior in fresh water and saltwater systems.

Both of these criteria control the production and input of radium in coastal systems. In coastal waters, Th is efficiently scavenged by colloids and particles, and rapidly removed to the seabed. In fresh water, radium is bound strongly onto particle surfaces; however, as the ionic strength of a water mass increases during mixing into seawater, desorption occurs and Ra is released. Therefore, Ra exhibits strongly nonconservative behavior in estuaries, with data generally falling above a mixing line between river and seawater (Li and Chan 1979; Ghose et al. 2000). This same process of ^{226}Ra release from underlying sediments permits the use of Ra as a useful water mass tracer, if one can identify and constrain water column re-distribution processes (Cochran 1982). For example, Ra behavior in estuaries can be complicated by involvement in biological cycles (Moore and Dymond 1991; Carroll et al. 1993). Estuarine sediments, enriched in Th outside crystal lattices of detrital material, provide a continuous source to coastal waters of Ra isotopes, which are supplied according to production rates that are defined directly by parent isotope decay constants (Burnett et al. 1990; Moore and Todd 1993; Webster et al. 1995; Hancock et al. 2000). Source functions for Ra in an estuary may thus include the following components: (1) riverine, (2) oceanic, (3) estuarine sediments and (4) groundwater (Miller et al. 1990; Bollinger and Moore 1993; Rama and Moore 1996; Moore 1996; Torgersen et al. 1996; Sun and Torgersen 1998). The relative significance of each of these source terms is defined by the particular environmental constraints of an estuary.

The distribution of Ra and Th isotopes in the Kalix Sea estuary and the surrounding Baltic Sea was studied by Porcelli et al. (2001). The activities of ^{226}Ra in the Baltic are generally greater than either bulk river inputs, as represented by the Kalix River, or Atlantic surface seawater. Therefore, a source of ^{226}Ra in addition to the inflow of these sources is required. Interestingly, Ba/^{226}Ra ratios are below a conservative mixing line between Kalix River water and Atlantic surface seawater; this requires that the additional source of ^{226}Ra has a relatively low Ba/^{226}Ra ratio. The (^{228}Ra/^{226}Ra) ratios in the Baltic fall within a narrow range (1.2-1.7) and are comparable to that of the Kalix (1.3) and that of average crust. This actually poses a problem, since the residence time of water (and so of conservative elements like Ra) in the Baltic (35 years) is substantially greater than the half-life of ^{228}Ra. Therefore, a supply of ^{228}Ra relative to ^{226}Ra is required to balance ^{228}Ra decay. While the absolute fluxes required for Ba and the Ra isotopes can be satisfied from sediments (or even by other rivers to the Baltic that may have higher concentrations), the fortuitous balance between Ra isotope ratios in the relatively long-lived waters of the Baltic and the surrounding continental rocks, is difficult to explain.

4.2. Ra as submarine groundwater tracer

The estuarine behavior of Ra has recently received much attention as coastal scientists begin to include a submarine groundwater discharge component in their mass balance models and estuarine studies. Much of this work has been pioneered by Moore (1996), who suggested that as much as 40% of the ^{226}Ra in the coastal waters of the South Atlantic Bight might be derived from exchange processes across the sediment-water interface, including submarine groundwater discharge. In a recent series of conceptual papers, Moore (1996, 1999) emphasized that it is necessary to extend the estuarine reaction zone to include "subterranean estuaries," a region at the land-sea margin where interstitial fluids, regardless of origin, are readily exchanged by a suite of physico-chemical processes with coastal bottom water. Moore and others concluded that many of the well-known reactions and processes are quite similar for surface and subterranean estuaries, and that a vigorous subsurface flux could introduce significant amounts of nutrients and other reactive constituents into the coastal ocean (Bokuniewicz 1980; Johannes 1980; Giblin and Gaines 1990; Lapointe et al. 1990; Valiela et al. 1990; Reay et al. 1992; Simmons 1992; Bugna et al. 1996; Cable et al. 1996a,b; Shaw et al. 1998;

Corbett et al. 1999; Krest et al. 2000; Swarzenski et al. 2001). By omitting such a subterranean component, the overall delivery of radionuclides during their transport out to sea might be significantly underestimated.

There is concern that nutrient-laden submarine groundwater discharge, whether from natural or anthropogenic sources, may contribute to coastal eutrophication (Burnett et al. 2002). Large-scale submarine groundwater discharge (SGD) is generally a widespread coastal phenomenon that can occur wherever hydrogeologic gradients enable lateral and upward groundwater transport to coastal bottom waters (Johannes 1980). There is an abundance of historic and recent evidence for global submarine groundwater discharge (Zektzer et al. 1973). "Leaky" land-sea margins where the discharge of submarine groundwater is likely to be enhanced include: karstic terrain (Back et al. 1979; Paull et al. 1990; Spechler 1994; Swarzenski et al. 2001), buried river channels (Chapelle 1997), geopressured and/or geothermal aquifers (Kohout 1967), and lagoons (Simms 1984; Martin et al. 2000).

Ra isotopes have proven to be one of the most effective tracers of large-scale groundwater flow into estuaries (Moore 1996; Moore and Shaw 1998; Yang et al. 2002). In natural groundwaters, the isotopic composition of Ra by various processes, including weathering and recoil from U- and Th-bearing phases, and adsorption (see Porcelli and Swarzenski 2003). Due to continuous inputs, the inventories of even very short-lived nuclides are maintained. Therefore, groundwaters may develop Ra isotopic compositions that are distinct from surface waters, and so provide a convenient tracer of groundwater supply into estuaries.

Submarine groundwater discharge, which can consist of recycled marine water or fresh water or a mixture thereof, may enhance the diffusive/advective flux of dissolved Ra from the seabed to a coastal water column wherever the hydraulic gradients and sediment transmissivities are favorable (Moore 1996). In coastal systems where groundwater is discharged to coastal waters either continuously or ephemerally due to tidal forcing, distinctive ^{228}Ra, ^{228}Th and ^{224}Ra activities can develop. This disequilibrium occurs because dissolved radium will be rapidly advected into the water column while thorium remains attached to bottom sediments (Webster et al. 1995). In surficial sediments that are flushed by the upward movement of groundwater, (^{224}Ra/^{228}Ra) activity ratios can become quite large as a result of this process and can be used to model the flux across the sediment/water interface (Rama and Moore 1996). Below this diagenetically active surface sediment layer, secular equilibrium is expected to develop from ^{232}Th down to ^{224}Ra. Such radioactive equilibrium in the ^{232}Th decay series requires approximately 20 years of sediment storage. Resuspension, bioturbation, and chemical dissolution, as well as groundwater flow, will allow these two sediment layers to interact. Moore and his colleagues have used such disequilibria to derive a ground-water flux rate in a South Carolina salt marsh (Bollinger and Moore 1993; Rama and Moore 1996).

To use the activity of excess ^{224}Ra in a water sample as a geochronometer for water movement or transport (i.e., residence times), one would write a mass balance equation as follows:

$$^{224}Ra_{obs} = {}^{224}Ra_i f_{EM} e^{-\lambda_{224} t} \qquad (2)$$

and solving for time, t:

$$t = -\frac{\ln\left[\dfrac{^{224}Ra_{obs}}{^{224}Ra_i f_{EM}}\right]}{\lambda_{224}} \qquad (3)$$

where λ_{224} is the decay constant for ^{224}Ra, 0.189 days^{-1}, ^{224}Ra$_i$ is the initial activity of ^{224}Ra in the water sample and f_{EM} is the fraction of the endmember remaining in the sample. Age determinations calculated in such a manner reflect the time elapsed since the water sample became enriched in Ra by the discharge of groundwater. ^{224}Ra is regenerated on the order of days. The fraction f_{EM} can be estimated either from salinity signal or from the distribution of ^{228,226}Ra isotopes, but this term can be difficult to constrain in non-two endmember systems. There are three basic assumptions that must be upheld to correctly apply this Ra model: 1) we can define a single value for the ^{224}Ra activity over the time of interest; 2) the endmembers can not change over the time period of interest; 3) there can be no inputs/sinks for Ra except for mixing and radioactive decay.

One can similarly write an equation to describe the change in activity ratios as a result of mixing and decay for ^{223}Ra and ^{224}Ra which yields (f_{EM} drops out):

$$\left[\frac{^{223}Ra}{^{224}Ra}\right]_{obs} = \left[\frac{^{223}Ra}{^{224}Ra}\right]_i \frac{e^{-\lambda_{223}t}}{e^{-\lambda_{224}t}} \tag{4}$$

This equation is particularly useful to derive apparent estuarine water mass ages (Fig. 6) because the term f_{EM} is removed. Using (^{223}Ra/^{224}Ra) isotope ratios in this manner is based on the assumption that the initial (^{223}Ra/^{224}Ra) activity ratio must remain constant. This conclusion is reasonable as the long-lived parent isotopes (^{231}Pa and ^{232}Th) have relatively constant activity ratios in sediments, and the intermediate Th isotopes (^{227}Th and ^{228}Th) are scavenged efficiently in the near-shore water column. The utility of Ra as

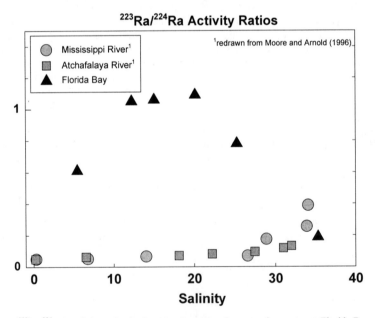

Figure 6. (^{223}Ra/^{224}Ra) activity ratios in three vastly contrasting estuarine systems: Florida Bay, and the Atchafalaya and Mississippi Rivers. Florida Bay waters, which overlie U-rich sediments, contain much higher (^{223}Ra/^{224}Ra) activity ratios than other estuaries. The increased (^{223}Ra/^{224}Ra) values observed at high salinities in the Mississippi/Atchafalaya systems indicate preferential decay of the shorter-lived ^{224}Ra over ^{223}Ra during estuarine mixing.

an accurate coastal groundwater discharge tracer has to be carefully evaluated to assure realistic flux estimates (see Porcelli and Swarzenski 2003). This validation is best accomplished by incorporating a suite of complementary tracers (e.g., ^{222}Rn, CH$_4$) and other techniques (e.g., streaming resistivity surveys) into coastal Ra mass balance studies.

5. RADON-222

Although there are three Rn isotopes in the U- and Th-decay series, only ^{222}Rn is sufficiently long lived ($t_{1/2}$= 3.8 days) to be a useful estuarine tracer. Radioactive decay of ^{226}Ra continuously produces ^{222}Rn, which because of its short half-life is generally in secular equilibrium in seawater. Being chemically non-reactive except for very weak Van der Waals bonding makes this isotope a unique marine tracer in that it is not directly involved in biogeochemical cycles.

There are two main sources of ^{222}Rn to the ocean, (1) the decay of sediment-bound ^{226}Ra and (2) decay of dissolved ^{226}Ra in the water column. Radon can enter the sediment pore water through alpha recoil during decay events (see discussion in Porcelli and Swarzenski 2003). Since radon is chemically inert, it readily diffuses from bottom sediments into overlying waters. The diffusion of radon from sediments to the water column gives rise to disequilibrium between ^{222}Rn and ^{226}Ra, with ratios of (^{222}Rn/^{226}Ra) >1 due to the addition of excess ^{222}Rn. This excess ^{222}Rn is unsupported and so is rapidly diminished by decay. Therefore, the excess ^{222}Rn signal is only resolvable where significant transport of ^{222}Rn away from the sediment can occur over timescales that do not significantly exceed the half-life of ^{222}Rn.

Loss of radon in the ocean occurs typically through radioactive decay (producing 4 short-lived daughters before decaying to ^{210}Pb) or loss to the atmosphere at the air-sea interface. Loss of radon due to turbulence or diffusion at the air-sea interface leads to a depletion of radon activity with respect to that of ^{226}Ra, allowing for studies on gas exchange at this interface (Broecker and Peng 1982). Radon can also be used as a tracer for studying the movement of other inert gaseous tracers.

Based upon these considerations, Rn has been used extensively for determining gaseous exchange rates across the air-sea interface as well as for studying exchange of pore waters into coastal bottom waters (Berelson et al. 1987). The observed depletion of ^{222}Rn in porewaters relative to the sedimentary production rate from its parent, ^{226}Rn has been widely used to examine exchange processes at the sediment/water interface (Broecker and Peng 1982). In addition, ^{222}Rn has also been used as a tracer for evaluating coefficients of isopycnal and diapycnal mixing in the deep sea (Broecker and Peng 1982; Sarmiento and Rooth 1980), and more recently, for tracing submarine groundwater discharge into the coastal zone (Cable et al. 1996a; Corbett et al. 1999; see Porcelli and Swarzenski 2003).

6. LEAD AND POLONIUM

The majority of ^{210}Pb studies address the utility of ^{210}Pb as a recent geochronological tool (e.g., Paulsen et al. 1999; Santschi et al. 2001) rather than as an element that is involved in complex biogeochemical cycles. Nonetheless, some of these studies do provide insight into the geochemical behavior of ^{210}Pb. The distribution of stable Pb to the coastal and estuarine waters also provides information regarding behavior of ^{210}Pb in this environment. Nearly all of the lead in the world's surface oceans is believed to be of anthropogenic origin—derived from combustion of leaded gasoline and a major portion of this has already been removed from the surface waters (Broecker and Peng 1982). Even the most remote areas of the world—the polar caps—show a very clear effect of lead pollution. Recent measurements of polar ice cores show that man's activities have caused a 300-fold increase in Pb since the beginning of the industrial revolution. ^{210}Pb and its indirect daughter ^{210}Po are both important radionuclides used in geochemical research. ^{210}Pb is delivered to the water column by three sources: 1) atmospheric fallout from the decay of gaseous ^{222}Rn that has been released mainly from the continents, 2) the *in-situ* decay of ^{226}Ra (via ^{222}Rn) in the water column, 3) river inputs, which are supplied by interactions with rocks and soils during weathering (see Chabaux et al. 2003; Porcelli and Swarzenski 2003) as well from the leaching of atmospherically-delivered ^{210}Pb-laden soils (Gascoyne 1982), and 4) lateral transport (boundary scavenging) from continental margins, which could also contribute to coastal waters (Olsen et al. 1989) but which is often a minor source. ^{210}Po is a decay product of ^{210}Pb and is produced mainly by this source in the water column, but this isotope may also have an atmospheric source term within surface waters due to dust deposition. In broad terms, the marine cycle of ^{210}Pb is characterized by supply or production of dissolved ^{210}Pb, which is then scavenged on to solid surfaces and eventually removed from the water column (Santschi et al. 1983). Therefore, ^{210}Pb has been used extensively as a tracer for particle-reactive elements (Benninger 1978), as well as for geochronological studies of lake, estuarine and coastal sediments (Fuller et al. 1999). ^{210}Po appears to be somewhat more reactive than ^{210}Pb in seawater (Kadko 1993).

Like Th isotopes, ^{210}Pb is rapidly removed in estuaries by various processes in the water column (Ravichandran et al. 1995) and deposited onto the seafloor during particle deposition and accumulation (Fuller et al. 1999). The first study by Rama and Goldberg (1961) identified deficiencies of ^{210}Pb with respect to its grandparent, ^{226}Ra and the residence time of ^{210}Pb in the water column in coastal and estuarine regions was calculated to be <1 year, in contrast to the deep ocean where it is 30-100 years (Rama and Goldberg 1961; Li et al. 1981; Baskaran and Santschi 1993; Baskaran et al. 1997). In shallow sediments, the profiles of organic carbon and ^{210}Pb are often well correlated (Paulsen et al. 1999). In the coastal ocean, sinking particles are adsorption sites for ^{210}Pb and even here iron and manganese oxides are important removal phases. In the Baltic Sea near the mouth of the Kalix River and at a salinity of ~3, activities were found to be substantially depleted compared to a bulk river water value (Porcelli et al. 2001). This indicates that ^{210}Pb is rapidly removed at low salinities, with a considerable fraction of the remaining ^{210}Pb on particles. Even lower concentrations were found further into the Baltic. In anoxic environments, ^{210}Pb may be readily removed as a lead sulfide precipitate, causing this mineral component to be of importance in sulfide-rich sediments (Swarzenski et al. 1999a). In organic-rich estuaries, the residence time of dissolved ^{210}Pb could be considerably longer and where the hydraulic residence time is comparable to the residence time of ^{210}Pb, a major portion of the dissolved ^{210}Pb could be directly exported to the coastal zone, without significant removal in the estuarine zone (Baskaran et al. 1997; Baskaran and Santschi 2002). Note that in shallow environments, the atmospheric flux of ^{210}Pb may be the dominant input, rather than production within the water column (e.g., Gustafsson et al. 1998). In rivers with large watershed, the erosional input of ^{210}Pb could be significant and could be compared to the direct atmospheric fallout (Fig. 7), such as in the estuarine systems of the Texas-Louisiana border (Baskaran and Santschi 1993; Baskaran et al. 1996, 1997).

Involvement in biological cycles may affect the distribution of both ^{210}Pb and ^{210}Po. Santschi et al. (1979) noticed large seasonal variations of ^{210}Pb and ^{210}Po in Narragansett Bay. This seasonal trend was ascribed to two factors: remobilization out of the sediments during the spring and early summer, or formation of organic complexes. In a summary paper, Nozaki (1991) suggested that the removal of ^{234}Th, ^{210}Pb and ^{210}Po from marine waters may be accelerated by biological activity. One consequence of this is that

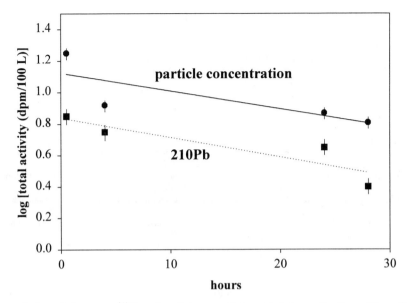

Figure 7. Log of the excess ^{210}Pb and particle concentrations plotted as a function of time after a prolonged rain event in Galveston Bay. Data from Baskaran and Santschi (1993).

preferential biological uptake of ^{210}Po relative to ^{210}Pb in some sediments may cause local disequilibria, which can affect the interpretation of ^{210}Pb-derived geochronologies. Hodge et al. (1979) demonstrated that Po, U and Pu have different uptake kinetics onto solid phases. They exposed different solid phases to seawater to determine if Po, U or Pu could be removed by inorganic material. Their results suggest that organic surface coatings may collect particles, with their associated elements, at differing rates. Such variable uptake rates led to the inference that the particulate Pu, Po, and U may be associated with different particulate phases.

7. CONCLUSIONS

Studies of U- and Th-series nuclides have discovered a range of interactions, fluxes, and removal mechanisms that can affect relatively soluble elements such as U and Ra, and particle-reactive elements such as Th and Pb. However, because of the great complexity within individual estuaries, and the tremendous differences between them, considerable work is still required. Now that the broad outline of mechanisms has been identified, more detailed, integrated studies are required to document the time-dependence of estuarine processes, as well as that of river input characteristics. This includes focusing on the composition and reactive properties of particles in rivers and estuaries, biogeochemistry of compounds that serve as ligands or surface-reactive flocs, and the hydrodynamics of sediment transport and resuspension. There is limited data regarding the role of colloids in different environments, and understanding colloid behavior will require advances in separating and characterizing a wide range of possible compositions. An important goal will be to predict transport properties to locations where less data is available, both in order to make broad calculations regarding global fluxes, and to make inferences about estuarine transport in the past.

REFERENCES

Aller RC, Cochran JK (1976) ^{234}Th/^{238}U disequilibrium and diagenetic time scales. Earth Planet Sci Lett 29:37-50

Andersson PS, Wasserburg GJ, Chen JH, Papanastassiou DA, Ingri J (1995) ^{238}U-^{234}U and ^{232}Th-^{230}Th in the Baltic Sea and in river water. Earth Planet Sci Lett 130:217-234

Andersson PS, Porcelli D, Wasserburg GJ, Ingri J (1998) Particle transport of ^{234}U – ^{238}U in the Kalix River and in the Baltic Sea. Geochim Cosmochim Acta 62:385-392

Andersson PS, Porcelli D, Gustafsson Ö, Ingri J, Wasserburg GJ (2001) The importance of colloids for the behavior of uranium isotopes in the low-salinity zone of a stable estuary. Geochim Cosmochim Acta 65:13-25

Anderson RF (1987) Redox behavior of uranium in an anoxic marine basin. Uranium 3:145-164

Anderson RF, Fleisher MQ, LeHuray AP (1989) Concentration, oxidation state, and particulate flux of uranium in the Black Sea. Geochim Cosmochim Acta 53:2215-2224

Back W, Hanshaw BB, Pyler TE, Plummer LN, Weiede AE (1979) Geochemical significance of groundwater discharge in Caleta Xel Ha, Quintana Roo, Mexico. Water Res 15:1521-1535

Barnes CE, Cochran JK (1990) Uranium removal in oceanic sediments and the oceanic U balance. Earth. Planet. Sci. Lett 97:94-101

Barnes CE, Cochran JK (1993) Uranium geochemistry in estuarine sediments: Controls on removal and release processes. Geochim Cosmochim Acta 57:555-569

Baskaran M, Santschi PH (1993) The role of particles and colloids in the transport of radionuclides in coastal environments of Texas. Marine Chem 43:95-114

Baskaran M, Santschi PH (2002) Particulate and dissolved Pb-210 activities in the shelf and slope regions of the Gulf of Mexico waters. Cont Shelf Res 22:1493-1510

Baskaran M, Santschi PH, Benoit G, Honeyman BD (1992) Scavenging of thorium isotopes by colloids in seawater of the Gulf of Mexico. Geochim Cosmochim Acta 56:3375-3388

Baskaran M, Santschi PH, Guo LD, Bianchi TS, Lambert C (1996) ^{234}Th:^{238}U disequilibria in the Gulf of Mexico: The importance of organic matter and particle concentration. Cont Shelf Res 16:353-380

Baskaran M., Ravichandran M, Bianchi TS (1997) Cycling of ^{7}Be and ^{210}Pb in a high DOC, shallow, turbid estuary of Southeast Texas. Estuarine Coastal Shelf Sci 45: 165-176

Baskaran M, Swarzenski PW, Porcelli D (2003) Role of colloidal material in the removal of ^{234}Th in the Canada Basin of the Arctic Ocean. Deep-Sea Res (in press)

Benninger LK (1978) ^{210}Pb balance in Long Island Sound. Geochim Cosmochim Acta 42:1165-1174

Berelson WM, Buchholtz MR, Hammond DE, Santschi PH (1987) Radon fluxes measured with the MANOP bottom lander. Deep-Sea Res 34:1209-1228

Berner EK, Berner RA (1987) The global water cycle: Geochemistry and environment. Prentince-Hall, New Jersey

Bertine KK, Chan LH, Turekian KK (1970) Uranium determinations in deep-sea sediments and natural waters using fission tracks. Geochim Cosmochim Acta 34:641-648

Bhat SG, Krishnaswami S, Lal D, Rama, Moore WS (1969) ^{234}Th/^{238}U ratios in the ocean. Earth Planet Sci Lett 5:483-491

Bokuniewicz H (1980) Groundwater seepage into Great South Bay, New York. Estuar Coast Mar Sci 10:437-444

Bollinger MS, Moore WS (1993) Evaluation of salt marsh hydrology using radium as a tracer. Geochim Cosmochim Acta 57:2203-2212

Borole DV, Krishnaswami S, Somayajulu BLK (1977) Investigations on dissolved uranium, silicon and on particulate trace elements in estuaries. Estuarine Coastal Mar Sci 5:743-754

Borole DV, Krishnaswami S, Somayajulu BLK (1982) Uranium isotopes in rivers, estuaries and adjacent coastal sediments of western India: their weathering, transport and oceanic budget. Geochim Cosmochim Acta 46:125-137

Bourdon B, Turner S, Henderson GM, Lundstrom CC (2003) Introduction to U-series geochemistry. Rev Mineral Geochem 52:1-21

Boyle EA, Edmonds JM, Sholkovitz ER (1977) The mechanism of iron removal in estuaries. Geochim Cosmochim Acta 41:1313-1324

Broecker WS, Kaufman A, Trier R (1973) The residence time of thorium in surface water and its implications regarding rate of reactive pollutants. Earth Planet Sci Lett 20:35-44.

Broecker WS, Peng T-H (1982) Tracers in the Sea. Lamont-Doherty Geological Observatory, New York

Buesseler KO, Bauer J, Chen R, Eglinton T, Gustafsson Ö, Landing W, Mopper K, Moran SB, Santschi P, Vernon Clark R, Wells M (1996) Sampling marine colloids using cross-flow filtration: Overview and results from an intercomparison study. Marine Chem 55:1-31

Buffle J, van Leeuwen H (eds) (1992) Environmental Particles. Lewis Publishers, Boca Raton Florida

Bugna GC, Chanton JP, Cable JE, Burnett WC, Cable PH (1996) The importance of groundwater discharge to the methane budgets of nearshore and continental shelf waters of the northeastern Gulf of Mexico. Geochim Cosmochim Acta 60:4735-4746

Burnett WC, Cowart JB, Deetae S (1990) Radium in the Suwannee River and estuary. Biogeochem 10:237-255

Burnett WC, Chanton J, Christoff J, Kontar E, Krupa S, Lambert M, Moore W, O'Rourke D, Paulsen R, Smith C, Smith L, Taniguchi M (2002) Assessing methodologies for measuring groundwater discharge to the ocean. EOS Trans Am Geophys Un 83:117-123

Burton JD (1965) Radioactive nuclides in seawater, marine sediments and marine organisms In: Chemical Oceanography. Vol 2 Riley JP, Skirrow G (eds) Academic Press, London, p 425-475

Cable JE, Bugna GC, Burnett WC, Chanton JP (1996a) Application of ^{222}Rn and CH_4 for assessment of groundwater discharge to the coastal ocean. Limnol Oceanogr 41:1347-1353

Cable JE, Burnett WC, Chanton JP, Weatherly GL (1996b) Estimating groundwater discharge into the northeastern Gulf of Mexico using radon-222. Earth Planet Sci Lett 144:591-604

Carroll J, Falkner KK, Brown ET, Moore WS (1993) The role of the Ganges-Brahmaputra mixing zone in supplying barium and ^{226}Ra to the Bay of Bengal. Geochim Cosmochim Acta 57:2981-2990

Carroll J, Moore WS (1994) Uranium removal during low discharge in the Ganges-Brahmaputra mixing zone. Geochim Cosmochim Acta 58:4987-4995

Chabaux F, Riotte J, Dequincey O (2003) U-Th-Ra fractionation during weathering and river transport. Rev Mineral Geochem 52:533-576

Chapelle FH (1997) The Hidden Sea. Geoscience Press, Tucson, Arizona

Chen JH, Edwards RL, Wasserburg GJ (1986) ^{238}U, ^{234}U and ^{232}Th in seawater. Earth Planet Sci Lett 80:241-251

Chen JH, Edwards RL, Wasserburg GJ (1992) Mass spectrometry and applications to uranium-series disequilibrium. In: Uranium-series Disequilibrium: Applications to Earth, Marine and Environmental Sciences. Ivanovich M, Harmon RS (eds) Clarendon Press, Oxford, p 174-206

Cheng H, Edwards RL, Hoff J, Gallup CD, Richards DA, Asmerom Y (2000) The half lives of uranium-234 and thorium -230. Chem Geol 169:17-33Church TM, Sarin MM, Fleisher MQ, Ferdelman TG (1996) Salt marshes: An important coastal sink for dissolved uranium. Geochim Cosmochim Acta 60:3879-3887

Cochran JK (1982) The oceanic chemistry of the U- and Th-series nuclides. In: Uranium Series Disequilibrium: Applications to Environmental Problems. Ivanovich M, Harmon RS (eds), Clarendon Press, Oxford, p 334-395

Cochran JK (1984) The fates of U and Th decay series nuclides in the estuarine environment. In: The Estuary as a Filter. Kennedy VS (ed) Academic Press, London, p 179-220

Cochran JK (1992) The oceanic chemistry of the uranium - and thorium – series nuclides. In: Uranium-series Disequilibrium: Applications to Earth, Marine and Environmental Sciences. Ivanovich M, Harmon RS (eds) Clarendon Press, Oxford, p 334-395

Cochran JK, Masqué P (2003) Short-lived U/Th-series radionuclides in the ocean: tracers for scavenging rates, export fluxes and particle dynamics. Rev Mineral Geochem 52:461-492

Cochran JK, Carey AE, Sholkovitz ER, Surprenant LD (1986) The geochemistry of uranium and thorium in coastal marine-sediments and sediment pore waters. Geochim Cosmochim Acta 50:663-680

Corbett DR, Chanton J, Burnett W, Dillon K, Rutkowski C. (1999) Patterns of groundwater discharge into Florida Bay. Limnol Oceanogr 44:1045-1055

Dearlove JPL, Longworth G, Ivanovich M, Kim J, Delakowitz B, Zeh P (1991) A study of groundwater-colloids and their geochemical interactions with natural radionuclides in Gorleben aquifer systems, Radiochim Acta 52-3:83-89

DeMaster DJ, Kuehl SA, Nittrouer CA (1986) Effects of suspended sediments on geochemical processes near the mouth of the Amazon river - examination of biological silica uptake and the fate of particle-reactive elements. Cont Shelf Res 6:107-125

Djogic R, Branica M (1991) Dissolved uranyl complexed species in artificial seawater. Marine Chem 36:121-135

Dyer KR (1973) Estuaries - A Physical Introduction. John Wiley and Sons, New York

Edmonds HN, Moran SB, Hoff JA, Smith JN, Edwards RL (1998) Protactinium-231 and thorium-230 abundances and high scavenging rates in the western Arctic Ocean. Science 280:405-407

Fairbridge RW (1980) The estuary: its definition and geodynamic cycle. In: Chemistry and biogeochemistry of estuaries. Olausson E, Cato I (eds) John Wiley and Sons, New York, p 1-36

Feng H, Cochran JK, Hirschberg DJ (1999) ^{234}Th and ^{7}Be as tracers for the transport and dynamics of suspended particles in a partially mixed estuary. Geochim Cosmochim Acta 63:2487-2505

Föyn E, Karlik B, Pettersson H, Rona E (1939) The radioactivity of seawater. Göteborgs Kungl. Vetenskaps- och Vitterhets – Samhälles Handlingar, Ser. B, 6, No. 12

Fuller CC, van Geen A, Baskaran M, Anima R (1999) Sediment chronology in San Francisco Bay, California, defined by ^{210}Pb, ^{234}Th, ^{137}Cs, and ^{239}Pu, ^{240}Pu. Marine Chem 64:7-27

Gascoyne M (1982) Geocgemcistry of the actinitdes and their daughters. *In:* Uranium Series Disequilibrium: Applications to Environmental Problems. Ivanovich M, Harmon RS (eds) Clarendon Press, Oxford, p 33-55

Ghose, S, Alam, MN, Islam, MN (2000) Concentrations of ^{222}Rn, ^{226}Ra and ^{228}Ra in surface seawater of the Bay of Bengal. J Environ Radio 47:291-300

Giblin, AE, Gaines AG (1990) Nitrogen inputs to a marine embayment: the importance of groundwater. Biogeochem 10:309-328

Goldstein SJ, Stirling CH (2003) Techniques for measuring uranium-series nuclides: 1992-2002. Rev Mineral Geochem 52:23-57

Guo LD, Santschi PH, Baskaran M, Zindler A (1995) distribution of dissolved and particulate ^{230}Th and ^{232}Th in seawater from the Gulf of Mexico and off Cape Hatteras as measured by SIMS. Earth Planet Sci Lett 133:117-128

Guo LD, Santschi PH (1996) A critical evaluation of the cross-flow ultrafiltration technique for sampling colloidal organic carbon in seawater. Marine Chem 55:113-127

Guo LD, Wen LS, Tang DG, Santschi PH (2000) Re-examination of cross-flow ultrafiltration for sampling aquatic colloids: evidence from molecular probes. Marine Chem 69:75-90

Guo LD, Hunt BJ, Santschi PH (2001) Ultrafiltration behavior of major ions (Na, Ca, Mg, F, Cl, and SO$_4$) in natural waters. Water Res 35:1500-1508

Gustafsson Ö, Gschwend PM (1996) Aquatic colloids: concepts, definitions, and current challenges. Limnol Oceanogr 42:519-528

Gustafsson Ö, Buesseler KO, Gschwend PM (1996) On the integrity of cross-flow filtration for collecting marine organic colloids. Marine Chem 55:93-111

Gustafsson Ö, Buesseler KO, Geyer WR, Moran SB, Gschwend PM (1998) An assessment of the relative importance of horizontal and vertical transport of particle-reactive chemicals in the coastal ocean. Cont Shelf Res 18:805-829

Gustafsson Ö, Düker A, Larsson J, Andersson P, Ingri J (2000a) Functional separation of colloids and gravitoids in surface waters based on differential settling velocity: coupled cross-flow filtration-split flow thin-cell fractionation (CFF-SPLITT). Limnol Oceanogr 45:1731-1742

Gustafsson Ö, Widerlund A, Andersson P, Ingri J, Roos P, Ledin A (2000b) Colloid dynamics and transport of major elements through a boreal river-brackish bay mixing zone. Marine Chem 71:1-21

Halliday AN, Lee D-C, Christensen JN, Rehkämper M, Yi W, Lou X, Hall CM, Ballentine CJ, Pettke T, Stirling C (1998) Applications of multiple collector-ICPMS to cosmochemsitry, geochemistry and paleoceanography. Geochim Cosmochim Acta 62:919-940

Hancock GJ (2000) Identifying resuspended sediment in an estuary using the ^{228}Th/^{232}Th activity ratio: the fate of lagoon sediment in the Bega River estuary, Australia. Mar Freshwater Res 51:659-667

Hancock GJ, Webster IT, Ford PW, Moore WS (2000) Using Ra isotopes to examine transport processes controlling benthic fluxes into a shallow estuarine lagoon. Geochim Cosmochim Acta 21:685-3699

Henderson GM (2002) Seawater (^{234}U/^{238}U) during the last 800 thousand years. Earth Planet Sci Lett 199:97-110

Hernegger F, Karlik B (1935) Uranium in seawater. Göteborgs Kungl Vetenskaps- och Vitterhets – Samhälles Handlingar, Ser. B4, No. 12

Hodge VF, Koide M, Goldberg GD (1979) Particulate uranium, plutonium and polonium in the biogeochemistries of the coastal zone. Nature 277:206-209

Holland HD, Kulp JL (1954) The transport and deposition of uranium, ionium and radium in rivers, oceans and ocean sediments. Geochim Cosmochim Acta 5:197-213

Honeyman BD, Santschi PH (1989) A Brownian pumping model for oceanic trace-metal scavenging - evidence from Th isotopes. J Mar Res 47:951-992

Huh C-A, Moore WS, Kadko DC (1989) Oceanic ^{232}Th: a reconnaissance and implications of global distribution from manganese nodules. Geochim Cosmochim Acta 53:1357-1366

Ingri J (1996) Hydrochemistry of the Kalix River. Luleå University of Technology. Report. 126p (in Swedish)

Johannes RE (1980) The ecological significance of the submarine discharge of groundwater. Mar Ecol Prog Ser 3:365-373

Joly J (1908) On the radium content of deep-sea sediments. Phil Mag 16:190

Kadko D (1993) Excess ^{210}Po and nutrient recycling within the California coastal transition zone. J Geophys Res 98:857-864

Kaufman A, Li Y-H, Turekian KK (1971) The removal rates of ^{234}Th and ^{228}Th from waters of the New York Bight. Earth Planet Sci Lett 54:385-392

Kennedy VC, Zellweger GW, Jones BF (1974) Filter pore-size effects on the analysis of Al, Fe, Mn, and Ti in water. Water Resource Res 10:785-790

Kersten M, Thomsen S, Priebsch W, Garbe-Schonberg CD (1998) Scavenging and particle residence times determined from Th-234/U-238 disequilibria in the coastal waters of Mecklenburg Bay. Appl Geochem 13:339-347

Kineke GC, Sternberg RW, Trowbridge JH, Geyer WR (1996) Fluid-mud processes on the Amazon continental shelf. Cont Shelf Res 16:667-696

Klinkhammer GP, Palmer MR (1991) Uranium in the oceans: Where it goes and why. Geochim Cosmochim Acta 55:1799-1806

Koczy FF, Tomic E, Hecht F (1957) On the geochemistry of U in Ostseebecken. Geochim Cosmochim Acta 11:86-102 (in German)

Kohout FA (1967) Groundwater flow and the geothermal regime of the Floridian Plateau. Gulf Coast Assoc Geol Soc 17:339-354

Koide M, Goldberg ED (1964) $^{234}U/^{238}U$ ratios in seawater. Progress in Oceanography. Vol 3. Sears M (ed) Pergamon Press, Oxford p 173-177

Krest JM, Moore WS, Gardner LR, Morris JT (2000) Marsh nutrient export supplied by groundwater discharge: Evidence from radium measurements. Global Biogeochem Cycles 14:167-176

Krishnaswami S, Graustein WC, Turekian KK (1982) Radium, thorium and radioactive isotopes in groundwaters: Application to the in situ determination of adsorption-desorption rate constants and retardation factors. Water Resource Res 18:1633-1675

Ku T-L, Knauss K and Mathieu GG (1977) Uranium in open ocean: concentration and isotopic composition. Deep Sea Res 24:1005-1017

Langmuir D (1978) Uranium solution-mineral equilibria at low temperatures with applications to sedimentary ore deposits. Geochim Cosmochim Acta 42:547-569

Langmuir D, Herman JS (1980) The mobility of thorium in natural waters at low temperatures. Geochim Cosmochim Acta 44:1753-1766

Lapointe BE, O'Connell JD, Garrett GS (1990) Nutrient couplings between on-site sewage disposal systems, groundwaters, and nearshore surface waters of the Florida Keys. Biogeochem 10:289-307

Li Y-H, Chan L-H (1979) Desorption of Ba and ^{226}Ra from river-borne sediments in the Hudson estuary. Earth Planet Sci Lett 43:343-350

Li Y-H, Santschi PH, Kaufman A, Benninger LK, Feely HW (1981) Natural radionuclides in waters of the New-York Bight. Earth Planet Sci Lett 55:217-228

Lienert C, Short SA, Von Gunten HR (1994) Uranium infiltration from a river to shallow groundwater. Geochim Cosmochim Acta 58:5455-5463

Loder TC, Reichard RP (1981) The dynamics of conservative mixing in estuaries. Estuar 4:64-69

Löfvendahl R (1987) Dissolved uranium in the Baltic Sea. Marine Chem 21:213-227

Maeda M, Windom H L (1982) Behavior of uranium in two estuaries of the southeastern United States. Marine Chem 11:427-436

Mann DK, Wong GTF (1993) "Strongly bound" uranium in marine waters: occurrence and analytical implications. Marine Chem 42:25-37

Martin JB, Cable JE, Swarzenski PW (2000) Quantification of groundwater discharge and nutrient loading to the Indian River Lagoon. St. Johns River Water Management District Report, Palatka Fl

Martin J-M, Meybeck M, Pusset M (1978a) Uranium behavior in the Zaire estuary. Netherlands J Sea Res 12:338-344

Martin J-M, Nijampurkar V, Salvadori F (1978b) Uranium and thorium isotope behavior in estuarine systems. In: Biogeochemistry of Estuarine Sediments. Goldberg ED (ed) UNESCO, Paris p 111-127

McKee BA, Todd JF (1993) Uranium behavior in a permanently anoxic fjord: microbial control? Limnol Oceanogr 38:408-414

McKee BA, DeMaster DJ, Nittrouer CA (1983) Concepts of sediment deposition and accumulation applied to the continental shelf near the mouth of the Yangtze River. Geol 11:631-633

McKee BA, DeMaster DJ, Nittrouer CA (1986a) Temporal variability in the partitioning of thorium between dissolved and particulate phases on the Amazon shelf - implications for the scavenging of particle-reactive species. Cont Shelf Res 6:87-106

McKee BA, DeMaster DJ, Nittrouer CA (1986b) The use of $^{234}Th/^{238}U$ disequilibrium to examine the fate of particle-reactive species on the Yangtze continental. Shelf. Earth Planet Sci Lett 68:431-442

McKee BA, DeMaster DJ, Nittrouer CA (1987) Uranium geochemistry on the Amazon shelf: evidence for uranium release from bottom sediments. Geochim Cosmochim Acta 51:2779-2786

Miller RJ, Kraemer, TF, McPherson BF (1990) Radium and radon in Charlotte Harbor estuary, Florida. Estuar Coastal Shelf Sci 31:439-457

Millero FJ, Sohn ML (1992) Chemical Oceanography. CRC Press, Boca Raton Florida

Moore WS (1967) Amazon and Mississippi River concentrations of uranium, thorium and radium isotopes. Earth Planet Sci Lett 2:231-234

Moore WS (1992) Radionuclides of the uranium and thorium decay series in the estuarine environment. *In*: Uranium-series Disequilibrium: Applications to Earth, Marine and Environmental Sciences. Ivanovich M, Harmon RS (eds) Clarendon Press, Oxford, p 396-422

Moore WS (1996) Large groundwater inputs to coastal waters revealed by ^{226}Ra enrichments. Nature 380:612-614

Moore WS (1997) High fluxes of radium and barium from the mouth of the Ganges-Brahmaputra during low river discharge suggest a large groundwater source. Earth Planet Sci Lett 150:141-150

Moore WS (1999) The subterranean estuary: a reaction zone of groundwater and seawater. Marine Chem 65:111-125

Moore WS, Dymond J (1991) Fluxes of ^{226}Ra and barium in the Pacific Ocean: the importance of boundary processes. Earth Planet Sci Lett 107:55-68

Moore WS, Shaw TJ (1998) Chemical signals from submarine fluid advection onto the continental shelf. J Geophys Res 103:21543-21552

Moore WS, Todd JF (1993) Radium isotopes in the Orinoco estuary and the eastern Caribbean Sea. J Geophys Res 98:2233-2244

Moore WS, Demaster DJ, Smoak JM, McKee BA, Swarzenski PW (1996) Radionuclide tracers of sediment-water interactions on the Amazon shelf. Cont Shelf Res 16:645-665

Moran SB, Moore RM (1989) The distribution of colloidal aluminum and organic carbon in coastal and open ocean waters off Nova Scotia. Geochim Cosmochim Acta 53:2519-2527

Nozaki Y (1991) The systematics and kinetics of U/Th decay series nuclides in ocean water. Rev Aquatic Sci 4:75-105

Officer CB, Lynch DR (1981) Dynamics of mixing in estuaries. Estuar Coast Shelf Sci 12:525-533

Olsen CR, Thein M, Larsen IL, Lowry PD, Mulholland PJ, Cutshall NH, Byrd JT, Windom HL (1989) Plutonium, ^{210}Pb, and carbon isotopes in the Savannah Estuary - Riverborne versus marine sources. Environ Sci Technol 23:1475-1481

Osmond J, Ivanovich M (1992) Uranium – series mobilization and surface hydrology. *In*: Uranium-series Disequilibrium: Applications to Earth, Marine and Environmental Sciences. Ivanovich M, Harmon RS (eds) Clarendon Press, Oxford, p 259-289

Palmer MR, Edmond JM (1993) Uranium in river water. Geochim Cosmochim Acta 57:4947-4955

Paull CK, Speiss F, Curry J, Twichell D (1990) Origin of Florida canyon and the role of spring sapping on the formation of submarine box canyons. Geol Soc Am Bull 102:502-515

Paulsen SC, List EJ, Santschi PH (1999) Modeling variability in ^{210}Pb and sediment fluxes near the Whites Point Outfalls, Paleo Verdes Shelf, California. Environ Sci Tech 33:3077-3085

Ponter C, Ingri J, Burman JO, Boström K (1990) Temporal variations in dissolved and suspended iron and manganese in the Kalix River, northern Sweden. Chem Geol 81:121-131

Porcelli D, Swarzenski PW (2003)The behavior of U- and Th- series nuclides in groundwater. Rev Mineral Geochem 52:317-361

Porcelli D, Andersson PS, Wasserburg GJ, Ingri J, Baskaran M (1997) The importance of colloids and mires for the transport of uranium isotopes through the Kalix River watershed and Baltic Sea. Geochim Cosmochim Acta 61:4095-4113

Porcelli D, Andersson PS, Baskaran M, Wasserburg GJ (2001) Transport of U- and Th-series nuclides in a Baltic Shield watershed and the Baltic Sea. Geochim Cosmochim Acta 65:2439-2459

Quigley MS, Santschi PH, Guo LD, Honeyman BD (2001) Sorption irreversibility and coagulation behavior of ^{234}Th with marine organic matter. Marine Chem 76:27-45

Rama MK, Goldberg ED (1961) Lead-210 in natural waters. Science 134:98-99

Rama, Moore WS (1996) Using the radium quartet for evaluating groundwater input and water exchange in salt marshes. Geochim Cosmochim Acta 60:4645-4652

Ravichandran M, Baskaran M, Santschi PH, Bianchi TS (1995) Geochronology of sediments in the Sabine-Neches estuary, Texas, USA. Chem Geol 125:291-306

Ray SB, Mohanti M, Somayajulu BLK (1996) Uranium isotopes in the Mahandi River – Estuarine System, India. Estuar Coastal Shelf Sci 40:634-645

Reay WG, Gallagher DL, Simmons GM (1992) Groundwater discharge and its impact on surface water quality in a Chesapeake Bay inlet. Water Res Bull 28:1121-1134

Santschi PH, Li YH, Bell J (1979) Natural radionuclides in the water of Narragansett Bay. Earth Planet Sci Lett 45:201-213

Santschi PH, Li YH, Adler DM, Amdurer M, Bell J, Nyffeler UP (1983) The relative mobility of natural (Th, Pb and Po) and fallout (Pu, Am, Cs) radionuclides in the coastal marine-environment - results from model-ecosystems (MERL) and Narragansett Bay. Geochim Cosmochim Acta 47:201-210

Santschi PH, Guo LD, Asbill S, Allison M, Kepple AB, Wen LS (2001) Accumulation rates and sources of sediments and organic carbon on the Palos Verdes shelf based on radioisotopic tracers (^{137}Cs, ^{240}Pu, ^{240}Pu, ^{210}Pb, ^{234}Th, ^{238}U and ^{14}C). Marine Chem 73:125-152

Sarin MM, Church TM (1994) Behavior of uranium during mixing in the Delaware and Chesapeake Estuaries. Estuar Coastal Shelf Sci 39:619-631

Sarmiento JL, Rooth CG (1980) A comparison of vertical and isopycnal mixing models in the deep-sea based on ^{222}Rn measurements. J Geophys Res 85:1515-1518

Shaw TJ, Moore WS, Kloepfer J, Sochaski MA (1998) The flux of barium to the coastal waters of the Southeastern United States: the importance of submarine groundwater discharge. Geochim Cosmochim Acta 62:3047-3052

Shiller AM (1996) The effect of recycling traps and upwelling on estuarine chemical flux estimates. Geochim Cosmochim Acta 60:3177-3185

Shiller AM, Boyle EA (1991) Trace elements in the Mississippi River Delta outflow region: behavior at high discharge. Geochim Cosmochim Acta 55:3241-3251

Shiller AM, Mao L (2000) Dissolved vanadium in rivers: Effects of silicate weathering. Chem Geol 165:13-22

Sholkovitz ER (1977) The flocculation of dissolved Fe, Mn, Al, Cu, Ni, Co, and Cd during estuarine mixing. Earth Planet Sci Lett 41:77-86

Simmons GM Jr (1992) Importance of submarine groundwater discharge (SGWD) and seawater cycling to material flux across sediment/water interfaces in marine environments. Mar Ecol Prog Ser 84:173-184

Simms BS (1984) Dolomitization by groundwater flow systems in carbonate platforms. Trans. Gulf Coast Assoc Geol Soc 34: 411-420

Smoak JM, DeMaster DJ, Kuehl SA, Pope RH, McKee BA (1996) The behavior of particle-reactive tracers in a high turbidity environment: ^{234}Th and ^{210}Pb on the Amazon continental shelf. Geochim Cosmochim Acta 60:2123-2137

Somayajulu BLK (1994) Uranium isotopes in the Hooghly Estuary, India. Marine Chem 47:291-296

Spechler RM (1994) Saltwater intrusion and the quality of water in the Floridian Aquifer system, northeastern Florida. US Geol Surv Water Resource Invest Rept 92-4174

Sun Y, Torgersen T (1998) Rapid and precise measurement for adsorbed ^{224}Ra on sediments. Marine Chem 61:163-171

Swarzenski PW, McKee BA (1998) Seasonal uranium distributions in the coastal waters off the Amazon and Mississippi Rivers. Estuar 21:379-390

Swarzenski PW, McKee BA, Booth JG (1995) Uranium geochemistry on the Amazon shelf: Chemical phase partitioning and cycling across a salinity gradient. Geochim Cosmochim Acta 59:7-18

Swarzenski PW, McKee BA, Sørensen K, Todd JF (1999a) ^{210}Pb and ^{210}Po, manganese and iron cycling across the O$_2$/H$_2$S interface of a permanently stratified Fjord: Framvaren, Norway. Marine Chem 67:199-217

Swarzenski PW, McKee BA, Skei JM, Todd JF (1999b) Uranium biogeochemistry across the redox transition zone of a permanently stratified fjord: Framvaren, Norway. Marine Chem 67:181-198

Swarzenski PW, Reich CD, Spechler RM, Kindinger JL, Moore WS (2001) Using multiple geochemical tracers to characterize the hydrogeology of the submarine spring off Crescent Beach, Florida. Chem Geol 179:187-202

Swarzenski PW, Sutton P, Porcelli D, McKee BA (2003) ^{234,238}U isotope systematics in two large tropical estuaries: The Amazon and Fly River outflow regions. Cont Shelf Res (in press)

Thurber D L (1962) Anomalous ^{234}U/^{238}U in nature. J Geophys Res 67:4518-4520

Toole J, Baxter M S and Thomson J (1987) The behavior of uranium isotopes with salinity change in three U.K. Estuaries. Estuarine Coastal Shelf Sci. 25:283-297

Torgersen T, Turekian KK, Turekian VC, Tanaka N, DeAngelo E, O'Donnell JO (1996) ^{224}Ra distribution in surface and deep water of Long Island Sound: Sources and horizontal transport rates. Cont Shelf Res 16:1545-1559

Valiela I, Costa J, Foreman K, Teal JM, Howes B, Aubrey D (1990) Transport of groundwater-borne nutrients from watersheds and their effects on coastal waters. Biogeochem 10:177-197

Viers J, Dupre B, Polve M, Schott J, Dandurand JL, Braun JJ (1997) Chemical weathering in the drainage basin of a tropical watershed (Nsimi-Zoetele site, Cameroon): Comparison between organic-poor and organic-rich waters. Chem Geol 140:181-206

Webster IA, Hancock GJ, Murray AS (1995) Modeling the effect of salinity on radium desorption from sediments. Geochim Cosmochim Acta 59:2469-2476

Wen LS, Santschi PH, Tang DG (1997) Interactions between radioactively labeled colloids and natural particles: Evidence for colloidal pumping. Geochim Cosmochim Acta 61:2867-2878

Whitehouse BG, Yeats PA, Strain PM (1990) Cross-flow filtration of colloids from aquatic environments. Limnol Oceanogr 35:1368-1375

Windom H, Smith R, Niencheski F, Alexander C (2000) Uranium in rivers and estuaries of globally diverse smaller watersheds. Marine Chem 68:307-321

Yang H-S, Hwang D-W, Kim G (2002) Factors controlling excess radium in the Nakdong River estuary, Korea: submarine groundwater discharge versus desorption from riverine particles. Marine Chem 78:1-8

Zektzer LS, Ivanov VA, Meskheteli AV (1973) The problem of direct groundwater discharge to the seas. J Hydrol 20:1-36

15 U-series Dating and Human Evolution

A. W. G. Pike

*Research Laboratory for Archaeology
University of Oxford, 6 Keble Rd.
Oxford, OX1 3QJ, United Kingdom*

P. B. Pettitt

*Keble College
University of Oxford
Oxford, OX1 3QJ, United Kingdom*

1. INTRODUCTION

Some of the major events in human evolution occurred during the Middle Pleistocene. These include the evolutionary rise of the Neanderthals in Eurasia and of anatomically modern humans either in Africa or on a broader geographical front, in addition to major biogeographic changes among mammals including the hominine colonization of Europe. Prior to the Middle Pleistocene, the application of Potassium-Argon and Argon-Argon dating has yielded a relatively detailed chronological picture for the earlier phases of hominid evolution, and from *c.* 50 ky radiocarbon dating provides the backbone of our chronology for the extinction of the Neanderthals, colonization of Europe by anatomically modern humans and all subsequent events. Between these, however, the Middle Pleistocene has only been amenable to precise chronometric dating for the past few decades. Before the advent of radiometric dating, as Glynn Isaac (1975) famously observed, the Middle Pleistocene was "the muddle in the middle."

U-series dating is one of a range of dating methods employed by archaeologists to provide a chronology for human evolution (Fig. 1). Other applicable methods include optically stimulated luminescence (OSL), which can be used to date sediments perhaps back to about 0.5 My, measuring the time since the last exposure to sunlight. It is increasingly employed as a dating method especially since suitable sediments are common on archaeological sites (e.g., Aitken 1992). A related technique, Thermoluminescence (TL), can be used to date burnt flint artifacts (e.g., Aitken 1985), which provide a more direct date for specific human activity on a site, although material that is suitably burnt is relatively rare. Potassium-Argon and Argon-Argon dating have been the most useful techniques for calibrating early hominid evolution because its limit for dating extends considerably beyond the origins of the earliest hominids at around 5 My (e.g., Curtis and Hay 1972). Their application, however, is limited by the requirement of pyroclastic deposits with K-bearing phases that must be related stratigraphically to the hominid bearing layers. Electron Spin Resonance dating (ESR) can date the enamel and dentine of fossil teeth directly (e.g., Grün and Stringer 1991), and is used widely to date hominid remains themselves back to several My (e.g., Grün and Stringer 2000; Curnoe et al. 2001). There are problems specific to each of these methods and it is usual that an important archaeological site is dated by as many methods as possible to provide independent checks on the results. Although the main focus of this chapter is U-series dating, other dating methods will be considered.

In the book edited by Ivanovich and Harmon (1992), Schwarcz and Blackwell (1992) provided an extensive review of U-series dating in archaeology, which at that time focused on the dating of calcite deposits using ^{234}U/^{238}U and ^{230}Th/^{234}U disequilibria. U-series dates on calcite are generally considered both accurate and precise but the

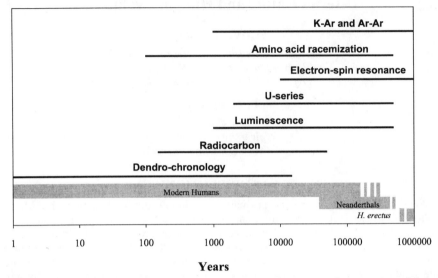

Figure 1. Dating methods employed by archaeologists (e.g., for an up-to-date overview of dating methods applied to human evolution see Grün et al., in press). Also shown is the rudimentary chronology of human evolution for the last million years.

occurrence of hominid remains and datable calcite is rare. In the words of Schwarcz (1992), "we are somewhat in the position of the drunken man who has lost his keys on a dark stretch of road but who is searching for them a few yards away, under a lamp post, because the light is better there. Thus, we must sometimes content ourselves with dating sites or strata in sites which are far removed from critical hominid loci, but which are better suited for dating and can, hopefully, be correlated to hominid sites." The dating of typologically diagnostic lithic technologies, where they are interstratified with calcite deposits, provides an indirect chronology for later human evolution. But given that the archaeological signatures left by late archaic and early modern humans are so similar, the direct dating of hominid fossils themselves is the only unambiguous way to investigate their evolutionary relationships. Occasionally, human remains are found with calcite coatings (e.g., Shen and Yokoyama 1984; Schwarcz et al. 1991; McDermott et al. 1996), but these only provide an *ante quem* date for the fossil. Thus, considerable effort has been made over the last decade to date directly the hominid fossils themselves (i.e., the bones and teeth), and this endeavor provides the focus of this chapter.

By 1992, the application of U-series dating to bones and teeth had not yet come of age. It was generally considered unreliable, and results were certainly controversial. It probably hasn't come of age even today, but a number of new developments, which are discussed in the first half of this paper have given it a new maturity. An increasingly large proportion of new U-series dating applications in archaeology involve the direct dating of fossil bones and teeth. A review of the major U-series studies, and their implications to our understanding of human evolution is presented in the second half of this chapter.

2. U-SERIES DATING OF BONES AND TEETH

While U-series dating of calcite deposits undoubtedly gives the most reliable dates (see Richards and Dorole 2003) with which to construct archaeological chronologies beyond the *c.* 50 ky limit of radiocarbon, such deposits are by no means ubiquitous on archaeological sites. Even where, for example, a stalagmitic floor seals an archaeological deposit, its U-series date will provide only a *terminus ante quem* (a minimum age estimate) for the archaeology. In some cases the archaeology may be fortuitously sandwiched between layers of datable calcite, thereby providing both *termini ante* and *post quem*, but only rarely is the date range narrow enough to provide a useful high resolution chronology, and often the layers are separated by 10s to 100s of ky. Furthermore, site formation processes are complex, and since calcite itself is not an archaeological material, considerable effort must be made to ensure the stratigraphic and chronological relationship between the actual archaeology and the dated calcite is known. In the case of some museum collections, particularly from old excavations, stratigraphic information is so poor that the archaeology cannot easily be related to the resulting dates.

Bones and teeth, however, are primary archaeological materials and are common to many archaeological sites. Bones bearing cut marks from stone tools are a clear proxy for human occupation of a site, and in the study of human evolution, hominid remains provide the primary archive material. Hence, many attempts have been made to directly date bones and teeth using the U-series method. Unlike calcite, however, bones and teeth are open systems. Living bone, for example, contains a few parts per billion (ppb) of Uranium, but archaeological bone may contain 1-100 parts per million (ppm) of Uranium, taken up from the burial environment. Implicit in the calculation of a date from $^{230}Th/U$ or $^{231}Pa/^{235}U$ is a model for this Uranium uptake, and the reliability of a U-series date is dependent on the validity of this uptake model.

The most commonly employed model is the *early uptake* (EU) model, where U is deemed to have been taken up sufficiently shortly after burial for the bone to approximate to a closed system. Justification for the validity of early uptake seems to have stemmed from Szabo's (1979) suggestion, later elaborated on by Rae and Ivanovich (1986), that Uranium is fixed in the bone in the U^{IV} oxidation state, facilitated by the reduction of U^{VI} by decay products of the organic phase of bone, collagen. Since the bulk of collagen is lost rapidly from the bone (on the U-series time-scale at least), it is assumed Uranium will be taken up rapidly, and then uptake will cease.

An alternative to the early uptake assumption, *linear uptake*, assumes that bones and teeth continue to take up Uranium at a constant rate (Ikeya 1982; Bischoff et al. 1995), giving a U-series date something over twice that calculated using the EU assumption. Although more common for teeth than for bone, both EU and LU dates are often quoted for the same sample, with the implication that the true age of the sample probably lies somewhere in between.

Recently, however, the validity of the EU and LU assumptions has been challenged, for bone at least, when applied *a priori* with no additional evidence for a particular mode of uptake for a particular sample. Figure 2 shows the results of a meta-analysis undertaken by Pike and Millard (forthcoming) of 103 published U-series dates on bones for which reasonable independent age constraints exist. The analysis shows that only about 25% of the dates calculated using an EU assumption agree with their control dates to within 10% of the measured U-series date. Even fewer bones agree using the LU assumption. For 65% of dates to agree using EU, we are required to assume about a 50% error on the U-series date, and for 95% to agree the error is in excess of 100%! If Pike and Millards' data-set is representative, we are forced to conclude that the *a priori* assumption of EU for bone gives about a 1 in 4 chance of giving a reasonably accurate date (i.e., within 10%). Considerably fewer dates will actually agree to within the quoted precision on a TIMS measurement (generally of the order of 1%). We would not like to stake our archaeological theory against such long odds, yet dates for bones based on the EU assumption are still appearing in the literature.

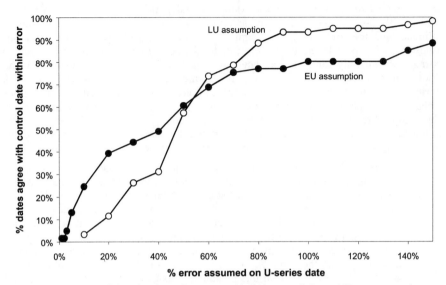

Figure 2. Percentage of U-series dates that agree with their control date within an assumed error. Results from dates calculated for each bone using both the EU and LU assumptions are shown (see forthcoming paper by Pike and Millard).

EU and LU are not the only uptake models that have been proposed. More mathematically sophisticated models have been developed (e.g., Szabo and Rosholt 1969; Hille 1979; Chen and Yuan 1988), in some cases using both the ^{238}U and ^{235}U decay chains (see Ivanovich 1982; Millard and Pike 1999 for summaries). While there have been some apparently successful applications of these models, none have been found to be universally applicable, and the search for a reliable method of U-series dating of bones and teeth continues. In the last decade or so, two important new approaches to the modeling of U uptake in bones and in teeth have been developed. These are discussed in detail below.

2.1. The diffusion-adsorption (D-A) model

The previously proposed uptake models were mathematical assumptions and had no physical or chemical basis. Millard and Hedges, on the other hand, considered the chemistry of bone-uranium interactions. With the D-A model, they proposed that U was diffusing into bone as uranyl complexes, and adsorbing to the large surface area presented by the bone mineral hydroxyapatite (Millard and Hedges 1996). Laboratory experiments showed a partition coefficient between uranyl and hydroxyapatite under oxic conditions of 10^4-10^5, demonstrating U uptake in the U^{VI} state without the need for reduction by protein decay products as proposed by Rae and Ivanovich (1986).

The D-A model predicts the distribution of uranium and U-series isotopes across a bone section (Figs. 3 and 4). Under constant conditions Uranium is diffusing from the inner and outer surfaces of the bone, giving a ∪-shaped Uranium concentration profile that gradually flattens with time to a uniform uranium distribution when the bone reaches equilibrium with the uranium in the groundwater. Because the uranium is equilibrating with the outer portions of the bone section first, closed system U-series dates approach the true age of the bone towards the surfaces, but are underestimated towards the centre. Further details of the D-A model are given in the Appendix.

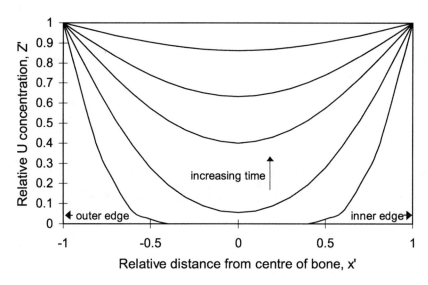

Figure 3. The development of U concentration profiles over time according to the D-A model. The model assumes constant conditions (e.g., water content of the soil, U concentration of the groundwater, and diagenetic state of the bone). [Used by permission of Elsevier Science, from Pike et al. (2002), *Geochim Cosmochim Acta*, Vol. 66, Fig. 1, p. 4275.]

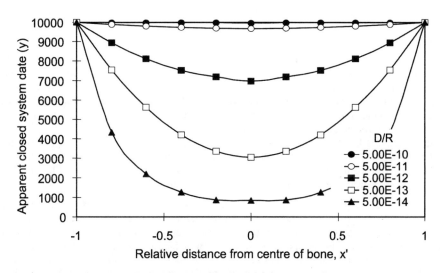

Figure 4. Modeled U-series date profiles across a 10 ky bone according to the D-A model under constant conditions. The dates are calculated using the closed system assumption. The parameter D/R is the diffusion-adsorption parameter and is related to the water content of the soil, the state of preservation of the bone and aspects of the geochemistry of the burial environment. After Pike et al. (2002). [Used by permission of Elsevier Science, from Pike et al. (2002), *Geochim Cosmochim Acta*, Vol. 66, Fig. 2, p. 4275.]

Pike et al. (2002) extended the D-A model by looking at how uranium uptake changes with changes in the burial environment. For example, a decrease in the uranium concentration in the groundwater will lead to desorption and diffusion of uranium out of the bone. The uranium can be leached out of a bone in this way, but the insoluble Th would remain, increasing the ^{230}Th/U and giving a falsely high U-series date calculated using the closed system assumption. Figures 5 and 6 show the effect of a drop in uranium concentration in the groundwater as predicted by the D-A model. Because the uranium is leached from the outer and inner surfaces of the bone first, the uranium concentration profile becomes a characteristic M-shape at first, becoming ∩-shaped as leaching progresses, and will eventually become uniform at the new equilibrium concentration. In the scenario shown, the apparent closed-system date at the surface jumps from 10 ky to about 240 ky after just 1 ky of leaching! Similarly, an increase in uranium concentration in the groundwater causes increased uptake of uranium by the bone, reducing the measured ^{230}Th/U, thereby giving underestimated closed system ages (this phenomenon is known as "recent uptake"). The U-series dates for such bones also exhibit characteristic profiles (Fig. 7).

Cases of gross over- and under-estimation of the age of a bone by U-series, attributable to leaching and recent uptake, are common in the literature. For example, Rae et al. (1989) give U-series measurements which have a ^{230}Th/^{238}U greater than the theoretical equilibrium value, because of the leaching of uranium. In other cases, the U-series dates come out far too young. The majority of Rae's et al. dates on bones from the British site of Hoxne (Oxygen Isotope Stage 9 or 11, 300-430 ky) fell in the range 10-30 ky, a massive underestimation, due to recent uptake of uranium. Grün et al. (1988) presented U-series dates on teeth from the same site in the range 27-56 ky, suggesting that teeth and bone have a similar response to changes in the burial environment. In another example, dental samples from the Oxygen Isotope Stage 7 (*c.* 190-250 ky) British site of Stanton Harcourt gave EU U-series dates of 61-147 ky prompting Zhou et al. (1997) to suggest between one and two thirds of the uranium in the teeth had been acquired very recently. U-series dates on bone from the same site gave EU dates in the range 7-90 ky (Pike 2000).

As well as using the shape of measured U concentration and U-series isotope profiles to identify and reject leached and recent uptake bones, Pike et al. (2002) identify the mechanisms by which bones can undergo early uptake. When a bone has reached equilibrium with uranium in the groundwater, no further uranium is taken up, provided the burial conditions remain constant. If the time taken to reach equilibrium is fast, compared to the age of the bone, the uptake can be considered early. The speed of equilibration is largely controlled by the diagenetic state of the bone. Pristine bones may take as long as 600 ky to equilibrate, but severely diagenetically altered bone (which has lost a large portion of its internal surface area) may equilibrate in as little as 12 ky, provided the burial environment is close to water saturation.

Pike's et al. (2002) application of the D-A model to U-series dating relies on identifying two groups of bones; those that have undergone early uptake, and those that fit the diffusion-adsorption method under constant conditions (for which the D-A model can be used to calculate an open system date). Bone samples are first screened using ICP-MS to measure the uranium concentration profiles. Bones that show evidence for leaching (e.g., Fig. 8) or that show irregular profiles, are rejected as unsuitable for dating. U-series isotopes are then measured on selected samples across the bone section using TIMS. Early uptake bones can be identified by their uniform distribution of dates across the section, although some recent uptake bones are rejected at this stage (e.g., those similar to Fig. 7). Bones that fit the D-A model under constant conditions, show ∪-

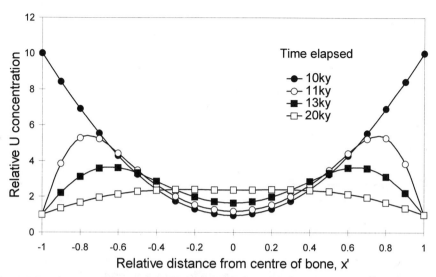

Figure 5. Development of U concentration profiles during a leaching scenario, $D/R=10^{-13}$. Here diffusive uptake occurs for 10 ky at a relative groundwater concentration of 10 after which the concentration is dropped to 1. U is lost initially from the outer portions of the bone, leading to distinctive M- and ∩-shaped profiles. [Used by permission of Elsevier Science, from Pike et al. (2002), *Geochim Cosmochim Acta*, Vol. 66, Fig. 3f, p. 4276.]

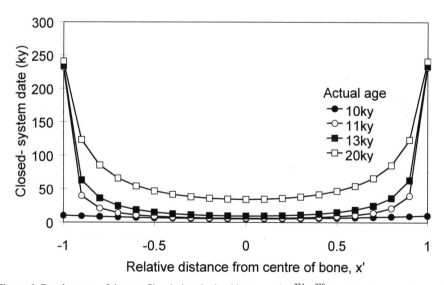

Figure 6. Development of date profiles during the leaching scenario ($^{234}U/^{238}U = 1.0$). Because U but not Th is being lost from the bone, the apparent closed system dates get older with leaching. The effect is marked, with dates at the surfaces of the bone giving dates of about 240 ky after just 1 ky of leaching for an 11 ky bone. [Used by permission of Elsevier Science, from Pike et al. (2002), *Geochim Cosmochim Acta*, Vol. 66, Fig. 3f, p. 4276.]

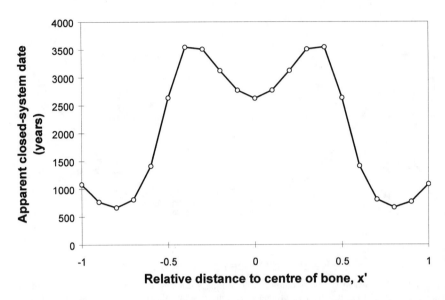

Figure 7. Date profile predicted by the D-A model for a recent uptake scenario. Here, the bone takes up U for 9 ky, at a notional groundwater U concentration of 1 after which the concentration is increased to 100 for a further 1 ky. Because the surfaces experience increased U uptake, the apparent closed system dates increase towards the centre of the bone. [Used by permission of Elsevier Science, from Pike et al. (2002), *Geochim Cosmochim Acta*, Vol. 66, Fig. 3e, p. 4276.]

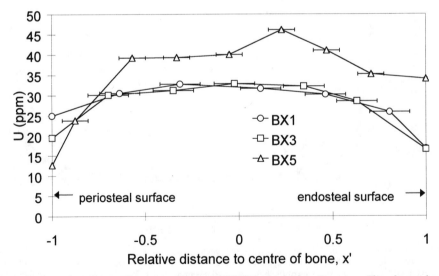

Figure 8. U concentration profiles measured using ICP-MS for bones from Boxgrove. These bones show characteristic leached profiles (compare with Fig. 5), and are rejected as unsuitable for dating. [Used by permission of Elsevier Science, from Pike et al. (2002), *Geochim Cosmochim Acta*, Vol. 66, Fig. 5f, p. 4280.]

shaped date profiles, and the D-A model can be used to calculate an open system date (e.g., Fig. 9). Preliminary results have shown a good agreement between the calculated U-series dates, and control dates (Pike et al. 2002).

There are a few drawbacks to this method. Using 4 or 5 TIMS measurements to produce a U-series date profile across a bone is time consuming, although a single reliable U-series date is surely worth hundreds where the accuracy is not known. In future, the application of Laser-Ablation ICP-MS to measuring profiles will significantly reduce the analytical effort required to obtain a date.

Correcting for detrital Thorium is also a problem. Because of the inhomogeneous distribution of U-series dates across a bone, an isochron cannot be constructed. Careful cleaning of the bones is certainly required, and in some cases a correction using an assumed $^{230}Th/^{232}Th$ of the detrital component is necessary, which leads to greater uncertainty in the dates. Because the ability to date a site depends on the geochemical history of that site, some sites may be "undateable." In some cases, bones from different stratigraphic units, often separated by many meters show similar patterns of leaching or recent uptake, and the geochemical changes that cause such phenomena may affect the whole site, leaving no bones suitable for dating. Nevertheless, bones are many times more common on archaeological sites than dateable calcite, and continued work on dating bone will improve the scope of our archaeological chronologies beyond the limit of radiocarbon.

2.2. U-series combined with electron spin resonance dating

Electron spin resonance (ESR) measures the trapped electron population in a lattice, which is directly related to the amount of ionizing radiation received by the sample since its formation. The total radiation dose received by the sample is estimated from the ESR

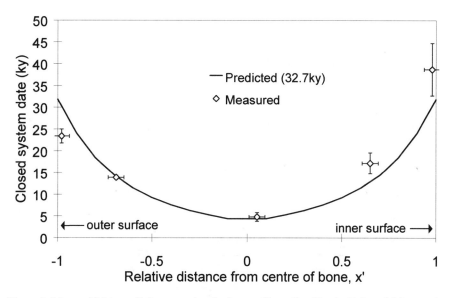

Figure 9. Measured "date profile" compared to the date profile predicted by the D-A model for sample P8008 from Combe Sauniere, giving a date of 32.7ky. The uncalibrated ^{14}C date for this bone is 28.7 ky which is in good agreement. The large errors on the U-series dates stem from a correction for detrital Th using a $^{230}Th/^{232}Th$ of 1.7 ± 0.7.

signal and measurements of the sensitivity of the ESR signal by further irradiation of the sample. A date can be calculated from the total radiation dose received by the sample in antiquity if the annual radiation dose can be calculated. ESR signals can be measured on a number of materials (e.g., see Ikeya 1993), but of primary relevance to archaeology is ESR dating of tooth enamel. The ambient radiation dose rate of the sediment can be estimated either from measurements of radionuclides in a sediment sample, or from γ-spectrometric measurements made on site and combined with estimations of the cosmic dose rate. An additional component of the annual dose will come from the decay of uranium (and daughter nuclides) taken up by both the enamel and dentine of the tooth. Thus, for teeth with significant uranium concentrations, a uranium uptake model needs to be applied to calculate an ESR date. Traditionally the EU or LU (or both) models were applied, but as for U-series dating, there were many cases where dates calculated using these assumptions disagreed with the known age of the samples, probably because similar phenomena such as recent uptake and leaching affect teeth as well as bone (e.g., Zhou et al. 1997)

Grün et al. developed an approach which combines a U-series date and an ESR date from the same sample (Grün et al. 1988). U uptake is assumed to take the form:

$$U_t = U_T \left[\frac{t}{T}\right]^{p+1} \tag{1}$$

where U_t is the uranium concentration at time t, U_T is the measured uranium concentration, T is the age of the tooth and p is a parameter that takes a value > -1. When $p = -1$ uptake is instantaneous and early, when $p = 0$ uptake is linear and when $p > 0$, uptake is sublinear or "recent uptake" (Fig. 10). A value for p is found iteratively such that both the ESR and U-series dates agree when calculated using uranium uptake according to Eqn 1.

This method had several advantages over traditional uptake assumptions. It can predict uptake somewhere between early and linear ($-1 < p < 0$), giving greater precision

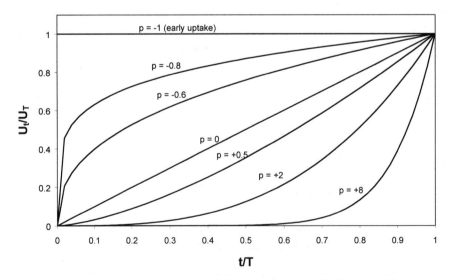

Figure 10. Uptake according to Grün's et al. (1988) "p-parameter model." When $p = -1$ the system is one of early uptake, when p = 0, uptake is linear, becoming increasingly sub-linear as p is increased above 0.

than the traditional presentation of EU and LU dates as maxima and minima which often span a broad range. It can also be used to calculate a date when the bulk of the uranium has been taken up late in the burial history of the tooth (i.e., when p > 0). The authors initial example was from the site of Hoxne, where U-series results on teeth and bone had given extremely young dates (see above). Using combined ESR and U-series measurements Grün et al. (1998) calculated values for p of 3.1-8.7 for dentine and 3.4-4.9 for enamel, showing sublinear, "recent" uptake in all cases. Their average calculated date of 319 ± 38 ky places Hoxne within Oxygen Isotope Stage 9, and should be contrasted with the EU U-series dates of just 27-56 ky. Of course if Hoxne is Oxygen Isotope Stage 11 then this date is still an underestimate, although considerably closer.

2.3. Non-destructive U-series dating by gamma spectrometry

Although not a very recent development, measurement of U-series isotopes by gamma spectrometry is worth a mention here because of its increasing application to the dating of bones. About 50 mg, or less, of bone is required for a TIMS U-series measurement, although a bone section is required if profiles are to be measured for the application of the D-A model. Understandably, museum curators are reluctant to allow the most valuable specimens to be cut or drilled, so the non-destructive measurement of a U-series date is extremely advantageous.

γ-spectrometric determination of U-series isotopes was first proposed to date marine samples in 1980, and applied to the dating of bone shortly after (Yokoyama and Nguyen 1981). The technique measures gamma emissions of U-series isotopes of interest or their short-lived daughters as a proxy, and its application to bone is described in detail in Simpson and Grün (1998). The measurement of $^{231}Pa/^{235}U$ is the most straight-forward, and suitable for dating samples up to about 100 ky. Measurement of $^{234}U/^{238}U$ and $^{230}Th/^{234}U$ is more problematic. There are only two usable γ-emissions from ^{234}U decay and both have interference from other more intense γ-rays. The contribution of interfering gammas can be subtracted but gives a poor precision. In general, the precision of a γ-spectrometric U-series date is rarely better than 10% and can be as poor as 30%. Further details of measurement techniques for U-series isotopes are given in Goldstein and Stirling (2003).

2.4. Future developments

There are a few developments on the horizon that will increase our ability to date bones and teeth reliability. Both γ- and α-spectrometric methods can measure $^{231}Pa/^{235}U$ and $^{230}Th/U$ and concordance between dates calculated using the two can provide a measure of reliability. However, the discordance between the two is not very sensitive to different uptake regimes, and it is difficult to resolve, for example, bones that have undergone EU from those that have undergone LU with the analytical errors commonly encountered in measurements by γ- and α-spectrometry. On the other hand, it has been shown recently that TIMS can measure both isotopic ratios with a precision usually better than 1% (Edwards et al. 1997). TIMS measurements of $^{231}Pa/^{235}U$ and $^{230}Th/U$ have yet to be routinely applied to dating fossil remains, but in the future, concordance between the two decay series will provide further evidence of the validity of a particular uptake model to a particular sample.

The measurement of U-series isotopes using laser ablation will potentially reduce the damage inflicted on specimens to almost zero. The increased spatial resolution will allow profiles, as required by the D-A model, to be measured quickly and simply in the smallest of bone samples and across thin layers of enamel (e.g., Fig. 11). Early studies (e.g., Belshaw et al. 2002; Eggins et al. 2002) have demonstrated the potential of the technique to provide U-series profiles for bones and teeth, but consistent and fully quantitative results are still a little way off.

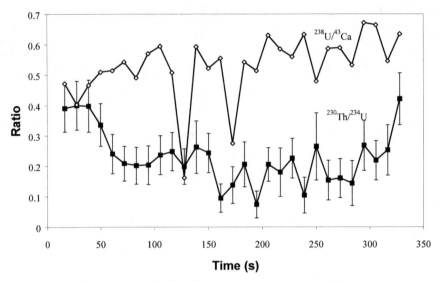

Figure 11. U concentration profile and $^{230}Th/^{234}U$ profile measured using laser ablation ICP-MS across a bone using the method of Eggins et al. (in press). $^{238}U/^{43}Ca$ is expressed as an atomic ratio, $^{230}Th/^{234}U$ as activity. The laser is scanned at 1 mm/min and the graph shows the profiles from the outer to the inner surfaces of the bone. U is expressed as a ratio with Ca to correct for fluctuations in laser intensity and surface topography.

3. APPLICATIONS

3.1. The issue of chronology in hominid evolution

Theories of human evolution are based on genetic and morphological data read in spatial and chronological context. In particular, the chronological relationship of different grades of fossils is used to make inferences about extinctions, speciation events and the relationship between morphologically distinct hominid taxa. The development of a precise chronology for hominid fossils is the only way to test competing hypotheses of human evolution, and to calibrate models of evolution based on genetic studies. The chronology for the last archaic and earliest modern humans is still in a state of flux, and because of this it has not been possible to resolve convincingly the conflict between the two main hypotheses for modern human origins, i.e., whether it occurred in a single or multiple regions.

The multi-regional evolution hypothesis proposes that the evolution of modern humans from late archaics occurred as a gradual process in many old world regions, with sufficient gene flow between these to ensure taxonomic integrity (e.g., Wolpoff 1989, who would contend that archaic and modern humans do not exist as real entities, but represent only older and younger forms of *Homo sapiens*). By contrast, the single centre hypothesis proposes that such an evolution occurred in only one region (e.g., Africa) and archaic populations elsewhere played little or no role in the evolution of our own species, ultimately being replaced by the expansion (from its single origin) of the new species (e.g., Stringer 1989).

From a chronological point of view, the single centre hypothesis is easier to test, as it predicts that the earliest fossils will be found in only one area. In addition, the chronological contemporaneity of two distinct hominid taxa will inevitably rule out an ancestor-descendent relationship between the two. By contrast, the multi-regional hypothesis would be supported, or at least not refuted, in a given region if hypothetical ancestor and descendent populations are chronologically distinct.

Chronology is not the only contentious issue: debates over the interpretation of morphological and genetic data are as contentious as that of dating. While the anatomical classification of hominids is often problematic and can take a subjective slant, the dating is in theory objective and absolute. The following examples, which have been selected to illustrate these issues are exemplars of the current debate, and cover the major issues noted above specifically from the perspective of U-series dating.

3.2. Neanderthals and modern humans in Israel

The caves of Skhul, Qafzeh, Kebara, Tabun and Amud in Israel are critical in our understanding of later human evolution and the relationship between anatomically modern human and Neanderthals. Early non-radiometric chronologies put the Neanderthal layers (e.g., of Tabun, Fig. 12) at 50-60 ky (Jelinek 1982), with anatomically modern humans at Skhul and Qafzeh succeeding them at around 40 ky. This chronology was in keeping with the multi-regionalist view that Neanderthals evolved into modern humans in the Levant and subsequently dispersed into Europe. ESR dating (Grün et al. 1991; Stringer et al. 1989), and U-series dating on teeth (McDermott et al. 1993) from these sites had a major impact on this view.

McDermott et al. (1993) produced EU U-series dates and both EU and LU ESR dates on teeth from the hominid bearing layers of Tabun, Qafzeh and Skhul. Their results suggest both the Neanderthals and Modern Humans from Tabun, Skhul and Qafzeh are much older that previously thought, at around 100 ± 5 ky. Some faunal dates from Skhul were younger at around 43 ky, but were possibly the result of recent accumulation of uranium. Not only do these dates represent some of the earliest modern human skeletal material found anywhere, but the dates of the Neanderthal layers were broadly

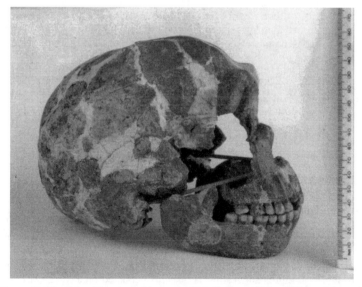

Figure 12. The Tabun Neanderthal (courtesy Chris Stringer, National History Museum).

contemporaneous with the modern human layers, casting doubt on the ancestor/descendent evolution model of Neanderthals and modern humans. The dates fit with a model of evolution of modern humans in Africa perhaps as early as 150-200 ky (e.g., Grün and Beaumont 2001) who then spread via the Near East to the rest of Asia (e.g., Stringer 1989).

The interpretation of the significance of the Tabun Neanderthal is somewhat hampered by the uncertainty as to which layer the hominid came from, layer B or C (Bar Yosef 2000). Despite the method's lack of precision, Schwarcz et al. (1998) dated elements of the Neanderthal skeleton directly and non-destructively using γ-spectrometry. Using an EU assumption they calculated a date of 34 ± 5 ky for the mandible, and 19 ± 2 ky for the femur (which they state would give a date of 33 ± 4 ky if a LU model were employed). They claim these dates show the burial to be intrusive to layer B or C, and "may be one of the youngest Neanderthals known" (Schwarcz et al. 1998). Their result generated considerable discussion, since it disagreed with the ESR chronology (Grün et al. 1991) which would place it between 69 ± 13 (latest date from layer B) and 134 ± 36 (earliest date from layer C), the thermo-luminescence dates on burnt flint which date layer C to at least 144 ± 14 ky (Mercier et al. 1995), and the U-series dates on layers B and C of 51-105 ky (McDermott et al. 1993). Millard and Pike (1999) urged caution in accepting these young dates, and dates by γ-spectrometry in general since, by the very nature of the non-destructive measurements, they lack any evidence to suggest that EU or LU or, for that matter, any U uptake model is appropriate for a particular bone. While in some cases γ-spectrometric U-series has given dates in agreement with other dating methods (e.g., Yokoyama et al. 1997), such young dates as for Tabun could be the result of recent accumulation of U. Both $^{230}Th/^{238}U$ and $^{231}Pa/^{235}U$ can be estimated using γ-spectrometry and could provide independent evidence for the mode of U uptake, but the degree of discordance between dates calculated using the two ratios is relatively insensitive, showing a maximum difference of only 8% between the $^{231}Pa/^{235}U$ for the EU and LU assumptions despite EU and LU dates differing by more than a factor of 2 (e.g., see Ku 1982; Cheng et al. 1998; Pike and Millard, forthcoming). Further criticism of Schwarcz et al. (1998) came from Alperson et al. (2000) who see an incongruity between the young dates, and the lithic technologies found at Tabun. If the Neanderthal were really as young as 34 ky, one might expect it to be associated with an Upper Palaeolithic lithic assemblage rather than a Levantine Middle Palaeolithic one. Such controversy is common with U-series dating of bone in archaeology, because of the uncertainty in U uptake. The consensus view, based on the ESR, U-series and luminescence chronologies is probably that the Neanderthal is in the region of 100 ky (e.g., Grün and Stringer 2000), although there is still considerable disagreement between dates from the various techniques. The authors of the paper presenting the γ-spectrometric dating of Tabun have since compared their results for the Neanderthal with new measurements of faunal material from the same site. Using the discordance between $^{230}Th/^{238}U$ and $^{231}Pa/^{235}U$ they have now concluded that the Neanderthal probably experienced some later uptake of U, which would account for the younger than expected date (H.P. Schwarcz pers. comm.).

3.3. *Homo erectus* and *Homo sapiens* in Java

The dating of bovid teeth from the *H. erectus* bearing layers from Ngandong and Sambungmacan, Central Java, has caused considerable controversy. Mean EU U-series and ESR dates produced dates of 27 ± 2 and 53 ± 4 ky for the two sites (Swisher III et al., 1996). Furthermore, Swisher III et al. suggest the sample had lost U through leaching, making these *maximum* ages. Such dates are very much younger than previous age estimates for these hominids suggesting *H. erectus* persisted on Java 250 ky longer than on the Asian mainland, and perhaps 1 My longer than in Africa. If these dates are to believed, it makes Javanese *H. erectus* contemporary with the later Neanderthals in Europe, and with anatomically modern humans. The authors argue that the unilineal transformation of *H. erectus* to *H. sapiens* in Southeast Asia is no longer chronologically plausible, since the two overlap, and since such late dates for *H. erectus* are not paralleled elsewhere, it is unlikely that *H. erectus* would evolve into *H. sapiens* everywhere else except Java. By implication they favor a replacement model for the decline of *H. erectus* rather than an evolution of *H. erectus* to *H. sapiens*. In their interpretation, it is simply that the Java *erectus* was not replaced until relatively recently.

The criticisms of these results centre on the provenance of the dated faunal material, and the U uptake models employed by Swisher III et al.(1996). In a reply to the paper, Grün and Thorne (1997) suggest the river terrace where the material was found had been reworked many times, mixing older with younger material, and that the state of preservation between the hominid remains and the faunal remains was sufficiently different to suggest they had fossilized in a different environment. In addition, they doubt whether the U uptake models employed are appropriate. In the future, direct dating of the hominid remains themselves will address the first problem, a more sophisticated U uptake model would be required to address the second.

Others interpret the relationship between *H. erectus* and *H. sapiens* differently for Java. Wolpoff (2001), for example, suggests that the recently discovered SM3 (*"Java Gal"*) calvaria lies morphologically between *H. erectus* and *H. sapiens* and interprets this as indicating a local evolution from *H. erectus* to *H. sapiens*. However, given the lack of unambiguous dating for this fossil (and arguably the latest *H. erectus* in Java as a whole), this remains at present a very open issue.

3.4. *Homo erectus* and *Homo sapiens* in China

In 1988, Chen and Yuan published U-series dates on bones and teeth for more than 20 Chinese sites, providing the first extensive Palaeolithic chronology for China. They used α-spectrometric measurements of ^{230}Th/U, and ^{231}Pa/^{235}U, and used concordance between dates calculated from the two series to demonstrate the reliability of a date. Where there was discordance between the two calculated dates, they used a two-stage model of initial U uptake, followed by a second uptake or loss event to calculate a date. These results, and others, prompted Chen and Yinyun (1991) to suggest a possible contemporaneity between *H. erectus* and *H. sapiens, sensu lato,* in China.

Zhoukoudian Locality I has provided the largest sample of *H. erectus* remains from any single locality in the world. Chen and Yuan (1988) provided dates on faunal bones of 220-460 ky for the hominid bearing strata at the site, which is in broad agreement with ESR, TL and palaeomagnetic dating of the same levels (e.g., see Brown 2001). In 1980-81, a number of cranial and dental fossils were recovered from the site of Hexian. While these are clearly of *H. erectus* grade, a number of characteristics contrast with Zhoukoudian, and it may be that the Hexian population was morphologically closer to Javan *H. erectus* than other Chinese populations. Chen and Yuan's chronology provides dates of 150-190 ky, for these fossils with a possible maximum of 270 ky. Overall, it would appear that the Chinese *H. erectus* populations date to 200-400 ky, which it is claimed is a similar age range to fossils interpreted as *H. sapiens* found at other Chinese sites such as Chaohu, Dali, and Jinniushan.

The site of Chaohu, 50 km south of the site of Hexian, yielded a few fragments of hominids classified as archaic *H. sapiens* which were dated by U-series to 160-200 ky (Chen and Yuan 1988). The site of Dali yielded a reasonably complete cranium which has been assigned to archaic *H. sapiens*. An EU U-series date of 209 ± 23 ky was obtained from an Ox tooth apparently associated with the hominid cranium (Chen et al. 1994). A number of U-series dates exist from the Jinniushan cave, which range from 200-310 ky (Chen and Yuan 1988). Layer 7 yielded a partial skeleton which, like Dali, appears to be of archaic *H. sapiens*. ESR dating of teeth from layer 7 suggest that the skeleton may be at the younger end of this age range (Chen et al. 1993). Many people have interpreted the existing dates as indicating a degree of contemporaneity between the two hominid taxa (e.g., Chen and Yinyun 1991), and as such appear inconsistent with the notion of a local evolution from *H. erectus* to *H. sapiens*.

However, the date ranges given are themselves too broad to confirm unambiguous contemporaneity. Instead, we suggest that they could be taken to suggest chronological separation, with *H. erectus* pre-dating and *H. sapiens* post-dating about 200 ky. Recent work has added more weight to this latter interpretation. Using U-series on calcite from Zhoukoudian Locality 1, Shen et al. (2001) dated the youngest *H. erectus* fossil to >400 ky. A similar age (412 ± 25 ky) was obtained by Grün and colleagues using combined ESR-U-series on teeth from the *H. erectus* bearing layers at Hexian (Grün et al. 1998). U-series dating of calcite deposits from immediately above the *H. erectus* bearing layers at Nanjing, similarly showed how previous ESR and U-series measurements on fossil teeth were probably underestimating the dates of the latest *H. erectus* fossils in China (Zhao et al. 2001). The calcite dates gave a youngest age of about 510 ky, some 270 ky older than previous ESR and U-series measurements. These latest dates increase significantly the gap between the earliest *H. sapiens* and latest *H. erectus*, forcing a re-evaluation of the hypothesis of co-occurrence of these two species.

Note that this debate is quite distinct from the question of the timing for the appearance of "modern" *H. sapiens* in China. The *H. sapiens* fossils discussed above have been variously assigned to archaic *H. sapiens*, advanced *H. erectus* or *H. heidlebergesis*, and their role in the origins of modern humans is hotly debated (e.g., see Wolpoff et al. 1984; Stringer and McKie 1996). Until recently there was no unequivocal evidence for anatomically modern humans in China prior to 30 ky, and the traditional view was for the appearance of modern humans after 40-50 ky (Wu 1989). Recently, however, tantalizing, if ambiguous, evidence may back-date the appearance of modern humans in China to >100 ky, which would be in line with those from Qafzeh and Skhul (see above). Shen et al. (in press) measured U-series dates on speleothem layers at the site of Lujing, from where an almost complete modern human cranium was discovered by workers digging for fertilizer in 1958. The exact layer from which the fossil came was not recorded, but Shen et al. were able to provide date ranges for the layers which gives 3 distinct possibilities for the age of the skull: 20-61 ky; 68-153 ky, >153 ky. They prefer the date of >68 ky, and with additional dates on fragments of calcite, give a most likely date of 111-139 ky. Such an early date would have extremely important implications for the origins of modern humans.

The problem here, however, is that the evidence that the skull came from the layer dating between 68-153 ky is inconclusive. Contemporary reports of the appearance, texture and mineralogy of the hominid bearing layer are contradictory, and the possibility that the fossil was from a grave cut into lower layers is not considered. Direct dating of the skull would seem to be the only way to solve these problems.

3.5. The "Pit of the Bones" and a new species of hominid in Spain

The Sierra de Atapuerca in Burgos, Spain, has recently yielded two impressive and important hominid bearing sites, the Gran Dolina, and the Sima de los Huesos.

At the Sima de los Huesos ("Pit of the Bones") the well preserved remains of at least 32 individuals, classified as *H. heidelbergensis* and apparently ancestral to Neanderthals, were found along with bear and fox bones within mud breccias enigmatically deep within the cave system and far from the modern day entrance. In an extensive study, Bischoff et al. (1997) attempted to date the bear and human bones using U-series along with ESR dates on selected bones. The U-series dates for the humans show a large scatter, 80-220 ky, probably attributable to differential uranium uptake by different bones. In addition, a relationship between sample geometry and apparent date might be suggested by the 8 repeat sub-samples of the same bone that produced dates from 121-162 ky. The diffusion-adsorption model predicts in most cases an inhomogeneous distribution of U-series dates across a bone (e.g., see Fig. 4). Samples removed by drilling deeper into the bone should produce different results to those from the surface. As a general rule, the samples towards the surface will tend to appear older, although this is not the case if there has been recent uptake of uranium (Pike et al. 2002). A similar but more pronounced effect is predicted for the distribution of U-series dates in teeth (Pike and Hedges 2001). Nevertheless, Bischoff et al. (1997) went on to suggest the results provide a firm minimum age of about 200 ky for the deposition of the human fossils in the shaft. They also undertook ESR measurements on 9 of the bear bones and calculated combined U-series–ESR dates using the *p*-parameter model, in the range 195-323 ky. They argue that a minimum date of 320 ky for the human can be inferred because they appear stratigraphically to pre-date the bear bones. However, bone is a less than ideal material for ESR dating since it is vulnerable to diagenetic change, including recrystallization which resets the ESR signal (e.g., see Grün and Schwarcz 1987). Given this, and the uncertainty in U-series dates (even those sub-sampled from a single bone), the dates proposed by this study for the hominids should be taken as provisional.

In the 2001 excavation season, however, a 14 cm thick *in situ* speleothem was discovered immediately above two hominid phalanges, providing the opportunity to obtain a *terminus ante-quem* date for the *H. heidelbergensis* fossils. The upper layers of the speleothem gave finite dates in stratigraphic order from 153-281 ky, but for the lower layers, samples were at or near isotopic equilibrium, yielding a minimum age of 350 ky (Bischoff et al. in press). By extrapolating the growth rate of the speleothem calculated from the three finite U-series dates, Bischoff et al. (in press) suggest that the hominids are perhaps 400-500 ky old. This would fit the context of hominids of this date from Europe as being *H.heidelbergensis*—which some view as an ancestor common to both Neanderthals and modern humans (e.g., Stringer and Hublin 1999). However, Bischoff et al. (in press) view the Sima de los Huesos and other European fossils of this date as already exhibiting Neanderthal traits and therefore as ancestral only to Neanderthals. Their proposed common ancestor to modern humans and Neanderthals is older, and evidence for it was found in the vicinity of the Sima de los Huesos at the collapsed cave site of Gran Dolina.

The Gran Dolina deposits contained remains of a minimum of 6 individuals associated with lithic tools, and a rich fauna characteristic of a later Early Pleistocene age (Laplana and Cuenca-Bescos 1997). Although some scholars would view these fossils as falling within the broad morphological range of *H. heidelbergensis*, it has been proposed that the hominid remains from level 6 represent a new species, *H. antecessor*, a possible common ancestor of modern humans and Neanderthals (Bermudez de Castro et al. 1997). The hominid bearing layer lies below a layer containing the Matuyama-Brunhes magnetic reversal boundary which is dated to *c.* 780 ky providing a *terminus ante quem* date for the hominid material. Falguères et al. (1999) attempted to confirm this date using combined U-series and ESR dating and the *p*-parameter model on faunal teeth. For materials that have behaved as a closed system, that are older than 780 ky, the ^{230}Th/U would be indistinguishable from equilibrium and only a minimum age (probably in the region of 500 ky) could be calculated. For systems that continue to take up uranium, however, the effective limit of Th/U dating is extended as the influx of new U maintains the disequilibrium. U-series dates were measured using α-spectrometry on dentine, enamel and cementum. Despite the supposed great antiquity of the teeth, the U-series dates for layer 6 were 176-280 ky for the enamel, and 151-374 ky for the dentine, but ESR gave EU dates of 626-685 ky. When combined using the p-parameter model, dates of 676-762 ky were calculated. The error on these calculated dates are large ($\pm$ c. 110 ky), primarily because of the relative imprecision of the α-spectrometry. Nevertheless, these dates agree within error of the palaeomagnetic evidence, and confirm the hominids as some of the earliest in Europe.

3.6. The earliest Australian human remains

The timing of the initial colonization of Australia has long been debated. Only a few sites pre-date c. 40 ky. Initial OSL dates of between c. 100-170 ky for archaeology at the Jinmium rock shelter were eventually refuted as they had not been completely zeroed by exposure to sunlight, with the real age of the shelter's occupation probably closer to c. 3 ky (Spooner 1998). At the Malakunanja rock shelter the lowest artifact bearing horizon has been dated by TL to 41-63 ky (Roberts et al. 1990), although the artifacts may have been stratigraphically mobile and the resulting age therefore an overestimation. Given this, and the large error on the date, the site is not convincing evidence of a pre-40 ky human occupation of Australia.

Recently, a study was published claiming to have provided the earliest dated human in Australia. The LM 3 human skeleton was discovered in a shallow grave lined with red ochre cut into a Pleistocene lunette fringing Lake Mungo in New South Wales. Morphologically it is gracile and is claimed to fit within the range of living Aboriginal Australians. It was the subject of an extensive dating study with U-series dates on the bone by γ-spectrometry and by TIMS, an ESR date on a tooth, optical dating (OSL) of the underlying sediments and U-series dating of the carbonate fraction of the sediment (Thorne et al. 1999). Results from all the dating techniques were in general agreement. The EU TIMS U-series results on the bone fell in the range 51-70 ky, but the true age may be younger because the bones had probably lost some uranium. The mean of the γ-spectrometric dates gave 69.5 ± 2.9 ky, the U-series on the leached fraction of the sediment 82 ± 21 ky, ESR of the tooth 63 ± 6 (EU) and 78.7 ± 7 ky (LU), combined U-series-ESR 62 ± 6 ky and the OSL date on the underlying sediment 61 ± 2 ky. These dates are about 20 ky older than previous age estimates of the sediments containing the burial, but are in general agreement, and would make LM 3 the earliest Australian human remains yet discovered.

Bowler and Magee (2000) published a comment refuting the results of Thorne et al. (1999). Bowler and Magee note that Thorne et al. did not stress that the OSL samples were in fact taken 350 m and 500 m away from the LM3 burial site, and the relationship between the burial and the dates may not be as close as perhaps perceived. Furthermore, they suggest that the OSL date of c. 61 ky was not from the layer into which the grave was cut, but from a lower layer. Bowler and Magee maintain that the burial was dug into a higher layer, formed during an arid transitional phase characterized by pelletal clays, fine sand and a significant increase in Wüstenquartz (regional desert-derived dust) which were found in the burial infill. The age of this phase, estimated from dates for other locations in the region, is placed at about 45 ky. Bowler and Magee also present a new OSL date from this transitional phase at 43.3 ± 1.7 ky. Furthermore, they challenge the "good agreement" between dating methods by Thorne et al., drawing attention to the sediment-carbonate date, which at the 2σ range gives 40-124 ky. Similarly, they cite Thorne and coworkers own comment that because there is evidence for loss of U, the true age of the bones may be less than 50 ky.

Gillespie and Roberts (2000) also draw attention to the identified loss of uranium from the bones. They suggest that the U-series date of the sample with the highest uranium concentration (by implication the least affected by uranium loss) should be taken as a maximum age, making the burial <50.7 ky. They also suggest that the Thorne et al. (1999) dosimetry calculations for the ESR dates do not take into account the inhomogeneity of the surrounding sediment which would lead to considerable uncertainty in the date. Furthermore, Thorne et al. (1999) made no attempt to control for partial bleaching (incomplete resetting of the luminescence signal) in their OSL dates, which if present would lead to over-estimated dates. Gillespie and Roberts concur with Bowler and Magee (2000) that the likely date for LM3 is around 43 ky.

Stringer (1999) has recently reviewed the debate surrounding the peopling of Australia. The dual-origin hypothesis proposes that there were two original colonizations. An early one, with a robust population originating from the *H. erectus* population of Java arriving at around 50 ky, and represented by fossils such as that from Willandra Lakes (WLH-50). A second, more gracile population, supposedly derived from China, then arrived around 30 ky. An alternative hypothesis, of a single origin of the Australian population sees the gracile and robust forms as morphological variation that occurred within Australia after the initial colonization.

The LM3 skeleton is gracile, and with such an early date, if accurate, suggests the gracile population arrived first, and this order of events according to Stringer fits better with the Single Origin model. The early date of LM3 also has wider implications for our understanding of modern human evolution. The LM3 human would have had needed to use a sea-going craft to reach Australia, and the symbolic use of red ochre is cognitively sophisticated. It is generally perceived that the transition from the Middle to Late Stone age lithic industries in Africa at around 50 ky marked cognitive and behavioral changes resulting in culturally "modern" humans. However, LM3 predates this transition by 10 ky, suggesting that the Middle Palaeolithic humans of Africa may have been behaviorally "modern" (Stringer 1999).

4. CONCLUSION

It is rare for a new date for hominid remains to be published without significant controversy and debate. The extent of such discussion, exemplified above, reveals how important a single date can be to our understanding of human evolution. The date of a single hominid fossil has the potential to force re-evaluation of our evolutionary theories. Consider the impact, if, for example, human remains from Australia were dated by U-series to 200 ky. We might be talking about the Out of Australia model of modern human evolution! Yet there is sufficient evidence to suggest that the blind application of an EU assumption in U-series dating of bone can lead to dates so erroneous as to give just such a result. Leaching of uranium can lead to grossly overestimated dates, and recent uptake can lead to gross under-estimations using an EU assumption. With the exception of sites where calcite deposits are present, most of the U-series dates presented above were calculated using the EU model on bones and teeth, and if not of doubtful accuracy, then certainly of unknown accuracy.

This is the most fundamental challenge for U-series dating in archaeology. With the advent of TIMS, the precision of our measurements has improved more then an order of magnitude, but precision is nothing without accuracy. It is generally accepted that U-series dates on calcite are accurate (accurate enough to calibrate ^{14}C beyond the tree ring record). However, the challenge is to produce accurate dates on more archaeologically relevant and ubiquitous materials such as bones and teeth. Approaches have been developed (e.g., the diffusion-adsorption uptake model, and combined U-series-ESR dating) that clearly have the potential to increase the reliability of our chronologies, but require more widespread adoption before the chronology will be reliable enough to resolve even some of the most basic debates in the study of later human evolution. The persistence of both single origin and multi-regional interpretations for nearly all the Middle Pleistocene hominid fossils is testament to the current inadequacy of our chronology. The science has certainly progressed in the last decade, and many more attempts to date fossil remains directly are being made, but we are still a long way from bringing order to the *muddle in the middle*.

ACKNOWLEDGMENTS

The application of the Diffusion-Adsorption model to dating bone (by AP) was funded by a NERC grant to Robert Hedges at the Research Laboratory for Archaeology, University of Oxford. The U-series date profiles shown here were measured at the NERC U-series dating facility at Open University, and the laser ablation U-series profile was measured at the Research School for Earth Sciences, Australian National University, Canberra in collaboration with Steve Eggins and Rainer Grün.

REFERENCES

Aitken MJ (1985) Thermoluminescence dating. Academic Press, London
Aitken MJ (1992) Optical dating. Quat Sci Rev 11:127-132
Alperson N, Barzilai O, Dag D, Matskevich Z (2000) The age and context of the Tabun I skeleton: a reply to Schwarcz et al. J Human Evol 38:849-853
Bar-Yosef O (2000) The Middle and Early Upper Palaeolithic in southwest Asia and neighbouring regions. *In:* The Geography of Neanderthals and Modern Humans in Europe and the Greater Mediterranean Vol. 8, Bar-Yosef O, Pilbeam D (eds), Peabody Museum Bulletin: Cambridge, MA, p 107-156
Belshaw NS, Pike AWG, Henderson GM (2002) U-series dating of archaeological bone material by laser ablation multiple-ion-counter ICP-MS. Poster presented at Golschmidt 2002 conference. 18th-25th August 2002, Davos, Swtizerland
Bermudez de Castro JM, Arsuaga JL, Carbonell E, Rosas A, Martinez I, Mosquera M (1997) A hominid from the Lower Pleistocene of Atapuerca, Spain: possible ancestor to Neanderthals and modern humans. Science 176:1392-1395
Bischoff JL, Shamp DD, Aramburu A, Arsuaga JL, Carbonell E, Bermudez de Castro JM (in press) The Sima de Los Huesos Hominids date to beyond U/Th equilibrium (>350ky) and perhaps to 400-500 ky: new radiometric dates. J Arch Sci
Bischoff JL, Fitzpatrick JA, León L, Arsuaga JL, Falgueres C, Bahain JJ, Bullen T (1997) Geology and prelimiary dating of the hominid-bearing sedimentary fill of the Sima de los Huesos Chamber, Cueva Mayor of the Sierra de Atapuerca, Burgon, Spain. J Human Evol 33:129-154
Bischoff JL, Rosenbauer RJ, Moench AF, Ku T-L (1995) Uranium series age equations for uranium assimilation by fossil bone. Radiochim Acta 69:127-135
Bowler JM, Magee JW (2000) Redating Australia's oldest human remains: a sceptic's view. J Human Evol 38:719-726
Brown P (2001) Chinese Middle Pleistocene Hominids and modern human origins in East Asia. *In:* Human Roots: Africa and Asia in the Middle Pleistocene, Barham L, Robson-Brown K (eds) Weston Academic and Specialist Press: Bristol, p. 135-147
Chen T, Yang Q, Wu E (1992) ESR dating of teeth enamel samples from Jinniushan palaeoanthropological site. Acta Anthro Sinica 12:337-346
Chen T, Yang Q, Wu E (1994) Antiquity of *Homo Sapiens* in China. Nature 368:55-56
Chen T, Yinyun Z (1991) Paleolithic chronology and possible coexistence *of Homo erectus* and *Homo spaiens* in China. World Arch 23:147-154
Chen T, Yuan S (1988) Uranium-series dating of bones and teeth from Chinese Paleolithic sites. Archaeometry 30:59-76
Cheng H, Lawrence Edwards R, Murrell MT, Benjamin TM (1998) Uranium-thorium-protactinium dating systematics. Geochim Cosmochim Acta 62:3437-3452
Crank J (1975) The Mathematics of Diffusion. Clarendon Press, Oxford

Curnoe D, Grün R, Thackeray F (2001), Direct ESR dating of a Pliocene hominid from Swartkrans. J Human Evol 40:379-391

Curtis GH, Hay RL (1972) Further geological studies and potassium-argon dating at Oludvai Gorge and Ngorongoro Crater. In: Calibration of Hominid Evolution. Bishop WW, Miller JA (eds), Scottish Academic Press, Edinburgh

Edwards RL, Cheng JH, Murrell MT, Goldstein SJ (1997) Protactinium-231 dating of carbonates by thermal ionization mass spectrometry: implications for quaternary climate change. Science 276:782-786

Eggins S, Grün R, Pike AWG, Shelly M, Taylor L (in press) ^{238}U, ^{232}Th profiling and U-series isotope analysis of fossil teeth by laser ablation-ICPMS. Quat Sci Rev

Falgueres C, Bahain J-J, Yokoyama Y, Arsuaga JL, Bermudez de Castro JM, Carbonell E, Bischoff JL, Dolo J-M (1999) Earliest Humans in Europe: the age of TD6 Gran Dolina, Atapuerca, Spain. J Human Evol 37:343-352

Gillespie R, Roberts RG (2000) On the reliability of age estimates for human remains at Lake Mungo. J Human Evol 38:727-732

Goldstein SJ, Stirling CH (2003) Techniques for measuring uranium-series nuclides: 1992-2002. Rev Mineral Geochem 52:23-57

Grün R, Beaumont P (2001) Border Cave revisited: a revised ESR chronology. J Human Evol 40:467-482

Grün R, Huang PH, Huang W, McDermott F, Thorne A, Stringer CB, Yan G (1998) ESR and U-series analyses of teeth from the palaeoanthropological site of Hexian, Anhui Province, China. J Human Evol 34:555-64

Grün R, Richards M, Davies W, Stringer C (in press) The chronology of modern human evolution, Part I: Methods. Quat Sci Rev

Grün R, Schwarcz H (1987) Some remarks on "ESR dating of bones." Ancient TL 5:1-9

Grün R, Schwarcz HP, Chadham J (1988) ESR dating of tooth enamel: coupled correction for U-uptake and U-series disequilibrium. Nuclear Tracks Rad Meas 14:237-241

Grün R, Stringer CB (1991) Electron spin resonance dating and the evolution of modern humans. Archaeometry 33:153-199

Grün R, Stringer CB (2000) ESR and U-series analyses of dental material from Tabun C1. J Human Evol 39:601-612

Grün R, Stringer CB, Schwarcz HP (1991) ESR dating of teeth from Garrod's Tabun cave collection. J Human Evol 20:231-248

Grün R, Thorne A (1997) Dating the Ngandong humans. Science 276:1575

Hille P (1979) An open system model for uranium series dating. Earth Planet Sci Lett 42:138-142

Ikeya M (1982) A model of linear uranium accumulation for ESR age of Heidelberg (Mauer) and Tautavel bones. Jap J App Phys (Lett) 21:690-692

Ikeya M (1993) New Applications of Electron Spin Resonance, Dating, Dosimetry and Microscopy. World Scientific Publishing, Singapore

Isaac GL (1975) Sorting out the muddle in the middle: an anthropologist's post-conference appraisal. In: After the Australopithecines. Butzer KW, Isaac GL (eds) Mouton de Gruyter, The Hague, p 875-88

Ivanovich M (1982) Uranium series disequilibrim applications in geochronology. In Uranium Series Disequilibrium: Applications to Environmental Problems. Ivanovich M, Harmon RS (eds), Clarendon Press, Oxford, p 56-78.

Jelinek AJ (1982) The Tabun Cave and Palaeolithic man in the Levant. Science 216:1369-75

Ku T-L (1982) Progress and Perspectives. In: Uranium-series disequiliubrium: Application to Environmental Problems. Ivanovich M, Harmon RS, (eds), Clarendon Press, Oxford, p 268-301.

Laplana C, Cuenca-Bescos G (1997) Los arvicolidos (Arvicolidae, Rodentia) del limite Pleistoceno Inferior-Medio en al relleno carstico Trinchera Dolina (Sierra de Atapuerca, Burgos, Espana) J Paleontologia XIII:192-195

McDermott F, Grün R, Stringer CB, Hawkesworth CJ (1993) Mass-spectrometric U-series dates for Israeli Neandertal/early hominid sites. Nature 363:252-255

McDermott F, Stringer C, Grün R, Williams CT, Din VK, Hawkesworth CJ (1996) New Late-Pleistocene Uranium-Thorium and ESR dates for the Singa hominid (Sudan) J Human Evol 31:507-516

Mercier N, Valladas H, Reyss J-L, Jelinek A, Meignen L, Joron J-L (1995) TL dates of burnt flints from Jelinek's excavations at Tabun and their implications. J Arch Sci 22:495-509

Millard AR, Hedges REM (1996) A diffusion-adsorption model of uranium uptake by archaeological bone. Geochim Cosmochim Acta 60:2139-2152

Millard AR, Pike AWG (1999) Uranium-series dating of the Tabun Neanderthal: a cautionary note. J Human Evol 36:581-585

Pike AWG (2000) U-series dating of Archaeological Bone using TIMS. PhD dissertation, University of Oxford, Oxford, UK

Pike AWG, Hedges REM (2001) Sample geometry and U-uptake in archaeological teeth: implications for U-series and ESR dating. Quat Sci Rev 20:1021-1025

Pike AWG, Hedges REM (in press) U uptake in boxgrove bones: implications for U-series and ESR dating. *In* Boxgrove A Middle Pleistocene hominid site at Eartham Quarry, Boxgrove, West Sussex. Volume II. Roberts MB, Parfitt SA (eds) English Heritage, London

Pike AWG, Hedges REM, Van Calsteren P (2002) U-series dating of bone using the diffusion-adsorption model. Geochim Cosmochim Acta 66(24):4273-4286

Pike AWG, Millard AR (forthcoming) An empirical assessment of the accuracy and precision of U-series dating of bone.

Rae AM, Hedges REM, Ivanovich M (1989) Further studies for uranium-series dating of fossil bones. Appl Geochem 4:331-337

Rae AM, Ivanovich M (1986) Successful application of uranium series dating of fossil bone. Appl Geochem 1:419-426

Richards DA, Dorale JA (2003) U-series chronology and environmental applications of speleothems. Rev Mineral Geochem 52:407-460

Roberts RG, Jones R, Smith MA (1990) Thermoluminescence dating of a 50,000 year old human occupation site in northern Australia. Nature 345:153-156

Schwarcz HP (1992) Uranium-series dating and the origin of modern man. *In:* The Origin of Modern Humans and the Impact of Chronometric Dating. Aitken MJ, Stringer CB, Mellars PA (eds) Princeton University Press, Princeton, p. 12-16

Schwarcz HP, Bietti A, Buhay WM, Stiner M, Grün R, Segre E (1991) On the reexamination of Grotta Guattari: Uranium series and ESR dates. Current Anthro 32:313-316

Schwarcz HP, Blackwell BA (1992) Archaeological applications. *In* Uranium Series Disequilibrium Applications to Earth, Marine and Environmental Sciences. Ivanovich M, Harmon RS (eds) Clarendon, Oxford. p. 513-552

Schwarcz HP, Simpson JJ, Stringer CB (1998) Neanderthal skeleton from Tabun: U-series data by gamma-ray spectrometry. J Human Evol 35:635-645

Shen G, Ku T-L, Cheng H, Edwards RL, Yuan Z, Wang Q (2001) High-precision U-series dating of Locality 1 at Zhoukoudian, China. J Human Evol 41:679-688

Shen G, Wang W, Wang Q, Zhao J, Collerson K, Zhou C, Tobias PV (in press) U-series dating of Lujing hominid site in Guangxi, southern China. J Human Evol

Shen G, Yokoyama Y, (1984) Th-230/U-234 dating of Petralona speleothems. Anthropos (Athens) 11:23-32

Simpson JJ, Grün R (1998) Non-destructive gamma spectrometric U-series dating. Quat Geochron 17:1009-1022

Spooner N (1998) Human occupation at Jimnium, N. Australia, 116,000 years ago, or much less? Antiquity 72:173-177

Stringer CB (1989) The origin of early modern humans: a comparison of the European and non-European evidence. *In:* The Human Revolution. Mellars P, Stringer CB (eds) Edinburgh University Press, Edinburgh, p. 232-244

Stringer CB (1999) Has Australia backdated the Human Revolution? Antiquity 73:876-879

Stringer CB, Grün R, Schwarcz HP, Goldberg P (1989) ESR dates for the hominid burial site of Es Skhul in Israel. Nature 338:756-758

Stringer CB, Hublin J-J (1999) New age estimates for the Swanscombe hominid and their significance for human evolution. J Human Evol 37:873-877

Stringer CD, McKie R (1996) African Exodus, The Origins of Modern Humanity. Jonathan Cape, London.

Swisher III CC, Rink WJ, Antoin SC, Schwarcz HP, Curtis GH, Suprijo A, Widiasmoro (1996) Latest *Homo erectus* of Java: Potential contemporaneity with *Homo sapiens* in Southeast Asia. Science 274:1870-1874

Szabo BJ (1979) Dating fossil bone from Orange Free State, South Africa. Journal of Archaeological Science 6:201-203

Szabo BJ, Rosholt JN (1969) Uranium-series dating of Pleistocene molluscan shells from southern California - an open system model. J Geophys Res 74:3253-3260

Thorne A, Grün R, Spooner NA, Simpson JJ, McCulloch M, Curnoe D (1999) Australia's oldest human remains: age of the Lake Mungo 3 skeleton. J Human Evol 36:591-612

Wolpoff M (1989) Multiregional evolution: the fossil alternative to Eden. *In* The Human Revolution. Mellars P, Stringer CB (eds), Edinburgh University Press, Edinburgh. p 62-108

Wolpoff M (2001) Modern human ancestry at the peripheries. Science 291:293

Wolpoff MH, Wu X, Thorne A (1984) Modern *Homo Sapiens* origins: a general theory of hominid evolution involving the fossil evidence from East Asia, *In:* The Origins of Modern Humans: A World Survey of the Fossil Evidence. Smith F, Spencer F (eds), Alan Liss, New York, p 411-83

Wu RK (1989) Palaeoanthropology. Cultural Relics Publishing House, Beijing
Yokoyama Y, Nguyen HV (1981) Direct dating by non-destructive gamma-ray spectrometry of fossil human skull Arago XXI, fossil animal bones and stalagmites of the Caune de l'Arago at Tautavel. *In:* Absolute Dating and Isotope Analysis in Prehistory - Methods and Limits. DeLumley H, Labeyrie J (eds), Proceedings Pretirage, p. 355-375
Yokoyama Y, Falgueres C, Luumley MA (1997) Datation direct d'un crane proto-cro-magnon de Qafzeh par la spectrometrie gamma non destructive. Comptes Rendues de l'Academie de Science, Paris, Series 2, 324:773-79
Zhao J, Hu K, Collerson KD, Xu H (2001) Thermal ionization mass spectrometry U-series dating of a hominid site near Nanjing, China. Geology 29(1):27-30
Zhou LP, McDermott F, Rhodes EJ, Marseglia EA, Mellars PA (1997) ESR and mass-spectrometric uranium-series dating studies of a mammoth tooth from Stanton Harcourt, Oxfordshire, England. Quat Sci Rev (Quat Geochron) 16:445-454

APPENDIX:
FURTHER DETAILS OF THE D-A MODEL

Assuming a linear adsorption isotherm Millard and Hedges (1996) give a form of Fick's Second Law as the equation for diffusion in an infinite plane slab:

$$\frac{\partial C}{\partial t} = \frac{D}{(R+1)} \frac{\partial^2 C}{\partial x^2} \quad (A1)$$

where D is the diffusion coefficient, reduced for diffusion through a porous media, and R is the volumetric equilibrium coefficient, and is related by the specific porosity of bone, p, to K_d, the partition coefficient between groundwater and bone by $R = K_d/p$.

Drawing on Crank (1975) the give the U concentration, Z, at a distance into the bone, x, at time t as:

$$Z_{(x,t)} = pRC_1 \left(1 - \frac{4}{\pi} \sum_{n=0}^{\infty} \frac{(-1)^n}{2n+1} e^{\left[\frac{-D(2n+1)^2 \pi^2 t}{(R+1)4l}\right]} \cos\left[\frac{(2n+1)\pi x}{2l}\right] \right) \quad (A2)$$

Where C_1 is the groundwater U concentration and l is half the thickness of the bone. Using the following dimensionless parameters Millard and Hedges were able to produce generalized diffusion profiles (Z' vs. x') for the parameter t' (e.g., Fig. 3):

$$Z' = \frac{Z}{pRC_1}, \quad x' = \frac{x}{l}, \quad t' = \frac{Dt}{(R+1)l^2} \quad (A3)$$

The distribution of ^{230}Th/^{234}U can be calculated using a finite differences method:

$$\frac{\delta\left(^{230}\text{Th}/^{234}\text{U}\right)}{\delta t} \approx \lambda_{230}\left(1 - \frac{^{230}\text{Th}}{^{234}\text{U}}\right) - \frac{^{230}\text{Th}}{^{234}\text{U}}\left[\frac{1}{Z}\frac{\partial Z}{\partial t} - \lambda_{234}\left(1 - \frac{^{234}\text{U}}{^{238}\text{U}}\right)\right] \quad (A4)$$

where λ_{230} and λ_{234} are the decay constants of ^{230}Th and ^{234}U respectively, and isotope ratios are expressed as activities.

Pike and co-workers (Pike 2000, Pike and Hedges 2001, Pike et al. 2002) used a finite element method incorporating Equations (A1) and (A4) to model the effect of changes in the groundwater U concentration, C_1, on the development of U concentration and U-series date profiles (e.g., Figs. 5-7). They went on to use the D-A model to calculate open-system dates for bones where the uptake appeared to have taken place under constant geochemical conditions. In order to calculate a U-series date from a measured date profile, Pike et al. (2002) estimate the parameter $t'l^2$ from the measured U concentration profile (e.g., Fig. 3). Since $t'l^2=Dt/(R+1)$, see Equation (A3), a value of $D/(R+1)$ can be calculated for a model time, t, that keeps the modeled U concentration profile consistent with the measured profile. The D-A date is the calculated by forward modeling to find the value of t that gives the best fit to the measured U-series date profile (e.g., Fig. 9).

16   Mathematical–Statistical Treatment of Data and Errors for ^{230}Th/U Geochronology

K. R. Ludwig
Berkeley Geochronology Center
2455 Ridge Rd.
Berkeley, California, 94709, U.S.A.

1. INTRODUCTION

The various methods of geochronology have, as a rule, evolved along a characteristic path whereby the earliest work lays out the physical and mathematical foundation and demonstrates its practicality for a few cases using optimal samples. Later work confronts the problems and failures of early assumptions on more general problems, and as the method matures and becomes more generally applied, a standard toolkit of more-or-less rigorous protocols for data interpretation and error estimation is developed. The ^{230}Th/U dating method has evolved along broadly similar lines, but with a discontinuity arising from the revolutionary application by Chen et al. (1986) and Edwards et al. (1987) of TIMS (thermal ionization mass spectrometry), which not only improved the precision of the analyses by an order of magnitude, but also changed the "culture" and mind-set of the practitioners from a small community trained largely in AS (alpha spectrometry) to a larger community that included the TIMS and ICP-MS (inductively-coupled polarization mass-spectrometry) laboratories as well. At this writing, these two cultures have not completely integrated their approaches, so that there is little practical consensus on a "standard" way of evaluating common types of ^{230}Th/U data-sets for geochronology, in contrast to geochronology by the U-Pb, Rb-Sr, or Ar-Ar methods. The viewpoint taken in this chapter is to outline the basic tools of data handling and evaluation for ^{230}Th/U geochronology, while at the same time pointing out the limitations of these tools and possible directions of improvement.

Though this chapter is focused on the ^{230}Th/U method, most of the concepts are directly applicable to other uranium-series chronometers, such as ^{231}Pa/^{235}U, ^{231}Pa/^{230}Th or ^{226}Ra/^{230}Th. However, because the details of individual applications of these methods are more varied than for the ^{230}Th/U method, and in some cases can be expected to evolve significantly (e.g., coupled ^{231}Pa-^{230}Th evolution; Cheng et al. 1998), they will not be specifically addressed here.

2. WHY ERROR ESTIMATION IS IMPORTANT

Geochronology is one of the most quantitative sub-disciplines within the earth sciences, at least in the sense that the "product" of geochronology—the age of a rock, mineral, or process—is unusable in the absence of an estimate of its uncertainty. For example, if a sample of coral is reported as having ^{230}Th/U date of 135 ka, but with no uncertainty given, what use could be made of such a number? Is the date highly precise and accurate (say to better than 1 ka), so that it might directly bear on the contentious history of Marine Isotope Stage 5 sea-level rise? Or perhaps it is one of the many such dates with 2σ precisions of ±15-20 ka, in which case the date at best constrains the sample to a Substage 5*e-d*? Or perhaps the ratios were measured with a technique such as direct gamma-ray counting, where the 95%-confidence errors on the calculated age might be as great as +100/−60 ka? Or, suppose the date is not on a coral but from an isochron for an impure aragonite cement from a sediment, with a quoted error of, say, ±1.0 ka. Could this apparently precise date have been given with a "1 sigma" error of an isochron with only 3 points and with excess scatter, so that a far more realistic uncertainty would be the 95%-confidence uncertainty of ±13 Ma?

This example does not exaggerate the importance of knowing the uncertainty on a ^{230}Th/U date. Indeed, for most studies, *the uncertainty of a date is no less significant than the date itself.* The exceptions are those cases where, once the age-uncertainty is below a certain limit, its precise value is not particularly important since the problem which the date addresses is itself imperfectly constrained.

3. ERRORS OF THE MEASURED ISOTOPIC RATIOS

As a starting point, the analyst must be able to assign accurate uncertainties to the isotope ratios required by the ^{230}Th/U age equation (Kaufman and Broecker 1965)[1],

$$\frac{^{230}\text{Th}}{^{238}\text{U}} = 1 - e^{-\lambda_{230}t} - \left(\frac{^{234}\text{U}}{^{238}\text{U}} - 1\right)\left(\frac{\lambda_{230}}{\lambda_{234} - \lambda_{230}}\right)\left[1 - e^{(\lambda_{234}-\lambda_{230})t}\right] \qquad (1)$$

For measurements by AS, the errors of the isotope ratio will be dominated by counting statistics for each isotope. For measurements by TIMS or ICP-MS, the counting-statistic errors set a firm lower limit on the isotopic measurement errors, but more often than not contribute only a part of the total variance of the isotope-ratio measurements. For these techniques, other sources of (non-systematic) error include:

1. Instability of the ion beam (if measured with a single collector)
2. Dark noise of the ion counter or analog electron multiplier (especially important for the latter, for which the 2σ dark noise is typically 10-15 ions/second)
3. Error in measuring the gain or efficiency of the electron multiplier or ion counter
4. Amplifier noise on large peaks measured via Faraday cups[2]
5. Correction for tails from nearby large peaks (for instruments with abundance sensitivities worse than 0.1 ppm)
6. Variable mass-discrimination (generally a problem only for TIMS analyses of Th or of U when not double-spiked and not normalized via ^{238}U/^{235}U)
7. Unrecognized isobaric interferences
8. Correction for residual ion beams from previous runs or instrumental background (ICP-MS only)

Because of these complications, regardless of the very high within-run precision attainable via TIMS or ICP-MS, the true precision of the runs (as opposed to the "internal" or "within run" precision provided by the TIMS or ICP-MS operating software) can only be reliably established by replicate analyses of natural samples. One useful approach is to establish the "external" variance of a measurement technique by subtracting the "internal" variance from the total (= run-to-run) variance from replicate analyses, e.g.,

---

[1] All isotopic ratios in this chapter are activity rather than atomic.
[2] Even for charge collection (Esat 1995) instead of current amplification, though in the former case the magnitude is much less.

$$\sigma^2_{external} = \sigma^2_{total} - \sigma^2_{internal}$$

The true variance of an analysis can then be calculated as the sum of $\sigma^2_{external}$ and the square of the standard error of the mean provided by the mass spectrometer software ($=\sigma^2_{internal}$).

Systematic errors (e.g., from decay-constant uncertainties, spike-calibration bias ...) affecting the *accuracy* of U-series dates only approach or exceed the precision errors when the analyses are very precise (at the permil level or better). Systematic errors must be treated separately from random measurement errors, and can generally be ignored when comparing apparent ages from the same laboratory or (in the case of decay constants) decay systems. When measurement precision is very high and comparison with other geochronometers is possible, however, systematic errors arising from uncertainties in the ^{230}Th and ^{234}U decay constants should be taken into account, as well as laboratory biases arising from spike calibration or integrity of secular-equilibrium standards.

4. ERROR CORRELATIONS

Error correlations are often ignored both mathematically and graphically in the treatment of ^{230}Th/U data. Nonetheless, error correlations can be as important a parameter in error propagation and evaluation as the errors themselves, and must be considered in all cases. Though sometimes approached with trepidation, the significance of error correlations is easily understood at an intuitive level and demonstrated visually (Fig. 1). Basic examples of procedures for calculating error correlations are given in Appendix I.

The important, intuitively understandable properties of error correlations are that:

- The correlation coefficient, ρ_{xy}, between the errors of two variables x and y is a measure of the tendency of the errors of any measured x, y pair to be in the same direction.

- The correlation coefficient ranges between −1 and +1. An error correlation of +1 indicates that the errors in both x and y for any particular measurement of an x, y pair will always be in the same direction (either both greater than, or both less than the true values). An error correlation of −1 indicates that the x error will always be in the opposite direction than the y error. An error correlation of zero indicates that the x- and y-errors are independent of one another.

The analyst may wish to avoid the complications of significant error correlations, since their estimation via analytical derivations may require tracking cross-terms in the multiplication of complex algebraic expressions, and if significant rules out most types of error symbols (error ellipses being the exception) for data plotting. In many cases, error-correlation avoidance can be achieved by recasting the data in a form such that error correlations are minimized, as error correlations whose absolute values are less than about 0.3 can usually be neglected without significant consequences.

For example, one of the common forms of ^{230}Th/U isochron diagrams involves plotting and regressing ^{230}Th/^{232}Th versus ^{234}U/^{232}Th and ^{234}U/^{232}Th versus ^{238}U/^{232}Th. If samples analyzed via AS have high ^{238}U/^{232}Th, however, the ^{232}Th measurement errors are generally much greater than those of the other U-Th isotopes, resulting in high error correlations which could not be justifiably ignored in any mathematical or graphical evaluation. However, changing the denominator isotope to one of the more-precisely measured isotopes—say ^{238}U, so that the isochrons become ^{230}Th/^{238}U-^{232}Th/^{238}U and ^{234}U/^{238}U-^{232}Th/^{238}U—often reduces the error correlations sufficiently that they can be neglected. In contrast, for TIMS or ICP-MS measurements the ^{234}U and (especially) ^{230}Th ion beams are almost always much larger than the other U-Th isotopes, so than any

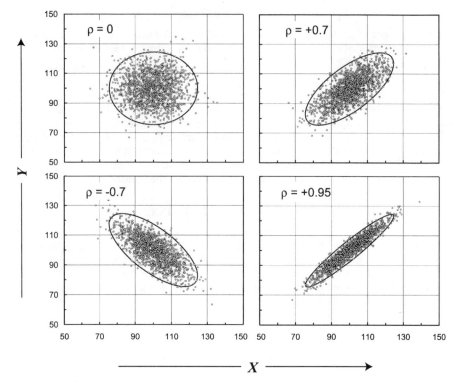

Figure 1. Results of Monte Carlo simulations for 1500 pairs of x, y points with a mean of 100, Gaussian errors of ±10 (1σ), and four different x-y error correlations (ρ). Ellipses show 95% confidence limits for the joint x-y distribution. Note that the ellipses extend farther than the 2σ range of either the x or y errors themselves—a non-intuitive characteristic of joint distributions that arises because an x- (or y-) value deviating *less* than expected "permits" a y- (or x-) value that deviates *more* than expected.

isochron diagram which does not have ^{230}Th or ^{234}U as the common-denominator isotope can be considered (for practical purposes) free of error correlations.

5. FIRST ORDER ESTIMATION OF ERRORS

The simplest method for propagation of the analytical errors into the ^{230}Th/U age equation (Eqn. 1) involves linear expansion (Albarède 1995, ch. 4.3) of the effect on the calculated age of very small perturbations of the measured ratios (in effect, a Taylor expansion using only the first term). As long as the effect of the errors of the measured ratios on the age is not a large fraction of the age itself, this method will yield acceptably accurate age-errors with minimal effort.

For TIMS or ICP-MS measurements, where the ^{230}Th/^{238}U and ^{234}U/^{238}U errors are almost entirely uncorrelated (see discussion later in this chapter), this first-derivative expansion method can be thought of as calculating approximate error-magnification factors for the effect of each input-error on the age. Thus, for a reasonably small error, say σ_φ, in the measured ^{230}Th/^{238}U (= φ) the resulting error of the ^{230}Th/U age, t will be close to $(\partial t/\partial \varphi)\sigma_\varphi$, and similarly for the measured ^{234}U/^{238}U (= γ). The total error in t and the initial ^{234}U/^{238}U (= γ_0), can therefore be estimated by differentiation of Eqn. (1) as

$$\sigma_t^2 = \left[\sigma_\varphi^2 + k_1^2 \sigma_\gamma^2 + 2k_1 \text{cov}(\varphi,\gamma)\right]/D^2 \qquad (2)$$

and

$$\sigma_{\gamma_0}^2 = \left(k_2 \sigma_t^2\right) + \left(e^{\lambda_{234} t} \sigma_\gamma\right)^2 + \left(\frac{2k_2 e^{\lambda_{234} t}}{D}\right)\left[k_1 \sigma_\gamma^2 + \text{cov}(\varphi,\gamma)\right] \qquad (3)$$

respectively (Ludwig and Titterington 1994), where

$$\gamma_0 = 1 + (\gamma - 1)e^{\lambda_{234} t} \qquad (4)$$

$$k_1 = \frac{\lambda_{230}}{\lambda_{\text{diff}}}\left[1 - e^{-\lambda_{\text{diff}} t}\right] \qquad (5)$$

$$\lambda_{\text{diff}} = \lambda_{230} - \lambda_{234} \qquad (6)$$

$$k_2 = \lambda_{234}(\gamma - 1)e^{\lambda_{234} t} \qquad (7)$$

$$D = \lambda_{230}\left[e^{-\lambda_{230} t} + (\gamma - 1)e^{-\lambda_{\text{diff}} t}\right] \qquad (8)$$

$$\rho_{t,\gamma_0} = \frac{k_2 \sigma_t^2 + \dfrac{e^{\lambda_{234} t}}{D}\left[k_1 \sigma_\gamma^2 + \text{cov}(\varphi,\gamma)\right]}{\sigma_t \sigma_{\gamma_0}} \qquad (9)$$

and $\text{cov}(\varphi,\gamma)$ is the covariance between the two measured ratios and is defined by the identity $\text{cov}(\varphi,\gamma) \equiv \rho_{\varphi,\gamma}$.[3] The equations for estimation of the error correlation between the ^{230}Th/U age and back-calculated initial ^{234}U/^{238}U are important, as the age–γ_0 diagram is often very useful.

If errors in the ^{230}Th and ^{234}U decay constants are included (necessary for any comparison with dates from another method), then two additional terms must be added to the numerator of Eqn. (2),

$$\sigma_t^2 = \left(\sigma_\varphi^2 + k_1^2 \sigma_\gamma^2 + 2k_1 \text{cov}(\varphi,\gamma) + k_3^2 \sigma_{\lambda_{230}}^2 + k_4^2 \sigma_{\lambda_{234}}^2\right)/D^2 \qquad (10)$$

where

$$k_3 = te^{-\lambda_{230} t} - \frac{\gamma - 1}{\lambda_{\text{diff}}}\left[\frac{\lambda_{234}}{\lambda_{\text{diff}}}\left(1 - e^{-\lambda_{\text{diff}} t}\right) - \lambda_{230} t e^{-\lambda_{\text{diff}} t}\right] \qquad (11)$$

$$k_4 = \frac{\lambda_{230}}{\lambda_{\text{diff}}}(\gamma - 1)\left(\frac{1 - e^{-\lambda_{\text{diff}} t}}{\lambda_{\text{diff}}} - te^{-\lambda_{\text{diff}} t}\right) \qquad (12)$$

provided that the ratios are calculated by AS or by normalization to a secular equilibrium standard (if the activity ratios are calculated from the product of the atomic abundances and decay constants, the expressions are somewhat different; Cheng et al. 1998, p. 3439).

6. WHEN FIRST ORDER ERROR ESTIMATION IS INADEQUATE

At the extreme limits of the ^{230}Th/U method—somewhere between 5 to 10 half-lives

---

[3] $\text{cov}(\varphi,\gamma)$ is substituted for the cross-term $d\varphi \times d\gamma$ in the derivatives in the same way that σ_φ is substituted for the differential $d\varphi$

of ^{230}Th for TIMS or ICP-MS, or 3 to 5 half-lives with AS—the error distribution of the ^{230}Th/U age becomes distinctly asymmetric and non-Gaussian, with a much larger tail towards the older part of the distribution than the younger. Similar problems occur when estimating the error of the regressions used in isochrons, especially as the isochron errors become large and/or the distance between the data-points along the isochron becomes comparable. First order methods of error estimation are then inadequate, and a more accurate method should be used.

6.1. Improving the first order estimate analytically

The accuracy of any expression for first order errors can be improved as much as desired (so long as the expression is differentiable) by including additional terms of its Taylor expansion. For example, the error magnification factor in the age equation for the ^{230}Th/^{238}U term, φ, can be improved to include the second derivative, so the estimate of the effect of a given error in φ on the age becomes:

$$\left(\frac{\partial t}{\partial \varphi} + \frac{1}{2}\frac{\partial^2 t}{\partial \varphi^2}\sigma_\varphi\right)\sigma_\varphi \tag{13}$$

Generally, however, increasing the number of Taylor-expansion terms beyond the first derivative results in such a plethora of terms to be squared and cross-multiplied that analysts are often reluctant to proceed with the algebra, much less the coding and subsequent debugging to implement the solution.

6.2. Error estimation by Monte Carlo

Though inelegant and computationally expensive, one of the easiest and most flexible methods of error estimation to understand and implement, regardless of the complexity of the problem, is the Monte Carlo method (Anderson 1976; Albarède 1995, p. 233). Monte Carlo estimates of errors of a function or algorithmic procedure are accomplished by repeatedly perturbing (say N times) the input data according to their assigned errors, error correlations, and error distribution (generally but not necessarily assumed to be Gaussian), then numerically sorting the results into a ranked $N \times 1$ array. The plus- and minus-errors at any confidence limit of the output parameters can then be obtained by subtracting its calculated value from the ith and j^{th} elements of the array, respectively, where $i = 0.5 \times (100 - C) \times N/100$, $j = N - i + 1$ and C is the confidence limit in percent. Thus for 9,999 trials, the 95% confidence-limits on the solution are approximated by the 250th and 9,999th elements of the sorted array. Monte Carlo results can also be used to construct a visual impression of the error distribution, by plotting the sorted array as a histogram. For example, consider two ^{230}Th/U analyses (Table 1) with the following values and Gaussian errors[4]:

Table 1. Two synthetic ^{230}Th/U analyses for a ~330 ka sample.

^{230}Th/^{238}U	1σ error	^{234}U/^{238}U	1σ error	Error correlation
1.130	2.5%	1.140	2.0%	+0.5
1.130	0.30%	1.140	0.25%	+0.1

---

[4] If the errors were greater than ~10% and dominated by counting statistics, a Poisson distribution would yield slightly more accurate results.

The first analysis is one with AS-level precision, the second with TIMS-level precision. The first order 2σ error for the resulting 331 ka age is ±96 ka, but examination of the distribution of a Monte Carlo simulation (Fig. 2) shows that the actual age distribution is strongly asymmetric, with 95% confidence limits of +158/−79 ka. For either younger ages or more-precise analyses, however, the first-order age errors are more than adequate, as shown by the Monte Carlo results for the same data, but with TIMS-level precision (Fig. 2B).

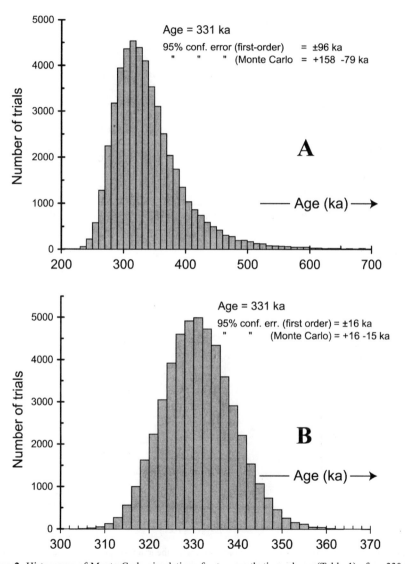

Figure 2. Histograms of Monte Carlo simulations for two synthetic analyses (Table 1) of a ~330 ka sample. The lower precision analysis (A) has a distinctly asymmetric, non-Gaussian distribution of age errors and a misleading first-order error calculation. The higher precision analysis (B) yields a nearly symmetric, Gaussian age distribution with confidence limits almost identical those of the first-order error expansion.

Monte Carlo simulations can also be used to deal with more subtle questions of ^{230}Th/U data-evaluation. Thus if there is prior knowledge about the true age of a sample, this knowledge can be incorporated into the Monte Carlo routine as a prior constraint (Bernardo and Smith 2000). For example, ^{230}Th/U analyses with near-secular equilibrium ^{230}Th/^{234}U may yield "infinite" ages (that is, the isotope ratios may not correspond to any possible closed-system age), though the possible range of the true ratios (as indicated by their measurement uncertainties) may overlap with ratios corresponding to finite ages (Table 2).

Table 2. Synthetic ^{230}Th/U analysis for a sample whose age is close to the limit of ^{230}Th/U dating.

^{230}Th/^{238}U	1σ error	^{234}U/^{238}U	1σ error	Error correlation
1.65	2.5%	1.43	2.0%	+0.5

A Monte Carlo simulation (Fig. 3) can be made as usual (that is, without constraints on the output age), in which case only about 24% of the trials will yield ratios corresponding to a finite age, and a younger limit of >821 ka (95% confidence) or >531 ka (68% conf.) is indicated. If, however, the *a priori* assumption of a closed system with no initial ^{230}Th is made, the "failed" trials can be ignored (since they violate the *a priori* constraints), and solution of both age and age-error (630 +370/−210 ka at 95% conf., or +150/−140 ka at 68% conf.) can be obtained from the Monte Carlo simulations.

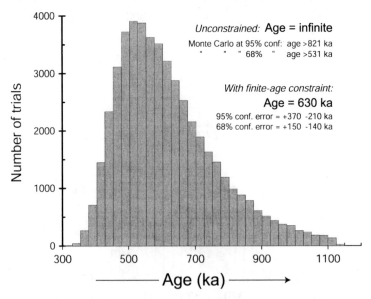

Figure 3. Histogram of Monte Carlo simulation for a synthetic alpha-spectrometric analysis (Table 1) of a sample with near-secular equilibrium ^{230}Th/^{234}U. No age can be calculated for the "measured" ratios, but the unconstrained Monte Carlo simulation (for which 76% of the trials yielded ratios with no age solution—i.e., "infinite" ages) restricts the minimum age to >821 ka at 95% confidence. If the sample is assumed to be a closed system, however, trials with no age solutions can be ignored, and the simulation defines an age and 95% conf. error of 630 +370/-210 ka.

More subtle use of the probability distributions obtained from Monte Carlo simulations of ^{230}Th/U analyses. For example, Rowe et al. (1999) were able to incorporate geologic and stratigraphic information which constrained the age of two samples two within the same interglacial period, permitting a significant improvement in the age uncertainty for both of the two samples.

7. CORRECTING A SINGLE ANALYSIS FOR DETRITAL THORIUM AND URANIUM

Because there is no natural mechanism for producing isotopically pure ^{232}Th, any measurable ^{232}Th in a sample for ^{230}Th/U dating implies the presence of at least some ^{230}Th when the sample (mineral or mineral assemblage) was formed. This initial ^{230}Th must then be subtracted from the total measured ^{230}Th before a valid ^{230}Th/U age can be calculated. In addition, a correction for initial U isotopes is required when the material being dated is a mechanical mixture of an isotopically homogeneous detrital material with an initially thorium-free mineral such as chemically precipitated carbonate. In these cases we are usually interested only in the time of formation of the Th-free material, and so must also subtract the detrital ^{234}U and ^{238}U before calculating the ^{230}Th/U age.

For such mechanical mixtures, we may be able to isolate and analyze material whose U-Th isotopic composition is arguably representative of the detrital component—for example, volcanic rock from which the sand-sized material contaminating a pedogenic carbonate was derived, or clays believed to carry the ^{232}Th of a macroscopically pure carbonate. Or, we might decide that the isotopic composition of the detrital component is known within broad limits. In such cases, so long as the correction is small and the uncertainty of the assigned detrital ratios propagated, we can subtract the composition of the assumed detrital material as a first-order correction. It could be argued that that latter sort of correction degrades rather than enhances the accuracy of the resulting ages, since the assumed isotopic compositions might be very wide of the true mark. Perhaps, but it does not take much geochemical insight to improve upon the default assumption of a detrital ^{230}Th/^{232}Th of zero (inherent in making no correction at all for detrital U and Th).

Correction for detrital ^{230}Th can be visualized with a plot of ^{230}Th/^{238}U versus ^{232}Th/^{238}U (Fig. 4)—effectively, as a 2-point isochron (see later discussion) where one of the points can be measured, assumed, or calculated in some way. The corrected ^{230}Th/^{238}U and ^{234}U/^{238}U (if not in secular equilibrium) is then obtained by the extrapolation of the mixing line between the sample and the detrital component to zero ^{232}Th. The errors arising from these initial-isotope corrections must then be propagated when calculating the error of the ^{230}Th/U age, especially when the correction approaches or exceeds the analytical errors on the uncorrected isotope ratios, or when the errors assigned to the detrital ratios are large. The relevant equations for the detrital corrections to ^{230}Th/^{238}U and ^{234}U/^{238}U can be written as

$$\varphi_a = \varphi_m - b\omega_m \qquad (14)$$

$$\gamma_a = \gamma_m - B\omega_m \qquad (15)$$

where

m = measured or mix a = authigenic d = detrital

$\varphi = {}^{230}\text{Th}/{}^{238}\text{U}$ $\gamma = {}^{234}\text{U}/{}^{238}\text{U}$ $\omega = {}^{232}\text{Th}/{}^{238}\text{U}$

$b = \dfrac{\varphi_d - \varphi_m}{\omega_d - \omega_m}$ $B = \dfrac{\gamma_d - \gamma_m}{\omega_d - \omega_m}$

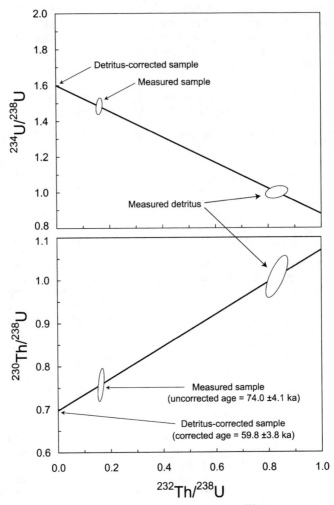

Figure 4. Graphical representation of detrital-isotope correction for ^{230}Th/U dating. Apparent ages (in parentheses) age shown for both the measured and detritus-corrected isotope ratios.

(Ludwig and Titterington 1994), from which one can calculate the expressions for the corresponding (first order) errors and error correlations,

$$\sigma^2_{\varphi_a} = b^2\left(r_2^2\sigma^2_{\omega_d} + r_1^2\sigma^2_{\omega_m}\right) + r_2^2\sigma^2_{\varphi_d} + r_1^2\sigma^2_{\varphi_m} - 2r_2 b\sigma_{\omega_d}\sigma_{\varphi_d}\rho_{\omega_d\varphi_d} - 2r_1^2 b\sigma_{\omega_m}\sigma_{\varphi_m}\rho_{\omega_m\varphi_m} \quad (16)$$

$$\sigma^2_{\gamma_a} = B^2\left(r_2^2\sigma^2_{\omega_d} + r_1^2\sigma^2_{\omega_m}\right) + r_2^2\sigma^2_{\gamma_d} + r_1^2\sigma^2_{\gamma_m} - 2r_2 B\sigma_{\omega_d}\sigma_{\gamma_d}\rho_{\omega_d,\gamma_d} - 2r_1^2\sigma_{\omega_m}\sigma_{\gamma_m}\rho_{\omega_m,\gamma_m} \quad (17)$$

$$\mathrm{cov}(\varphi_a,\gamma_a) = bB\left(r_2^2\sigma^2_{\omega_d} + r_1^2\sigma^2_{\omega_m}\right)$$
$$+ r_2^2\left[\mathrm{cov}(\varphi_d,\gamma_d) - B\,\mathrm{cov}(\omega_d,\varphi_d) - b\,\mathrm{cov}(\omega_d,\gamma_d)\right] \quad (18)$$
$$+ r_1^2\left[\mathrm{cov}(\varphi_m,\gamma_m) - B\,\mathrm{cov}(\omega_m,\varphi_m) - b\,\mathrm{cov}(\omega_m,\gamma_m)\right]$$

with the covariances converted to error correlations as usual, using

$$\rho_{x,y} \equiv \frac{\text{cov}(x,y)}{\sigma_x \sigma_y} \tag{19}$$

A worked example of the detrital-component correction and error propagation for the data of Figure 4 is given in Appendix II.

8. ISOCHRONS

If analyses for 3 or more cogenetic samples are available and believed to share the same initial or detrital ^{230}Th/^{238}U and ^{234}U/^{238}U, both the age and initial isotopic ratios are overdetermined. A set of simultaneous equations could be set up to solve for the combination of age and initial ^{234}U/^{238}U that is the "best match" to the analyses, where "best match" could be defined as "most consistent with the assigned analytical errors" or any other arguably appropriate criteria. However, this purely mathematical approach is never used. Instead, a mathematically equivalent but graphically-based method which permits ready visualization of both the solution and its reliability is always employed—the isochron[5].

In the original and most common form of isochrons (first applied to Rb-Sr geochronology; Nicolaysen 1961), the stable radiogenic daughter-isotope of interest is plotted against its radioactive parent-isotope, both normalized to a stable isotope of the daughter elements—for example, ^{87}Sr/^{86}Sr versus ^{87}Rb/^{86}Sr. If analyses for samples having the same age and same initial isotopic composition of the daughter element are plotted on this diagram, they will form a linear trend (the isochron) whose slope defines the present-day ratio of the daughter isotope to its parent nuclide and whose y-intercept defines the initial isotopic composition of the daughter element. This isochron will exist regardless of whether the samples are a physical mixture of minerals with different ratios of parent/daughter nuclides, or internally homogeneous minerals with variable parent/daughter ratios at the time of their formation. The samples may display more or less scatter from the linear trend, but as long as they fulfill the three fundamental assumptions of an isochron—the samples formed at the same time, their parent and daughter isotopes were initially homogeneous, and they have behaved as a closed-system—this scatter must arise entirely from analytical error.

For the ^{230}Th/U system of igneous rocks (^{234}U/^{238}U assumed to be at secular equilibrium), the only difference is that the daughter isotope is itself radioactive; the graphical representation is essentially the same, with ^{238}U being the parent isotope, ^{230}Th the daughter isotope, and ^{232}Th the (essentially) stable isotope of the daughter element. The isochron diagram can then be a plot of ^{230}Th/^{232}Th versus ^{238}U/^{232}Th (Fig. 5), with the slope yielding the ^{230}Th/^{238}U corrected for initial ^{230}Th. The age is then calculated from the slope, using

$$t = -\frac{1}{\lambda_{230}} \ln\left(1 - \frac{^{230}\text{Th}}{^{238}\text{U}}\right) \tag{20}$$

obtained by including the constraint of secular equilibrium ^{234}U into Equation (1). The slope itself can be determined in a variety of ways, including fitting by eye with a

---

[5] For other philosophies of isochron regressions and errors, see Schwarcz and Latham 1989 or Luo and Ku 1991.

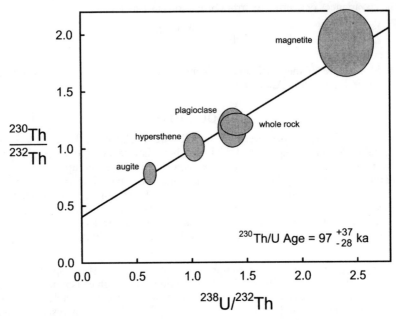

Figure 5. Internal ^{230}Th/U isochron for a volcanic rock, with assumed secular equilibrium ^{234}U/^{238}U. Slope of regression line yields the initial-isotope corrected ^{230}Th/^{238}U for the age equation (Eqn. 2). Data from Allègre and Condomines (1976); error ellipses are 2σ.

straightedge, performing a conventional *y*-on-*x* least-squares regression, and performing a least-squares regression where each of the data points is weighted according to its analytical errors.

8.1. Isochron representations for the general ^{230}Th/U system

Depending on the denominator isotope shared between the axes of the isochron(s), the initial- or detrital-corrected isotope ratio (and thus the age) will be given by the slope or the intercept of the isochron. At present, most laboratories use the "Rosholt" isochron pairs (Rosholt 1976; Fig. 6A), with the slope of the ^{234}U/^{232}Th-^{230}Th/^{232}Th isochron giving the detrital-corrected ^{230}Th/^{234}U ratio, and the slope of the ^{238}U/^{232}Th-^{234}U/^{232}Th isochron giving the detrital-corrected ^{234}U/^{238}U ratio. The alternative diagram (Fig. 6B), wherein the detrital-corrected ratios are given by the *y*-intercepts, e.g., ^{232}Th/^{234}U-^{230}Th/^{234}U plus ^{232}Th/^{238}U-^{234}U/^{238}U (Osmond et al. 1970) or ^{232}Th/^{238}U-^{230}Th/^{238}U plus ^{232}Th/^{238}U-^{234}U/^{238}U (Ludwig and Titterington 1994), is less popular, but in fact has several advantages in both calculation and visualization, especially when the goal is to accurately convey the balance of strengths and weakness of a data set.

The tendency of Rosholt plots to give a misleading impression of an well-defined isochron for data of little statistical power arises from three factors. First, the *x-y* variables of both of the Rosholt plots tend to reflect, to a greater or lesser degree, the U/Th elemental ratio (since the abundances of both ^{234}U and ^{230}Th for nonzero-age materials are generally highly correlated with the concentration of U), so that the usual dispersion in ^{238}U/^{232}Th will greatly amplify the appearance of even a dubious isochron trend. This arithmetical fact almost guarantees that Rosholt isochrons for identical data will have a much higher r^2

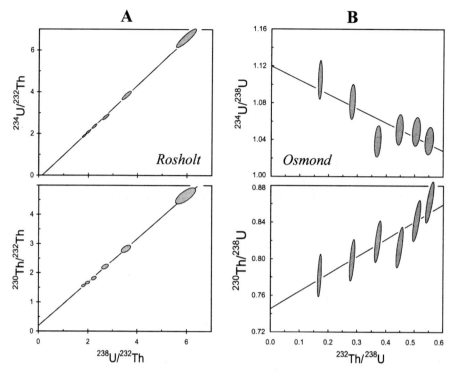

Figure 6. Rosholt (A) and Osmond (B) isochron diagrams for the same, synthetic AS data, shown as their 1σ error ellipses. Apparent age is 115.7 ± 9.7 ka (*MSWD* = 1.7), calculated from either the slopes of the "Rosholt" isochrons or the intercepts of the "Osmond" isochrons.

statistic than an Osmond isochron (Fig. 7). Second, for AS data the errors of the Rosholt-isochron variables must be highly correlated for samples with relatively low ^{232}Th concentrations (because ^{232}Th is always the denominator isotope)[6]. As a result, analytical error will tend to disperse the data points along an oblique trend sub-parallel to most isochrons, thus paradoxically strengthening the visual impression of an isochron by the *imprecision* of the data points. And third (again, for AS data only), if calculation of the relevant error correlations is neglected but the data points are plotted as error symbols, the data will appear to fall more-precisely on the regression line than the error symbols imply, even if there is much more scatter than the true errors and error correlations can support.

These drawbacks of the Rosholt isochron largely disappear when used for igneous systems, where ^{232}Th, being the most-abundant actinide, is measured with approximately equal or better precision than the other U-Th isotopes, thus minimizing error correlations. In fact, for data where all analyses have relatively high ^{232}Th/^{238}U, Osmond diagrams can be *less* informative, since the long extrapolation to the *y*-intercept is difficult to visually appreciate unless the plot extends to zero ^{232}Th/^{238}U.

---

[6] For TIMS or ICP-MS analyses, error correlations in the measured isotope ratios are almost always negligible, as the ^{238}U ion beam is always much larger that of ^{234}U, and the ^{232}Th ion-beam is almost always much larger than that of ^{230}Th (if the latter is not the case, recourse to isochrons is unnecessary).

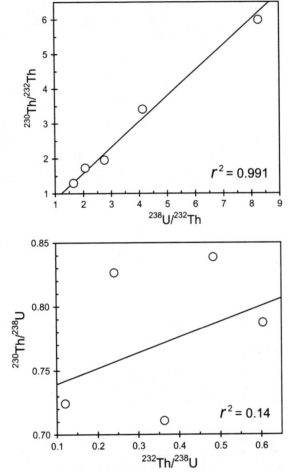

Figure 7. Identical data plotted on ^{230}Th/^{238}U isochron diagrams with different choices of axes. Note the marked difference in the *apparent* conformity to an isochron line and in the magnitude of the often-quoted but geochronologically irrelevant r^2 statistic.

8.2. Error-weighted regressions and isochrons for x-y data

Assumptions and characteristics. Because evaluating the validity of the three isochron assumptions is essential in evaluating the reliability of the isochron age, EWLS (error-weighted least-squares[7]) regression should be regarded as an essential starting point for isochron dating. EWLS regressions assume that (1) all of the observed scatter about the regression line arises from data-point analytical error, (2) data-point analytical errors are known and conform to a Gaussian distribution. Given these two preconditions,

---

[7] Albarède (1995), p. 294-307 gives a modern summary of EWLS isochron mathematics and error propagation. Bevington (1969), chapter 6.3 presents the classical approach, limited to cases with *y*-error only.

EWLS regression is by statistical definition the optimum method, and has the added benefits that (a) uncertainties of the slope and intercept of the regression are easily estimated, and (b) the probability that the data support assumption (1) above can be readily calculated.

Several algorithms have been published for EWLS *x-y* regressions (York 1966, 1969; McIntyre et al. 1966; Williamson 1968; Cumming et al. 1972; Titterington and Halliday 1979), all of which yield the same results for the regression slope and intercept. Those algorithms requiring data with uncorrelated *x-y* errors (York 1966; McIntyre et al. 1966, Williamson 1968) are more restricted in their application. The errors of the regression slope/intercept of such regressions are generally estimated by propagating the errors and error correlations of the input data using first-order expansion, and may differ slightly from one algorithm to another; however, so long as the assumptions of an isochron are met and the errors of the regression slope/intercept are not too large, these differences will be slight.

The MSWD and probability of fit. All EWLS algorithms calculate a statistical parameter from which the observed scatter of the data points about the regression line can be quantitatively compared with the average amount of scatter to be expected from the assigned analytical errors. Arguably the most convenient and intuitively accessible of these is the so-called *MSWD* parameter (Mean Square of Weighted Deviates; McIntyre et al. 1966; Wendt and Carl 1991), defined as:

$$MSWD = \frac{1}{\nu} \sum_{i}^{N} \left(\frac{Resid_i}{\sigma_{y_i}} \right)^2 \tag{21}$$

where $Resid_i$ is the difference between the measured y_i and the y_i predicted from the regression slope and intercept, for each x_i. N is the number of data points, and ν is the degrees of freedom of the regression ($\nu = N-2$ for an *x-y* regression, or $2N-4$ for an linear *x-y-z* regression). The average value expected for the *MSWD* is 1 (e.g., Albarède 1995 p. 291), as suggested intuitively by the fact that it is essentially the mean value of the ratio of the observed data-point scatter to their predicted scatter. Values much greater indicate the presence of more scatter than predicted by the analytical errors, and so imply that either the analytical errors were underestimated or that some non-analytical source of scatter exists. Values much less than 1 imply only that the analytical errors were overestimated.

Terms such as "much less than" are not of much practical use, of course. However, the *MSWD* can be used for two additional calculations that are very useful. First, suppose a *MSWD* much different than 1 raises concerns regarding possible under- or overestimation of the analytical errors. To evaluate such a hypothesis, the square root of the *MSWD* can be used as a correction factor to increase or decrease the assumed analytical errors such that the observed scatter becomes precisely compatible with the "corrected" analytical errors. That is, an *MSWD* of 0.25 is consistent with the assigned analytical errors all being too high by a factor of two, and an *MSWD* of 4 is consistent with errors that are too low by a factor of two.

Second, the probability that the assigned analytical errors would yield at least the observed amount of scatter (usually referred to as the "probability of fit") can be calculated from the chi-squared distribution[8] of $\nu \times MSWD$ about ν. For example,

---

[8] Indeed, the *MSWD* parameter is often referred to as the "reduced Chi-square," and sometimes just as "chi square," though in fact the *MSWD* alone has an *F* distribution (McIntyre et al. 1966).

consider an *x-y* regression with 6 data points yielded an *MSWD* of 1.85. Entering = CHIDIST(1.85*4, 4) in an *Excel* spreadsheet[9] would give the number 0.1162, indicating that these data and errors should produce the observed amount of scatter or more about 12% of the time. By convention, probabilities of fit greater than 0.05 are generally considered as arguably satisfying the mathematical assumptions of an isochron, while lower probabilities are generally taken as indicating the presence of "geological" scatter, and hence a significant possibility of bias in the isochron age. Note that the *MSWD* itself is an inadequate parameter for evaluating the presence of "geological" scatter, since the highest permissible value depends greatly on the number of data-points in the isochron (for example, though an *MSWD* as high as 3.8 for a 3-point isochron is arguably acceptable, an *MSWD* of 1.9 for a 12-point isochron is not). This lack of power of the *MSWD* parameter to detect non-analytical scatter for isochrons with few data points is a serious and under-appreciated limitation which qualitatively diminishes the reliability of such isochrons (Powell et al. 2002).

How one should estimate the errors of low probability-of-fit isochrons (sometimes referred to as an "errorchrons") is a matter of some dispute, and will be discussed later in this chapter.

8.3. 3-dimensional error-weighted regressions and isochrons

Isochrons in the general ^{230}Th-^{234}U-^{238}U system (that is, with disequilibrium ^{234}U/^{238}U) are more complex than classical *x-y* isochrons. Because the ^{230}Th/U age is a function of two coupled decay systems, wherein the daughter of one (^{234}U in the ^{238}U $\rightarrow$ ^{234}U system) is the parent of the other (^{234}U in the ^{234}U $\rightarrow$ ^{230}Th system), an isochron treatment must involve at least 3 isotopic ratios. As discussed earlier, this complexity has traditionally been addressed using two separate isochrons – one to infer the detrital- or initial-isotope corrected ^{234}U/^{238}U, and the other the corrected ^{230}Th/^{238}U, ^{230}Th/^{234}U, or ^{230}Th/^{232}Th. However, this "paired isochron" approach has an inherent statistical flaw.

Because both isochrons of the paired isochrons (regardless of the choice of axes) share some of the same analytical information, both the regression slopes/intercepts and their calculated errors must be correlated in complex ways. These correlations, however, are inherently ignored in the mathematics of the regressions, so that not only are the slopes and intercepts of the regressions sub-optimal, their errors are to some degree inaccurate as well. Fortunately, the mathematical solution to the generalized ^{230}Th/U isochron can be readily visualized by recognizing that the data satisfying the isochron assumptions for any possible combination of ^{230}Th-^{232}Th-^{234}U-^{238}U isotopic ratios with one common-denominator isotope must define a mixing line in three dimensions. The problem then reduces to one of deriving an error-weighted, *x-y-z* linear regression (Fig. 8; Ludwig and Titterington 1994).

The visually most convenient choice of axes for this regression are the ones whose common denominator is ^{238}U, so that projection of the 3-D regression line to the ^{232}Th-free plane will give the detrital-corrected ^{230}Th/^{238}U and ^{234}U/^{238}U from which the ^{230}Th/U age is calculated. Mathematically, however, the choice of ratios is unimportant, provided that there is a common-denominator isotope to all three axes, and the errors and error correlations for the data points are correctly calculated. Note that, unlike paired *x-y* isochrons, not only can the errors of the detrital-corrected ^{230}Th/^{238}U and ^{234}U/^{238}U be calculated using this algorithm, but also their error correlations (required for accurate estimation of age errors).

---

[9] CHIDIST is the *Excel* function for the one-tailed probability of the chi-squared distribution.

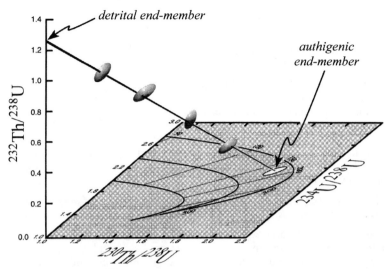

Figure 8. The 3-dimensional ^{230}Th/U isochron (Ludwig and Titterington 1994). Intersection with the ^{232}Th-free plane gives the age and initial ^{234}U/^{238}U of the authigenic end member. Curved lines on the ^{232}Th-free plane show the evolution for systems with different initial ^{234}U/^{238}U; straight lines are isochrons, labeled in ka.

8.4. Isochrons with excess scatter

Error-weighted, least-squares regressions have several strong points. First, they are mathematically and statistically on firm ground, in that their few assumptions have relatively simple mathematical consequences with extensive predictive power. Second, they make the best conceivable use of data which fulfill the simple requirements of a classical isochron. Third, they offer a range of quantitative predictions that can be of great value for not only the geochronologist, but also the scientist attempting to build or test quantitative models from such isochron dates. It is thus disheartening to confess that these elegant mathematical constructs are often irrelevant to real-world systems of interest, especially with the advent of the very small errors produced by TIMS or ICP-MS analyses.

Part of the problem, paradoxically, lies in the sensitivity with which EWLS regressions can test for their own reliability (at least for isochrons with more than 4 or 5 data points), using the "probability of fit" calculation as outlined above. This is seldom a matter of arguing about whether a probability of fit of, say, 0.045 is or is not a red flag indicating an "errorchron." It is much more-commonly a matter of dealing with a probability of fit of something like 0.000023 and an *MSWD* of 56. In the latter case, there is simply no arguing that the assumptions of the isochron approach are grossly violated, and that at least one of the following failures have affected the samples: 1) multiple detrital components, 2) multiple ages, or 3) an open-system history.

There are several common methods in geochronology to deal with "excess scatter" isochrons (that is, isochrons whose data points scatter significantly more than indicated by their analytical errors), none of which are particularly satisfactory. These are to:

1. Ignore the problem completely and simply quote the errors propagated from the analytical errors of the data points.

This approach will always give the smallest errors, but is scientifically (and ethically) devoid of merit. Devotees of this method tend to avoid discussion of *MSWD* or related parameters, as well as probabilities of fit.

2. *Expand the slope/intercept errors by $\sqrt{MSWD}$ and possibly also by a Student's-t factor for a confidence level of interest.*

This is a common approach, and is equivalent to saying either "I think the analytical errors may be more-or-less uniformly underestimated by some unknown amount, so I'm going to correct for this underestimate using the actual scatter of the data about the isochron as a guide", or "There is some 'geologic' scatter in this data, but I'm going to assume that it's more or less proportional to the analytical errors and that this scatter has a Gaussian distribution about the regression line". Not perfect, and with rather shaky assumptions, but certainly much better than (1) above and much more likely to yield realistic errors.

3. *Abandon error weighting entirely and revert to a pair of simple y-on-x least-squares regressions.*

Unweighted least-squares regressions have a variety of flaws, including different results depending on which ratio is chosen as the dependent variable, the invalid assumption of errors and scatter in only one, arbitrarily chosen axis, and unreliable calculated errors with unknown error correlations amongst the two isochrons. For any particular data set, whether these flaws will result in significant errors in estimation of either the isochron parameters themselves or their uncertainties depends is difficult to predict. In any case, the risk of bias and error underestimation with this approach increases markedly for data-sets that are either small or have marginal dispersion along the isochron.

8.5. Beyond error-weighted least-squares isochrons

Explicit inclusion of geologic complexities in the weighting approach. Part of the power of EWLS isochrons is that, for a "true" isochron, the algorithms pay precisely as much attention to each data point as it deserves. When excess scatter becomes statistically obvious, analytical-error weighting can no longer be optimum. However, if the "geologic" source of scatter is, say, a multi-component detrital end-member with roughly Gaussian variation of its Th-U isotopic ratios that could be characterized via either reasonable assumptions or actual analyses on candidate materials, one should be able to quantitatively model this situation along the lines of the "Model 3" McIntyre et al. (1966) or Ludwig (1984, 2002) *x-y* regression algorithms. Such a weighting algorithm would place heavy emphasis on the data points with low ^{232}Th/^{238}U, which points are least affected by variations in the detrital component's isotopic composition, and the resulting probability of fit would be a good test of the validity of the assigned variation. Further, if one could reliably know the true (non-Gaussian) distribution of the detrital-isotope ratios, one could use more-general methods such as Maximum Likelihood Estimation (e.g., Titterington and Halliday 1979) to estimate the regression-line parameters, making possible a less-restrictive set of assumptions that nonetheless involves only slight modifications to the fundamental isochron requirements. These are promising approaches, but are as yet untried.

8.6. Robust and resistant isochrons

Methods. Perhaps the best way of dealing with this thorny problem (common to not only ^{230}Th/U geochronology, but also the more "classical" methods of isotope geochronology as well) is to abandon the reliance on a strictly Gaussian distribution of residuals, whether arising from analytical error or "geologic" complexities. "Robust" in the statistical sense implies insensitivity to departure of the data from the initial assumptions (such as a Gaussian distribution of residuals), generally exacting only a minimal "penalty" in the form of only slight increase in error estimates. "Resistant" methods (for example, the median as opposed to the mean) tend to be even less sensitive to gross deviations from expected values by a minority of the data, at the expense of generally larger uncertainties in the resulting statistics (Hoaglin et al. 1983, p. 2). A major advantage of both robust and resistant isochrons is that subjective decisions as to whether or not a data-point should be "accepted" or "rejected" are eliminated. However, neither type of isochron abandons the requirements that most of the data conform to the fundamental characteristics (a closed system of cogenetic samples) of a true isochron, and neither can give *accurate* results in the presence of large departures from these requirements.

Robust and resistant isochrons can have very different characteristics than traditional least-squares or error-weighted least squares regressions. Some methods ignore analytical errors entirely, others infer them from the observed scatter of the data, and still others make use of analytical errors only to the extent that they are validated by their observed scatter.

For *x-y* isochrons, several robust/resistant algorithms have been proposed (see especially Vugrinovich 1981, Rock and Duffy 1986, and Rock et al. 1987). Three of the most interesting are the Theil's median of pairwise slopes (Theil 1950; Vugrinovich 1981), Siegel's method of repeated medians of pairwise slopes (Siegel 1982) and Powell's method (Powell et al. 2002). With the Theil-Vugrinovich method (Fig. 9), the slope of the regression line is taken as the median of the slopes of all of the possible 2-point lines defined by the data, with errors based on the order statistics of the resulting arrays of slopes (Vugrinovich 1981). Siegel's method (Siegel 1982) is even more

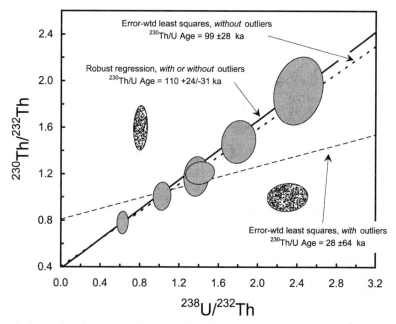

Figure 9. Comparison between performance of error-weighted least-squares isochron (dashed lines) with a robust isochron (solid line) for a "dirty" data-set (^{234}U at secular equilibrium). Both age and errors are greatly affected by inclusion of the two outliers (stippled error ellipses), whereas the robust isochron (Theil 1950; Vugrinovich 1981) is unchanged. Errors are 95% confidence.

resistant to outliers, as the slope is taken as the median of the medians of all pairwise slopes from each data point in turn, but has the disadvantage that error estimation is not straightforward, especially for small data-sets.

The method proposed by Powell et al. (2002) is more sophisticated, in that it is based on an algorithm that self-weights the individual data points according to their scatter from the regression line. Outlier points are deweighted to zero and thus effectively ignored. Analytical errors are not entirely discounted, in that when calculating the errors of the isochron, the data points are never assumed to be "better" (in a statistical sense) than implied by their analytical errors. Regression errors are calculated using a bootstrap method (Davison and Hinkley 1997) on the residuals, with the resulting disadvantage that their reliability becomes doubtful for data sets much smaller than 8-10 points (Porter et al. 1997).

Advantages and disadvantages of existing robust/resistant isochron methods. The precision penalty (compared to EWLS) imposed on regressions by most robust methods for well-behaved, Gaussian data sets of at least moderate size (say $N > 8$) is generally rather small—often in the range of 10%-40%. Such a penalty is arguably a small price to pay for abandoning the assumption of analytical error-related scatter when that assumption is objectively discredited by very low probabilities-of-fit. Regardless of enthusiasm for robust approaches, however, one should not entirely abandon EWLS regressions, because their associated probability-of-fit parameter signals the researcher with more or less clarity whether the data *might permissibly* represent a "true" isochron. For isochrons with few data-points or little dispersion (compared to their analytical errors), an acceptably high probability of fit is only a weak verification (Powell et al. 2002). However, an unacceptably low probability of fit is a convincing argument for the existence of non-analytical scatter, thus at the same time justifying a robust regression and diminishing, in an unquantifiable way, its age significance

Further development of robust methods tailored for isochron data can be anticipated, with the ultimate goal being a method that yields the same errors as EWLS for well-behaved data (i.e., strictly Gaussian, with accurate estimates of data-point errors), and only modest loss of precision for increasingly "dirty" data (R. Powell, written commun. 2002).

9. PITFALLS IN DATA PRESENTATION

There is no standard set of specifications for the presentation of ^{230}Th/U data in scientific publications; however, there are a number of ways in which the data can be presented so as to be at best ambiguous, and at worst misleading. These include:

- Not specifying the sigma or confidence level of isotopic ratios or isochron results;
- Using the standard error from replicates or excess-scatter isochrons with small N (say <6 or so), instead of a confidence level;
- Not including error correlations where significant (say >0.3) in isotope-ratio data tables;
- Using error crosses as isochron data-point symbols where significant x-y error correlations exist;
- Not giving the isotope ratios and errors actually used in age or isochron calculations (for example, data tables with only ^{230}Th/^{232}Th, ^{230}Th/^{234}U, and ^{234}U/^{238}U ratios but using ^{230}Th/^{232}Th-^{234}U/^{232}Th and ^{234}U/^{232}Th-^{238}U/^{232}Th isochrons) so that the reader cannot reproduce the calculations;

- Using "internal" or "within-run" TIMS/ICP-MS isotope-ratio errors without evaluating the magnitude of the "external" (run to run) error;
- Quoting only first order (linear) age errors when the age plus age-error approaches the secular-equilibrium limit;
- Presenting excess-scatter isochrons without discussion of the cause and significance to the validity of the apparent age.
- Showing and appealing to high r^2 statistics as a justification for the reliability of an isochron (especially for Rosholt-type isochron axes).

Like any method of dating, the ^{230}Th/U method can yield apparent ages that range from almost iron-clad to extremely suspect, depending on the nature of the samples and the complexity of their history. The hurdles provided by natural processes should provide sufficient complexity for the scientist attempting to evaluate the significance of the dates, without having to worry about careless or deceptive data-handling on the part of their provider.

10. NOTE ON IMPLEMENTATION OF ALGORITHMS

Most of the algorithms and formulae discussed in this chapter can be implemented as expressions in computer spreadsheets, and the rest as simple computer programs. Most are also incorporated into the Microsoft *Excel* spreadsheet program by the *Isoplot* add-in (Ludwig 1999, in press) as user-available functions and graphical routines (Appendix III).

ACKNOWLEDGMENTS

I am indebted to Roger Powell for a constructive review and trenchant observations from a real expert in statistical examination of geochemical data, and to Pyramo Marianelli for pointing out the relevance of prior constraints in Monte Carlo simulations of ^{230}Th/U ages.

REFERENCES

Albarède F (1995) Introduction to Geochemical Modeling. Cambridge University Press, Cambridge

Allègre CJ, Condomines M (1976) Fine chronology of volcanic processes using ^{238}U-^{230}Th systematics. Earth Planet Sci Lett 28:395-406

Anderson GM (1976) Error propagation by the Monte Carlo method in geochemical calculations. Geochim Cosmochim Acta 40:1533-1538

Bevington PR (1969) Data Reduction and Error Analysis for the Physical Sciences. McGraw-Hill, New York

Bernardo JM, Smith AFM (2000) Bayesian Theory. John Wiley & Sons, New York, p 353-356

Bischoff JL, Fitzpatrick JA (1991) U-series dating of impure carbonates: An isochron technique using total-sample dissolution. Geochim Cosmochim Acta 55:543-554

Chen JH, Edwards RL, Wasserburg GJ (1986) ^{238}U, ^{234}U, and ^{232}Th in seawater. Earth Planet Sci Lett 80:241-251

Cheng H, Edwards RL, Murrell MT, Benjamin TM (1998) Uranium-thorium-protactinium dating systematics. Geochim Cosmochim Acta 62:3437–3452

Cumming GL, Rollett JS, Rosotti FJC, Whewell RJ (1972) Statistical methods for the computation of stability constants, I. Straight-line fitting of points with correlated errors. J Chem Soc Dalton Trans 23:2652-2658

Davison AC, Hinkley DV (1997) Bootstrap Methods and their Application. Cambridge University Press, Cambridge

Edwards RL, Chen JH, Wasserburg GJ (1987) ^{234}U-^{238}U-^{230}Th-^{232}Th systematics and precise measurement of time over the past 500,000 years. Earth Planet Sci Lett 81:175-192

Esat TM (1995) Charge collection thermal ion mass spectrometry of thorium. Int J Mass Spectrom Ion Proc 148:159-170

Hoaglin DC, Mosteller F, Tukey JW (1983) Understanding Robust and Exploratory Data Analysis. John Wiley and Sons, New York

Kaufman A, Broecker W (1965) Comparison of ^{14}C and ^{230}Th ages for carbonate minerals from Lakes Lahontan and Bonneville. J Geophys Res 70:4039-4054

Ludwig KR, Titterington DM (1994) Calculation of ^{230}Th/U Isochrons, ages, and errors: Geochim Cosmochim Acta 58:5031-5042

Ludwig KR (1999) Using Isoplot/Ex, Version 2.01, A Geochronological Toolkit for Microsoft Excel: Berkeley Geochronology Ctr. Spec. Pub. 1a

Ludwig KR (in press) Using Isoplot/Ex, Version 3, A Geochronological Toolkit for Microsoft Excel: Berkeley Geochronology Ctr. Spec. Pub. 4

Luo S, Ku T-L (1991) U-series isochron dating: a generalized method employing total sample dissolution. Geochim Cosmochim Acta 55:555-564

McIntyre GA, Brooks C, Compston W, Turek A (1966) The statistical assessment of Rb-Sr Isochrons. J Geophys Res 71:5459-5468

Nicolaysen NO (1961) Graphic interpretation of discordant age measurements of metamorphic rocks. Ann NY Acad Sci 91:198-206

Osmond JK, May JP, Tanner WF (1970) Age of the Cape Kennedy barrier-and-lagoon complex. J Geophys Res 75:5459-5468

Porter PS, Rao ST, Ku J-Y, Poirot RL, Dakins M (1997) Small sample properties of nonparametric bootstrap t confidence intervals. J Air Waste Management Assoc 47:1197-1203

Powell R, Hergt J, Woodhead J (2002) Improving isochron calculations with robust statistics and the bootstrap. Chem Geol 185:191–204

Rickmers AD, Todd HN (1967) Statistics: An Introduction. McGraw-Hill, New York

Rock NMS, Duffy TR (1986) REGRES: A FORTRAN-77 program to calculate nonparametric and "structural" parametric solutions to bivariate regression equations. Comp Geosci 12:807-818

Rock NMS, Webb JA, McNaughton NJ, Bell GD (1987) Nonparametric estimation of averages and error for small data-sets in isotope geoscience: a proposal. Chem Geol 66:163-177

Rosholt JN (1976) ^{230}Th/U dating of travertine and caliche rinds. Geol Soc Amer Abstr Prog 8:1076

Rowe PJ, Atkinson TC, Turner C (1999) U-series dating of Hoxnian interglacial deposits at Marks Tey, Essex, England. J Quat Sci 14: 693–702

Schwarcz HP, Latham AG (1989) Dirty calcites 1. Uranium series dating of contaminated soils using leachates alone. Chem Geol (80) 35–43

Siegel AF (1982) Robust regression using repeated medians. Biometrika 69:242-244

Theil H (1950) A rank-invariant method of linear and polynomial regression analysis, I, II, and III. Proc. Koninklijke Nederlands Akadamie van Wetenschappen 53:386-392, 521-525, 1397-1412

Titterington DM, Halliday AN (1979) On the fitting of parallel isochrons and the method of maximum likelihood. Chem Geol 26:183-195

Vugrinovich RG (1981) A distribution-free alternative to least-squares regression and its application to Rb/Sr isochron calculations. J Math Geol 13:443-454

Wendt I, Carl C (1991) The statistical distribution of the mean squared weighted deviation. Chem Geol 86:275-285

Williamson JH (1968) Least-squares fitting of a straight line. Can J Phys 46:1845-1847

York D (1966) Least-squares fitting of a straight line. Can J Phys 44:1079-1086

York D (1969) Least-squares fitting of a straight line with correlated errors. Earth Planet Sci Lett 5:320-324

APPENDIX I:
ESTIMATING ERROR CORRELATIONS

The error correlation between two quantities can be determined empirically, from a number (N) of replicate analyses of pairs of the two quantities (say x and y), and evaluating the expression for the linear correlation (Rickmers and Todd 1967)

$$\rho_{xy} = \frac{\sum_{i=1}^{N}(x_i - \bar{x})(y_i - \bar{y})}{\sqrt{\sum_{i=1}^{N}(x_i - \bar{x})^2 \sum_{i=1}^{N}(y_i - \bar{y})^2}} \tag{22}$$

where $\bar{x}$ and $\bar{y}$ are the average of the N x_i and y_i. Alternatively, if the x_i and y_i are derived values from precursor measurements whose (Gaussian) errors are known, the error correlation can be derived analytically by differentiation. For example if $x = a/c$ and $y = b/c$, where a, b, c are measured quantities with known errors, differentiation yields

$$\frac{dx}{x} = \frac{da}{a} - \frac{dc}{c} \tag{23}$$

$$\frac{dy}{y} = \frac{db}{b} - \frac{dc}{c} \tag{24}$$

so

$$\left(\frac{dx}{x}\right)^2 = \left(\frac{da}{a}\right)^2 + \left(\frac{dc}{c}\right)^2 - 2\frac{(da)(dc)}{ac} \tag{25}$$

$$\left(\frac{dy}{y}\right)^2 = \left(\frac{db}{b}\right)^2 + \left(\frac{dc}{c}\right)^2 - 2\frac{(db)(dc)}{bc} \tag{26}$$

and

$$\frac{(dx)(dy)}{xy} = \frac{(da)(db)}{ab} - \frac{(da)(dc)}{ac} - \frac{(db)(dc)}{bc} + \left(\frac{dc}{c}\right)^2 \tag{27}$$

If the errors in a, b, c are not too large, a linearized error approximation (see later discussion) can then be made by expanding the squares and cross-products of the infinitesimally small derivatives to variances and covariances. Using S_x to indicate the coefficient of variation, σ_x/x, we obtain

$$S_x^2 = S_a^2 + S_c^2 - 2\frac{\text{cov}(a,c)}{ac} \tag{28}$$

$$S_y^2 = S_b^2 + S_c^2 - 2\frac{\text{cov}(b,c)}{bc} \tag{29}$$

and

$$\frac{\text{cov}(x,y)}{xy} = \frac{\text{cov}(a,b)}{ab} - \frac{\text{cov}(a,c)}{ac} - \frac{\text{cov}(b,c)}{bc} + S_c^2 \tag{30}$$

By definition, the x-y error correlation is

$$\rho_{x,y} \equiv \frac{\text{cov}(x,y)}{\sigma_x \sigma_y} \tag{31}$$

so, if the covariances among the parameters a, b, and c are small (for example, if a, b, c were AS-measured peak heights, or (neglecting mass fractionation), TIMS ion-beams for different isotopes), the covariances among a, b, and c can be neglected, giving

$$\frac{\text{cov}(x,y)}{xy} = S_c^2 \tag{32}$$

thus

$$\rho_{x,y} = \frac{S_c^2}{S_x S_y} \tag{33}$$

In practice, the expressions to be differentiated and squared or cross-multiplied can be much more complex. However, as long as the expressions are differentiable, the procedure is the same.

APPENDIX II:
WORKED EXAMPLE OF DETRITAL CORRECTION AND ERROR PROPAGATION

Synthetic data for detrital correction.

	$^{232}Th/^{238}U$	$\pm 1\sigma\%$	$^{230}Th/^{238}U$	$\pm 1\sigma\%$	$^{234}U/^{238}U$	$\pm 1\sigma\%$	$\rho\,^{232}Th/^{238}U\text{-}^{230}Th/^{238}U$	$\rho\,^{232}Th/^{238}U\text{-}^{2230}Th/^{238}U$	$\rho\,^{230}Th/^{238}U\text{-}^{234}U/^{238}U$
Measured Detritus	0.833	2.05	1.008	2.00	0.9985	1.42	0.733	0.345	0.354
Measured Sample	0.160	3.04	0.757	2.09	1.485	1.29	0.475	0.254	0.372
Corrected Sample			0.697	2.71	1.601	1.61			0.408

Uncorrected sample age:	74.0 ±4.1 ka (2σ)
Corrected sample age:	**59.8 ±3.8 ka (2σ)**

The data for the above table (shown graphically in Fig. 4) were generated via synthetic alpha-spectrometric measurements, wherein ratios involving both Th and U isotopes involved not only the natural isotopes of the sample or detritus, but also the artificial spike isotopes ^{228}Th and ^{236}U. Errors and error correlations were then derived assuming perfectly-resolved alpha-counting peaks with negligible background and the formulae given in Ludwig and Titterington (1994):

$$S^2_{^{232}Th/^{238}U} = S^2_{^{228}Th} + S^2_{^{232}Th} + S^2_{^{236}U} + S^2_{^{238}U}$$

$$S^2_{^{230}Th/^{238}U} = S^2_{^{228}Th} + S^2_{^{230}Th} + S^2_{^{236}U} + S^2_{^{238}U}$$

$$S^2_{^{234}U/^{238}U} = S^2_{^{234}U} + S^2_{^{238}U}$$

$$\rho_{^{232}Th/^{238}U\text{-}^{230}Th/^{238}U} = \frac{S^2_{^{228}Th} + S^2_{^{236}U} + S^2_{^{238}U}}{S_{^{232}Th/^{238}U}\,S_{^{230}Th/^{238}U}}$$

$$\rho_{^{232}Th/^{238}U\text{-}^{234}U/^{238}U} = \frac{S^2_{^{238}U}}{S_{^{232}Th/^{238}U}\,S_{^{234}U/^{238}U}}$$

$$\rho_{^{230}Th/^{238}U\text{-}^{234}U/^{238}U} = \frac{S^2_{^{238}U}}{S_{^{230}Th/^{238}U}\,S_{^{234}U/^{238}U}}$$

Where $S_x = \sigma_x/x$ and ρ_{xy} is the *x-y* error correlation.

The resulting corrected age of 59.8 ± 3.8 ka can be calculated either from the two-point isochron defined by the data, or by application of Equations (16)-(18) of this chapter.

APPENDIX III:
FUNCTIONS/ROUTINES FOR ^{230}Th/U
DATING PROVIDED BY *ISOPLOT*

As a convenience to readers, the editors have asked that the following index to relevant *Isoplot* functionality be included. This Excel add-in program is available from the author upon request (Ludwig 1999, in press).

Worksheet functions

AlphaMS — Converts Rosholt-type (^{238}U/^{232}Th-^{230}Th/^{232}Th-^{234}U/^{232}Th) isochron ratios, errors, and error correlations to Osmond-type (^{232}Th/^{238}U-^{230}Th/^{238}U-^{234}U/^{238}U) ratios, errors, and error correlations, and vice versa.

CorrThU — Corrects measured ^{232}Th/^{238}U-^{230}Th/^{238}U-^{234}U/^{238}U ratios, errors, and error correlations for a detrital component, using ^{232}Th as the index isotope.

InitU234U238 — Returns the initial ^{234}U/^{238}U for a system of specified age and present-day ^{234}U/^{238}U.

Th230238ar — Returns the present-day ^{230}Th/^{238}U for a system of specified age and initial ^{234}U/^{238}U.

Th230Age — Returns the age of a system with specified present-day ^{230}Th/^{238}U and ^{234}U/^{238}U.

Th230AgeAndInitial — Returns the ^{230}Th/U age and initial ^{234}U/^{238}U, together with their errors (with or without ^{230}Th-^{234}U decay-constant errors) and error correlation, for a system with specified present-day ^{230}Th/^{238}U and ^{234}U/^{238}U, errors, and error-correlation.

U234age — Returns the ^{234}U/^{238}U age for a specified present-day ^{234}U/^{238}U and initial ^{234}U/^{238}U.

U234U238ar — Returns the present-day ^{234}U/^{238}U for a specified age and initial ^{234}U/^{238}U.

Graphical-numerical routines

- ^{238}U/^{232}Th-^{230}Th/^{232}Th 2-D isochrons, ages, and errors.
- ^{232}Th/^{238}U-^{230}Th/^{238}U-^{234}U/^{238}U 3-D isochrons (graphics are 2-D projections), ages, and errors.
- ^{230}Th/^{238}U-^{234}U/^{238}U evolution diagrams, with evolution curves, isochron lines, and data-point error ellipses.
- ^{230}Th/U age and error calculation, with or without decay-constant errors, via first-order error expansion or Monte Carlo (including age-histogram), with or without finite-age constraint.